OXFORD

GCSE Maths

For Edexcel

HIGHER PLUS

Marguerite Appleton
Dave Capewell
Derek Huby
Jayne Kranat
Peter Mullarkey

OXFORD
UNIVERSITY PRESS

Great Clarendon Street, Oxford OX2 6DP
Oxford University Press is a department of the University of Oxford.
It furthers the University's objective of excellence in research, scholarship,
and education by publishing worldwide in

Oxford New York
Auckland Cape Town Dar es Salaam Hong Kong Karachi
Kuala Lumpur Madrid Melbourne Mexico City Nairobi
New Delhi Shanghai Taipei Toronto

With offices in

Argentina Austria Brazil Chile Czech Republic France Greece
Guatemala Hungary Italy Japan South Korea Poland Portugal
Singapore Switzerland Thailand Turkey Ukraine Vietnam

Oxford is a registered trade mark of Oxford University Press
in the UK and in certain other countries

© Oxford University Press 2010

The moral rights of the author have been asserted

Database right Oxford University Press (maker)

First published 2010

All rights reserved. No part of this publication may be reproduced,
stored in a retrieval system, or transmitted, in any form or by any means,
without the prior permission in writing of Oxford University Press,
or as expressly permitted by law, or under terms agreed with the appropriate
reprographics rights organization. Enquiries concerning reproduction
outside the scope of the above should be sent to the Rights Department,
Oxford University Press, at the address above

You must not circulate this book in any other binding or cover
and you must impose this same condition on any acquirer
British Library Cataloguing in Publication Data
Data available

ISBN 9780199139439

10 9 8 7 6 5 4 3 2

Printed in Singapore by KHL Printing Co. Pte Ltd.

Paper used in the production of this book is a natural, recyclable product made from wood
grown in sustainable forests. The manufacturing process conforms to the environmental
regulations of the country of origin.

Acknowledgements
The Publisher would like to thank Edexcel for their kind permission to reproduce past exam
questions.
Edexcel Ltd, accepts no responsibility whatsoever for the accuracy or method of working in the answers given

The Publisher would like to thank the following for permission to reproduce photographs:
p2-3: Godrick/Dreamstime.com; p18-19: Ivan Kmit/Dreamstime.com p36-37: OUP/Photodisc; p54-55: Slobodan Djajic/Dreamstime; p68-69: Alex Segre/Alamy; p82-83: Juan Fuertes/Shutterstock; p102-103: Dreamstime Agency/Dreamstime.com; p116-117: Saniphoto/Dreamstime.com; p132-133: Monkey Business Images/Dreamstime.com; p137: Raja Rc/Dreamstime.com; p146-147: SSPL via Getty Images; p160-161: Diman Oshchepkov/Dreamstime.com; p174-175: OUP/Corbis; p192-193: Dgareri/Dreamstime; p206-207: Rex Features; p222-223: Theo Gottwald/Dreamstime.com; p236-237: Martin Fischer/Shutterstock; p256-257: Creative Commons/Wikipedia; p274: Raja Rc/Dreamstime.com; p278-279: SHOUT/Alamy; p293-294: Idrutu/Dreamstime.com; p308-309: Vandystadt/Michel Hans/Allsport; p324-325: Bbbar/Dreamstime.com; p364-365: European Space Agency/Science Photo Library; p378-379: NTERFOTO/Alamy; p396-370: Sharpshot/Dreamstime.com; M. Dykstra/Shutterstock.

The Publisher would also like to thank Anna Cox for her work in creating the case studies.
Figurative artwork is by Peter Donnelly

About this book

This book has been specifically written to help you get the best possible grade in your Edexcel GCSE Mathematics examinations. It is designed for students who have achieved a secure level 7 at Key Stage 3 and are looking to progress to a grade A* at GCSE, Higher tier.

The authors are experienced teachers and examiners who have an excellent understanding of the Edexcel specification and so are well qualified to help you successfully meet your objectives.

The book is made up of chapters that are based on Edexcel specification B, and is organised clearly into the three units that will make up your assessment.

Functional maths and **problem solving** are flagged in the exercises throughout.

- In particular there are **case studies**, which allow you apply your GCSE knowledge in a variety of engaging contexts.
- There are also **rich tasks**, which provide an investigative lead-in to the chapter – you may need to study some of the techniques in the chapter in order to be able to complete them properly.

Also built into this book are the new **assessment objectives:**

AO1 recall knowledge of prescribed content.
AO2 select and apply mathematical methods in a range of contexts.
AO3 interpret and analyse problems and select strategies to solve them.

AO2 and AO3 are flagged throughout, particularly in the regular **summary assessments,** as these make up around 50% of your examination.

Finally, you will notice an icon that looks like this:

This shows opportunities for **Quality of Written Communication,** which you will also be assessed on in your exams.

Best wishes with your GCSE Maths – we hope you enjoy your course and achieve success!

Contents

Finding your way around this book

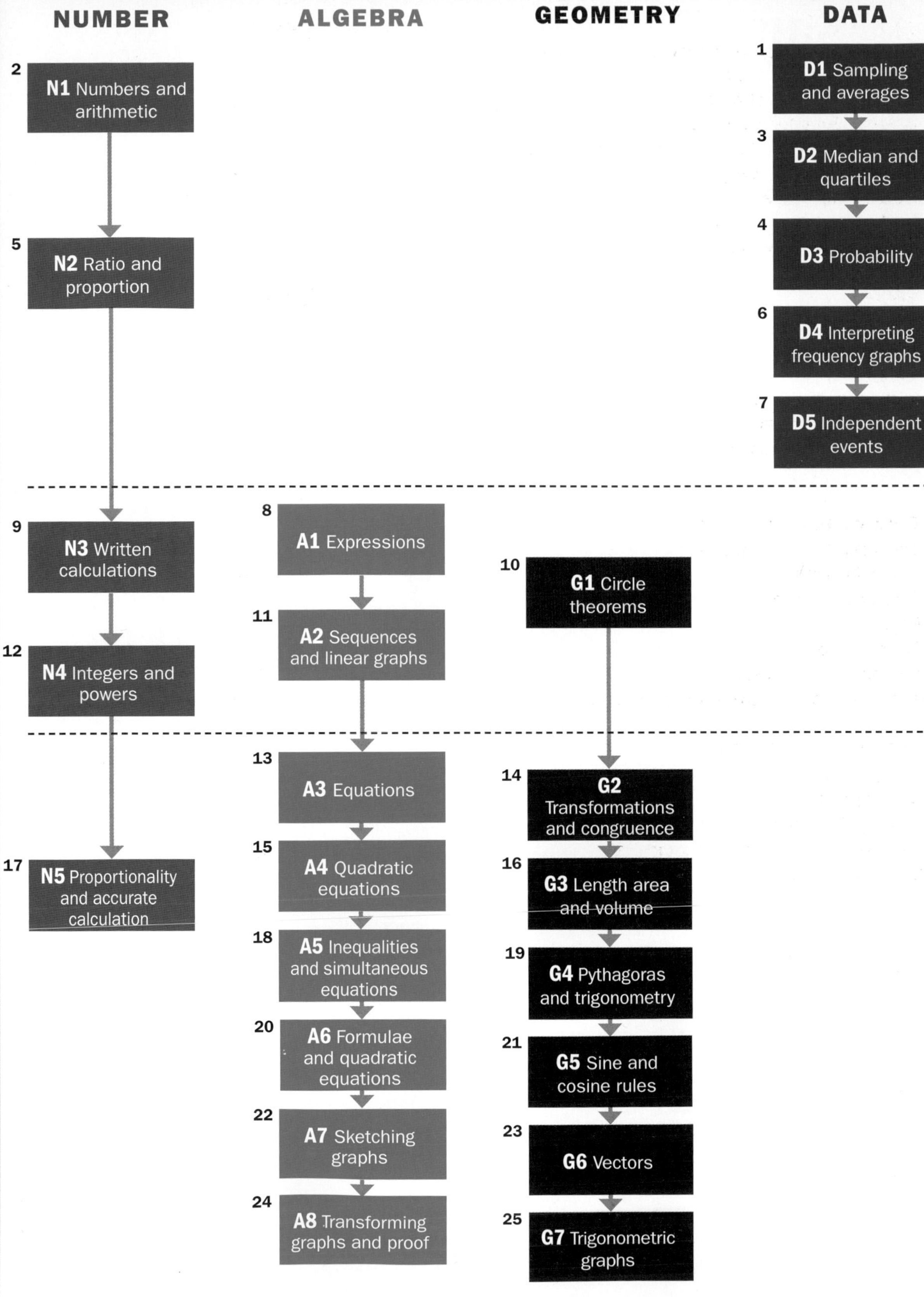

D1

Sampling and averages

In the run up to a general election, opinion polls are taken of which political party people are likely to vote for. The results of just a 1000 peoples' voting intentions are taken very seriously by the media and the politicians.

What's the point?

By selecting a representative sample, surveys allow statisticians to obtain reliable results from manageable amounts of data. The technique is used by polling organisations and throughout industry to monitor quality and productivity *etc*.

Check in

You should be able to

- **appreciate the issues involved in collecting data**

1 Janine wanted to find out who had school dinners. In her survey, Janine asked everyone at her school. This is called a census.

a Explain what a census is.

b How is a census different from a sample?

2 Nicky and Mike wanted to find out how many people had MMR inoculation. Nicky asked 100 people in the town centre one morning. Mike looked up MMR statistics on the internet.

a Who collected primary data and who collected secondary data?

b Explain the difference between primary and secondary data.

3 Oliver asked this question in a questionnaire:

'You like to eat cucumber in a salad, don't you?'

Explain why this is a leading question.

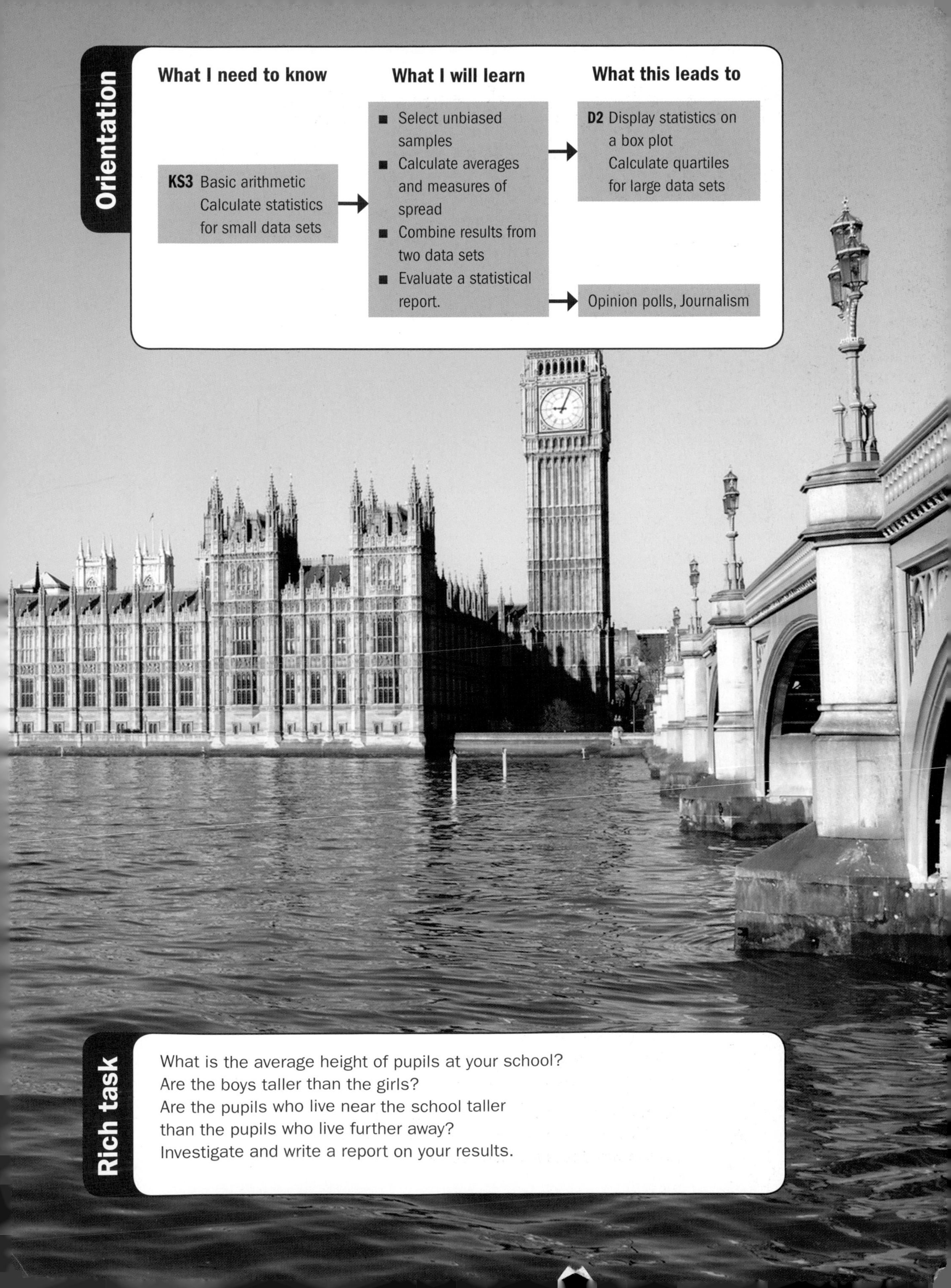

Orientation

What I need to know	What I will learn	What this leads to
KS3 Basic arithmetic Calculate statistics for small data sets	■ Select unbiased samples ■ Calculate averages and measures of spread ■ Combine results from two data sets ■ Evaluate a statistical report.	**D2** Display statistics on a box plot Calculate quartiles for large data sets Opinion polls, Journalism

Rich task

What is the average height of pupils at your school?
Are the boys taller than the girls?
Are the pupils who live near the school taller than the pupils who live further away?
Investigate and write a report on your results.

D1.1 Random and systematic sampling

This spread will show you how to:

- Plan to minimise bias in a survey
- Use suitable methods of sampling to investigate a population, including random sampling

Keywords
Bias
Data
Random
Sample
Systematic sample

In a statistical survey, the larger the **sample**, the more reliable the results.

Your sample should represent the whole population you are studying.

- You need to plan your survey to minimise **bias**; an unrepresentative sample will distort your results.

The most reliable **data** comes from a census – a survey of the whole population.

- In a **random sample**:
  - Each member of the population has the same chance of being included.
  - The person choosing the sample has no control over who is included.

To select a random sample:
- Assign a number to every member of the population.
- Use a calculator to generate random numbers.
- Choose the members with these numbers

'Picking names from a hat' is another method of random sampling.

- In a **systematic sample** you use a system to choose sample members from a list.

To select a systematic sample:
- List every item in the population.
- Choose a starting point at random, then select every *n*th item after it.

Every 10th calculator produced by a defective machine is faulty. Sampling the 10th, 20th, etc would pick every faulty component. Sampling the 1st, 11th, 21st would pick none of them. Each result is biased.

A systematic sample could be unrepresentative if there is a pattern in the data.

Example

A bus company wants to carry out a survey on local transport. They plan to ask 100 out of 8000 households in the vicinity.
Suggest
a a suitable random sampling method
b a suitable systematic sampling method.

Design the survey method to minimise bias.

---

a Assign a number to each household.
Generate a series of random numbers.
Read these in fours, to give one hundred 4-digit numbers.
Choose the households these 4-digit numbers represent.

b Assign a number to each household.
Generate two random digits to give a number between 1 and 80.
This is the first member of the sample.
Choose every 80th number after this member.

0603 represents household number 603 on the list.

100 : 8000 = 1 : 80.
To spread the sample over the population, select one from each 'batch' of 80.

Exercise D1.1

Grade C

Unit 1

A03 Functional Maths

1 Tristan is doing a survey to find out how often people go bowling.
He writes this question:

Do you go bowling often?

He stands outside a bowling alley and asks people as they go in.
Write two reasons why this is not a good way to find out how often people go bowling.

2 Dervla asked people at a netball club this question:

Do you agree that tennis is the most exciting sport?

Write two reasons why this is not a good way to find out which sport people find most exciting.

3 The school cook wants to know the favourite baked potato fillings.
She plans to carry out a random sample of Year 11 students.

a Explain why the results could be biased.

There are 1000 students in the school. Describe how the cook could select a sample of 50 students, using

b random sampling

c systematic sampling.

4 A market researcher wants to find out how far people would travel to watch their football team.
He goes to a match and asks every tenth person who goes in through the gates.

a What sampling method is he using?

b Write a reason why his sample could be biased.

5 A machine producing rivets develops a fault.
Every 20th item it produces is sub-standard.
The quality control officer takes a sample of rivets produced by the machine. None are found to be substandard.
Which type of sampling method do you think was used – random or systematic? Give a reason for your answer.

6 Pritesh is organising a film show for Years 7–9. He wants to find out which film he should show.
There are 120 students in each year group. He wants to ask a sample of 90 students.

a Describe how he could take a random sample.

b Describe how he could take a systematic sample.

c Explain why either sampling method could give a biased sample.

D1.2 Stratified sampling

This spread will show you how to:
- Plan to minimise bias in a survey
- Use suitable methods of sampling to investigate a population, including random sampling

Keywords
Bias
Stratified sample

When a population is made up of different groups you can reduce **bias** by ensuring each group is represented in your sample.

- In a **stratified sample**, each group (or strata) is represented in the same proportion as in the whole population.

If the population has $\frac{2}{3}$ men and $\frac{1}{3}$ women, men should make up $\frac{2}{3}$ of the sample.

To choose a stratified sample

- Divide the population into groups.
- Work out the proportion, sample size as a fraction of population.
- Multiply the number in each group by this fraction, to find the number needed in the sample.
- Use random sampling to choose the correct sized sample from each group.

The groups could be by age, gender or some other criterion.

Example

A sports centre has 600 school-aged members.
Work out the number from each group needed for a stratified sample of 90 members.

Age	5–15	16–18
Girls	140	96
Boys	205	159

Total number: $140 + 96 + 205 + 159 = 600$
Proportion of sample size in population $= \frac{90}{600} = \frac{3}{20}$
Group sample sizes

p.70

Girls 5–15: $\frac{3}{20} \times 140 = 21$

Boys 5–15: $\frac{3}{20} \times 205 = 30.75 \approx 31$

Girls 16–18: $\frac{3}{20} \times 96 = 14.4 \approx 14$

Boys 16–18: $\frac{3}{20} \times 159 = 23.85 \approx 24$

For a sample of 32 taken from a population of 192, the fraction is $\frac{32}{192} = \frac{1}{6}$.

30.75 boys is not possible. Round up to 31.

Check: $21 + 31 + 14 + 24 = 90$

Example

A sixth form college has 400 students aged 16 to 18.
A sample is chosen, stratified by age and gender, of 50 of the 400 students.
The sample includes six 16-year-old female students.
Find the least possible total number of 16-year-old female students.

Sample size as a fraction of population $= \frac{50}{400} = \frac{1}{8}$
6 students in the sample means $\frac{1}{8} \times n \geqslant 5.5$
where n is the number of 16-year-old females in the college
so $n \geqslant 5.5 \times 8$
$n \geqslant 44$
Least total number of 16-year-old females is 44.

5.5 is the lowest number that would round to 6.

Exercise D1.2

Grade C/B

1 Shona carries out a survey of favourite sports stars at a gym club which has 178 girls and 42 boys as members.

She selects a random sample of 50. In her sample there are 20 girls and 30 boys.

Explain why this sample could be biased.

2 A company has a database of 4000 customers, of which 2750 are women and the rest men. The company wants to survey 800 customers. Describe how to select a sample stratified by gender.

3 A tennis club wants to find out what facilities to offer.

The club's membership is:

	18–30	31–50	over 50
Male	100	97	83
Female	140	133	147

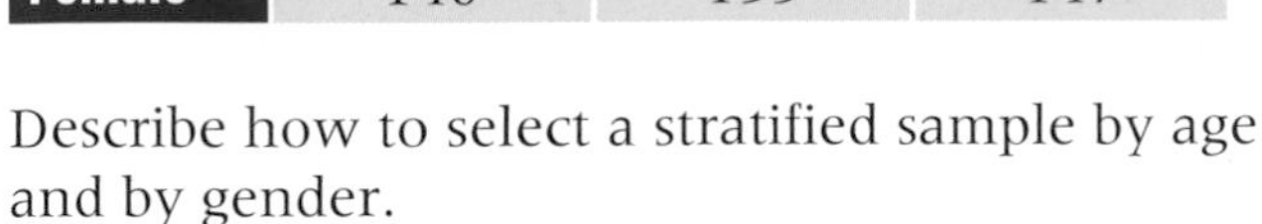

Describe how to select a stratified sample by age and by gender.

4 There are 600 students in Years 6, 7 and 8 at a middle school. This incomplete table shows information about the students.

	Boys	Girls
Year 6	96	81
Year 7	87	102
Year 8		

A sample is chosen, stratified by both age and gender, of 60 of the 600 students.

a Calculate the number of Year 6 boys and Year 6 girls to be sampled.

In the sample there are nine Year 8 boys.

b Work out the least possible number of Year 8 boys in the middle school.

5 Of 200 people on an activity holiday, 30 are adults, 85 are boys and 85 girls.

A representative sample is to be chosen.

Explain why it would be difficult to choose a stratified sample of 60.

D1.3 Averages and spread

This spread will show you how to:

- Find the mean, median, mode and range of a small data set

Keywords
Interquartile range
Lower quartile
Mean
Median
Mode
Range
Upper quartile

The three common types of average are

- **mode** – the value that occurs most often
- **median** – the middle value when the data are arranged in order
- **mean** – the total of all the values divided by the number of values.

Two measures of spread are

- **range** = highest − lowest value
- p.42 the **interquartile range** is the spread of the middle half of the data interquartile range (IQR) = **upper quartile** (UQ) − **lower quartile** (LQ).

For data arranged in ascending order:

- Lower quartile: value $\frac{1}{4}$ of the way along
- Upper quartile: value $\frac{3}{4}$ of the way along

Example

The table shows the lengths of words in a paragraph of writing.

For these data, work out the

a mode **b** median
c mean **d** range
e IQR.

Word length	Frequency
4	4
5	5
6	8
7	7
8	2
9	1

a Mode = 6 letters — Highest frequency = 8, for words with 6 letters.

b Total number of words = 27 — Add the values in the frequency column.

Middle value is the $\frac{1}{2}(27 + 1) = 14$th

This occurs in the 'Word length 6' category. — Adding frequencies, 4 + 5 = 9, 9 + 8 = 17, so the 14th is in the 3rd group, that is, 6 letter words.

Median = 6 letters

c Mean

Word length	Frequency	Word length × frequency
4	4	$4 \times 4 = 16$
5	5	$5 \times 5 = 25$
6	8	$6 \times 8 = 48$
7	7	$7 \times 7 = 49$
8	2	$8 \times 2 = 16$
9	1	$9 \times 1 = 9$
Total	**27**	**163**

Total of word length × frequency = 163 = total number of letters in paragraph.
Total number of words = 27

The mean value does not have to be an integer or a member of the original data set.

Mean = 163 ÷ 27 = 6.03 letters

d Range: Longest − shortest word length = 9 − 4 = 5 letters

e IQR = UQ − LQ

LQ = 5 $\frac{1}{4}(27 + 1) = 7$th value, in the category 'word length 5'.

UQ = 7 $\frac{3}{4}(27 + 1) = 21$st value, in the category 'word length 7'.

IQR = 7 − 5 = 2 letters

You can use the statistical mode of your scientific calculator to calculate the mean of a data set.

Exercise D1.3 — Grade C

1 Anya counted the contents of 12 boxes of paperclips.
Her results are shown in the table.
Work out the mean number of paperclips in a box.

Number of paperclips	Frequency
44	1
45	5
46	4
47	2

2 The tables give information about the length of words in four different paragraphs.
Copy each table, add an extra working column and find the
i mode **ii** median **iii** mean **iv** range **v** interquartile range.

a

Word length	Frequency
4	8
5	4
6	9
7	5
8	5

b

Word length	Frequency
3	4
4	6
5	9
6	6
7	4

c

Word length	Frequency
4	7
5	3
6	4
7	2
8	3
9	6
10	2

d

Word length	Frequency
3	5
4	5
5	7
6	8
7	6
8	4
9	2

3 A rounders team played 20 matches. The numbers of rounders scored in these matches are given in the table.

Number of rounders	Frequency
0	4
1	7
2	5
3	3
4	1

Work out the mean number of rounders scored.

AO3 Problem

4 Reuben counted the raisins in some mini-boxes. Five boxes contained 13 raisins, nine boxes contained 14 raisins and the remainder had 15 raisins.
The mean number of raisins per box was 14.1 (to 1 dp).
Find the number of boxes that contained 15 raisins.

Mean of combined data sets

This spread will show you how to:

- Calculate the mean for large and combined sets of data

Keywords
Mean

- **Mean** = $\frac{\text{Total of all values}}{\text{Number of values}}$

If you combine two data sets with known means, you can use these means to calculate the mean for the combined set.

Example

A group of 14 men and 22 women took their driving theory test.
The men's mean mark was 68%.
The women's mean mark was 74%.
Work out the mean mark for the whole group.

Men: Total of all marks: $68 \times 14 = 952$
Women: Total of all marks: $74 \times 22 = 1628$

Total of all marks for men and women: $952 + 1628 = 2580$
Mean mark for whole group $= 2580 \div 36 = 71.66666...$
$= 72\%$

Number in group: $14 + 22 = 36$

Example

In a survey, 50 bus passengers were asked how long they had had to wait for their bus.
30 passengers were asked on daytime buses, 20 on evening buses.

The mean waiting time for all 50 passengers was 14 minutes.
The mean waiting time for the daytime buses was 12 minutes.

Compare the mean waiting times for daytime and evening buses.

Total waiting time for all 50 passengers $50 \times 14 = 700$ minutes
Total waiting time for daytime buses $30 \times 12 = 360$ minutes
Total waiting time for evening buses $700 - 360 = 340$ minutes
Mean waiting time for evening buses $340 \div 20 = 17$ minutes
The mean waiting time for evening buses (17 minutes) was longer than for daytime buses (12 minutes).

Example

42 scuba divers took a diving exam.
16 divers were under 18 and 26 were adults.
The mean mark for the adults was p.
The mean mark for the under 18s was q.
Find an expression for the mean mark of all 42 divers.

Total of all adult marks $= 26p$
Total of all under 18s marks $= 16q$

Mean for all divers $= \frac{26p + 16q}{42}$

Exercise D1.4

Grade C/B

AO2 Functional Maths

1 A swimming club has 180 members of which 110 are male and 70 are female.
The mean daily training time for males is 86 minutes.
The mean daily training time for females is 72 minutes.
Find the mean daily training time all 180 members.

2 A college has 240 A-level students of which 135 are girls and 105 are boys.
A survey of home study time for one week found that the mean time for boys was 6.8 hours and the mean time for girls was 8.2 hours.
Find the mean time spent on home study for all 240 A-level students.

3 A driving school calculated that 40 of their pupils passed the driving test after a mean number of 24.5 lessons. Of the 40 pupils, the six that already had a motorcycle driving licence had an average of 16 lessons.
Find the mean number of driving lessons that the remaining pupils had before passing their driving test.

4 500 people were asked how many times they had visited a museum in the past year.
290 of them lived in a large cities, and the remainder lived in small towns.
The mean number of visits for the whole group was 4.8.
For the city people, the mean number of visits was 6.3.
Compare the mean number of museum visits for people in cities and small towns.
Suggest a reason for the difference.

5 From a survey of 800 in store and online customers, a retailer calculates the mean amount spent per customer as £74.63.
For the 575 in store customers, the mean was £71.44.
Compare the mean amounts spent by in store and online customers.
Suggest a reason for the difference.

6 36 women and 42 men take a fitness test.
The mean score for the women is c.
The mean score for the men is d.
Write an expression for the mean fitness score for the whole group.

AO3 Problem

7 A squash club has m male and f female members.
The average age for the males is x.
The average age for the females is y.
Write an expression for the average age for all the members of the squash club.

D1.5 Averages and spread for grouped data

This spread will show you how to:

- Use grouped frequency tables
- Calculate the modal class, class containing the median, and estimated mean and range for large data sets

Keywords
Estimate
Grouped frequency
Median
Midpoint
Modal class

You can present a large set of data in a **grouped frequency** table.

Lengths of runner beans in cm
3.9, 5.2, 7.6, 10.6, 12.4, 14.2

Length, x cm	Frequency
$0 < x \leqslant 5$	I
$5 < x \leqslant 10$	II
$10 < x \leqslant 15$	III

The classes must not overlap.

- For a grouped frequency table, the **modal class** is the class with thehighest frequency.
- You can work out which class contains the median.
- You can **estimate** the mean and range.

You do not know the actual data values, so you can only estimate the mean and range.

Example

The table shows the times taken, to the nearest minute, for commuters to solve a sudoku puzzle.

Find

a the modal class
b the class containing the median.

Calculate an estimate for

c the mean
d the range.

p.84

Time, t, minutes	Frequency
$5 < t \leqslant 10$	4
$10 < t \leqslant 15$	18
$15 < t \leqslant 20$	11
$20 < t \leqslant 25$	5
$25 < t \leqslant 30$	2

a Modal class is $10 < t \leqslant 15$ (10 to 15 minutes). The class with the highest frequency.

b Total number of commuters = 40 Add the frequency column.

Median is the $\frac{1}{2}(40 + 1) = 20\frac{1}{2}$th value Look for 20th and 21st values.

which is in the class $10 < t \leqslant 15$ (10 to 15 minutes).

c

Time, t, minutes	Frequency	Midpoint	Midpoint × frequency
$5 < t \leqslant 10$	4	7.5	30
$10 < t \leqslant 15$	18	12.5	225
$15 < t \leqslant 20$	11	17.5	192.5
$20 < t \leqslant 25$	5	22.5	112.5
$25 < t \leqslant 30$	2	27.5	55
Totals	40		615

Add two columns to your table. Use the **midpoint** as an estimate of the mean value for each class.

Estimate of mean = 615 ÷ 40 = 15.375 = 15 minutes

d Range = 30 − 5 = 25 minutes. Longest possible time = 30 minutes. Shortest possible time = 5 minutes.

Exercise D1.5 Grade C

1 The tables give information about the times taken for visitors to find their way through different mazes.
Copy each table, add extra working columns and find

i the modal class
ii the class containing the median
iii an estimate of the mean
iv an estimate of the range.

a **Hampton Court Maze**

Time, t, minutes	Frequency
$5 < t \leqslant 10$	5
$10 < t \leqslant 15$	8
$15 < t \leqslant 20$	6
$20 < t \leqslant 25$	4
$25 < t \leqslant 30$	2

b **Marlborough Maze**

Time, t, minutes	Frequency
$0 < t \leqslant 10$	1
$10 < t \leqslant 20$	9
$20 < t \leqslant 30$	7
$30 < t \leqslant 40$	5
$40 < t \leqslant 50$	2

c **The Maize Maze**

Time, t, minutes	Frequency
$5 < t \leqslant 10$	6
$10 < t \leqslant 15$	7
$15 < t \leqslant 20$	5
$20 < t \leqslant 25$	0
$25 < t \leqslant 30$	2
$30 < t \leqslant 35$	1

d **Leeds Castle Maze**

Time, t, minutes	Frequency
$5 < t \leqslant 15$	12
$15 < t \leqslant 25$	9
$25 < t \leqslant 35$	8
$35 < t \leqslant 45$	5
$45 < t \leqslant 55$	4
$55 < t \leqslant 65$	2

Functional Maths **A02**

2 Alfie kept a record of his monthly food bills for one year.
a Find the class interval that contains the median.
b Calculate an estimate for Alfie's mean monthly food bill.

Food bill, B, pounds	Frequency
$100 < B \leqslant 150$	5
$150 < B \leqslant 200$	4
$200 < B \leqslant 250$	2
$250 < B \leqslant 300$	1

3 A computer game and video store gives its staff a discount.
The table shows the amount spent by staff members in one month.

a Calculate an estimate for the mean amount of money spent.
b Write down the class interval that contains the median.
c The manager of the shop spent £250, which was not included in the table. If this amount is included, would your answer to **b** change? Explain your answer.

Monies spent, M, pounds	Frequency
$0 < M \leqslant 40$	11
$40 < M \leqslant 80$	8
$80 < M \leqslant 120$	8
$120 < M \leqslant 160$	9
$160 < M \leqslant 200$	3
$200 < M \leqslant 240$	1

D1.6 Statistical reports

This spread will show you how to:

- Understand statistical reports

Keywords
Average
Measure of spread
Modal class
Range

A report of a statistical investigation summarises the main findings from the data.
A report could include

- a summary of the data, using **averages** and **measures of spread**
- graphs of the data
- observations on the data
- interpretation of the data
- comparisons between data sets.

For example,
Observation: 'The **modal class** for the basketball players is higher than for the jockeys'
Interpretation: 'In general, the basketball players are taller than the jockeys.'

p.46

Example

This is an excerpt from a news report on the final football Premiership scores.

Teams in the Premiership this season gained an average of $52\frac{1}{2}$ points with a range of 67

These two graphs summarise the Premiership points score for two different years.

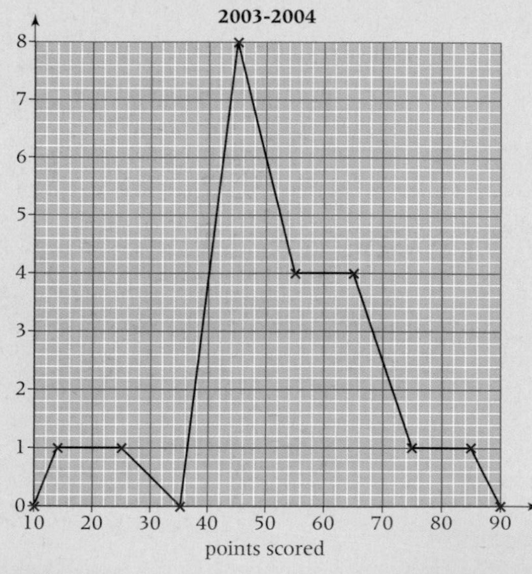

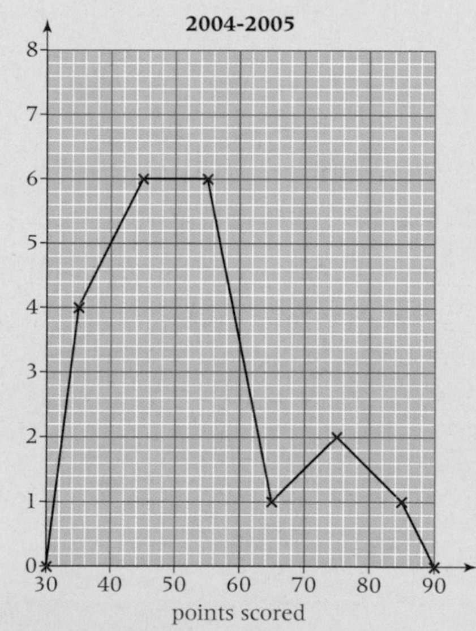

Which year was the news report commenting on?

Give a reason for your choice.

You can only estimate the range from a frequency polygon, as you don't have the exact data values.

Range of values for 2004–2005 cannot be 67 as $90 - 30 = 60 < 67$.

For the 2003–2004 graph the range could be 67.

Ranges could be between $90 - 10 = 80$ and $80 - 20 = 60$.

Exercise D1.6 Grade B

1 This news report was written about sales representatives of a small firm.

On average, sales representatives at the firm travel 77km per day. The range of distances travelled is 116km.

These two graphs were drawn to summarise the distances travelled by representatives at two different firms.

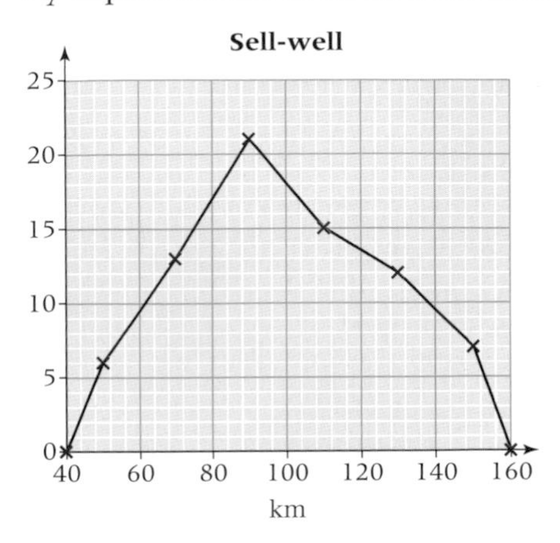

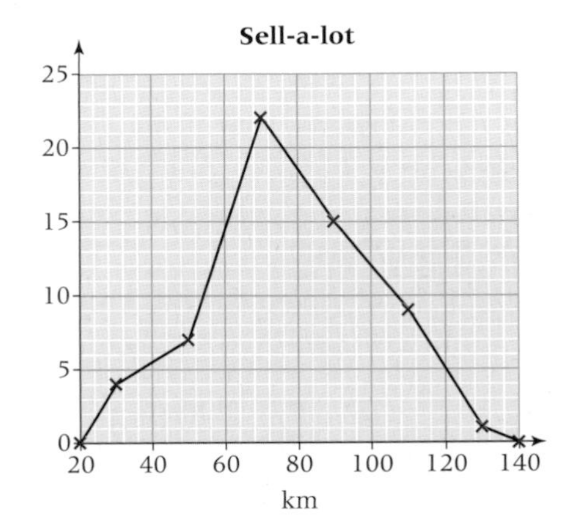

Use the graphs to identify which firm was being reported on. Give a reason for your choice.

2 The school magazine carried this report about a homework survey.

Year 7 students are doing twice as much homework every night as Year 10 students.

The two graphs were drawn with the report.

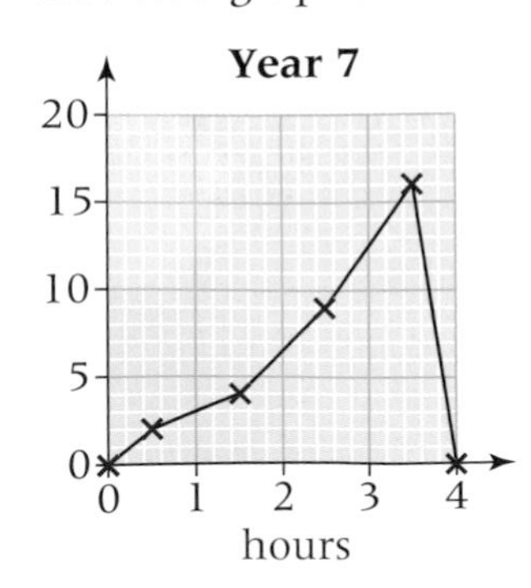

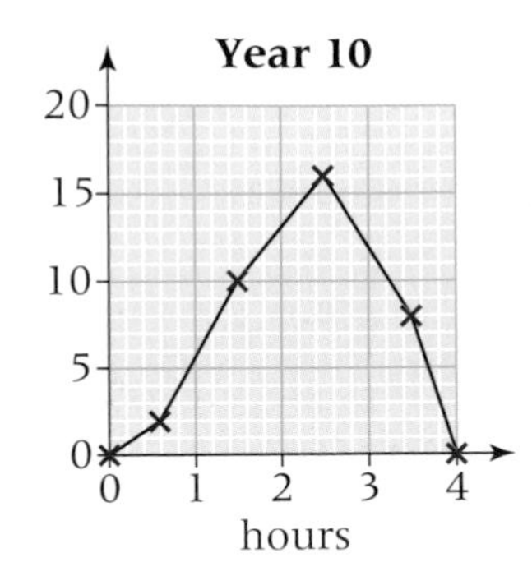

Use the graphs to comment on whether the report was valid or not. Give a reason for your choice.

D1 Summary

Check out

You should now be able to:

- Select and justify methods of sampling to investigate a population, including random and stratified sampling
- Identify possible sources of bias
- Calculate averages and range of data sets with discrete and continuous data
- Calculate the modal class, class containing the median and estimated mean for large sets of grouped data
- Interpret graphs and diagrams and draw conclusions

Worked exam question

258 students each study one of three languages.
The table shows information about these students.

Language studied

	German	French	Spanish
Male	45	52	26
Female	25	48	62

A sample is taken, stratified by the language studied and by gender, of 50 of the 258 students is taken.

a Work out the number of male students studying Spanish in the sample. (2)

b Work out the number of female students in the sample. (2)

(Edexcel Limited 2009)

a

$\frac{26}{258} \times 50 = 5$ male students

Show these calculations and round to the nearest whole student.

b

$25 + 48 + 62 = 135$

$\frac{135}{258} \times 50 = 26$ female students

OR

$\frac{25}{258} \times 50 = 5$ female German students

$\frac{48}{258} \times 50 = 9$ female French students

$\frac{62}{258} \times 50 = 12$ female Spanish students

$5 + 9 + 12 = 26$ female students

Show these calculations and round to the nearest whole student.

Exam questions

1 Bill recorded the times, in minutes, taken to complete his last 40 homeworks. This table shows information about the times.

Time (t minutes)	Frequency	
$20 \le t < 25$	8	
$25 \le t < 30$	3	
$30 \le t < 35$	7	
$35 \le t < 40$	7	
$40 \le t < 45$	15	

a Find the class interval in which the median lies. (1)

b Calculate an estimate of the mean time it took Bill to complete each homework. (4)

(Edexcel Limited 2006)

2 In a class of nine, the mean mark in a test for both boys and girls was 32. The three girls had a mean mark of 26.

What is the mean mark for the boys? (3)

A02

3 The table shows the number of boys and the number of girls in each year group at Springfield Secondary School.

There are 500 boys and 500 girls in the school.

Year group	Number of boys	Number of girls
7	100	100
8	150	50
9	100	100
10	50	150
11	100	100
Total	500	500

Azez took a stratified sample of 50 girls, by year group.

Work out the number of Year 8 girls in his sample. (2)

(Edexcel Limited 2007)

N1

Numbers and arithmetic

Many systems in the real world are governed by equations that allow you to predict their future behaviour. However in chaotic systems, such as double pendulums, electric circuits, the weather, *etc*., tiny changes in how you describe the initial state of the system can lead to dramatic changes in its future behaviour. For example, could the flap of a butterfly's wing in Brazil set off a tornado in Texas?

What's the point?

Nature has shown how important small errors in numbers can be. So mathematicians take great care to write numbers accurately and have developed special ways to write them without losing 'exactness'.

Check in

You should be able to

- **write numbers to a given degree of accuracy**

1 Write these numbers to **i** one decimal place (1dp)
ii two significant Figures (2sf)

a 38.5 **b** 16.08 **c** 103.88 **d** 0.082 **e** 0.38

- **do basic arithmetic**

2 For each calculation **i** write down a mental estimate
ii use a calculator to give an exact answer

a 18×53 **b** 3.77×89.5 **c** $3870 \div 79$

d $642 \div 28.7$ **e** $\dfrac{101 \times 23}{17.1 + 4.9}$ **f** $\dfrac{37 - 84}{0.13 \times 8}$

Orientation

What I need to know	What I will learn	What this leads to
KS3 Round numbers Basic arithmetic with decimals and fractions	■ Understand the effect of rounding ■ Make estimates and exact calculations with decimals and fractions ■ Convert between fractions and (recuring) decimals	**N3** Written methods for arithmetic **N5** Use a calculator efficiently understand the effects of rounding

Rich task

Continued fractions provide an alternative way to write any number, For example,

$$\sqrt{2} = 1.414213562\ldots = 1+\cfrac{1}{2+\cfrac{1}{2+\cfrac{1}{2+\ldots}}} = [1; 2, 2, 2, \ldots]$$

$$\frac{7}{30} = 0.2\dot{3} = \cfrac{1}{4+\cfrac{1}{3+\cfrac{1}{2}}} = [0; 4, 3, 2]$$

Investigate.

What are [1; 1, 2, 1, 2, 1, 2,...], [1; 1, 1, 1,...], [3; 7, 15, 1, 292, 1, 1, 2, 3,...]?

N1.1 Rounding

This spread will show you how to:

- Round to a given number of significant figures
- Estimate answers to calculations
- Use appropriate degrees of accuracy for solutions

Keywords
Digit
Round
Significant figures

Numbers are rounded to make them easier to handle.

- Numbers **round** up if the 'next' **digit** is a 5 or more.

Example

Round 104.458 to

a 2 dp **b** 1 dp **c** the nearest 10
d the nearest 100 **e** the nearest 1000.

a 104.46 **b** 104.5 **c** 100
d 100 **e** 0

- When rounding to a given number of **significant figures**, start counting at the first non-zero digit.

Example

Round to 2 significant figures

a 16.668 **b** 4.923 **c** 14 559
d 105 **e** 0.000 675

a 16.668
= 17 to 2 sf

b 4.923
= 4.9 to 2 sf

c 14 559
= 15 000 to 2 sf

d 105
= 110 to 2 sf

e 0.000 675
= 0.000 68 to 2 sf

You should round measurements to a realistic degree of accuracy. For example, if you are using a metre rule your measurements will be accurate to the nearest half centimetre.

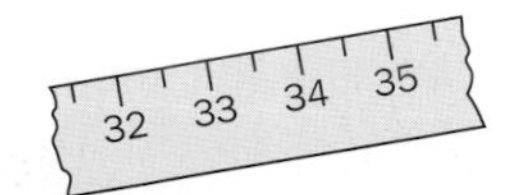

Example

A block of glass weighs 46 grams, and has a volume of 18.3 cm^3.
Estimate the density of the glass in g/cm^3.

Density = mass ÷ volume
Estimate: $46 \div 18.3 \approx 50 \div 20 = 2.5$
Density of glass ≈ 2.5 g/cm^3

p.26

To estimate the result of a calculation, round the numbers to 1 significant figure.

An exact calculation gives $46 \div 18.3 = 2.513661202...$ However since 46 g is only given to 2 sf it is only sensible to quote the answer to 2 sf: density = 2.5 g/cm^3.

Exercise N1.1 Grade D

1 Round each of these numbers to the nearest 10.
a 306 **b** 445 **c** 534.5 **d** 2174.9 **e** 56 685

2 Round each of these numbers to the nearest whole number.
a 43.475 **b** 0.508 **c** 23.486 **d** 31.503 **e** 44.499

3 Round each of these numbers
i to the nearest 100 **ii** to the nearest 1000.
a 1286 **b** 1094 **c** 49 **d** 508 **e** 41 450

4 Round each of these numbers to the accuracy shown.
a 0.31 (1 dp) **b** 0.735 (2 dp) **c** 0.1505 (3 dp) **d** 0.675 (2 dp)

5 Round each of these numbers to two significant figures.
a 0.0564 **b** 3.175 **c** 14.67 **d** 948

6 Round these numbers to the accuracy shown.
a 0.518 (2 sf) **b** 34 591 (3 sf) **c** 72 736 (3 sf) **d** 0.004 49 (2 sf)

7 Round these numbers to one significant figure.
a 352 **b** 0.632 **c** 0.0045 **d** 651 330

8 Evaluate these expressions giving your answers correct to two significant figures.
a $5 \div 12$ **b** $6 \div 7$ **c** 37×49 **d** 239×18

9 **a** Round all the numbers to one significant figure and write a calculation that you could do in your head to estimate the answer to each of these calculations.
i $451 \div 18$ **ii** $17 + 38 \div 4$ **iii** $1736 + 33 \times 48$ **iv** $1.9173 + 3.2013$
b Now write the answers to the calculations you wrote in part **a**.
c Use a calculator to find an exact answer for each calculation in part **a**. For each one, write a sentence to say how well the calculator result agrees with the estimate that you wrote in part **b**.

10 **a** Write a calculation that you can do in your head to estimate the answers to these calculations.
i $258 + 362$ **ii** $64 \div 27$ **iii** 62.7×211.8 **iv** $96.7 - 64.8$
b Explain carefully whether each of the calculations that you wrote in part **a** will produce an overestimate or an underestimate of the actual result.
c Use a calculator to find the exact answers, and check your answers to part **b**.

N1.2 Upper and lower bounds

This spread will show you how to:

- Understand that data and measurements are not always exact
- Calculate the upper and lower bounds of data

Keywords
Lower bound
Upper bound

- Measurements are not exact. Their accuracy depends on the precision of the measuring instrument and the skill of the person making the measurement.

The height of a tree is given as 5 metres, correct to the nearest metre.

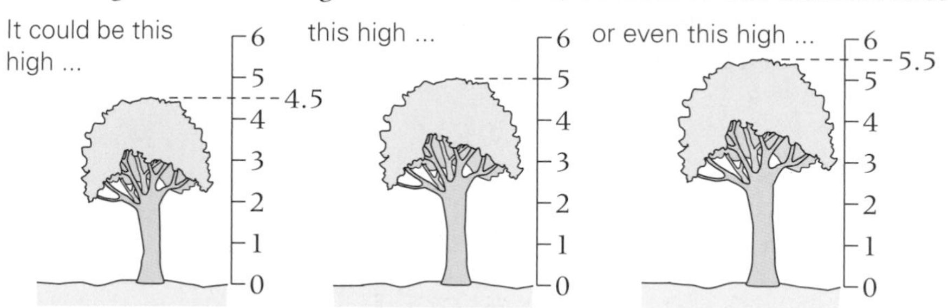

- 4.5 m is the **lower bound**.
- 5.5 m is the **upper bound**.

Compare this to data obtained by counting, which are exact.

The upper bound is not actually included in the range of possible values.

You always use the lower bound and the upper bound when stating the range of possible values of measurements or data.

Example

The mass of a meteorite is given as 235.6 g.
Find the lower and upper bounds of the mass.

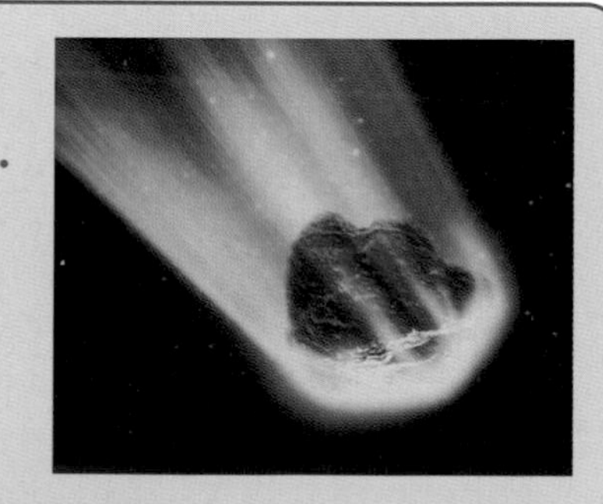

The mass is given correct to the nearest 0.1 g.
Lower bound of the actual mass is 235.55 g.
Upper bound is 235.65 g.
If the mass of the meteorite is m then

$$235.55 \leq m < 235.65$$

Example

A student was asked to give the upper and lower bounds for a measurement given as 4.5 seconds. He gave the lower bound as 4.45 seconds, and the upper bound as 4.5499999 seconds.
What is wrong with his answer?

The lower bound is correct, but the upper bound should be 4.55 seconds. The student was trying to show that the maximum possible value is 'a bit less than 4.55 seconds'. However, this is never necessary; the upper bound is not a possible value of the measurement.

Exercise N1.2 Grade B/A

1 Each of these measurements was made correct to one decimal place. Write the upper and lower bounds for each measurement.

a 5.8 m **b** 16.5 litres **c** 0.9 kg **d** 6.3 N
e 10.1 s **f** 104.7 cm **g** 16.0 km **h** 9.3 m/s

2 Find the upper and lower bounds of these measurements, which were made to varying degrees of accuracy.

a 6.7 m **b** 7.74 litres **c** 0.813 kg **d** 6 N
e 0.001 s **f** 2.54 cm **g** 1.162 km **h** 15 m/s

3 Find the maximum and minimum possible total weight of

a 12 boxes, each of which weighs 14 kg, to the nearest kilogram
b 8 parcels, each weighing 3.5 kg.

4 Find the upper and lower bounds of these measurements, which are correct to the nearest 5 mm.

a 35 mm **b** 40 mm **c** 110 mm **d** 4.5 cm

5 A box contains 38 nails, each weighing 12 g, to the nearest gram. Alex calculated the maximum and minimum possible total weight of the nails. He wrote

- Maximum $= 38.5 \times 12.5\,\text{g} = 481.25\,\text{g}$
- Minimum $= 37.5 \times 11.5\,\text{g} = 431.25\,\text{g}$

These calculations are incorrect. Explain why, and correct them.

AO2 Functional Maths

6 A lift can hold up to six people, with a maximum safe load of 460 kg. A group of six people are waiting for the lift. Their weights are 85 kg, 96 kg, 63 kg, 73 kg, 68 kg and 73 kg (all measured to the nearest kilogram). Is it possible that the total weight of the group exceeds the limit for the lift? Show your working.

7 A completed jigsaw puzzle measures 24 cm by 21 cm (both to the nearest centimetre). Find the maximum and minimum possible area of the puzzle. Show your working.

8 A bag contains 98 g of sugar, then 34 g of sugar are taken out. Find the upper and lower bounds for the amount of sugar remaining in the bag, if the measurements are all correct to the nearest gram.

AO3 Problem

9 Unladen, a lorry weighs 2.3 tonnes, measured to the nearest 100 kg. The lorry is loaded with crates, each weighing 250 kg, correct to the nearest 10 kg. On its journey the lorry crosses a bridge with a maximum safe load of 15 tonnes. What is the maximum number of crates that the driver can load onto the lorry?

N1.3 Multiplying and dividing

This spread will show you how to:

- Multiply and divide by a number between 0 and 1

Keywords
Divide
Multiply
Reciprocal

Multiplying doesn't always make numbers bigger.
Similarly, **dividing** doesn't always make them smaller.

p.138

Multiplying by 1
No change

Multiplying by **numbers < 1** make positive numbers smaller	Multiplying by **numbers > 1** make positive numbers bigger
0 Dividing by **numbers < 1** make positive numbers bigger	1 Dividing by **numbers > 1** make positive numbers smaller

Dividing by 1
No change

Dividing by a number is the same as multiplying by its **reciprocal**, for example

$$9 \div 4 = 9 \times \frac{1}{4} = 2\frac{1}{4}$$

A number multiplied by its reciprocal is equal to 1.

Multiplying by a number is the same as dividing by its reciprocal, for example

$$2 \times 4 = 2 \div \frac{1}{4} = 8$$

There are 8 quarters in 2.

Example

State whether a positive number will be larger, smaller or the same size when it is

a multiplied by 1.3 **b** divided by 0.3 **c** multiplied by 0.4
d divided by 1.5 **e** multiplied by 1.

a Larger **b** Larger **c** Smaller
d Smaller **e** The same

Example

Use reciprocals to give an equivalent division for each multiplication.

a 7×2 **b** 1.8×5 **c** 29×1 **d** 15×0.5 **e** 46×0.2

a $7 \div 0.5$ **b** $1.8 \div 0.2$ **c** $29 \div 1$ **d** $15 \div 2$ **e** $46 \div 5$

Example

Use reciprocals to give an equivalent multiplication for each division.

a $17 \div 4$ **b** $5.9 \div 8$ **c** $44 \div 1$ **d** $2.9 \div 0.25$ **e** $78 \div 0.4$

a 17×0.25 **b** 5.9×0.125 **c** 44×1 **d** 2.9×4 **e** 78×2.5

$0.4 = \frac{2}{5}$, so the reciprocal of $0.4 = \frac{5}{2} = 2.5$

Exercise N1.3 Grade D

1 State whether a positive number will give you an answer that is larger, smaller or the same size as the original number, when you

a multiply by 8 **b** divide by 7 **c** multiply by $\frac{1}{4}$
d multiply by 0.01 **e** divide by 0.75 **f** divide by $\frac{3}{5}$
g multiply by 0 **h** divide by 1.

2 Use reciprocals to write each of these multiplications as a division.

a 8×0.5 **b** 10×0.2 **c** 12×0.25 **d** 18×0.1

3 Write an equivalent division for each of these multiplications.

a $15 \times \frac{1}{5}$ **b** $28 \times \frac{1}{4}$ **c** $10 \times \frac{2}{5}$ **d** $72 \times \frac{1}{8}$

4 Write an equivalent multiplication for each of these divisions.

a $18 \div \frac{1}{2}$ **b** $24 \div 0.25$ **c** $8 \div \frac{2}{3}$ **d** $5.5 \div 0.1$
e $5.9 \div 1$ **f** $66 \div \frac{3}{5}$ **g** $7 \div \frac{7}{10}$ **h** $8 \div 1.25$

5 Calculate these, without using a calculator. Show your method.

a $12 \div 0.5$ **b** $36 \div 0.25$ **c** $13 \div 0.2$ **d** $7.2 \div 0.1$
e 8×0.25 **f** 45×0.2 **g** $24 \times \frac{1}{3}$ **h** $20 \div \frac{2}{3}$

6 Evaluate these without using a calculator. Show your method.

a $9 \div \frac{1}{3}$ **b** $12 \div \frac{1}{5}$ **c** $8 \div 0.5$ **d** 84×0.25
e 35×0.2 **f** $60 \div \frac{2}{3}$ **g** $16 \div 0.4$ **h** $15 \div 2.5$

7 Find the value of each expression. Do not use a calculator.

a $24 \div \frac{3}{2}$ **b** $30 \div \frac{5}{4}$ **c** $15 \div 1.25$ **d** $24 \div 1.5$
e $35 \div \frac{7}{6}$ **f** $28 \div 1\frac{1}{6}$ **g** $45 \div 1\frac{1}{4}$ **h** $32 \div 1\frac{1}{7}$

8 For any pair of numbers, the greater number is the one further to the right on the number line. Write the greater number in each of these.

a 2 and 2.5 **b** 2 and −2 **c** −3 and 3.5 **d** 0 and −4.2
e 1 and −25 **f** −7 and −23 **g** 4.8 and −38 **h** −3.9 and −3.85

9 Explain whether each statement is true or false, giving examples to support your argument.

a When you divide a number by 2, the answer is always smaller than the number you started with.
b Dividing by $\frac{1}{2}$ is always the same as doubling a number.
c When you multiply by 5, the answer is always bigger than the number you started with.
d When you multiply by 10, the answer will always be different from the original number.

AO3 Problem

10 a Explain the effect of repeatedly multiplying a number by +0.9.
b Explain the effect of repeatedly multiplying a number by −0.9.

N1.4 Estimation

This spread will show you how to:

- Check and estimate answers to problems
- Give answers to an appropriate degree of accuracy

Keywords
Approximation
Estimate
Significant figures
Standard form

- You can use **approximations** to one **significant figure** to make **estimates**.

You need to be careful when estimating powers.
For example, 1.3 is quite close to 1, but 1.3^7 is not close to 1^7.

Example

Estimate the value of these calculations.

a $\dfrac{563 + 1.58}{327 - 4.72}$ **b** $\dfrac{3.27 \times 4.49}{1.78^2}$ **c** $\dfrac{\sqrt{2485}}{1.4^3}$ **d** $\dfrac{2.45^3}{2.5 - 2.4}$

a Estimate: $600 \div 300 = 2$

b 1.78^2 is 'a bit more than 3', so it cancels with 3.27 to give an estimate of 4.5.

c $2485 \approx 2500$ and $\sqrt{2500} = 50$
$1.4^2 \approx 2$, so $1.4^3 \approx 1.4 \times 2 = 2.8 \approx 3$
Estimate: $50 \div 3 \approx 17$

d $2.45 \approx 2.5$ and $25^2 = 625$, so $2.5^2 \approx 6$
$2.45^3 \approx 6 \times 2.5 = 15$
$2.5 - 2.4 = 0.1$, so estimate is: $15 + 0.1 = 150$

In part **a**, ignore the relatively small amounts added and subtracted.

In part **b**, $\sqrt{3} = 1.73$ to 2 dp.

In part **d**, to 1 sf the denominator is $2 - 2 = 0$, which is not possible.

p.182

- You can use **standard form** to estimate calculations involving very large or very small numbers. For example,

$$\begin{aligned} 5130 \times 0.000\,178 &\approx (5 \times 10^3) \times (2 \times 10^{-4}) \\ &= 5 \times 2 \times 10^3 \times 10^{-4} \\ &= 10 \times 10^{-1} = 1 \end{aligned}$$

Example

Estimate the value of $\dfrac{4217 \times 0.0625}{23\,563}$.

Writing the calculation in standard form.

$(4.217 \times 10^3) \times (6.25 \times 10^{-2}) \div (2.3563 \times 10^4)$
$\approx (4 \times 10^3) \times (6 \times 10^{-2}) \div (2 \times 10^4)$
$= 12 \times 10^{-3}$
$= 1.2 \times 10^{-2}$
$= 0.012$

For standard form, the multiplier must be between 1 and 10.

Exercise N1.4 Grade C/B

1 Estimate answers to these calculations.

a $4.88 + 3.07$ **b** $216 + 339$

c $0.0049 + 0.00302$ **d** $43.89 - 28.83$

e 3.77×0.85 **f** $44.66 \div 0.89$

2 Estimate these square roots mentally, to 1 decimal place.

a $\sqrt{2}$ **b** $\sqrt{8}$

c $\sqrt{10}$ **d** $\sqrt{15}$

e $\sqrt{20}$ **f** $\sqrt{26}$

g $\sqrt{32}$ **h** $\sqrt{45}$

i $\sqrt{70}$ **j** $\sqrt{85}$

Use a calculator to check your estimates.

3 Explain why approximating the numbers in these calculations to 1 significant figure would *not* be an appropriate method for estimating the results of the calculations.

a $\dfrac{5.39 + 4.72}{0.53 - 0.46}$ **b** $(2.45 - 0.96)^8$

c $(1.52 - 1.49)^2$

4 Use approximations to estimate the value of each of these calculations. You should show all your working.

a $\dfrac{317 \times 4.22}{0.197}$ **b** $\dfrac{4.37 \times 689}{0.793}$

c $\dfrac{4.75 \times 122}{522 \times 0.38}$ **d** $4.8^3 - 8.5^2$

e $\dfrac{9.32 - 3.85}{0.043 - 0.021}$ **f** $7.73 \times \left(\dfrac{0.17 \times 234}{53.8 - 24.9}\right)$

5 Find approximate values for these calculations. Show your working.

a $\dfrac{48.75 \times 4.97}{10.13^2}$ **b** $\sqrt{\dfrac{305.3^2}{913}}$

c $\dfrac{\sqrt{9.67 \times 8.83}}{0.087}$ **d** $\dfrac{6.8^2 + 11.8^2}{\sqrt{47.8 \times 52.1}}$

e $\dfrac{(23.4 - 18.2)^2}{3.2 + 1.8}$ **f** $\sqrt{\dfrac{2.85 + 5.91}{0.17^2}}$

6 Use standard form approximations to find an estimate for each of these calculations. Show your working.

a $4800 \div 465$ **b** $7326 \div 0.069$

c $\dfrac{83550 \times 0.039}{4378}$ **d** $\dfrac{653 \times 0.415}{0.07 \times 0.38}$

e $\dfrac{735 + 863}{0.06 \times 0.85}$ **f** $\dfrac{3400 \times 475}{(28.5 + 36.9)^2}$

Arithmetic with numbers written in standard form is examined in unit 3.

N1.5 Fraction calculations

This spread will show you how to:

- Add, subtract, multiply and divide with fractions

Keywords
Cancel
Common denominator
Common factors
Multiplicative inverse

p.138
You need to be able to:

- Add and subtract fractions.
  You can add or subtract fractions if they have the same denominator.

p.130

$$\frac{3}{5} + \frac{1}{4} = \frac{12}{20} + \frac{5}{20} = \frac{17}{20}$$

$$3\frac{1}{4} - 2\frac{5}{8} = 1 + \frac{1}{4} - \frac{5}{8} = 1 + \frac{2}{8} - \frac{5}{8} = 1 - \frac{3}{8} = \frac{5}{8}$$

Subtract the whole numbers first.

Use a **common denominator**.

- Multiply fractions.

p.128

$$\frac{2}{3} \times \frac{5}{8} = \frac{10}{24} = \frac{5}{12}$$

Multiply numerators together and multiply denominators together.

- Divide fractions.
  Dividing by any number is the same as multiplying by its **multiplicative inverse**.

$$\frac{3}{4} \div \frac{2}{5} = \frac{3}{4} \times \frac{5}{2} = \frac{15}{8} = 1\frac{7}{8}$$

$\frac{5}{2}$ is the inverse of $\frac{2}{5}$.

Example

Work out each of these calculations.

a $\frac{3}{5} + \frac{11}{16}$ **b** $\frac{2}{7} - \frac{1}{4}$ **c** $\frac{3}{5} \div 8$ **d** $\frac{4}{9} \div \frac{1}{3}$ **e** $\frac{3}{8} \times \frac{5}{9}$ **f** $\frac{3}{5} \div \frac{7}{10}$

a $\frac{3}{5} + \frac{11}{16} = \frac{48}{80} + \frac{55}{80} = \frac{48 + 55}{80} = \frac{103}{80} = 1\frac{23}{80}$

The LCM of 5 and 16 is 80.

b $\frac{2}{7} - \frac{1}{4} = \frac{8}{28} - \frac{7}{28} = \frac{1}{28}$

The LCM of 7 and 4 is 28.

c $\frac{3}{5} \div 8 = \frac{3}{5} \times \frac{1}{8} = \frac{3 \times 1}{5 \times 8} = \frac{3}{40}$

The multiplicative inverse of 8 is $\frac{1}{8}$.

d $\frac{4}{9} \div \frac{1}{3} = \frac{4}{9} \times \frac{3}{1} = \frac{4 \times 3}{9 \times 1} = \frac{12}{9} = \frac{4}{3} = 1\frac{1}{3}$

The multiplicative inverse of $\frac{1}{3}$ is $\frac{3}{1}$ (which is 3).

e $\frac{{}^1\cancel{3}}{8} \times \frac{5}{\cancel{9}_3} = \frac{1}{8} \times \frac{5}{3} = \frac{5}{24}$

Notice how you can **cancel common factors** before multiplying.

f $\frac{3}{5} \div \frac{7}{10} = \frac{3}{{}_1\cancel{5}} \times \frac{\cancel{10}^2}{7} = \frac{6}{7}$

Again, cancel common factors before multiplying.

Exercise N1.5 — Grade D/C

1 Add these fractions.

a $\frac{1}{2}+\frac{1}{2}$ **b** $\frac{1}{2}+\frac{1}{4}$ **c** $\frac{1}{2}+\frac{1}{3}$ **d** $\frac{1}{5}+\frac{1}{10}$ **e** $\frac{1}{3}+\frac{1}{4}$

2 Subtract these fractions.

a $\frac{2}{3}-\frac{1}{3}$ **b** $\frac{1}{2}-\frac{1}{6}$ **c** $\frac{1}{3}-\frac{1}{4}$ **d** $\frac{2}{5}-\frac{1}{3}$ **e** $\frac{2}{3}-\frac{3}{7}$

3 Do these calculations with mixed numbers.

a $1\frac{1}{2}+\frac{1}{4}$ **b** $2\frac{1}{3}-\frac{2}{3}$

c $1\frac{1}{5}+2\frac{1}{10}$ **d** $3\frac{1}{4}-\frac{1}{8}$

4 Do these multiplications and divisions.

a $\frac{3}{8}\times 4$ **b** $2\times\frac{2}{5}$

c $\frac{7}{8}\div 2$ **d** $\frac{15}{16}\div 3$

5 Calculate

a $\frac{2}{3}\times\frac{1}{3}$ **b** $\frac{5}{8}\div\frac{1}{4}$

c $\frac{5}{9}\times\frac{1}{5}$ **d** $\frac{9}{20}\div\frac{1}{5}$

6 Calculate

a $\frac{2}{5}\div\frac{2}{3}$ **b** $\frac{3}{7}\times\frac{7}{8}$

c $\frac{3}{7}\times\frac{2}{5}$ **d** $\frac{4}{9}\div\frac{5}{6}$

7 Evaluate

a $2\frac{1}{4}\div\frac{3}{4}$ **b** $3\frac{1}{2}\div\frac{5}{8}$

c $2\frac{1}{4}\times 3\frac{2}{3}$ **d** $1\frac{1}{8}\times 2\frac{3}{4}$

8 Calculate

a $3\frac{7}{8}+2\frac{1}{4}$ **b** $3\frac{7}{8}-3\frac{1}{4}$

c $5\frac{1}{2}\times 1\frac{7}{8}$ **d** $2\frac{1}{2}\div 3\frac{3}{4}$

9 Calculate

a $\dfrac{\frac{3}{4}+\frac{1}{2}}{\frac{5}{8}-\frac{1}{4}}$ **b** $\left(\frac{3}{4}-\frac{1}{8}\right)\left(\frac{3}{4}+\frac{1}{8}\right)$

c $\dfrac{1\frac{1}{2}-\frac{7}{8}}{2\frac{3}{4}-1\frac{1}{6}}$ **d** $\left(2\frac{3}{5}-1\frac{3}{4}\right)\left(1\frac{7}{8}+3\frac{1}{4}\right)$

Functional Maths — A02

10 Tom planted 48 daffodils. On Easter Day, $\frac{2}{3}$ of them were in bloom.
How many daffodils were blooming?

11 Ed irons his shirts every Sunday, and it takes him $5\frac{3}{4}$ minutes to do one shirt.
Last week he washed all his shirts and spent 46 minutes on his ironing. How many shirts does he have?

N1.6 Fractions and decimals

This spread will show you how to:

- Recognise terminating and recurring decimals when written as fractions
- Convert between fractions, decimals and percentages

Keywords
Recurring
Terminating

- To convert a fraction to a decimal divide the numerator by the denominator.

Example

Write these fractions as decimals.

a $\frac{7}{8}$ **b** $\frac{2}{3}$

a $\frac{7}{8} = 7 \div 8 = 8\overline{)7.000}$ giving 0.875

0.875 is a **terminating** decimal.

b $\frac{2}{3} = 2 \div 3 = 3\overline{)2.000\ldots}$ giving $0.666\ldots$

$0.666\ldots = 0.\dot{6}$ is a **recurring** decimal.
The dot shows the recurring digit.

To decide if a fraction will be a terminating or a recurring decimal:

- If the only factors of the denominator are 2 and/or 5 or combinations of 2 and 5 then the fraction will be a terminating decimal.
- If the denominator has any factors other than 2 and/or 5 then the fraction will be a recurring decimal.

$\frac{7}{8}$ is a terminating decimal as $8 = 2 \times 2 \times 2$.

$\frac{5}{6}$ is a recurring decimal as $6 = 2 \times 3$.

To convert a terminating decimal to a fraction write the decimal as a fraction with the denominator as a power of ten.

$$0.385 = \frac{385}{1000}$$

Then cancel common factors: $\frac{385}{1000} = \frac{77}{200}$

Example

Write these as fractions. **a** 0.3 **b** 0.56 **c** 0.625

a $0.3 = \frac{3}{10}$ **b** $0.56 = \frac{56}{100} = \frac{14}{25}$ **c** $0.625 = \frac{625}{1000} = \frac{5}{8}$

To convert a recurring decimal to a fraction use this method.

Write $0.\dot{3}\dot{6}$ as a fraction. Put $0.\dot{3}\dot{6} = x$

$$\begin{aligned} \text{Then } 100x &= 36.\dot{3}\dot{6} \\ x &= 0.\dot{3}\dot{6} \\ 99x &= 36 \quad \text{Subtract one from the other.} \\ x &= \frac{36}{99} = \frac{4}{11} \end{aligned}$$

If there is one recurring digit, find 10x. If there are three, find 1000x.

Example

Write these as fractions. **a** $0.\dot{3}$ **b** $0.\dot{1}\dot{6}$ **c** $0.\dot{1}2\dot{3}$

a
$$\begin{aligned} 10x &= 3.\dot{3} \\ x &= 0.\dot{3} \\ 9x &= 3 \\ x &= \frac{3}{9} = \frac{1}{3} \end{aligned}$$

b
$$\begin{aligned} 100x &= 16.\dot{1}\dot{6} \\ x &= 0.\dot{1}\dot{6} \\ 99x &= 16 \quad \text{(Subtract)} \\ x &= \frac{16}{99} \end{aligned}$$

c
$$\begin{aligned} 1000x &= 123.\dot{1}2\dot{3} \\ x &= 0.\dot{1}2\dot{3} \\ 999x &= 123 \\ x &= \frac{123}{999} = \frac{41}{333} \end{aligned}$$

The dots tell you that the group of digits, 123, recurs.

Exercise N1.6 Grade B/A

1 Write the decimal equivalents of these fractions.

a $\frac{1}{2}$ **b** $2\frac{2}{5}$ **c** $\frac{3}{20}$ **d** $\frac{1}{25}$

2 Use a written method to convert these fractions to decimals. Show your working.

a $\frac{3}{8}$ **b** $\frac{4}{5}$ **c** $\frac{1}{16}$ **d** $\frac{3}{25}$

3 Use a calculator to check your answers to questions **1** and **2**.

4 Write these fractions as recurring decimals, using the 'dot' notation. Show your working – do not use a calculator.

a $\frac{1}{3}$ **b** $\frac{2}{3}$ **c** $\frac{1}{6}$ **d** $\frac{1}{9}$ **e** $\frac{5}{6}$

5 Write each of the fractions $\frac{1}{7}, \frac{2}{7}, \frac{3}{7}, \frac{4}{7}, \frac{5}{7},$ and $\frac{6}{7}$ as a recurring decimal, using 'dot' notation. Describe any patterns that you see in your results.

6 Convert these decimals to fractions.

a 0.5 **b** 0.3 **c** 0.75 **d** 0.95 **e** 0.65

7 Convert each of these recurring decimals to a fraction in its simplest form. Show your working.

a $0.1\dot{1}$ **b** $0.2\dot{2}$ **c** $0.\dot{1}\dot{5}$ **d** $0.\dot{1}2\dot{5}$ **e** $0.\dot{2}1\dot{6}$

8 Write these decimals as fractions. Show your working.

a $0.2\dot{1}$ **b** $0.7\dot{2}$ **c** $0.8\dot{2}\dot{7}$ **d** $0.6\dot{3}2\dot{1}$ **e** $0.81\dot{7}\dot{5}$

9 **a** Which one of these is a recurring decimal?

$\frac{18}{25}$ $\frac{19}{20}$ $\frac{8}{11}$ $\frac{9}{18}$

b Write $\frac{7}{9}$ as a recurring decimal.

c You are told that $\frac{1}{54} = 0.0\dot{1}8\dot{5}$

Write $\frac{4}{54}$ as a recurring decimal.

AO3 Problem

10 **a** Prove that $0.\dot{5}\dot{7} = \frac{19}{33}$.

b Hence, or otherwise, write the decimal number $0.3\dot{5}\dot{7}$ as a fraction.

N1.7 Fractions, decimals and percentages

This spread will show you how to:
- Convert between fractions, decimals and percentages
- Order fractions, decimals and percentages

Keywords
Ascending
Denominator
Numerator
Order

- To convert a decimal to a percentage, multiply by 100%.
  $0.65 = 0.65 \times 100\% = 65\%$
- To convert a percentage to a decimal, divide by 100.
  $18.\dot{3}\% = 18.\dot{3} \div 100 = 0.18\dot{3}$
- To convert a fraction to a percentage, divide the **numerator** by the **denominator**, and then multiply by 100%.
  $\frac{5}{8} = 5 \div 8 \times 100\% = 0.625 \times 100\% = 62.5\%$

Example

a Convert these decimals to percentages.

i 0.74 **ii** 1.315 **iii** $0.\dot{8}$ **iv** $0.2\dot{8}1\dot{5}$

b Convert these percentages to decimals.

i 67.5% **ii** 255% **iii** 0.1%

c Convert these fractions to percentages.

i $\frac{3}{7}$ **ii** $\frac{4}{9}$ **iii** $\frac{9}{40}$ **iv** $\frac{13}{25}$

a **i** $0.74 = 0.74 \times 100\% = 74\%$ **ii** $1.315 = 1.315 \times 100\% = 131.5\%$

iii $0.\dot{8} = 0.\dot{8} \times 100\% = 88.\dot{8}\% = 88.9\%$ to 1 dp

iv $0.2\dot{8}1\dot{5} = 0.2815815\ldots \times 100\% = 28.158158\ldots\% = 28.2\%$ to 1 dp

b **i** $67.5\% = 0.675$ **ii** $255\% = 2.55$ **iii** $0.1\% = 0.001$

c **i** $\frac{3}{7} = 3 \div 7 = 0.\dot{4}2857\dot{1} = 0.\dot{4}2857\dot{1} \times 100\% = 42.\dot{8}5714\dot{2}\% = 43\%$

ii $\frac{4}{9} = 4 \div 9 = 0.\dot{4} = 0.\dot{4} \times 100\% = 44.\dot{4}\% = 44\%$

iii $\frac{9}{40} = 9 \div 40 = 0.225 = 0.225 \times 100\% = 22.5\% = 23\%$

iv $\frac{13}{25} = 13 \div 25 = 0.52 = 0.52 \times 100\% = 52\%$

In part **c**, answers are given to the nearest whole number.

Example

Write these quantities in **ascending** order.

a $\frac{7}{20}, \frac{3}{8}, \frac{1}{5}, \frac{3}{10}$ **b** $35\%, \frac{2}{7}, \frac{4}{15}, 0.347$

a Rewrite the list as decimals: 0.35, 0.375, 0.2, 0.3

The original list, in ascending order, is: $\frac{1}{5}, \frac{3}{10}, \frac{7}{20}, \frac{3}{8}$

b The decimal equivalents are $0.35, 0.\dot{2}8571\dot{4}, 0.2\dot{6}, 0.347$

The original list, in ascending order, is: $\frac{4}{15}, \frac{2}{7}, 0.347, 35\%$

Exercise N1.7 — Grade C/B

1 Convert these decimals to percentages.

a 0.35 **b** 0.607 **c** 0.995 **d** 1.00
e 2.15 **f** 0.00056 **g** 17 **h** 0.101

2 Convert these recurring decimals to percentages.

a $0.5\dot{5}$ **b** $0.3\dot{4}$ **c** $0.\dot{7}\dot{5}$ **d** $0.51\dot{2}\dot{8}$
e $0.4\dot{3}\dot{7}$ **f** $0.83\dot{8}$ **g** $0.\dot{8}3\dot{8}$ **h** $1.0\dot{5}$

3 Convert these percentages to decimals.

a 22% **b** 18.5% **c** $55.5\dot{5}\%$ **d** $35.\dot{5}\%$
e $6.\dot{5}\dot{6}\%$ **f** $61.\dot{4}\dot{9}\%$ **g** $54.\dot{4}\dot{6}\%$ **h** $152.\dot{2}\%$

4 Convert these fractions to percentages. You should be able to do these without a calculator.

a $\frac{4}{5}$ **b** $\frac{3}{20}$ **c** $\frac{7}{8}$ **d** $\frac{3}{4}$
e $\frac{7}{25}$ **f** $\frac{3}{16}$ **g** $\frac{9}{20}$ **h** $\frac{7}{50}$

5 Convert these fractions to percentages. You may use a calculator.

a $\frac{2}{3}$ **b** $\frac{1}{15}$ **c** $\frac{3}{7}$ **d** $\frac{5}{6}$
e $\frac{7}{9}$ **f** $\frac{4}{11}$ **g** $\frac{1}{12}$ **h** $\frac{2}{15}$

6 Rewrite each set of fractions with a common denominator.
Then use your answers to rewrite each list in ascending order.

a $\frac{2}{3}, \frac{1}{2}, \frac{3}{5}, \frac{1}{6}$ **b** $\frac{3}{8}, \frac{1}{12}, \frac{7}{24}, \frac{1}{2}, \frac{1}{3}$ **c** $\frac{3}{4}, \frac{1}{20}, \frac{3}{5}, \frac{7}{40}, \frac{5}{8}$ **d** $\frac{7}{36}, \frac{4}{9}, \frac{1}{2}, \frac{2}{3}, \frac{3}{4}, \frac{5}{18}$

7 Rewrite each list of fractions as a list of equivalent decimals.
Use your answers to rewrite each list in descending order.

a $\frac{2}{5}, \frac{1}{4}, \frac{3}{8}, \frac{4}{10}$ **b** $\frac{3}{10}, \frac{1}{3}, \frac{2}{9}, \frac{2}{5}, \frac{3}{11}$ **c** $\frac{2}{7}, \frac{7}{20}, \frac{1}{5}, \frac{4}{9}, \frac{1}{3}$ **d** $\frac{4}{13}, \frac{3}{11}, \frac{1}{3}, \frac{5}{9}, \frac{3}{8}, \frac{2}{7}$

8 Rewrite these sets of numbers in ascending order.
Show your working.

a 33.3%, 0.33, 33, $33\frac{1}{3}\%$ **b** 0.45, 44.5%, 0.454, $0\dot{4}$

c $0.2\dot{3}$, 0.232, 22.3%, 23.22%, 0.233

d $\frac{2}{3}$, 0.66, $0.\dot{6}\dot{5}$, 66.6%, 0.6666

e $\frac{1}{7}$, 14%, 0.142, $\frac{51}{350}$, $14.\dot{1}\%$ **f** 86%, $\frac{5}{6}$, $0.8\dot{6}$, 0.866, $\frac{6}{7}$

Problem **AO3**

9 Jodie says that the recurring decimal $0.\dot{9}$ is a little smaller than 1. Abby says that $0.\dot{9}$ is equal to 1. Who is correct? Explain your reasoning.

N1 Summary

Check out

You should now be able to:

- Multiply and divide by a number between 0 and 1
- Add, subtract, multiply and divide fractions
- Round numbers to a number of decimal places and significant figures
- Give answers to an appropriate degree of accuracy using upper and lower bounds
- Calculate the upper and lower bounds of measurements
- Check and estimate answers to calculations
- Use the correct order of operations, including brackets, in a calculation
- Convert between fractions, decimals and percentages
- Convert a recurring decimal to a fraction and vice-versa

Worked exam question

Work out an estimate for $\frac{6.8 \times 191}{0.051}$ (3)

(Edexcel Limited 2009)

$6.8 \longrightarrow 7$

$191 \longrightarrow 200$

$0.051 \longrightarrow 0.05$

Write an approximation for each number.

$$\frac{6.8 \times 191}{0.051} \approx \frac{7 \times 200}{0.05} = \frac{1400}{0.05} = 28000$$

OR

$$\frac{6.8 \times 191}{0.051} \approx \frac{7 \times 200}{0.05} = 7 \times 4000 = 28000$$

OR

$$\frac{6.8 \times 191}{0.051} \approx \frac{7 \times 200}{0.05} = 7 \times 200 \times 20 = 28000$$

You should show your working whichever method you choose.

Exam questions

1 Estimate the value of $\dfrac{21 \times 3.86}{0.207}$ (3)

(Edexcel Limited 2006)

2 A tomato weighs 73 grams, correct to the nearest gram.

a Write down the least possible weight of the tomato. (1)

b Write down the greatest possible weight of the tomato. (1)

A02

3. The distance from Bristol to Leeds is 216 miles.

a Cara drove the 216 miles in 4 hours 30 minutes.
Calculate her average speed.
State the units of your answer. (2)

b The amount of petrol Cara's car used for the journey was 23 litres, correct to the nearest litre.

i Write down the least possible amount of petrol used.

ii Write down the greatest possible amount of petrol used. (2)

(Edexcel Limited 2006)

4 Prove that the recurring decimal $0.\dot{4}\dot{5} = \dfrac{15}{33}$ (3)

(Edexcel Limited 2007)

5 Express the recurring decimal $0.2\dot{1}\dot{3}$ as a fraction. (3)

(Edexcel Limited 2008)

Median and quartiles

Statistics are vital in medicine were they are used to test the safety and performance of new drugs. Tests are performed on large groups of people and the analysis of the results is used to evaluate the safety and reliability of the new drug.

What's the point?

When data is analysed it is essential for that analysis to be correct. Statisticians use a range of techniques to analyse and compare large data sets. It is the use of their statistical techniques that ensures that a drug is safe to be on sale to the general public.

Check in

You should be able to

- **find fractions of an amount**

1 Work out:

a $\frac{1}{2}$ of 124 **b** 50% of 120 **c** $\frac{1}{2}$ of (36 + 1) **d** 25% of 60

e $\frac{1}{4}$ of (27 + 1) **f** 25% of 120 **g** $\frac{3}{4}$ of 144 **h** 75% of 200

- **interpret a line graph**

2 Here is a graph showing the cost per day to hire a power tool.

How much does it cost to hire the power tool for

a 3 days **b** 5 days?

c Mike has £40.

What is the maximum number of days he can hire the power tool?

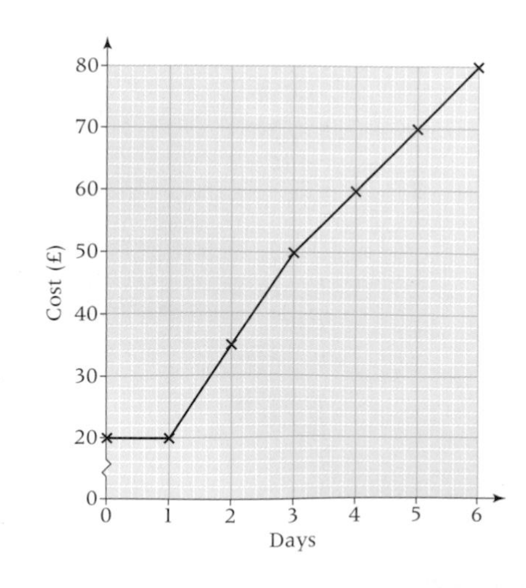

Orientation

What I need to know	What I will learn	What this leads to
KS3 Plot points and draw accurate graphs →	■ Construct and interpret box diagrams, cumulative frequency diagrams and stem-and-leaf diagrams →	**D4** Further displaying and comparing data sets
D1 Calculate quartiles and ranges →	■ Compare distributions →	**A-level** Maths, Biology, Economics, Geography

Rich task

As you get older you are more likely to become heavier.
Investigate if this statement is true by looking at different age groups.

D2.1 Box plots

This spread will show you how to:

- Draw and interpret box plots

Keywords
Box-and-whisker diagram
Box plot
Interquartile range
Lower quartile
Median
Upper quartile

You can represent ungrouped data on a **box plot**.

p.8 First you need to calculate the **median** and **upper** and **lower quartiles** of the data.

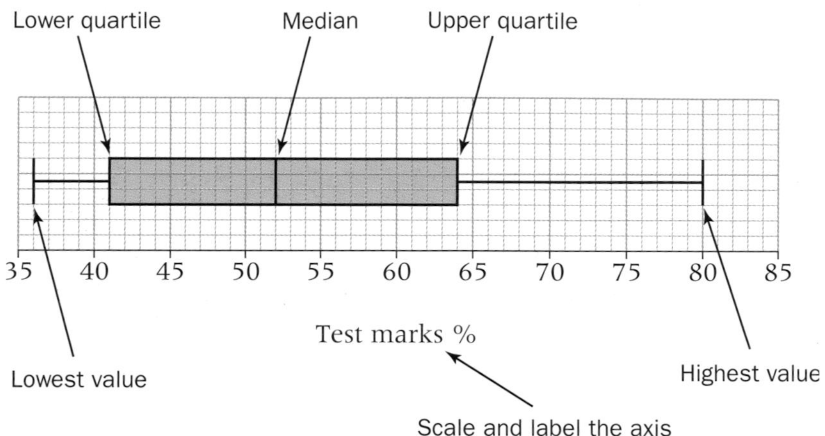

Box-and-whisker diagram is another name for box plot.

- The box width is the **interquartile range**.
- The distance between the lowest and highest value is the range.

Example

The weekly wages for 11 employees, to the nearest pound, are:
168, 144, 207, 432, 156, 316, 189, 224, 193, 97, 191.

a Draw a box plot to represent these results.
b Find the interquartile range.

a Write the data in order.
97, 144, 156, 168, 189, 191, 193, 207, 224, 316, 432
Median is $\frac{1}{2}(11 + 1)$th value = 6th value
Median = 191
LQ = $\frac{1}{4}(11 + 1)$th value = 3rd value
LQ = 156
UQ = $\frac{3}{4}(11 + 1)$th value = 9th value
UQ = 224

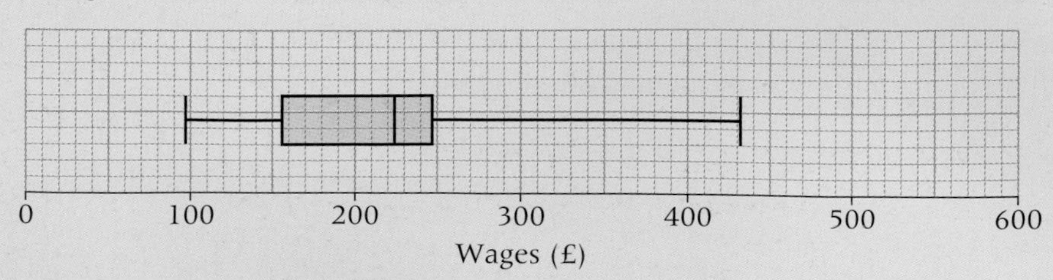

b IQR = UQ − LQ = 224 − 156 = 68

Exercise D2.1 Grade B

1 Mr Cheong summarised the test results for a group of students.

Lowest mark 41%   Lower quartile 54%   Median 60%
Highest mark 84%   Upper quartile 70%

Draw a box plot for these results.

2 In a science experiment, Davinder recorded reaction times to the nearest tenth of a second.
He summarised the data.

Quickest time 3.2   Lower quartile 3.7   Median 4.4
Slowest time 9.6   Upper quartile 7.5

Draw a box plot to represent these results.

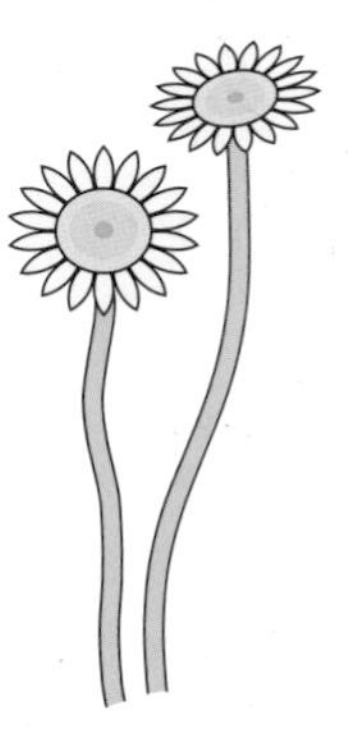

3 Daisy summarised the heights of sunflowers in her garden.

Smallest 148 cm   Lower quartile 154 cm   Median 157 cm
Tallest 177 cm   Upper quartile 162 cm

Draw a box plot to represent these results.

4 Marcus recorded how long, to the nearest second, it took 11 people to complete a calculation.

44 47 49 53 55 58 63 67 75 79 88

a Find the median.
b Find the lower quartile.
c Find the upper quartile.
d Draw a box plot for the data.

5 These are the marks for a drama assessment.

36 21 64 29 35 44 56 55 68 27 37 23 49 61 46

a Find the median, lower and upper quartiles.
b Draw a box plot for the data.

A03 Problem

6 Draw box plots to represent the weights of three consignments of cargo containers to the nearest tonne.
Consignment A
95 161 152 173 144 98 128 156 121 125 176 123 164
Consignment B
201 157 169 92 188 168 92 57 222 143
Consignment C
137 175 101 224 186 77 123 132 166 168 113 99

D2.2 Cumulative frequency diagrams

This spread will show you how to:

- Draw and interpret cumulative frequency diagrams for grouped data

Keywords
Cumulative frequency
Upper bound

You can represent grouped data on a **cumulative frequency** diagram by:

- calculating the cumulative frequencies and recording them in a cumulative frequency table
- plotting the cumulative frequency against the **upper bound** for each class
- joining the points with a smooth curve.

The cumulative frequencies are the running totals.

Example

The table shows the heights of 120 men.

Height, h cm	$160 \leqslant h < 165$	$165 \leqslant h < 170$	$170 \leqslant h < 175$	$175 \leqslant h < 180$	$180 \leqslant h < 185$
Frequency	9	26	47	33	5

a Draw a cumulative frequency table for these data.
b Draw a cumulative frequency diagram for these data.
c Estimate the number of men over 168 cm tall.

a

Height, h cm	<165	<170	<175	<180	<185
Cumulative frequency	9	35	82	115	120

$9 + 26 = 35$

Upper bound of each class.

Add frequencies to get cumulative frequency (CF).

b

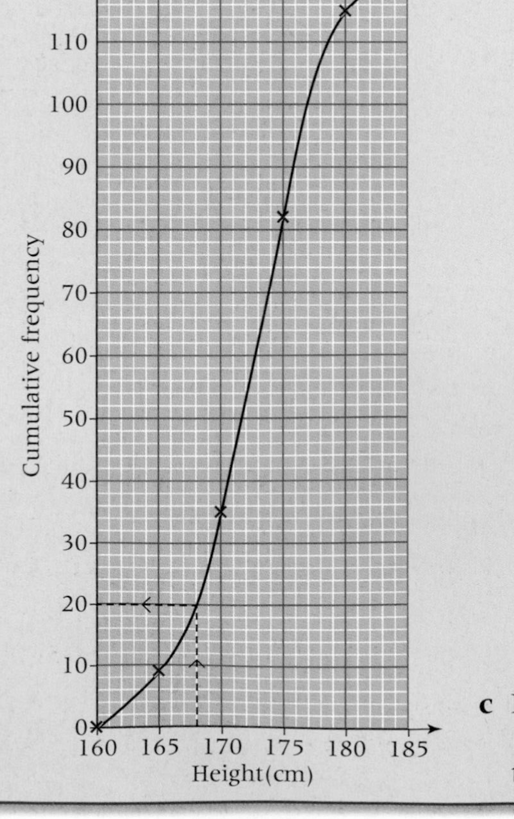

Plot: (upper bound, CF).

Lower bound of 1st class is 160, so plot (160, 0).

If the points are joined by straight lines the graph is called a frequency polygon.

c From the graph,
$120 - 20$ men are over 168 cm,
that is, 100 men.

Read up from 168 on the horizontal axis to the curve. Read across to the vertical axis to find the number of men *less than* 168 cm.

Exercise D2.2 Grade C

You will need the cumulative frequency tables and diagrams you draw in this exercise in **D2.3** and **D2.4.**

1 The heights of 100 women are given in the table.

Height, h cm	$150 \leqslant h < 155$	$155 \leqslant h < 160$	$160 \leqslant h < 165$	$165 \leqslant h < 170$	$170 \leqslant h < 175$
Frequency	9	27	45	16	3

a Draw up a cumulative frequency table for these data.
b Draw a cumulative frequency diagram for these data.

2 The table gives information about the ages of staff in a large store.

Age, A	$20 \leqslant A < 30$	$30 \leqslant A < 40$	$40 \leqslant A < 50$	$50 \leqslant A < 60$	$60 \leqslant A < 70$
Frequency	20	36	51	27	11

a Draw a cumulative frequency table for these data.
b Draw a cumulative frequency diagram for these data.

3 The table gives information about journey times to work.

Time, t minutes	$0 \leqslant t < 10$	$10 \leqslant t < 20$	$20 \leqslant t < 30$	$30 \leqslant t < 40$	$40 \leqslant t < 50$	$50 \leqslant t < 60$
Frequency	6	18	29	35	21	11

Draw a cumulative frequency diagram to represent this data.

4 The table gives information about the birth weights of babies.

Weight, w grams	$2500 \leqslant w < 3000$	$3000 \leqslant w < 3500$	$3500 \leqslant w < 4000$	$4000 \leqslant w < 4500$	$4500 \leqslant w < 5000$
Frequency	7	23	34	25	11

Draw a cumulative frequency diagram for these data.

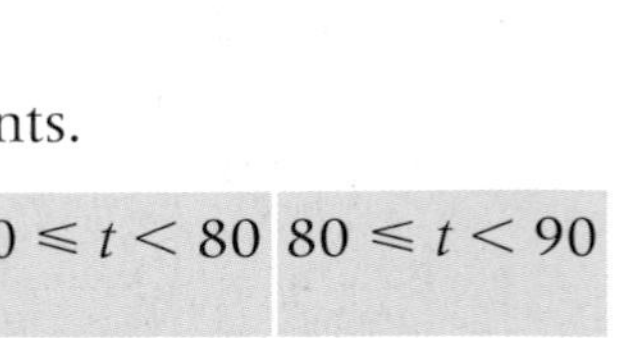

5 The table gives information about height of maize plants in a crop trial.

Height, h cm	$40 \leqslant h < 60$	$60 \leqslant h < 80$	$80 \leqslant h < 100$	$100 \leqslant h < 120$	$120 \leqslant h < 140$	$140 \leqslant h < 160$
Frequency	3	19	29	40	21	8

a Draw a cumulative frequency diagram for these data.
b Estimate how many maize plants were taller than 130 cm.

6 The table gives information about test results of a group of students.

Test, result, t %	$30 \leqslant t < 40$	$40 \leqslant t < 50$	$50 \leqslant t < 60$	$60 \leqslant t < 70$	$70 \leqslant t < 80$	$80 \leqslant t < 90$
Frequency	5	16	25	33	18	3

a Draw a cumulative frequency diagram for these data.
b Marks over 65% are awarded grade A. Estimate how many students were awarded grade A.

Using a cumulative frequency diagram

This spread will show you how to:

- Draw and interpret cumulative frequency diagrams for grouped data
- Estimate the averages and range for grouped data using a cumulative frequency diagram

Keywords
Cumulative frequency
Interquartile range
Median
Quartiles

For grouped data:

- You can estimate the mean
- You can state which class contains the median.

p.8

- You can estimate the **median** and **quartiles** from a **cumulative frequency** diagram.

This gives a more accurate estimate of the median.

Example

The weights of 100 rats are recorded in the table:

Weight, w grams	$100 \leqslant w < 120$	$120 \leqslant w < 140$	$140 \leqslant w < 160$	$160 \leqslant w < 180$	$180 \leqslant w < 200$	$200 \leqslant w < 220$
Frequency	8	19	23	27	18	5

a Draw a cumulative frequency diagram to represent the data.

b Use your diagram to estimate

i the median

ii the **interquartile range**.

c Use your diagram to estimate the number of rats that weigh between 165 g and 190 g.

a

Weight, w, grams	<120	<140	<160	<180	<200	<220
Cumulative Frequency	8	27	50	77	95	100

For large data sets, use $\frac{1}{2}n$ (not $\frac{1}{2}(n + 1)$) for the median.

b From the graph

i Median = 160 g

ii IQR = 177 − 138
= 39 g

c From the graph:
56 rats weigh < 165 g
88 rats weigh < 190 g
So, 88 − 56 = 32 rats weigh between 165 g and 190 g.

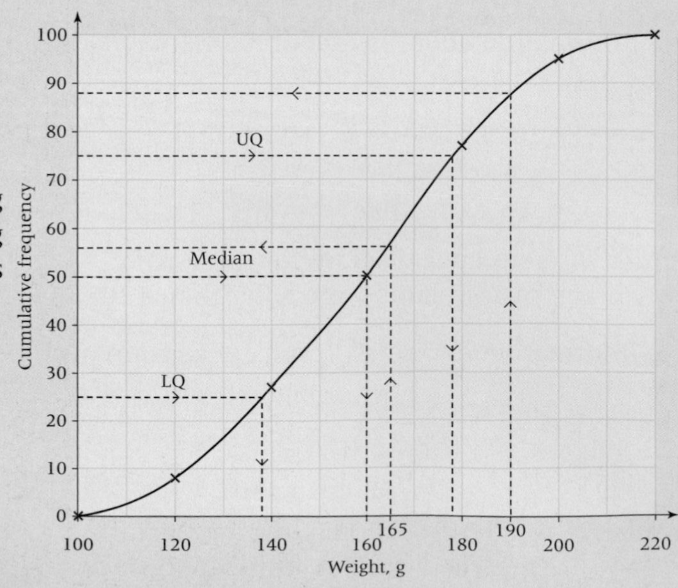

For 100 pieces of data:

- median is the 50th
- LQ is the 25th
- UQ is the 75th

Exercise D2.3

Grade B

You will need some of your results from this exercise in **D2.4**

1 Use your table and graph from Exercise **D2.2** question **1** to estimate
 a the median
 b the interquartile range
 c the number of women taller than 163 cm.

2 Use your table and graph from Exercise **D2.2** question **2** to estimate
 a the median
 b the interquartile range
 c the number of staff under 25.

3 Use your table and graph from Exercise **D2.2** question **3** to estimate
 a the median
 b the interquartile range
 c the number of journeys that take between 15 and 35 minutes.

4 Use your table and graph from Exercise **D2.2** question **4** to estimate
 a the median
 b the interquartile range
 c the number of babies that weighed less than 2600 g or more than 3300 g.

5 Use your table and graph from Exercise **D2.2** question **5** to estimate
 a the median
 b the interquartile range
 c the number of maize plants between 75 and 110 cm tall.

6 The cumulative frequency graph shows the IQ scores of a sample of 100 adults.
 Estimate from the graph
 a the median
 b the interquartile range
 c the number in the sample with an IQ score
  i less than 85
  ii greater than 115.

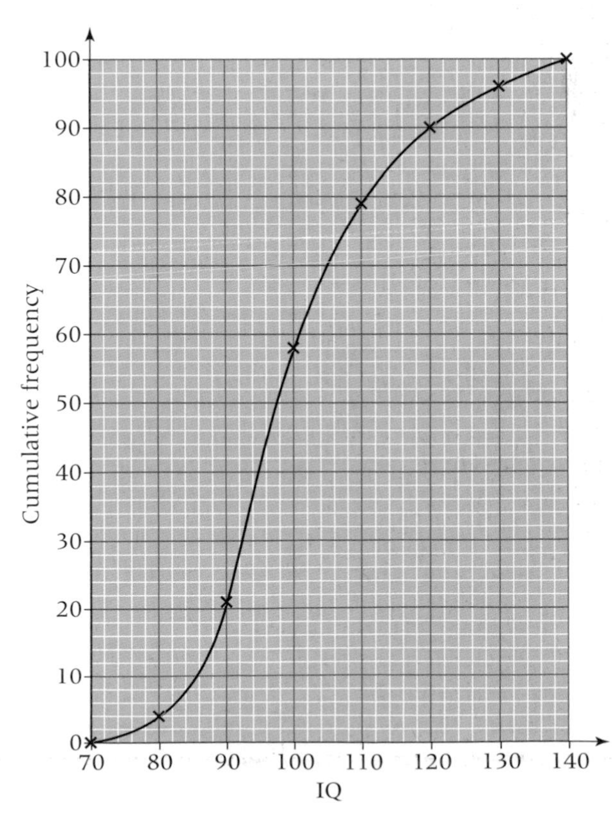

D2.4 Box plots for large data sets

This spread will show you how to:

- Draw and interpret box plots for grouped data
- Find the median, quartiles and interquartile range for large sets of data

Keywords
Box plot
Cumulative frequency
Median
Quartiles

To draw a **box plot** for grouped data:

– draw a **cumulative frequency** graph
– estimate the **median** and **quartiles** from your graph.

Example

Del spends his summer holidays working at his Aunt Jo's fruit farm. He weighs 100 apples from the new orchard and creates a cumulative frequency graph of the results.
Draw a box plot for the data.

If you are not given the actual highest and lowest values, use:

- lower bound of first class as lowest value
- upper bound of last class as highest value.

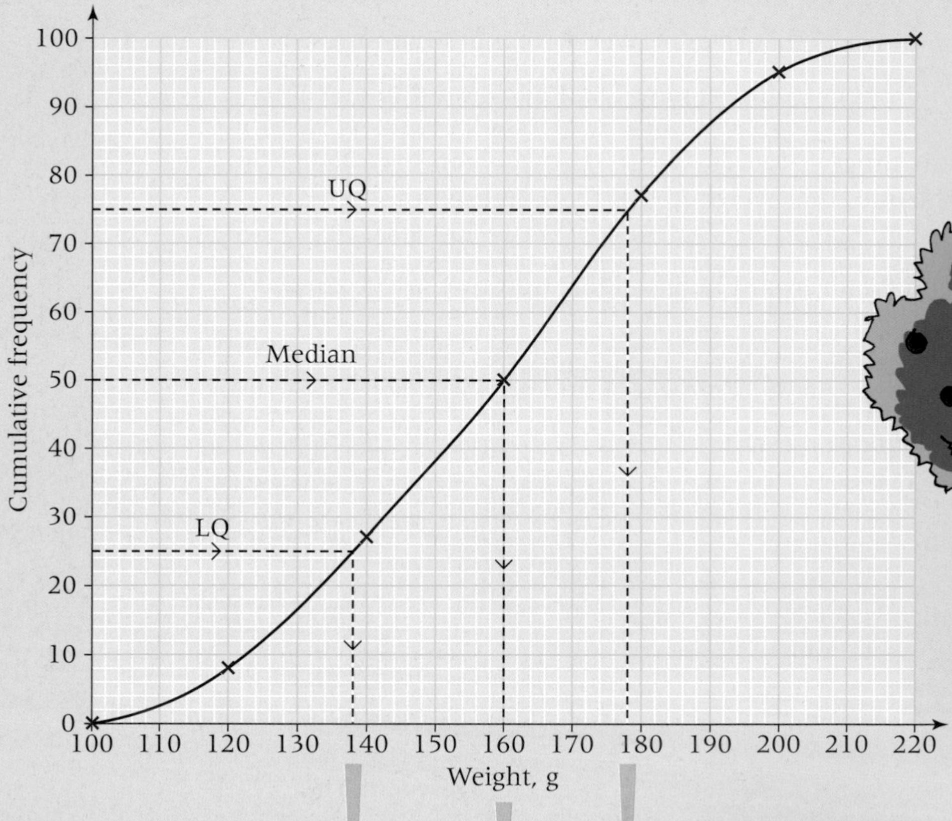

Cumulative frequency
100
90
80
70
60
50
40
30
20
10
0
UQ
Median
LQ
100 110 120 130 140 150 160 170 180 190 200 210 220
Weight, g

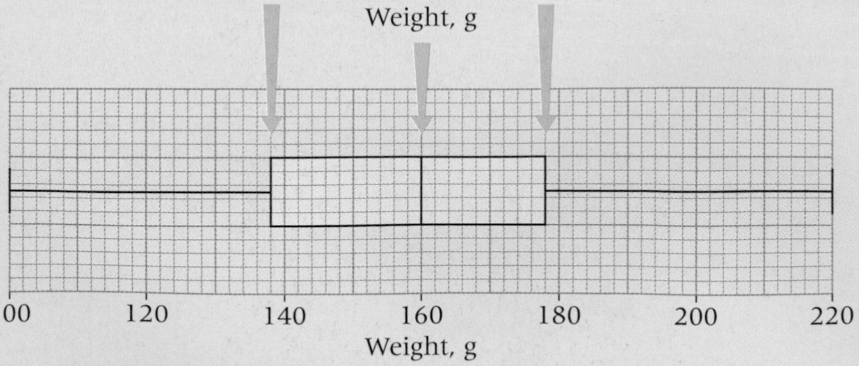

100 120 140 160 180 200 220
Weight, g

Exercise D2.4

Grade B

1–5 Use your table, graph and values from questions **1–5** in Exercises **D2.2** and **D2.3** to draw box plots for the data.

6 For the survey in question **6** of Exercise **D2.3**, the lowest IQ score in the sample was 72 and the highest IQ score was 136.
Use this information and your answer to question **6** in Exercise **D2.3** to draw a box plot.

7 Draw box plots for the data shown in these graphs.

a Waiting times of 100 patients at a dentist's surgery

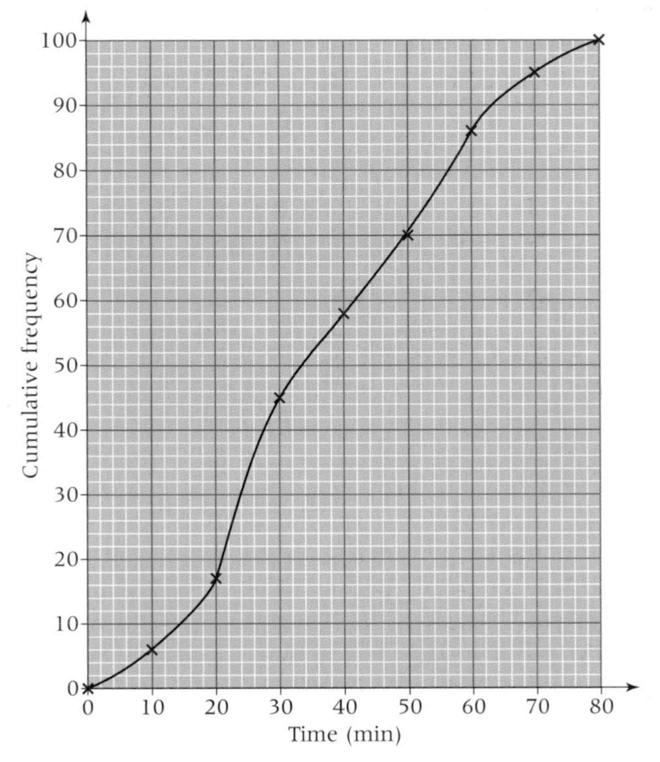

b Waiting times of 80 patients at a doctor's surgery

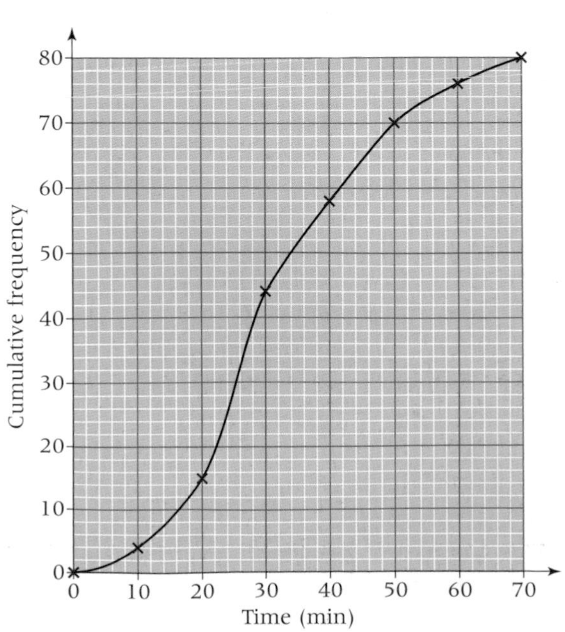

D2.5 Comparing data sets

This spread will show you how to:

- Compare distributions and make inferences, using shapes of distributions and measures of average and spread, including median and quartiles

Keywords
Skewness
Statistics

- You can compare data sets using summary **statistics**.

Summary statistics are averages and measures of spread.

Box plots show if the data is skewed.

- Skewed data is not evenly grouped about the median.

The median is nearer to the lower quartile.
The third quarter of the data is more spread out than the second quarter.
The data is positively skewed.

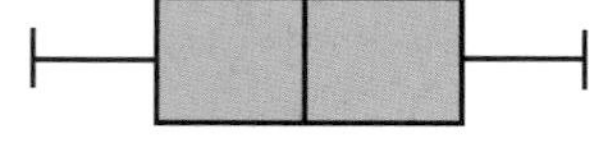

The median is central.
The data is not skewed.

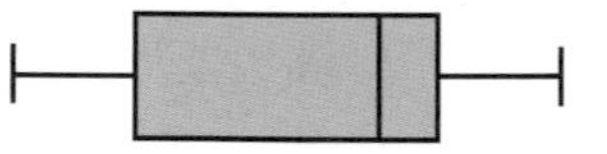

The median is nearer to the upper quartile.
The second quarter of the data is more spread out than the third quarter.
The data is negatively skewed.

- For two or more distributions, compare:
  - a measure of average (mean, median or mode)
  - a measure of spread (range or interquartile range)
  - **skewness**.

Compare like with like, for example:
- two medians
- two means
- two IQRs

Example

The box plots summarise the heights of samples of two strains of sunflower, Strain A and Strain B.

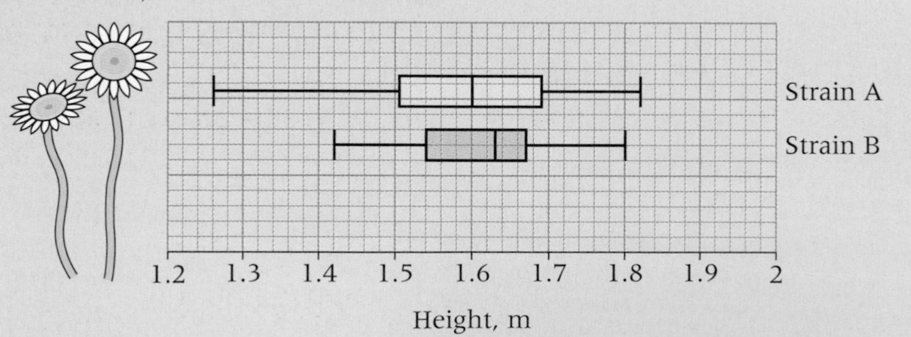

Write four comparisons between the heights of the two strains.

a Strain B median is greater than Strain A, so on average Strain B is taller than Strain A.
b Range for Strain A is greater than for Strain B, so the heights of Strain A plants are more varied than the heights of Strain B plants.
c IQR for Strain A is greater than IQR for Strain B, so middle 50% of heights vary more for Strain A than Strain B.
d Strain A heights are not skewed. Strain B heights are negatively skewed, so more of the middle 50% of heights are less than the median.

Exercise D2.5

Grade B/A

Functional Maths

AO3

1 The box plots summarise the durations of phone calls to a computer helpline and a mobile phone helpline.
Write four comparisons between the durations of calls to the two helplines.

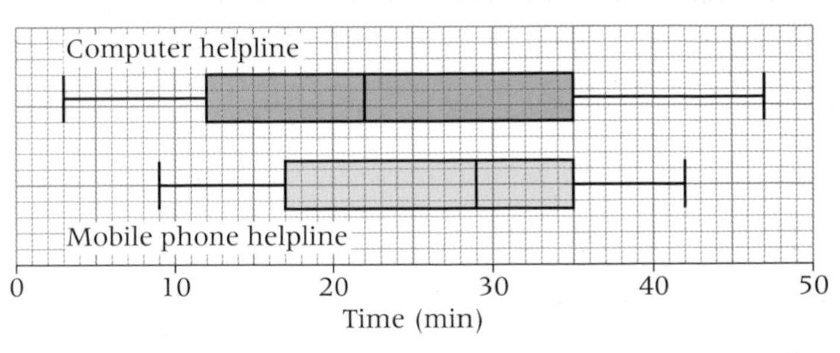

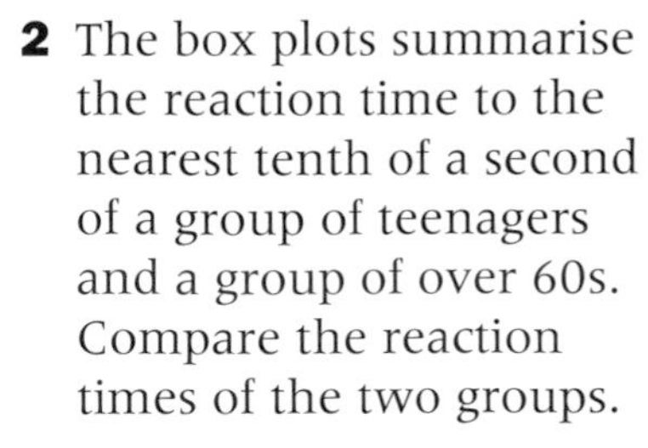

2 The box plots summarise the reaction time to the nearest tenth of a second of a group of teenagers and a group of over 60s.
Compare the reaction times of the two groups.

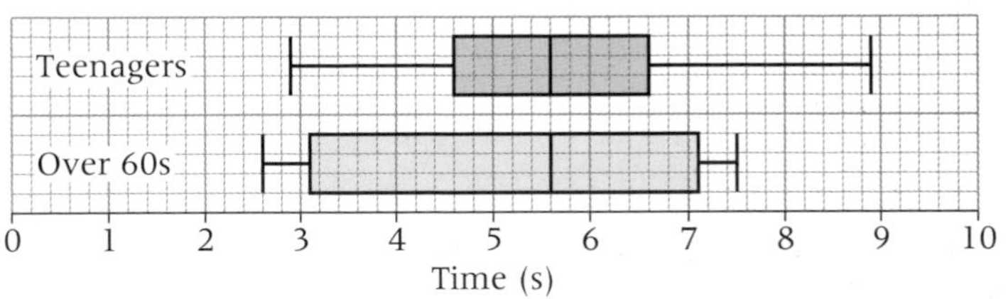

3 Write three comparisons between the test results of a group of girls and boys summarised in the cumulative frequency graph.

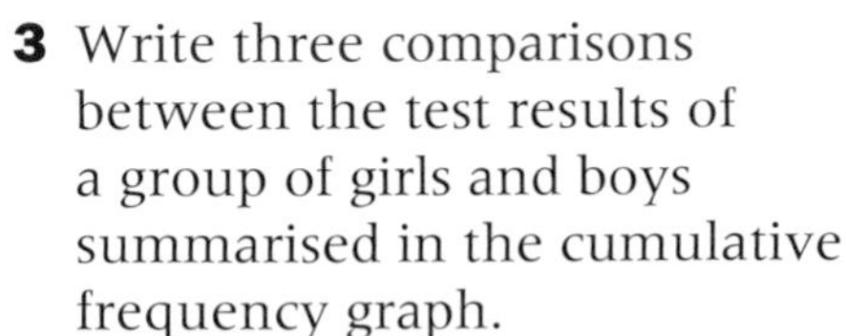

Consider the medians, the ranges and the interquartile ranges.

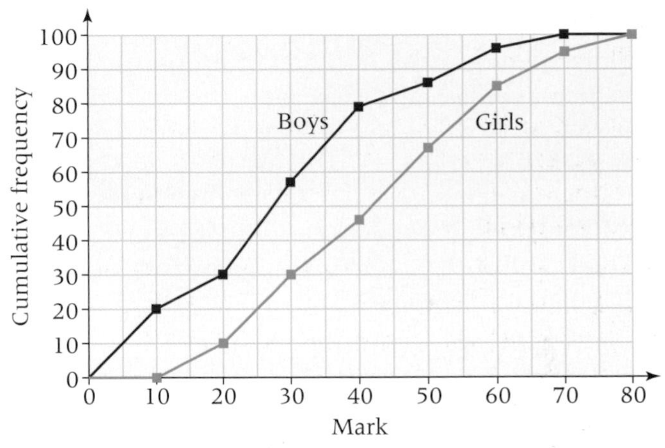

4 Compare the heights of samples of sunflowers grown by two farmers summarised in the cumulative frequency graph.

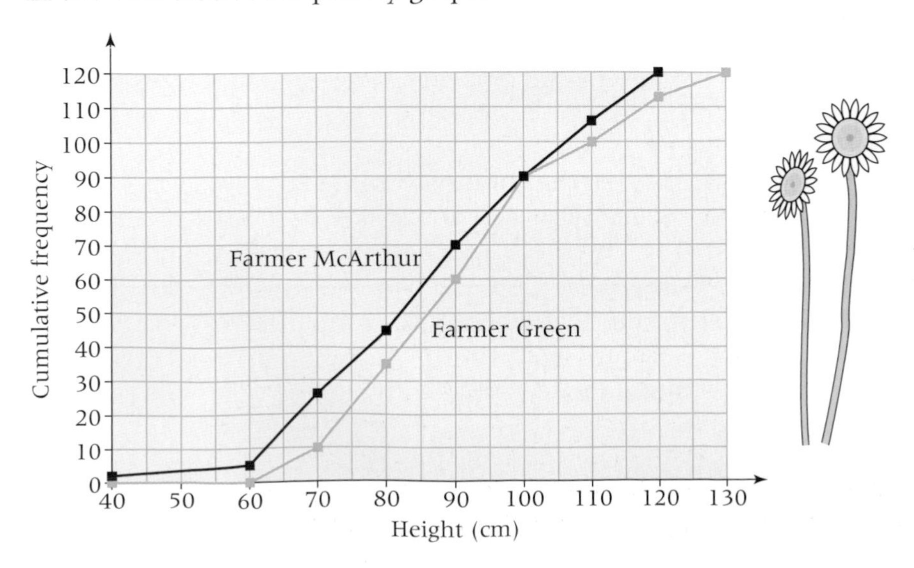

D2.6 Stem-and-leaf diagrams

This spread will show you how to:
- Draw and interpret stem-and-leaf diagrams

Keywords
Interquartile range
Median
Mode
Stem-and-leaf diagram

You can represent small sets of data on a **stem-and-leaf diagram**. The diagram shows the trend of the distribution and all the individual values.

You can find the **mode, median**, upper quartile (UQ) and lower quartile (LQ) from a stem-and-leaf diagram.

- **Interquartile range** (IQR) = UQ − LQ

The IQR represents the middle 50% of the data.

Example

These are the times, in seconds, for 21 athletes who ran a 100 m race.

9.7	12.2	11.4	10.9	10.1	9.6	11.1	11.8	10.3	9.9	10.7
11.4	10.1	11.4	10.4	9.8	10.2	12.1	10.6	9.7	10.7	

a Draw a stem-and-leaf diagram for this data.
b Describe the distribution.
c Find the mode, median and interquartile range.
d Which average best represents the data, the mode or the median?

a

Stem	Leaves
9	7 6 9 8 7
10	9 1 3 7 1 4 2 6 7
11	4 1 8 4 4
12	2 1

Write the 'leaves' in order to see the distribution. ⟶

Stem	Leaves
9	6 7 7 8 9
10	1 1 2 3 4 6 7 7 9
11	1 4 4 4 8
12	1 2

Key: 10 | 1 represents 10.1 s

b Most runners completed the race in under 11 seconds.
c Mode = 11.4 seconds
Median is the $\frac{1}{2}(21 + 1)$th = 11th term = 10.6 seconds
IQR = UQ − LQ
UQ is $\frac{3}{4}(21 + 1)$th = $16\frac{1}{2}$th term = 11.4
LQ = $\frac{1}{4}(21 + 1)$th = $5\frac{1}{2}$th term = 10.0
IQR = 11.4 − 10.0 = 1.4 seconds
d The median (10.6 seconds), as most of the runners completed the race in under 11 seconds. The mode (11.4 seconds) is slower than the majority of the runners.

The median is the $\frac{1}{2}(n + 1)$th term.

Data is in order, so count up to find 11th term.

16th term = 11.4, 17th term = 11.4

5th term = 9.9, 6th term = 10.1
10.0 is halfway between 9.9 and 10.1.

- You can draw a back-to-back stem-and-leaf diagram to compare two data sets.

Puzzle A		Puzzle B
9 9 8	0	9
7 6 4 4 2 1 1	10	1 2 4
5 5 4 2 2	20	1 1 2 3 3 5 7 9
	30	1 2 2

Key: 2 | 20 represents 22 min   Key: 30 | 1 represents 31 min

Time to solve puzzle A is shorter on average. Puzzle A is probably easier.

Exercise D2.6 Grade B/A

For these sets of data

a Draw a stem-and-leaf diagram.

b Describe the trend.

c Find the mode, median and interquartile range.

d State, with reasons, whether the mode or median best represents the data.

1 IQ scores

132 120 105 118 111 128 109 104 117 90 95
130 89 93 132 127 119 115 107 106 92
102 113 126 110 99 108 107 116 120 115

2 Journey times to work (to the nearest minute)

12 21 9 25 23 18 33 8 32 15 11 22
11 16 4 18 23 26 15 10 5 31 23

3 Weights, in kg, of sacks of compost

15.3 15.5 16.2 14.9 15.1 14.8 16.6 17.0 14.5 15.3 14.7
15.7 14.8 16.5 15.4 16.7 14.8 15.6 15.7 15.6 14.9

For these sets of data

a Draw back-to-back stem-and-leaf diagrams.

b Work out the median, interquartile range and range for each.

c Compare the two sets of data.

4 Heights to the nearest cm

Men: 177 181 179 184 183 173 189 169 172 174 177 182
157 158 164 168 176 173 168 159 166 172 181

Women: 158 156 149 172 152 157 168 163 175 167 166 148
168 177 157 153 152 151 154 149 160 157 162

5 Marks in two tests, for the same group of students

Test A: 103 121 129 115 116 102 96 94 101 110 108
119 128 131 125 111 110 117 101 91 123

Test B: 112 105 129 97 104 122 131 117 124 112 120 104
121 106 123 128 120 111 99 116 131 129 130

6 Scores for a gymnastics routine

Club X: 4.3 5.7 5.1 3.2 3.3 3.7 3.9 5.7 6.5 4.6 3.9 4.4
4.4 6.2 6.1 5.8 6.9 5.2 7.2 5.7 6.5 3.7 7.1

Club Y: 7.7 6.2 5.6 7.6 6.2 8.1 6.5 4.1 5.7 8.1 6.6 7.3
6.1 6.8 7.4 4.4 7.5 5.1 5.3 4.7 5.8 4.3 4.7

7 Compare and contrast

a a stem-and-leaf diagram and a bar chart

b a back-to-back stem-and-leaf diagram and a comparative bar chart.

D2 Summary

Check out

You should now be able to:

- Draw and interpret cumulative frequency diagrams
- Use cumulative frequency graphs to find median, quartiles and interquartile range
- Draw and interpret box plots for small and large sets of data
- Compare distributions using box plots and cumulative frequency graphs
- Draw and interpret stem-and-leaf diagrams

Worked exam question

The box plot gives information about the distribution of the weights of bags on a plane.

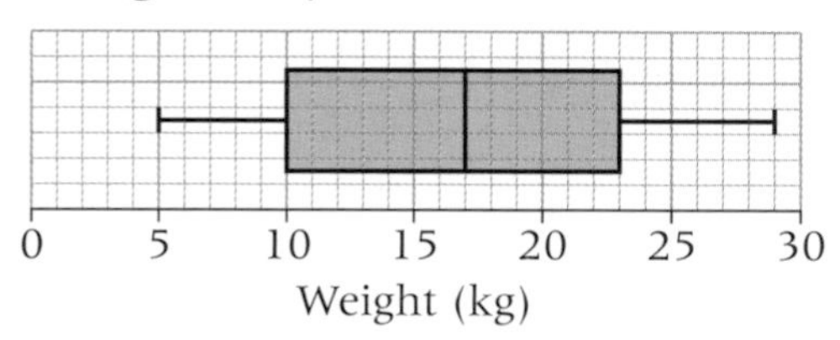

a Jean says the heaviest bag weighs 23 kg.
She is wrong.
Explain why. (1)

b Write down the median weight. (1)

c Work out the interquartile range of the weights. (1)

There are 240 bags on the plane.

d Work out the number of bags with a weight of 10 kg or less. (2)

(Edexcel Limited 2009)

---

a The heaviest bag is 29 kg

b 17 kg

c $23 - 10 = 13$ kg

d 10 kg is the Lower Quartile

25% of 240 $= \frac{1}{4} \times 240$

$= 60$ bags

State that 10 kg is the Lower Quartile and so 25% of the bags weigh 10 kg or less. Show this calculation.

Exam question

AO3

1 Here are four cumulative frequency diagrams.

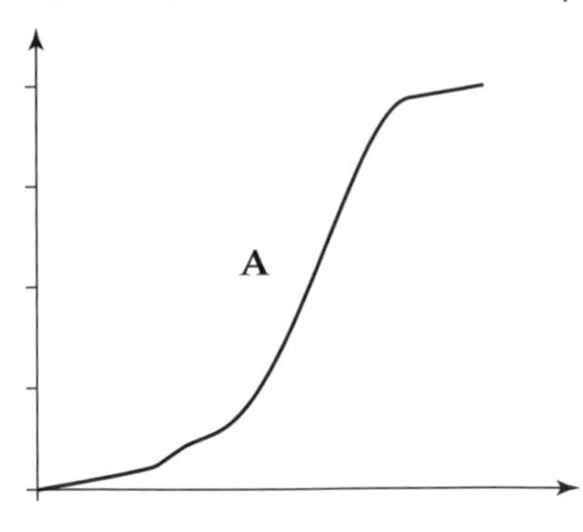

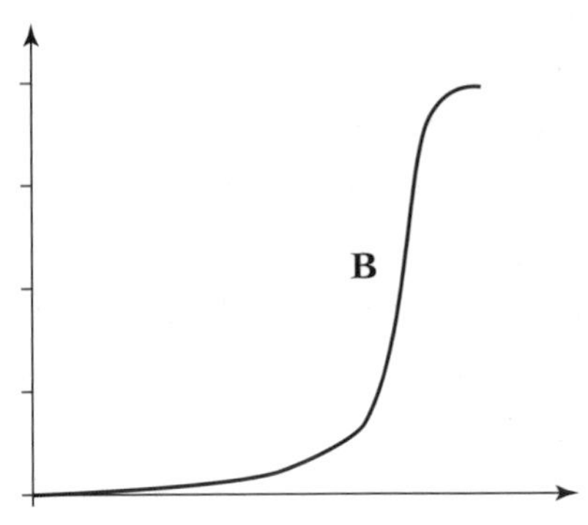

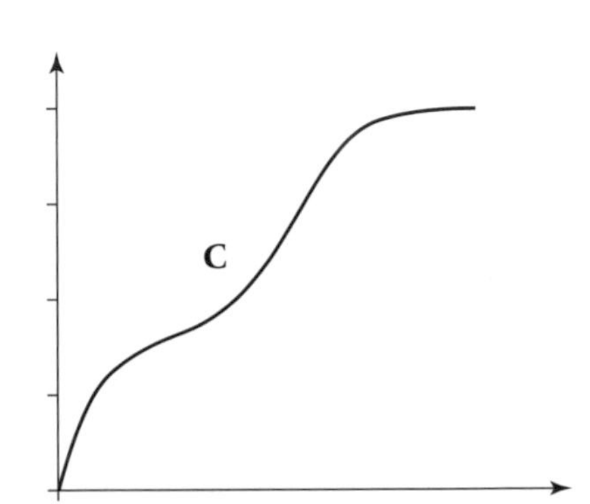

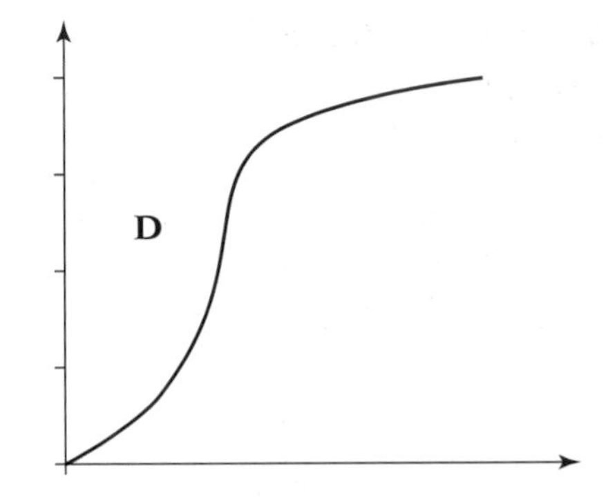

Here are four box plots.

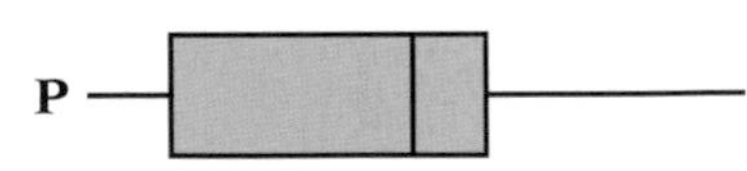

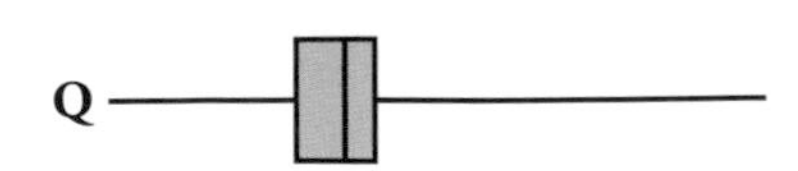

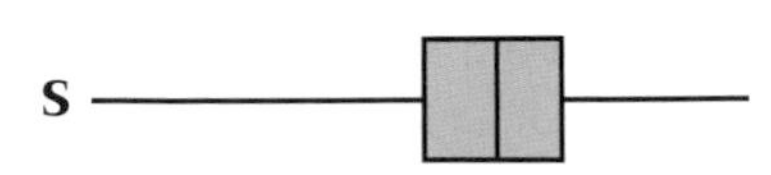

For each box plot, write down the letter of the appropriate cumulative frequency diagram.

P and ....
Q and ....
R and ....
S and ....

(2)

(Edexcel Limited 2006)

Functional Maths 1: Weather

Before creating a weather forecast, data is collected from all over the world to give information about the current conditions. A supercomputer and knowledge about the atmosphere, the Earth's surface and the oceans are then used to create the forecast.

Write down the temperature shown on each of these thermometers:

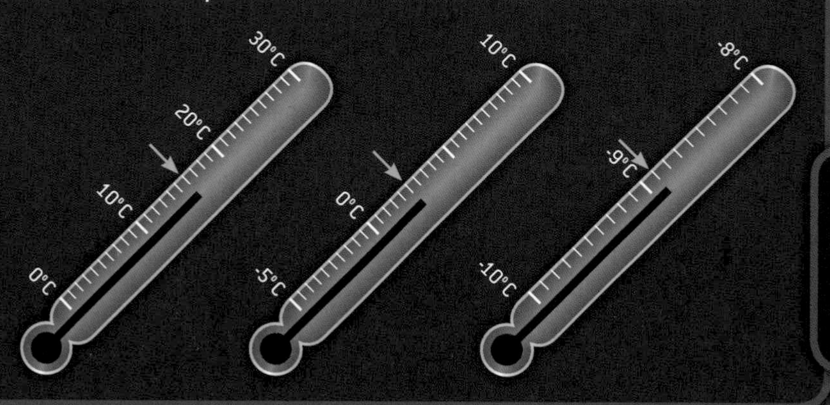

Each day, the Met Office receives and uses around half a million observations.

These tables show the national and UK weather records, last updated 24th November 2008:

HIGHEST DAILY TEMPERATURE RECORDS			
England	38.5 °C	10 August 2003	Faversham (Kent)
Wales	35.2 °C	2 August 1990	Hawarden Bridge (Flintshire)
Scotland	32.9 °C	9 August 2003	Greycrook (Scottish Borders)
Northern Ireland	30.8 °C	30 June 1976 12 July 1983	Knockarevan (County Fermanagh) Shaw's Bridge, Belfast (County Antrim)

HIGHEST DAILY TEMPERATURE RECORDS			
Scotland	-27.2 °C	11 February 1895 10 January 1982 30 December 1995	Braemar (Aberdeenshire) Braemar (Aberdeenshire) Altnaharra (Highland)
England	-26.1°C	10 January 1982	Newport (Shropshire)
Wales	-23.3 °C	21 January 1940	Rhayader (Powys)
Northern Ireland	-17.5 °C	1 January 1979	Magherally (County Down)

Use the internet to find out if these records have since been broken.

Calcuate the difference (in °C) between the maximum and minimum

Wind direction is measured in tens of degrees relative to true North and is always given from where the wind is blowing. In the UK, wind speed is measured in knots, where 1knot = 1.15mph, or in terms of the Beaufort Scale.

Look up the Beaufort Scale on the internet. Use it to describe the weather shown on this map.

Describe the wind speed (in knots) and direction (in tens of degrees and in words) shown by each of the arrows shown on this map:

20
30
38
56
35
53
9
21
10
18
36
9
9
8
14
25
20
31
10
14
12
25
24
37
13
23
7
18
18
22
38
18
35
13
25

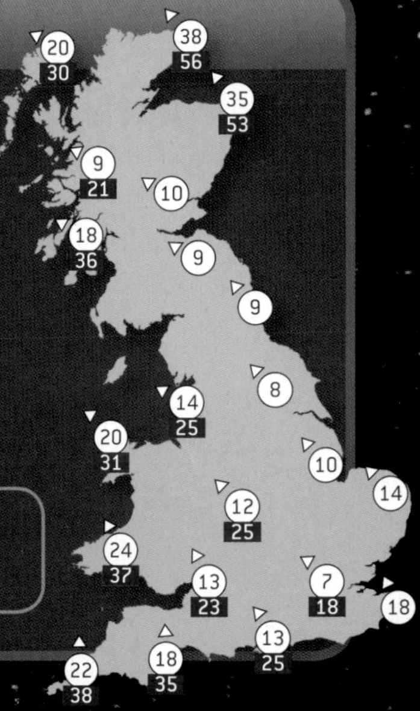

12
24
indicates a mean wind of 12 m.p.h., coming from the south, gusting 24 m.p.h.

Observed data can be used to make predictions, but there is always some level of uncertainty. This graph shows the range of uncertainty in temperature in Exeter with some indication of the most probable values:

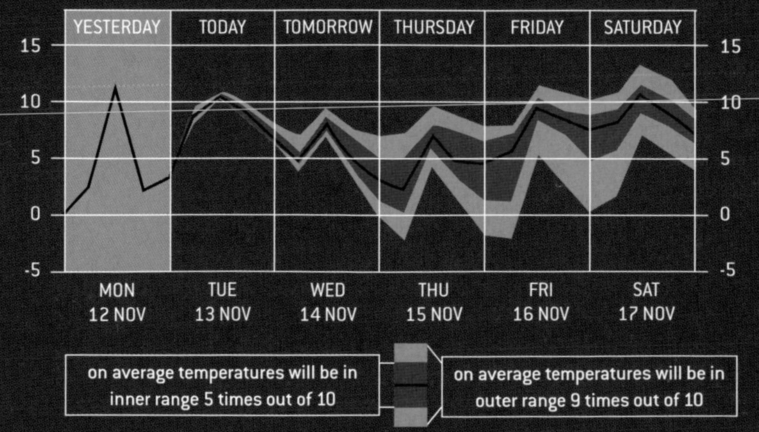

YESTERDAY
TODAY
TOMORROW
THURSDAY
FRIDAY
SATURDAY
15
10
5
0
-5
MON 12 NOV
TUE 13 NOV
WED 14 NOV
THU 15 NOV
FRI 16 NOV
SAT 17 NOV
on average temperatures will be in inner range 5 times out of 10
on average temperatures will be in outer range 9 times out of 10

What predictions do you think a weather forecaster would have made about the temperature in Exeter during the week shown?

Justify your response by referring to the graph.

D3

Probability

There are many things in life which are uncertain. Will it be sunny tomorrow? Will my football team win the Premier League? Will I be able to afford a house in the future? The mathematics used to deal with uncertainty is called probability.

What's the point?

Probability is a way of quantifying the uncertainty of an event, which is usually expressed as a fraction, decimal or percentage. When the Met Office gives a weather forecast, they use a complex mathematical model of the earth's climate with many variables, to predict the probability of sunshine in a particular region which they often express as a percentage.

Probability

Check in

You should be able to

- **add and subtract fractions**

1 Word out each of these.

a $1 - \frac{2}{5}$ **b** $1 - \frac{4}{7}$ **c** $1 - \frac{3}{8}$

d $\frac{2}{3} + \frac{1}{6}$ **e** $\frac{1}{5} + \frac{1}{4}$ **f** $\frac{1}{3} + \frac{5}{8}$

- **multiply fractions**

2 Word out each of these.

a $\frac{1}{4} \times \frac{2}{3}$ **b** $\frac{1}{5} \times \frac{3}{4}$ **c** $\frac{1}{3} \times \frac{2}{5}$

- **convert between fractions and decimals**

3 Change these fractions to decimals.

a $\frac{7}{10}$ **b** $\frac{3}{4}$ **c** $\frac{3}{8}$

d $\frac{2}{5}$ **e** $\frac{1}{3}$ **f** $\frac{1}{7}$

Orientation

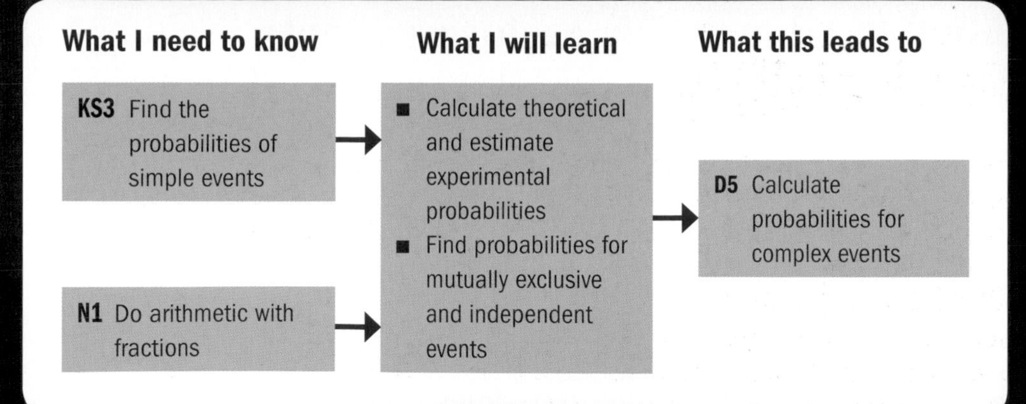

Rich task

Inside a bag are three cards with the letters A, N and D written on them.
You are allowed to pick one card from the bag and then replace the card in the bag.
How many cards will you have picked before you have all the letters to spell the word AND?

D3.1 Probability and mutually exclusive events

This spread will show you how to:

- Understand the probability scale
- Understand mutually exclusive events and their related probabilities
- Know that if A and B are mutually exclusive, then the probability of A or B occurring is P(A) + P(B)

Keywords
Event
Mutually exclusive
Outcome

Probability is a measure of the likelihood that an **event** will happen.

- An event is one or more of the possible **outcomes**.
- The probability of an event lies between 0 and 1.
- Probability of an event happening $= \dfrac{\text{number of favouable outcomes}}{\text{total number of all possible outcomes}}$

You can write probability as a decimal, fraction or percentage.

- Events are **mutually exclusive** if they cannot happen at the same time.
- For two mutually exclusive events A and B:
  P(A or B) = P(A) + P(B)
- For an event A,
  P(not A) = 1 − P(A) or P(A) = 1 − P(not A)

P(A) is the probability of event A happening.

A and not A are mutually exclusive events.

Example

A spinner has 10 equal sections:
3 green, 2 blue, 4 red and 1 white.
What is the probability that the spinner lands on

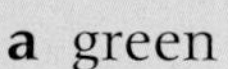

a green **b** green or blue
c blue or white **d** blue or red or white
e not green?

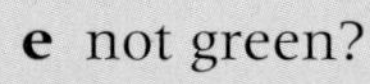

a P(green) $= \frac{3}{10}$

b P(green or blue) = P(green) + P(blue)

p.28

$= \frac{3}{10} + \frac{2}{10}$

$= \frac{5}{10} = \frac{1}{2}$

c P(blue or white) = P(blue) + P(white)

$= \frac{2}{10} + \frac{1}{10}$

$= \frac{3}{10}$

d P(blue or red or white) = P(blue) + P(red) + P(white)

$= \frac{2}{10} + \frac{4}{10} + \frac{1}{10}$

$= \frac{7}{10}$

e P(not green) = 1 − P(green)

$= 1 - \frac{3}{10}$

$= \frac{7}{10}$

Exercise D3.1 Grade C

1 A bag contains 6 green, 10 blue and 9 red cubes.
8 of the cubes are large, 17 are small.
One cube is chosen at random.
What is the probability that the cube is

a red **b** not red **c** large
d small **e** blue **f** not green?

2 Four teams are left in a hockey tournament.
Their probabilities of winning are

Colts	Falcons	Jesters	Hawks
0.17	0.38	x	$2x$

a Compare the Hawks' chance of winning with the Jesters' chance of winning.
b Work out the value of x.

3 The probabilities that a player will win, draw or lose at snap are

Win	Draw	Lose
x	$2x$	x

Work out the value of x.

4 A nine-sided spinner has sections numbered 1, 2, 3, 4, 5, 6, 7, 8, 9.
What is the probability that it lands on

a 6 **b** not the number 6 **c** an even number
d a multiple of 5 **e** a multiple of 3 **f** not a multiple of 3
g a factor of 24 **h** not a factor of 24 **i** a prime number?

5 The table shows how many TVs are owned by some families.

Number of TVs	0	1	2	3	more than 3
Number of families	8	12	5	1	4

a How many families were surveyed?
b What is the probability that a family chosen at random from the sample will own

i exactly 1 TV **ii** no TV **iii** either 1 or 2 TVs
iv 2 or more TVs **v** 3 or more TVs **vi** 4 TVs?

6 The table shows staff absences over a two-month period.

Number of days off	0	1	2	3	more than 3
Number of staff	5	1	9	11	6

Find the probability that a member of staff chosen at random had

a 1 day off **b** 1 or 2 days off
c 2 or more days off **d** 2 or fewer days off.

D3.2 Theoretical and experimental probability

This spread will show you how to:

- Understand theoretical and experimental probability and how estimates can be used in probability
- Calculate the expected frequency of events

Keywords
Expected frequency
Experimental probability
Theoretical probability

If all the outcomes are equally likely, you can calculate the **theoretical probability**.

- Theoretical probability $= \dfrac{\text{number of favourable outcomes}}{\text{total number of outcomes}}$

You use theoretical probability for fair activities, for example rolling a fair dice.

You can use data from a survey or experiment to estimate **experimental probability**.

- Experimental probability $= \dfrac{\text{number of successful trials}}{\text{total number of trials}}$

For unfair or biased activities, or where you cannot predict the outcome, you use experimental probability.

Example

A fair spinner has 12 sides:
5 yellow, 2 red, 4 blue and 1 white.
Work out the probability that the spinner lands on

a yellow **b** red or blue **c** not white.

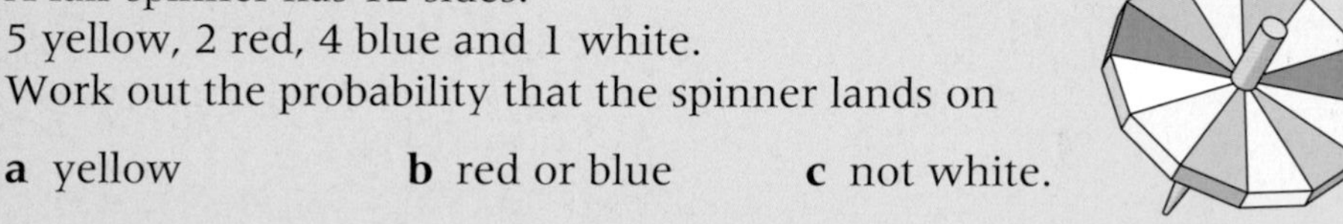

a P(yellow) $= \frac{5}{12}$ **b** P(red or blue) $= \frac{6}{12} = \frac{1}{2}$
c P(not white) $= 1 - \frac{1}{12} = \frac{11}{12}$

P(red or blue) = P(red) + P(blue)

You can use probability to calculate the **expected frequency** of an event.

- Expected frequency = number of trials × probability

Example

Finn carries out a survey on left-handedness. He asks 50 students at his school. Six of them are left-handed.
a Estimate the probability that a student at the school is left-handed.
b Estimate how many of the 1493 students at the school are left-handed.

p.70

a P(left-handed) $= \frac{6}{50}$
b Expected frequency $= 1493 \times \frac{6}{50} = 179.16$
Estimate: 179 students are left-handed.

'Number of trials' = 'number of students' in this context.

Example

A biased coin is thrown 150 times. It lands on heads 105 times.
a Estimate the probability of the coin landing on heads next time.
b What number of heads would you expect in 400 throws?

a P(Head) $= \frac{105}{150} = 0.7$
b Estimated number of heads in 400 throws $= 400 \times 0.7 = 280$

Exercise D3.2 Grade C

1 The probability that a biased dice will land on a six is 0.24. Estimate the number of times the dice will land on a six if it is rolled 500 times.

2 A biased coin is thrown 140 times. It lands on tails 28 times. Estimate the probability that this coin will land on tails on the next throw.

3 A tetrahedral dice is rolled 100 times. The table shows the outcomes.

Score	1	2	3	4
Frequency	16	28	33	23

a The dice is rolled once more. Estimate the probability that the dice will land on
i 4 **ii** 3 **iii** 1 or 2 **iv** not 2.

b If the dice is rolled another 200 times, how many times would you expect the dice to land on
i 2 **ii** 3 or 4?

A02 Functional Maths

4 There are 240 adults in the village of Hollowdown.

a In a survey of 30 adults from Hollowdown, 19 owned a bicycle. How many adults in the village would you expect to own bicycle?

b In a survey of 40 adults, 21 owned a dog. How many adults in the village would you expect to own a dog?

c In a survey of 48 adults, 43 felt the speed limit through the village was too high. Estimate the number of adults in the village who think the speed limit is too high.

A03 Problem

5 On a holiday, people chose one activity.

	Canoeing	Abseiling	Potholing	Total
Male	13		5	22
Female		9		
Total	20			50

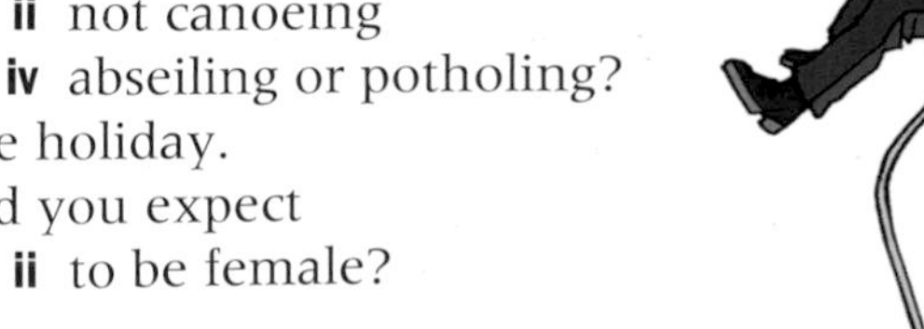

a Copy and complete the table.

b What is the probability that a person chosen at random chose
i canoeing **ii** not canoeing
iii canoeing or potholing **iv** abseiling or potholing?

c There are 300 people in total on the holiday. How many of the total group would you expect
i to choose canoeing **ii** to be female?

D3.3 Relative frequency and best estimate

This spread will show you how to:

- Calculate the expected frequency and relative frequency of events
- Understand that more trials give a more reliable estimate of probability

Keywords
Relative frequency

Experimental probability is also called **relative frequency**.
Relative frequency is the proportion of successful trials in an experiment.

- Relative frequency $= \frac{\text{number of successful trials}}{\text{total number of trials}}$

The more trials you carry out, the more reliable the relative frequency will be as an estimate of probability.

Example

A double glazing representative telephones people to make appointments to demonstrate his product. One week he makes 50 calls each morning and creates a table to record the number of appointments.

Day	Mon	Tues	Wed	Thurs	Fri
Number of appointments	4	6	9	7	5

a Work out the relative frequency of the number of appointments for each day.
b Write down the best estimate of the relative frequency of setting up an appointment.

a

Day	Mon	Tues	Wed	Thurs	Fri
Number of appointments	4	6	9	7	5
Relative frequency	$4 \div 50 = 0.08$	$6 \div 50 = 0.12$	$9 \div 50 = 0.18$	$7 \div 50 = 0.14$	$5 \div 50 = 0.1$

b Total number of calls made = 250
Best estimate of relative frequency $= \frac{4 + 6 + 9 + 7 + 5}{250} = \frac{31}{250} = 0.124$

For the best estimate of relative frequency, pool all the results.

You can compare theoretical probability with relative frequency.

Example

Kaseem rolled a dice 50 times and in 14 of those he scored a 2.
a What is the relative frequency of rolling a 2?
b Is the dice biased towards 2? Explain your answer.

a Relative frequency of rolling a 2 $= \frac{14}{50} = \frac{28}{100} = 0.28$
b Theoretical probability of rolling a 2 $= \frac{1}{6} = 0.166...$
In 50 rolls expected number of 2s $= 50 \times 0.166... = 8.333$
The relative frequency of rolling 2 is 0.28, which is much higher than the theoretical probability, 0.17.
The actual number of 2s is 14, which is much higher than the expected number, 8.
The dice does appear to be biased towards 2.

Exercise D3.3 Grade C

1 In a statistical experiment, a coin is flipped 360 times and lands on tails 124 times.
Is the coin fair? Explain your answer.

2 Surjeet picks a card from a deck of playing cards and records whether it is black or red.
She returns the card to the pack and shuffles it, before picking the next card.
In total she picks a card 420 times and records 213 black.
Is the deck of cards complete? Explain your reasons.

In a complete deck, there are equal numbers of black and red cards.

3 The results when a dice is rolled 200 times are

Score	1	2	3	4	5	6
Frequency	32	33	34	32	35	34

Is the dice fair? Explain your answer.

4 A circular spinner is divided into four sections: red, blue, yellow and green.
The table shows the results from 100 spins.

Colour	Red	Blue	Yellow	Green
Frequency	19	18	47	16

Do you think the sections of the spinner are equally sized?
Explain your answer.

5 A group of students carry out an experiment to investigate the probability that a dropped drawing pin lands point up.
Each student drops a drawing pin 10 times and records the number of times it lands point up.

Point up

Point down

Student	1	2	3	4	5	6	7	8	9	10
Point up	2	3	4	1	1	2	3	2	1	3
Relative frequency										

a Copy the table and complete the relative frequency row.
b Write the best estimate of the drawing pin landing point up.

A02 Functional Maths

6 A four-sided dice is rolled 36 times. These are the results.

2	1	4	2	2	3	2	1	3	4	4	1
3	1	3	2	3	2	1	1	3	1	2	3
2	4	4	1	1	4	1	4	4	1	3	4

a Find the relative frequency of the dice landing on 1, 2, 3 and 4.
b Is the dice biased? Explain your answer.

D3.4 Independent events

This spread will show you how to:

- Recognise independent events
- Know that if A and B are independent events, the probability of both A and B occurring is P(A and B) = P(A) × P(B)

Keywords
Event
Independent

- Two or more events are **independent** if when one **event** occurs it has no effect on the other event(s) occurring.

For example, when you flip two coins, the result for one coin does not affect the result for the other.

- For two independent events A and B
  P(A and B) = P(A) × P(B)

This is called the 'multiplication rule' or the 'AND rule'.

Example

A spinner has 9 equal sides:
2 green, 3 blue, 1 red and 3 white.
A fair coin is thrown and the spinner is spun.

a What is the probability of
  i head and green **ii** tail and white
  iii tail and red **iv** head and red?

b Comment on your answers to **iii** and **iv**.

p.28

a **i** P(head and green) = P(H) × P(green) $= \frac{1}{2} \times \frac{2}{9} = \frac{1}{9}$

ii P(tail and white) = P(T) × P(white) $= \frac{1}{2} \times \frac{3}{9} = \frac{1}{6}$

iii P(tail and red) = P(T) × P(red) $= \frac{1}{2} \times \frac{1}{9} = \frac{1}{8}$

iv P(head and red) = P(H) × P(red) $= \frac{1}{2} \times \frac{1}{9} = \frac{1}{18}$

b The answers to **iii** and **iv** are the same, because the coin is fair, so P(H) = P(T).

Example

Find the probability of rolling three consecutive sixes on a fair dice.

$P(6) = \frac{1}{6}$

$P(6 \text{ and } 6 \text{ and } 6) = P(6) \times P(6) \times P(6)$
$= \frac{1}{6} \times \frac{1}{6} \times \frac{1}{6}$
$= \frac{1}{216}$

Rolling a dice three times will give independent results.

Exercise D3.4 Grade B

1 **a** Copy and complete the table to list all the outcomes, when a fair coin is thrown and a fair dice is rolled. One has been done for you.

		Dice					
		1	2	3	4	5	6
Coin	Head						
	Tail				Tail and 4		

b Find the probability of
i head and 3 **ii** tail and 6.
Comment on your answers.

2 A red and a blue dice are rolled.
a Draw a table to show all the possible outcomes.
b Find the probability of
i 1 on the red dice and 1 on the blue dice
ii 2 on the red dice and 5 on the blue dice
iii 3 on the red dice and an even number on the blue dice
iv 5 or greater on the red dice and 5 on the blue dice.

3 A fair dice is rolled twice.
Find the probability that the dice shows an odd number on the first roll and a number less than 3 on the second roll.

4 A spinner has ten equal sides, 4 show squares, 3 pentagons, 2 hexagons and 1 circle. The spinner is spun twice.
Find the probability of
a square on the first and second spins
b square on first and circle on second spin
c pentagon on first and hexagon on second spin
d hexagon on first and circle on second spin
e circle on first and pentagon on second spin.

5 A 10 pence and a 2 pence coin are spun.
a Draw a table to show all the possible outcomes.
b Find the probability that the 10 p shows tails and the 2 p shows heads.

A03 Problem

6 The probability that Richard will catch a bus to work is 0.24.
The probability that Kerry will catch a bus to work is 0.15.
The probability that Richard and Kerry will catch the bus to work is 0.036.
Are the events 'Richard catches a bus to work' and 'Kerry catches a bus to work' independent? Explain your reasons.

D3.5 Probability of two events

This spread will show you how to:

- Calculate the probabilities for combinations of two events

Keywords
Event
Outcome

To calculate the probability of two **events** both occurring, you need to identify all the possible **outcomes** systematically.

Example

In a game of backgammon you can win or lose.
The probability that Joshua will win any game of backgammon against Reuben is 0.4.
Joshua and Reuben play two games of backgammon.
Work out the probability that Joshua wins at least one game.

The only outcome where Joshua *does not* win at least one game is 'lose, lose'.

P(win at least one game) = 1 − P(lose, lose)
$= 1 - 0.6 \times 0.6 = 1 - 0.36 = 0.64$

The results of the two games are independent.
P(lose, lose) = P(lose) × P(lose)

Example

Two boxes of chocolates contain caramel, nut and cream centres.
Box X has 4 caramel, 7 nut and 9 cream centres.
Box Y has 8 caramel, 5 nut and 2 cream centres.

X

Y

p.106

Jodie chooses one chocolate at random from box X and one from box Y.
Work out the probability that she picks two chocolates of different types.

P(same types) = P(caramel, caramel) + P(nut, nut) + P(cream, cream)

$= (\frac{4}{20} \times \frac{8}{15}) + (\frac{7}{20} \times \frac{5}{15}) + (\frac{9}{20} \times \frac{2}{15})$

$= \frac{32}{300} + \frac{35}{300} + \frac{18}{300} = \frac{85}{300} = \frac{17}{60}$

P(different types) $= 1 - \frac{17}{60} = \frac{43}{60}$

The events 'from box X' and 'from box Y' are independent.

Example

The table shows the number of boys and the number of girls in Years 12 and 13 at a school.
Two students are to be chosen at random.

	Year 12	Year 13
Boys	48	96
Girls	72	54

a One student is chosen from Year 12 and one from Year 13.
Calculate the probability that both students will be girls.

b One student is chosen from all the girls and one from all the boys.
Calculate the probability that both students are in Year 13.

a P(girl Y12 and girl Y13) $= \frac{72}{120} \times \frac{54}{150} = \frac{27}{125}$

b P(Y13 girl and Y13 boy) $= \frac{54}{126} \times \frac{96}{144} = \frac{2}{7}$

The denominator for each fraction depends on the group from which the student is chosen.

Exercise D3.5

Grade B/A

1 In a game of Armageddon chess you can win or lose.
The probability that Edmund will win any game of Armageddon chess against Kevin is 0.9.
Kevin and Edmund play two games of chess.
Work out the probability that Edmund will win at least one game.

2 James has two bags of marbles.
Bag A contains 4 red, 3 green and 5 blue marbles.
Bag B contains 7 red, 6 green and 2 blue marbles.
James chooses one marble at random from each bag.
Find the probability that the two marbles he chooses are different colours.

3 Penny has two folders full of photos.
Folder X has 7 pictures of her family, 11 of friends and 6 of her cat.
Folder Y has 9 pictures of her family, 12 of friends and 15 of her cat.
Penny chooses one picture at random from each folder.
Work out the probability that the pictures she chooses are of different subjects.

4 The table shows the number of male and female staff in two branches of a company.

	London	Manchester
Male	76	74
Female	84	66

Two people are to be chosen at random.

a One person is chosen from London and one from Manchester.
Calculate the probability that both are women.

b One woman and one man are chosen.
Calculate the probability that both are from Manchester.

A03 Problem

5 The table shows how many boys and girls in Class 10Z wear glasses.

	Wears glasses	Does not wear glasses
Boys	6	10
Girls	4	13

Two students are to be chosen at random from class 10Z.

a One student is chosen who wears glasses and one student is chosen who does not wear glasses. Calculate the probability that both students will be boys.

b One student is chosen from all the girls and one student is chosen from all the boys. Calculate the probability that neither student wears glasses.

Summary

Check out

You should now be able to:

- Understand and use the probability scale and calculate theoretical probabilities
- Identify different mutually exclusive outcomes and their related probabilities
- Know that if two events, A and B, are mutually exclusive, then the probability of A or B occurring is P(A) + P(B)
- Know that if two events, A and B, are independent, then the probability of A and B occurring is P(A) × P(B)
- Compare experimental data and theoretical probabilities
- List all outcomes for events in a systematic way

Worked exam question

Martin has a pencil case which contains 4 blue pens and 3 green pens.
Martin picks a pen at random from the pencil case.
He notes its colour, and then replaces it.
He does this two more times.
Work out the probability that when Martin takes three pens, exactly two are the same colour. (3)

(Edexcel Limited 2007)

BBB

BBG ⟶ $\frac{4}{7} \times \frac{4}{7} \times \frac{3}{7} = \frac{48}{343}$

BGB ⟶ $\frac{4}{7} \times \frac{3}{7} \times \frac{4}{7} = \frac{48}{343}$

BGG ⟶ $\frac{4}{7} \times \frac{3}{7} \times \frac{3}{7} = \frac{36}{343}$

GGG

GBG ⟶ $\frac{3}{7} \times \frac{4}{7} \times \frac{3}{7} = \frac{36}{343}$

GGB ⟶ $\frac{3}{7} \times \frac{3}{7} \times \frac{4}{7} = \frac{36}{343}$

GBB ⟶ $\frac{3}{7} \times \frac{4}{7} \times \frac{4}{7} = \frac{48}{343}$

$\frac{48}{343} + \frac{48}{343} + \frac{36}{343} + \frac{36}{343} + \frac{36}{343} + \frac{48}{343} = \frac{252}{343}$

Show the calculations for the probabilities of the 6 outcomes, that give 2 pens with the same colour.

OR

BBB ⟶ $\frac{4}{7} \times \frac{4}{7} \times \frac{4}{7} = \frac{64}{343}$

GGG ⟶ $\frac{3}{7} \times \frac{3}{7} \times \frac{3}{7} = \frac{27}{343}$

$\frac{64}{343} + \frac{27}{343} = \frac{91}{343}$

$1 - \frac{91}{343} = \frac{252}{343}$

There are 2 outcomes that give 3 pens with the same colour. Show the subtraction from 1.

Exam questions

AO3

1 Jeremy designs a game for a school fair.

He has two 5-sided spinners.
The spinners are equally likely to land on each of their sides.

One spinner has 2 red sides, 1 green side and 2 blue sides.
The other spinner has 3 red sides, 1 yellow side and 1 blue side.

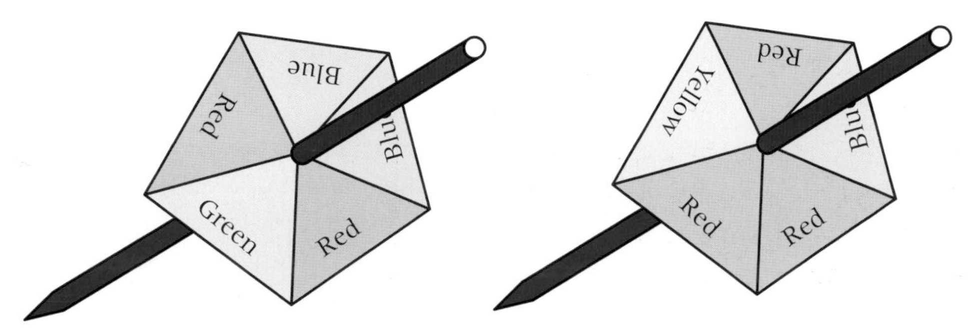

The game consists of spinning each spinner once.
It costs 20p to play the game.

To win a prize both spinners must land on the same colour.
The prize for a win is 50p.

100 people play the game.

Work out an estimate of the profit that Jeremy should expect to make. (5)

(Edexcel Limited 2005)

2 The probability that any piece of buttered toast will land buttered side down when it is dropped is 0.62

Two pieces of buttered toast are to be dropped, one after the other.

Calculate the probability that exactly one piece of buttered toast will land buttered side down. (4)

(Edexcel Limited 2006)

N2

Ratio and proportion

In modern society people want to buy their own house, own a new car and go to university. To do this they need to borrow money from a bank or building society. These organisations lend the money but charge a fee (called 'interest') that is calculated as a percentage (or fraction) of the amount borrowed.

What's the point?

Being able to solve problems involving percentages gives people greater control of their finances. It allows them to budget properly and be aware of the risks involved in borrowing too much money. In real life not being able to pay back enough money each month can lead to debt and bankruptcy.

Ratio and proportion

Check in

You should be able to

- **recognise equivalent fractions, decimals and percentages**

1 Carry out these conversions.

a Convert $\frac{3}{5}$ to a percentage. **b** Convert 0.35 to a fraction.
c Convert 65% to a fraction. **d** Convert $\frac{7}{20}$ to a decimal.

- **calculate fractions of an amount**

2 Calculate each of these.

a $\frac{2}{3}$ of 120 **b** 0.35×60 **c** 28% of 70 **d** $\frac{3}{5}$ of 280

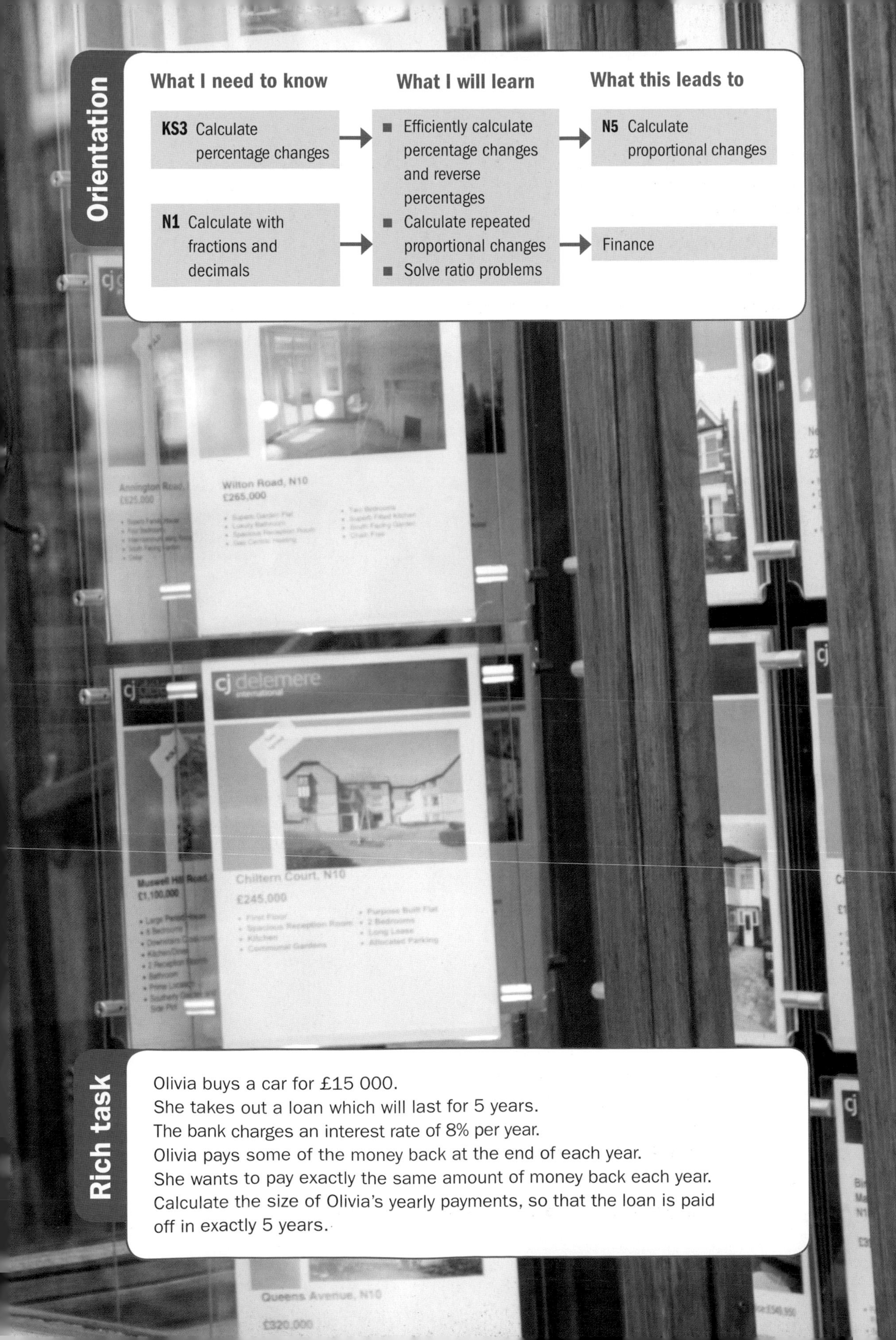

Orientation

What I need to know	What I will learn	What this leads to
KS3 Calculate percentage changes	■ Efficiently calculate percentage changes and reverse percentages	**N5** Calculate proportional changes
N1 Calculate with fractions and decimals	■ Calculate repeated proportional changes ■ Solve ratio problems	Finance

Rich task

Olivia buys a car for £15 000.
She takes out a loan which will last for 5 years.
The bank charges an interest rate of 8% per year.
Olivia pays some of the money back at the end of each year.
She wants to pay exactly the same amount of money back each year.
Calculate the size of Olivia's yearly payments, so that the loan is paid off in exactly 5 years.

Proportion problems

This spread will show you how to:

- Calculate a proportion of an amount, writing it as a fraction of the total
- Solve problems involving proportion

Keywords
Part
Proportion
Total

- To calculate a **proportion**, write the **part** as a fraction of the **total**.

  In a cooking class with 14 girls and 16 boys, the proportion of girls is $\frac{14}{30} = \frac{7}{15} = 0.467 = 46.7\%$.

- If you know what proportion of the total a part is, you can find the actual size of the part by multiplying the proportion by the total.

  If a 450 g pie contains 12% chicken, the actual weight of chicken is $450\,\text{g} \times \frac{12}{100} = 54\,\text{g}$.

Example

a Find: **i** $\frac{3}{5}$ of 205 **ii** 31% of 490 **iii** 135% of 27 **iv** $\frac{15}{16}$ of 95

b What proportion is **i** 5 of 35 **ii** 7 of 31 **iii** 18 of 24?
Give your answers as fractions and percentages.

a **i** Estimate: $\frac{3}{5}$ of $205 \approx \frac{3}{5}$ of $200 = 120$

$\frac{3}{\cancel{5}_1} \times \cancel{205}^{41} = \frac{3}{1} \times 41 = 123$

ii Estimate: 31% of 490 ≈ 30% of 500 = 150
By long multiplication, $31 \times 49 = 1519$
So 31% of 490 = 151.9

iii Estimate: 135% of 27 ≈ 100% + 30% of 30 = 30 + 9 = 39
By long multiplication, $27 \times 135 = 3645$
So 135% of 27 = 36.45

iv Estimate: $\frac{15}{16}$ of $95 \approx 90$

$\frac{15}{16} \times 95 = \frac{15 \times 95}{16} = 89.0625$ (by calculator)

b **i** $\frac{\cancel{5}^1}{\cancel{35}_7} = \frac{1}{7}$
$\frac{1}{7} = 14.3\%$
5 is $\frac{1}{7}$ or 14.3% of 35

ii $\frac{7}{31}$
$\frac{7}{31} = 22.6\%$
7 is $\frac{7}{31}$ or 22.6% of 31

iii $\frac{\cancel{18}^3}{\cancel{24}_4} = \frac{3}{4}$
$\frac{3}{4} = 75\%$
18 is $\frac{3}{4}$ or 75% of 24

You can use these techniques to solve proportion problems.

Example

Tom has £2800. He gives $\frac{1}{5}$ to his son and $\frac{1}{4}$ to his daughter.
How much does Tom keep himself? You must show all of your working.

$\frac{1}{5}$ of £2800 = 2 × £280 = £560
$\frac{1}{4}$ of £2800 = £700
Tom keeps £2800 − (£560 + £700) = £1540

Exercise N2.1 Grade D

1 Work out these proportions, giving your answers as
i fractions in their lowest terms **ii** percentages.

a 7 out of every 20 **b** 8 parts in a hundred
c 6 out of 20 **d** 75 in every 1000
e 18 parts out of 80 **f** 9 parts in every 60

2 A 250 ml glass of fruit drink contains 30 ml of pure orange juice. What proportion of the drink is pure juice? Give your answer as

a a fraction in its lowest terms
b a percentage.

3 A soft drink comes in two varieties. The 'Regular' variety contains 8.4% sugar by weight, and the 'Lite' version contains 2.5% sugar by weight. Find the weight of sugar in these amounts of soft drink.

a 200 g of Regular **b** 350 g of Lite **c** 360 g of Lite
d 800 g of Regular **e** 750 g of Lite **f** 580 g of Regular

4 Calculate

a 90% of 50 kg **b** 105% of 80 g **c** 80% of 60 cm
d 95% of 300 m **e** 55% of 29 cc **f** 65% of 400 mm

Do these mentally.

5 Use an appropriate method to work out these percentages. Show your method each time.

a 50% of 270 kg **b** 27.9% of 115 m **c** 37.5% of £280
d 25% of 90 cm^3 **e** 19% of 2685 g **f** 27.5% of £60.00

6 Calculate

a $\frac{4}{5}$ of £800 **b** $\frac{3}{7}$ of €420 **c** $\frac{3}{8}$ of 640 kg
d $\frac{2}{5}$ of 760 cc **e** $\frac{2}{3}$ of 2460 kg **f** $\frac{5}{6}$ of 28 m

7 Jo wins £3600 in a competition. She gives $\frac{1}{3}$ to her mother, and $\frac{1}{4}$ to her sister.

a How much does Jo keep? Show your working.
b What proportion of the prize money does she give away? Give your answer as
i a fraction
ii a percentage.

N2.2 Percentage problems

This spread will show you how to:

- Find percentages of quantities
- Calculate percentage increases and decreases

Keywords
Percentage decrease
Percentage increase

- To work out 72% of a quantity, multiply by 0.72
  72% of £18 = 0.72 × £18 = £12.96

0.72 is the decimal equivalent of 72%.

- You can find a **percentage increase** in the same way.
  For example, to find a 2% increase, multiply by 1.02

  10 million increased by 2% = 1.02 × 10 000 000
  = 10 200 000

The new amount will be 102% of the original amount.

- Use the same method to find a **percentage decrease**.
  For example, to find a 15% decrease, multiply by 0.85

  £50 reduced by 15% = 0.85 × £50 = £42.50

The new amount will be 85% of the original amount.

Example

What decimal equivalent would you use to find
a 37.5% of a quantity **b** a 12% increase **c** a 42% decrease?

a 37.5% = 0.375
b For a 12% increase, multiply by 1.12
c For a 42% decrease, multiply by 0.58

100% − 42% = 58%

Example

a Increase 500 g by 25%. **b** Reduce £265 by $\frac{1}{3}$.

a 500 g × 1.25 = 625 g
b The new price is $\frac{2}{3}$ or $66\frac{2}{3}$% of the original price.
265 × $0.\dot{6}$ = £176.67 to the nearest penny

The new mass is 125% of the original mass. 125% = 1.25

$66\frac{2}{3}\% = 0.\dot{6}$

Example

A 740 ml bottle of shampoo costs £2.45. In a special offer the bottle size is increased by 15% and the price is reduced by 15%.

Find the original and the new cost per litre of the shampoo.

original cost per litre = $\frac{£2.45}{0.74}$ = £3.31 per litre
Size of new bottle = 740 × 1.15 = 851 ml
New Price = £2.45 × 0.85 = £2.08
New cost per litre = $\frac{2.08}{0.851}$ = £2.45 per litre

Exercise N2.2 Grade D/C

1 Use a mental method to find these amounts.

a 25% of £48 **b** 40% of 600 m **c** 90% of 58 kg
d 5% of 3.60 km **e** 110% of €64.00 **f** 30% of 60 seconds
g 105% of 82 cm **h** 95% of 36 grams

2 **a** Write a decimal equivalent for each of the percentages in question **1**.
b Use a calculator to check your answers to part **a**.

3 Write a decimal equivalent that you would multiply by, to find these percentages of a quantity.

a 30% **b** 32% **c** 32.5%
d 1.25% **e** 112% **f** 0.006%

4 Write the decimal multipliers for these percentage increases.

a 15% **b** 2.5% **c** 22.5%
d 87.5% **e** 108% **f** 0.045%

5 Write the decimal multipliers for these percentage decreases.

a 5% **b** 8.25% **c** 22.75%
d 38.25% **e** 98% **f** 100%

6 The price of petrol is increased by 10%.

a Write the decimal multiplier equivalent to 10%.

This new price is then increased by a further 10%.

b Work out the decimal equivalent you would use to find, in one step, the result of the two successive 10% increases.

7 Find the single decimal equivalent for these percentage changes.

a A 5% increase, followed by a 6% increase.
b An 8% increase, followed by a 10% decrease.
c A 22% decrease, followed by a 30% increase.

AO3 Problem

8 Mandy buys a sweater over the internet. She has to add VAT at 17.5% to the price shown. Her loyalty club card entitles her to a 7.5% discount on the order.

- Mandy's Dad tells her to deduct the discount before adding the VAT.
- Her Mum tells her it would be better to add the VAT first, then subtract the discount.

What should Mandy do? Explain your answer.

9 The rate of VAT is increased from 17.5% to 20%.
What is the percentage increase in prices?

N2.3 Reverse percentage problems

This spread will show you how to:

- Solve reverse percentage problems

Keywords
Reverse percentage

- To find one quantity as a percentage of another, simply divide one quantity by the other, and then multiply by 100%.

If 36 people out of 400 wear contact lenses, this represents $(36 \div 400) \times 100\% = 9\%$

- To find the original amount before a percentage change, first decide what multiplication was needed to find the new amount.
For example

A train fare is £31.72 after a **22%** increase, so the original fare was multiplied by **1.22**

New price is 122% of original price.

If a freshly-baked cake weighs 2200 grams, and this is a **12%** reduction on the uncooked weight, then the uncooked weight was multiplied by **0.88**

Cooked weight is 88% of uncooked weight.

- To find the original amount before the percentage change, simply 'undo' the multiplication by dividing the new amount by the appropriate number.
For example

This is a **reverse percentage**.

The original train fare $= 31.72 \div 1.22 =$ £26.00

The uncooked weight of the cake $= 2200 \div 0.88 = 2500$ g

Example

a A shop buys DVD recorders at a wholesale price of £85 each, and sells them at £155. Find the percentage profit.
b A computer costs £998.75 including VAT at 17.5%. Find the cost before VAT.
c The price of a dress is reduced by 30% in a sale. If the sale price is £77, find the original price.

a Profit = £155 − £85 = £70.
Profit as a percentage of the wholesale price $= \frac{70}{85} \times 100\% = 82.4\%$
b Original cost × 1.175 = £998.75
⇒ original cost = £998.75 ÷ 1.175 = £850
c Original price × 0.7 = £77
⇒ original price = £77 ÷ 0.7 = £110

Example

A restaurant sells a bottle of wine for £10.35, making a profit of 130%. How much did the restaurant pay for the wine?

The restaurant adds 130% to the original cost, so the selling price is 230% of the original cost.
Original cost × 2.3 = £10.35
⇒ original cost = £10.35 ÷ 2.3 = £4.50

Exercise N2.3

Grade B

Unit 1

Reverse percentage problems are examined in unit 3.

1 Find

a 20 as a percentage of 800 **b** 25 as a percentage of 750

c 16 as a percentage of 29 **d** 22 as a percentage of 1760.

2 A suit is normally sold for £350. In a sale, the price is reduced to £205. What percentage reduction is this?

3 Copy and complete the table to show the percentage reduction for each original price.

Original price	Sale price	Percentage reduction
£25.00	£21.00	
£37.50	£32.75	
£1500.00	£950.50	
£2850.75	£2200.00	

DID YOU KNOW?

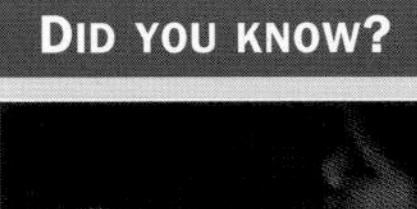

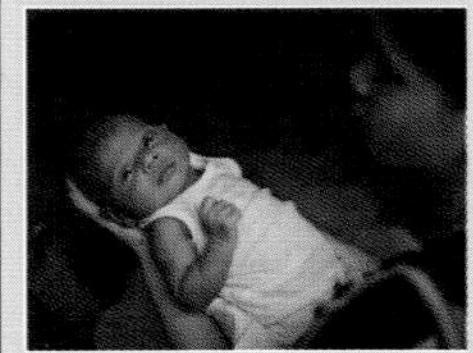

World population is growing at a rate of 1.14% per year.

AO2 Functional Maths

4 A car manufacturer increases the price of a Sunseeker sports car by 6%. The new price is £8957. Calculate the price before the increase.

5 During 2005 the population of Camtown increased by 5%. At the end of the year the population was 14 280. What was the population at the beginning of the year?

6 In a clearance sale at a jewellery shop, the price of all unsold items is reduced by 10% at the end of each day.

a A brooch is on sale for £250 on Monday morning. It is still unsold on Wednesday evening. Find the sale price of the brooch on Thursday morning.

b A pair of earrings is on sale for £280.67 on Thursday morning. Find the price of the earrings on the previous Monday morning.

AO3 Problem

7 The Retail Price Index (RPI) is a government statistic that shows the percentage increase in the price of a representative 'basket' of goods and services over the last 12 months. The table shows the RPI for the month of July for selected years.

Year	1975	1976	1977	1978	1979	1980
July RPI	26.3	13.8	17.7	7.4	11.4	21.0

a The total household expenditure for one family in July 1975 was £104.75. Estimate the same family's expenditure in July 1974.

b Another family had a monthly expenditure of £216.50 in July 1977. Estimate their monthly expenditure in July 1980. Show your working, and state any assumptions you have made.

c A third family had a monthly expenditure of £198.45 in July 1979. Estimate their monthly expenditure in July 1976.

N2.4 Repeated proportional change

This spread will show you how to:
- Calculate proportional and percentage changes
- Solve problems involving proportional change and interest

Keywords
Compound interest
Depreciation
p.a. (per annum)
Proportional change
Simple interest

- To work out a **proportional change**, first work out the size of the change, and then add to, or subtract from, the original amount.

  If fuel charges go up by $\frac{1}{4}$ and original charge = £48,

  increase = $\frac{1}{4}$ × £48 = £12 and new charge = £48 + £12 = £60.
- Percentage changes can be found using a decimal multiplier.

  To increase £20 by 15%, multiply £20 by 1.15.
  To reduce 18 kg by 12.5%, multiply 18 kg by 0.875.

100% − 12.5% = 87.5%

Example

Find the new total when
a £340 is increased by a third **b** 210 cm is decreased by 45%
c VAT at 17.5% is added to a price of £297.

a £340 ÷ 3 = £113.33
New total £340 + £113.33 = £453.33
b 210 cm × 0.55 = 115.5 cm — To decrease an amount by 45%, multiply by 0.55.
c £297 × 1.175 = £348.98 — To increase an amount by 17.5%, multiply by 1.175.

- Repeated percentage changes can be found using powers of a decimal multiplier. A common application of this is in **compound interest**.

Example

Find the final amount when an initial sum of £1000 is invested for 12 years at
a 5% p.a. simple interest **b** 5% p.a. compound interest.

a In **simple interest**, the original amount invested (the principal) earns interest, but the interest is not added to the principal.
The principal earned £50 per year for 12 years = £600.
Final amount = £1000 + £600 = £1600.
b In **compound interest**, the interest payments are added to the principal, and will themselves earn interest. The sum in the account after 12 years will be £1000 × 1.05^{12} = £1795.86.

p.a. (per annum) means 'per year'.

Some items, like cars, decrease in value as they age. This is called **depreciation**.

Example

Sue has had her car for 3 years. Each year, the car has depreciated in value by 15%. Now the car is worth £5220.
How much was the car worth when Sue bought it?

Original cost = C
C × 0.85^3 = £5220
so C = £5220 ÷ 0.85^3 = £8500 (to the nearest pound)

Exercise N2.4 — Grade C/B

Unit 1

1 Copy and complete the table to show the result of some proportional changes.

Original quantity	Proportional change	Result
235 mm	Decrease by $\frac{1}{5}$	
38 litres	Increase by $\frac{1}{8}$	
295 cm^2	Decrease by $\frac{2}{9}$	
£96.55	Increase by $\frac{5}{6}$	

2 Calculate these percentage changes.

a £375 decreased by 12% **b** £290 decreased by 15%

c £885 decreased by 9% **d** £439 increased by 22%

3 Find the simple interest earned on

a £860 at 5.3% p.a. for 4 years **b** £350 000 at 4.75% for 18 years.

4 Find the compound interest earned on

a £720 at 3.8% p.a. for 5 years **b** £410 000 at 5.15% for 12 years.

AO2 Functional Maths

5 Charlene invests £950 for 5 years in an account that pays 4.5% per year simple interest.

a How much interest will Charlene earn over 5 years?

b Charlene's Dad tells her that it would be better to invest the money in an account that paid 3.5% compound interest. Is he right? Explain your answer.

c Would your answer for part **b** have been different if the money was invested for 10 years instead of 5? Explain your reasoning.

6 £6500 is invested at 3.9% compound interest per annum. How many years will it take for the investment to exceed £8000?

7 During the 21 years that Pat owned a house, it increased in value at an average rate of 3.5% per year. When she sold the house, it was worth £245 000. Pat's sister says, 'Your house must have doubled in value since you bought it.' Is she right? Show your working.

AO3 Problem

8 A special savings account earns 7% per annum compound interest.

a Liz invests £2400 in the special account. How much will she have in her account after 2 years?

b Paul invests some money in the same account. After earning interest for 1 year, he has £1669.20 in his account. How much money did Paul invest?

N2.5 Ratio problems

This spread will show you how to:

- Use ratio notation, including reduction to its simplest form and its various links to fraction notation
- Divide a quantity in a given ratio
- Solve problems involving ratio

Keywords
Ratio
Simplify

- **Ratios** are used to compare quantities.

If there are 18 girls and 14 boys in a class, the ratio of girls to boys = 18 : 14.
This simplifies to 9 : 7.

Simplify a ratio by cancelling common factors, for example 8 : 12 simplifies to 2 : 3.

- To divide a quantity in a given ratio
  - first find the total number of parts
  - find the size of one part
  - multiply to find each share.

Example

Share £400 between Tom and Ed in the ratio 3 : 2.

Total number of parts = 3 + 2 = 5
Size of one part = £400 ÷ 5 = £80
3 × £80 = £240
2 × £80 = £160
Tom gets £240, Ed gets £160.

Example

Divide
a 460 cm in the ratio 3 : 2
b £2500 in the ratio 3 : 2 : 5.

a 3 + 2 = 5
460 ÷ 5 = 92
2 × 92 = 184
3 × 92 = 276
The two parts are 276 cm and 184 cm.

b 3 + 2 + 5 = 10
2500 ÷ 10 = 250
250 × 3 = 750
250 × 2 = 500
250 × 5 = 1250
The three parts are £750, £500 and £1250.

Example

Andy and Brenda share a bingo prize of £65 in the ratio of their ages. Andy is 38, and Brenda is 42. How much does each get?

Ratio of ages = 38 : 42 = 19 : 21
19 + 21 = 40
Andy receives (£65 ÷ 40) × 19 = £30.88
Brenda receives (£65 ÷ 40) × 21 = £34.12.

One part is £65 ÷ 40.

Exercise N2.5 Grade C

1 Simplify these ratios.

a 8 : 4 **b** 15 : 3 **c** 6 : 4
d 2 : 4 : 6 **e** 45 : 18 **f** 14 : 28 : 84

2 Divide each of these amounts in the ratio given. Show your working.

a £55 in the ratio 3 : 2 **b** 120 cm in the ratio 5 : 3
c 96 seats in the ratio 4 : 3 : 1 **d** 42 tickets in the ratio 9 : 3 : 2
e 144 books in the ratio 8 : 3 : 1 **f** 160 hours in the ratio 8 : 5 : 3

3 Copy and complete the table to show how the quantities can be divided in the ratios given. Show your working.

Quantity	Ratio	First share	Second share
£140	4 : 3		
85 cm	3 : 2		
18 hours	5 : 4		
49 cc	1 : 6		
45 minutes	7 : 8		
€720	19 : 5		

4 Mrs Jackson wins £800 in a competition. She decides to keep half of the money, and share the rest between her two children, Amber (who is 8 years old) and Benny (who is 10) in the ratio of their ages. Work out how much each child receives. Show your working.

5 Prize money of £9600 is shared between Peggy, Grant and Mehmet in the ratio 8 : 9 : 7 respectively. How much do they each receive?

6 Steven, Will and Phil divide prize money of £12 000 between them in the ratio 10 : 6 : 3. Find the amount that each person receives, giving your answers to the nearest penny.

7 Copy and complete the table to show how each quantity can be divided in the ratio given. Give your answers to a suitable degree of accuracy.

Quantity	Ratio	First share	Second share
200 cm^2	12 : 4 : 1		
38 cm	3 : 2 : 1		
450 m	6 : 3 : 4		
720 mm	2 : 5 : 4		
$95	8 : 3 : 5		

A02 Functional Maths

8 John and Janine entered a quiz. John came first with 32 points, and Janine came second with 27 points. They shared prize money of £45 in the ratio of their scores. How much did they each receive?

9 In the 2010 U.K. general election the Conservatives won 307 seats, Labour won 258 seats, the Liberal Democrates won 57 seats and other parties won 28 seats.
Show this information on a pie chart.

N2 Summary

Check out

You should now be able to:

- Describe and calculate proportions using fractions, decimals or percentages
- Solve problems involving ratio, proportion and proportional change
- Divide a quantity in a given ratio
- Calculate percentage increase and decrease
- Calculate the original amount after a percentage increase or decrease
- Understand and find reciprocals

Worked exam question

Toby invested £4500 for 2 years in a saving account.
He was paid 4% per annum compound interest.

a How much did Toby have in his savings account after 2 years? (3)

Jaspir invested £2400 for n years in a savings account.
He was paid 7.5% per annum compound interest.
At the end of the n years he had £3445.51 in the savings account.

b Work out the value of n. (2)

(Edexcel Limited, 2009)

a

104% of £4500 = 1.04 × £4500
= £4680
104% of £4680 = 1.04 × £4680
= £4867.20

Show a method to find 104% of £4500, then 104% of £4680

b

1.075 × £2400 = £2580
1.075 × £2580 = £2773.50
1.075 × £2773.50 = £2981.5125
1.075 × £2981.5125 = £3205.1259
1.075 × £3205.1259 = £3445.5103
So $(1.075)^5 \times £2400$ = £3445.51 i.e. 5 years
$n = 5$

Show that 1.075 is used for the repeated multiplications.

State the value of n.

Exam questions

1 Three women earned a total of £36
They shared the £36 in the ratio 7 : 3 : 2
Donna received the largest amount.
a Work out the amount Donna received. (3)

A year ago, Donna weighed 51.5 kg.
Donna now weighs $8\frac{1}{2}$% less.
b Work out how much Donna now weighs.
Give your answer to an appropriate degree of accuracy. (4)
(Edexcel Limited 2005)

A02

2 James invested £2000 for three years in an Internet Savings Account.
He is paid 5.5% per annum compound interest.
Work out the total interest earned after three years. (3)
(Edexcel Limited 2007)

3 Gwen bought a new car.
Each year, the value of her car depreciated by 9%.
Calculate the number of years after which the value of her car was 47% of its value when new. (3)
(Edexcel Limited 2006)

4 The value of a car depreciates by 35% each year.
At the end of 2007 the value of the car was £5460
Work out the value of the car at the end of 2006 (3)
(Edexcel Limited 2008)

Interpreting frequency graphs

Over 30% of the numbers in everyday use begin with the digit 1 whilst less that 5% begin with a 9. ‘Benford’s law’, as it is called, makes it possible to detect when a list of numbers has been falsified. This is particularly useful in fraud investigations for detecting ‘made-up’ entries on claim forms and expense accounts.

What’s the point?

One of the tasks statisticians work on is finding ways to display and characterise different data sets using statistical ‘fingerprints’. This is so that they can first identify and then measure any possible differences to decide if they are significant or not.

Check in

You should be able to

- **carry out basic arithmetic**

1 Find the midpoint of each pair.

a 5 and 15 **b** 20 and 30 **c** 0 and 40

d 10 and 15 **e** 25 and 30

2 Work out each of these.

a $(20 \times 0.2) + (10 \times 1.4) + (5 \times 0.8)$

b $(30 \times 0.6) + (20 \times 1.2) + (5 \times 0.6)$

3 Work out each of these.

a $13 \div (50 - 40)$ **b** $12 \div (10 - 5)$

c $9 \div (40 - 15)$ **d** $9 \div (0.5 - 0.2)$

Orientation

What I need to know	What I will learn	What this leads to
N1 Basic arithemetic	■ Construct and interpret frequency polygons, histogram and scatter diagrams	**A-level** Maths, Biology, Economics, Geography
D1 Calculate summary statistics **D2** Compare data sets		Business

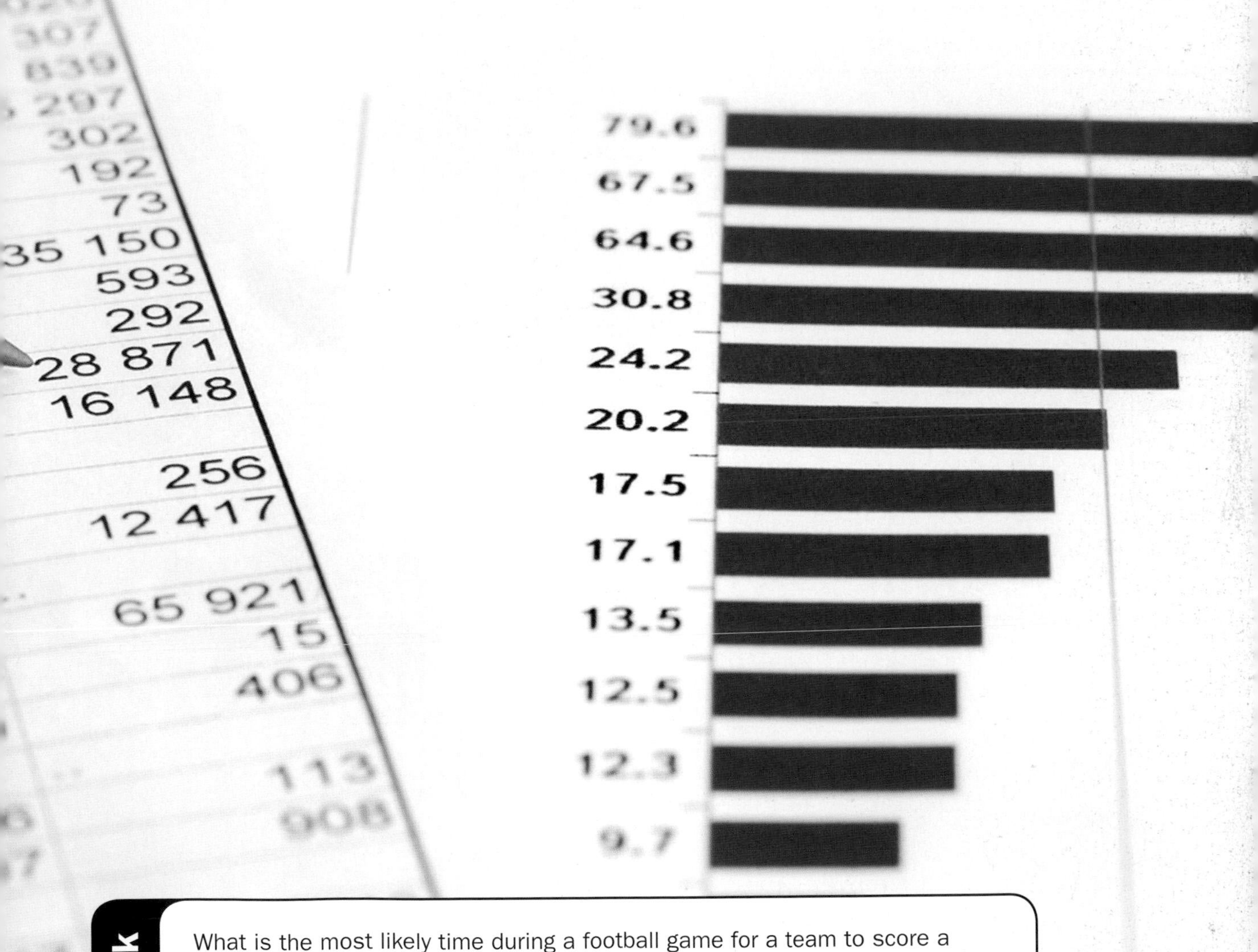

Rich task

What is the most likely time during a football game for a team to score a goal?
It is frequently stated by football commentators that teams are most likely to concede a goal within five minutes of scoring a goal themselves. Is this true?
Investigate and write a report on your results

D4.1 Frequency polygons

This spread will show you how to:

- Draw frequency polygons to represent grouped data
- Compare data sets using summary statistics

Keywords
Frequency polygon
Mean
Midpoint
Modal
Range

- You can represent grouped data in a **frequency polygon**.
- You can use frequency polygons to compare data sets.

Example

The tables show the age distribution of people in a gym and a bowls club.
Draw frequency polygons for this data.
Compare the ages of the people in the two clubs.

The data needs to be grouped in equal sized class intervals. It can be discrete or continuous.

Gym

Age, a, years	Frequency
$0 \quad a < 10$	32
$10 \leqslant a < 20$	56
$20 \leqslant a < 30$	23
$30 \leqslant a < 40$	14
$40 \leqslant a < 50$	5

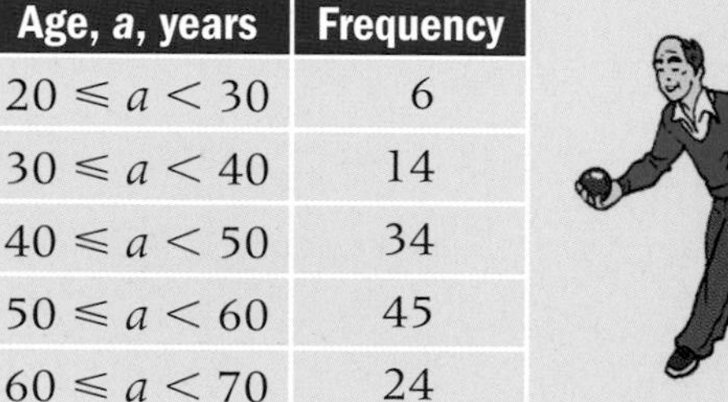

Bowls

Age, a, years	Frequency
$20 \leqslant a < 30$	6
$30 \leqslant a < 40$	14
$40 \leqslant a < 50$	34
$50 \leqslant a < 60$	45
$60 \leqslant a < 70$	24

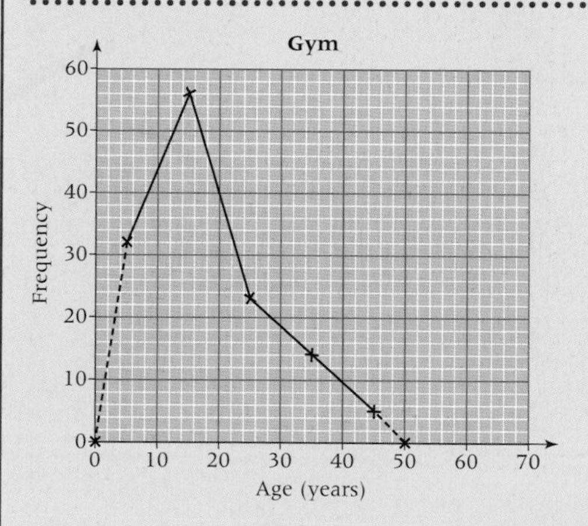

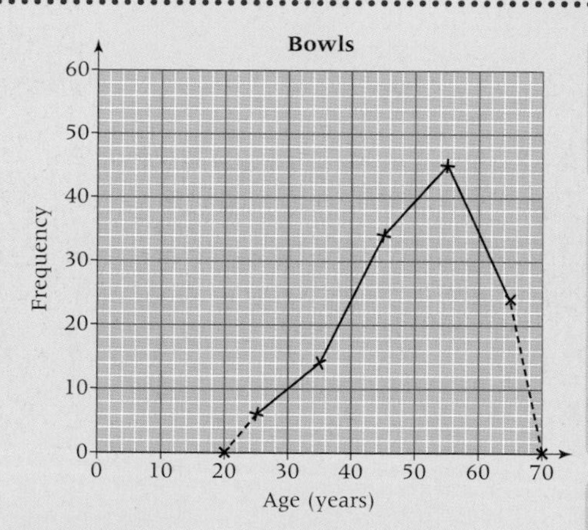

For each class you plot (**midpoint**, frequency).

For the end points you plot (0, lower bound of first class) and (0, upper bound of last class).

Join the points with straight lines.

p.12

The **modal** age is greater in the bowls club.
Highest frequency for gym is in age range 10–20 years.
Highest frequency for bowls is in age range 50–60 years.

The **range** of ages is the same at both clubs.
Estimated range for gym: $50 - 0 = 50$
Estimated range for bowls: $70 - 20 = 50$

To estimate the range, use the highest and lowest possible data values.

The estimated **mean** age for bowls is much higher than that for gym.

Gym: $\frac{(5 \times 32) + (15 \times 56) \div (25 \times 23) + (35 \times 14) + (45 \times 5)}{130}$

$= 2290 \div 130 = 17.6$ years (1 dp)

Bowls: $\frac{(25 \times 6) + (35 \times 14) + (45 \times 34) + (55 \times 45) + (65 \times 24)}{123}$

$= 6205 \div 123 = 50.4$ years (1 dp)

Use midpoint frequency to estimate the means.

1 The frequency tables show the ages of the first 100 people to visit a shopping centre on a Monday and a Saturday.

Monday

Age, a, years	Frequency
$0 \leqslant a < 10$	11
$10 \leqslant a < 20$	3
$20 \leqslant a < 30$	14
$30 \leqslant a < 40$	12
$40 \leqslant a < 50$	15
$50 \leqslant a < 60$	19
$60 \leqslant a < 70$	26

Saturday

Age, a, years	Frequency
$0 \leqslant a < 10$	7
$10 \leqslant a < 20$	18
$20 \leqslant a < 30$	31
$30 \leqslant a < 40$	20
$40 \leqslant a < 50$	12
$50 \leqslant a < 60$	9
$60 \leqslant a < 70$	3

Draw frequency polygons for these data.
Compare the ages of the shoppers on the two days.

2 The frequency polygons show the time taken by 120 teachers and 120 office workers to travel home from work.

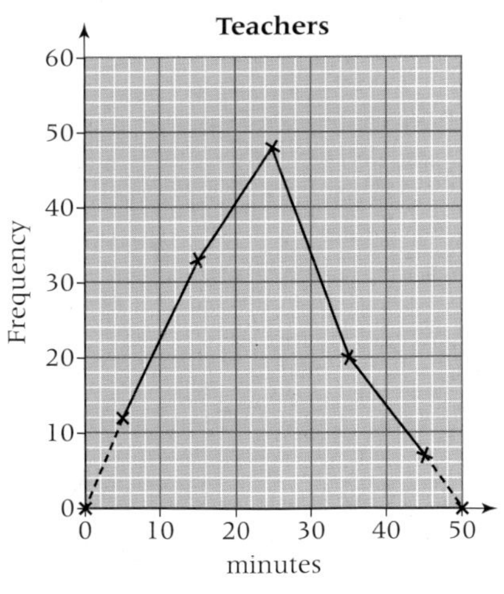

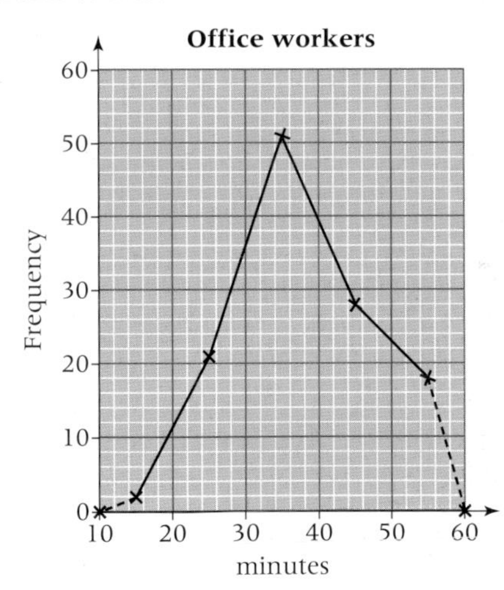

Compare the two sets of journey times.

3 The frequency polygons show the number of miles Jayne travelled each day in her car during December and January.

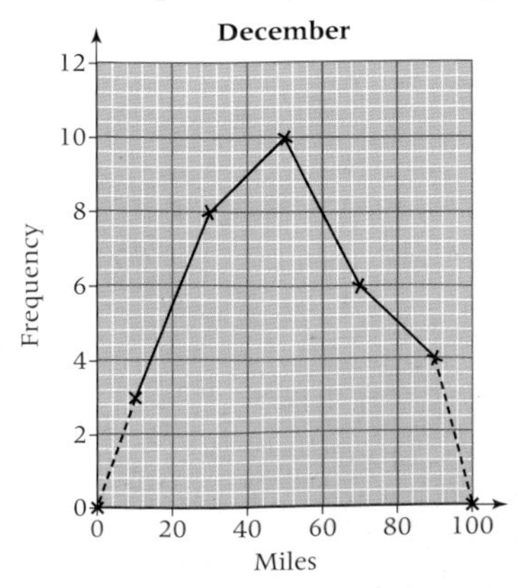

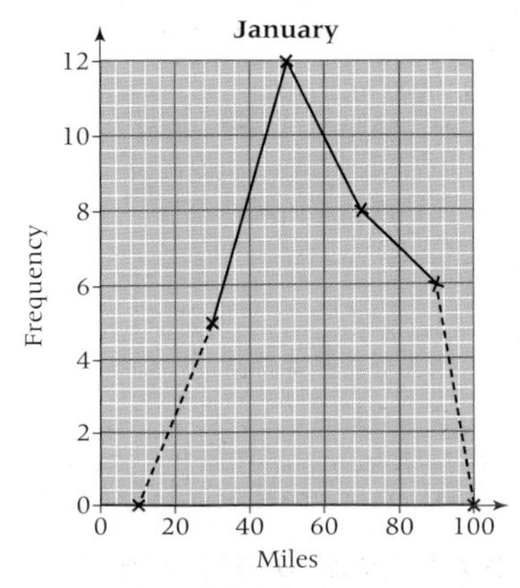

Compare the daily mileages for the two months.
Suggest reasons for any differences or similarities.

D4.2 Histograms

This spread will show you how to:

- Understand frequency density
- Represent grouped data on a histogram

Keywords
Class width
Frequency
Frequency density
Histogram

You can represent grouped continuous data in a **histogram**.

- In a histogram, the area of each bar represents the **frequency**.

The data can be in equal or unequal sized class intervals.
The vertical axis represents **frequency density**.

- Frequency density $= \dfrac{\text{frequency}}{\text{class width}}$

Area = frequency = **class width** × bar height
So bar height $= \dfrac{\text{frequency}}{\text{class width}}$

Example

Ursula collected data on the time taken to complete a simple jigsaw.
Draw a histogram to represent these data.

Time, t seconds	$40 \leqslant t < 60$	$60 \leqslant t < 70$	$70 \leqslant t < 80$	$80 \leqslant t < 90$	$90 \leqslant t < 120$
Frequency	6	6	10	7	6

Time, t seconds	$40 \leqslant t < 60$	$60 \leqslant t < 70$	$70 \leqslant t < 80$	$80 \leqslant t < 90$	$90 \leqslant t < 120$
Class width	20	10	10	10	30
Frequency	6	6	10	7	6
Frequency density	0.3	0.6	1	0.7	0.2

Add rows to the table to calculate class width and frequency density.

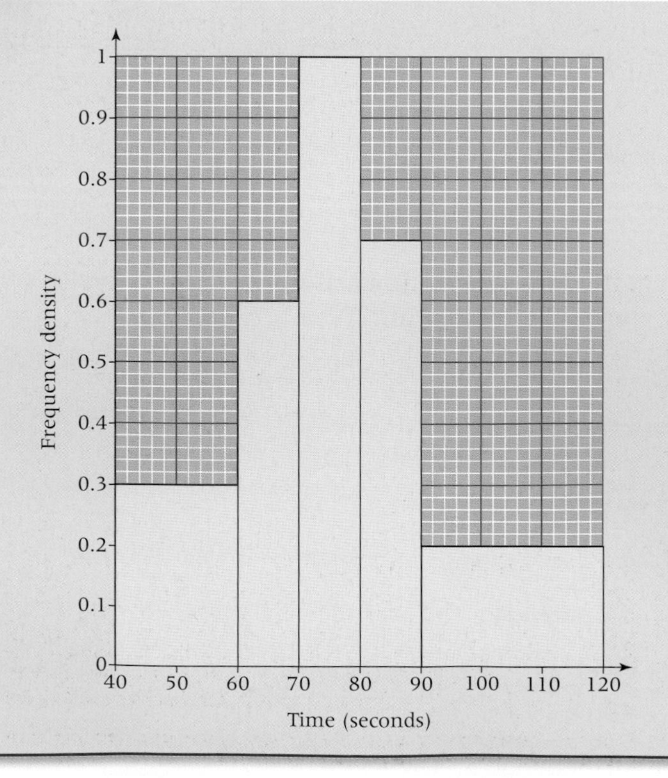

Exercise D4.2 Grade A

For each set of data

a Copy and complete the table.

b Draw a histogram to represent the data.

1 Reaction times of a sample of students.

Time, t seconds	$1 \leqslant t < 3$	$3 \leqslant t < 4$	$4 \leqslant t < 5$	$5 \leqslant t < 6$	$6 \leqslant t < 9$
Class width					
Frequency	12	17	19	11	18
Frequency density					

2 Amounts spent by the first 100 customers in a shop one Saturday.

Amount spent, £a	$0 \leqslant a < 5$	$5 \leqslant a < 10$	$10 \leqslant a < 20$	$20 \leqslant a < 40$	$40 \leqslant a < 60$	$60 \leqslant a < 100$
Class width						
Frequency	6	10	23	29	24	8
Frequency density						

3 Distance travelled to work by 100 office workers.

Distance, d miles	$0 \leqslant d < 2$	$2 \leqslant d < 5$	$5 \leqslant d < 10$	$10 \leqslant d < 20$	$20 \leqslant d < 30$
Class width					
Frequency	8	15	27	44	6
Frequency density					

4 Times of goals scored in Premiership football matches one Saturday.

Time, t minutes	$0 \leqslant t < 10$	$10 \leqslant t < 40$	$40 \leqslant t < 45$	$45 \leqslant t < 55$	$55 \leqslant t < 85$	$85 \leqslant t < 90$
Class width						
Frequency	12	48	18	11	30	22
Frequency density						

5 Distances swum by children in a sponsored swim.

Distance, d km	$0.1 \leqslant d < 0.2$	$0.2 \leqslant d < 0.5$	$0.5 \leqslant d < 1$	$1 \leqslant d < 2$	$2 \leqslant d < 5$
Class width					
Frequency	3	12	22	25	18
Frequency density					

A03 Problem

6 Times dog owners spend on daily walks.

Time, t minutes	$10 \leqslant t < 20$	$20 \leqslant t < 40$	$40 \leqslant t < 60$	$60 \leqslant t < 90$	$90 \leqslant t < ?$
Class width					30
Frequency	8				9
Frequency density		0.8	1.4	1.3	

D4.3 Interpreting histograms

This spread will show you how to:

- Understand frequency density
- Calculate frequencies from a histogram

Keywords
Class width
Frequency
Frequency density
Histogram

You can calculate frequencies from a **histogram**.
For each bar, the area represents the **frequency**.

- Frequency = frequency density × class width

Frequency density = $\frac{\text{frequency}}{\text{class width}}$

Example

The histogram shows the times a sample of students spent on the internet one evening.

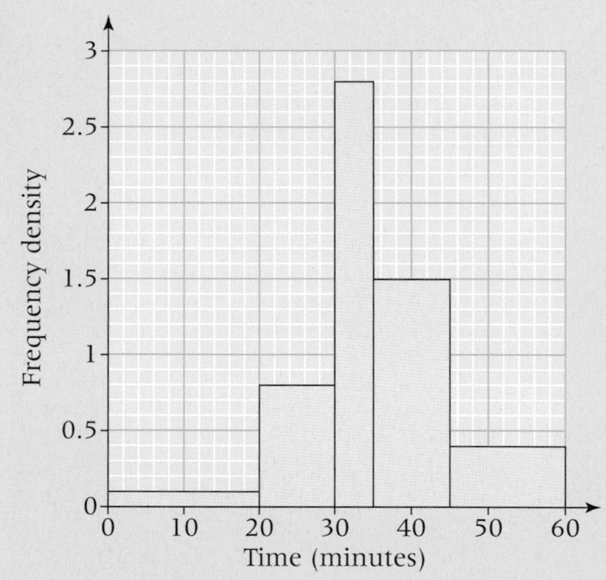

a Estimate how many students spent longer than 50 minutes on the internet?
b Complete the frequency table for these data.

Time, t minutes	$0 \leq t < 20$	$20 \leq t < 30$	$30 \leq t < 35$	$35 \leq t < 45$	$45 \leq t < 60$
Frequency					

c How many students were included in the sample?

a The area in the histogram that represents >50 minutes is only part of the last bar.

Area $50 \leq t < 60$ = height × width of 50 − 60 class interval
= 0.4 × 10 = 4

Four students spent longer than 50 minutes on the internet.

b

Time, t minutes	$0 \leq t < 20$	$20 \leq t < 30$	$30 \leq t < 35$	$35 \leq t < 45$	$45 \leq t < 60$
Frequency	0.1 × 20 = 2	0.8 × 10 = 8	2.8 × 5 = 14	1.5 × 10 = 15	0.4 × 15 = 6

c 2 + 8 + 14 + 15 + 6 = 45
45 students were included in the sample.

The area of each bar gives the frequency.

Exercise D4.3 Grade A/A*

1 The histogram shows the times a sample of students spent watching TV one evening.

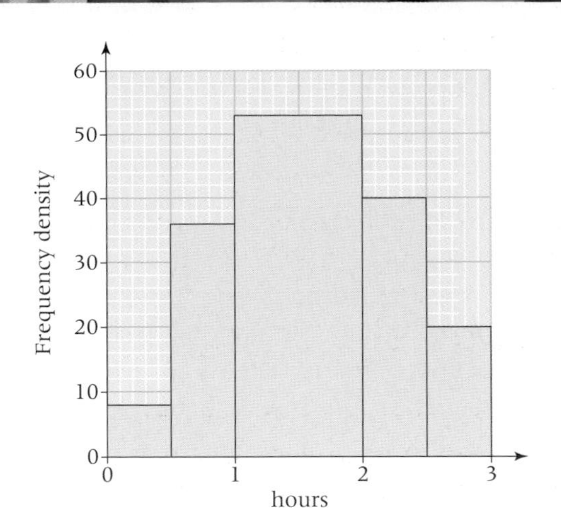

a How many students spent longer than $2\frac{1}{2}$ hours watching TV?

b Copy and complete the frequency table for these data.

c How many students were in the sample?

Time, *t* hours	$0 \leqslant t < 0.5$	$0.5 \leqslant t < 1$	$1 \leqslant t < 2$	$2 \leqslant t < 2.5$	$2.5 \leqslant t < 3$
Frequency					

2 The histogram shows the distances a sample of teachers travel to work each day.

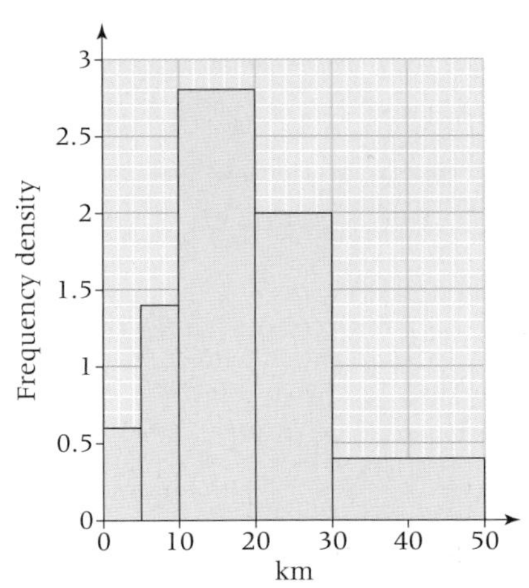

a How many teachers travel between 10 and 30 kilometres?

b Copy and complete the frequency table for these data.

c How many teachers were in the sample?

Distance, *d* km	$0 \leqslant d < 5$	$5 \leqslant d < 10$	$10 \leqslant d < 20$	$20 \leqslant d < 30$	$30 \leqslant d < 50$
Frequency					

3 The histograms show the heights of some boys aged 11 and 16. For each histogram, draw a frequency table and calculate the number of boys in each sample **a** aged 11 **b** aged 16.

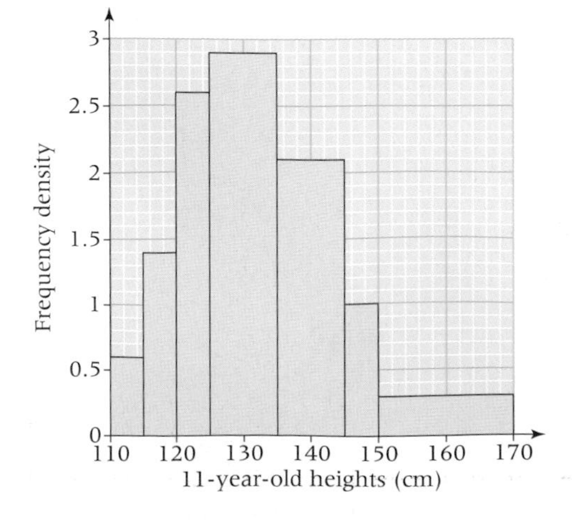

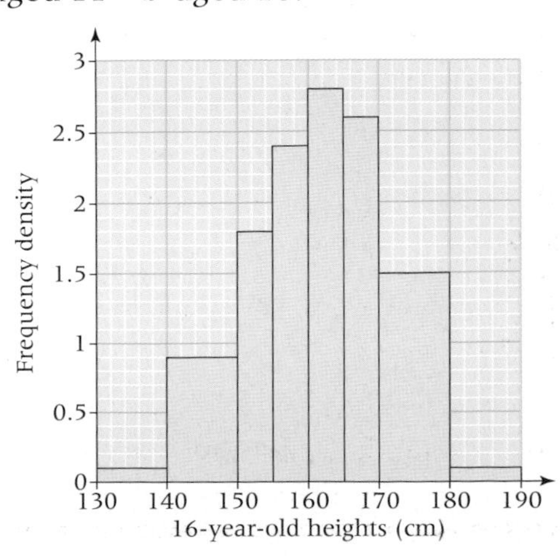

D4.4 Further histograms

This spread will show you how to:

- Use the information given in tables and histograms to deal with problems such as missing data

Keywords
Class width
Frequency
Frequency density
Histogram

You can use information from tables and **histograms** to fill in gaps in data.

Example

The incomplete table and histogram give some information about the lengths of phone calls Wendy made at work one day.

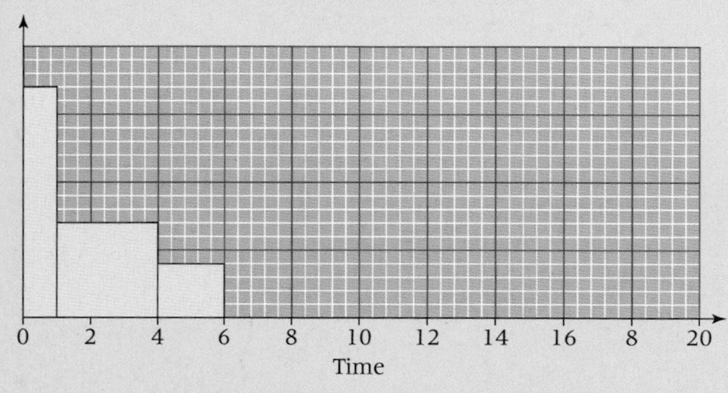

Time, t minutes	Frequency
$0 \leqslant t < 1$	17
$1 \leqslant t < 4$	
$4 \leqslant t < 6$	
$6 \leqslant t < 10$	12
$10 \leqslant t < 20$	10

a Use the information in the histogram to complete the table.
b Complete the histogram.

a The class $0 \leqslant t < 1$ has **frequency density** $= \dfrac{\text{frequency}}{\text{class width}} = \dfrac{17}{1} = 17$

On the histogram, $0 \leqslant t < 1$ bar has height 3.4 cm,
so vertical scale $= 17 \div 3.4 = 5$ per cm

From the histogram the class, $1 \leqslant t < 4$ has
frequency = frequency density × **class width** $= 7 \times 3 = 21$ calls.
The class $4 \leqslant t < 6$ has frequency $= 4 \times 2 = 8$ calls.
The completed table is:

Time, t min	Frequency	Class width	Frequency density
$0 \leqslant t < 1$	17	1	17
$1 \leqslant t < 4$	21	3	7
$4 \leqslant t < 6$	8	2	4
$6 \leqslant t < 10$	12	4	3
$10 \leqslant t < 20$	10	10	1

Use the information in the table to work out the scale on the vertical axis.

Calculate the frequency densities:
$6 \leqslant t < 10$:
fd = 12/4 = 3
$10 \leqslant t < 20$:
fd = 10/10 = 1

b

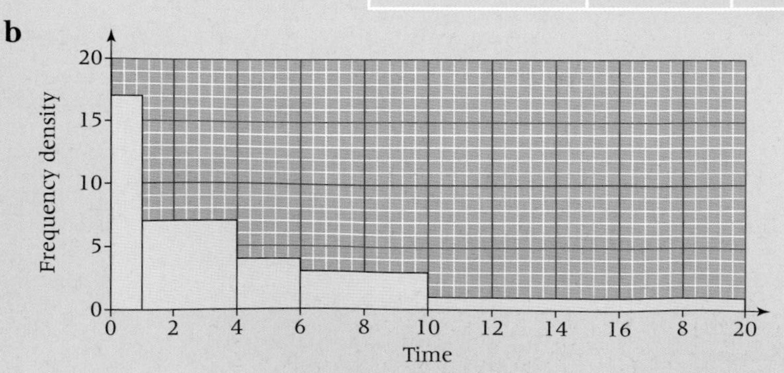

Remember to scale and label the vertical axis.

Exercise D4.4 Grade A*

1 The incomplete table and histogram give some information about the weights, in grams, of a sample of apples.

Weight, g grams	Frequency
$120 \leqslant g < 140$	8
$140 \leqslant g < 150$	6
$150 \leqslant g < 155$	
$155 \leqslant g < 160$	
$160 \leqslant g < 165$	
$165 \leqslant g < 175$	16
$175 \leqslant g < 185$	12
$185 \leqslant g < 200$	6

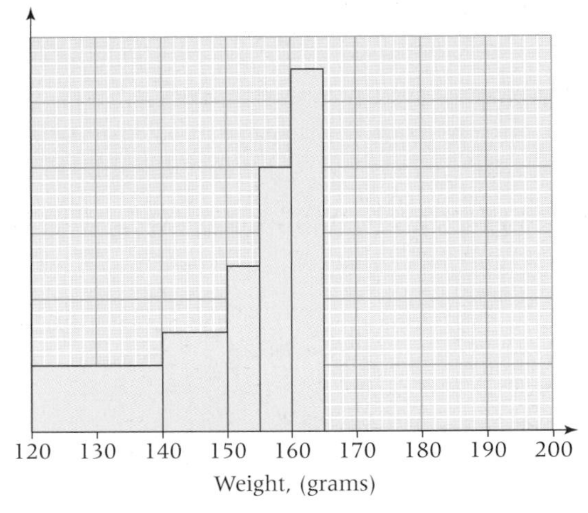

a Use the information in the histogram to work out the missing frequencies in the table.

b Copy and complete the histogram.

2 The incomplete table and histogram give some information about distances travelled by sales representatives on one day.

Miles travelled, m	Frequency
$0 \leqslant m < 80$	32
$80 \leqslant m < 100$	24
$100 \leqslant m < 120$	
$120 \leqslant m < 140$	
$140 \leqslant m < 160$	44
$160 \leqslant m < 200$	28

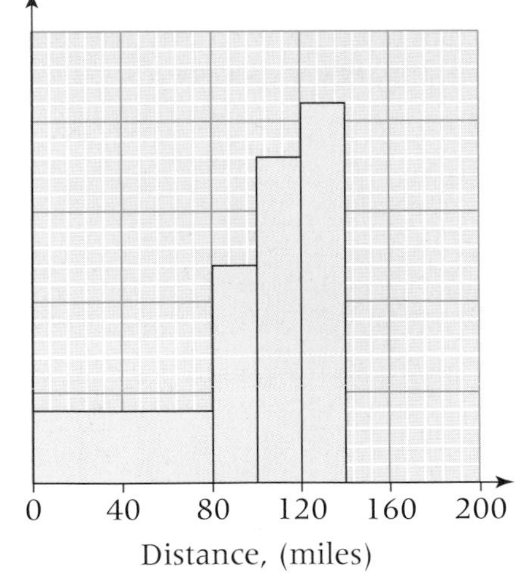

a Copy and complete the table.

b Copy and complete the histogram.

Problem A03

3 Copy the tables and work out the height, in terms of f, for the second bar in each of these histograms.

a

Time, t minutes	Frequency	Bar height	Frequency density
$0 \leqslant t < 10$	8	3.2 cm	$8 \div 10 = 0.8$
$10 \leqslant t < 40$	f		

b

Time, t minutes	Frequency	Bar height	Frequency density
$0 \leqslant t < 25$	10	2 cm	
$25 \leqslant t < 30$	f		

D4.5 Using histograms to compare data sets

This spread will show you how to:

- Use histograms to compare two or more data sets, considering the modal class, range and skewness

Keywords
Frequency density
Histogram
Modal class
Skewness

You can use **histograms** to compare data sets.

p.46

- The highest bar on a histogram represents the **modal class**.
- You can estimate the range.

This is the class with the highest **frequency density.** It **may not** be the class with the highest frequency.

The range is an estimate, as you do not have the actual data values.

The shape of a histogram shows whether the data is skewed.

Positive skew — Heights of jockeys; No skew – symmetrical — Heights of random sample of men; Negative skew — Heights of basketball players

- To compare histograms for two or more data sets, consider the modal class, range and **skewness**.

Example

The histograms show the times taken by a sample of boys and a sample of girls to complete the same puzzle.
Compare the times taken by the boys and the girls.

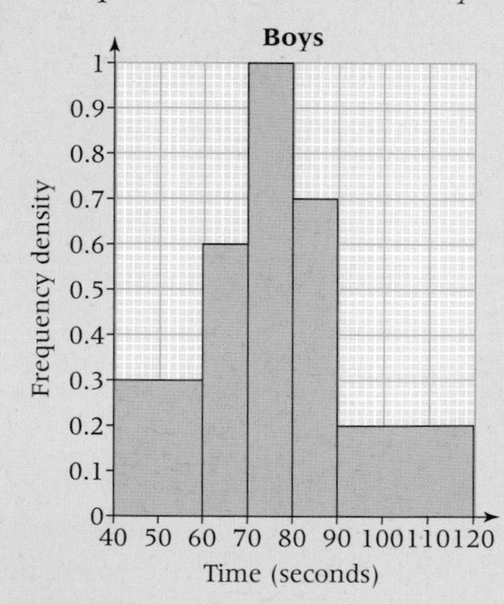

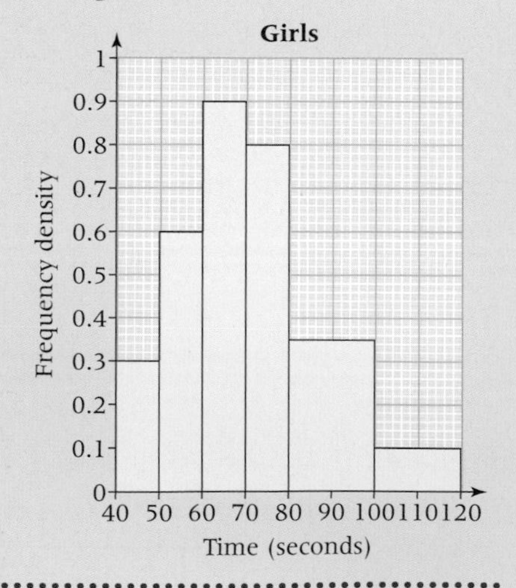

The range of times is the same for both boys and girls: $120 - 40 = 80$.

The modal class for boys (70–80 seconds) is a slower time than the modal class for girls (60–70 seconds), so the boys were generally slower.

The girls' times are more positively skewed than the boys' times indicating that girls' times were shorter, that is, the girls were quicker.

Exercise D4.5

Grade A*

1 Compare the heights of two samples of boys of different ages summarised in these histograms.

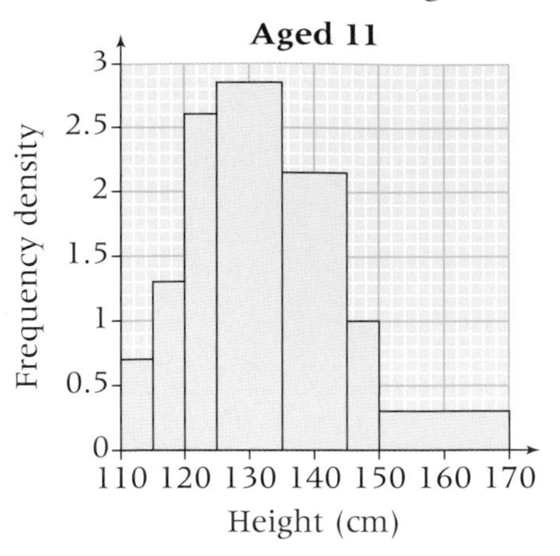

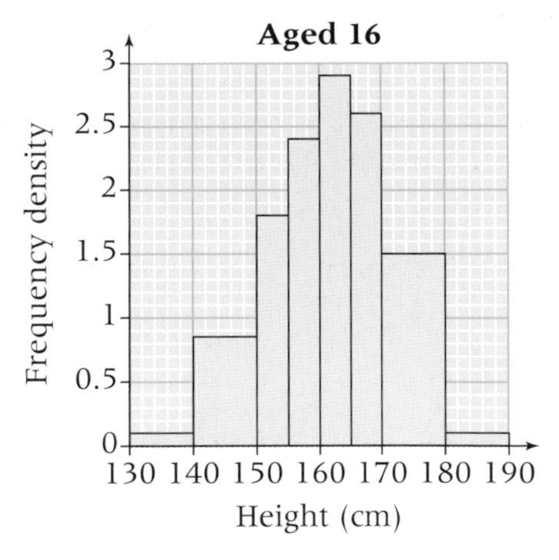

2 Compare the weights of samples of apples and pears summarised in these histograms.

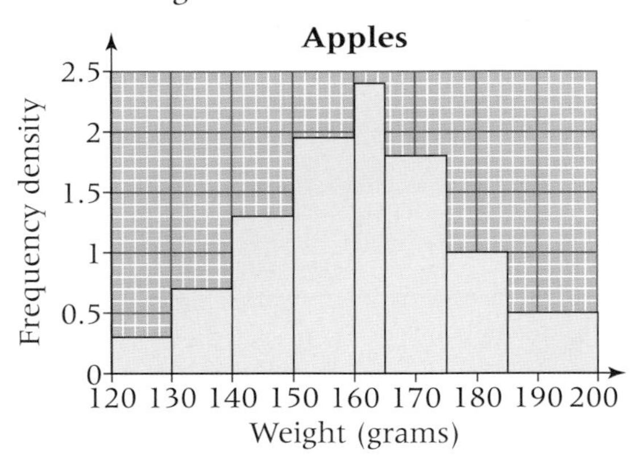

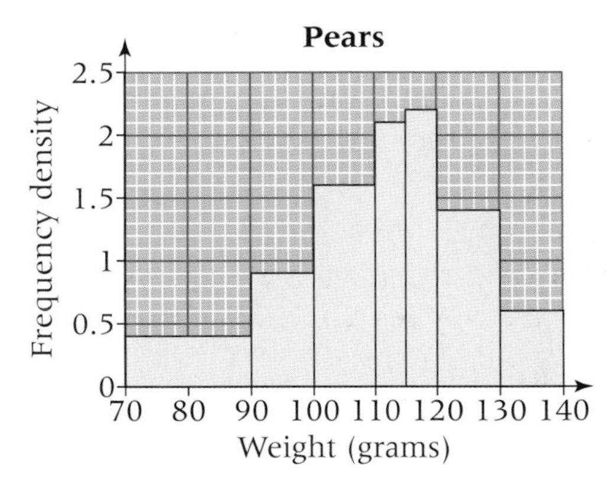

3 Compare the reaction times of girls and boys summarised in these histograms.

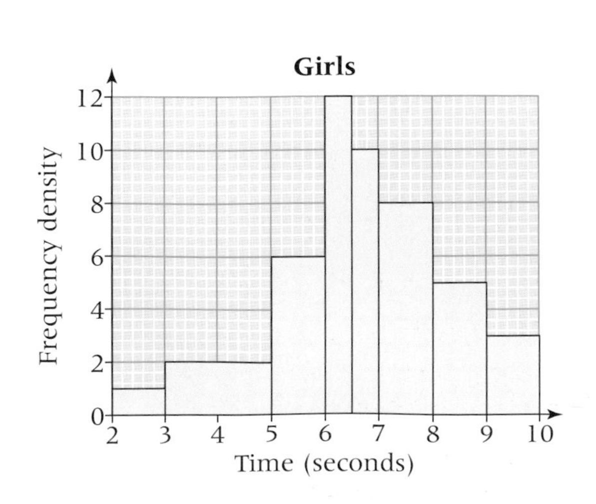

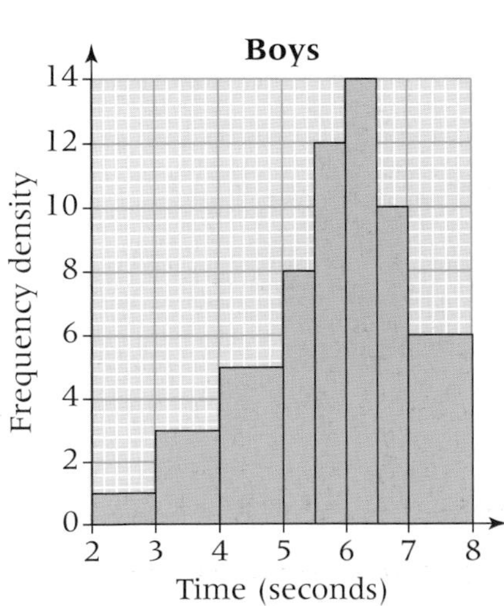

D4.6 Scatter graphs

This spread will show you how to:

- Draw and use scatter graphs
- Draw lines of best fit, understanding positive, negative and zero correlation

Keywords
Correlation
Line of best fit
Relationship

A scatter graph shows the **correlation** between two sets of data.

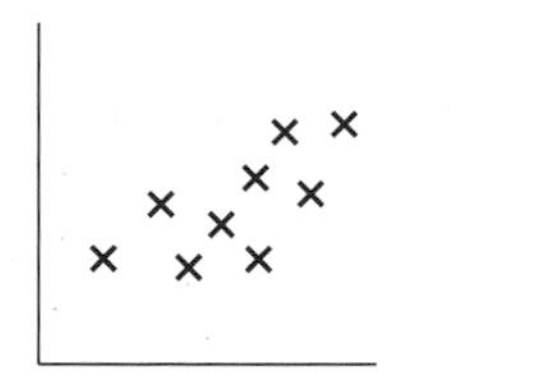
Positive correlation

Zero correlation

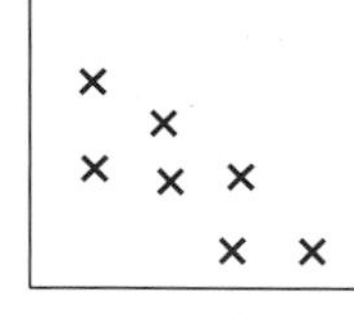
Negative correlation

A **line of best fit**
- Passes through the point (mean of set A, mean of set B).
- Has a similar number of points above and below the line.

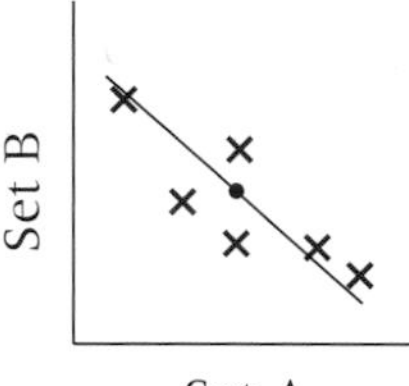

- The stronger the correlation, the closer the points lie to the line of best fit.
- Positive or negative correlation implies variable A increases or decreases, respectively, as variable B increases.
- Zero correlation implies no linear **relationship**.

There could be a non-linear relationship, for example $y = x^2$, or $y = 2^x$.

You can use a line of best fit to predict data values within the range of data collected.

Example

The table gives the marks earned in two exams by 10 students.

Maths %	70	76	61	70	89	65	59	58	73	82
Statistics %	78	82	74	75	93	70	66	62	77	89

a Draw a scatter graph for the data.
b Describe the correlation and the relationship shown.
c Draw a line of best fit.
d Predict the Statistics mark for a student who scored 62% in Maths.
e Could you use your graph to predict the Maths mark for a student who scored 32% in Statistics? Give your reasons.

a, c See graph.
b Positive correlation. Students who scored higher in Maths also tended to score higher in Statistics.
c Maths mean: $703 \div 10 = 70.3$
Statistics mean: $766 \div 10 = 76.6$
d Predicted Statistics mark = 70%
e No. This Statistics mark is outside the range of the data.

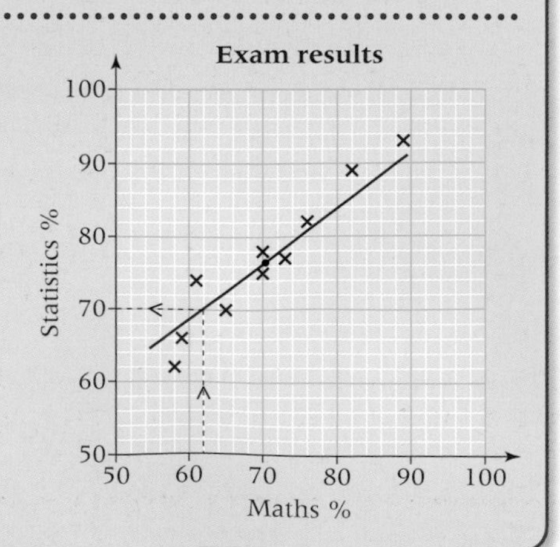

Read up from the horizontal axis to the line of best fit, then across to the vertical axis.

Exercise D4.6

Grade C/B

1 Louise recorded the average number of minutes per day spent playing computer games and the reaction times of nine students.

Minutes per day spent playing computer games	40	60	75	40	35	20	80	50	45
Reaction time, seconds	5.2	4.3	3.9	5.5	6.0	7.2	3.6	4.8	5.0

a Draw a scatter diagram of these data.
b Describe the correlation shown.
c Describe the relationship between the average number of minutes per day spent playing computer games and reaction time.

2 Bob caught nine fish during one angling session.
He recorded the weights and lengths of the fish he caught in this table.

Weight, g	500	560	750	625	610	680	600	650	580
Length, cm	30	32	50	44	39	48	40	45	36

a Draw a scatter diagram of these data.
b Describe the correlation shown.
c Describe the relationship between the weights and lengths of the fish.

Functional Maths **A02**

3 The table shows the number of hot water bottles sold per month in a chemist's shop and the average temperature for each month.

Month	Jan	Feb	Mar	April	May	June	July	Aug	Sept	Oct	Nov	Dec
Average monthly temperature °c	2	4	7	10	14	19	21	20	18	15	11	5
Sales of hot water bottles	32	28	10	4	6	0	2	3	7	15	22	29

a Draw a scatter graph for the data.
b Describe the correlation shown and the relationship between the two sets of results.
c Draw in a line of best fit.
d Predict the average temperature in a month when the shop sold 20 hot water bottles.
e The weather forecast predicts an average January temperature of −5°C. Could you use your graph to find how many hot water bottles would be sold?

Problem **A03**

72 52 54

4 For the data in question **3** find the equation of the line of best fit.
How should you interpret this equation?

5 Suggest three pairs of variables whose scatter graphs have curves of best fit in the shape of quadratic, cubic or reciprocal functions.

D4.7 Line graphs

This spread will show you how to:

- Draw and interpret line graphs

Keywords
Line graph
Seasonal variation
Trend

Some types of data are collected over an extended period of time.
For example,

Electricity and gas bills are produced every quarter (three monts).
Mobile phone bills are generated each month.
Unemployment reates arre published each month.

- Plotting the data on a line graph makes it easier to see any patterns.

Data often shows a short term, **seasonal variation**

For example, icecream sales are higher in Summer than in Winter .

and a longer term **trend**.

For example, inflation measures the annual increase or decrease in prices.

Example

Jenny's quarterly gas bills over a period of two years are shown in the table.

	Jan-March	April-June	July-Sept	Oct-Dec
2003	£65	£38	£24	£60
2004	£68	£42	£30	£68

Plot the data on a graph and comment on any pattern in the data.

Draw axes on graph paper with time on the horizontal axis.
Plot the coordinates as crosses on the grid.
Join them up with straight lines.

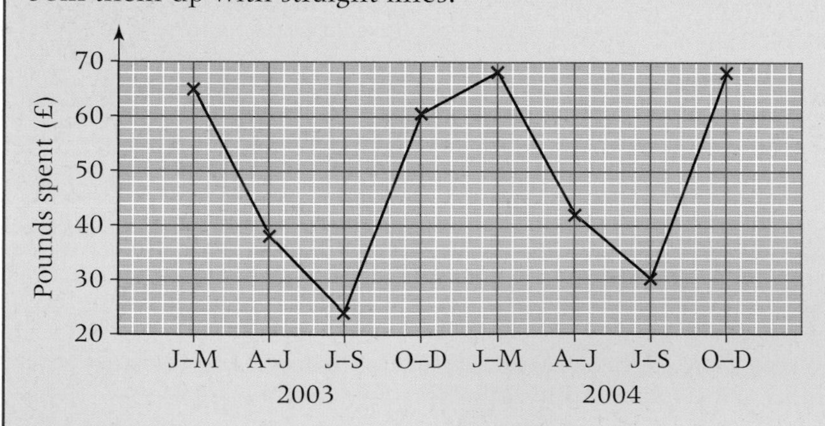

Plot time on the horizontal axis, J–M means Jan–March.

Gas bills are highest in the Winter months and lowest in the Summer months. This annual pattern appears to repeat itself.

There is a slight trend for the bills to rise from year-to-year.

Exercise D4.7

Grade C

Unit 1

For each of questions **1–6**

a Plot the data on a graph

b Comment on any patterns in the data.

1 The table shows Ken's monthly mobile phone bills.

Jan	Feb	Mar	April	May	June	July	Aug	Sept	Oct	Nov	Dec
£16	£12	£15	£18	£16	£18	£12	£10	£12	£15	£16	£20

2 The table shows Mary's quarterly electricity bills over a two-year period.

	Jan–March	April–June	July–Sept	Oct–Dec
2004	£45	£20	£15	£48
2005	£54	£24	£18	£50

3 The table shows monthly ice-cream sales at Angelo's shop during one year.

Jan	Feb	Mar	April	May	June	July	Aug	Sept	Oct	Nov	Dec
£16	£12	£15	£18	£38	£48	£52	£58	£18	£15	£16	£40

4 A town council carried out a survey over a number of years to find the percentage of local teenagers who used the town's library. The table shows the results.

year	1998	1999	2000	2001	2002	2003	2004	2005
%	14	18	24	28	25	20	18	22

5 Christabel kept a record of how much money she had earned from babysitting during three years.

	Jan–April	May–August	Sept–Dec
2001	£12	£18	£30
2002	£21	£33	£60
2003	£39	£42	£72

6 Steve kept a record of his quarterly expenses over a period of two years.

	Jan–March	April–June	July–Sept	Oct–Dec
2003	£35	£56	£27	£12
2004	£39	£68	£29	£18

D4 Summary

Check out

You should now be able to:

- Use frequency polygons to compare two sets of data
- Draw and interpret histograms using frequency density
- Use histograms to compare two or more sets of data
- Draw and use scatter graphs
- Recognise correlation and draw and use lines of best fit
- Look at data to find patterns and exceptions
- Interpret line graphs

Worked exam question

The histogram gives information about the times, in minutes, 135 students spent on the Internet last night.

Use the histogram to complete the table. (2)

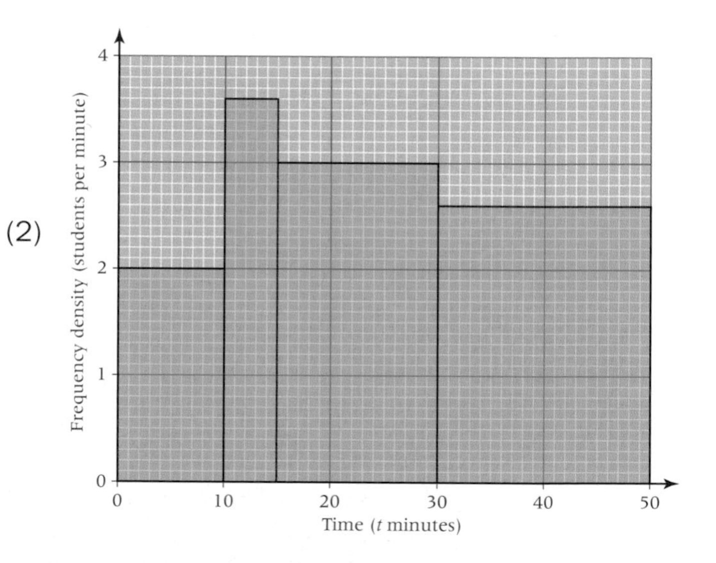

Time (t minutes)	Frequency
$0 < t \leq 10$	
$10 < t \leq 15$	
$15 < t \leq 30$	
$30 < t \leq 50$	
TOTAL	135

(Edexcel Limited 2004)

Time (t minutes)	Frequency
$0 < t \leq 10$	20
$10 < t \leq 15$	18
$15 < t \leq 30$	45
$30 < t \leq 50$	52
TOTAL	135

$10 \times 2 = 20$
$5 \times 3.6 = 18$
$15 \times 3 = 45$
$20 \times 2.6 = 52$ +
135

Show these calculations.

Exam question

1

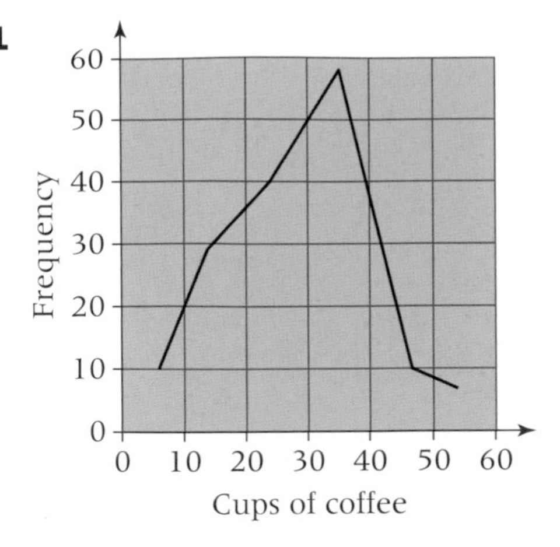

A café collected information about the number of cups of coffee it sold each day.
The frequency polygon shows the information about these sales.
Each class interval is the width of 10 cups of coffee.

a Write down the modal class interval. (1)

The café also collected information about the number of cups of tea it sold each day.
The frequency table gives the information about these sales.

Number of cups of tea sold each day	Frequency
0 – 10	20
11 – 20	46
21 – 30	50
31 – 40	38
41 – 50	31
51 – 60	18

b Using the same scale as in the grid above, draw a frequency polygon to show this information. (2)

Kath says "The range of sales of coffee is the same as the range of sales of tea."

c Explain why this may not be true. (1)

Functional Maths 2: Recycling

The focus on protecting the environment from further damage is now stronger than ever. Recycling and reusing waste materials have become an important part of everyday life both for manufacturers and consumers.

This time-series chart shows the amounts of different materials recycled from households in England between 1997/98 and 2007/8.

What can you say about the different types of materials being recycled by households in England during this time? Do you notice any trends? Justify your response by referring to the data.

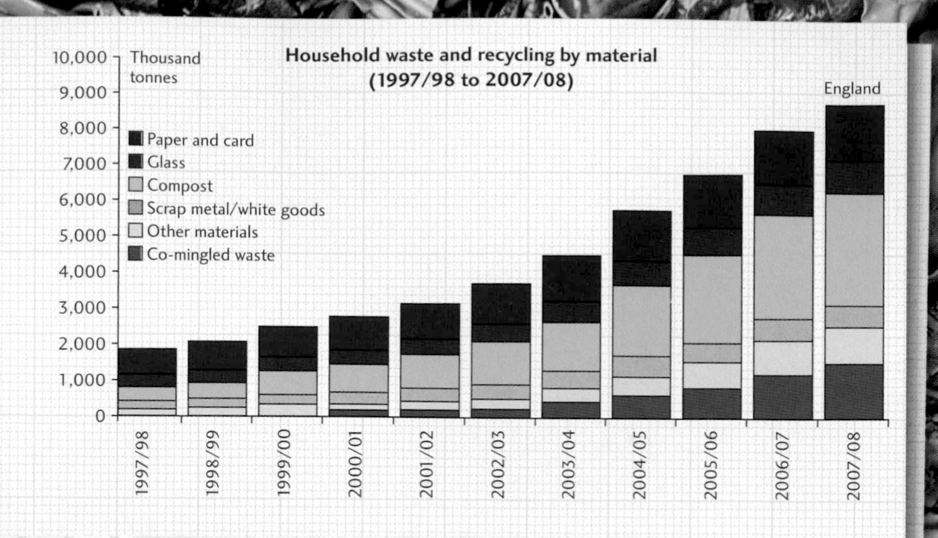

Copy and complete this table, using the data in the chart.

Year	1997/98	1998/99	1999/00	2000/01	2001/02	2002/03	2003/04	2004/05	2005/06	2006/07	2007/08
Co-mingled waste (amount)	0	0	0		221	268	469			1241	1563
Co-mingled waste (% of total)	0.00	0.00		7.33	6.94	7.16			12.65	15.39	
Other materials (amount)	230	257	355		235	269	385	516			989
Other materials (% of total)	12.31	12.27		5.94	7.38			8.92	10.74		
Scrap metal/white goods (amount)	231	253	265		369	419	465	577		601	598
Scrap metal/white goods (% of total)	12.36	12.08		11.02	11.58	11.20		9.97	7.83	7.45	
Compost (amount)	383		668		954	1189	1362	1960		2895	3189
Compost (% of total)	20.49	21.68		28.38	29.94	31.78		33.88	35.89	35.90	
Glass (amount)	335	347	383		426	470	568	670	760		902
Glass (% of total)	17.92	16.57		14.12	13.37	12.56		11.58	11.18	10.42	
Paper and card (amount)	690	783	842			1126	1272	1406		1535	1599
Paper and card (% of total)	36.92	37.39		33.21		30.10		24.30	21.70	19.04	
Total recycled (1000 tonnes)			2513	2812	3186					8063	

Compare the three largest components of recycled waste in 1997/98 and 2007/08. Can you think of any explanation for the difference? Justify your response, referring to the data.

A total of 25.3 million tonnes of household waste was collected in England in 2007/08, what percentage of this collected waste was re-used, recycled or composted?

The amount of household waste NOT re-used, recycled or composted was 7.0% lower in 2007/08 than in 2006/07. What was the total amount of household waste collected in tonnes in 2006/07?

Can you think of any reason for the trend shown by this data?

In 2007 the government set a target to reduce the amount of household waste in England not re-used, recycled or composted to 15.8 million tonnes. Do you think that this was a realistic target? Justify your response by referring to the data.

Manufacturers are responsible for designing packaging that is as environmentally friendly as possible while also protecting the product.

A company sells its own brand of baked beans in cans made of steel. The weight of these cans has been reduced by 13% every 10 years over the past 50 years. 50 years ago a can weighed 112g. What is the weight of a new can? By what proportion has the weight of a can changed over the last 50 years?

Glass milk bottles are 50% lighter than they were 50 years ago.

As well as reducing the consumption of raw materials, lighter packaging also saves money in other ways such as transport costs.

A supermarket sells tomatoes in packs of six. The packaging consists of a plastic tray with a lid as shown.

Given that on average this variety of tomato is spherical with a radius of 3cm, on average what percentage of the available volume of each package is empty?

Do you think that not having packaging would risk the quality of the tomatoes?

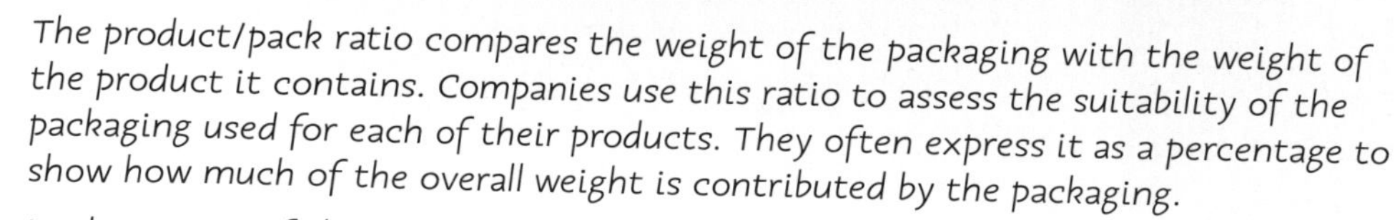

The product/pack ratio compares the weight of the packaging with the weight of the product it contains. Companies use this ratio to assess the suitability of the packaging used for each of their products. They often express it as a percentage to show how much of the overall weight is contributed by the packaging.

Look at some of the packaging you have at home. Could it be adapted to use less material without increasing the risk of damage to the product? If so, how?

How does the packaging used for perishable goods (e.g. food) differ from that used for non-perishable goods (e.g. electrical items)?

Research some well-known manufacturing companies on the Internet to find out about their packaging guidelines. Do they have different rules for different products (e.g. perishable/non-perishable goods)?

D5 Independent events

How much will I be charged to insure my car? Insurance companies work out the likelihood of you having an accident in your car, based on age, type of vehicle, driving experience and other factors, and then calculate your risk. This determines the cost of your insurance premium.

What's the point?

Mathematicians need to know how to combine the probabilities of simple events so that more complex probability questions, such as the risk of a car accident, can be calculated.

Check in

You should be able to

- **do arithmetic with decimals**

1 Work out each of these.

a $1 - 0.45$ **b** $1 - 0.96$ **c** $1 - 0.28$
d $1 - 0.375$ **e** $0.2 + 0.4$ **f** $0.3 + 0.04$
g $0.65 + 0.25$ **h** 0.5×0.36 **i** 0.25×0.68
j 0.64×0.3 **k** $1 - 0.125 - 0.64$ **l** $1 - 0.125 \times 0.64$

- **do arithmetic with fractions**

2 Work out each of these.

a $1 - \frac{5}{6}$ **b** $1 - \frac{1}{5}$ **c** $1 - \frac{7}{9}$ **d** $\frac{1}{5} + \frac{2}{3}$
e $\frac{3}{4} + \frac{1}{6}$ **f** $\frac{2}{3} \times \frac{5}{6}$ **g** $\frac{2}{9} \times \frac{4}{5}$ **h** $1 - \frac{3}{4} \times \frac{4}{5}$

Orientation

What I need to know	What I will learn	What this leads to
N1 Do arithmetic with decimals and fractions	■ Draw and use tree diagrams ■ Calculate with conditional probabilities	**A-level** Maths, Biology, Geography
D3 Calculate probabilities for independent and mutually exclusive events		Finance, Insurance, Meteorology, Quality assurance

Rich task

A bag contains twice as many red balls as blue balls. What is the probability of drawing out a red and a blue ball on two successive draws?

D5.1 Drawing tree diagrams

This spread will show you how to:

- Show the possible outcomes of two or more events on a tree diagram
- Use tree diagrams to calculate probabilities of combinations of independent events

Keywords
Event
Independent
Outcome
Replaced
Tree diagram

- You can show the possible **outcomes** of two or more **events** on a **tree diagram**.

p.56

Example

A box contains 4 red and 6 blue counters.
A counter is chosen at random from the box and its colour noted.
It is then replaced in the box.
The box is shaken and then a second counter is chosen at random.
Draw a tree diagram to show all the possible outcomes.

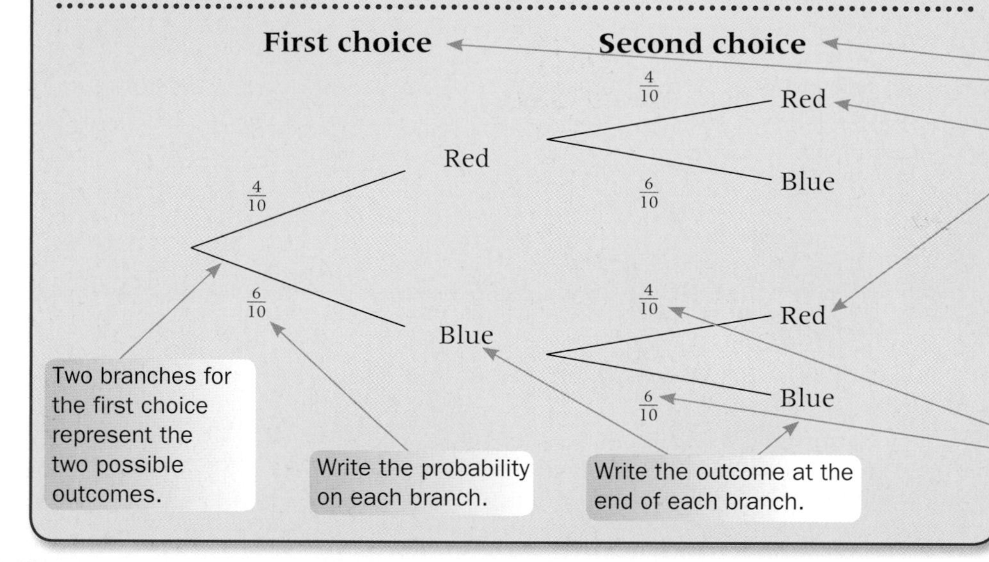

The counter is **replaced** in the box each time and the box is shaken, so the two events are **independent**.

Label the sets of branches.

For each branch of the first choice, there are two branches/possible outcomes for the second choice.

For each pair of branches, the probabilities should sum to 1.

Example

Records show that there are 10 wet days in every August. The other days are dry. The weather on any day is independent of the weather on the previous day.
Draw a tree diagram to show the possible types of weather on the weekend of 19 and 20 August one year.

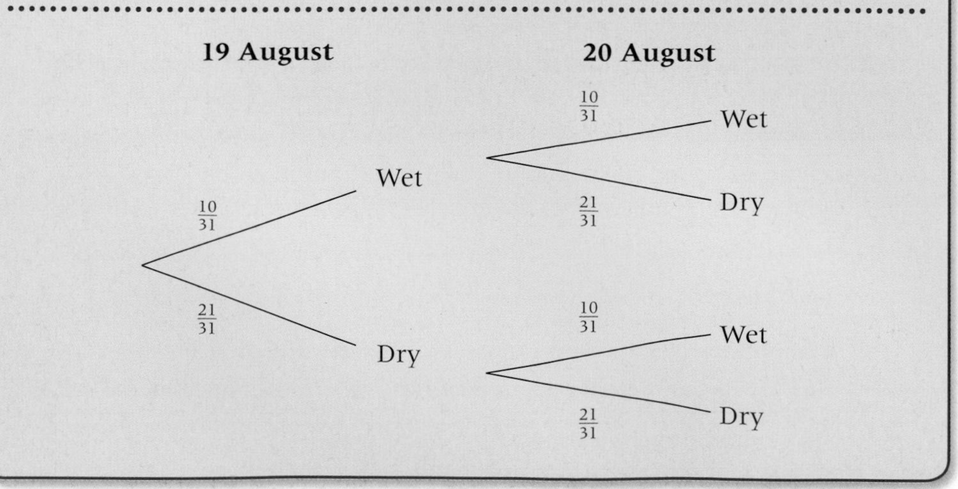

Exercise D5.1 Grade C

1 A bag contains 8 purple cubes and 9 orange cubes.
A cube is chosen at random from the bag and its colour noted.
It is then replaced in the bag.
The bag is shaken and then a second cube is chosen at random.
Copy and complete the tree diagram to show all the possible outcomes.

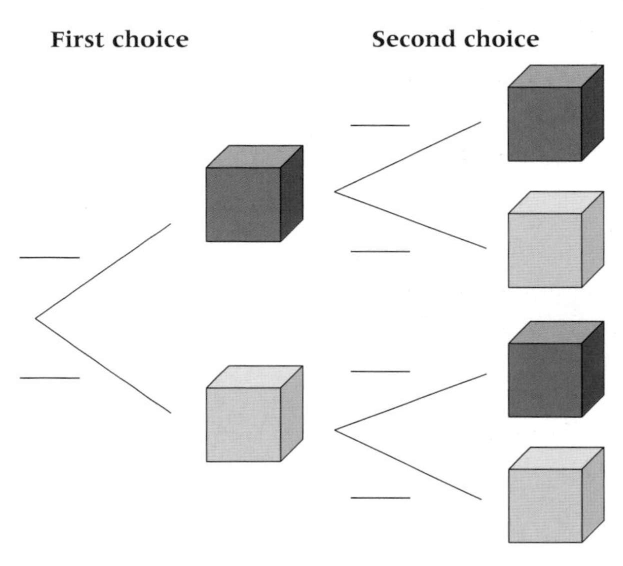

2 A set of cards contains 6 number cards and 5 picture cards.
A card is chosen at random from the set.
It is then replaced and the cards are shuffled.
Then a second card is chosen at random.
Draw a tree diagram to show all the possible outcomes.

3 A 50 pence coin and a 20 pence coin are tossed.
Draw a tree diagram to show all the possible outcomes.

4 A bag contains 7 red and 2 blue marbles.
A marble is chosen from the bag and a fair coin is tossed.
Draw a tree diagram to show all the possible outcomes.

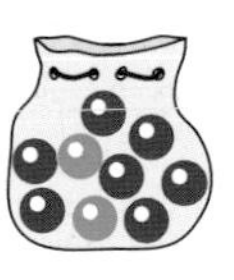

5 Kaz has 16 DVDs. Three of them are Star Wars films.
She chooses one of the DVDs at random, notes whether it is a Star Wars film, and then replaces it.
She then chooses another DVD at random.
Copy and complete this tree diagram for the outcomes.

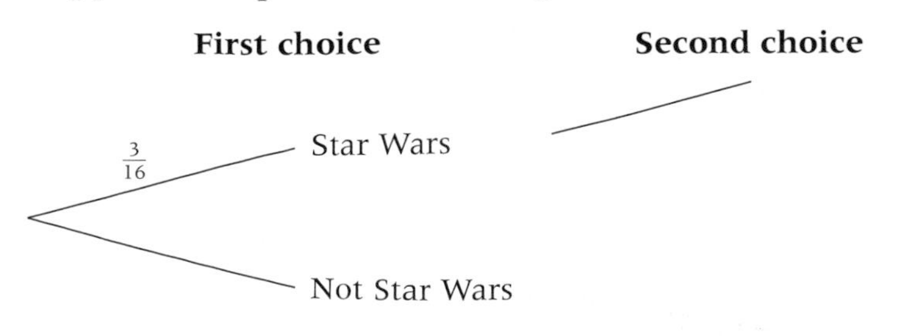

D5.2 Using tree diagrams to find probabilities

This spread will show you how to:

- Use tree diagrams to calculate probabilities of combinations of independent events
- Calculate probabilities based on the given conditions

Keywords
Event
Independent
Mutually exclusive
Outcome
Tree diagram

You can use a **tree diagram** to calculate probabilities of combinations of **independent events**.

- To calculate the probability of independent events, multiply along the branches.

Example

Shireen makes two chocolate rabbits by pouring melted chocolate into moulds and leaving it to set. She then removes the moulds.
The probability that a chocolate rabbit is cracked is 0.1

a Draw a tree diagram to show all the **outcomes** of the two chocolate rabbits being cracked or not cracked.

b Calculate the probability that both chocolate rabbits will be cracked.

c Work out the probability that one of the chocolate rabbits will be cracked.

---

a **First chocolate rabbit** **Second chocolate rabbit**

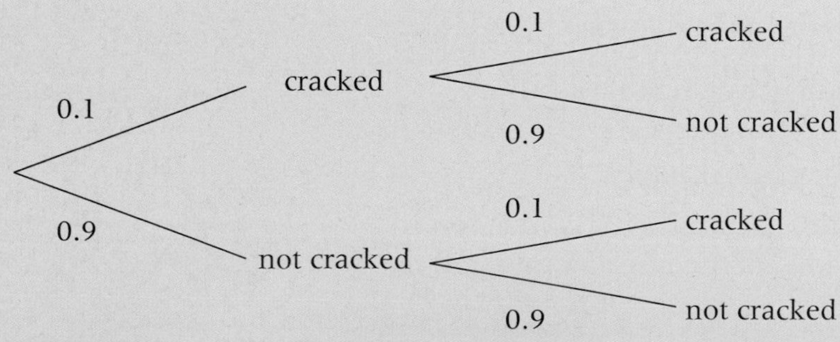

The two events are independent.

'Cracked' and 'not cracked' are **mutually exclusive** events, so P(not cracked) = 1 − P(cracked)

The four possible outcomes are: (cracked, cracked), (cracked, not cracked), (not cracked, cracked) and (not cracked, not cracked).

b P(cracked, cracked) $= 0.1 \times 0.1 = 0.01$

c Two outcomes result in one cracked rabbit:
(not cracked, cracked) or (cracked, not cracked).

P(one rabbit cracked)
= P(not cracked, cracked) + P(cracked, not cracked)
$= 0.9 \times 0.1 + 0.1 \times 0.9$
$= 0.09 + 0.09$
$= 0.18$

When A and B are mutually exclusive P(A or B) = P(A) + P(B).

- If you can take more than one route through the tree diagram:
  - Multiply the probabilities along each route.
  - Add the resultant probabilities for each route.

Exercise D5.2 Grade B

1 Two torches, one red and one black, are fitted with new batteries.
The probability that a battery lasts for more than 20 hours is 0.7.

a Draw a tree diagram to show the probabilities of the torch batteries lasting more or less than 20 hours.

b Find the probability that the battery lasts more than 20 hours
i in both torches **ii** in only one torch **iii** in at least one torch.

2 Rajen has five 10p and three 2p coins in his pocket.
He takes a coin at random from his pocket, notes what it is, and replaces it.
He then picks a second coin at random.

a Draw a tree diagram to show all the possible outcomes for picking two coins.

b Find the probability that he picks
i two 10p coins **ii** one coin of each type **iii** at least one 10p coin.

3 A bag contains 7 yellow and 3 green marbles.
A marble is chosen at random from the bag, its colour is noted and it is replaced.
A second marble is chosen at random.

a Draw a tree diagram to show all the possible outcomes for choosing two marbles.

b Find the probability of choosing
i one marble of each colour
ii two marbles the same colour
iii no green marbles.

4 Two vases are fired in a kiln.
The probability that a vase breaks in the kiln is $\frac{1}{50}$.

a Draw a tree diagram to show all the possible outcomes.

b Calculate the probability that in the kiln
i neither vase breaks **ii** at least one vase breaks.

5 A spinner has 10 equal sectors, 6 green and 4 white.
The spinner is spun twice.

a Draw a tree diagram to show all the possible outcomes of two spins on the spinner.

b Find the probability that on two spins the spinner lands on
i green both times
ii green at least one time
iii one of each colour.

D5.3 Sampling without replacement

This spread will show you how to:

- Use tree diagrams to calculate probabilities in 'without replacement' sampling

Keywords
Independent
Tree diagram

When an item is chosen and *not* replaced, the probability of choosing a second item changes.

A bag contains 4 white and 3 black counters.

First choice
A counter is chosen at random.
P(black) $= \frac{3}{7}$
P(white) $= \frac{4}{7}$

A black counter is chosen and not replaced.
Bag now contains 4 white and 2 black counters.

Second choice
A counter is chosen at random.
P(black) $= \frac{2}{6} = \frac{1}{3}$
P(white) $= \frac{4}{6} = \frac{2}{3}$
The probability of choosing a white counter has increased.

- You can use a **tree diagram** to show the probability changes for each choice for 'without replacement' sampling.

Example

A bag contains 7 yellow and 3 blue marbles.
A marble is chosen at random from the bag and its colour noted.
It is *not* replaced in the bag.
The bag is shaken and then a second marble is chosen at random.

a Draw a tree diagram to show all the possible outcomes.

b Find the probability that

i both marbles are blue

ii one marble of each colour is chosen.

a

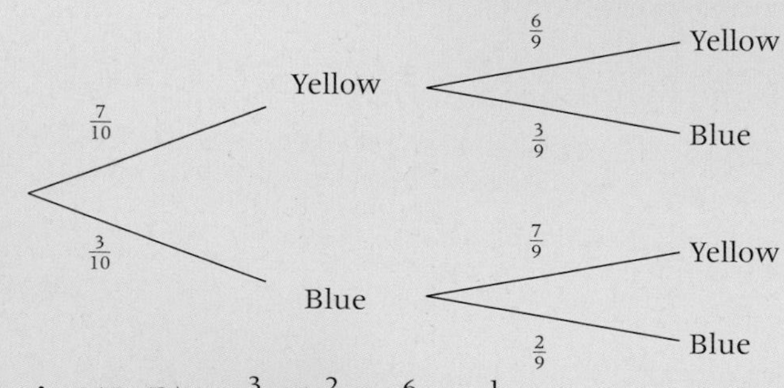

b i $P(B, B) = \frac{3}{10} \times \frac{2}{9} = \frac{6}{90} = \frac{1}{15}$

ii $P(Y, B) + P(B, Y) = \left(\frac{7}{10} \times \frac{3}{9}\right) + \left(\frac{3}{10} \times \frac{7}{9}\right)$

$= \frac{21}{90} + \frac{21}{90} = \frac{42}{90} = \frac{7}{15}$

For the second choice there is one less marble in the bag, so the denominator decreases by 1. The numerator depends upon the first choice.

For each pair of branches, the probabilities add to 1.

Exercise D5.3 Grade A

1 A bag contains 3 yellow and 4 blue marbles.
A marble is chosen at random from the bag and its colour noted.
It is *not* replaced in the bag.
The bag is shaken and then a second marble is chosen at random.
The tree diagram shows all the possible outcomes.

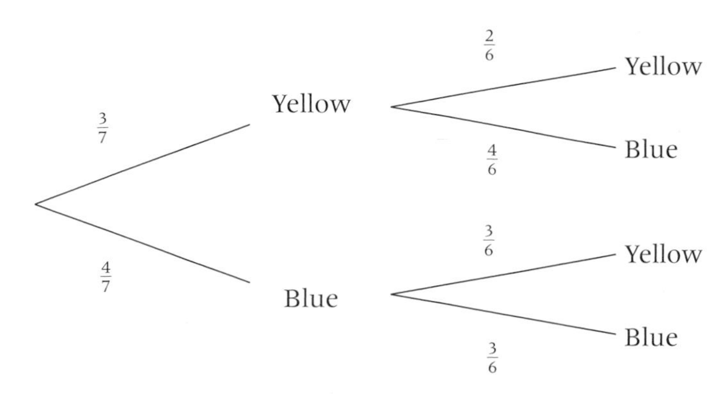

Use the tree diagram to find the probability of choosing
a two yellow marbles
b two blue marbles
c one marble of each colour.

2 A bag contains 3 white counters and 8 black counters.
A counter is chosen at random from the bag and its colour noted. It is *not* replaced in the bag.
The bag is shaken and then a second counter is chosen at random.
a Draw a tree diagram to show all the outcomes and their probabilities.
b Find the probability of choosing
i two white counters
ii one counter of each colour
iii at least one white counter.

3 A bag contains 12 lemon and 4 orange sweets.
Reuben chooses a sweet at random and eats it.
He then chooses a second sweet and eats it.
a Draw a tree diagram to show all the outcomes and probabilities.
b Find the probability of choosing
i two orange sweets
ii one sweet of each flavour
iii at least one orange sweet.

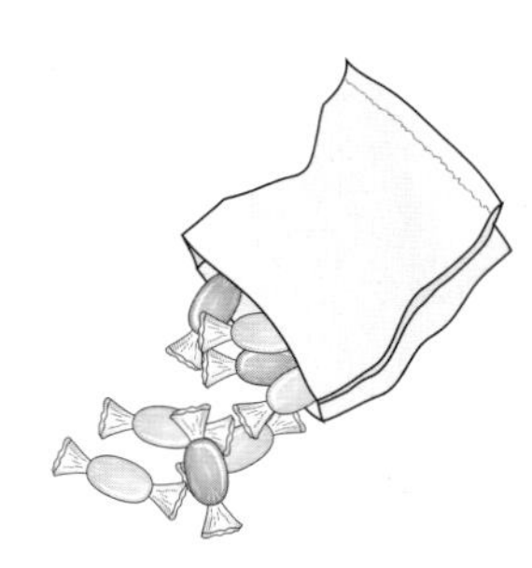

D5.4 Tree diagrams – conditional probability

This spread will show you how to:

- Use tree diagrams to calculate probabilities in 'without replacement' sampling
- Understand conditional probability

Keywords
Conditional
Event

- In **conditional** probability, the probability of subsequent **events** depends on previous events occuring.

Example

Gareth travels through two sets of traffic lights on his way to work.
The probability that the first set of traffic lights is on red is 0.6.
If the first set of lights is on red, then the probability that the second set of lights will be on red is 0.9.
If the first set of lights is not on red then the probability that the second set of lights is on red is 0.25.

The conditional statement 'If ... then ... ' describes a conditional probability.

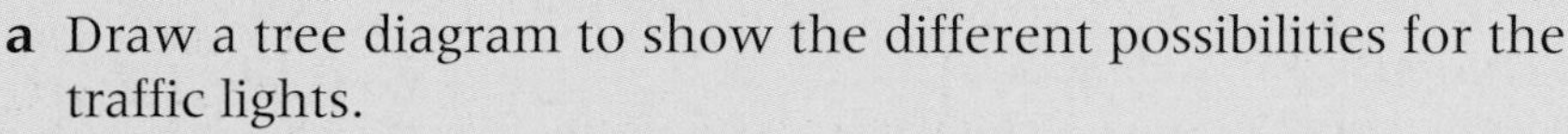

a Draw a tree diagram to show the different possibilities for the traffic lights.

b Work out the probability that on Gareth's way to work

i both sets of lights will be on red

ii only one set of traffic lights will be on red.

a

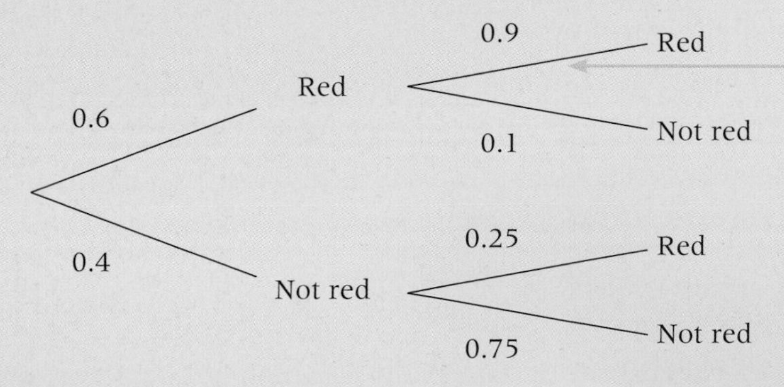

If the first set is red, P(second set red) = 0.9.

Each pair of branches has different probabilities.

On each pair of branches the probabilities add to 1.

b **i** P(red, red) $= 0.6 \times 0.9 = 0.54$

ii P(only one on red) = P(red, not red) + P(not red, red)

$= (0.6 \times 0.1) + (0.4 \times 0.25)$

$= 0.06 + 0.1$

$= 0.16$

Exercise D5.4 Grade A

1 Richard either drives or cycles to work.
The probability that he drives to work is 0.25.
If Richard drives, the probability that he is late is 0.2.
If Richard cycles, the probability that he is late is 0.4.

a Draw a tree diagram to show the probability of Richard being late for work.

b Work out the probability that on any one day Richard will *not* be late for work.

2 Julie's college course offers lectures and accompanying tutorials.
The probability that Julie attends any one lecture is 0.8.
If she attends the lecture the probability that she attends the accompanying tutorial is 0.9.
If she does not attend the lecture the probability that she attends the accompanying tutorial is 0.6.

a Draw a tree diagram to show the probabilities that Julie attends lectures and tutorials.

b Work out the probability that Julie attends
i only one tutorial **ii** at least one tutorial.

AO3 Problem

3 40 girls and 60 boys completed a questionnaire.
The probability that a girl completed the questionnaire truthfully was 0.8.
The probability that a boy completed the questionnaire truthfully was 0.3.

a Copy and complete the tree diagram.

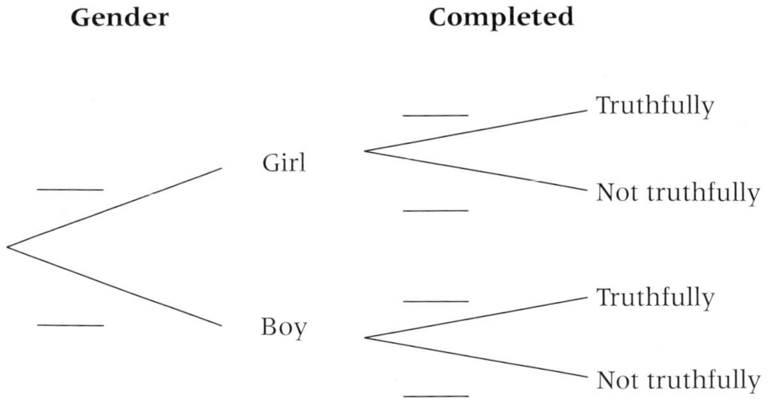

Gender **Completed**

Girl: ____ Truthfully; ____ Not truthfully
Boy: ____ Truthfully; ____ Not truthfully

b One questionnaire is chosen at random.
Work out the probability that
i the questionnaire has not been completed truthfully
ii the questionnaire has been completed truthfully by a girl.

You could rephrase this as: If a girl completes the questionnaire, then the probability that she completes it truthfully is 0.8.

D5.5 Further conditional probability

This spread will show you how to:

- Understand conditional probability
- Calculate probabilities based on the given conditions

Keywords
Conditional probability

When solving probability problems, look out for **conditional probabilities**.

- You need to calculate probabilities based on the conditions you are given.

Example

The table shows the number of boys and the number of girls in Years 12 and 13 at a school.

	Year 12	Year 13	Total
Boys	48	96	144
Girls	72	54	126
Total	120	150	270

a One student is to be chosen at random.
What is the probability that it is a boy?

b One of the boys is chosen at random.
What is the probability that he is in Year 13?

c One of the Year 13 students is chosen at random.
What is the probability that the student is a boy?

a 144 boys and 270 students in total

$P(\text{boy}) = \frac{144}{270} = \frac{8}{15}$

Parts **a** and **b** involve conditional probability

b Total 144 boys, 96 boys in Year 13

$\text{Probability} = \frac{96}{144} = \frac{2}{3}$

P(student is in Y13 *given* they are a boy)

c Total 150 in Year 13, 96 are boys

$\text{Probability} = \frac{96}{150} = \frac{16}{25}$

P(student is a boy *given* in Y13)

Example

At a party, the menu choice is meat or fish.
The table shows the choices made.

	Meat	Fish	Total
Men	8	2	10
Women	4	11	15
Total	12	13	25

a One of the women is chosen at random.
What is the probability that she ate fish?

b One person who ate fish is chosen at random.
What is the probability that it was a man?

a There are 15 women of whom 11 ate fish. $\text{Probability} = \frac{11}{15}$

b 13 people chose fish, of whom 2 are men. $\text{Probability} = \frac{2}{13}$

Exercise D5.5 Grade A*

1 The table shows the numbers of boys and girls in Years 10 and 11 at a school.

	Year 10	Year 11	Total
Boys	57	60	117
Girls	51	52	103
Total	108	112	220

a One student is chosen at random. What is the probability that the student is a girl?

b One of the girls is chosen at random. What is the probability that she is in Year 10?

c One of the Year 10 students is chosen at random. What is the probability that the student is a girl?

2 The table shows the ages of men and women enrolling for a course at an adult education centre.

	Aged under 40	Age 40 and over	Total
Men	24	30	54
Women	42	54	96
Total	66	84	150

a One person is chosen at random. What is the probability that the person is a man?

b One of the men is chosen at random. What is the probability that he is under 40?

c One person under 40 is chosen at random. What is the probability that this person is male?

3 After the main course, people chose to have tea or coffee with their dessert. The table shows their choices.

	Tea	Coffee	Total
Men	11	1	12
Women	4	16	20
Total	15	17	32

a One of the men is chosen at random. What is the probability that he drank coffee?

b One of the coffee drinkers is chosen at random. What is the probability that the person is a man?

4 A spinner has 24 equal sides. Each side is coloured red or black and has a circle or triangle on it.
The table shows the colour and the shape drawn on each side.
The spinner is spun once. What is the probability that the spinner lands on

	Circle	Triangle	Total
Red	4	12	16
Black	6	2	8
Total	10	14	24

a red

b red, given that it shows triangle

c a triangle, given that it shows red?

Summary

Check out

You should now be able to:

- Recognise when two events are independent
- Show the possible outcomes of two or more events on a tree diagram
- Use a tree diagram to calculate the probability of combinations of independent events
- Understand and calculate conditional probabilities
- Use a tree diagram to calculate conditional probability

Worked exam question

In a game of chess, a player can either win, draw or lose.
The probability that Vishi wins any game of chess is 0.5
The probability that Vishi draws any game of chess is 0.3

Vishi plays 2 games of chess.

a Complete the probability tree diagram. (2)

b Work out the probability that Vishi will win both games. (2)

(Edexcel Limited 2009)

a

$1 - (0.5 + 0.3) = 0.2$

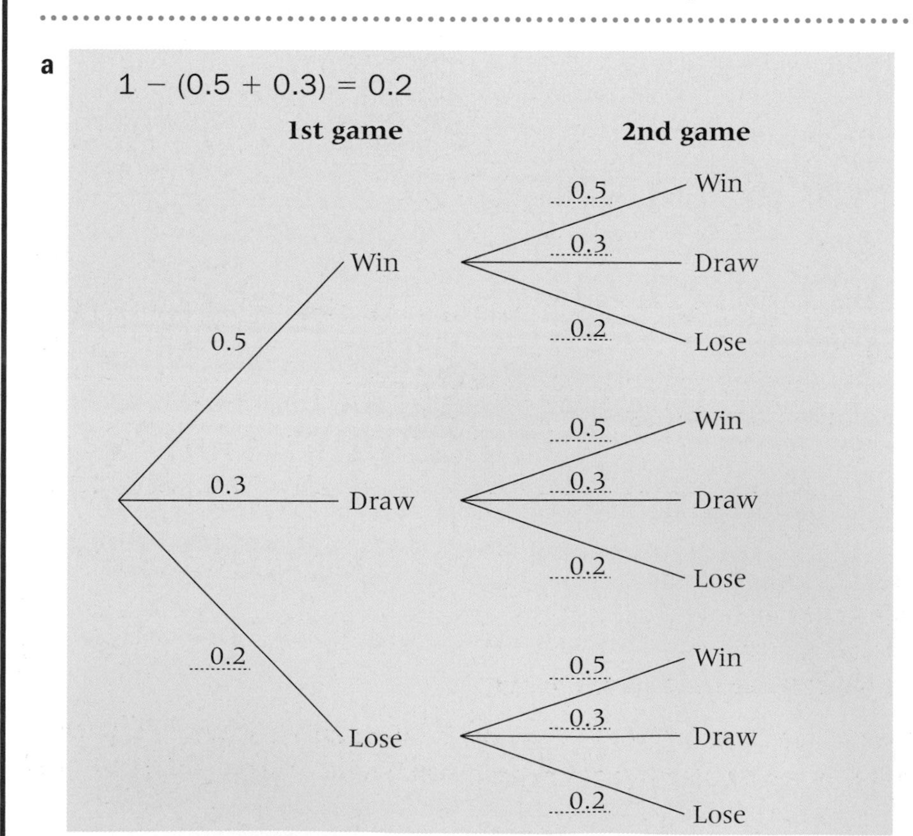

Write a probability on each branch of the tree.

b

$0.5 \times 0.5 = 0.25$

Show this multiplication.

Exam questions

1 Julie has 100 music CDs.
58 of the CDs are classical.
22 of the CDs are folk.
The rest of the CDs are jazz.

On Saturday, Julie chooses one CD at random from the 100 CDs.
On Sunday, Julie chooses one CD at random from the 100 CDs.

a Complete the probability tree diagram.

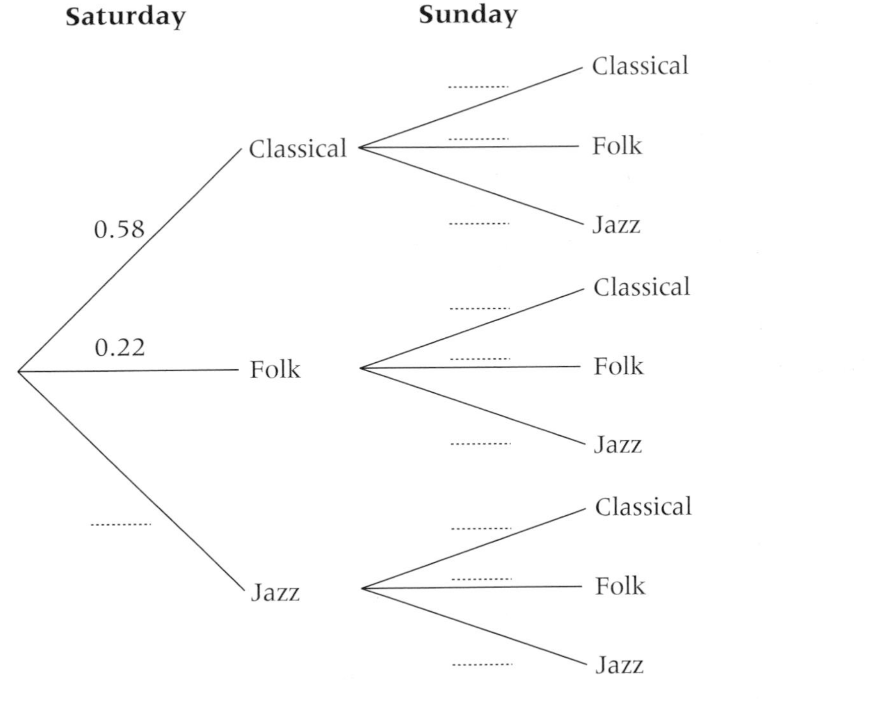

(2)

b Calculate the probability that Julie will choose a jazz CD on **both** Saturday and Sunday. (2)

c Calculate the probability that Julie will choose at least one jazz CD on Saturday and Sunday. (3)

(Edexcel Limited 2007)

AO3

2 Jim has 20 balls in a bag.

5 of the balls are red.
6 of the balls are blue.
9 of the balls are green.
Jim takes at random **two** balls from the bag.

Work out the probability that the balls will **not** be the same colour. (4)

Expressions

Engineers and scientists use quadratic expressions to model and explain the behaviour of events and activities in real life. Without quadratic expressions there wouldn't be aircraft, mobile phones or satellite TV.

What's the point?

Algebra provides the language of modern science and engineering and at the heart of algebra lie expanding brackets and the reverse factorising into brackets.

Check in

You should be able to

■ **use indices with numbers**

1 Find the value of these without a calculator.

a 14^2 **b** 1^{50} **c** $(-2)^5$ **d** $\left(\frac{2}{3}\right)^3$

e 0.02^3 **f** 10^{10} **g** $(-5)^1$ **h** $\sqrt{\frac{4}{25}}$

■ **write and interpret algebraic expressions**

2 Write these statements using the rules of algebra.

a I think of a number, multiply it by 4 and add 7.

b I think of a number, subtract 6 and then multiply by 3.

c I think of a number, multiply it by itself then subtract this from 10.

d I think of a number, treble it, take away 6 then divide this by two.

e I think of a number and multiply it by itself five times.

■ **find the HCF of a set of numbers**

3 a Find the highest common factor of these sets of numbers.

i 24 and 36 **ii** 30 and 75 **iii** 56 and 72

iv 90, 180 and 225 **v** 17, 28 and 93

b What is the highest common factor of $2x$, x^2 and $5x^3$?

Orientation

What I need to know	What I will learn	What this leads to
KS3 Collect like terms Use indices with numbers Find the HCF of two numbers	■ Use the index laws ■ Expand single and double brackets ■ Factorise into double brackets ■ Manipulate algebraic fractions	**A3** Use factorisation and expansion to simplify expressions **A4** Solve quadratic equations by factorisation **N4** Further index laws

Rich task

A square grid is numbered from 1 to 100.
A 2 × 2 square is shaded in as shown on the grid.
The numbers in the opposite corners of the 2 × 2 square are multiplied together.
Investigate.

1	2	3	4	5	6	7	8	9	10
11	12	13	14	15	16	17	18	19	20
21	22	23	24	25	26	27	28	29	30
31	32	33	34	35	36	37	38	39	40
41	42	43	44	45	46	47	48	49	50
51	52	53	54	55	56	57	58	59	60
61	62	63	64	65	66	67	68	69	70
71	72	73	74	75	76	77	78	79	80
81	82	83	84	85	86	87	88	89	90
91	92	93	94	95	96	97	98	99	100

A1.1 Simplifying using index laws

This spread will show you how to:

- Use index notation and simple laws of indices

Keywords
Base
Coefficient
Index
Indices

An **index** is a power. The **base** is the number which is raised to this power.

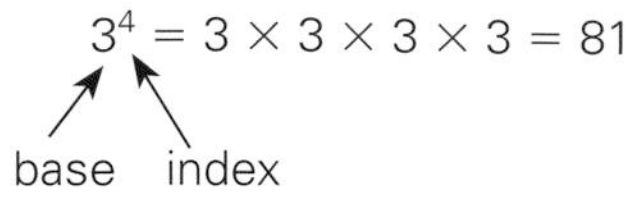

The plural of index is **indices**.

You can simplify expressions using the three index laws.

p.180

- when multiplying, add the indices $3^2 \times 3^3 = (3 \times 3) \times (3 \times 3 \times 3) = 3^5$
- when dividing, subtract the indices $5^4 \div 5^2 = (5 \times 5 \times 5 \times 5) \div (5 \times 5) = 5^2$
- with brackets, multiply the indices $(4^2)^3 = (4 \times 4) \times (4 \times 4) \times (4 \times 4) = 4^6$

To use the index laws, the bases must be the same.

When terms have numerical **coefficients**, deal with these first.

$$5p^7 \times 8p^{-3} = 40p^4$$

$7 + (-3)$

5×8

Example

Simplify each of these using the index laws.

a $p^9 \times p^{-7}$ **b** $\dfrac{q^{-3}}{q^{-5}}$

c $(w^6)^{-4}$ **d** $p^5 \times q^3$

e $\dfrac{(k^3 + k^2)^7}{k}$ **f** $(5p^2)^3 \times 2p^{-7}$

a $p^9 \times p^{-7} = p^2$

b $\dfrac{q^{-3}}{q^{-5}} = q^2$

c $(w^6)^{-4} = w^{-24}$

d $p^5 \times q^3 = p^5q^3$

e $\dfrac{(k^3 \times k^2)^7}{k} = \dfrac{(k^5)^7}{k} = \dfrac{k^{35}}{k} = k^{34}$

f $(5p^2)^3 \times 2p^{-7} = 125p^6 \times 2p^{-7} = 250p^{-1} = \dfrac{250}{p}$

Take care with negatives:
$9 + (-7) = 2$
$(-3) - (-5) = 2$
$6 \times (-4) = -24$.

Remember k is really k^1.

$\frac{1}{p} = \frac{p^0}{p} - p^{0-1} = p^{-1}$.

Remember $p^0 = 1$.

Example

Expand $(2p^2q)^3$

$(2p^2q)^3 = 8p^6q^3$

Everything inside the bracket is cubed:
$2^3 = 8$, $(p^2)^3 = p^6$, $(q^1)^3 = q^3$.

Exercise A1.1 — Grade B

1 Simplify

a $y^3 \times y^9$ **b** $k^9 \div k^5$ **c** $(m^3)^4$ **d** $g^8 \times g^{-5}$

e $\frac{h^{-2}}{h^4}$ **f** $(b^{-4})^3$ **g** $j^{-4} \times j^{-2}$ **h** $(t^{-5})^{-2}$

i $n^{-8} - n^{-6}$

2 Simplify fully

a $5h^7 \times 3h^6$ **b** $\frac{15p^3}{5p}$ **c** $(2p^8)^2$ **d** $10r^3 \times 6r^{-4}$

e $(3h^{-3})^3$ **f** $9b^3 \div 3b^{-5}$ **g** $(3m^3 \times 2m^{-7})^2 \div 18m$

h $18(f^{-4})^4 \div 9f^{-16}$

3 Write a simplified expression for the area of this triangle.

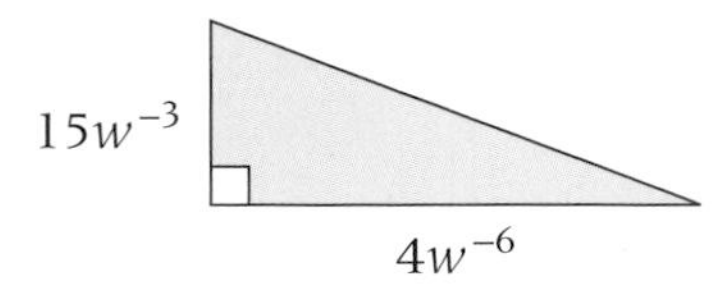

4 Show that the expression $(4p^4)^3 \div (8p^7)^2$ simplifies to $\frac{1}{p^2}$.

5 True or false? $3^x \times 3^y$ simplifies to give 9^{x+y}.
Explain your answer.

6 If $x = 3$ and $y = 4$, evaluate these expressions.

a x^2y **b** $3(x - y)^2$ **c** y^x

d $(x + y)(x - y)$ **e** $(x + y)^2$

7 Which expression is the odd one out?
Explain your answers.

$5t^2 \times 10t^{-4} \div (5t^{+3})^2$ | $\frac{2}{t^6}$ | $\left(\frac{4t^2}{16t}\right) \times 8t^{-7}$

8 Find the value of x in this equation.

$$(2^2)^x \times 2^{3x} = 32$$

AO3 Problem

9 **a** If $u = 3^x$, show that $9^x + 3^{x+1}$ can be written as $u^2 + 3u$.

b Write an expression in terms of x for $u^3 + 9u$.

c Write an expression in terms of x for $u^2 - \frac{1}{u}$.

d Write $81^x - 9^{x-1}$ in terms of u.

A1.2 Expanding single and double brackets

This spread will show you how to:

- Expand single and double brackets

Keywords
Expand
F.O.I.L.

To **expand** a single bracket, you multiply all terms in the bracket by the term outside.

$$3(2x - 9) = 6x - 27$$

To expand double brackets, you multiply each term in the second bracket by each term in the first bracket.

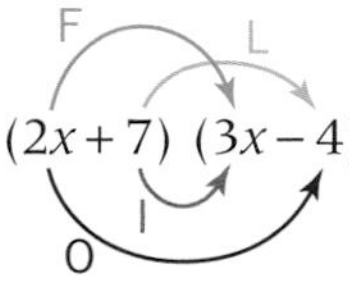

$$(2x+7)(3x-4) = 6x^2 - 8x + 21x - 28$$
$$= 6x^2 + 13x - 28$$

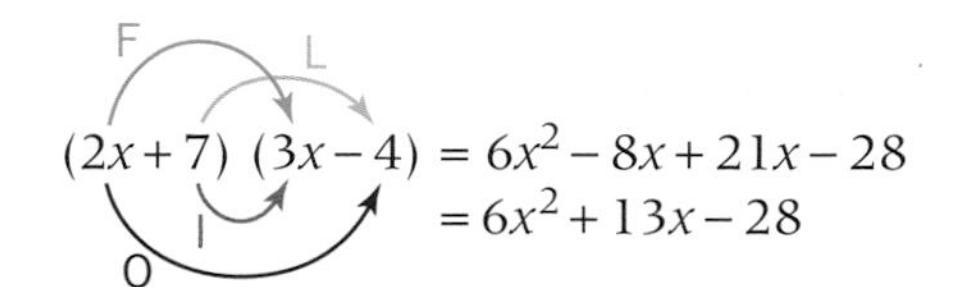

F ... **F**irsts
O ... **O**uters
I ... **I**nners
L ... **L**asts

Example

Expand and simplify each of these.

a $6(2y - 5) - 3(2 - 2y)$ **b** $(2m - 7)(4m - 2)$

a $6(2y - 5) - 3(2 - 2y)$
$= 12y - 30 - 6 + 6y$
$= 18y - 36$

b $(2m - 7)(4m - 2)$
$= 8m^2 - 4m - 28m + 14$
$= 8m^2 - 32m + 14$

$-3 \times 2 = -6$
$-3 \times -2y = +6y$

Use the index laws:
$2m^1 \times 4m^1 = 8m^2$

Example

Expand $(3x - 1)^2$.

$$(3x - 1)^2 = (3x - 1)(3x - 1)$$
$$= 9x^2 - 3x - 3x + 1$$
$$= 9x^2 - 6x + 1$$

This question is a double bracket in disguise!

Example

Given that the length and width of a rectangle are $x + 5$ and $x - 2$ respectively and its area is 15, show that $x^2 + 3x - 25 = 0$.

Sketch a diagram:

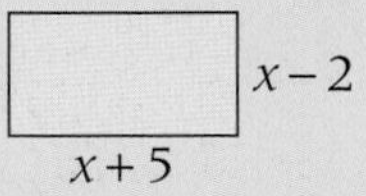

Area of rectangle = length × width
$15 = (x + 5)(x - 2)$
$15 = x^2 + 5x - 2x - 10$
$15 = x^2 + 3x - 10$
Hence, $x^2 + 3x - 25 = 0$

Collect like terms. Subtract 15 from each side.

Exercise A1.2 Grade B

1 Expand and simplify

a $3(5x + 9)$ **b** $2p(4p - 8)$

c $3m(5 - 2m)$ **d** $3(2y + 9) + 5(3y - 2)$

e $5x(2x + 2y - 9)$ **f** $4(t + 9) - 3(2t - 7)$

g $(7h + 9) - (3h - 7)$ **h** $x(3x^2 + x^3)$

2 Expand and simplify

a $(x + 7)(x + 6)$ **b** $(2x - 8)(x + 3)$

c $(3p + 2)(4p + 5)$ **d** $(3m - 7)(2m - 6)$

e $(5y - 9)(2y + 7)$ **f** $(3t - 2)^2$

g $(x + 4)(x - 6) + (x + 3)^2$ **h** $(4 + 8b)(2 - 3b) - (3 - b)^2$

3 Write an expression for the area of each of these shapes.

a

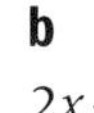

b

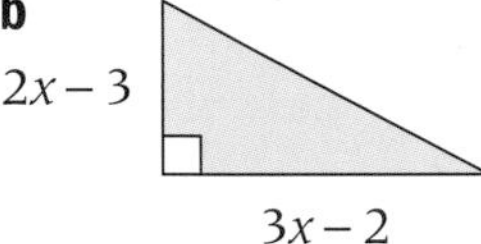

42

4 Write an expression, in terms of p, for the length of the hypotenuse of this right-angled triangle.

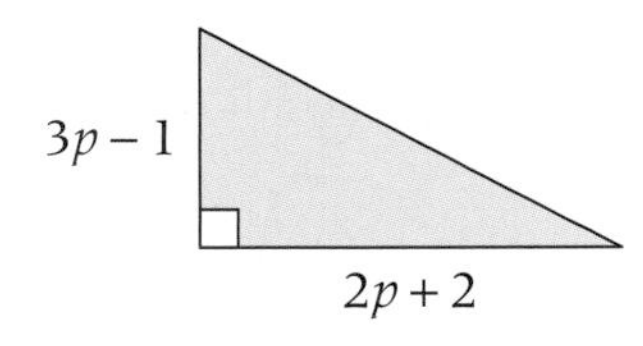

5 Given that the perimeter of this rectangle is equal to 24 cm, find an expression, in terms of x, for its area.

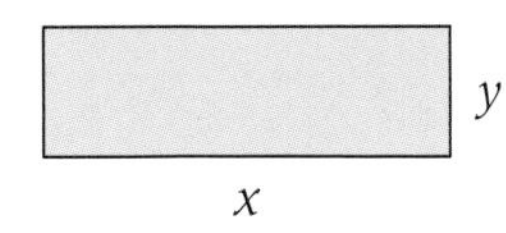

6 Given that two consecutive odd numbers can be written as $2n - 1$ and $2n + 1$ respectively, show that the difference between their squares is $8n$.

A03 Problem

7 Expand each of these.

a $(x + 2)^3$

b $(2y - 1)(y + 5)(y - 3)$

To expand a triple bracket,
- first multiply two of the brackets together
- multiply each term in your answer with each term in the remaining bracket.

A1.3 Factorisation

This spread will show you how to:

- Manipulate algebraic expressions by taking out common factors and factorising quadratic expressions

Keywords
Common factors
Factorise
Quadratic

p.228

- Factorising is the 'reverse' of expanding.

To **factorise** into single brackets, remove the **common factors** of the terms.
Find the HCF of the coefficients first, then the HCF of the algebraic terms.

$$9xy + 6x^2 = 3x(3y + 2x)$$

HCF of 9 and 6 is 3.

HCF of xy and x^2 is x.

Example

Factorise

a $12mn + 18m$ **b** $5p - 10p^3$

a $12mn + 18m = 6m(2n + 3)$

b $5p - 10p^3 = 5p(1 - 2p^2)$

HCF of 12 and 18 is 6. HCF of mn and m is m.

$5p \div 5p$ is 1. A common mistake is to write $5p(0 - 2p^2)$

To factorise into double brackets, look for two numbers that add to give the coefficient of x and multiply to give the constant.

Quadratic expressions often factorise into double brackets

EXPAND

$$(x + 4)(x + 7) \rightleftarrows x^2 + 11x + 28$$

$4 + 7$ 4×7

FACTORISE

Example

Factorise

a $x^2 + 9x + 18$ **b** $x^2 - 5x - 36$ **c** $3(a - b) + (a - b)^2$

a $x^2 + 9x + 18 = (x + 3)(x + 6)$

Two numbers that multiply to 18 and add to 9 are +6 and +3.

b $x^2 - 5x - 36 = (x - 9)(x + 4)$

List the factor pairs of -36, then check if they add to -5:
-1 and 36 -2 and 18 -3 and 12 -4 and 9 -6 and 6
$-9 \times 4 = -36$; $-9 + 4 = -5$

c $3(a - b) + (a - b)^2 = (a - b)(3 + (a - b))$
$= (a - b)(3 + a - b)$

The terms have $(a - b)$ in common.

Exercise A1.3 Grade B/A

1 Factorise each of these by removing common factors.

a $3x + 6y + 9z$ **b** $10p - 15$ **c** $5xy + 7x$
d $6mn + 9mt$ **e** $16x^2 - 12xy$ **f** $3p + 9pq$
g $7xy - 56x^2$ **h** $3x^2 + 12x^3 - 6x$ **i** $3(m + n) + (m + n)^2$
j $4(p - q) + (p - q)^3$ **k** $ax + bx + ay + by$ **l** $ac - ad - bc + bd$

2 Factorise each of these using double brackets.

a $x^2 + 7x + 10$ **b** $x^2 + 8x + 15$ **c** $x^2 + 8x + 12$
d $x^2 + 12x + 35$ **e** $x^2 - 3x - 10$ **f** $x^2 - 2x - 35$
g $x^2 - 8x + 15$ **h** $x^2 - x - 20$ **i** $x^2 - 8x - 240$
j $x^2 + 3x - 108$ **k** $x^2 - 25$ **l** $x^2 - 6 - x$

3 **a** Factorise $x^2 + 2xy + y^2$.
b Use your answer to **a** to find, without a calculator, the value of $12.3^2 + 2 \times 12.3 \times 7.7 + 7.7^2$.

4 **a** Given that the length of a rectangle is $x + 4$ and its width is $2x - 3$, write a factorised expression for its perimeter.
b Repeat for a rectangle with length $x^2 + 7x$ and width $24 - 17x$.

5 Given that $x(x + 10) = -21$, write a fully factorised expression with a value of zero.

6 Some expressions can be factorised 'twice' by first removing common factors and then using double brackets.

For example

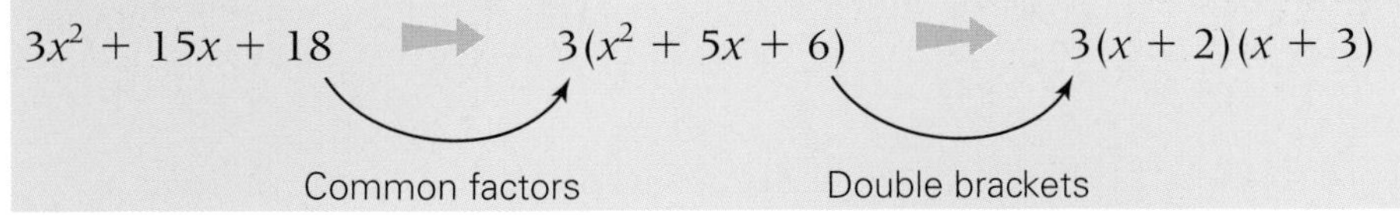

Factorise each of these twice.

a $2x^2 + 16x + 24$ **b** $3y^2 + 45y + 108$ **c** $4m^2 - 4m - 80$
d $x^3 + 8x^2 + 15x$ **e** $xy^2 - 3xy - 108x$ **f** $x^2y - 16y$

7 Factorise these using common factors, double brackets or both.

a $p^2 - p - 12$ **b** $3p^2 + 6p$ **c** $10x^2 + 70x + 120$
d $x^2y - 63y - 2xy$ **e** $a^3 + ab^2 - 2a^2b$ **f** $3am + 3an + 3ab + 3ac$

A1.4 Further factorisation

This spread will show you how to:

- Manipulate algebraic expressions by factorising quadratic expressions

Keywords
Coefficient
Factorise
Quadratic

To **factorise** quadratics with more than one x^2, you need to adapt your method.

$2x^2 + 5x + 6 = (2x + \ )(x + \ )$
Two numbers that multiply to give 6
and add to give 5 are +2 and +3
$(2x + 3)(x + 2) = 2x^2 + 7x + 6$ ✗
or
$(2x + 2)(x + 3) = 2x^2 + 8x + 6$ ✗

Not all quadratics will factorise easily.

To factorise a **quadratic** where the coefficient of x^2 is not 1:

- Multiply the **coefficient** of x^2 and the constant. — $2x^2 + 11x + 12 \rightarrow 2 \times 12 = 24$
- Find two numbers that multiply to give this value and add to give the coefficient of x. — Find two numbers that multiply to give +24 and add to give +11 → +3, +8
- Write the quadratic with the x-term split into two x-terms, using these numbers. — $2x^2 + 3x + 8x + 12$
- Factorise the pairs of terms. — $x(2x + 3) + 4(2x + 3)$
- Factorise again, taking the bracket as the common factor. — $(2x + 3)(x + 4)$
- Check by expanding. — $(2x + 3)(x + 4) = 2x^2 + 8x + 3x + 12$
  $= 2x^2 + 11x + 12$ ✓

This is a suggested method but you may find a method that works better for you.

Example

Factorise

a $3x^2 + 26x + 16$ **b** $5x^2 + 34x - 7$

a $3x^2 + 26x + 16$

$3 \times 16 = 48$ so find two numbers that multiply to +48 and add to +26. These are +2 and +24

Splitting the x term $3x^2 + 26x + 16 = 3x^2 + 2x + 24x + 16$

$= x(3x + 2) + 8(3x + 2)$

$= (3x + 2)(x + 8)$

b $5x^2 + 34x - 7$

$5 \times -7 = -35$

Two numbers that add to +34 and multiply to −35 are +35, −1.

Splitting the x term $5x^2 + 35x - 1x - 7 = 5x(x + 7) - 1(x + 7)$

$= (5x - 1)(x + 7)$

Try the factor pairs systematically.

Be careful with signs as you take out the factor −1.

Exercise A1.4 Grade A

1 Factorise fully

a $2x^2 + 5x + 3$
b $3x^2 + 8x + 4$
c $2x^2 + 7x + 5$
d $2x^2 + 11x + 12$
e $3x^2 + 7x + 2$
f $2x^2 + 7x + 3$
g $2x^2 + x - 21$
h $3x^2 - 5x - 2$
i $4x^2 - 23x + 15$
j $6x^2 - 19x + 3$
k $12x^2 + 23x + 10$
l $8x^2 - 10x - 3$
m $6x^2 - 27x + 30$
n $4x^2 - 9$
o $6x^2 + 7x - 3$
p $18x^2 + 21x - 4$

2 Explain why you cannot factorise $2x^2 + 4x + 3$.

3 Factorise these expressions, using common factors (single brackets) and double brackets, or both.

Sometimes you may need to factorise twice.

a $18x^2 - 9x$
b $4ab - 16ab^3$
c $3mn + 8m - m^3$
d $x^2 - 7x - 18$
e $2x^2 - x - 15$
f $x^3 + 7x^2 + 12x$
g $2px^2 + 11px + 12p$
h $50x^2 - 50x - 1000$
i $40p^2 - 230p + 150$
j $8x^2 + 16xy + 8y^2$

.8

4 Show that the mean average of these expressions is $(2x + 3)(x + 5)$.

$6x^2 + 10x + 23$

$2x^2 + 20x - 11$

$4x^2 + 13x + 21$

$9x - 4x^2 + 27$

5 If a quadratic expression has a negative x^2 term, you can factorise it by first taking out a common factor of '-1'.

Use this method to factorise these expressions.

a $3 - 7y - 6y^2$
b $10p + 3 - 8p^2$
c $11y - 3y^2 - 10$
d $27m - 6m^2 - 30$
e $5xy + 2y - 3x^2y$

$21 - x - 2x^2$
$= -(-21 + x + 2x^2)$
$= -(2x^2 + x - 21)$
$= -(x - 3)(2x + 7)$
$= (3 - x)(2x + 7)$

A1.5 The difference of two squares

This spread will show you how to:

- Manipulate algebraic expressions by factorising quadratic expressions, including the difference of two squares

Keywords
Factorise
Factors
Quadratic

When you expand brackets, sometimes the 'x' terms cancel each other out.

$$(x+3)(x-3) = x^2 - 9$$
$$(x+5)(x-5) = x^2 - 25$$
$$(2x-7)(2x+7) = 4x^2 - 49$$

$$\uparrow \quad \uparrow$$
$$(2x)^2 \quad 7^2$$

You call quadratic expressions of the form $x^2 - 16 = (x+4)(x-4)$ the 'difference of two squares'.

You can use what you know about expanding brackets to factorise DOTS expressions.

DOTS is shorthand for '**d**ifference **o**f **t**wo **s**quares'.

$$x^2 - 64 = (x+8)(x-8)$$
$$\sqrt{x^2} = x, \sqrt{64} = 8$$

- When you factorise an expression, check for:
  - common factors
  - double brackets
  - difference of two squares.

Example

Factorise

a $x^2 - 81$ **b** $16y^2 - 49$ **c** $25a^2 - 36b^2$ **d** $2x^2 - 50$.

a $x^2 - 81 = (x+9)(x-9)$ Using DOTS.
b $16y^2 - 49 = (4y-7)(4y+7)$ $\sqrt{16y^2} = 4y$, $\sqrt{49} = 7$.
c $25a^2 - 36b^2 = (5a-6b)(5a+6b)$
d $2x^2 - 50 = 2(x^2 - 25)$ Take out the common factor 2.
$= 2(x-5)(x+5)$ $x^2 - 25$ is DOTS.

Example

By writing 2491 as $50^2 - 3^2$, find the prime factors of 2491.

$$50^2 - 3^2 = (50-3)(50+3)$$
$$= 47 \times 53$$

47 and 53 are both prime.
The prime factors of 2491 are 47 and 53.

Exercise A1.5 Grade A

1 Factorise these expressions.

a $x^2 - 100$ **b** $y^2 - 16$ **c** $m^2 - 144$ **d** $p^2 - 64$

e $x^2 - \frac{1}{4}$ **f** $k^2 - \frac{25}{36}$ **g** $w^2 - 2500$ **h** $49 - b^2$

i $4x^2 - 25$ **j** $9y^2 - 121$ **k** $16m^2 - \frac{1}{4}$ **l** $400p^2 - 169$

m $x^2 - y^2$ **n** $4a^2 - 25b^2$ **o** $9w^2 - 100v^2$ **p** $25c^2 - \frac{1}{4}d^2$

q $x^3 - 16x$ **r** $50y - 2y^3$ **s** $\left(\frac{16}{49}\right)x^2 - \left(\frac{64}{81}\right)y^2$

2 Use factorisation to work out these, without a calculator.

a $101^2 - 99^2$ **b** $10\,006^2 - 9994^2$

c $100^2 - 99^2$ **d** $407^2 - 93^2$

3 Rewrite each of these numbers as the difference of two squares in order to find their prime factors.

a 851 **b** 9991 **c** 627 **d** 319

AO3 Problem

4 Without using a calculator, find the missing side in this right-angled triangle in surd form.

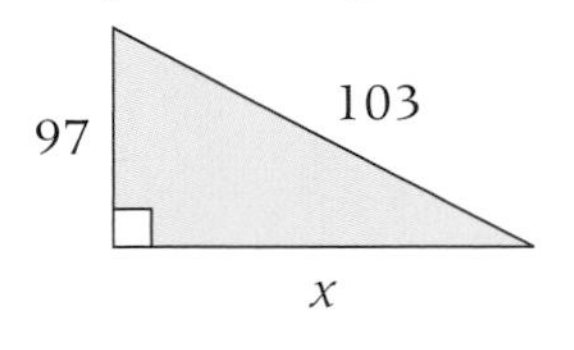

A surd is an expression like $2\sqrt{3}$ which contains a square root.

5 Factorise these algebraic expressions.

a $6x^2 - 15xy + 9y^2$ **b** $16a^2 - 9b^2$ **c** $x^2 - 11x + 28$

d $2x^2 + 11x - 21$ **e** $x^3 + 3x^2 - 18x$ **f** $5ab + 10(ab)^2$

g $10 - 3x - x^2$ **h** $10 - 10x^2$ **i** $2y + y^2 - 63$

j $2x^3 - 132x$ **k** $6x^2 + 6 - 13x$ **l** $x^4 - y^4$

AO2 Functional Maths

6 Show that the shaded area in the diagram is 6160 cm². Do not use a calculator.

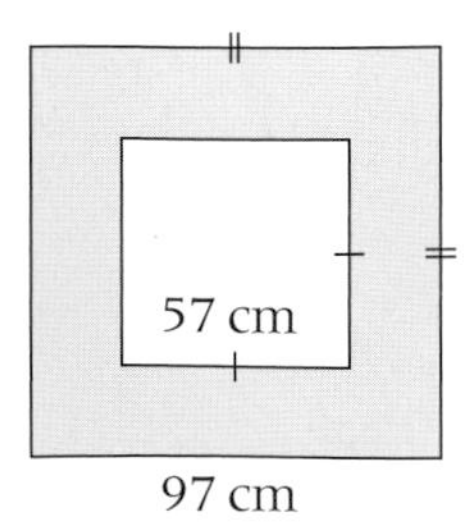

A1.6 Dealing with algebraic fractions

This spread will show you how to:

- Cancel common fractions in algebraic expressions

Keywords
Cancel
Denominator
DOTS
Factor
Numerator

- You can **cancel** common **factors** in algebraic expressions.

  $\frac{3p}{3} = \frac{\cancel{3} \times p}{\cancel{3}} = p$ 3 is a common factor of 3 and $3p$.

- You cannot cancel in the expression $\frac{3 + p}{3}$ because 3 is not a common factor of $3 + p$ and 3.

- You can factorise an expression and then cancel common factors.

  $\frac{x^2 + 5x + 6}{x + 2} = \frac{(x + 2)(x + 3)}{x + 2} = x + 3$

- You can cancel common factors in multiplication or division problems to make the calculation easier.

  $\frac{p}{7} \div \frac{3p}{4} = \frac{\cancel{p}}{7} \times \frac{4}{3\cancel{p}} = \frac{4}{21}$

p.28

Use common factors, double brackets or **DOTS** (Difference Of Two Squares).

$x + 2$ is a common factor of the **numerator** and **denominator**:

$\frac{\cancel{(x + 2)}(x + 3)}{\cancel{(x + 2)}} = (x + 3)$

p is a common factor of $p \times 4$ and $3p \times 7$.

Example

Cancel these fractions fully.

a $\frac{5p^3}{10p}$ **b** $\frac{3x + 6}{3}$ **c** $\frac{2x^2 - 5x - 12}{x^2 - 16}$

a $\frac{5p^3}{10p} = \frac{p^2}{2} = \frac{1}{2}p^2$ Divide numerator and denominator by the common factor $5p$.

b $\frac{3x + 6}{3} = \frac{3(x + 2)}{3} = x + 2$ Factorise first.

c $\frac{2x^2 - 5x - 12}{x^2 - 16}$

$= \frac{(2x + 3)(x - 4)}{(x + 4)(x - 4)}$

$= \frac{2x + 3}{x + 4}$

Numerator: factors of -24 that add to make -5 are 3 and -8.

Denominator: DOTS.

$2x^2 - 5x - 12$
$= 2x^2 - 8x + 3x - 12$
$= 2x(x - 4) + 3(x - 4)$
$= (2x + 3)(x - 4)$

Example

Simplify this expression

$\frac{x^2 - 9}{7} \times \frac{5}{x - 3}$

$\frac{x^2 - 9}{7} \times \frac{5}{x - 3} = \frac{(x + 3)(x - 3) \times 5}{7 \times (x - 3)}$

$= \frac{5(x + 3)}{7}$

Exercise A1.6 Grade B/A

1 Cancel these fractions fully.

a $\frac{15w}{5}$ **b** $\frac{3b}{9}$ **c** $\frac{10c^2}{5c}$ **d** $\frac{12bd}{3d^2}$

e $\frac{100(bd)^2}{25b}$ **f** $\frac{2x + 6}{2}$ **g** $\frac{x^2 + x}{x}$ **h** $\frac{5y - 10}{15}$

i $\frac{x^2 + 5x + 6}{x + 3}$ **j** $\frac{x^2 - 3x - 28}{x + 4}$

k $\frac{x - 5}{x^2 - 12x + 35}$ **l** $\frac{x^2 - 4}{x + 2}$

m $\frac{4y^2 - 25}{2y + 5}$ **n** $\frac{x - 9}{x^2 - 81}$

o $\frac{2x^2 - 7x + 5}{x - 1}$ **p** $\frac{3x^2 + 10x + 8}{x^2 - 4}$

q $\frac{x^3 - 16x}{x^2 + 4x}$

2 Explain why $\frac{x + 1}{x - 1}$ cannot be simplified, whereas $\frac{x + 1}{x^2 - 1}$ can be.

3 Simplify fully $\frac{a^3b^2 - a}{ab + 1}$.

4 By cancelling where possible, simplify these multiplication and division calculations.

a $\frac{4p}{3} \times \frac{9}{4p}$ **b** $\frac{6ab}{7} \times \frac{2}{b}$ **c** $\frac{4m^2}{8} \times \frac{2n}{5m^3}$ **d** $\frac{3}{g} \div \frac{g}{5}$

e $\frac{4w}{3} \div \frac{w}{2}$ **f** $\frac{2f^2}{p^3} \times \frac{p}{4f}$ **g** $\frac{y^2}{5} \div \frac{y^3}{25}$

h $\frac{(x^2 + 11x + 28)}{5} \times \frac{15}{(x + 4)}$ **i** $\frac{x^2 - 11x + 18}{12} \div \frac{x^2 - 17x + 18}{24}$

j $\frac{2x^2 - 7x - 15}{x^2 - 36} \times \frac{2x + 12}{2x^3 + 3x^2}$

5 A rectangle measures $\frac{x + 2}{8}$ by $\frac{7}{x^2 - 4}$ and its area is $\frac{1}{4}$ m². Set up an equation and solve it to find the dimensions of this shape.

A03 Problem

6 The product of these three expressions is 8.
Use this information to find the value of y.

$\frac{y^2 + 10y + 21}{2y + 8}$	$\frac{y^2 - 16}{15}$	$\frac{60}{y^2 - y - 12}$

A1.7 Adding and subtracting algebraic fractions

This spread will show you how to:

- Simplify equations involving fractions by using common denominators

Keywords
Denominator
Numerator

- To add (or subtract) numerical fractions, you convert them to equivalent fractions with a common **denominator**, and then add (or subtract) the **numerators**.

p.28

$$\frac{3}{4} + \frac{1}{6} = \frac{9}{12} + \frac{2}{12} = \frac{11}{12}$$

$$\frac{5}{7} - \frac{2}{3} = \frac{15}{21} - \frac{14}{21} = \frac{1}{21}$$

- You use the same method to add or subtract algebraic fractions.

To convert to an equivalent fraction, multiply numerator and denominator by the same number.

Example

Simplify these expressions.

a $\frac{4}{p} - \frac{3}{q}$ **b** $\frac{x+4}{6} - \frac{2x-1}{5}$

a $\frac{4}{p} - \frac{3}{q}$

$$= \frac{4q}{pq} - \frac{3p}{pq}$$

$$= \frac{4q - 3p}{pq}$$

b $\frac{x+4}{6} - \frac{2x-1}{5}$

$$= \frac{5(x+4)}{30} - \frac{6(2x-1)}{30}$$

$$= \frac{5x+20}{30} - \frac{(12x-6)}{30}$$

$$= \frac{5x + 20 - 12x + 6}{30}$$

$$= \frac{26 - 7x}{30}$$

In part **a**, convert to equivalent fractions with common denominator pq.

Example

Write each expression as a single fraction.

a $\frac{5}{x+2} + \frac{3}{x-4}$ **b** $\frac{3}{4x} + \frac{7}{4x^2}$

a $\frac{5}{x+2} + \frac{3}{x-4}$

$$= \frac{5(x-4)}{(x+2)(x-4)} + \frac{3(x+2)}{(x+2)(x-4)}$$

$$= \frac{5x - 20 + 3x + 6}{(x+2)(x-4)}$$

$$= \frac{8x - 14}{(x+2)(x-4)} = \frac{2(4x-7)}{(x+2)(x-4)}$$

b $\frac{3}{4x} + \frac{7}{4x^2}$

$$= \frac{3 \times x}{4x \times x} + \frac{7}{4x^2}$$

$$= \frac{3x + 7}{4x^2}$$

In part **a**, the common denominator is $(x + 2)(x - 4)$.

In part **b**, the common denominator is $4x^2$.

Factorise the numerator and denominator – sometimes you may be able to cancel further.

Exercise A1.7   Grade A

1 Simplify these expressions.

a $\frac{3p}{5} + \frac{p}{5}$ **b** $\frac{y}{7} + \frac{3y}{7}$

c $\frac{1}{3p} + \frac{8}{3p}$ **d** $\frac{5y}{4} + \frac{y}{8}$

e $\frac{2p}{5} - \frac{p}{3}$ **f** $\frac{6}{x} - \frac{7}{y}$

g $\frac{4}{x} + \frac{2}{x^2}$

2 Sort these expressions into equivalent pairs.
Which is the odd one out? Create its pair.

A $\frac{5x}{12} - \frac{3x}{12}$ **C** $\frac{2}{3}x - \frac{1}{3}x$ **D** $\frac{x}{3} - \frac{x}{4}$ **F** $\frac{x}{12}$

B $\frac{x}{6} + \frac{x}{4}$ **E** $\frac{x}{6}$ **G** $\frac{4x^2}{12x}$

3 Write each expression as a single fraction.

a $\frac{x+2}{5} + \frac{2x-1}{4}$ **b** $\frac{3x-2}{7} + \frac{5-3x}{11}$

c $\frac{2y-5}{3} - \frac{3y-8}{5}$ **d** $\frac{3(p-2)}{5} - \frac{2(7-2p)}{7}$

e $\frac{2}{x-7} + \frac{3}{x+4}$ **f** $\frac{5}{x-2} + \frac{3}{x+3}$

g $\frac{3}{y-2} - \frac{4}{y+1}$ **h** $\frac{2}{p+3} - \frac{5}{p-1}$

i $\frac{3}{w} + \frac{9}{w-8}$ **j** $\frac{4}{x-2} + \frac{5}{(x-2)^2}$

4 Here is a rectangle.

$\frac{5}{p+1}$

$\frac{3}{p-2}$

Find an expression for the perimeter of the rectangle.

Unit 2

A1.8 Deriving and using formulae

This spread will show you how to:

- Use formulae from mathematics and other subjects
- Substitute numbers into a formula
- Derive a formula

Keywords
Formula(e)
Substitute

- You can **substitute** numbers into **formulae** to work out the value of an unknown.

Example

The formula $v^2 = u^2 + 2as$ connects velocity (v) with initial speed (u), acceleration (a) and distance (s).
Find the final velocity of a race car that is stationary at the start line and then accelerates at 6 m/s^2 for 300 metres.

$u = 0$, $a = 6$ and $s = 300$.

The car is stationary, so initial speed (u) = 0.

$$v^2 = u^2 + 2as$$
$$= 0^2 + 2 \times 6 \times 300$$
$$= 3600$$
$$v = \sqrt{3600} = 60 \text{ m/s}$$

Substitute the values into the formula.

- You can use given information to generate a formula to represent a situation.

Example

A company sells three brands of mobile phones.
The company's profits on each are:
Brand A – £12 per phone
Brand B – £15 per phone
Brand C – £23 per phone
Write a formula for P, the company's profit in £s, for different numbers of phones sold.

If the company sold 7 phones of Brand A, its profit would be £(12 × 7). Suppose the company sells a of Brand A, b of Brand B and c of Brand C, then:

$$P = 12a + 15b + 23c$$

Use variables to represent unknowns – here, the numbers of each phone sold.

Example

Find the area, A, of these shapes

a

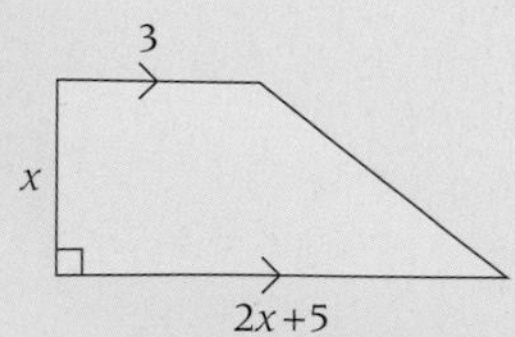

b

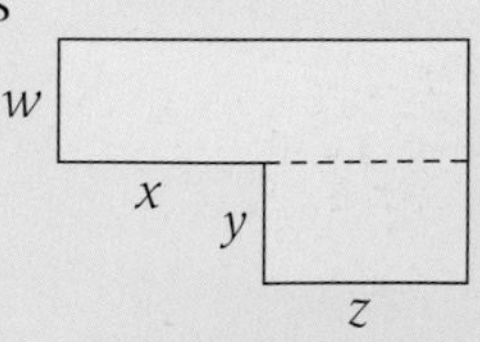

a $A = \frac{1}{2}(3 + 2x + 5) \times x$
$= \frac{1}{2}(2x + 8)\,x$
$= x\,(x + 4)$

b $A = w \times (x + z) + y \times z$
$= wx + wz + yz$

Exercise A1.8 — Grade B/A

Unit 2

AO2 Functional Maths

1 Use this formula to find the value of the required variable.

a $A = \sqrt{s(s-a)(s-b)(s-c)}$, where A is the area of a triangle with sides a, b and c, and s is half of the perimeter.
Find the area of the triangle:

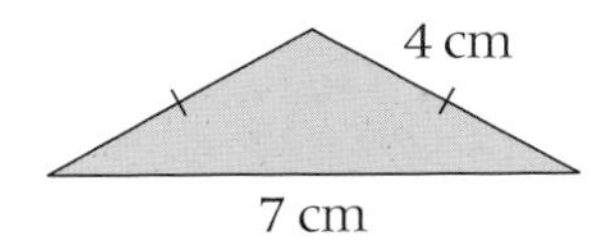

b $V = \frac{4}{3}\pi r^3$
Find the volume, V, of a sphere with diameter 6 mm.

c $T = 2\pi\sqrt{\frac{l}{g}}$
Find the length, l, of a pendulum that makes one complete swing in 8 seconds (T), given that gravity, g, is 9.8 m/s^2.

d $V = \frac{\pi^2 (r_1 + r_2)(r_1 - r_2)^2}{4}$
Find the volume of a torus ('doughnut') with outer radius $r_1 = 10$ cm and inner radius $r_2 = 4$ cm.

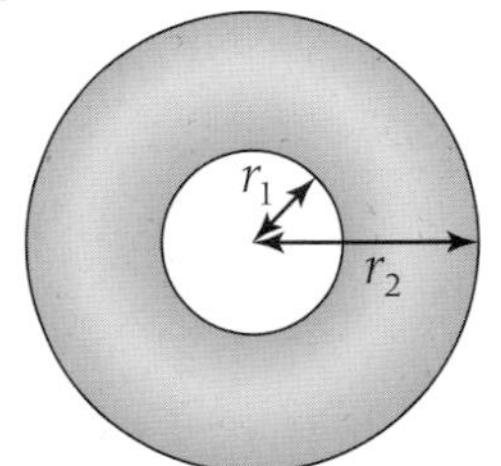

2 Write a formula to describe each of these situations.

a The taxi fare, F, with AB cabs is 130 p per mile plus a hire charge of £2.50 per mile.

b The total cost, P, of posting a bag of mail is based on the following prices: 39 p per small letter, 61 p per large letter, £1.28 per small packet and £4.45 per large packet.

c The cost of bricks, B, is based on a delivery charge plus the cost per 100 bricks.

D The cost, C, of an order of bags of chips and cans of cola at Dave's fish bar.

3 Show that the formula for each shaded area is

a $A = p^2\left(1 - \frac{\pi}{4}\right)$

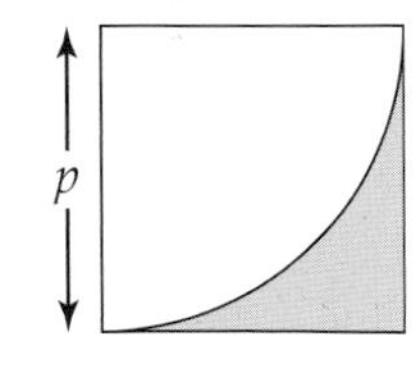

b $A = x^2 + 5x + 18$.

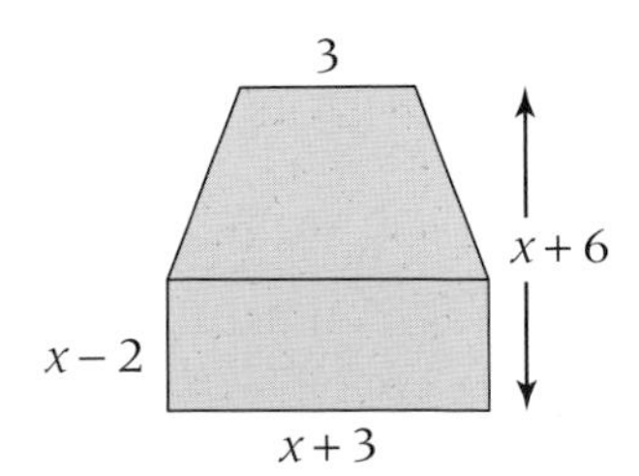

Summary

Check out

You should now be able to:

- Use index notation and simple laws of indices
- Multiply a single term over a bracket
- Expand the product of two linear expressions
- Factorise an algebraic expression using common factors
- Factorise quadratic expressions
- Distinguish in meaning between the words 'equation', 'formula', 'identity' and 'expression'
- Substitute numbers into a formula from mathematics and other subjects
- Derive a formula

Worked exam question

a Factorise fully $6x^2 + 9xy$ (2)

b Expand and simplify $(2x + 5)(x - 2)$ (2)

(Edexcel Limited 2008)

a

$$6x^2 + 9xy = 3(2x^2 + 3xy)$$
$$= 3x(2x + 3y)$$

The common factor is 3 then x.

OR

a

$$6x^2 + 9xy = x(6x + 9y)$$
$$= 3x(2x + 3y)$$

The common factor is x then 3.

OR

a

$$6x^2 + 9xy = 3x(2x + 3y)$$

The common factor is $3x$.

For part **a**, factorise **fully** suggests there are at least 2 common factors.

b

$$(2x + 5)(x - 2) = 2x^2 + 5x - 4x - 10$$
$$= 2x^2 + x - 10$$

There should be 4 terms.

These terms will not simplify any further.

Exam questions

1 **a** Expand $a(2a - 3b)$ (2)
b Factorise $a^2 - 49$ (1)

2 **a** Factorise fully $4x^2 - 6xy$ (2)
b Factorise $x^2 + 5x - 6$ (2)
(Edexcel Limited 2009)

3 **a** Factorise $x^2 - 3x$ (2)
b Simplify $k^5 \div k^2$ (1)
c Expand and simplify
i $4(x + 5) + 3(x - 7)$
ii $(x + 3y)(x + 2y)$ (4)
d Factorise $(p + q)^2 + 5(p + q)$ (1)
(Edexcel Limited 2004)

4 **a** Simplify $t^6 \times t^2$ (1)
b Simplify $\frac{m^8}{m^3}$ (1)
c Simplify $(2x)^3$ (2)
d Simplify $3a^2h \times 4a^5h^4$ (2)
(Edexcel Limited 2009)

5 Simplify fully

$$\frac{x^2 + x - 6}{x^2 - 7x + 10}$$ (3)

(Edexcel Limited 2008)

6 Simplify fully

$$\frac{x^2 - 7x + 12}{2x^2 - 5x - 12}$$ (3)

7 The diagram below shows a 6-sided shape.
All the corners are right angles.
All measurements are given in centimetres.

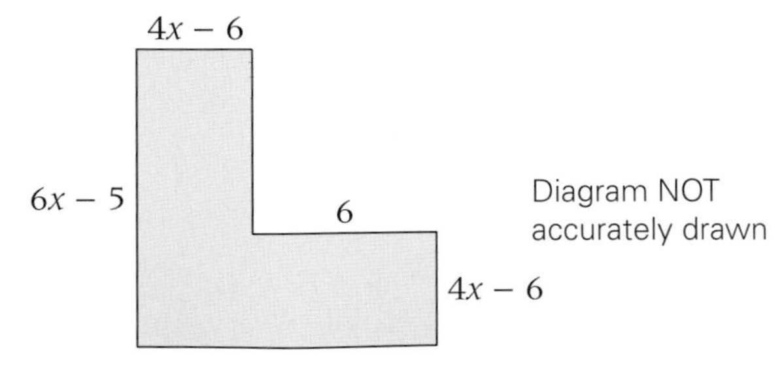

The area of the shape is A cm^2.
Show that $A = 24x^2 - 32x - 6$ (3)

N3

Written calculations

Written calculations

There are lots of professions in which it is vital to perform mental calculations as a check on the answer they have calculated either manually or by using a calculator. These include doctors working out the dose of medicine to give a patient, pilots checking the fuel required for a flight, or civil engineers calculating the amount of material required for the construction of a building.

What's the point?

Standard written methods are used to perform more complex mathematical calculations. However it is vital that at all times mental calculations (approximations) are used to check that any answers are sensible

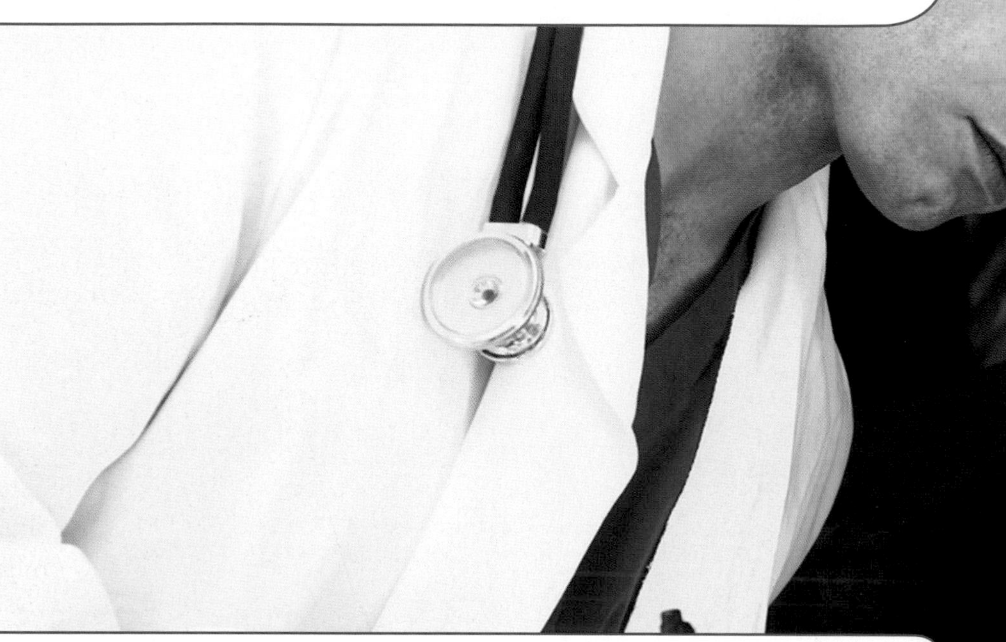

Check in

You should be able to

- **use mental and written methods of calculation**

1 For each of these calculations write down

i write down a mental estimate

ii an exact calculation, showing your method clearly

a $145 + 298$ **b** $775 - 402$ **c** $16.9 + 4.4$

d $973 - 708$ **e** $83.78 + 48.39$ **f** $315.7 - 48.89$

g $66.8 - 39.77$ **h** $102.65 + 35.88 + 17.7$ **i** 116×6

j $7728 \div 7$ **k** 38×19 **l** $104 \div 8$

m 3.7×4.2 **n** 238×8.5 **o** $809 \div 11$

2 Find the results of these additions and subtractions involving fractions.

a $\frac{1}{2} + \frac{1}{4}$ **b** $\frac{1}{2} - \frac{1}{4}$ **c** $\frac{3}{10} + \frac{1}{5}$ **d** $\frac{2}{3} + \frac{1}{2}$

3 Find the results of these multiplications and divisions involving fractions.

a $\frac{1}{2} \times \frac{2}{3}$ **b** $\frac{2}{5} \div \frac{1}{4}$ **c** $\frac{3}{7} \times \frac{7}{10}$ **d** $\frac{4}{5} \div \frac{1}{10}$

Orientation

What I need to know	What I will learn	What this leads to
KS3 Do mental and written arithmetic	■ Do arithmetic with decimals and fractions ■ Prove or refute simple statements	**A3** Calculate with algebraic fractions
N1 Do arithmetic with decimals, fractions and percentages		Personal finance

Rich task

How many hamsters would you need to make an elephant?
How much food would an elephant-sized hamster need to eat?
How big would the cage for an elephant-sized hamster need to be?

N3.1 Mental calculations

This spread will show you how to:

- Use mental methods for calculations involving fractions, decimals and percentages

Keywords
Equivalent calculation
Estimate
Product

You can use strategies such as finding **estimates** and **equivalent calculations** to help with mental calculations.

p.24,28

Example

Work out **a** $\frac{2}{3} + \frac{3}{4}$ **b** $\frac{2}{5} \times \frac{3}{8}$ **c** $0.3 - 0.23$
d 0.4×0.3 **e** $0.8 \div 0.02$ **f** 28% of 40.

a Start with an estimate. Both fractions are between $\frac{1}{2}$ and 1, so the total is between 1 and 2.
The lowest common denominator is 12, and the equivalent fractions are $\frac{8}{12}$ and $\frac{9}{12}$. The total is $\frac{17}{12} = 1\frac{5}{12}$.

b Both fractions are less than $\frac{1}{2}$, so the **product** is less than $\frac{1}{4}$.
Multiplying numerators and denominators gives $\frac{6}{40}$, which cancels down to $\frac{3}{20}$.

c $0.3 - 0.2 = 0.1$, so the answer must be a little smaller than 0.1.
$30 - 23 = 7$, so the answer is 0.07.

d Half of 0.3 is 0.15, so 0.4×0.3 must be a little smaller than 0.15.
$4 \times 3 = 12$, so the answer is 0.12.

e Multiply both numbers by 100, to give the equivalent calculation $80 \div 2 = 40$.

f The answer is a bit more than a quarter of 40, which is 10.
Work out $28 \times 4 = 56 \times 2 = 112$; the answer is 11.2.

Or you could work out 40% of 28.

Example

Work out $1\frac{1}{4} - \frac{3}{7}$.

Estimate first.
$\frac{3}{7}$ is nearly $\frac{1}{2}$, so the answer must be a little more than $\frac{3}{4}$.
The common denominator is 28 so work out

$$1\frac{7}{28} - \frac{12}{28} = 1 - \frac{5}{28}$$
$$= \frac{23}{28}$$

Exercise N3.1 Grade D

All of the calculations in this exercise should be done mentally, with jottings where necessary.

1 Write a fraction equivalent to each of these decimals.

a 0.5 **b** 0.3 **c** 0.25 **d** 0.4
e 0.6 **f** 0.05 **g** 0.45 **h** 0.375

2 Write the decimal equivalents of these fractions.

a $\frac{3}{4}$ **b** $\frac{2}{5}$ **c** $\frac{5}{8}$ **d** $\frac{3}{20}$
e $\frac{1}{25}$ **f** $\frac{4}{25}$ **g** $\frac{7}{20}$ **h** $\frac{3}{50}$

3 Add these fractions.

a $\frac{1}{2}+\frac{1}{4}$ **b** $\frac{1}{5}+\frac{1}{10}$ **c** $\frac{1}{3}+\frac{1}{6}$ **d** $\frac{1}{4}+\frac{1}{5}$
e $\frac{1}{8}+\frac{1}{2}$ **f** $\frac{1}{6}+\frac{1}{4}$ **g** $\frac{1}{3}+\frac{1}{5}$ **h** $\frac{3}{8}+\frac{1}{4}$

4 Subtract these fractions.

a $\frac{3}{4}-\frac{1}{2}$ **b** $\frac{5}{8}-\frac{1}{4}$ **c** $\frac{4}{5}-\frac{1}{10}$ **d** $\frac{3}{4}-\frac{1}{6}$
e $\frac{4}{9}-\frac{1}{3}$ **f** $\frac{5}{6}-\frac{2}{3}$ **g** $\frac{3}{8}-\frac{1}{6}$ **h** $\frac{5}{8}-\frac{2}{7}$

5 Evaluate these multiplications, giving your answers as fractions in their simplest form.

a $\frac{2}{3}\times\frac{1}{5}$ **b** $\frac{3}{4}\times\frac{1}{2}$ **c** $\frac{2}{9}\times\frac{3}{4}$ **d** $\frac{4}{5}\times\frac{5}{8}$
e $\frac{3}{8}\times\frac{4}{9}$ **f** $\frac{5}{12}\times\frac{3}{10}$ **g** $\frac{6}{7}\times\frac{1}{2}$ **h** $\frac{7}{24}\times\frac{3}{14}$

6 Work out these divisions and simplify your answers.

a $\frac{1}{2}\div\frac{1}{8}$ **b** $\frac{2}{3}\div\frac{1}{6}$ **c** $\frac{2}{5}\div\frac{1}{10}$ **d** $\frac{2}{3}\div\frac{3}{4}$
e $\frac{5}{8}\div\frac{1}{4}$ **f** $\frac{7}{10}\div\frac{2}{4}$ **g** $\frac{4}{9}\div\frac{2}{3}$ **h** $\frac{7}{20}\div\frac{1}{4}$

7 Do these additions and subtractions.

a 0.3 + 0.4 **b** 0.5 + 0.7 **c** 1.2 + 0.9 **d** 1.0 − 0.8
e 2.4 − 1.5 **f** 13.3 − 6.8 **g** 0.4 + 0.32 **h** 0.5 − 0.17
i 6.7 − 4.85 **j** 19 − 2.77 **k** 4.27 + 5.944 **l** 16.4 − 8.517

8 Do these multiplications and divisions.

a 4.8 × 0.2 **b** 6.2 ÷ 0.2 **c** 25.4 ÷ 0.25 **d** 4.3 × 0.3
e 2.8 × 0.4 **f** 2.8 ÷ 0.4 **g** 0.75 × 0.6 **h** 0.75 ÷ 0.05

9 Find these percentages. Show your method in each case.

a 10% of 480 **b** 15% of 340 **c** 19% of 800 **d** 41% of 600
e 32% of 75 **f** 53% of 740 **g** 65% of 90 **h** 79% of 4000

10 Work out the result of these percentage changes. Show your method in each case.

a £350 is increased by 12% **b** £840 is decreased by 12.5%
c £50 is increased by 17% **d** £390 is increased by 21%
e £780 is decreased by 5% **f** £245 is decreased by 15%

N3.2 Written calculations

This spread will show you how to:

- Use written methods for calculations involving fractions, decimals and percentages

Keywords
Decimal equivalent
Estimate
Place value

You need to know a range of efficient written techniques for calculating with fractions, decimals and percentages.
To divide by a fraction, multiply by its multiplicative inverse (invert the fraction).
Multiply fractions by multiplying numerators and denominators.

You can still use mental methods to estimate the answers, which gives you a check on the written calculations.

Example

Calculate
a 2.6×3.28 **b** $\frac{2}{3} \div \frac{3}{5}$ **c** 48% of 73 **d** 37 as a percentage of 120.

a

2.6 × 3.28
Estimate ≃ 3 × 3 = 9

```
  328
 × 26
 1968
 6560
 8528 ⇒ Ans = 8.528
```

Do the calculation with whole numbers, then use the **estimate** to adjust the **place value**.

b

$\frac{2}{3} \div \frac{3}{5}$
Estimate ≃ a bit more than 1

$\frac{2}{3} \div \frac{3}{5} = \frac{2}{3} \times \frac{5}{3}$
$= \frac{10}{9} = 1\frac{1}{9}$

Note the initial estimate, $\frac{2}{3}$, is slightly larger than $\frac{3}{5}$ (think of their **decimal equivalents**).

c

48% of 73
Estimate ≃ $\frac{1}{2} \times 72$ = 36

```
   73
 × 48
  584
 2920
 3504 ⇒ Ans = 35.04
```

Here you need to work out $73 \times 48 \div 100$. You can just do the whole-number multiplication, and adjust the place value.

d

Estimate $\frac{37}{120} \simeq 30\%$

```
      30833
  12)370000
     36
      100
       96
        40
        36
         4   Ans = 30.83̇%
```

Use long division. Notice the recurring digit, and how the place value is adjusted.

Exercise N3.2 — Grade D

Unless you are told otherwise, you should use a written method for the calculations in this exercise.

1 First write a mental estimate and then use a written method to calculate an exact answer for each of these.

a $134.6 - 7.859$ **b** 2.47×7
c 64.8×0.9 **d** $434.28 \div 7$
e $23.862 + 38.779$ **f** 5.7×8.9
g $84.66 \div 0.3$ **h** 13.2×0.74
i $234.2 - 166.89$

2 Use a calculator to check your answers to question **1**.

You will not be allowed to use a calculator in the unit 2 examination.

3 Evaluate

a $\frac{2}{7} + \frac{3}{5}$ **b** $\frac{4}{9} \div \frac{3}{5}$
c $\frac{3}{8} \times \frac{5}{6}$ **d** $\frac{7}{8} - \frac{3}{7}$
e $\frac{2}{3} + \frac{1}{6} + \frac{2}{5}$ **f** $\frac{4}{9} + \frac{2}{3} - \frac{3}{6}$
g $\frac{2}{5} \times \frac{5}{6} \times \frac{1}{2}$ **h** $\frac{4}{5} - \frac{3}{8}$
i $\frac{1}{2} \div \frac{3}{8}$

4 Use the fraction facility on a scientific calculator to check your answers for question **3**.

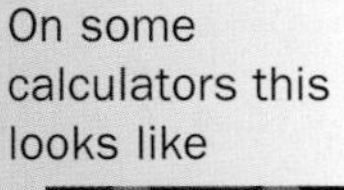
On some calculators this looks like

5 Calculate these percentages, giving your answers to 3 significant figures.

a 47% of 93 **b** 32% of 49
c 17% of 604 **d** 64% of 123
e 23% of 290 **f** 56% of 324
g 43.9% of 600 **h** 88.3% of 17
i 26.4% of 428

6 Use a calculator to check your answers to question **5**.

7 Find the results of these percentage changes.

a 480 is decreased by 4.75% **b** 725 is increased by 10.6%
c 285 is increased by 67.5% **d** 48.49 is decreased by 18.75%

8 Use a calculator to check your answers to question **7**.

9 For each pair of numbers, use long division (or an equivalent written method) to express one number as a percentage of the other. Give your answers to a suitable degree of accuracy.

a 15 as a percentage of 240 **b** 28.8 as a percentage of 120
c 81.9 as a percentage of 450 **d** 1.7 as a percentage of 130
e 48.75 as a percentage of 90 **f** 267.5 as a percentage of 1800

10 Use a calculator to check your answers to question **9**.

N3.3 Further written calculation methods

This spread will show you how to:

- Use written calculation methods

Keywords
BIDMAS
Estimate
Order of operations

- You can make an **estimate** before starting a written calculation. You can use it to check your answer or to adjust place value.
- In a multi-stage calculation, you need to remember the **order of operations**, **BIDMAS**.

BIDMAS
Brackets
Indices (or powers)
Division or
Multiplication
Addition or
Subtraction

Example

Evaluate these without a calculator.

a $\frac{2.3 \times (4^2 + 5^2)}{7}$ **b** $-2.1 + (-9 \times 2.4)$ **c** $\frac{2}{7} + \frac{3}{5}$ **d** 23% of 760

a
$23 \times 41 = 943$
$943 \div 7 = 134.7$ to 4sf
$\div 10$ to adjust place value: answer $= 13.5$ to 3sf

Estimate
$4^2 + 5^2 = 16 + 25 = 41$
$41 \div 7 \approx 6$, and $6 \times 2.3 \approx 14$.

b
$9 \times 24 = 216 \Rightarrow -9 \times 2.4 = -21.6$
$-2.1 - 21.6 = -23.7$

Estimate
$-9 \times 2.4 \approx -10 \times 2 = -20$, and $-2.1 - 20 \approx -22$.

c
$\frac{2}{7} + \frac{3}{5} = \frac{10}{35} + \frac{21}{35} = \frac{31}{35}$

Estimate
$\frac{2}{7} = \frac{1}{3}$ and $\frac{3}{5} \approx \frac{2}{3}$, so the answer should be about 1.

d
23% of 760 $\Rightarrow \frac{23}{100} \times 760$
$23 \times 76 = 1748$
adjust place value: answer $= 174.8$

Estimate
23% of 760 $\approx \frac{1}{4}$ of 760 $= \frac{1}{2}$ of 380 $= 190$.

Use your estimate to adjust place value.

Example

p.182

Evaluate these without a calculator, giving your answers in standard form to 2 significant figures.

a $(2.1 \times 10^4) \times (7.3 \times 10^{-6})$
b $(1.512 \times 10^{-2}) \div (9 \times 10^3)$

a $21 \times 73 = 1533$
adjust place value: 1.5×10^{-1}
b $1512 \div 9 = 168$
adjust place value: 1.68×10^{-6}

Estimate
$(2 \times 10^4) \times (7 \times 10^{-6}) = 14 \times 10^{-2} = 1.4 \times 10^{-1}$.

Estimate
$9 \times 10^3 \approx 10^4$
$\Rightarrow (1.512 \times 10^{-2}) \div (9 \times 10^3) \approx (1.5 \times 10^{-2}) \div 10^4 = 1.5 \times 10^{-6}$.

Exercise N3.3 Grade D/C

1 Use a written method to calculate

a 836×46 **b** 774×38
c 397×171 **d** 617×259

2 Use a written method to calculate

a $2496 \div 16$ **b** $4071 \div 23$
c $7942 \div 38$ **d** $21\,373 \div 67$

3 Without using a calculator, evaluate

a $\dfrac{-7 + (-18 \times +45)}{+3 \times -3}$ **b** $-6.7 + \dfrac{4.3 \times -2.1}{-2 \times +5}$

c $\dfrac{8.51 - 30.77}{1.8 + 0.3}$ **d** $\dfrac{25.19}{11} - 3.6^2$

4 Evaluate these without a calculator, giving your answers to 3 significant figures.

a $\dfrac{4.7 \times 3.8}{7}$ **b** $5.8^2 - \dfrac{3.2 \times 1.8}{9}$

c $\dfrac{5.2^2}{8} - \dfrac{2.1^2}{5}$ **d** $\dfrac{4.8 \times 6.3}{2.3 + 5.1}$

5 Use a written method to evaluate these. Show your working.

a 52% of 416 **b** 38% of 98
c 63% of 881 **d** 119% of 77
e 28.5% of 515 **f** 106.5% of 24.5
g 3.8% of 46.2 **h** 0.35% of 17
i 0.07% of 309.5

6 Use a written method for these calculations. Show your working.

a $\frac{2}{7} + \frac{1}{3}$ **b** $\frac{3}{4} - \frac{2}{5}$
c $\frac{3}{8} \times \frac{4}{9}$ **d** $\frac{7}{8} \div \frac{3}{4}$
e $3\frac{2}{3} - 1\frac{5}{8}$ **f** $15\frac{3}{4} + 7\frac{5}{9}$
g $4\frac{1}{2} \times 9\frac{3}{4}$ **h** $3\frac{2}{5} \div 2\frac{1}{10}$

7 Evaluate these without using a calculator. Give your answers in standard form.

a $(2.3 \times 10^3) + (4.2 \times 10^3)$ **b** $(2.7 \times 10^5) - (6.3 \times 10^4)$
c $(8.75 \times 10^{-8}) + (5.66 \times 10^{-8})$ **d** $(6.31 \times 10^7) + (6.09 \times 10^8)$
e $(1.82 \times 10^2) - (7.3 \times 10^1)$ **f** $(3.19 \times 10^{-6}) - (8.4 \times 10^{-7})$

Arithmetic with numbers written in standard form is examined in unit 3.

8 Evaluate these without using a calculator. Give your answers in standard form, correct to 3 significant figures.

a $(4.5 \times 10^5) \times (3.6 \times 10^2)$ **b** $(8.9 \times 10^{-4}) \times (7.7 \times 10^8)$
c $(5.85 \times 10^4) \div (5 \times 10^{-6})$ **d** $(2.2 \times 10^7)^2$
e $(3.77 \times 10^{-3}) \times (6.08 \times 10^{11})$ **f** $(4.82 \times 10^8) \div (2 \times 10^{-4})$

9 Use a calculator to check your answers to questions **1** to **8**.

N3.4 Number and algebra

This spread will show you how to:

- Use the commutative, associative and distributive laws of addition, multiplication and factorisation
- Understand that the rules of algebra obey and generalise the rules of arithmetic

Keywords
Associative
Commutative
Distributive

When you expand brackets you are using the **distributive** law.

- Multiplication is distributive over addition and subtraction.

$a(b + c) = ab + ac$ $\quad$ $3(2 + 6) = 3 \times 2 + 3 \times 6 = 6 + 18 = 24$

$a(b - c) = ab - ac$ $\quad$ $4(3 - 1) = 4 \times 3 - 4 \times 1 = 12 - 4 = 8$

- When you add or multiply two variables you can put the variables in any order. This is the **commutative** law.

Subtraction and division are not commutative.

For addition: $a + b = b + a$ $\quad$ $10 + 5 = 5 + 10$

For multiplication: $a \times b = b \times a$ $\quad$ $3 \times 6 = 6 \times 3$

- When you multiply together three variables you can use the **associative** law.

It does not matter which pair you add (or multiply) first.

Subtraction and division are *not* associative.

For addition:

$(a + b) + c = a + (b + c)$ $\quad$ $(6 + 5) + 7 = 11 + 7 = 18$

$\quad$ $6 + (5 + 7) = 6 + 12 = 18$

For multiplication:

$(a \times b) \times c = a \times (b \times c)$ $\quad$ $(4 \times 2) \times 3 = 8 \times 3 = 24$

$\quad$ $4 \times (2 \times 3) = 4 \times 6 = 24$

Example

Prove that, for any two fractions $\frac{a}{b}$ and $\frac{c}{d}$,

$$\frac{a}{b} \div \frac{c}{d} = \frac{ad}{bc}.$$

$\frac{a}{b} \div \frac{c}{d} = \frac{ad}{b} \div c$ $\quad$ Multiply both fractions by d

$\frac{ad}{b} \div c = ad \div bc = \frac{ad}{bc}$ $\quad$ Multiply both fractions by b

$b \neq 0$ and $d \neq 0$ (you cannot have a fraction with zero denominator).

Multiplying both terms in a division by any non-zero number leaves the result unchanged.

Example

p.126

a Calculate $83^2 - 17^2$.

b Without using a calculator, evaluate 350^2.

a $83^2 - 17^2 = (83 + 17)(83 - 17)$

$= 100 \times 66 = 6600$

b $x^2 - y^2 \equiv (x + y)(x - y)$

$x^2 \equiv (x + y)(x - y) + y^2$

$350^2 = (350 + 50)(350 - 50) + 50^2$

$= 400 \times 300 + 2500$

$= 120\,000 + 2500 = 122\,500$

Use DOTS: $x^2 - y^2 \equiv (x + y)(x - y)$

$x = 350$
You could choose any number for y. $y = 50$ gives 'easy' numbers in the brackets to multiply.

Exercise N3.4 Grade A

1 Charlie says, 'When I work out a multiplication like 7×17, I just do 7×10 and 7×7, and add the answers.' Which law is Charlie using? Explain your answer carefully.

2 Multiplication is distributive over subtraction. Show how this idea can be used to calculate 24×19 mentally.

$19 = 20 - 1$

3 Nancy says, 'I am working out $36 \div (2 + 3)$. I'll get the same result if I do $36 \div 2$ and $36 \div 3$, and then add the answers together.' Explain why Nancy is wrong.

4 Give numerical examples to show that

a subtraction is *not* commutative **b** division is *not* commutative

c subtraction is *not* associative **d** division is *not* associative.

5 Is the 'square root' operation distributive over addition? That is, is it true that $\sqrt{a+b} \equiv \sqrt{a} + \sqrt{b}$? Explain your answer carefully.

6 Evaluate these calculations, without using a calculator.

a $73^2 - 27^2$ **b** $6.4^2 - 3.6^2$

c $191^2 - 9^2$

Follow the method used to evaluate $83^2 - 17^2$ in the example.

7 Without using a calculator, work out these calculations.

a 35^2 **b** 29^2

c 4.1^2

Follow the method used to work out 350^2 in the example.

8 Given that $4.7^2 = 22.09$, show how you can find the value of 5.7^2 without further multiplication or squaring.

9 Given that $15.6^2 = 243.36$, show how you can use the identity $(x - 1)^2 = x^2 - 2x + 1$ to find the value of 14.6^2.

A03 Problem

10 Petra has noticed a pattern in the squares of decimal numbers. She writes:

$1.5^2 = 2.25$, $2.5^2 = 6.25$, $3.5^2 = 12.25$, and so on.
To find any square like this, for example 7.5^2,
multiply the unit number by the number +1, that's
$7 \times 8 = 56$. Then just add 0.25, so $7.5^2 = 56.25$.

Use algebra to prove that this pattern always works.

Summary

Check out

You should now be able to:

- Use calculators effectively and efficiently for complex calculations
- Use mental and written methods to calculate with fractions, decimals and percentages

Worked exam question

Hajra's weekly pay this year is £240
This is 20% more than her weekly pay last year.
Bill says 'This means Hajra's weekly pay last year was £192'.
Bill is wrong.

a Explain why. (1)

b Work out Hajra's weekly pay last year. (2)

(Edexcel Limited 2006)

a

Bill has calculated 20% of this year's pay.
He has calculated 20% of £240
He should have calculated 20% of last year's pay.

20% of £240 = £48 and
£240 − £48 = £192

b

120% = £240
1% = £240 ÷ 120
= £2
100% = 2 × £100
= £200

OR

b £240 ÷ 1.2 = £200

Show the calculation 240 ÷ 120 × 100 in your working.

You need to show your working whichever method you use.

Exam questions

1 Work out $4\frac{1}{4} - 2\frac{2}{5}$ (3)

AO3

2

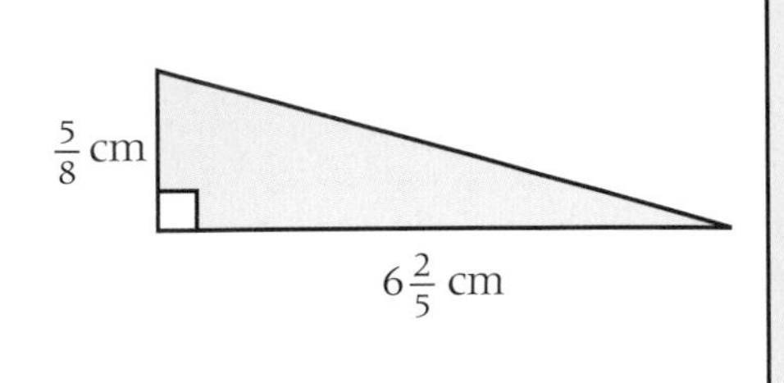

Diagrams **NOT** accurately drawn.

The area of the square is 18 times the area of the triangle.
Work out the perimeter of the square. (5)

(Edexcel Limited 2005)

AO2

3 In 2003 the population of Great Britain was 6.0×10^7
In 2003 the population of India was 9.9×10^8

a Work out the difference between the population of India and the population of Great Britain in 2003.
Give your answer in standard form. (2)

In 1933 the population of Great Britain was 4.5×10^7

b Calculate the percentage increase in population of Great Britain from 1933 to 2003.
Give your answer correct to one decimal place. (3)

(Edexcel Limited 2007)

Functional Maths 3: Sandwich shop

The manager of a catering company can use data about customer numbers in order to spot trends in customer behaviour and to plan for the future.

...mply Sandwiches

Simply ...andwiches

...ches, paninis, baguettes and salads

Simply Sandwiches

Customer numbers at 'Simply sandwiches' takeaway over a given two-week period were:

Day	Number of Customers	
	Week 1	Week 2
Monday	50	54
Tuesday	68	60
Wednesday	47	53
Thursday	58	57
Friday	52	56
Saturday	76	70
Total		

Simply Sandwiches

Work out an appropriate average number of customers for

a) each day of the week
b) the whole week in total.

How does your answer to b) differ if

c) you exclude Saturdays
d) a 24-person coach trip arrives on the second Wednesday?

Construct a pie chart to show what percentage of customers visited the sandwich shop on each day during this two-week period.

Comment on the spread of the data, referring to the data and your chart.

This frequency polygon shows customer numbers during each hour on the first Saturday, the busiest day during this two-week period:

What was the busiest/quietest time of day?

What can you say about the relationship between time of day and customer numbers? Would you expect every day of the week to have a similar pattern?

Justify your answers referring to the data.

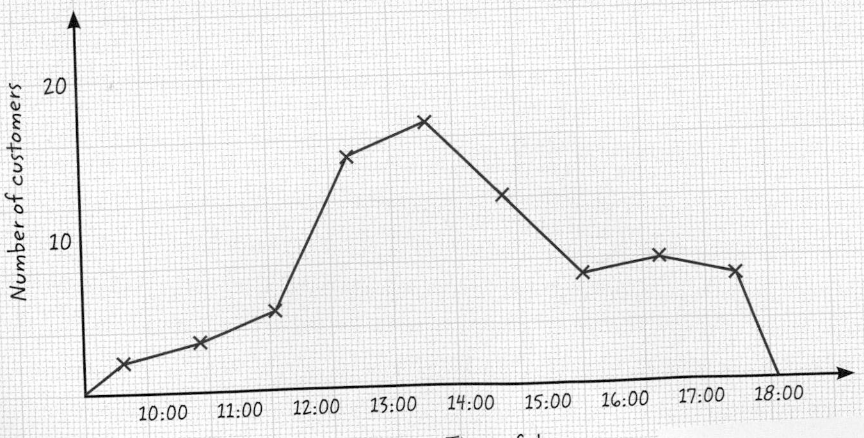

How do you think the manager could use such data about customer numbers?

A manager can use data about customer numbers to estimate how much stock to order each week. In reality, limitations due to space and the shelf life of products also apply.

In the second week of the two-week period at 'Simply sandwiches', total percentage sales of the different varieties of sandwiches were:

Variety	Ham	Cheese	Hummous	Tuna	Chicken
% sales	26	18	10	15	31

How many of each sandwich were sold during this week? (assuming every customer bought one sandwich)

The manager does a weekly stocktake every Sunday before placing the order for the following week.

The stocktake figures for this week were:

Product	Bread	Ham	Cheese	Hummous	Tuna	Chicken
Stock (packs)	6	2.5	3	2	1.5	1

Note that:

Each loaf of bread makes 20 sandwiches;

Each pack of ham, cheese and chicken contains 10 portions;

Each tub of hummous contains 8 portions;

Each can of tuna contains 14 portions.

Use the information given to estimate the amount of each product that the manager should order to last for the following week.

Record your estimations in a table.

The stock will be delivered on Wednesday.

A coach trip of 24 people arrives unexpectedly and places the following order on Tuesday, before the new stock arrives.

Sandwich	Ham	Cheese	Hummous	Tuna	Chicken
Quantity	4	7	5	3	5

Would 'Simply sandwiches' be able to cater for this order?

How would you advise the manager to prepare for such situations in the future?

Simply Sandwiches

Sandwiches, paninis, baguettes and salads

G1 Circle theorems

In everyday life you usually rely on experience to decide if something is true. In mathematics you start with something you already know is true and use steps of logical reasoning to show how it follows that something else is true. For example, using circle theorems you can prove that there is a best point from which to view a statue.

What's the point?

The skills used in being able to set out a logical train of thinking lie at the heart of every mathematician's work.

Circle theorems

Check in

You should be able to

- **use the sum of the angles on a line and around a point**

1 Work out the missing angles.

a

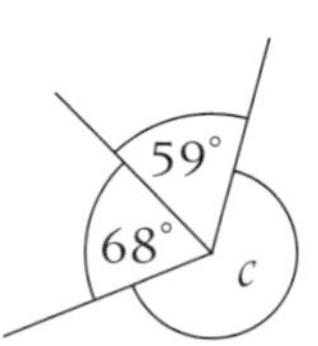

b

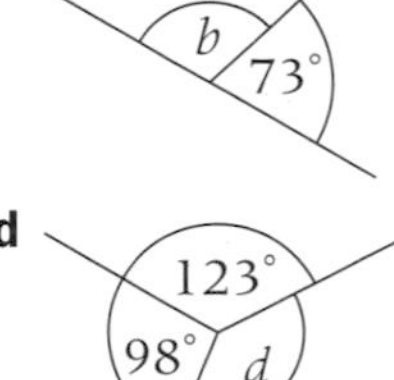

c

59°
68°
c

d

123°
98°
d

- **use the sum of the angles in a triangle and a quadrilateral**

2 Work out the missing angles in these shapes.

a

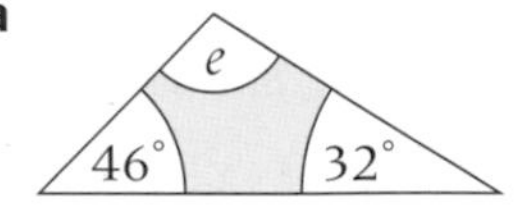

b

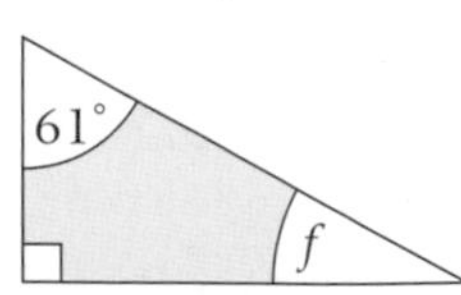

c

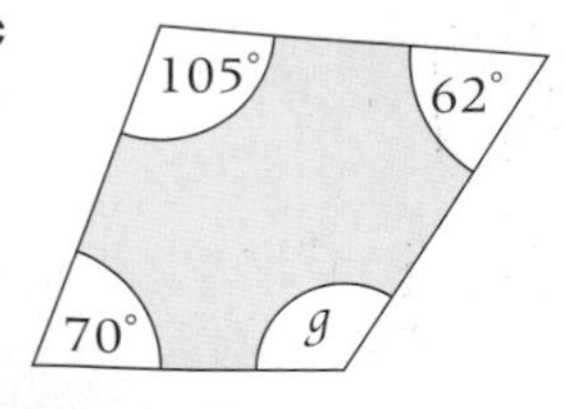

d

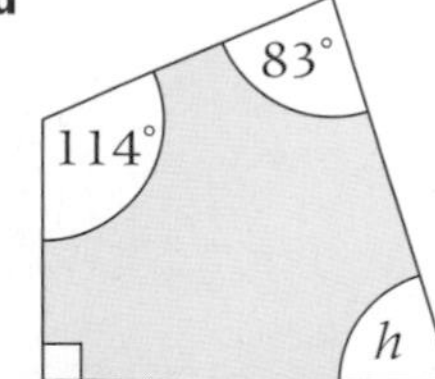

Orientation

What I need to know	What I will learn	What this leads to
KS3 Know and use the angle properties of lines and triangles	■ Apply circle theorems ■ Construct geometric proofs	**G2** Further geometric proof **A8** Algebraic proof

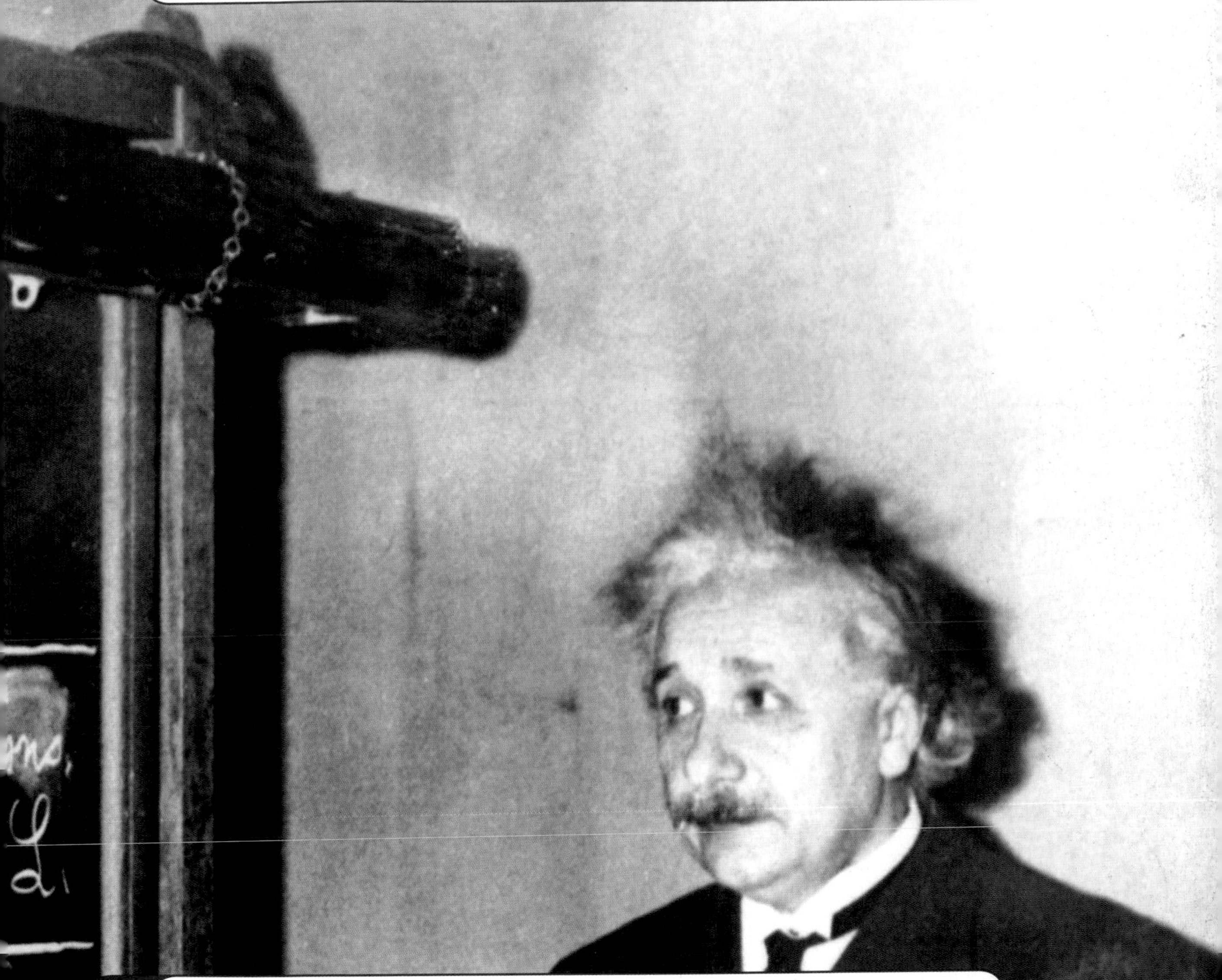

Rich task

Investigate the angles and lengths formed when a circle is cut by two lines.
Hint: Look for similar triangles.
Prove any results you find.

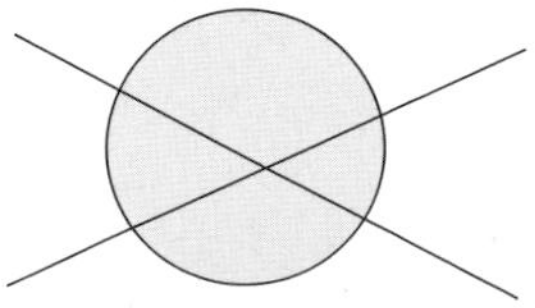

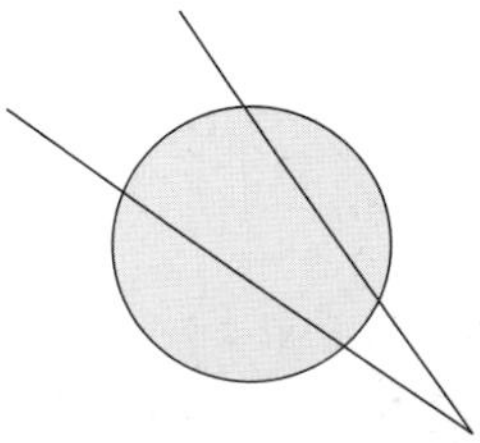

G1.1 Angle theorems in circles

This spread will show you how to:

- Use the correct vocabulary to describe parts of a circle
- Prove and use the theorems for angles in circles

Keywords
Arc
Chord
Diameter
Radii
Radius
Segment

You need to know the parts of a circle.

A straight line joining two points on the circumference is a **chord**.
A chord divides a circle into two **segments**.
The part of the circumference that joins two points is an **arc**.

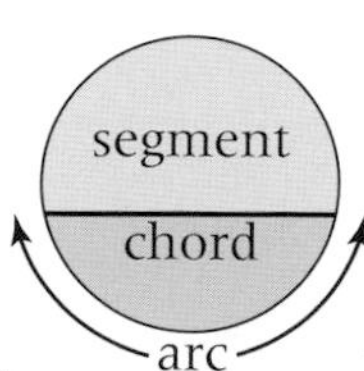

The longest chord is the **diameter**.

You need to be able to prove and use these circle theorems.

Proof

Prove the angle at the centre is twice the angle at the circumference from the same arc.

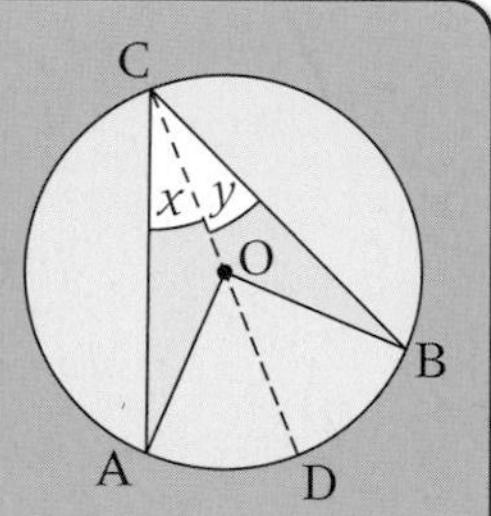

Draw the **radius** OC and extend it to D.

In $\triangle$ AOC, AO = OC — Both **radii**.

$\angle OAC = \angle OCA = x$ — Isosceles triangle.

$\angle COA = \angle 180° - 2x$

$\angle AOD = 2x$ — Angles on a straight line.

Similarly using $\triangle$ COB you can prove that

$\angle DOB = 2y$

Now ACB = $x + y$ and AOD = $2x + 2y$ as required.

Proof

Prove angles from the same arc in the same segment are equal.

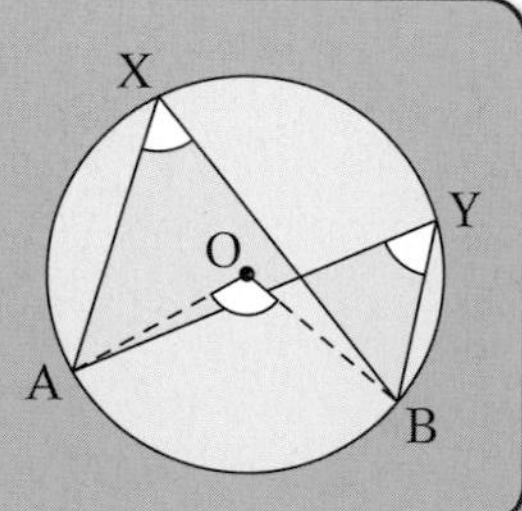

$\angle AOB = 2 \times \angle AXB$ — Angle at centre is twice

$\angle AOB = 2 \times \angle AYB$ — angle at circumference.

so $\angle AXB = \angle AYB$ as required.

Example

Find the missing angles. Give reasons.

a

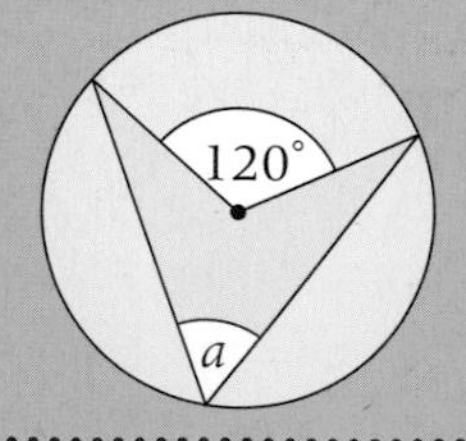

b

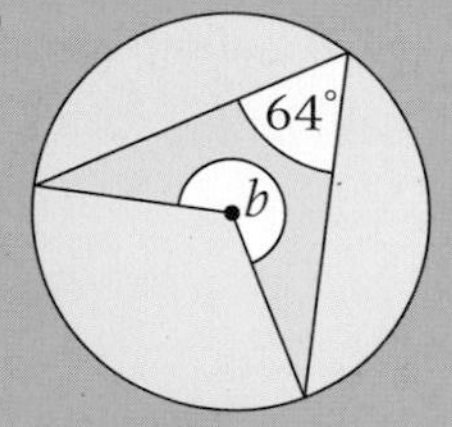

c

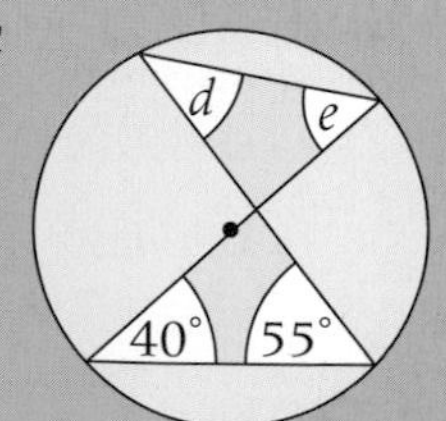

a $a = 60°$
Angle at centre is twice angle at circumference.

b Obtuse angle at centre = 128°
$b = 360° - 180° = 232°$
Angles at a point.

c $d = 40°$, $e = 55°$
Angles on same arc are equal.

Exercise G1.1 **Grade B**

Work out the missing angles.
Give a reason for each answer.

These circle theorems are examined in unit 3.

1

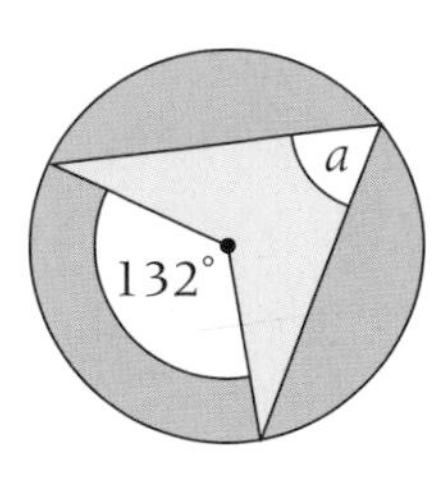

2

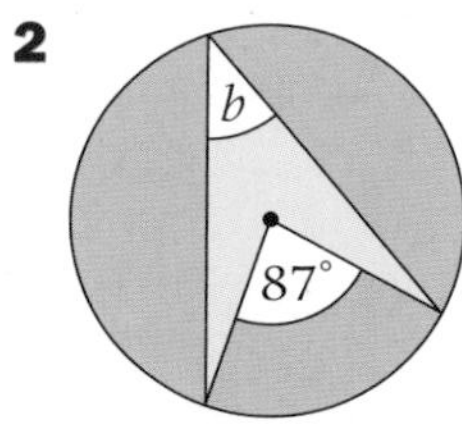

3

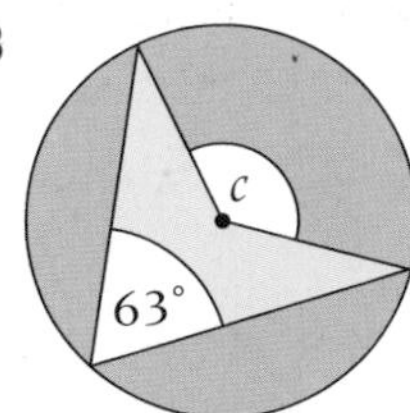

4

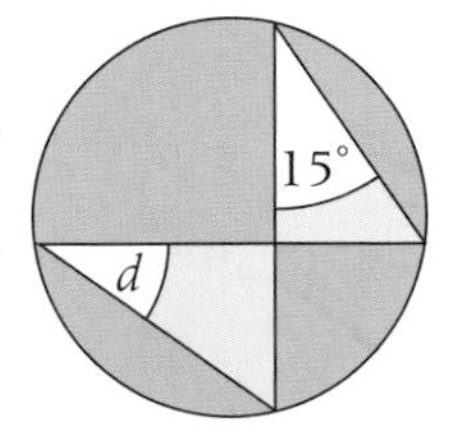

5

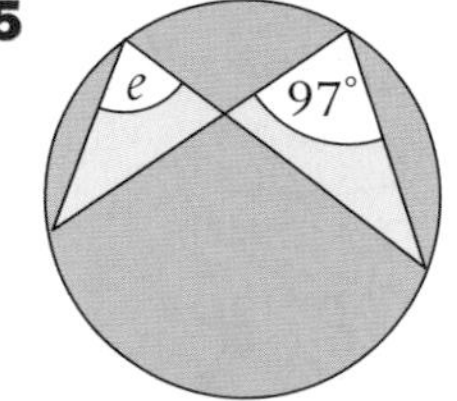

6

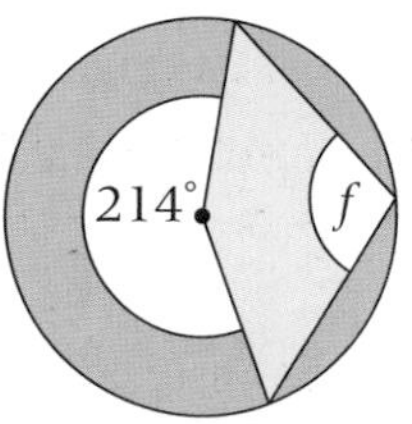

7

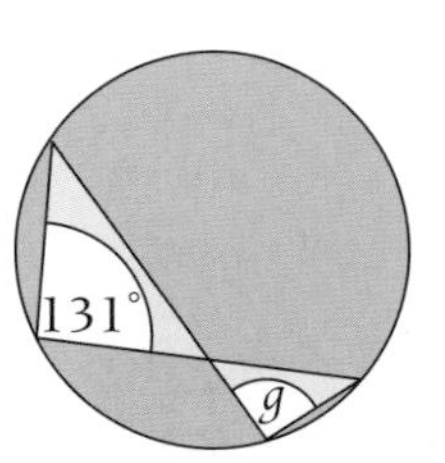

8

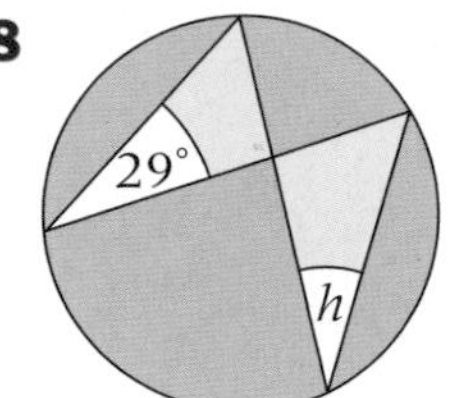

9

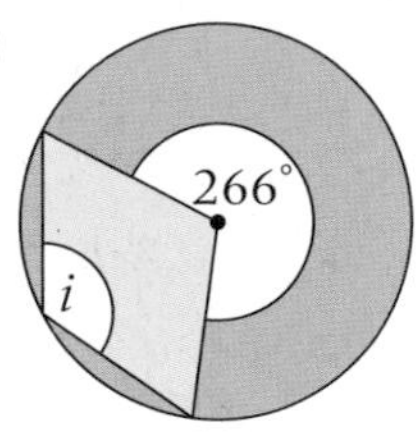

10

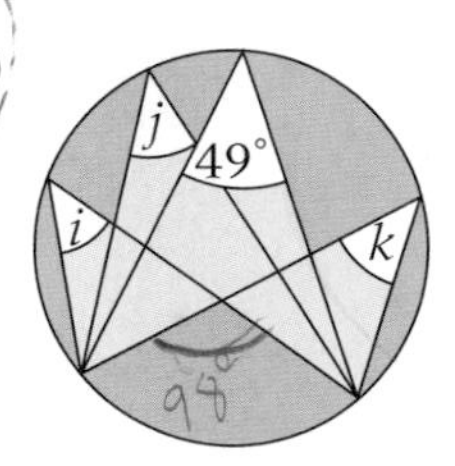

11

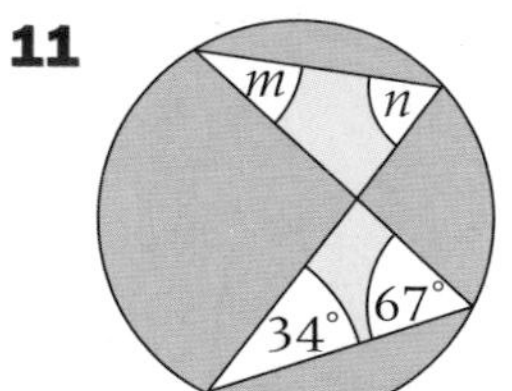

12

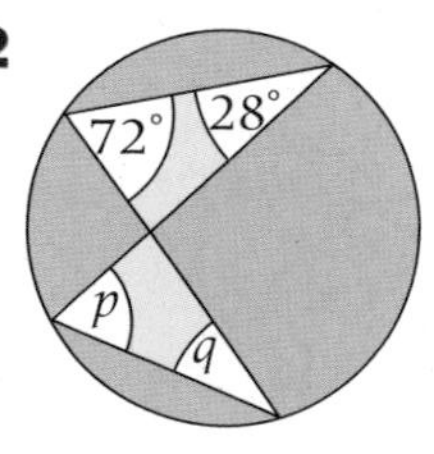

13

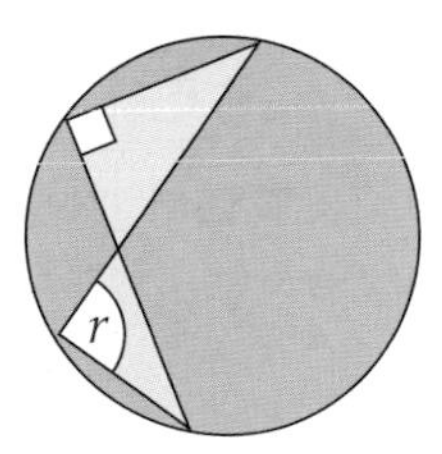

14

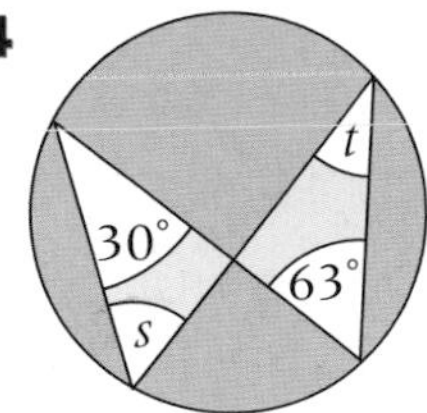

15

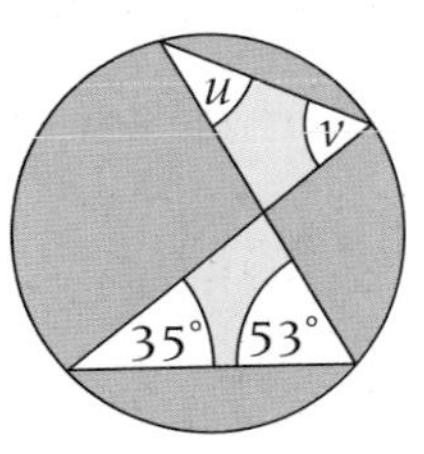

16

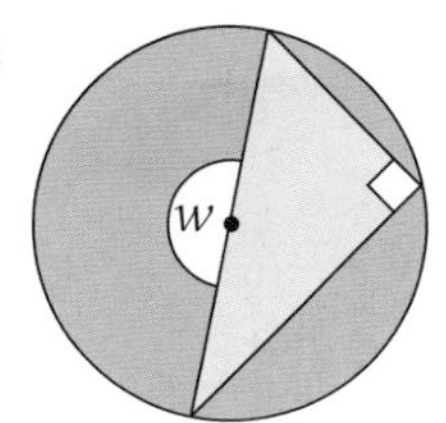

17

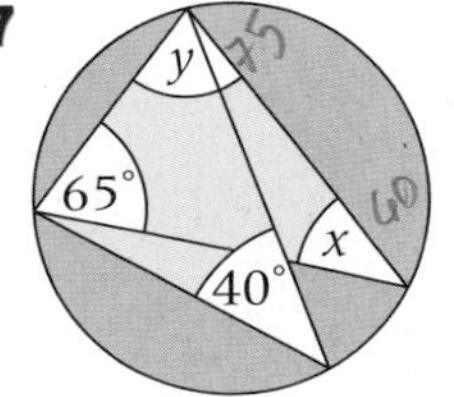

18

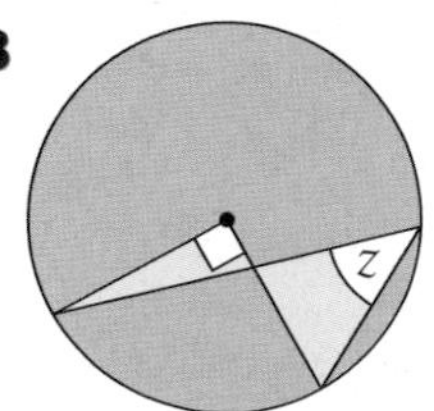

Further angle theorems in circles

This spread will show you how to:

- Use the correct vocabulary to describe parts of a circle
- Prove and use the theorems for angles in circles

Keywords
Cyclic quadrilateral
Semicircle

You need to be able to prove and use two more circle theorems.

Proof

Prove an angle in a **semicircle** is a right angle.

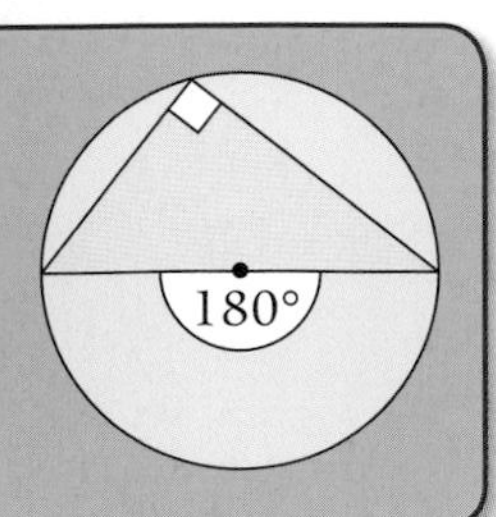

Angle at the centre = 180°
Angle at centre = 2 × angle at circumference
so angle at circumference = 90°

- All four vertices of a **cyclic quadrilateral** lie on the circumference of a circle.

Proof

Prove opposite angles of a cyclic quadrilateral add up to 180°.

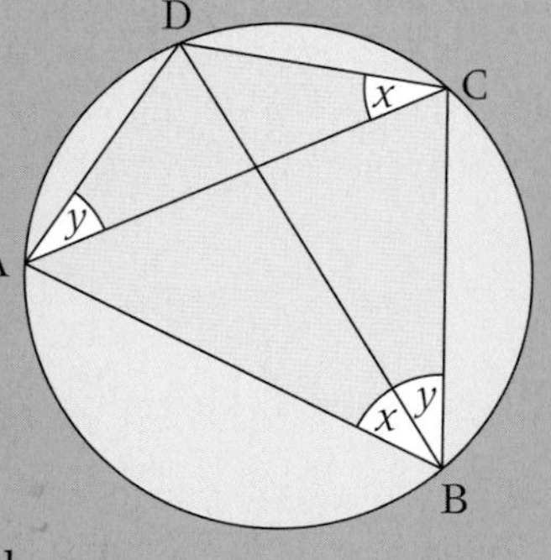

Draw the two diagonals AC and BD.

$\angle ABD = \angle ACD = x$ — Angles on same arc are equal.
$\angle CBD = \angle CAD = y$

$\angle ADC = 180° - x - y$ — Angles in a triangle.
$= 180° - (x + y)$

But $x + y = \angle ABC$
Therefore $\angle ADC + \angle ABC = 180°$ as required.

Example

Find the missing angles. Give reasons.

a

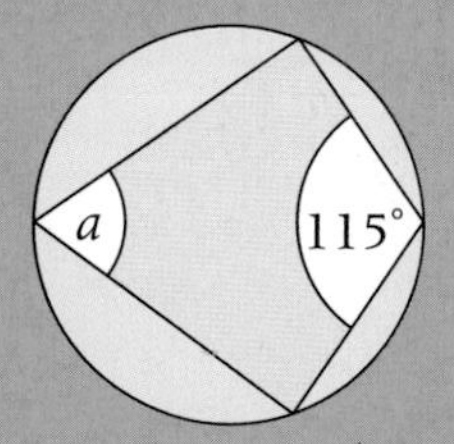

b

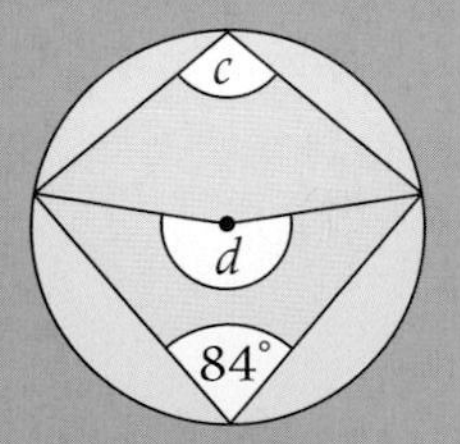

c

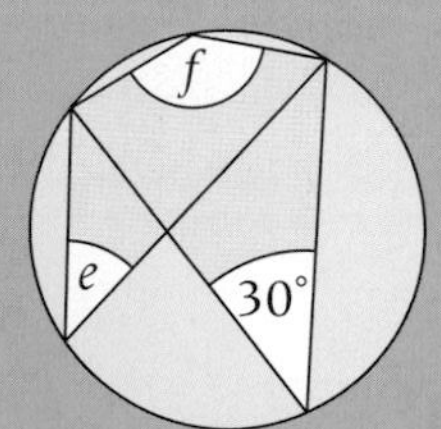

a $a = 65°$
Opposite angles of cyclic quadrilateral = 180°.

b $c = 96°$
Opposite angles of cyclic quadrilateral.

$d = 2 \times 96° = 192°$
Angle at centre = 2 × angle at circumference.

c $e = 30°$
Angles on same arc are equal.

$f = 150°$
Opposite angles of cyclic quadrilateral.

Exercise G1.2

Grade B/A

Work out the missing angles.
Explain how you worked out each answer.

These circle theorems are examined in unit 3.

1

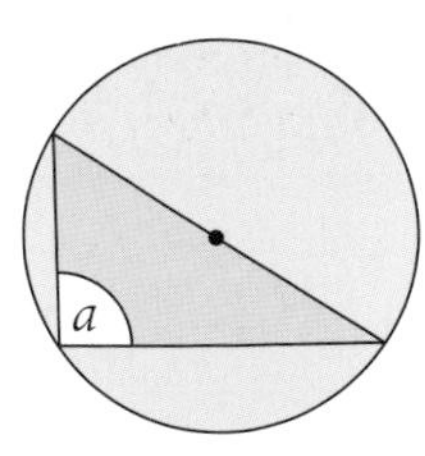

2

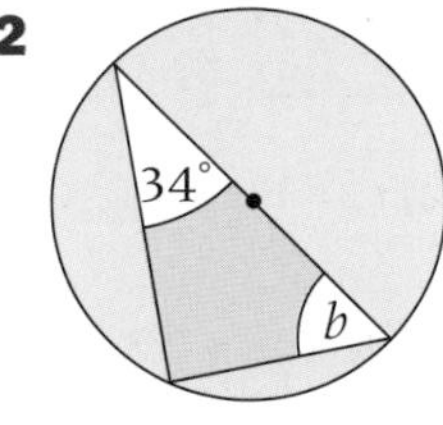

3

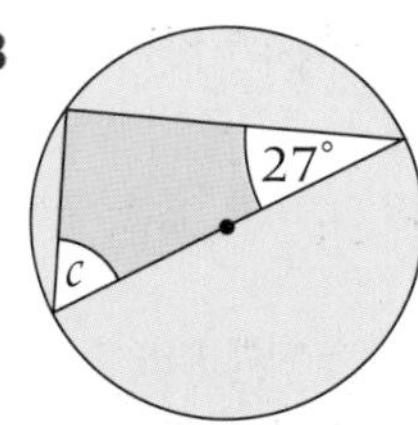

4

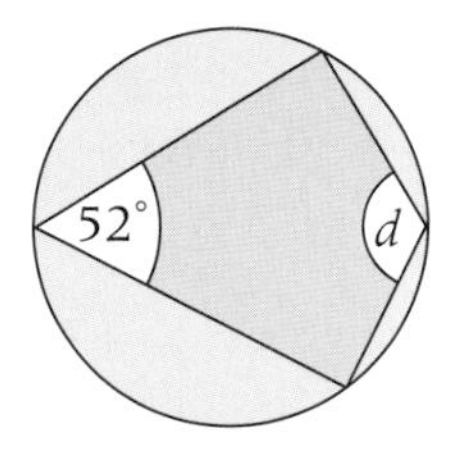

5

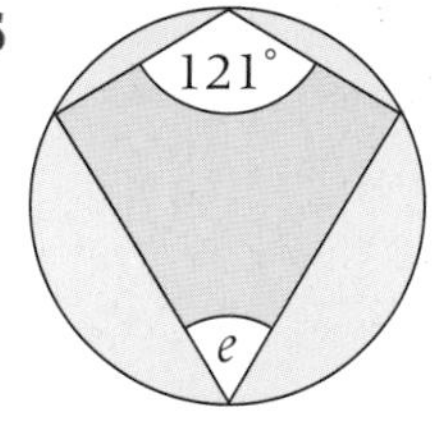

6

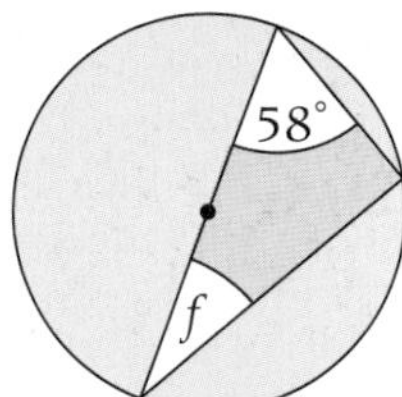

7

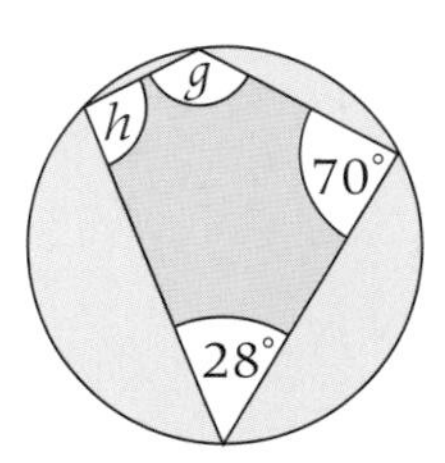

8

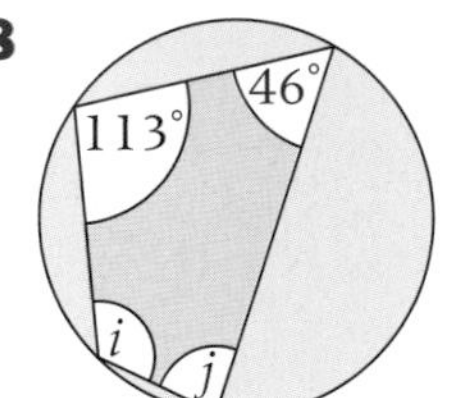

9

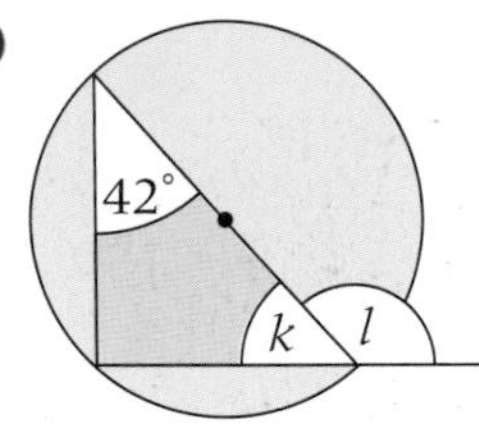

10

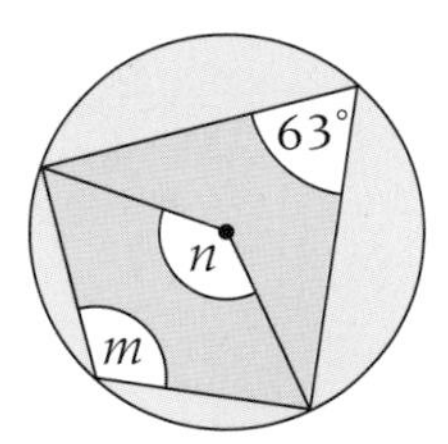

11

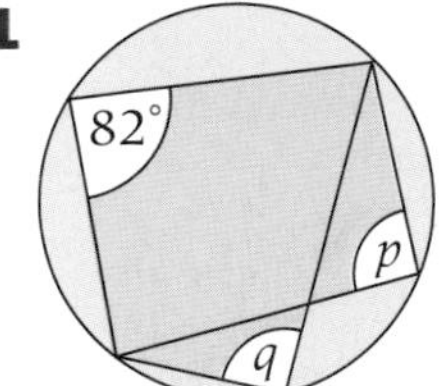

12

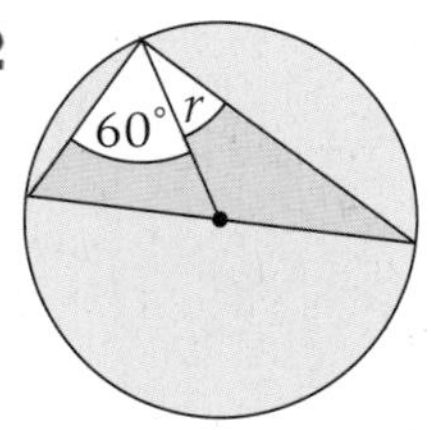

13

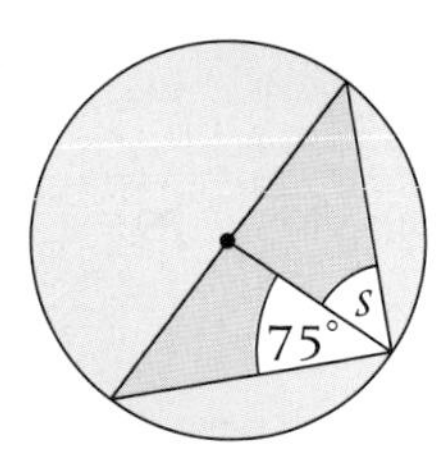

14

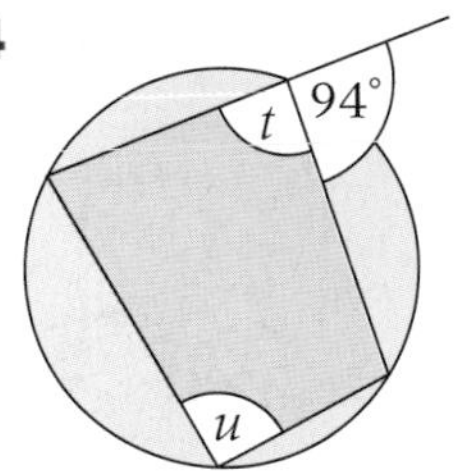

15

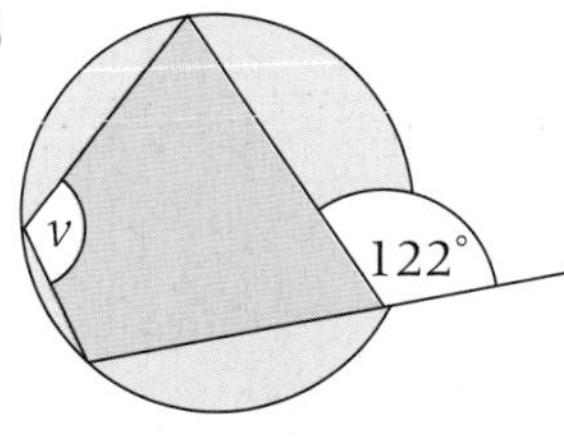

16

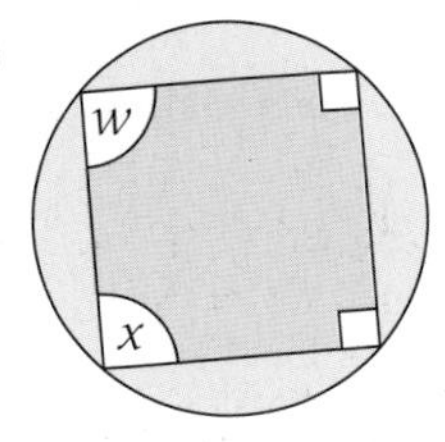

17

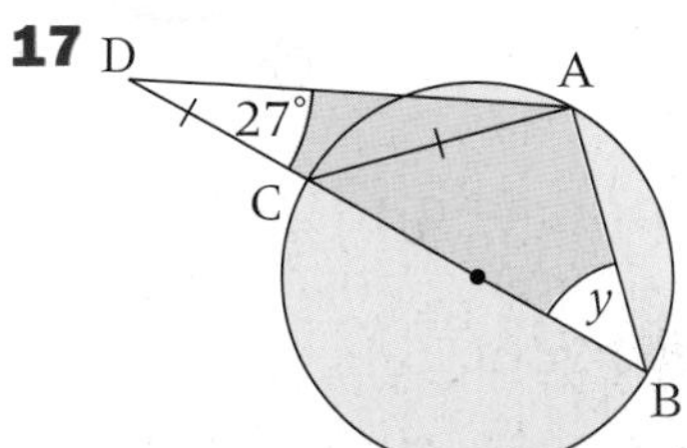

18

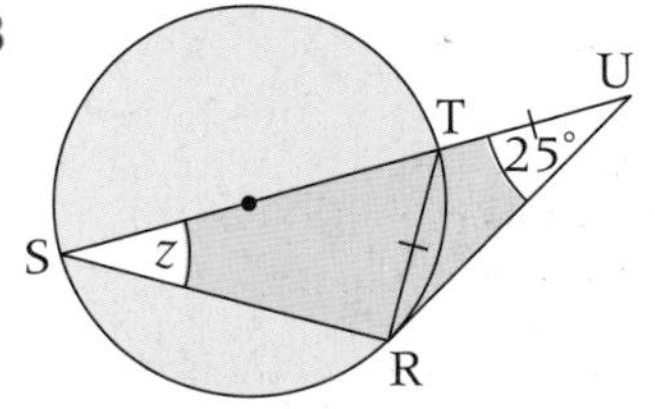

Unit 2

G1.3 Tangents to circles

This spread will show you how to:
- Use the correct vocabulary to describe parts of a circle
- Know and use the theorems for tangents at a point on a circle

Keywords
Chord
Perpendicular
Radius
Tangent

You need to know some facts about **tangents** to circles.

- The angle between a tangent and the **radius** at the point where the tangent touches the circle is a right angle.

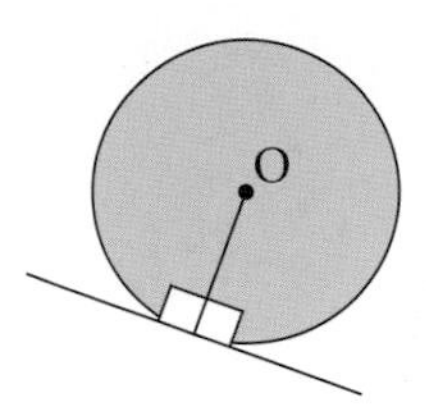

- Two tangents drawn from a point to a circle are equal.

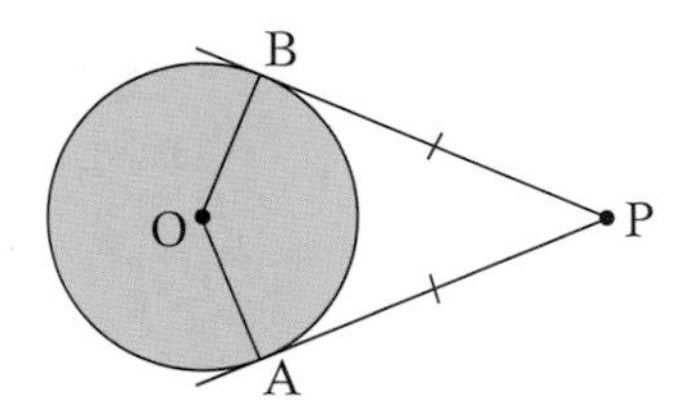

- The **perpendicular** line from the centre of a circle to a **chord** bisects the chord.

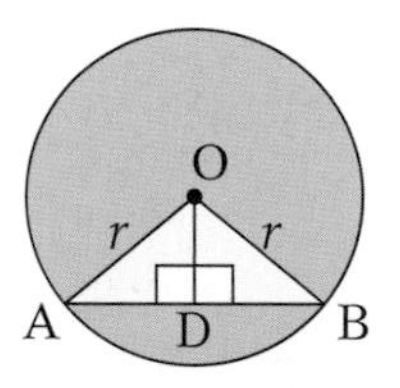

Example

Find the missing angles and lengths. Give reasons.

a

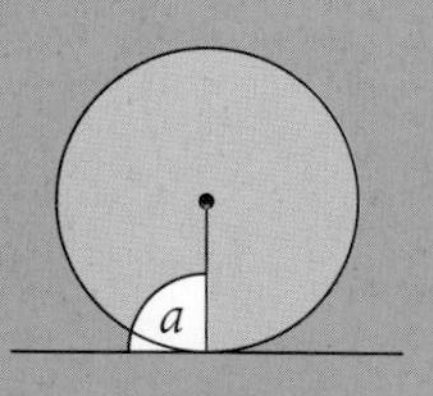

b

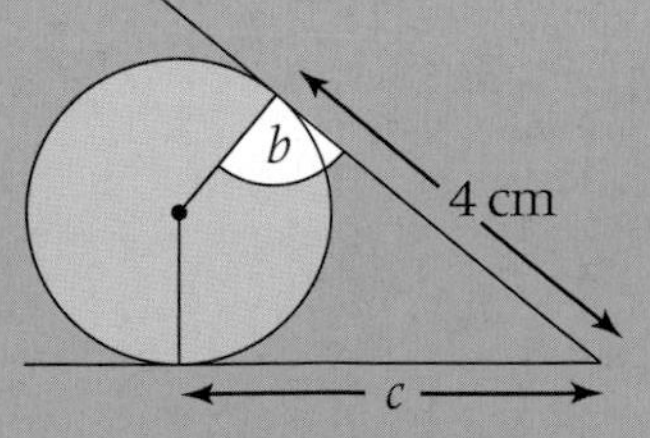

c

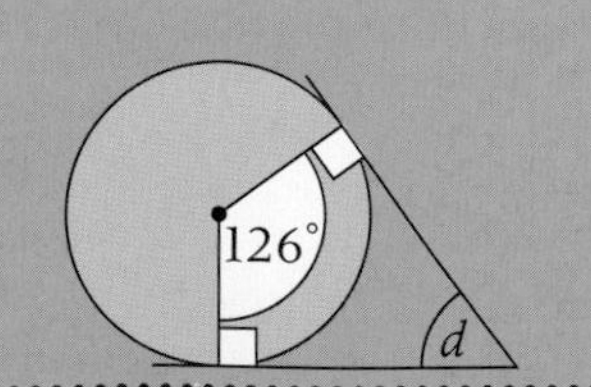

a $a = 90°$
Angle between tangent and radius.

b $b = 90°$
Angle between tangent and radius.
$c = 4\text{ cm}$
Tangents from a point.

c $d + 90° + 90° + 126° = 360°$
Angles of quadrilateral.
$d = 54°$

Exercise G1.3 — Grade B/A

Work out the missing angles and lengths.
Give a reason for each answer.

1

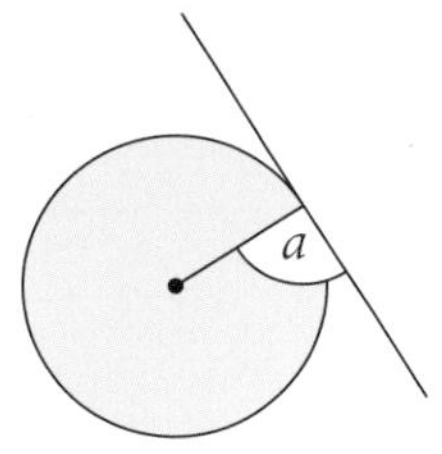

2

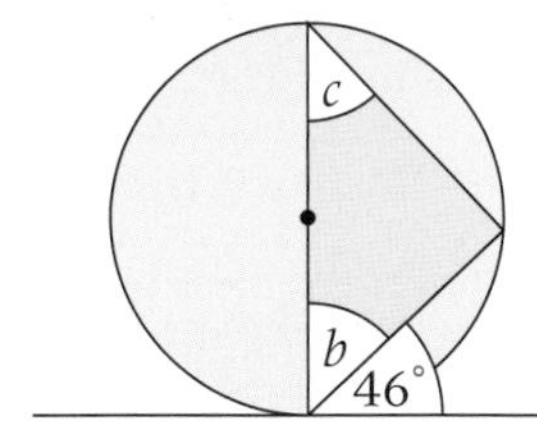

3

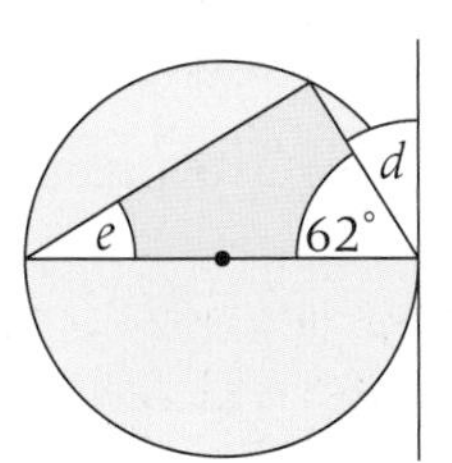

4

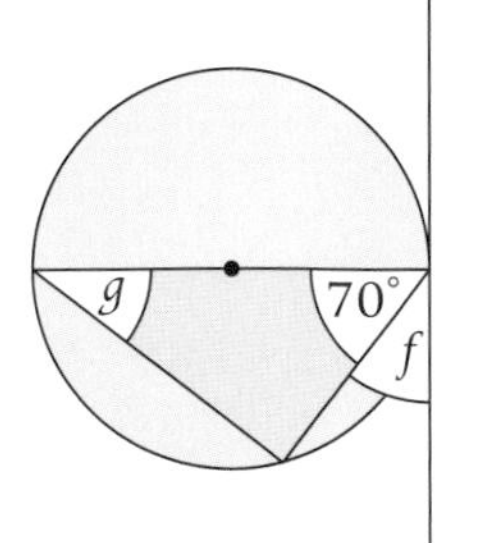

5

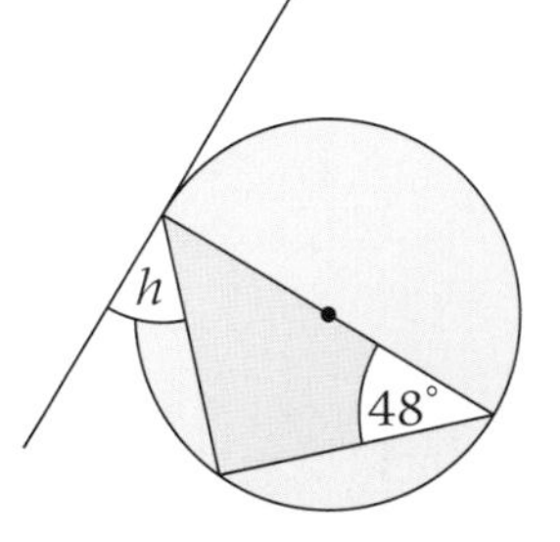

6

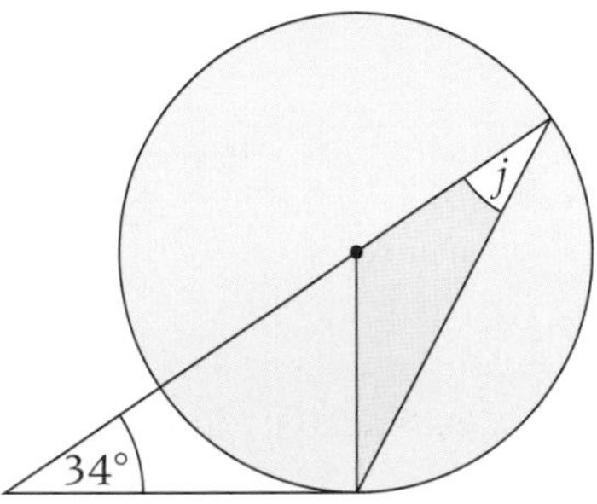

7

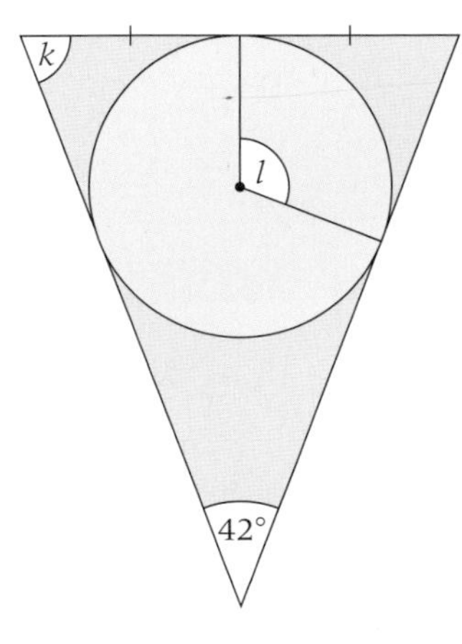

8

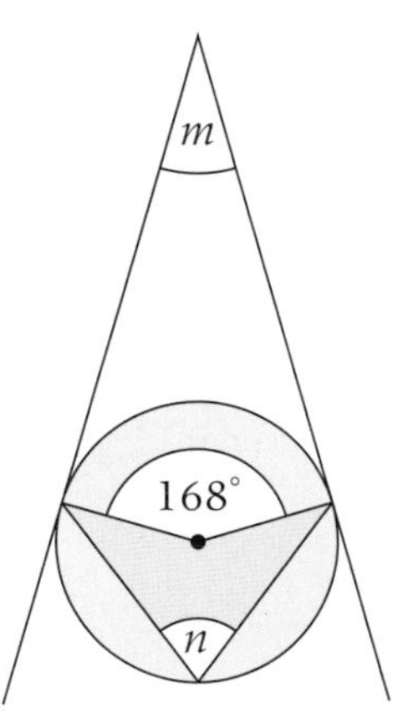

9

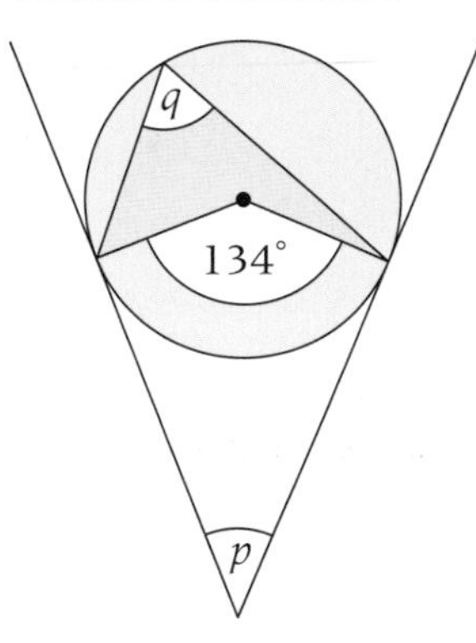

10

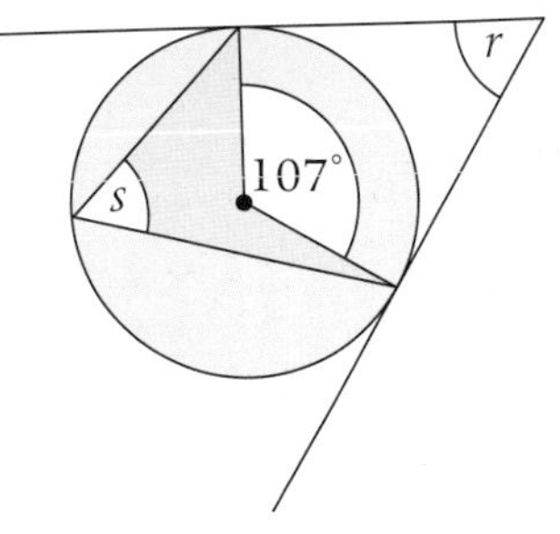

11

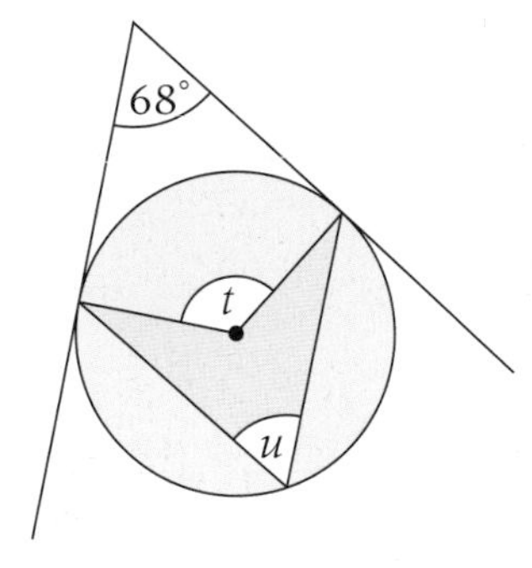

12

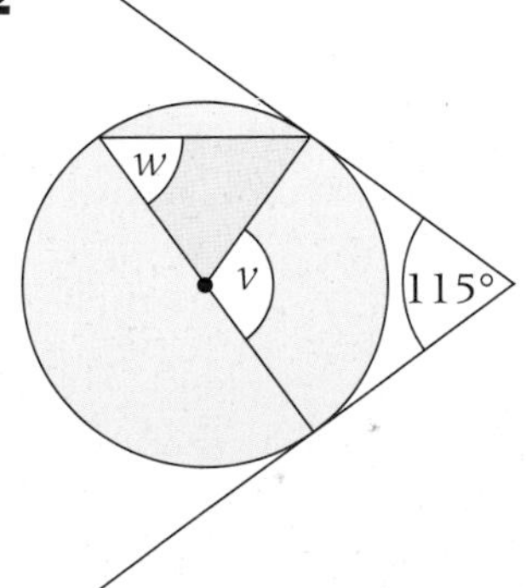

13

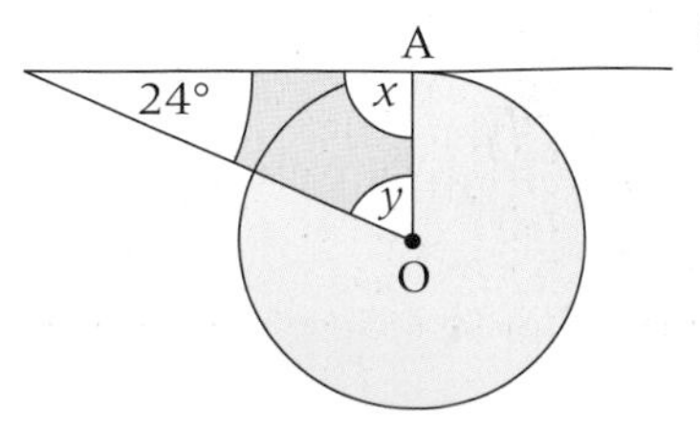

14

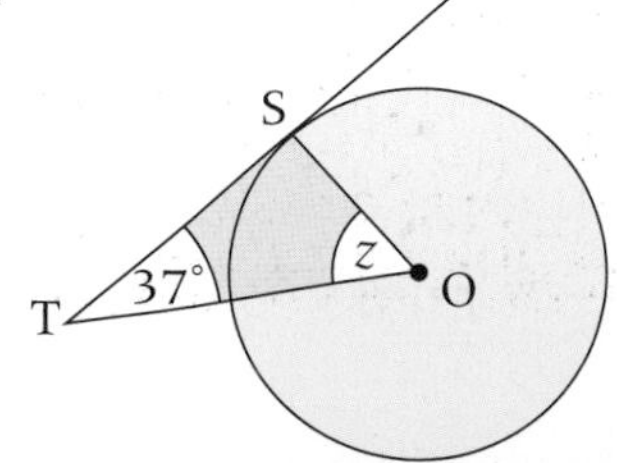

G1.4 Alternate segment theorem

This spread will show you how to:

- Know and use the theorems for tangents at a point on a circle

Keywords
Alternate segment
Chord
Segment
Tangent

You need to know the alternative segment theorem.

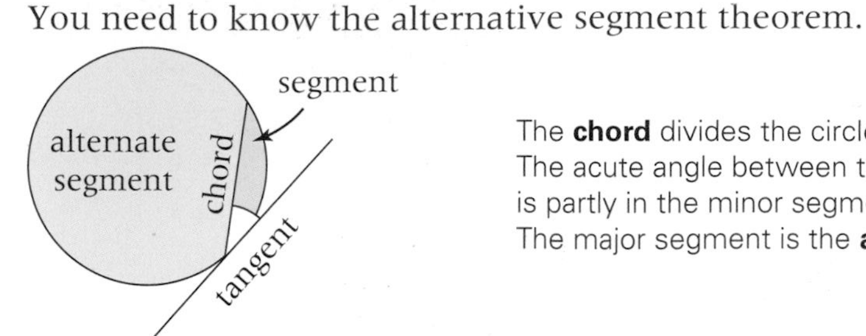

The **chord** divides the circle into two **segments**.
The acute angle between the **tangent** and the chord is partly in the minor segment.
The major segment is the **alternate segment**.

Proof

- Prove the angle formed between a tangent and a chord is equal to the angle from that chord in the alternate segment of the circle.

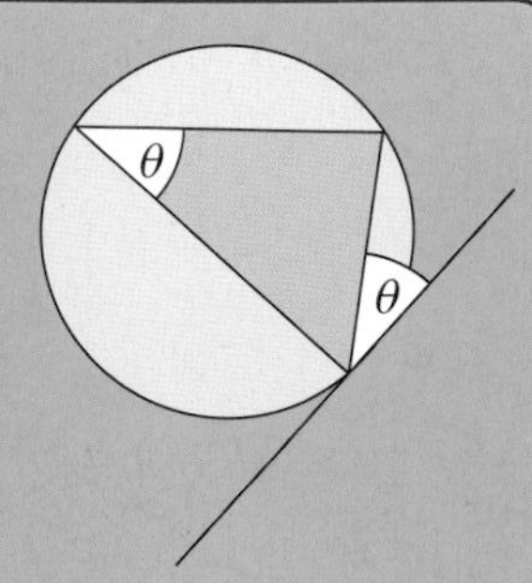

Draw a diameter where the tangent touches the circle (T).
Join this diameter (TX) to the other end of the chord (P) to form a triangle PTX.
$\angle XPT = 90°$ Angle in a semicircle.

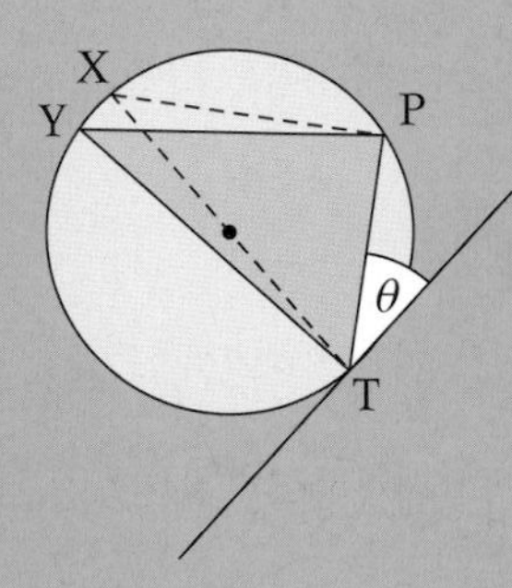

The angle between tangent and radius is a right-angle, so $\angle PTX = 90° - \theta$.

$\angle TXP + 90° + (90° - \theta) = 180°$ Angles in a triangle add up to 180°.
so $\angle TXP = 180° - 180° + \theta = \theta$

Angles from the same arc in the same segment are equal, so the angle in the alternate segment, $\angle PYT = \theta$.

Example

Find the missing angles. Give reasons.

a

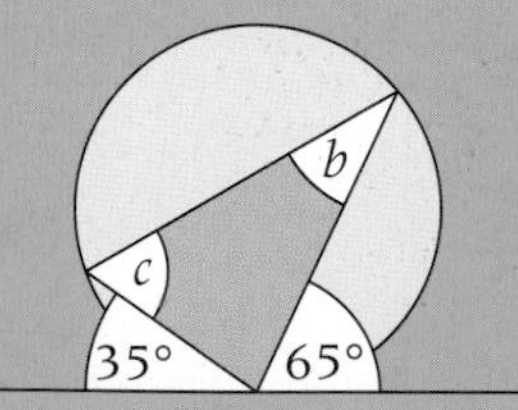

b

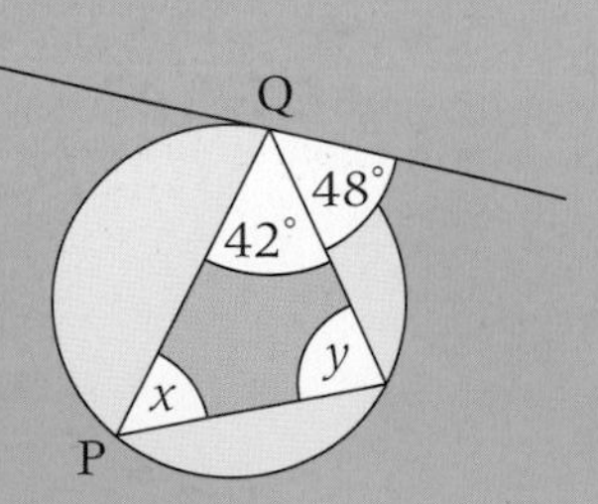

a $b = 35°$ Alternate segment.
$c = 65°$ Alternate segment.

b $x = 48°$ Alternate segment.
$y = 180° - 42° - 48°$ Angles in triangle.
$= 90°$

Since $y = 90°$, PQ must be a diameter.

Exercise G1.4 Grade A

Work out the missing angles. Explain how you worked out each answer.

These circle theorems are examined in unit 3.

Unit 2

1

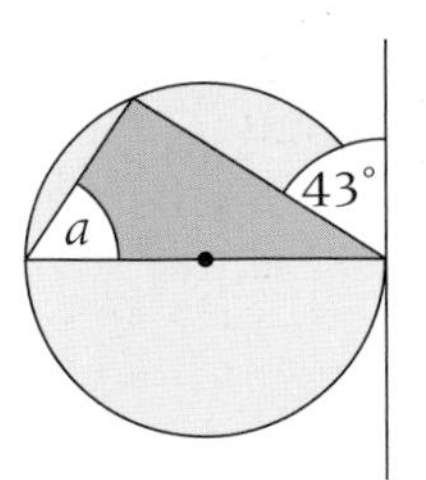

2

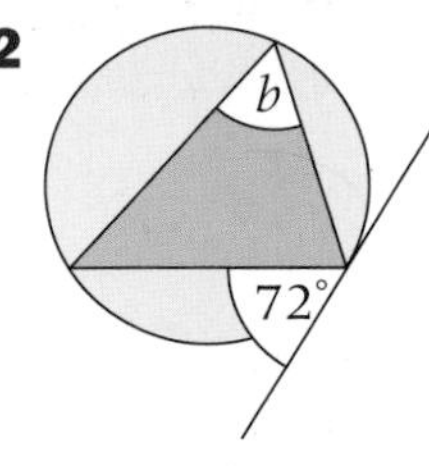

3

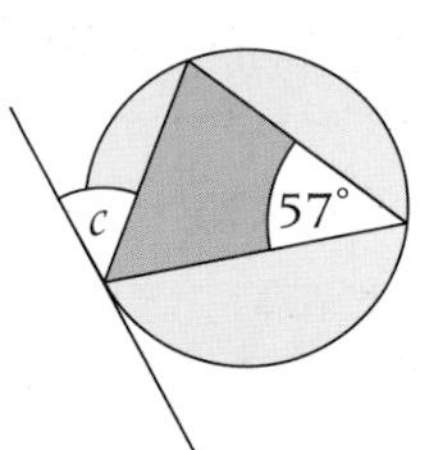

4

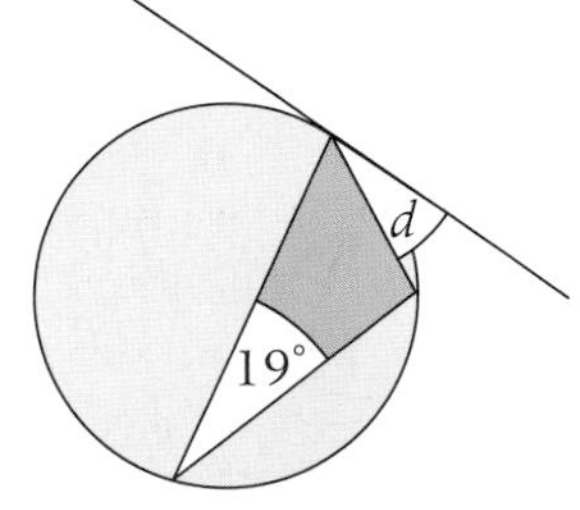

5

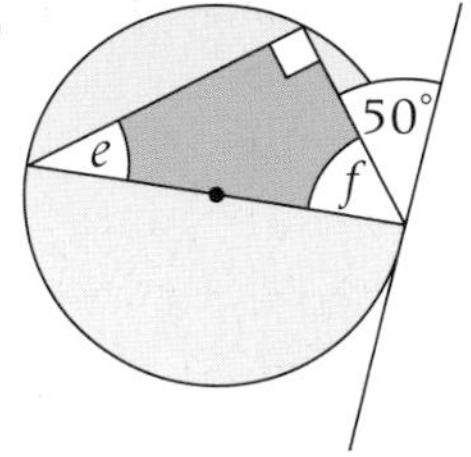

6

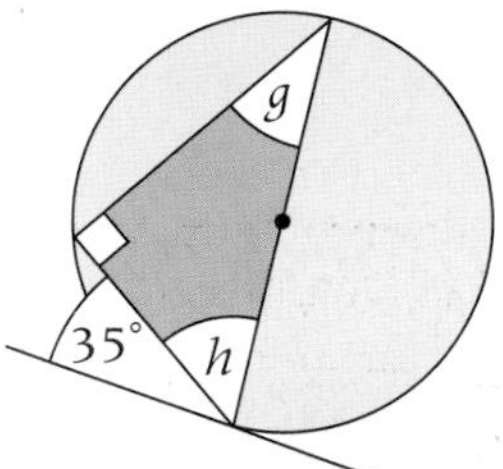

7

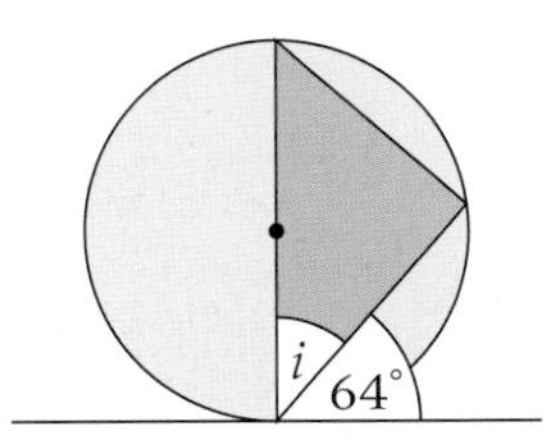

8

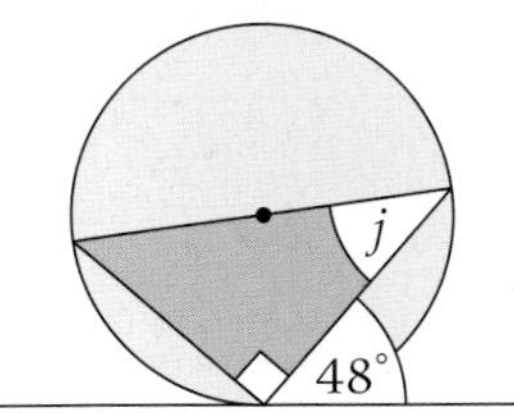

9

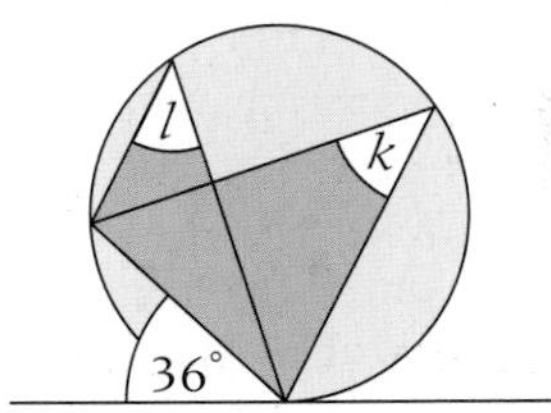

10

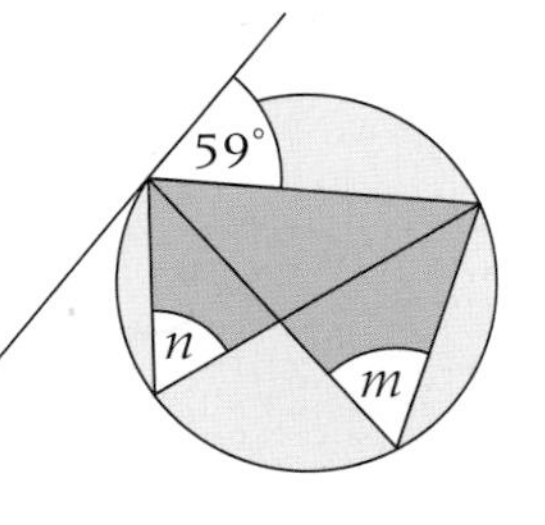

11

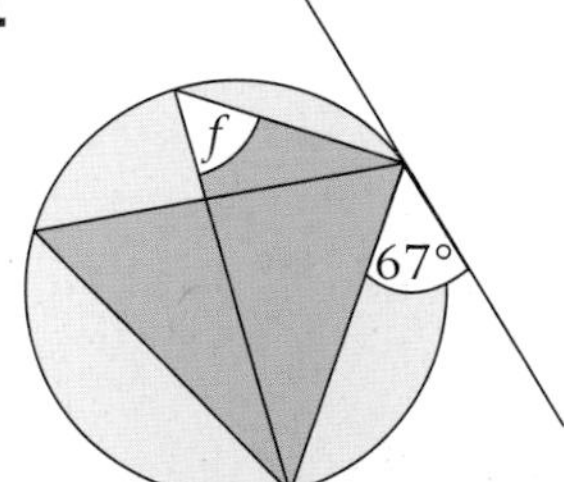

12

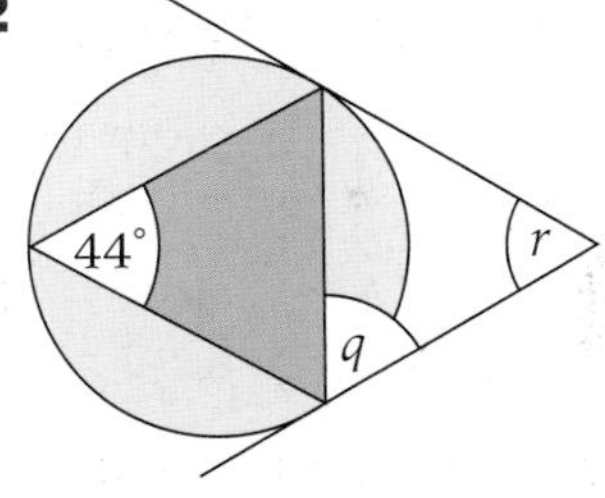

13

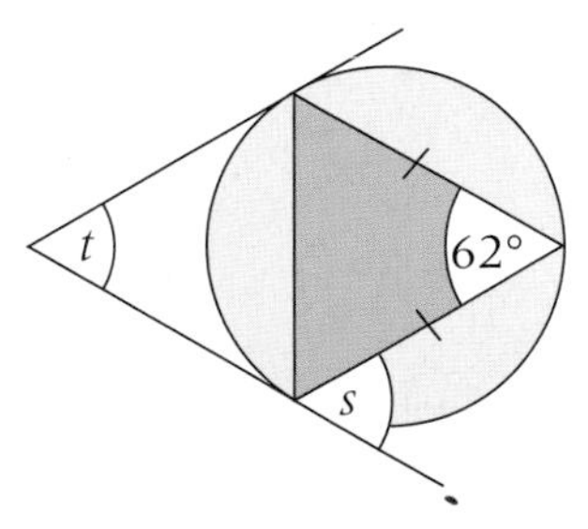

14

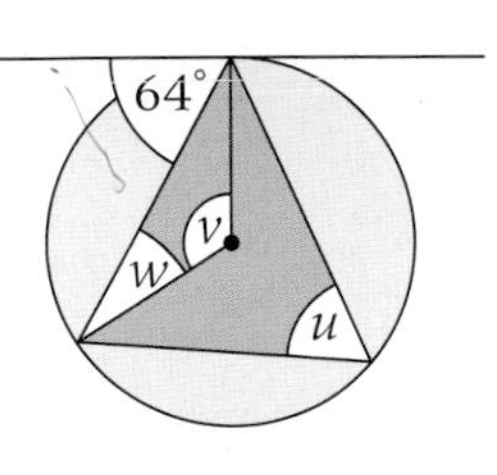

15

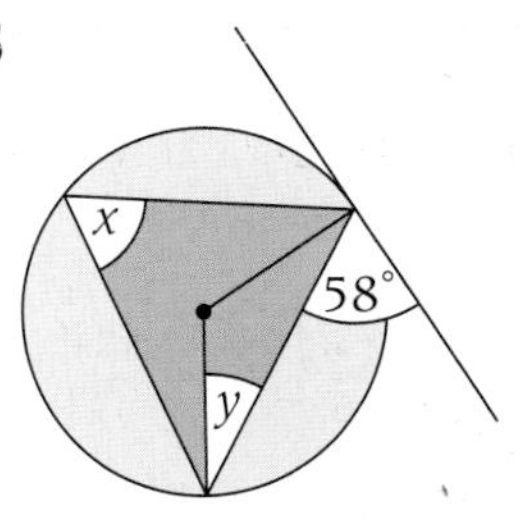

16

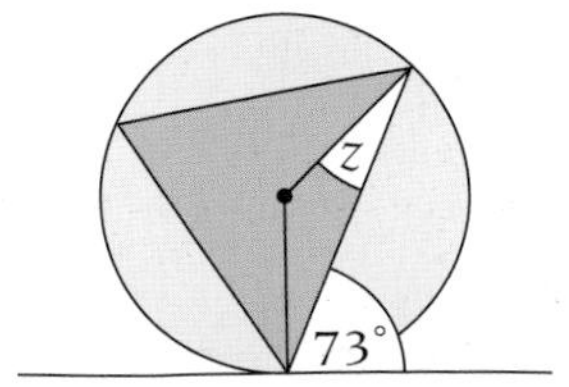

G1.5 Geometrical proof

This spread will show you how to:

- Derive proofs using the circle and tangent theorems

Keywords
Arc
Circumference
Cyclic quadrilateral

You can use circle theorems to explain and prove geometrical facts.

Proof

In a **cyclic quadrilateral** ABCD prove that an interior angle is equal to the opposite exterior angle, $\angle DAB = \angle BCE$.

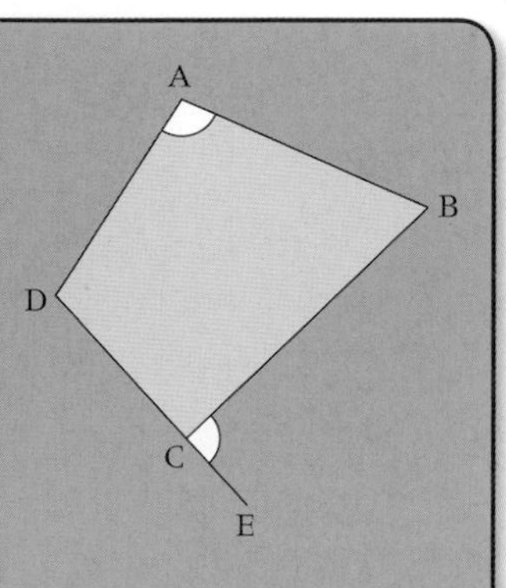

$\angle DCB = 180° - \angle DAB$ Opposite angles in a cyclic quadrilateral.

$\angle BCE = 180° - \angle DCB$ Angles on a line.

or

$\angle DCB = 180° - \angle BCE$

Therefore

$\angle DAB = \angle BCE$ as required.

You know that in a cyclic quadrilateral, opposite angles add up to 180°, so look at the opposite angles of ABCD.

Example

PQR is a triangle drawn inside a circle.
Angle PQR = 100°.
Is PR a diameter?

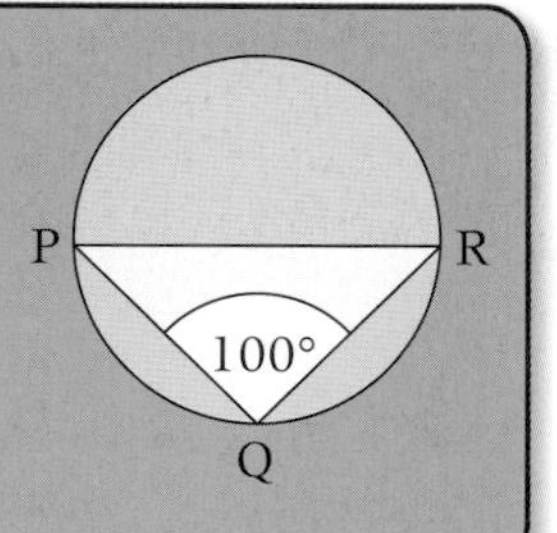

If PR were a diameter, angle PQR would be an angle in a semicircle, which is 90°.
Angle PQR $\neq$ 90°, so PR is not a diameter.

Suppose that PR is a diameter, and see if the rest of the information fits.
As it does not, your supposition must be wrong.

Example

In the diagram AB = ST.

Prove **a** $\angle AOB = \angle SOT$.
b Angles at the **circumference** from equal **arcs** are equal.

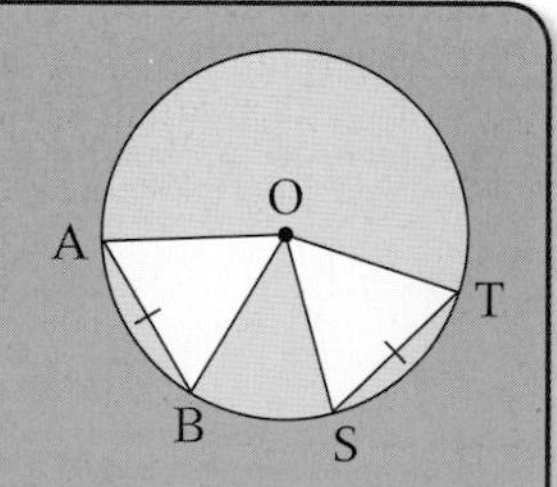

a AO = BO = SO = TO Radii.
AB = ST (given)
So $\triangle AOB$ and $\triangle SOT$ are congruent (SSS)
and $\angle AOB = \angle SOT$

b Arcs AB and ST are equal.
$AOB = 2 \times \angle APB$ Angle at centre is twice angle at circumference.
$SOT = 2 \times \angle SQT$
So $\angle APB = \angle SQT$

(given) means a fact you have been told.

Exercise G1.5

Grade A*

1 In the quadrilateral PQRS $\angle PQR = 38°$ and $\angle RSP = 138°$.
Explain why PQRS cannot be a cyclic quadrilateral.

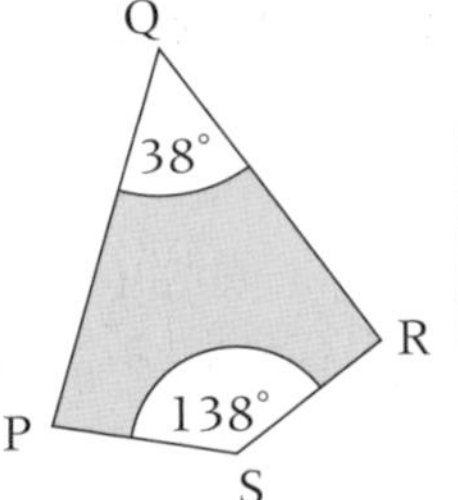

These proofs use circle theorems which are examined in unit 3.

Unit 2

2 ABCD is a parallelogram and it is also a cyclic quadrilateral.
Use circle theorems and properties of quadrilaterals to explain which special type of quadrilateral ABCD could be.

242

3 D, E and F are three points on the circumference of a circle of radius 68 mm.
DE = 120 mm and EF = 64 mm.
Show that DF is a diameter of the circle.

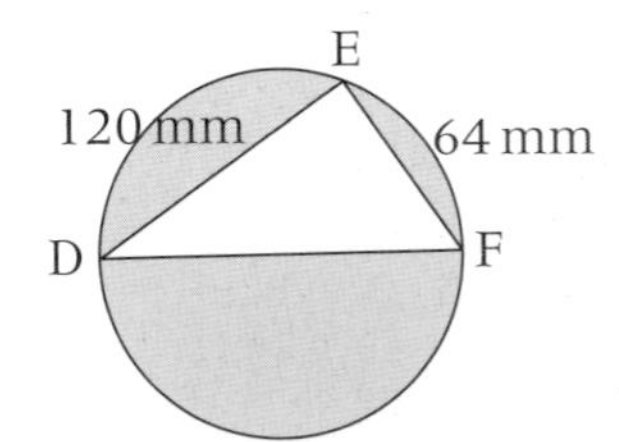

4 Explain why C is not the centre of the circle.

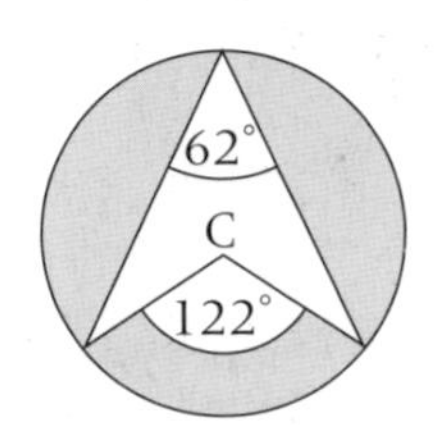

5 PQRS is a cyclic quadrilateral.
X lies outside the circle such that XQR and XPS are straight lines.
XQ and XP are equal in length.
Use the 'angles in cyclic quadrilaterals' theorem to show that PQRS is an isosceles trapezium.

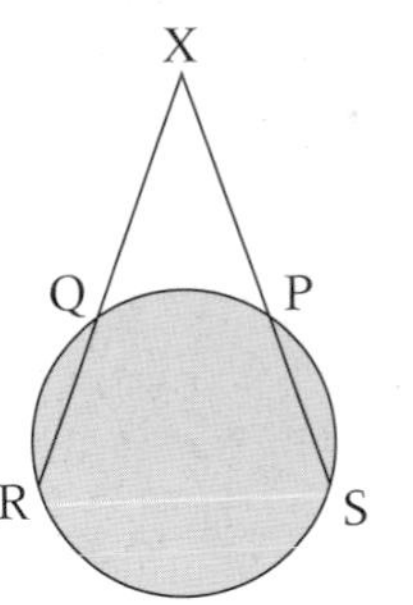

6 X and Y lie on the circumference of a circle with centre O.
P is a point outside the circle.
Angle PXY = 62°
Reflex angle XOY = 256°
Explain why PX is not a tangent to the circle.

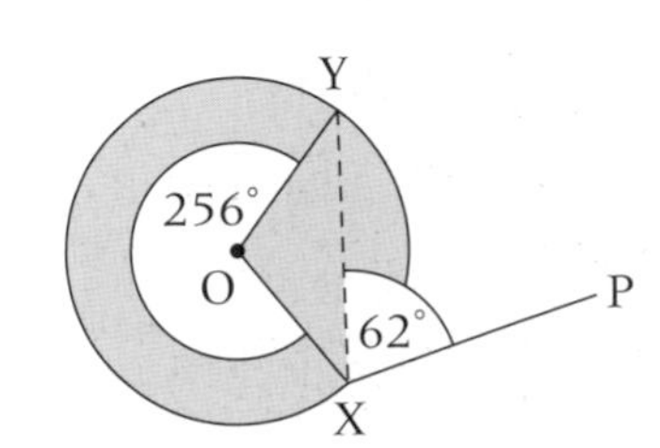

G1 Summary

Check out

You should now be able to:

- Use the correct vocabulary to describe parts of a circle
- Prove and use theorems for angles in circles
- Know and apply the tangent theorems to circle problems
- Derive proofs using the circle and tangent theorems

Worked exam question

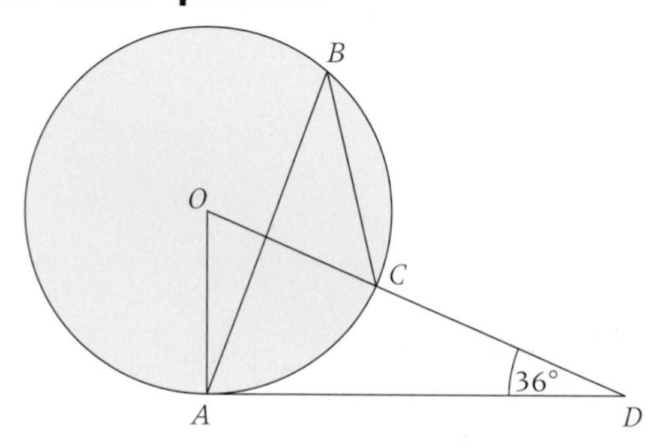

Diagrams NOT accurately drawn.

The diagram shows a circle centre *O*.
A, *B* and *C* are points on the circumference.
DCO is a straight line.
DA is a tangent to the circle.
Angle *ADO* = 36°

a Work out the size of angle *AOD*. (2)

b **i** Work out the size of angle *ABC*.

ii Give a reason for your answer. (3)

(Edexcel Limited 2009)

a

$90° + 36° = 126°$
$180° - 126° = 54°$

Either show this working or state angle *OAD* = 90°

b

i $\frac{1}{2}$ of $54° = 27°$

ii The angle at the centre is twice the angle at the circumference

Just state the appropriate circle theorem.

Exam questions

1

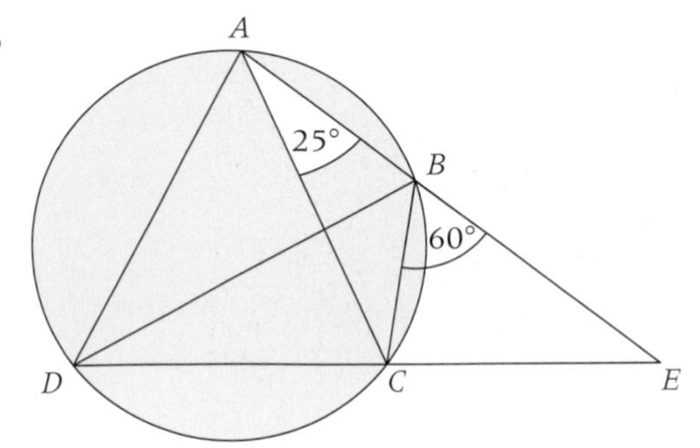

Diagram NOT accurately drawn.

A, *B*, *C* and *D* are four points on the circumference of a circle.
ABE and *DCE* are straight lines.
Angle $BAC = 25°$.
Angle $EBC = 60°$.

a Find the size of angle *ADC*. (1)

b Find the size of angle *ADB*. (2)

Angle $CAD = 65°$.
Ben says that *BD* is a diameter of the circle.

c Is Ben correct? You must explain your answer. (1)

(Edexcel Limited 2004)

2

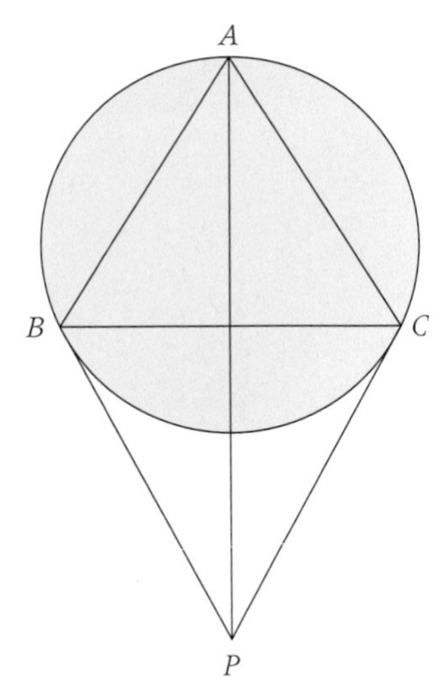

Diagram NOT accurately drawn.

A, *B* and *C* are three points on the circumference of a circle.
Angle *ABC* = angle *ACB*.
PB and *PC* are tangents to the circle from the point *P*.

a Prove that triangle *APB* and triangle *APC* are congruent. (3)

Angle $BPA = 10°$.

b Find the size of angle *ABC*. (4)

(Edexcel Limited 2004)

A2

Sequences and linear graphs

A sequence is simply an order list: it can be finite or infinite, random or have a pattern. As such there are many sequences studied in mathematics. For example, from the simple, even numbers, to the curious Fibonacci sequence used to try to describe the growth of rabbit populations, through to the elegant and complex world of continued fractions.

What's the point?

To really understand a sequence you need to be able give a formula for its nth term. Using the formula you can explain the patterns in a sequence and determine its behaviour for arbitrarily large n.

Sequences and linear graphs

Check in

You should be able to

- **identify patterns in sequences**

1 Find the next two terms of each sequence.

a 2, 5, 9, 14, ... **b** 2, 4, 8, 16, ... **c** $\frac{1}{1}, \frac{1}{4}, \frac{1}{9}, \frac{1}{16}, \ldots$

d 1, 8, 27, 64, ... **e** 5, 7, 12, 19, ... **f** 3, −9, 27, −81, ...

- **evaluate simple expressions**

2 Given that $n = 4$, find the value of each expression.

a $5n - 4$ **b** $2n^2$ **c** $n^3 - 2$

d $(5 - n)^3$ **e** $(-1)^n$ **f** $\frac{1}{2^n}$

- **plot straight line graphs**

3 On axes labelled from −8 to +8, plot these lines.

a $y = 6$ **b** $x = -2$ **c** $y = 2x + 1$ **d** $x + y = 8$

Orientation

What I need to know	What I will learn	What this leads to
KS3 Evaluate expressions using BIDMAS Recognise patterns in sequences Plot straight line graphs	■ Generate a sequence from an nth term ■ Find and explain the nth term of a linear sequence ■ Find and interpret the equation of a straight line graph ■ Find the equation of a perpendicular line.	**A5** Solving simultaneous linear equations **A7** Drawing and interpreting more complex graphs

Rich task

How many squares are there on an 8×8 chessboard?

How many squares are there on an $n \times n$ chessboard?

A2.1 Generating sequences

This spread will show you how to:

- Generate sequences from a general term
- Describe the behaviour of a sequence, using the relevant vocabulary

Keywords
Converge
Diverge
Limit
*n*th term
Oscillate

You can generate a sequence from a general or ***n*th term**.

If the general term is $T_n = n^2 + 3$,
to find T_1 you substitute $n = 1$: $T_1 = 1^2 + 3 = 4$
to find T_5 you substitute $n = 5$: $T_5 = 5^2 + 3 = 28$

T_1 is the 1st term, T_2 is the 2nd term, T_n is the nth term.

You can generate the terms of a sequence to observe its behaviour.
For example:

- 5, 8, 11, 14, 17, ... each term is larger than the one before.
  The sequence **diverges**.
- 2, −4, 6, −8, 10, −12, ... the terms alternate between positive and negative. The sequence **oscillates**. The difference between terms is increasing, so the sequence diverges.
- $1, \frac{1}{2}, \frac{1}{3}, \frac{1}{4}, \frac{1}{5}$, ... each term is smaller than the one before, but will never be equal to zero.
  The sequence **converges** towards a **limit** of zero.

Example

For each of these sequences
i find the first five terms
ii describe its behaviour.
a $T_n = 5n^2$ **b** $T_n = (-2)^n$

a **i** $T_n = 5n^2$
$T_1 = 5 \times 1^2 = 5$
$T_2 = 5 \times 2^2 = 20$
$T_3 = 5 \times 3^2 = 45$
$T_4 = 5 \times 4^2 = 80$
$T_5 = 5 \times 5^2 = 125$
First five terms are 5, 20, 45, 80, 125
ii The sequence is diverging.

b **i** $T_n = (-2)^n$
$T_1 = (-2)^1 = -2$
$T_2 = (-2)^2 = 4$
$T_3 = (-2)^3 = -8$
$T_4 = (-2)^4 = 16$
$T_5 = (-2)^5 = -32$
First five terms are −2, 4, −8, 16, −32
ii The sequence is diverging and oscillating.

BIDMAS: calculate indices before multiplying by 5.

$(-2)^2 = (-2) \times (-2) = +4$
$(-2)^3 = (-2) \times (-2) \times (-2) = -8$ etc.
Positive terms are getting larger, negative terms are getting smaller.

Exercise A2.1 Grade C

1 Generate the first five terms of each of these sequences.

a $T_n = 7n - 2$ **b** $T_n = n^2 - 4$

c $T_n = n^3$ **d** $T_n = 3n^2 + 5$

e $T_n = n^2 + 2n + 1$ **f** $T_n = 10 - 3n$

g $T_n = (-n)^2$ **h** $T_n = (n + 1)(n + 2)$

2 Generate the first five terms of each of these sequences.
Hence, name the sequences.
For example, $T_n = 2n$, so you get 2, 4, 6, 8, 10,... the even numbers.

a $T_n = 2n - 1$ **b** $T_n = 7n$

c $T_n = 2^n$ **d** $T_n = 10^n$

e $T_n = \dfrac{n(n + 1)}{2}$ **f** $T_n = (11 - n)^2$

g $T_n = n$ **h** $T_n = T_{(n-1)} + T_{(n-2)}$ where $T_1 = 1$ and $T_2 = 1$

3 Choose as many of the following words as necessary to describe these sequences. You will need to generate some terms of the sequence first in order to observe how it is behaving.

You will not be asked to describe the behaviour of a sequence in the exam.

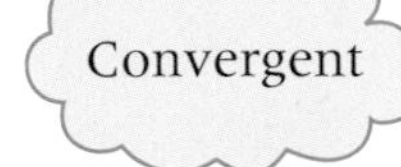

Divergent

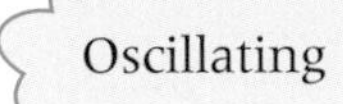

Has a limit

a $T_n = \dfrac{2n + 1}{2n}$ **b** $T_n = 3^n$

c $T_n = 2^{-n}$ **d** $T_n = \dfrac{1}{n+1}$

e $T_n = \sin(90n)°$ **f** $T_n = n!$

In question **3f**, *n*! is '*n* factorial'. You should find this function on your calculator. Try to work out what it does.

AO3 Problem

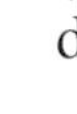

4 Match these sequences with the sketch graphs, where n (term number) is displayed on the x-axis and T_n (term) is displayed on the y-axis.

a $T_n = \frac{1}{n}$ **b** $T_n = 5^n$ **c** $T_n = 10 - n$ **d** $T_n = (n + 1)(n - 2)$

A

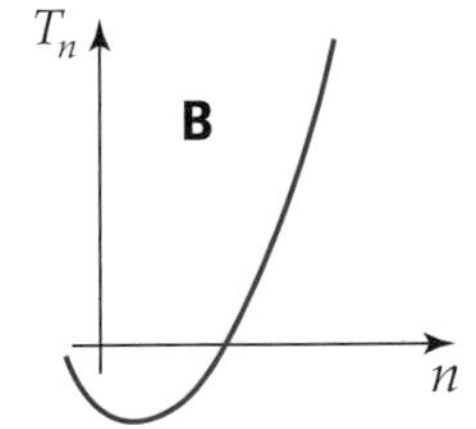

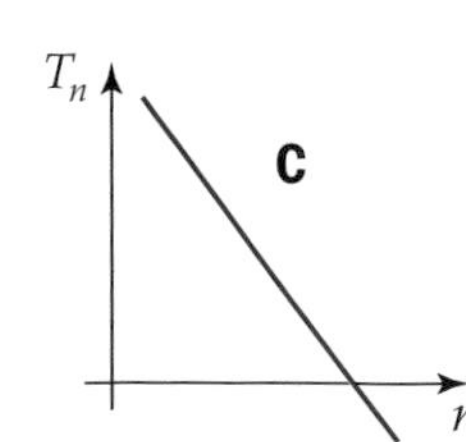

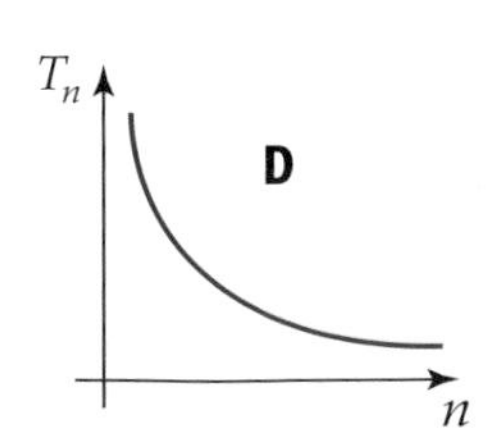

Try substituting some values for *n*, and see what happens to T_n.

Unit 2

A2.2 The *n*th term of a linear sequence

This spread will show you how to:

- Find the *n*th term of a linear sequence
- Justify the general term of a linear sequence by looking at the context from which it arises

Keywords
Arithmetic sequence
Linear sequence
*n*th term

- In an **arithmetic** or **linear sequence** the difference between successive terms is constant.

- You can find the ***n*th term** of a linear sequence by comparing the sequence to the multiples of its constant difference.

For example, for the sequence, 3, 10, 17, 24, 31, ...
Constant difference = 7, so compare to the multiples of 7.

Term number (n)	1	2	3	4	5
$7n$	7	14	21	28	35
Term	3	10	17	24	31

Each term is 4 less than $7n$.
The nth term is $7n - 4$.

The terms go up or down by the same amount each time.
12, 21, 30, 39, 48, ... goes up in 9s.
20, 18, 16, 14, 12, ... goes down in 2s.

Multiples of $7 = 7n$.

- Sometimes you can find a rule from the pattern that generates the sequence, for example

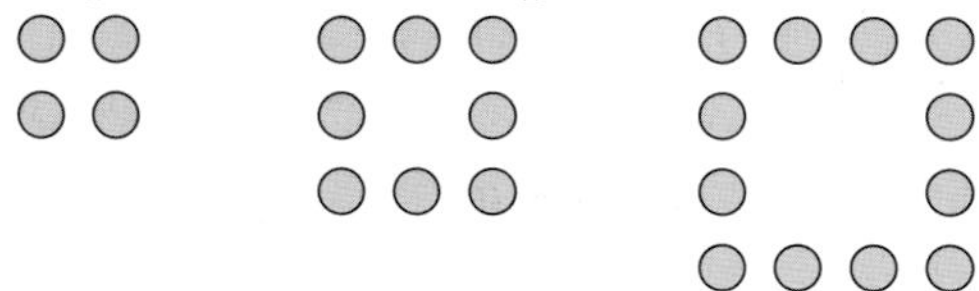

To find the number of dots (D) in a square of side L you add the number of dots along the top and along the bottom ($L + L$) to the number of dots in the sides ($(L - 2) + (L - 2)$).
$D = 2L + 2(L - 2) = 4L - 4$.

Thinking big will help you see the pattern. Number of dots in the 100th square would be: 100 across top + 100 across bottom + 98 down left + 98 down right $= 2 \times 100 + 2 \times (100 - 2)$.

Example

Find the nth term of the sequence: 100, 95, 90, 85, 80,...

Term number (n)	1	2	3	4	5
$-5n$	-5	-10	-15	-20	-25
Term	100	95	90	85	80

$T_n = 105 - 5_n$

Constant difference is -5, so compare to the multiples of -5.

Each term is 105 more than the multiple of -5.

Example

Find the rule for this pattern.

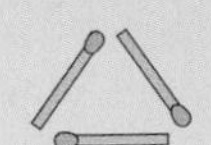
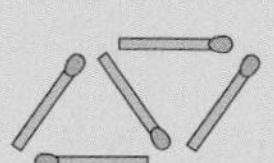
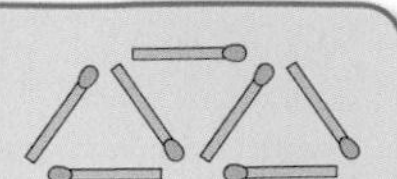

Term 1: 3 matches
Term 2: 3 + 2 matches
Term 3: 3 + 2 × 2 matches
Term 4: 3 + 3 × 2 matches
Term 100: 3 matches + 99 × 2 matches
Term n: $3 + 2(n - 1)$ matches

Number of matches $= M = 3 + 2(n - 1) = 3 + 2n - 2 = 2n + 1$
so $M = 2n + 1$.

Exercise A2.2 Grade C

1 Find the nth term formula for each of these linear sequences.

a 13, 16, 19, 22, 25,...
b 2, 5, 8, 11, 14,...
c 25, 30, 35, 40, 45,...
d 4, 5, 6, 7, 8,...
e 1, 1.2, 1.4, 1.6, 1.8,...
f 50, 47, 44, 41, 38,...
g Counting down in 4s from 10

2 Explain the rule for each pattern.

a

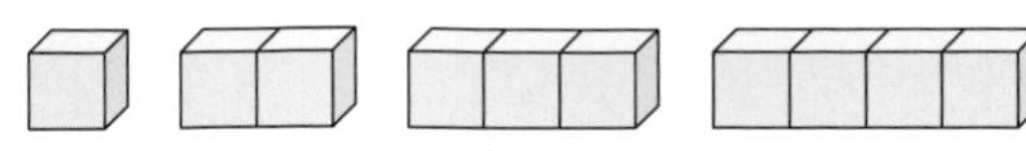

$G = 4L + 2$
G = number of green faces visible
L = length of strip

b

$A = H(H + 1)$
A = area
H = height of rectangle

c

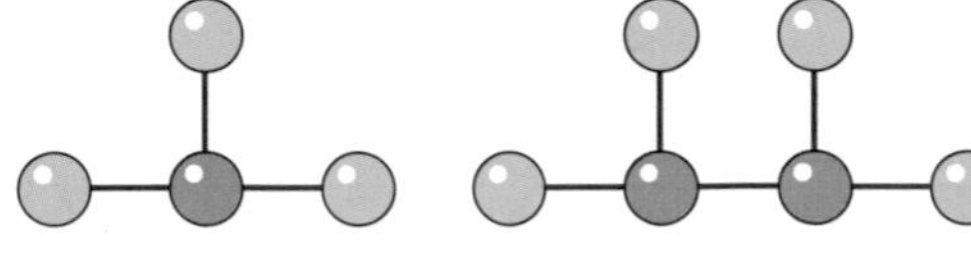

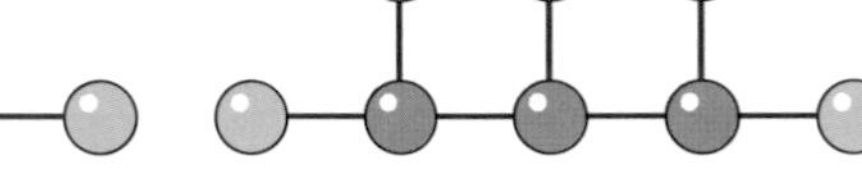

$B = R + 2$
B = number of blue beads
R = number of red beads

$L = 2R + 1$
L = number of links
R = number of red beads

3 By considering the sequence of patterns, write a formula to connect the quantities given.

a

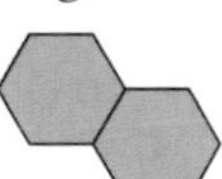

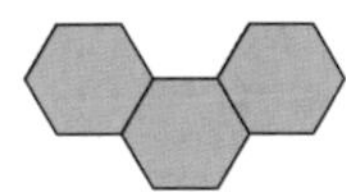

Relate the number of edges E to the number of hexagons H.

b

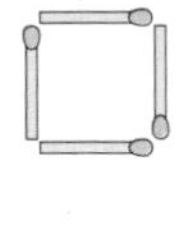

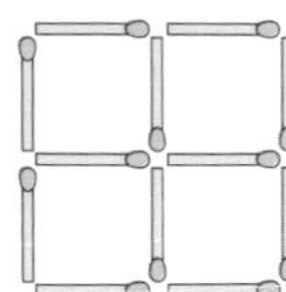

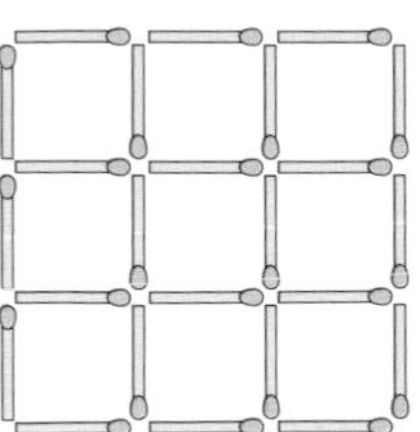

Relate the number of matches M with the length of the square L.

Account for the rows first, then the columns.

Problem AO3

4 Imagine a $3 \times 3 \times 3$ cube made from unit cubes stuck together. The outside is painted black. If this is then dismantled, how many cubes have 0, 1, 2, 3,... faces painted black?
Repeat for an $n \times n \times n$ cube and an $m \times n \times p$ cuboid.

A2.3 Line graphs

This spread will show you how to:

- Understand that the form $y = mx + c$ represents a straight line and that m is the gradient and c is the value of the y-intercept

Keywords
Gradient
Parallel
y-intercept

- The equation of a straight line is of the form $y = mx + c$, where m is the **gradient** and c is the ***y*-intercept**.

The y-intercept is the y-value where the graph cuts the y-axis.

Two lines that have the same gradient are **parallel**.

- A line with a positive gradient slopes upwards.
- A line with a negative gradient slopes downwards.

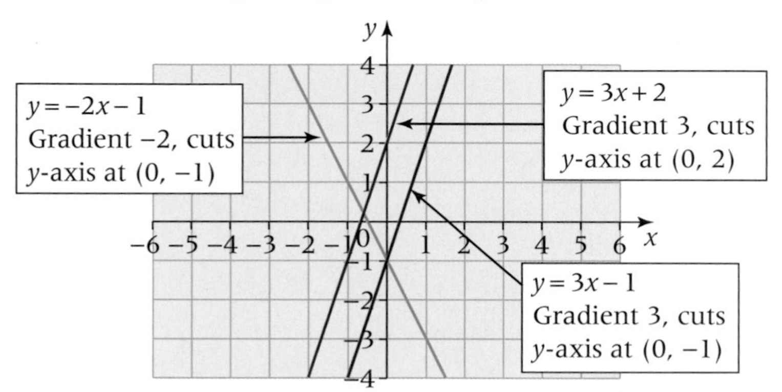

You can use a triplet of numbers (x, y, z) to specify points in 3-D. See **G5.4**.

- The coordinates of any point on a line satisfy the equation of the line.

Example

Describe the line $2y = 8 - 3x$.

$2y = 8 - 3x$

$y = 4 - \frac{3}{2}x$

$y = -\frac{3}{2}x + 4$

The line has gradient $-\frac{3}{2}$ and cuts the y-axis at $(0, 4)$.

First rearrange the equation into the form $y = mx + c$.

Example

The straight line L_1 has equation $y = 3x + 4$.
The straight line L_2 is parallel to L_1.
The straight line L_2 passes through $(3, 1)$.
Find an equation of the straight line L_2.

L_2 is parallel to L_1, so L_2 has gradient 3.

L_2: $y = 3x + c$

At $(3, 1)$: $1 = 3 \times 3 + c$

$-8 = c$

The equation of L_2 is $y = 3x - 8$.

Substitute $x = 3$ and $y = 1$ into the equation and solve to find c.

Exercise A2.3 Grade C

1 Describe each of these lines with reference to gradient and y-intercept.

a $y = 3x - 2$ **b** $y = 7 + \frac{1}{2}x$

c $3y = 9x - 6$ **d** $2y - 4x = 5$

2 Find the equations of these lines.

a Gradient 6, passes through $(0, 2)$

b Gradient -2, passes through $(0, 5)$

c Gradient -1, passes through $(0, \frac{1}{2})$

d Gradient -3, passes through $(0, -4)$

3 For each of these lines, give the equation of a line parallel to it.

a $y = 2x - 1$ **b** $y = -5x + 2$

c $y = -\frac{1}{4}x + 2$ **d** $y = 7 - 4x$

e $y = 6 + \frac{3}{4}x$ **f** $2y = 9x - 1$

4 Here are the equations of several lines.

A $y = 3x - 2$ **B** $y = 4 + 3x$ **C** $y = x + 3$

D $y = 5$ **E** $2y - 6x = -3$ **F** $y = 3 - x$

a Which three lines are parallel to one another?

b Which two lines cut the y-axis at the same point?

c Which a line has a zero gradient?

d Which a line passes through $(2, 4)$?

e Which a pair of lines are reflections of one another in the y-axis?

5 Find the equations of these lines.

a A line parallel to $y = -4x + 3$ and passing through $(-1, 2)$

b A line parallel to $2y - 3x = 4$ and passing through $(6, 7)$.

6 True or false?

86

a $y = 2x - 1$ and $y = 2 + 2x$ are parallel.

b $y = 3x - 4$ and $y = 6 - x$ both pass through $(2, 4)$.

56

c $x = 6$ and $y = 2x - 8$ never meet.

d $y = x + 1$ and $y = x^2$ meet once.

AO3 Problem

48

7 Find the coordinates of the points where the graphs of these lines meet.

a $y = 3x + 2$ and $2y + 3x = 13$

b $y = 2x - 2$ and $x^2 + y^2 = 25$

Did you know?

1:7

17%

Gradient road signs warn the driver when steep climbs or drops are near. In the UK they are given as either ratios or percentages.

A2.4 Finding the equation of a straight line

This spread will show you how to:

- Find the equation of a straight line by considering its gradient and axes intercepts

Keywords
Gradient
y-intercept

- The **gradient** of a line segment is calculated as $\dfrac{\text{Change in the } y\text{-direction}}{\text{Change in the } x\text{-direction}}$.

Gradient $= \frac{\text{rise}}{\text{run}}$

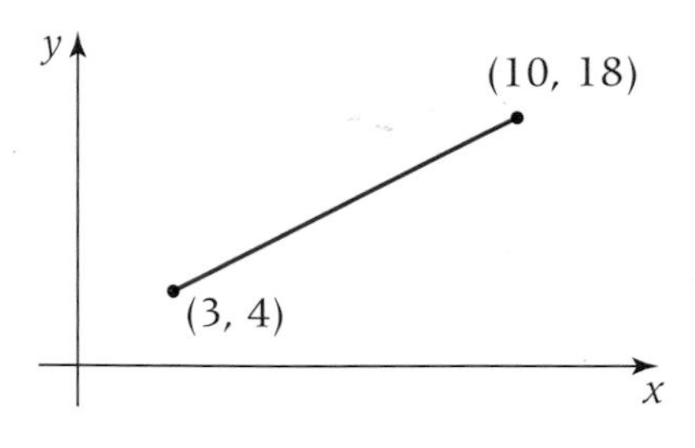

From (3, 4) to (10, 18):
Change in *y*-direction = +14
Change in *x*-direction = +7
Gradient $= \frac{14}{7} = 2$

- The gradient m of the line joining (x_1, y_1) to (x_2, y_2) is

$$m = \frac{y_2 - y_1}{x_2 - x_1}$$

- Once you know the gradient, you can use one of the points to find the value of the ***y*-intercept**, c.

Line joining (3, 4) to (10, 18):
$y = 2x + c$
Substitute $x = 3$, $y = 4$:
$4 = 6 + c \Rightarrow c = -2$
Equation: $y = 2x - 2$

Example

Find **a** the gradient **b** the equation of the line passing through (3, 5) and (9, 35).

a $m = \dfrac{35 - 5}{9 - 3} = \dfrac{30}{6} = 5$

b $y = 5x + c$

At the point (3, 5), $5 = 3 \times 5 + c$

$-10 = c$

Equation is $y = 5x - 10$.

$m = \frac{y_2 - y_1}{x_2 - x_1}$.
Learn this gradient formula.

Example

The line L is parallel to the graph of $y = 2x - 1$.
Find the equation of the line labelled L.

L is parallel to $y = 2x - 1$, so L has gradient 2.
For L: $y = 2x + c$
From the diagram, the *y*-intercept is -3.
Hence: $y = 2x - 3$.

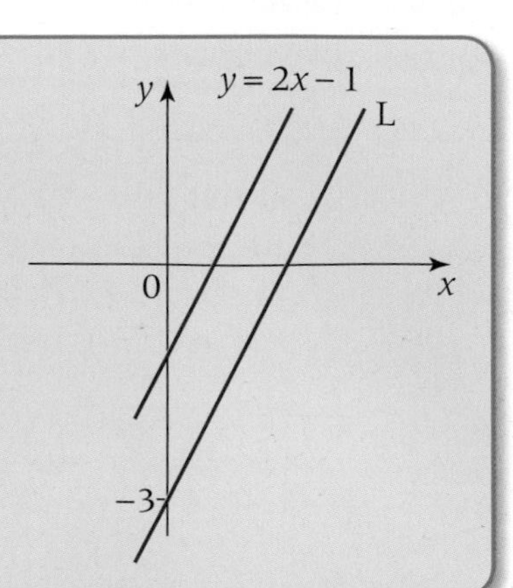

Exercise A2.4 Grade B

1 Find the gradients of the line segments joining these pairs of points.

a (4, 9) to (8, 25) **b** (5, 6) to (10, 16)

c (2, 1) to (5, 2)

2 Sketch each of these graphs, labelling the point where the graph crosses the y-axis.

a $y = 7x - 2$ **b** $y = 4 - 2x$

c $2y - x = 8$ **d** $2x + 3y = 4$

3 Find the equation of each line on this graph.

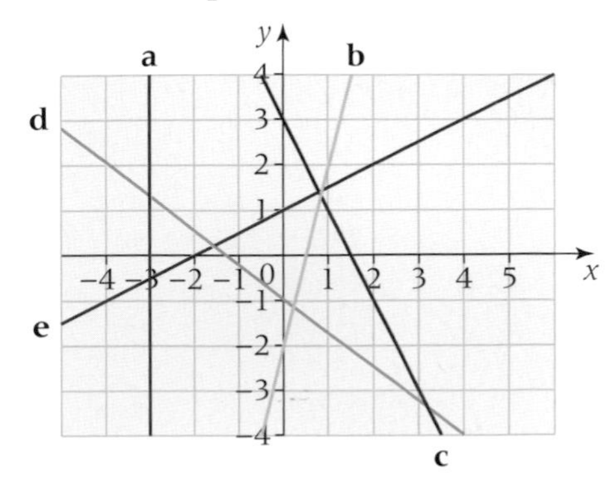

4 Find the equations of these line segments.

a A line joining (0, 5) to (3, 17)

b A line joining (2, 8) to (12, 13)

c

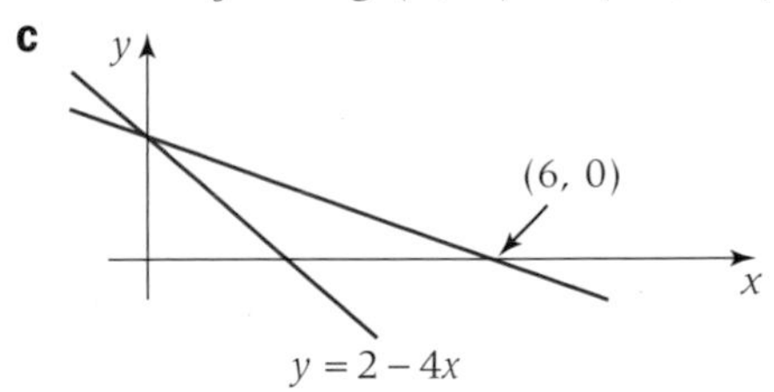

5 In questions **4**, parts **a** and **b** show that the midpoints of the two pairs of points lie on the lines joining them.

The midpoint of the points (x_a, y_a) and (x_b, y_b) is $\left(\frac{x_a + x_b}{2}, \frac{y_a + y_b}{2}\right)$

6 **a** The gradient of the line segment joining $(2, p)$ to $(6, 4p)$ is 4. Find the value of p.

b The gradient of the line segment joining (m, m) to $(5, 8)$ is 2. Find the value of m.

7 Use the information given in the diagram to find the equation of the line on each graph.

a

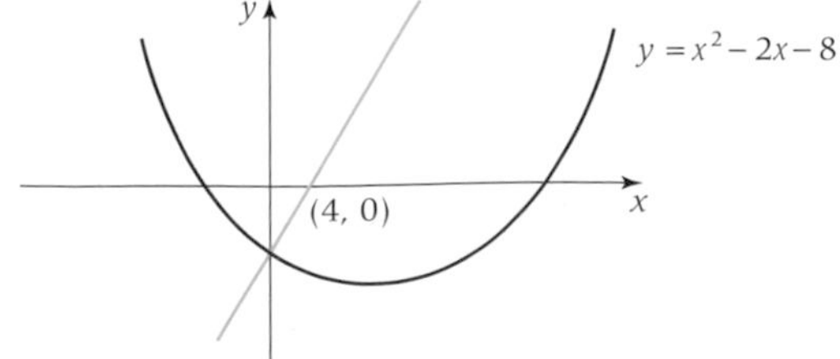

b

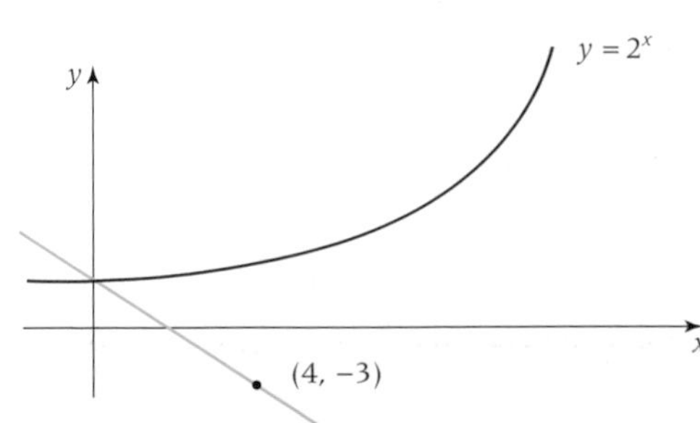

A2.5 Perpendicular lines

This spread will show you how to:

- Understand the relationship between perpendicular lines

Keywords
Bisector
Gradient
Perpendicular
Reciprocal

- Parallel lines have the same **gradient**.

The gradients of **perpendicular** lines are related.

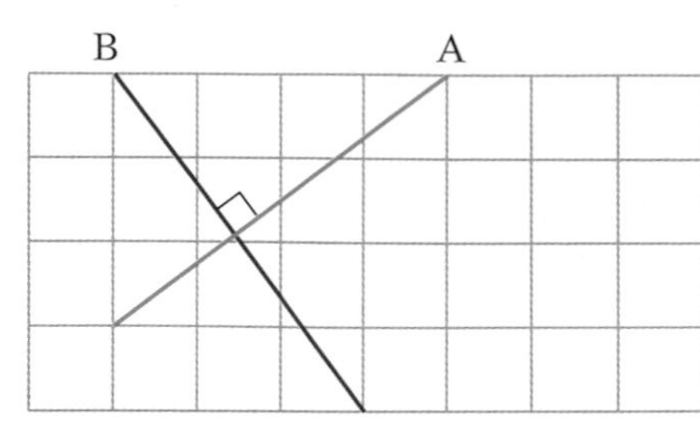

Lines A and B are perpendicular.
Line A has gradient $\frac{3}{4}$ and line B has gradient $-\frac{4}{3}$.
Notice that $\frac{3}{4} \times \frac{4}{3} = -1$.
The gradients of two perpendicular lines multiply to give -1.

- If line A has gradient m, any line perpendicular to line A has gradient $-\frac{1}{m}$.

$-\frac{1}{m}$ is the negative **reciprocal** of m.

Example

Find the equation of the line perpendicular to $y = 2x - 1$ that passes through (4, 5).

$y = 2x - 1$ has gradient 2, so a line perpendicular to it has gradient $-\frac{1}{2}$.

$y = -\frac{1}{2}x + c$

At (4, 5): $5 = \left(-\frac{1}{2}\right) \times 4 + c$

$c = 7$

The equation is $y = -\frac{1}{2}x + 7$.

Example

What is the equation of the perpendicular **bisector** of the line segment passing through (4, 8) and (6, 16)?

The midpoint of (4, 8) and (6, 16) is $\left(\frac{4+6}{2}, \frac{8+16}{2}\right) = (5, 12)$.

The gradient of this line segment is $\frac{16-8}{6-4} = 4$.

So, the perpendicular bisector has gradient $-\frac{1}{4}$ and passes through (5, 12).

$y = -\frac{1}{4}x + c$

At (5, 12): $12 = \left(-\frac{1}{4}\right) \times 5 + c$

$c = 13\frac{1}{4}$

The equation is $y = -\frac{1}{4}x + 13\frac{1}{4}$

A perpendicular bisector goes through the midpoint of a line segment and is at right angles to it.

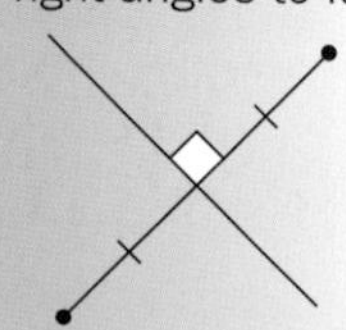

The midpoint is (mean of x-coordinates, mean of y-coordinates).

Exercise A2.5

Grade A/A*

1 Give the equation of a line perpendicular to each of these lines.

a $y = 2x - 1$ **b** $y = -5x + 2$ **c** $y = -\frac{1}{4}x + 2$

d $y = 7 - 4x$ **e** $y = 6 + \frac{3}{4}x$ **f** $2y = 9x - 1$

2 If these lines are arranged in perpendicular pairs, which is the odd one out?

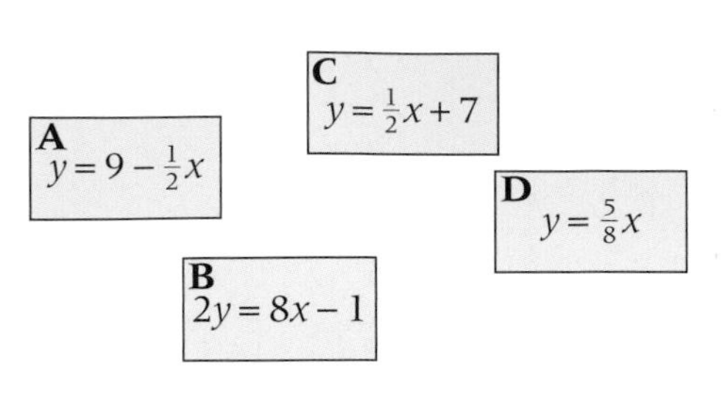

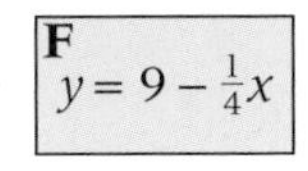

D $y = \frac{5}{8}x$

E $y = 2x + 4$

G $5y = 6 - 8x$

3 Find the equation of a line:

a parallel to $y = 3 - x$ and passing through (9, 10)
b perpendicular to $y = 2x + 4$ and passing through (3, 7)
c perpendicular to $2y = 8 - x$ and crossing the y-axis at the same point
d perpendicular to $y = -\frac{2}{3}x - 5$ and passing through (2, 8).

4 Find the equation of the perpendicular bisector of the line segment joining each pair of points.

a (3, 10) and (7, 12) **b** (2, 20) and (5, 18) **c** (−2, 7) and (4, −10)

5 Write an expression for the gradient of a line perpendicular to the line segment joining $(3t, 9)$ to $(2t, 12)$.

6 The triangle formed by joining the point (5, 12) to (14, 24) and (2, 40) is right-angled. True or false?

Use gradients to make your decision.

A03 Problem

7 Ben solved a pair of simultaneous equation graphically. His solution was $x = 2$ and $y = 3$. Use this diagram to help you find the second simultaneous equation that he solved.

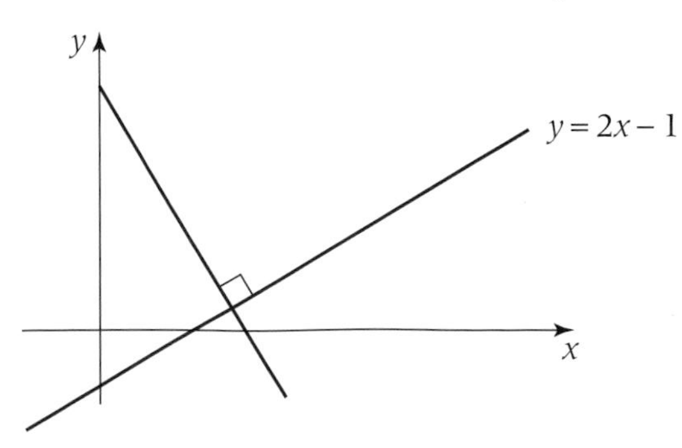

Unit 2

Summary

Check out

You should now be able to:

- Find and use the nth term of an arithmetic sequence
- Justify the nth term of an arithmetic sequence by looking at the context
- Understand that the form $y = mx + c$ represents a straight line and that m is the gradient and c is the value of the y-axis intercept
- Calculate the gradient of a line segment
- Find the equation of the straight line by considering its gradient and y-axis intercept
- Understand the gradients of parallel lines and lines perpendicular to each other

Worked exam question

The diagram shows three points $A\,(-1, 5)$, $B\,(2, -1)$ and $C\,(0, 5)$.
The line **L** is parallel to AB and passes through C.
Find the equation of the line **L**. (4)

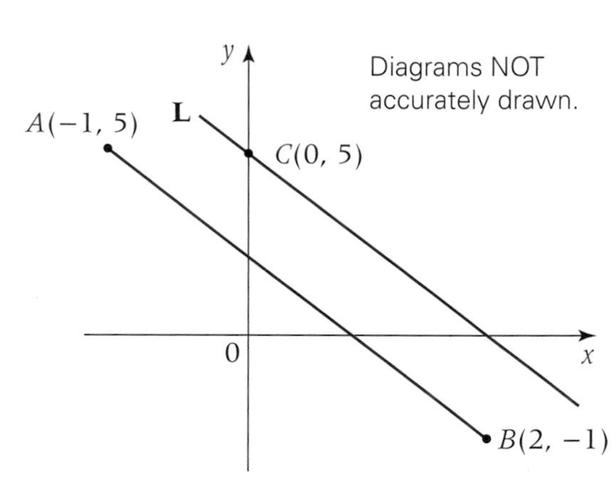

(Edexcel Limited 2005)

The equation of a straight line is $y = mx + c$

The gradient of the line AB is $\dfrac{y_2 - y_1}{x_2 - x_1} = \dfrac{-1 - 5}{2 - -1}$

$= \dfrac{-6}{3}$

$= -2$

The gradient of the line **L** is -2 and so $m = -2$
$y = -2x + c$ for line **L**

The intercept on the y-axis is 5 and so $c = 5$
$y = -2x + 5$ for line **L**

Calculate the gradient of the line AB.

Write down the equation $y = -2x + c$

Substitute $x = 0$ and $y = 5$ into the equation $y = -2x + c$ also gives $c = 5$

Exam questions

1 The table shows some rows of a number pattern.

a In the table, write down an expression, in terms of n, for Row n. (1)

b Simplify fully your expression for Row n.
You must show your working. (2)

Row 1	$1^2 - (0 \times 2)$
Row 2	$2^2 - (1 \times 3)$
Row 3	$3^2 - (2 \times 4)$
Row 4	$4^2 - (3 \times 5)$
Row n	

(Edexcel Limited 2006)

2

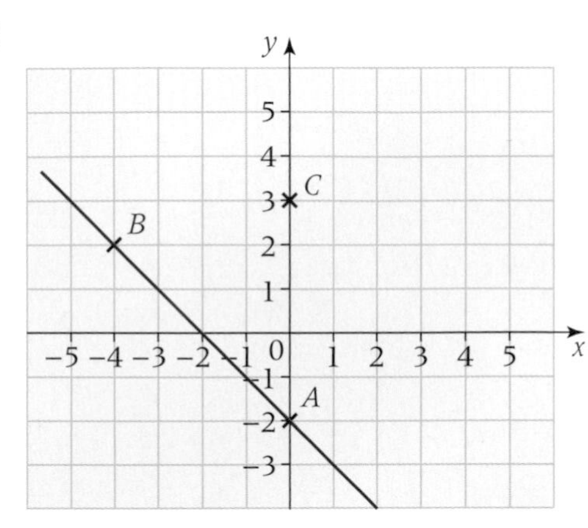

In the diagram
A is the point $(0, -2)$,
B is the point $(-4, 2)$,
C is the point $(0, 3)$.

Find an equation of the line that passes through C and is parallel to AB. (4)

(Edexcel Limited 2006)

3

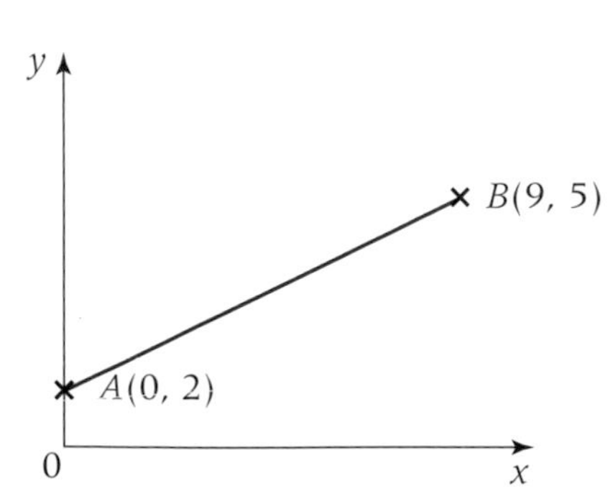

Diagram NOT accurately drawn.

A is the point $(0, 2)$ B is the point $(9, 5)$

The equation of the straight line through A and B is $y = \frac{1}{3}x + 2$

a Write down the equation of another straight line that is parallel to $y = \frac{1}{3}x + 2$ (1)

b Write down the equation of another straight line that passes through the point $(0, 2)$ (1)

c Find the equation of the line perpendicular to AB passing through B. (3)

(Edexcel Limited 2006)

N4

Integers and powers

When you use the internet to pay for goods you need to know that your financial details are safe. To make these details secure they are turned into a secret code, encrypted, using the product of two very large prime numbers. Whilst the product is publicly known, the difficulty in factorising it makes the message safe.

What's the point?

Prime numbers are those numbers that only have two factors: one and themselves. Since all numbers can be broken down into a unique product of prime numbers, mathematicians believe that studying prime numbers will lead to a better understanding of all numbers.

Integers and powers

Check in

You should be able to

- **find factors and recognise prime numbers**

1 Write all the factors of each number.
 a 12 **b** 30 **c** 120 **d** 360

2 Write a list of all of the prime numbers up to 100. (There are 25 of them.)

- **find HCFs and LCMs**

3 **a** Write the highest number that is a factor of both 12 and 30.
 b Write the smallest number that is a multiple of both 12 and 30.

- **use index laws and index notation**

4 Find the results of these calculations, giving the answers in index form.
 a $2^3 \times 2^4$ **b** $3^5 \div 3^2$ **c** $5^2 \times 5^3 \times 5^2$
 d $6^6 \div 6^4$ **e** $7^8 \div (7^2 \times 7^3)$ **f** $(4^6 \times 4^2) \div (4^3 \times 4)$

5 Write the value of each number.
 a 4^0 **b** 6^0 **c** 5^1 **d** 2^{-1}

- **simplify arithmetic expressions**

6 Simplify each expression.
 a $4\pi + 3\pi$ **b** $\pi(3^2 + 4)$ **c** $4\pi(5^2 - 4^2)$
 d $\sqrt{(5^2 - 3^2)}$ **e** $\sqrt{2} + 2\sqrt{2}$ **f** $5\sqrt{3} - \sqrt{3}(4^2 - 3 \times 4)$

Orientation

What I need to know	What I will learn	What this leads to
KS3 Identify the prime factors of a number	■ Use indices and standard form ■ Use prime factorisation and find HCFs and LCMs	**A7** The exponential funtion
N3 Do arithmetic with decimals **A1** Use the index laws	■ Use index laws for negative and fractional powers ■ Calculate with surds	

Rich task

Throughout history people have tried to find a mathematical formula that will generate prime numbers.

Here are some of the formulae that have been used,

$6n + 1$

$6n - 1$

$n^2 - n + 41$

$n^2 - 79n + 1601$

$2^n - 1$

Investigate these formulae and see if they really work.

N4.1 Powers of 10

This spread will show you how to:

- Use powers of 10 to represent large and small numbers
- Calculate with powers of 10 using the index laws

Keywords
Index form
Index laws
Power
Reciprocal

You can use **powers** of 10 to represent large and small numbers. Numbers written with powers are in **index form**.

Example

Write these numbers as powers of 10.

a 100 **b** 1000 **c** 1 000 000

a $100 = 10 \times 10 = 10^2$ **b** $1000 = 10^3$ **c** $1\,000\,000 = 10^6$

The number of zeros tells you the power of 10.

You can calculate with powers of 10, using the **index laws**.

p.118

- To multiply powers of 10 you add the indices: $10^a \times 10^b = 10^{a+b}$
- To divide powers of 10 you subtract the indices: $10^a \div 10^b = 10^{a-b}$

Example

Work out these calculations. Give your answers in index form.

a $10^2 \times 10^3$ **b** $10^2 \times 10^2$ **c** $10^3 \div 10^2$

a $10^2 \times 10^3 = 100 \times 1000 = 100\,000 = 10^5$
b $10^2 \times 10^2 = 10^{2+2} = 10^4$
c $10^3 \div 10^2 = 10^{3-2} = 10^1 = 10$

$10^1 = 10$

- Any number to the power 0 is equal to 1:

$3^0 = 1$ $12^0 = 1$ $10^0 = 1$

$x^0 = 1$ for any value of x except zero.

- A negative index represents a **reciprocal**:

$10^{-2} = \frac{1}{10^2} = 0.01$ $10^{-4} = \frac{1}{10^4} = 0.0001$

$x^{-n} = 1 \div x^n = \frac{1}{x^n}$

Example

a Write 0.000 001 as a power of 10.
b Work out $10^3 \div 10^3$
c Work out $10^0 \div 10^1$

a Count the number of digits from the decimal point to the digit 1, starting at the decimal point.
$0.000\,001 = 10^{-6}$
b $10^3 \div 10^3 = 10^{3-3} = 10^0 = 1$
c $10^0 \div 10^1 = 100^{0-1} = 10^{-1} = \frac{1}{10} = 0.1$

Exercise N4.1 Grade C

1 Write these numbers as powers of 10.

a 100 **b** 10 **c** 1000 **d** 1
e 10 000 **f** 1 000 000 **g** 100 000 **h** 100 000 000

2 Write these numbers as powers of 10.

a 0.01 **b** 0.1 **c** 0.001 **d** 0.000 01
e 0.0001 **f** 0.000 000 1 **g** 0.000 001 **h** 1.0

3 Write these expressions as ordinary numbers.

a 10^3 **b** 10^6 **c** 10^5 **d** 10^9
e 10^4 **f** 10^1 **g** 10^2 **h** 10^7

4 Write these expressions as ordinary numbers.

a 10^0 **b** 10^{-2} **c** 10^{-5} **d** 10^{-3}
e 10^{-7} **f** 10^{-1} **g** 10^{-4} **h** 10^{-6}

5 Work out these calculations and give your answers in index form.

a $10^2 \times 10^3$ **b** $10^4 \times 10^5$ **c** $10^5 \times 10^3$
d $10^6 \div 10^3$ **e** $10^8 \div 10^4$ **f** $10^6 \div 10^2$

6 Work out these calculations and give your answers in index form.

a $10^4 \div 10^6$ **b** $10^3 \div 10^7$ **c** $10^2 \div 10^{10}$
d $10 \div 10^9$ **e** $1 \div 10^8$ **f** $10^7 \div 10$

7 Work out these calculations and give your answers in index form.

a $10^4 \times 10^3 \div 10^5$ **b** $10^6 \div 10^{-6}$
c $10^3 \times 10^{-4} \div 10^5$ **d** $10^{-5} \div 10^{-2} \times 10^3$

8 Find the value of the letters in these equations.

a $10^4 \times 10^w \times 10^2 = 10^9$ **b** $10^x \div 10^2 \times 10^5 = 10^8$
c $10^{-2} \times 10^y \div 10^3 = 10^{-2}$ **d** $10^z \times 10^{-4} \times 10^z = 10$

9 Find the value of the letters in these equations.

a $300 = 3 \times 10^a$ **b** $0.000\,005 = b \times 10^{-6}$
c $0.15 = 1.5 \times 10^c$ **d** $130 = d \times 10^2$

10 Find the value of the letters in these equations.

a $0.004\,005 = m \times 10^{-3}$ **b** $10^n \times 10^{-5} \times 10^n = 10^n$
c $1 \div 10^{-7} = 10^p$ **d** $10^q \times 10^q \times 10^q = 10^{-12}$

N4.2 Standard form

This spread will show you how to:

- Understand and use standard index form to write large and small numbers

Keywords
Index
Standard form

You can write the numbers 44 506 as 4.4506×10^4 and 0.000 004 as 4×10^{-6}. This is called **standard form**.

Standard form makes it easier to compare and calculate with large and small numbers.

- You can write a number in standard form as $A \times 10^n$, where n is a positive or negative integer and $1 \leqslant A < 10$

- Write the number with the place values adjusted so that it is between 1 and 10.

  For 3667 write 3.667 — For 0.0876 write 8.76

- Work out the value of the **index**, n, the number of columns the digits have moved.

  In 3.667 the digits have moved 3 columns right so $n = 3$.
  $3667 = 3.667 \times 10^3$

  In 8.76 the digits have moved 2 columns left so $n = -2$.
  $0.0876 = 8.76 \times 10^{-2}$

To get from 8.76 to 0.0876 you divide by 100, or multiply by $\frac{1}{100} = 10^{-2}$.

Example

Express these numbers in standard form.

a 435 **b** 0.000 483
c 15.5 **d** 0.001 003

a $435 = 4.35 \times 10^2$ **b** $0.000483 = 4.83 \times 10^{-4}$
c $15.5 = 1.55 \times 10^1$ **d** $0.001003 = 1.003 \times 10^{-3}$

Example

Write these numbers in order, starting with the smallest.
6.35×10^4, 5.44×10^4, 6.95×10^3, 7.075×10^2, 9.9×10^{-1}

The correct order is:
9.9×10^{-1}, 7.075×10^2, 6.95×10^3, 5.44×10^4, 6.35×10^4

First order the powers of 10: $10^{-1} < 10^2 < 10^3 < 10^4$. Then compare numbers with the same powers of 10: $5.44 < 6.35$

You can use numbers written in standard form in calculations.

Example

Rosie wrote

$(7.2 \times 10^{-4}) \div (3.6 \times 10^{-8}) = 2 \times 10^{-12}$

Is she correct?

No. She has multiplied 10^{-4} and 10^{-8}, instead of dividing.
The correct answer is:
$(7.2 \div 3.6) \times (10^{-4} \div 10^{-8}) = 2 \times 10^{-4--8} = 2 \times 10^{4}$

Examiner's tip
A quick check shows Rosie's answer is incorrect. Since you are dividing one number by another smaller number, the answer must be **greater** than 1.

Exercise N4.2 — Grade B

1 Write these numbers in standard form.

a 1375 **b** 20 554 **c** 231 455 **d** 5.8 billion

2 Write these numbers in standard form.

a 0.000 34 **b** 0.1067 **c** 0.000 0091
d 0.315 **e** 0.000 0505 **f** 0.0182
g 0.008 45 **h** 0.000 000 000 306

1 billion = 1 000 000 000

Unit 2

3 Write these numbers in order of size, smallest first.

4.05×10^4 4.55×10^4 9×10^3 3.898×10^4 1.08×10^4 5×10^4

4 Write these numbers as ordinary numbers.

a 6.35×10^4 **b** 9.1×10^{17} **c** 1.11×10^2 **d** 2.998×10^8

5 Write these expressions as ordinary numbers.

a 4.5×10^{-3} **b** 3.17×10^{-5} **c** 1.09×10^{-6} **d** 9.79×10^{-7}

6 Write these numbers in standard form.

a 21.5×10^3 **b** 0.7×10^{14} **c** 122.516×10^{18} **d** 0.015×10^9

7 Work out these calculations, giving your answers in standard form.

a $2 \times 10^3 \times 3 \times 10^4$ **b** $(8 \times 10^{15}) \div (2 \times 10^3)$
c $7.5 \times 10^3 \times 2 \times 10^5$ **d** $(3.5 \times 10^3) \div (5 \times 10^2)$
e $5 \times 10^5 \times 3 \times 10^4$ **f** $4 \times 10^3 \times 5 \times 10^2 \times 6 \times 10^4$

8 Work out these calculations without using a calculator.

a $(4 \times 10^6) \div (2 \times 10^8)$ **b** $5 \times 10^4 \times 2 \times 10^{-6}$
c $(3 \times 10^{-3}) \div (2 \times 10^5)$ **d** $(4 \times 10^5) \div (2 \times 10^2)$
e $5 \times 10^{-3} \times 5 \times 10^{-4}$ **f** $(9.3 \times 10^{-2}) \div (3 \times 10^{-6})$

DID YOU KNOW?

The gravitational forces between Earth and the moon cause two high tides a day.

Functional Maths AO2

9 Light travels about 3×10^8 metres per second.

a Find the time it takes for light to travel 1 metre.
b Find the distance light travels in 1 year.
Give your answers in standard form.

10 A mass of 12 grams of carbon contains about 6.0×10^{23} carbon atoms.

a Write 12 grams as a mass in kilograms using standard form.
b Estimate the mass of one carbon atom, giving your answer in kilograms in standard form, correct to 2 significant figures.

Arithmetic with numbers written in standard form is examined in unit 3.

Problem AO3

11 As the moon orbits Earth the distance between them varies between 4.07×10^5 km and 3.56×10^5 km. Find the difference between these two distances.

N4.3 Prime numbers and factorisation

This spread will show you how to:

- Write any integer as a product of its factors
- Find the prime factor decomposition of any integer

Keywords
Factor
Integer
Prime
Prime factor decomposition

You can write any **integer** as the product of its **factors**.
Prime factor decomposition means writing a number as the product of **prime** factors only.

- The prime factor decomposition of a number can only be written in one way.

Example

Write 30 as the product of its prime factors.

$30 = 3 \times 10 = 3 \times 2 \times 5$
$= 2 \times 3 \times 5$

Write the prime factors in ascending order.

- Use index notation to show any repeated factors.
  $3 \times 2 \times 2 \times 2 = 2^3 \times 3$

With large numbers, work systematically through possible prime factors.

Example

Write 4095 as a product of its prime factors.

2 is not a factor
3 is a factor $\rightarrow 4095 = 3 \times 1365$
3 is a factor of 1365 $\rightarrow 4095 = 3 \times 3 \times 455$
5 is a factor of 455 $\rightarrow 4095 = 3 \times 3 \times 5 \times 91$
7 is a factor of 91 $\rightarrow 4095 = 3 \times 3 \times 5 \times 7 \times 13$
$4095 = 3^2 \times 5 \times 7 \times 13$

Examiner's tip
Working systematically and writing each line in full will help you keep track of the factors.
Check (by multiplying) that your prime factor decomposition is correct.

Example

A classroom has an area of 144 m^2.
a Write the prime factor decomposition of 144.
b Find the dimensions of the classroom if it is
i square **ii** 16 m long **iii** 18 m long.

a $144 = 2 \times 2 \times 2 \times 2 \times 3 \times 3$
$= 2^4 \times 3^2$

b **i** For a square room each side is $\sqrt{2^4 \times 3^2} = 2^2 \times 3$
$= 12\text{m}$

ii $144 \div 16 = (2^4 \times 3^2) \div 2^4$
$= 3^2$
So the classroom is 16 m × 9 m.

iii $144 \div 18 = (2^4 \times 3^2) \div (2 \times 3^2)$
$= 2^3$
So the classroom is 18 m × 8 m.

Exercise N4.3 Grade C

1 Sam thinks that each of these numbers is prime. Explain why he is wrong every time.

a 201 **b** 995 **c** 777 **d** 441 **e** 71 536

2 Find the prime factor decomposition of each number.

a 21 **b** 8 **c** 15 **d** 90 **e** 124

3 Write each number as a product of its prime factors. Check your answers by multiplying.

a 900 **b** 630 **c** 1001 **d** 2205
e 1371 **f** 891 **g** 2788 **h** 1431
i 3377 **j** 2 460 510

4 Copy and complete the table to show which numbers from 2000 to 2010 are prime. Give the prime factor decomposition of those that are not prime.

Number	Prime? (Yes/No)	Prime factor decomposition for non-primes
2000		
2001		
⋮		
2009		
2010		

5 **a** Write each number as a product of its prime factors.

i 1728 **ii** 423 **iii** 812 **iv** 23

b For each number in part **a**, find the smallest number you need to multiply by to give a square number.

6 A rectangular carpet has an area of 84 square metres. Its length and width are both whole numbers of metres greater than 1 m.

The prime factor decomposition of 84 is $2^2 \times 3 \times 7$.
One way of grouping the prime factors of 84 is $2^2 \times (3 \times 7) = 4 \times 21$.
The dimensions of the rectangle could be 4 m by 21 m.

By grouping the prime factors of 84 in different ways, find all the possible dimensions of the carpet.

AO3 Problem

7 A metal cuboid has a volume of $1815\,\text{cm}^3$. Each side of the cuboid is a whole number of centimetres, and each edge is longer than 1 cm.

a Find the prime factor decomposition of 1815.

b Use your answer to part **a** to find all the possible dimensions of the cuboid.

Volume of a cuboid = length × width × height.

Using prime factors: HCF and LCM

This spread will show you how to:

- Find the prime factor decomposition of any integer and the highest common factor and least common multiple of any two integers

Keywords
Highest common factor
Least common multiple
Prime factor

- The **highest common factor** (HCF) of two numbers is the largest number that is a factor of them both.
- The **least common multiple** (LCM) of two numbers is the smallest number that they both divide into.

You can find the HCF and LCM of two numbers by writing their **prime factors** in a Venn diagram.

- The HCF is the product of the numbers in the intersection.
- The LCM is the product of all the numbers in the diagram.

You do not need to use a Venn diagram method in your exam.

Example

Find the HCF of 54 and 84.

$54 = 2 \times 3^3 \qquad 84 = 2^2 \times 3 \times 7$

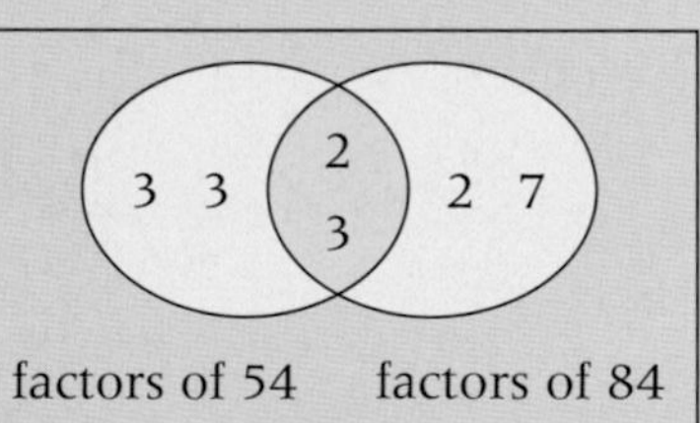

An alternative is to write:
$54 = (2) \times 3 \times (3) \times 3$
$84 = (2) \times 2 \times (3) \times 7$
Identify the common factors.
HCF $= 2 \times 3 = 6$

The prime factors are written in the diagram.
Common factors are in the intersection.
The HCF is the product of the numbers in the intersection:

$2 \times 3 = 6$, so the HCF of 54 and 84 is 6.

Example

Find the LCM of 60 and 280.

$60 = 2^2 \times 3 \times 5 \qquad 280 = 2^3 \times 5 \times 7$

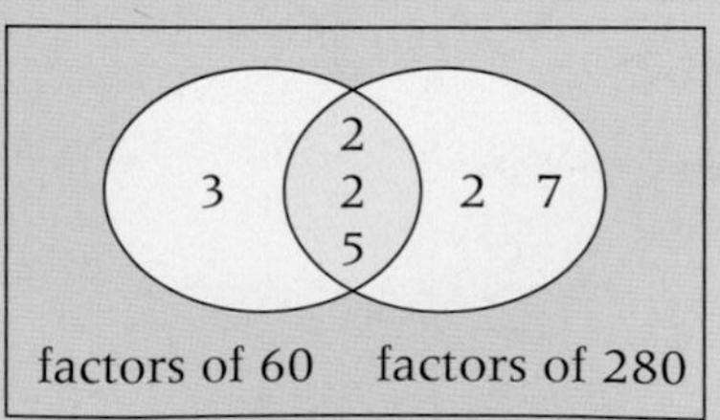

An alternative is to write:
$60 = 2^2 \times 3 \times 5$
$280 = 2^3 \times 5 \times 7$
Identify the highest power of each factor.
LCM $= 2^3 \times 3 \times 5 \times 7 = 840$

The LCM is the product of all the numbers in the diagram.
$2^3 \times 3 \times 5 \times 7 = 840$, so the LCM of 60 and 280 is 840.

Exercise N4.4 Grade C

1 Find the highest common factor (HCF) of each pair of numbers, by drawing a Venn diagram or otherwise.

a 35 and 20 **b** 48 and 16 **c** 21 and 24
d 25 and 80 **e** 28 and 42

2 Find the least common multiple (LCM) of each pair of numbers, by drawing a Venn diagram or otherwise.

a 24 and 16 **b** 32 and 100 **c** 22 and 33
d 104 and 32 **e** 56 and 35

3 Find the LCM and HCF of each pair of numbers.

a 180 and 420 **b** 77 and 735 **c** 240 and 336
d 1024 and 18 **e** 762 and 826

4 Write the prime factor decomposition of each number.

a 288 **b** 4725 **c** 67 375

5 **a** Find the LCM and HCF of each pair of numbers.

i 19 and 25 **ii** 36 and 23 **iii** 17 and 19 **iv** 11 and 9

b What can you say about the HCF of two numbers, if one (or both) of the numbers is prime?

c What can you say about the LCM of two numbers, if one (or both) of the numbers is prime?

6 Two numbers are co-prime if they have no common factors (other than 1). By drawing Venn diagrams or otherwise, decide whether each pair of numbers is co-prime.

a 105 and 429 **b** 63 and 715 **c** 121 and 175 **d** 455 and 693

7 **a** Find the HCF and LCM of each pair of numbers in question **6**.

b Use your answers to part **a** to explain how you can easily find the HCF and LCM of two numbers that are co-prime.

A03 Problem

8 Look at this statement.

> The product of any two numbers is equal to the product of their HCF and their LCM.

a Test the statement. Do you think it is true?

b Use a Venn diagram to justify your answer to part **a**.

9 Find the HCF and LCM of each set of three numbers.

a 30, 42, 54 **b** 90, 350 and 462 **c** 462, 510 and 1105

N4.5 Index laws

This spread will show you how to:

- Understand and use index notation and index laws

Keywords
Base
Index
Index laws
Power

You can write **powers** of a number in **index** form.

$10 \times 10 \times 10 \times 10 = 10^4$ (10 to the power 4)

In algebra, x^n means the **base** x raised to the power n.

In 10^4, the base is 10 and the index is 4.

- To multiply powers of the same base, add the indices.
  $x^m \times x^n = x^{m+n}$ $\quad$ $4^2 \times 4^3 = (4 \times 4) \times (4 \times 4 \times 4)$
  $= 4^{2+3} = 45$
- To divide powers of the same base, subtract the indices.
  $x^m \div x^n = x^{m-n}$ $\quad$ $7^5 \div 7^2 = \frac{7 \times 7 \times 7 \times 7 \times 7}{7 \times 7}$
  $= 7^{5-2} = 7^3$
- To raise a power, multiply the indices.
  $(x^m)^n = x^{m \times n}$ $\quad$ $(9^2)^3 = (9 \times 9) \times (9 \times 9) \times (9 \times 9)$
  $= 9^{2 \times 3} = 9^6$
- For all values of x except $x = 0$,
  $x^0 = 1$ $\quad$ $1 = x^n \div x^n = x^{n-n} = x^0$

You can only use the index laws with powers of the same base.

- $x^m \times x^n \times y^m = x^{m+n} \times y^m$

Example

Write the answers to these in index form.

a $5^5 \times 5^3$ **b** $8^2 \times 8$ **c** $4^7 \div 4^3$ **d** $7^5 \div 7$ **e** $(6^3)^4$

a $5^5 \times 5^3 = 5^{5+3} = 5^8$
b $8^2 \times 8 = 8^{2+1} = 8^3$
c $4^7 \div 4^3 = 4^{7-3} = 4^4$
d $7^5 \div 7 = 7^{5-1} = 7^4$
e $(6^3)^4 = 6^{3 \times 4} = 6^{12}$

You can write 8 as 8^1.

Example

For the calculation $2^3 \times 5^4$, Jack wrote:

$2 \times 5 = 10$, and $3 + 4 = 7$.
So, the answer is 10^7.

Is he correct?

No. The index laws only apply when both numbers have the same base. For this calculation you need to evaluate each term and then multiply them together:
$2^3 = 8$ and $5^4 = 625$ $\quad$ $2^3 \times 5^4 = 8 \times 625 = 5000$

Exercise N4.5 Grade C/B

1 Evaluate

a 4^3 **b** 5^4 **c** 2^6 **d** 3^3

e 9^2 **f** 7^4 **g** 12^0 **h** 8^0

2 Find the value of the letters in these equations.

a $u = 4^4$ **b** $32 = 2^v$ **c** $3^4 = w^2$ **d** $144 = 2^x \times y^2$

3 Simplify these expressions, giving your answers in index form.

a $3^2 \times 3^2$ **b** $5^2 \times 5^3$ **c** $6^3 \times 6^4$ **d** $7^2 \times 7^7$

e $2^8 \times 2^7$ **f** $4^4 \times 4^6$ **g** $10^4 \times 10^{10}$ **h** $9^8 \times 9^3 \times 9^2$

4 Copy and complete the multiplication table for powers of x.

$\times$	x	x^3		x^9
x^2			x^6	
	x^7			
x^3				
		x^8		

5 Simplify these expressions, giving your answer in index form.

a $4^4 \div 4^2$ **b** $5^7 \div 5^4$ **c** $9^5 \div 9^2$ **d** $6^8 \div 6^4$

e $7^6 \div 7^2$ **f** $8^5 \div 8^3$ **g** $9^8 \div 9^4$ **h** $3^7 \times 3^2 \div 3^4$

6 Simplify these expressions, giving your answers in index form.

a $\frac{3^4}{3^2}$ **b** $\frac{4^3 \times 4^4}{4^2}$ **c** $\frac{7^8}{7^3 \times 7^2}$ **d** $\frac{6^2 \times 6^7}{6^3 \times 6^4}$

e $5^2 \times \frac{5^7}{5^3 \times 5^2}$ **f** $\frac{2^3 \times 2^4}{2^5 \times 2^2} \times 2^5$ **g** $\frac{6^9 \div 6^3}{6^3 \times 6^2}$ **h** $\frac{4^5 \div 4}{4^8 \div 4^7}$

7 Use the relationship $(x^m)^n = x^{mn}$ to simplify these expressions, giving your answers in index form.

a $(2^3)^2$ **b** $(4^2)^5$ **c** $(7^2)^2$ **d** $(5^5)^3$

e $(3^4)^4$ **f** $(6^2)^2$ **g** $(5^7)^3$ **h** $(10^4)^4$

8 Find the value of the letter in each of these equations.

a $u = (4^3)^2$ **b** $64 = (v^2)^3$ **c** $81 = (3^w)^2$ **d** $x = (3^2)^3 + (2^3)^2$

9 Simplify these expressions, giving your answers in index form.

a $\left(\frac{4^4}{4^2}\right)^2$ **b** $\frac{(3^3)^2 \times 3^5}{3^7}$ **c** $\frac{6^3 \times (6^2)^4}{(6^5 \div 6^3)^2}$ **d** $\left(\frac{5^8 \div 5^2}{5^4 \div 5^2}\right)^3$

e $2^8 \times \left(\frac{2^4 \div 2^2}{2^9 \div 2^8}\right)^2$ **f** $\left(\frac{7^{15} \div 7^2}{7^8 \times 7^2}\right)^4$ **g** $\frac{(3^3)^5}{3^4 \times 3^2} \div (3^2)^3$ **h** $9^2 \times \left(\frac{9^7 \div 9}{9^2 \times 9^3}\right)^3$

N4.6 Further index laws

This spread will show you how to:

- Understand and use index notation and index laws, including integer, fractional and negative powers

Keywords
Index
Power
Reciprocal
Root

The **index** laws also apply for fractional and negative indices.

- Fractional indices represent **roots**.
  - $x^{\frac{1}{2}} = \pm\sqrt{x}$ for all values of x
  - $x^{\frac{1}{n}} = \sqrt[n]{x}$ the nth root of x
  - $x^{\frac{m}{n}} = (x^{\frac{1}{n}})^m = (\sqrt[n]{x})^m$, or $x^{\frac{m}{n}}$ 5 $(x^m)^{\frac{1}{n}} = \sqrt[n]{x^m}$
- Negative indices represent **reciprocals**.
  - $x^{-n} = \frac{1}{x^n}$

$5^{\frac{1}{2}} \times 5^{\frac{1}{2}} = 5^{\frac{1}{2}+\frac{1}{2}} = 5$, so $5^{\frac{1}{2}} = \sqrt{5}$

$x^{\frac{1}{3}} = \sqrt[3]{x}$

$\frac{1}{9^5} = 1 \div 9^5 = 9^0 \div 9^5 = 9^{0-5} = 9^{-5}$

The square root of a number can be positive or negative.

$4^{\frac{1}{2}} = \pm\sqrt{4} = \pm 2$ means the answer is $+2$ or -2.
$+2 \times +2 = 4$ $\quad$ $-2 \times -2 = 4$

The same is true for the square root of **any** positive number.

Example

Find the value of

a 4^{-1} **b** $4^{\frac{1}{2}}$ **c** $4^{-\frac{1}{2}}$ **d** $9^{\frac{3}{2}}$ **e** $8^{\frac{2}{3}}$ **f** $4^{-\frac{5}{2}}$

a $4^{-1} = \frac{1}{4^1} = \frac{1}{4}$

b $4^{\frac{1}{2}} = \pm\sqrt{4} = \pm 2$

c $4^{-\frac{1}{2}} = \frac{1}{4^{\frac{1}{2}}} = \pm\frac{1}{\sqrt{4}} = \pm\frac{1}{2}$

d $9^{\frac{3}{2}} = (9^{\frac{1}{2}})^3 = (\pm\sqrt{9})^3 = (\pm 3)^3 = \pm 27$

e $8^{\frac{2}{3}} = (8^{\frac{1}{3}})^2 = (\sqrt[3]{8})^2 = 2^2 = 4$

f $4^{-\frac{5}{2}} = \frac{1}{4^{\frac{5}{2}}} = \frac{1}{(4^{\frac{1}{2}})^5} = \frac{1}{(\pm\sqrt{4})^5} = \frac{1}{(\pm 2)^5} = \pm\frac{1}{32}$

Examiner's tip
In the exam, you are expected to give both $\pm$ answers where appropriate.

Example

To find the value of $16^{-\frac{1}{2}}$ Alex wrote:

$16^{\frac{1}{2}} = 4$, so the answer must be -4.

Is she correct?

No. $16^{\frac{1}{2}} = 4$ is correct, but it could also be -4.

$16^{-\frac{1}{2}} = \frac{1}{4}$ or $-\frac{1}{4}$. You could write $16^{-\frac{1}{2}} = \pm\frac{1}{4}$.

Examiner's tip
Rather than relying on learning the rules for fractional and negative indices 'by heart', make sure you understand the underlying relationships:

- $x^{\frac{1}{2}} \times x^{\frac{1}{2}} = x \Rightarrow x^{\frac{1}{2}} = \sqrt{x}$
- $x^{-n} = x^0 \div x^n = 1 \div x^n = \frac{1}{x^n}$

Exercise N4.6 Grade A

1 Find the value of

a $16^{\frac{1}{2}}$ **b** $9^{\frac{1}{2}}$ **c** $27^{\frac{1}{3}}$ **d** $0^{\frac{1}{2}}$ **e** $1000^{\frac{1}{3}}$

2 Find the value of the letter in each of these equations.

a $\sqrt{5} = 5^a$ **b** $\sqrt[3]{600} = 600^b$ **c** $\sqrt{100} = 1000^c$ **d** $2 = 16^d$

3 Evaluate

a $25^{\frac{1}{2}}$ **b** $25^{-\frac{1}{2}}$ **c** 25^0 **d** $25^{\frac{3}{2}}$ **e** 25^{-1}

4 Evaluate

a $4^{\frac{3}{2}}$ **b** $8^{\frac{2}{3}}$ **c** $9^{\frac{5}{2}}$ **d** $100^{-\frac{1}{2}}$ **e** $16^{-\frac{3}{2}}$

f $1000^{\frac{2}{3}}$ **g** $400^{-\frac{1}{2}}$ **h** $169^{\frac{3}{2}}$ **i** $81^{-\frac{3}{2}}$ **j** $4^{-\frac{3}{2}}$

5 Write these expressions as powers of 4.

a $\frac{1}{4}$ **b** 16 **c** $\frac{1}{16}$ **d** 8 **e** 32

6 Write each of the numbers given in question **5** as a power of 16.

7 Write these expressions as powers of 10.

a 100 **b** $\frac{1}{10}$ **c** $\sqrt{10}$

d $(\sqrt{10})^3$ **e** $\frac{1}{(\sqrt{10})^5}$

8 Write these expressions as powers of 5.

a $\frac{1}{25}$ **b** $\frac{1}{\sqrt[3]{5}}$ **c** $\sqrt[3]{25}$

d $\frac{1}{\sqrt{125}}$ **e** $\frac{1}{\sqrt[3]{625}}$

AO3 Problem

54

9 a Copy and complete the table to show the value of 9^x for various values of x.

x	-1	$-\frac{1}{2}$	0	$\frac{1}{2}$	1
9^x					

b Use your table to plot a graph to show the value of 9^x, for values of x in the range -1 to $+1$.

c On the same set of axes, plot the graph of 4^x for the same values of x. Comment on any similarities and differences between the two graphs.

N4.7 Irrational numbers

This spread will show you how to:

- Recognise the difference between rational and irrational numbers
- Rationalise a denominator, including when it involves a surd form

Keywords
Denominator
Irrational
Rational
Rationalise

p.30

- A number is **irrational** if it cannot be written as an exact fraction.

$\sqrt{2}$ and π are irrational.
If n is not a square number, then $\sqrt{n}$ is irrational.

- When approximated as decimals, irrational numbers have a non-recurring sequence of decimal digits.

$\sqrt{2} = 1.414\,213\ldots$ $\pi = 3.141\,592\ldots$ There are no repeating patterns in the decimal digits.

Example

Say whether each of these numbers is **rational** or irrational.

a $\frac{3}{17}$ **b** $\sqrt{16}$ **c** $\sqrt{17}$ **d** $\sqrt[3]{8}$ **e** $\frac{2\pi + 1}{3}$ **f** $\sqrt[3]{7}$

a $\frac{3}{17}$ rational $\frac{3}{17}$ is an exact fraction.
b $\sqrt{16} = 4$ rational 16 is a square number.
c $\sqrt{17}$ irrational 17 is not a square number.
d $\sqrt[3]{8} = 2$ rational 8 is a cube number.
e $\frac{2\pi + 1}{3}$ irrational π is irrational.
f $\sqrt[3]{7}$ irrational 7 is not a cube number.

Any multiple of π is irrational.

If you have an expression with an irrational number in the **denominator**, you should **rationalise** it.

- To rationalise the denominator in a fraction, multiply numerator and denominator by the denominator.

For example:
$\frac{2}{\sqrt{3}} = \frac{2}{\sqrt{3}} \times \frac{\sqrt{3}}{\sqrt{3}} = \frac{2\sqrt{3}}{3}$

Example

Rewrite each of these expressions without roots in the denominator.

a $\frac{5}{\sqrt{2}}$ **b** $\frac{7}{\sqrt{8}}$

a $\frac{5}{\sqrt{2}} = \frac{5}{\sqrt{2}} \times \frac{\sqrt{2}}{\sqrt{2}}$
$= \frac{5\sqrt{2}}{2}$

b $\frac{7}{\sqrt{8}} = \frac{7}{\sqrt{8}} \times \frac{\sqrt{8}}{\sqrt{8}}$
$= \frac{7\sqrt{8}}{8}$

Exercise N4.7 Grade A

Unit 2

1 Explain whether each of these numbers is rational or irrational.

a $\frac{1}{3}$ **b** 2π **c** $\frac{3}{19}$ **d** $\sqrt{3}$ **e** $\sqrt{25}$ **f** $\sqrt[3]{10}$

2 Nina says, 'I worked out $\frac{4}{19}$ on my calculator, and there was no repeating pattern. This means that $\frac{4}{19}$ is irrational.' Explain why Nina is wrong.

3 Give an example of

a an irrational number with a rational square

b a number with a rational square root

c a number with an irrational square, but a rational cube

d a number with an irrational square, cube, square root and cube root.

4 There are two whole numbers less than 100 for which both the square root and the cube root are rational. One of the numbers is 1. Find the other one.

5 Karla says, 'I am thinking of a rational number. When I halve it, I get an irrational number.' Can Karla be correct? Explain your answer.

6 Jim says, 'I am thinking of an irrational number. When I add one to my number, the answer is rational.' Can Jim be correct? Explain your answer.

7 Javier says, 'If a whole number is not a perfect cube, then its cube root is irrational.' Is Javier correct? Explain your answer.

8 Rationalise the denominator of each of these fractions.

a $\frac{1}{\sqrt{2}}$ **b** $\frac{1}{\sqrt{3}}$ **c** $\frac{1}{\sqrt{7}}$ **d** $\frac{1}{\sqrt{6}}$ **e** $\frac{1}{\sqrt{5}}$

9 Rewrite each of these fractions without roots in the denominator.

a $\frac{2}{\sqrt{8}}$ **b** $\frac{2}{\sqrt{10}}$ **c** $\frac{3}{\sqrt{12}}$ **d** $\frac{5}{\sqrt{30}}$ **e** $\frac{8}{\sqrt{40}}$

DID YOU KNOW?

According to legend, Pythagoras was so baffled by the concept that $\sqrt{2}$ was irrational that it eventually killed him.

A03 Problem

10 Lisa and Simone write down some decimal numbers.

Lisa writes: 0.121 121 112 111 121 111 12...
Simone writes: 0.454 545 454 545 454 545 45...

Simone says, 'My number is rational, because there is a repeating pattern in the digits.'
Lisa says, 'My number also has a pattern in the digits, so it is rational as well.'
Explain whether you agree with each of these statements.

N4.8 Calculations with surds

This spread will show you how to:

- Use surds and π in exact calculations, without a calculator
- Use algebraic techniques to expand and simplify brackets containing surds

Keywords
Irrational
Surds

In calculations, you can use approximate decimals for **irrational** numbers such as $\sqrt{2}$ or π.

$\pi \approx 3.142\ldots$
$\sqrt{2} \approx 1.414\ldots$

For more accurate results, you can carry out a calculation using **surds**.

A **surd** is a square root that is an irrational number, for example $\sqrt{2}$, $\sqrt{3}$, $\sqrt{7}$.

- The square root of a number is the product of the square roots of the number's factors.

For example, $\sqrt{20} = \sqrt{4 \times 5} = \sqrt{4} \times \sqrt{5} = 2\sqrt{5}$.

Example

a Simplify $\sqrt{12} + \sqrt{48}$.

b Write $6\sqrt{2}$ in the form $\sqrt{n}$, where n is an integer.

a $\sqrt{12} + \sqrt{48} = \sqrt{4} \times \sqrt{3} + \sqrt{16} \times \sqrt{3}$
$= 2\sqrt{3} + 4\sqrt{3} = 6\sqrt{3}$

b $6\sqrt{2} = \sqrt{6^2 \times 2} = \sqrt{36 \times 2} = \sqrt{72}$

- You can use algebraic techniques to expand and simplify brackets containing surds.

p.120

$(1 + \sqrt{3})^2 = 1 + 2\sqrt{3} + (\sqrt{3})^2 = 4 + 2\sqrt{3}$

For a combination of rational and irrational numbers, write the rational number first. To remove the surd, multiply by the surd with the opposite sign.

$(3 + \sqrt{5})(3 - \sqrt{5}) = 9 - 3\sqrt{5} + 3\sqrt{5} - 5 = 9 - 5 = 4$

$(a - x)(a + x) = a^2 - x^2$

Example

a Expand the brackets $(1 + \sqrt{2})(1 - \sqrt{2})$.

b Rewrite the fraction $\frac{1}{1 + \sqrt{3}}$ without surds in the denominator.

a $(1 + \sqrt{2})(1 - \sqrt{2}) = 1 - \sqrt{2} + \sqrt{2} - (\sqrt{2})^2 = 1 - 2 = -1$

b $\frac{1}{1 + \sqrt{3}} = \frac{1}{1 + \sqrt{3}} \times \frac{1 - \sqrt{3}}{1 - \sqrt{3}} = \frac{1 - \sqrt{3}}{1 - 3} = \frac{1 - \sqrt{3}}{-2}$
$= \frac{-1 + \sqrt{3}}{2}$

Multiply numerator and denominator by $(1 - \sqrt{3})$.

Exercise N4.8 Grade A/A*

1 Write each of these expressions as the square root of a single number.

a $\sqrt{2} \times \sqrt{3}$ **b** $\sqrt{5} \times \sqrt{3}$ **c** $\sqrt{3} \times \sqrt{7} \times \sqrt{11}$

2 Write these expressions in the form $\sqrt{a}\sqrt{b}$, where a and b are prime numbers.

a $\sqrt{14}$ **b** $\sqrt{33}$ **c** $\sqrt{21}$

3 Write these expressions in the form $a\sqrt{b}$, where b is a prime number.

a $\sqrt{20}$ **b** $\sqrt{27}$ **c** $\sqrt{98}$

4 Write these expressions in the form $\sqrt{n}$, where n is an integer.

a $4\sqrt{3}$ **b** $5\sqrt{2}$ **c** $4\sqrt{5}$

5 Without using a calculator, evaluate these expressions.

a $\sqrt{12} \times \sqrt{3}$ **b** $\sqrt{18} \times \sqrt{8}$ **c** $\sqrt{33} \times \sqrt{132}$

6 Simplify these expressions.

a $3\sqrt{5} + \sqrt{20}$ **b** $\sqrt{28} + 5\sqrt{7}$ **c** $7\sqrt{12} - 2\sqrt{27}$

7 **a** Use a calculator to evaluate your simplified expressions from question **6**, giving your answers to 3 decimal places.
b Use a calculator to evaluate the non-simplified expressions from question **6**.
c Compare your answers to parts **a** and **b**.

8 Expand the brackets and simplify.

a $\sqrt{5}(1 + \sqrt{5})$ **b** $\sqrt{3}(2 - \sqrt{3})$
c $(1 + \sqrt{5})(2 + \sqrt{5})$ **d** $(\sqrt{3} - 1)(2 + \sqrt{3})$
e $(4 - \sqrt{7})(6 - 2\sqrt{7})$ **f** $(5 + \sqrt{2})(5 - \sqrt{2})$
g $(6 + 2\sqrt{5})(6 - 2\sqrt{5})$ **h** $(5 - 2\sqrt{7})(5 + 3\sqrt{7})$
i $(7 + 5\sqrt{3})(7 - 5\sqrt{3})$

9 Rewrite these fractions without surds in the denominators.

a $\frac{1}{\sqrt{11}}$ **b** $\frac{1}{1 + \sqrt{2}}$ **c** $\frac{1}{1 - 2\sqrt{3}}$

d $\frac{\sqrt{5}}{1 + \sqrt{5}}$ **e** $\frac{1 - \sqrt{2}}{1 + \sqrt{2}}$ **f** $\frac{4 + \sqrt{2}}{3 - 2\sqrt{2}}$

Summary

Check out

You should now be able to:

- Use standard index form to represent large and small numbers
- Calculate with numbers written in standard index form
- Find the prime factor decomposition of positive integers
- Find the Highest Common Factor and Least Common Multiple of two or three numbers
- Understand and use index notation and index laws, including integer, fractional and negative powers

Worked exam question

a Write 4600 in standard form. (1)

b Write 255×10^{-5} in standard form. (2)

a

$4600 = 4.6 \times 10^3$

b

$255 \times 10^{-5} = 0.00255$

$= 2.55 \times 10^{-3}$

Write 0.00255 to show the working.

Worked exam question

a Write 5.7×10^{-4} as an ordinary number. (1)

b Work out the value of $(7 \times 10^4) \times (3 \times 10^5)$
Give your answer in standard form. (2)

(Edexcel Limited 2006)

a

0.00057

b

$(7 \times 10^4) \times (3 \times 10^5) = 7 \times 3 \times 10^5 \times 10^4$

$= 21 \times 10^{5+4}$

$= 21 \times 10^9$

$= 2.1 \times 10^{10}$

The final answer must be in standard index form.

Exam questions

1 a Write 84 000 in standard form.

b Write 0.0037 in standard form. (2)

2 The number of atoms in one kilogram of helium is 1.51×10^{26}

Calculate the number of atoms in 20 kilograms of helium.
Give your answer in standard form. (2)

(Edexcel Limited 2008)

3 a Find the Highest Common Factor (HCF) of 126, 210 and 294 (2)

b Work out the Least Common Multiple (LCM) of 12, 16 and 18 (2)

4 a Express 252 as a product of its prime factors. (3)

James thinks of two numbers.

He says "The Highest Common Factor (HCF) of my two numbers is 3
The Lowest Common Multiple (LCM) of my two numbers is 45"

b Write down two numbers that James could be thinking of. (3)

(Edexcel Limited 2008)

A02

5 Martin is organising a summer fair.
He needs bread buns and burgers for the barbecue.

Bread buns are sold in packs. Each pack contains 40 bread buns.
Burgers are sold in packs. Each pack contains 24 burgers.
Martin buys exactly the same number of bread buns as burgers.

What is the least number of each pack that Martin buys? (3)

(Edexcel Limited 2007)

6 a Find the value of

i 64^0 **ii** $64^{\frac{1}{2}}$ **iii** $64^{-\frac{2}{3}}$ (4)

b $3 \times \sqrt{27} = 3^n$

Find the value of n. (2)

(Edexcel Limited 2005)

7 a Rationalise the denominator of $\frac{1}{\sqrt{5}}$ (1)

b Expand $(1 + \sqrt{5})(3 + \sqrt{5})$

Give your answer in the form $a + b\sqrt{5}$, where a and b are an integers (2)

8 Expand and simplify $(\sqrt{3} - \sqrt{2})(\sqrt{3} + \sqrt{2})$ (2)

(Edexcel Limited 2007)

Functional Maths 4: Sport

Mathematics is used widely in sport, particularly when taking measurements and recording results.

Here are the results and wind speeds for the fastest thirteen all-time 100m Men's sprinters as of 20th September 2009:

Rank	Time (s)	Wind speed (m/s)	Athlete	Nation	Date
1	9.58	+0.9	Bolt	JAM	16/08/2009
2	9.69	+2.0	Gay	USA	20/09/2009
3	9.72	+0.2	Powell	JAM	02/09/2008
4	9.79	+0.1	Greene	USA	16/06/1999
5	9.84	+0.7	Bailey	CAN	27/07/1996
		+0.2	Surin	CAN	22/08/1999
7	9.85	+1.2	Burrell	USA	06/07/1994
		+0.6	Gatlin	USA	22/08/2004
		+1.7	Fasuba	NIG	12/05/2006
10	9.86	+1.2	Lewis	USA	25/08/1991
		−0.4	Fredericks	NAM	03/07/1996
		+1.8	Boldon	TRI	19/04/1998
		+0.6	Obikwelu	POR	22/08/2004

What level of accuracy is reported for
a) the result times b) the wind speeds?

Plot a time-series graph for these results. Can you see any trend? When (if ever) do you think the 9.50s barrier will be broken? Explain your answer referring to the data.

What is the a) fastest b) slowest actual time that each of the athletes could have run to give their reported result?

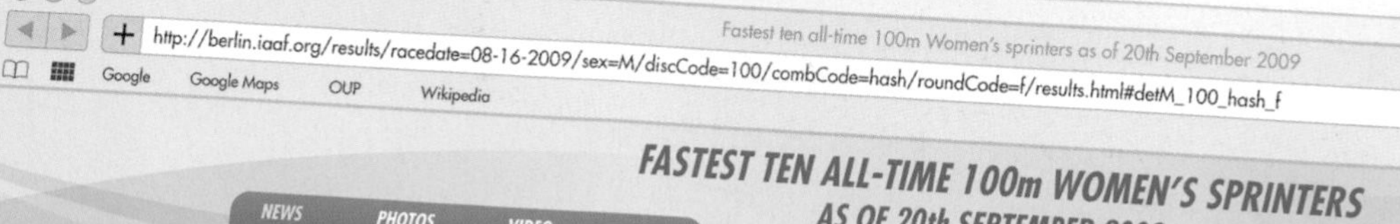

FASTEST TEN ALL-TIME 100m WOMEN'S SPRINTERS AS OF 20th SEPTEMBER 2009:

NEWS PHOTOS VIDEO AUDIO

By Date By Event Entry List Medal Table Placing Table Entry Standards

CHOOSE YOUR COUNTRY!
Select a country

Here are the results and wind speeds for the fastest ten all-time 100m Women's sprinters as of 20th September 2009:

Rank	Time (s)	Wind speed (m/s)	Athlete	Nation	Date
1	10.49	0.0	Griffith-Joyner	USA	16/07/1988
2	10.64	+1.2	Jeter	USA	20/09/2009
3	10.65	+1.1	Jones	USA	12/09/1998
4	10.73	+0.1	Fraser	JAM	17/08/2009
		+2.0	Arron	FRA	17/08/1998
6	10.74	+1.3	Ottey	JAM	07/09/1996
7	10.75	+0.4	Stewart	JAM	10/07/2009
8	10.76	+1.7	Ashford	USA	22/08/1984
9	10.77	+0.9	Privalova	RUS	06/07/1994
		+0.7	Lalova	BUL	19/06/2004

Florence Griffith-Joyner's World Record is quoted as being 'probably strongly wind assisted' because it is suspected that the wind speed measurer was faulty.

Comment on this, referring to the data and using your diagrams and statistics.

Use time-series diagrams and statistics to compare the Women's and Men's results. Refer to any trends you notice.

A maximum tail wind of +2.0m/s is allowed for 'wind legal' results. Head winds are not taken into account.

Tail winds follow the athlete and are recorded as +ve.

Head winds act against the athlete and are recorded as -ve.

Draw a scatter diagram of result time against wind speed for

a) the Men's results b) the Women's results.

Do you think there is any correlation between wind speed and time?

Justify your response by referring to the data.

A3

Equations

Equations

Mathematicians use algebra to turn real world problems into mathematical equations. Formula 1 engineers use complex mathematical equations to predict the effect on performance of their cars when they make technical modifications.

What's the point?

Learning how to write and solve equations allows you to understand the real world and make predictions.

Check in

You should be able to

- **collect like terms**

1 Simplify these expressions, by collecting like terms where possible.

a $5x + 7y - 4x + 9y$

b $11x + 9x^2 - 8x$

c $2ab + 5ab$

d $7p - 9$

e $11x^3 - 2x^3 + 5x^3$

f $3(2x + y) - x(x + 4)$

- **expand brackets**

2 Expand these sets of brackets, simplifying your answer where possible.

a $3(2x - 1) + 6(3x - 4)$

b $2(4y - 8) - 3(y - 9)$

c $x(3x - 2y)$

d $(x + 9)(x - 7)$

e $(2w - 8)(3w - 4)$

f $(p - q)^2$

- **do arithmetic with fractions**

3 Evaluate these, without a calculator, expressing your answers in their simplest form.

a $\frac{12}{30}$

b $\frac{1}{5} + \frac{2}{9}$

c $\frac{3}{4} - \frac{1}{4}$

d $2\frac{1}{6} + 3\frac{2}{5}$

e $\frac{5}{6} \times \frac{7}{10}$

f $\frac{5}{8} \div \frac{1}{5}$

Orientation

What I need to know	What I will learn	What this leads to
N3 Do arithmetic with fractions	■ Set up and solve linear equations	**A4** Quadratic equations
A1 Expand brackets and factorise expressions	■ Manipulate algebraic fractions	Engineering, Finance, Science

Rich task

$$\frac{3}{x+2} + \frac{2}{x-1} = \frac{5x-1}{x^2+x-2}$$

Find two algebraic fractions which add together to make $\frac{7x-6}{x^2-x-6}$

A3.1 Solving equations

This spread will show you how to:
- Set up simple equations
- Solve linear equations, including equations in which the term including the unknown is negative

Keywords
Equation
Inverse
Solve

- To **solve** an **equation** with the unknown on one side only, you use **inverse** operations.

$3x - 4 = 16$ Add 4 to both sides.
$3x = 20$ Divide both sides by 3.
$x = \frac{20}{3}$ or $6\frac{2}{3}$

You can 'read' the equation as: 'I think of a number, multiply it by 3 and subtract 4: this gives 16'. You can 'undo' the operations using the inverse operations 'add 4' then 'divide by 3'.

- To solve an equation when the unknown is in a negative term, first add this term to both sides to give a simpler equation.

$20 - 3x = 16$ Add $3x$ to both sides.
$20 = 16 + 3x$

- To solve an equation with the unknown on both sides, first subtract the smallest algebraic term from both sides to give a simpler equation.

$2x - 17 = 5x + 8 \xrightarrow{\text{Subtract } 2x \text{ from both sides}} -17 = 3x + 8$ Smallest algebraic term is $2x$.

$10 - 3x = 19 - 7x \xrightarrow{\text{Subtract } -7x \text{ from both sides}} 10 + 4x = 19$ Smallest algebraic term is $-7x$

Notice that subtracting $(-7x)$ is the same as adding $7x$, as $-(-7x) = +7x$.

Example

Solve **a** $25 - 8x = 5$ **b** $2(3 - 9x) = 4(10 - 2x)$

a $25 - 8x = 5$ Add $8x$ to both sides.
$25 = 5 + 8x$ Subtract 5 from both sides.
$20 = 8x$ Divide both sides by 8.
$x = \frac{20}{8}$ or $2\frac{1}{2}$ Change improper fractions to mixed numbers.

p.120

b $2(3 - 9x) = 4(10 - 2x)$ Expand the brackets.
$6 - 18x = 40 - 8x$ Subtract $-18x$ from both sides: $-(-18x) = +18x$.
$6 = 40 + 10x$ Subtract 40 from both sides.
$-34 = 10x$ Divide both sides by 10.
$x = -3.4$ Only use decimals if they are exact, otherwise, use fractions.

Inverse of $+5$ is -5
Inverse of $\times 8$ is $\div 8$

Example

Solve $6p + 3 = 2(p - 1)$

$6p + 3 = 2(p - 1)$
$6p + 3 = 2p - 2$
$4p + 3 = -2$
$4p = -5$
$p = -\frac{5}{4}$ or -1.25

Exercise A3.1 Grade C

1 Solve

a $5x - 4 = 17$

b $3(2x - 4) = 17$

c $4x - 9 = 27$

d $10 - 4x = 11$

e $3(2 - x) = 19$

f $15 = 9 - 5y$

g $x + 2x - 7 - 8 = 12$

h $3 + 7x - 9 - 15x = 18$

i $17\frac{1}{4} = 18\frac{3}{4} - \frac{1}{2}x$

2 Solve

a $10 + 2x = 7x - 9$

b $5x - 4 = 12 - 2x$

c $2 - 3y = 9 - 8y$

d $5 - y = 10 - 2y$

e $3w + 9 = -2w - 8$

f $4(1 - 2x) = 2(3 - x)$

g $2(3y - 1) = 3(y - 1)$

h $3z + 2(4z - 2) = 5(3 - z)$

i $6(p - 1) + 5(4 - p) = 6p$

j $10q - (2q + 4) = 15$

k $2(r - 7) - (2r - 3) = 9(r - 2)$

3 For each problem, set up an equation and solve it to find the unknown.

a Find the length of the rectangle:

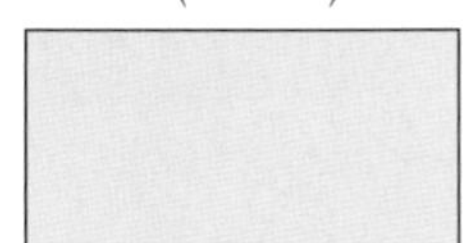

b Expression **A** is 10 more than twice expression **B**.
Find p.

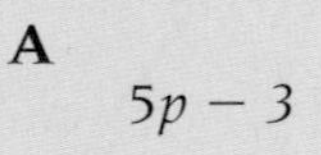

B
$3p - 4$

AO3 Problem

4 **a** In 10 years' time, my age will be double what it was 11 years ago. How old am I?

b Find the length of a square of side $2x + 3$ with an area equal to the area of a rectangle measuring $x + 6$ by $4x - 2$.

A3.2 Solving equations involving fractions

This spread will show you how to:

- Set up simple equations
- Solve linear equations, including equations in which the term including the unknown is negative or fractional

Keywords
Denominator
Equation
Solve

- To **solve equations** involving fractions, first clear the fractions by multiplying both sides of the equation by the denominator.

$$\frac{3x+2}{4} = 2x - 1$$ Multiply both sides by 4.

$$3x + 2 = 4(2x - 1)$$ $\cancel{4} \times \frac{3x+2}{\cancel{4}} = 3x + 2$

- For equations with fractions on both sides, multiply both sides of the equation by the product of the denominators.

$$\frac{2x-1}{5} = \frac{3x-4}{8}$$ Multiply both sides by 8 × 5.

$$8(2x - 1) = 5(3x - 4)$$ $8 \times \cancel{5} \times \frac{2x-1}{\cancel{5}} = \cancel{8} \times 5 \times \frac{3x-4}{\cancel{8}}$

This method is called 'cross-multiplying'.

- You can simplify equations involving several fractions by multiplying each term by every **denominator**.

Example

Solve

a $\frac{x}{5} - \frac{x}{6} = 10$

b $5 - \frac{5}{x} = 50$

a $\frac{x}{5} - \frac{x}{6} = 10$ Multiply each term by 5.

$x - \frac{5x}{6} = 50$ Multiply each term by 6.

$6x - 5x = 300$

$x = 300$

You could do this in one step by multiplying each term by 5 × 6 = 30.

b $5 - \frac{5}{x} = 50$ Add $\frac{5}{x}$ to both sides.

$5 = 50 + \frac{5}{x}$ Subtract 50 from each side.

$-45 = \frac{7}{x}$ Multiply each term by x.

$-45x = 5$

$x = -\frac{5}{45}$

Example

Solve $\frac{2m+1}{2} + \frac{3m-1}{3} = 2$

$\frac{2m+1}{2} + \frac{3m-1}{3} = 2$

$2m + 1 + \frac{2(3m-1)}{3} = 4$ Multiply by 2.

$3(2m + 1) + 2(3m - 1) = 12$ Multiply by 3.

$6m + 3 + 6m - 2 = 12$

$12m + 1 = 12$

$12m = 11 \Rightarrow m = \frac{11}{12}$

Exercise A3.2

Grade C

1 Solve these equations by clearing the fractions first.

a $\frac{x+7}{3} = \frac{2x-4}{5}$

b $\frac{5y-9}{2} = \frac{3-2y}{6}$

c $\frac{3z+1}{5} = \frac{2z}{3}$

d $\frac{p}{2} + \frac{p}{4} = 7$

e $\frac{4q}{3} - \frac{2q}{5} = 9$

f $\frac{3m}{4} + \frac{7m}{6} = \frac{1}{3}$

2 These equations have the same solution – true or false?

$\frac{x}{4} + \frac{2x}{5} = 10$ $\quad\quad$ $\frac{4x-2}{3} = \frac{3-8x}{5}$

3 These two rectangles have equal lengths. Find their areas.

5 mm — **Rectangle A** Area $= 2x + 3$

8 mm — **Rectangle B** Area $= 5x - 1$

4 A number is doubled and divided by 5. Three more than the number is divided by 8. Both calculations give the same answer. What is the number?

5 The sum of one-fifth of Lucy's age and one-seventh of her age is 12. Use algebra to work out Lucy's age.

6 These trapezia have equal areas. What are the lengths of the parallel sides in each?

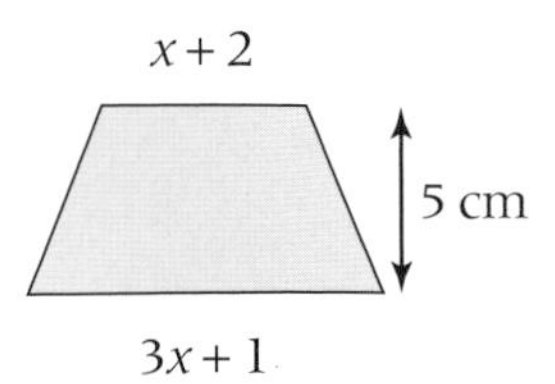

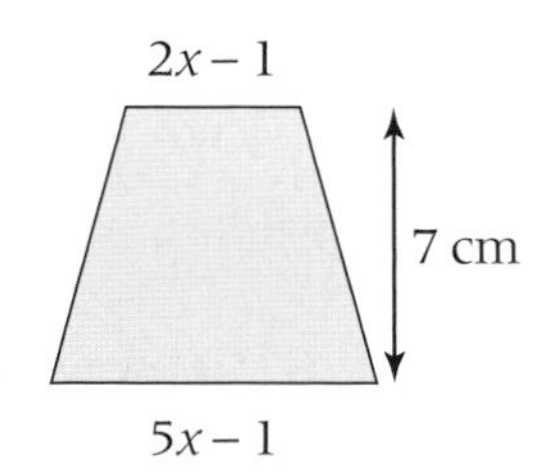

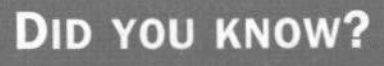

The word 'trapeze' originated from the trapezium shape formed by the ropes, crossbar and roof of the original trapeze.

Unit 3

A3.3 Further equations involving fractions

This spread will show you how to:

- Simplify equations involving fractions by using common denominators

Keywords
Cross-multiply
Solve

- You can **solve** equations involving fractions by simplifying first.

For equations of the form 'fraction = fraction' use cross-multiplication.

$$\frac{x+3}{4} = \frac{2x-1}{7}$$

$$7(x + 3) = 4(2x - 1)$$

For equations involving several fractions, multiply each term by the product of all the denominators.

$$\frac{x}{3} + \frac{x}{4} = 2$$

$$3 \times 4 \times \frac{x}{3} + 3 \times 4 \times \frac{x}{4} = 3 \times 4 \times 2$$

$$4x + 3x = 24$$

- You can simplify sums (or differences) of fractions to a single fraction and then **cross-multiply**.

Example

Solve the equation $\frac{x+3}{6} + \frac{x+5}{5} = 4$.

$$\frac{x+3}{6} + \frac{x+5}{5} = 4$$

$$5 \times 6 \times \frac{(x+3)}{6} + 5 \times 6 \times \frac{(x+5)}{5} = 5 \times 6 \times 4$$

$$5x + 15 + 6x + 30 = 120$$

$$11x + 45 = 120 \Rightarrow x = \tfrac{75}{11}$$

Example

By simplifying the equation $\frac{5}{x+2} - 16 = \frac{3}{x-4}$, show that $8x^2 - 17x - 51 = 0$.

$$\frac{5}{x+2} - 16 = \frac{3}{x-4}$$

$$\frac{5}{x+2} - \frac{3}{x-4} = 16$$

Collect the fractions on one side of the equation.

$$\frac{5(x-4)}{(x+2)(x-4)} - \frac{3(x+2)}{(x+2)(x-4)} = 16$$

Convert to fractions with a common denominator.

$$\frac{5x - 20 - 3x - 6}{(x+2)(x-4)} = 16$$

$$\frac{2x - 26}{(x+2)(x-4)} = 16$$

Cross-multiply.

$$2(x - 13) = 16(x + 2)(x - 4)$$

$$x - 13 = 8(x^2 - 2x - 8)$$

Divide both sides by 2.

$$x - 13 = 8x^2 - 16x - 64$$

$$8x^2 - 17x - 51 = 0$$

Rearrange into the required form.

Exercise A3.3 — Grade A*

1 Simplify the left-hand side of each original equation in the table and then show that it can be transformed to the new equation.

Original equation	Transform to
$\frac{2}{x} + \frac{2}{x+1} = 3$	$(3x + 2)(x - 1) = 0$
$\frac{3}{y-1} + \frac{3}{y+1} = 4$	$(y - 2)(2y + 1) = 0$
$\frac{4}{m+1} + \frac{2}{m-2} = 3$	$m^2 = 3m$
$\frac{3}{x-4} + \frac{5}{(x-4)(x+5)} = 10$	$10x^2 = 220 - 7x$
$\frac{6}{(x^2-4)} + \frac{3}{(x-2)} = 5$	$3x + 32 = 5x^2$
$\frac{5}{x^2+5x+6} + \frac{7}{x^2+8x+15} = 1$	$x^3 + 10x + 19x^2 - 9 = 0$

2 Solve these equations in two ways:

i by writing the left-hand side as a single fraction, then using cross-multiplication

ii by multiplying through by each denominator in turn and simplifying.

Check that both methods give the same solution.

a $\frac{(x+3)}{5} + \frac{(2x-1)}{4} = 12$ **b** $\frac{2y-3}{4} - \frac{3y-8}{2} = 20$

You need to find a **common denominator**.

AO3 Problem

3 Here is an equation:

$$\frac{2}{x+1} - \frac{5}{x+2} = \frac{3}{x+3}$$

Decide which of the following equations are possible as a next step in simplifying the original equation.

$$2(x + 2)(x + 3) - 5(x + 1)(x + 3) = 3(x + 1)(x + 2)$$

$$2x^2 + 10x + 12 - 5x^2 - 20x - 15 = 3x^2 + 9x + 6$$

$$\frac{x+1}{2} - \frac{x-2}{5} = \frac{x+3}{3}$$

A3

Summary

Check out

You should now be able to:

- Set up and solve simple equations
- Solve linear equations where the unknown appears on either side or on both sides of the equation
- Solve linear equations, including equations in which the term including the unknown is negative or fractional
- Simplify equations involving fractions by using common denominators
- Simplify algebraic expressions by cancelling factors

Worked exam question

a Solve $\frac{40 - x}{3} = 4 + x$ (3)

b Simplify fully $\frac{4x^2 - 6x}{4x^2 - 9}$ (3)

(Edexcel Limited 2004)

a

$$\frac{40 - x}{3} = 4 + x$$
$$40 - x = 3(4 + x)$$
$$40 - x = 12 + 3x$$
$$40 = 12 + 3x + x$$
$$40 = 12 + 4x$$
$$40 - 12 = 4x$$
$$28 = 4x$$
$$x = 7$$

Multiply both sides of the equation by 3.

Show each line of working out.

b

$$\frac{4x^2 - 6x}{4x^2 - 9} = \frac{2x(2x - 3)}{(2x - 3)(2x + 3)}$$
$$= \frac{2x}{(2x + 3)}$$

Write down the factorisation of each quadratic expression.

Exam questions

1 $A = \frac{h(12 + b)}{2}$

$A = 54$

$h = 8$

Work out the value of b. (3)

2 Solve the equation

$$\frac{3}{x + 3} - \frac{4}{x - 3} = \frac{5x}{x^2 - 9}$$ (4)

(Edexcel Limited 2007)

3 Solve the equation

$$\frac{x}{2x - 3} + \frac{4}{x + 1} = 1$$ (5)

(Edexcel Limited 2007)

G2

Transformations and congruence

When you play a computer game your character is able to move around the screen because of mathematical transformations. Combinations of these transformations, often taking place in 3D worlds, allow the characters to move in many different ways.

What's the point?

Transformations move points and shapes from one place to another. Mathematicians use transformations not just to move shapes but also to move graphs and statistics. This helps them match the power of mathematics to real life situations.

Transformations and congruence

Check in

You should be able to

- **apply basic transformations and recognise congruent shapes**

1 a Copy the diagram and reflect shape A in the line $y = 1$. Label the image B.

b Reflect shape B in the line $y = x + 1$. Label the image C.

c Translate the shape C by $\begin{pmatrix} -3 \\ 1 \end{pmatrix}$. Label the image D. What single transformation maps shape A to shape D?

d Enlarge the original shape A by scale factor 2, centre of enlargement (–1, –1). Label the image E.

e Which of the shapes are congruent?

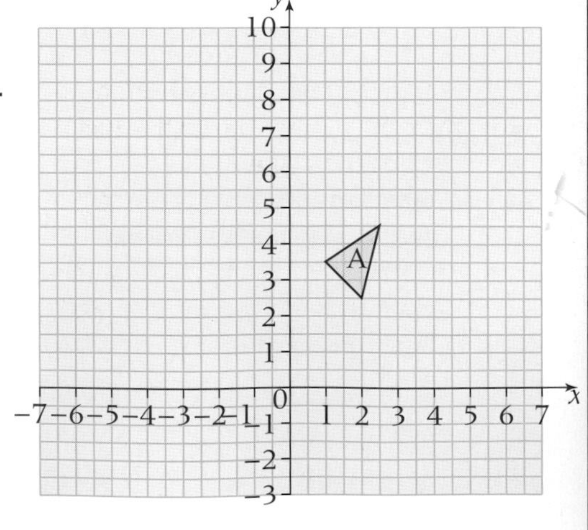

Orientation

What I need to know

KS3 Knowledge of basic transformation, constructions and ratio

G1 How to construct a geometric proof

What I will learn

- Describe and combine transformations including enlargements
- Use congruence in proofs
- Recognise and use similarity

What this leads to

G3 Similarity for areas and volumes
G4 Trigonometric ratios
G6 Vectors
A8 Transforming graphs

Computer graphics

Rich task

A shape (object) is reflected in the line $y = x$. To find the new coordinates of each vertex of the shape (image), simply swap the x and y coordinates around. Investigate rules for finding the image coordinates for this and other transformations.

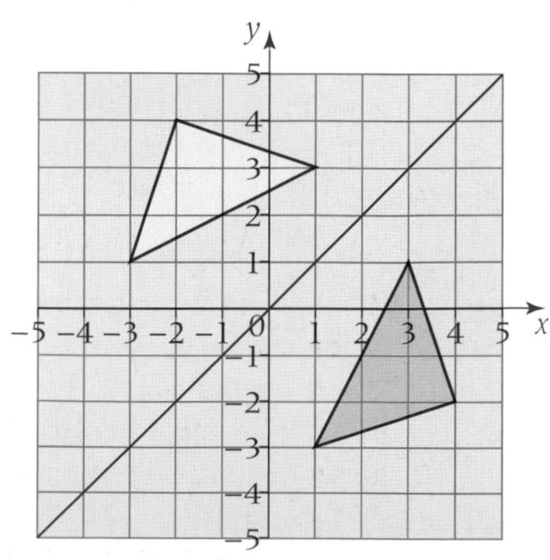

G2.1 Describing transformations

This spread will show you how to:

- Describe reflections, using mirror lines
- Understand that rotations are specified by a centre and an (anticlockwise) angle
- Describe translations by giving a distance and direction (or vector)

Keywords
Maps
Reflection
Rotation
Transformation
Translation

- To describe a **reflection**, you give the equation of the line.
- To describe a **rotation**, you give the centre and the angle of rotation.
- To describe a **translation**, you give the distance and direction or you specify the vector.

p.366

Reflections, rotations and translations are all **transformations**.

Example

Describe the transformation that **maps** shape A on to

a shape B **b** shape C **c** shape D.

Maps means changes.

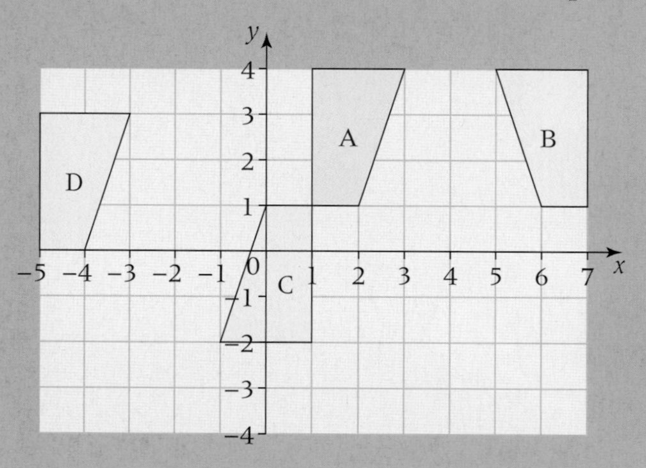

a In a reflection the mirror line bisects the line joining corresponding points on the object and image.
Shape B is a reflection of shape A in the line $x = 4$.

b The vertex (1, 1) does not move during the rotation, so it must be the centre of rotation.
Shape C is a rotation of shape A through 180°.

c Shape D is a translation of shape A by the vector $\begin{pmatrix} -5 \\ -1 \end{pmatrix}$.

Exercise G2.1 Grade C

1 Describe fully the transformation that maps
- **a** shape A onto shape B
- **b** shape A onto shape C
- **c** shape A onto shape D
- **d** shape B onto shape D
- **e** shape C onto shape D
- **f** shape D onto shape C.

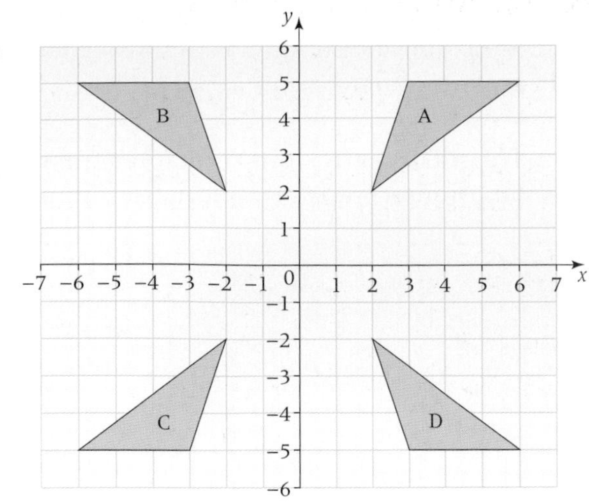

2 Describe fully the transformation that maps
- **a** shape J onto shape K
- **b** shape L onto shape K
- **c** shape M onto shape K
- **d** shape L onto shape M
- **e** shape J onto shape M
- **f** shape M onto shape J.

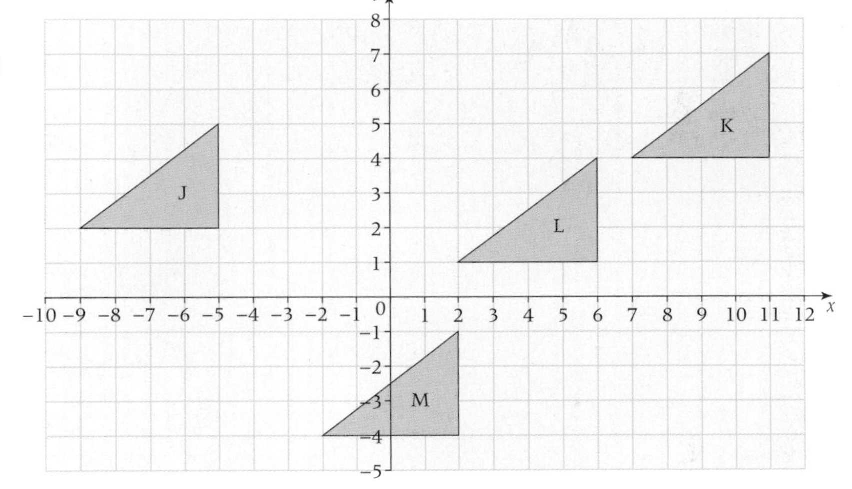

3 Describe fully the transformation that maps
- **a** shape W onto shape X
- **b** shape W onto shape Y
- **c** shape W onto shape Z
- **d** shape Z onto shape W
- **e** shape X onto shape Y
- **f** shape Z onto shape X.

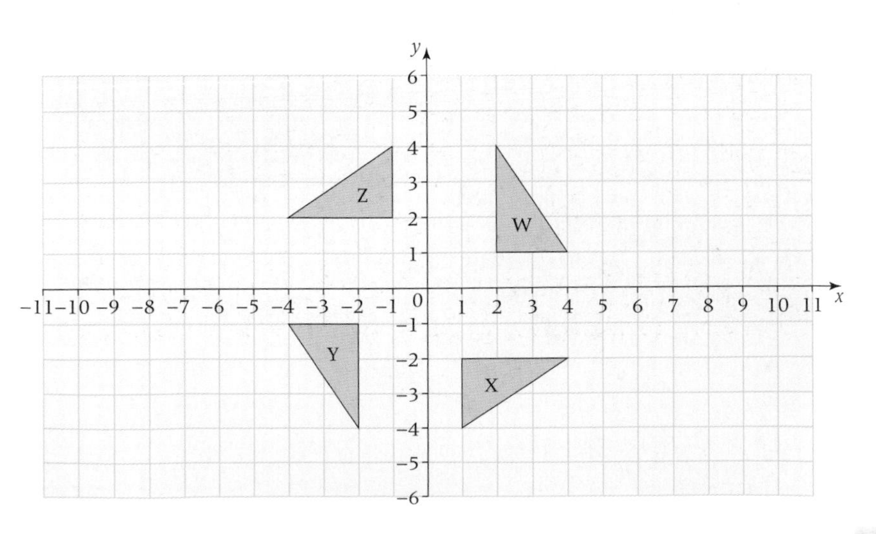

Combining transformations

This spread will show you how to:

- Transform 2-D shapes by translation, rotation and reflection and combinations of these transformations

Keywords
Reflection
Rotation
Transformations
Translation

You can combine **transformations** by doing one after the other.
You can describe a combination of transformation as a single transformation.

Example

In this diagram, triangle A undergoes three pairs of transformations.

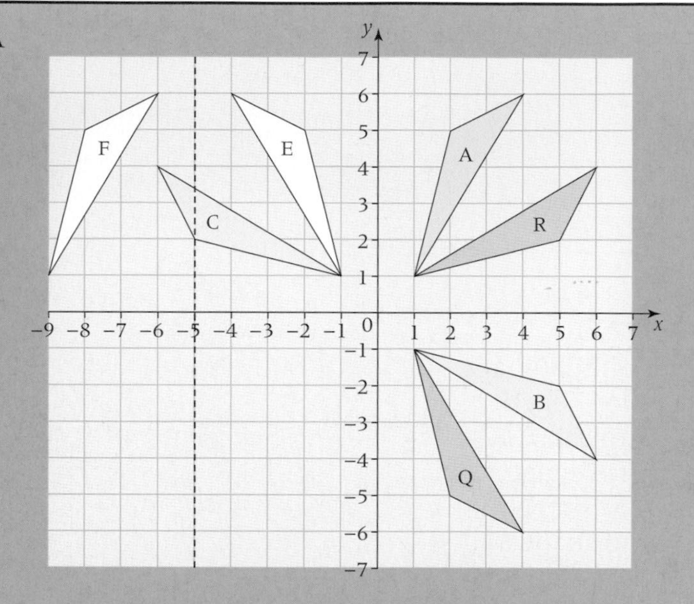

a Triangle A is rotated 90° clockwise about (0, 0) to triangle B.
Then triangle B is rotated 180° about (0, 0) to triangle C.
What single transformation maps triangle A onto triangle C?

b Triangle A is reflected in the line $y = 0$ (the x-axis) to triangle Q.
Then triangle Q is rotated through 90° anticlockwise about (0, 0) to triangle R.
What single transformation maps triangle A onto triangle R.

c Triangle A is reflected in the line $x = 0$ (the y-axis) to triangle E.
Then triangle E is reflected in the line $x = -5$ to triangle F.
What single transformation maps triangle A onto triangle F?

---

a A rotation of 90° anticlockwise about (0, 0) maps A onto C.

- A combination of rotations that have the same centre is equivalent to a single **rotation**.

b A reflection in the line $y = x$ maps A onto R.

- A combination of a reflection and a rotation is equivalent to a single **reflection**.

c A translation by the vector $\begin{pmatrix} -10 \\ 0 \end{pmatrix}$ maps A onto F.

- A combination of reflections is a **translation** when the mirror lines are parallel otherwise it is equivalent to a single rotation.

Exercise G2.2 Grade C

1 Copy this diagram.

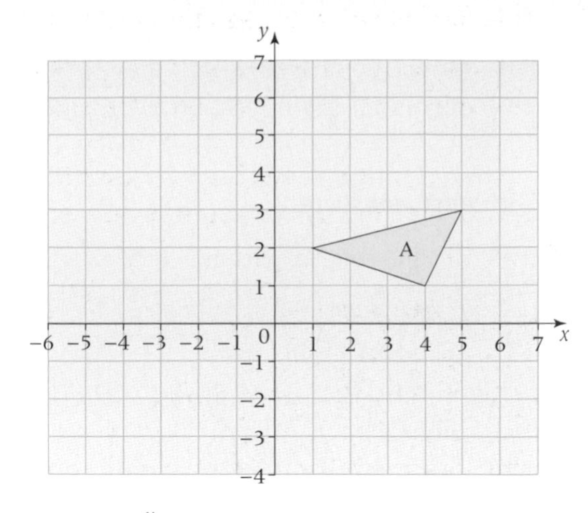

a Rotate triangle A 90° anticlockwise about centre (0, 0). Label the image B.

b Rotate triangle B 90° anticlockwise about centre (0, 0). Label the image C.

c Describe fully the single transformation that takes triangle A to triangle C.

d Rotate triangle B 90° clockwise about centre (2, 1). Label the image D.

e Describe fully the single transformation that takes triangle A to triangle D.

2 Copy this diagram.

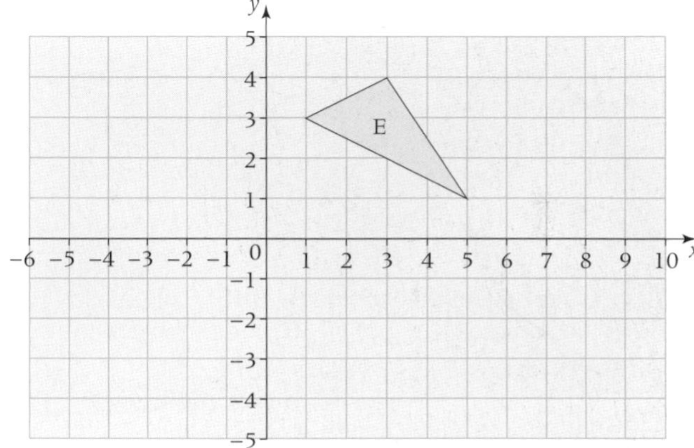

a Reflect triangle E in the y-axis. Label the image F.

b Reflect triangle F in the x-axis. Label the image G.

c Describe fully the single transformation that takes triangle G to triangle E.

d Reflect triangle F in the line $x = 2$. Label the image H.

e Describe fully the single transformation that takes triangle E to triangle H.

3 Copy this diagram.

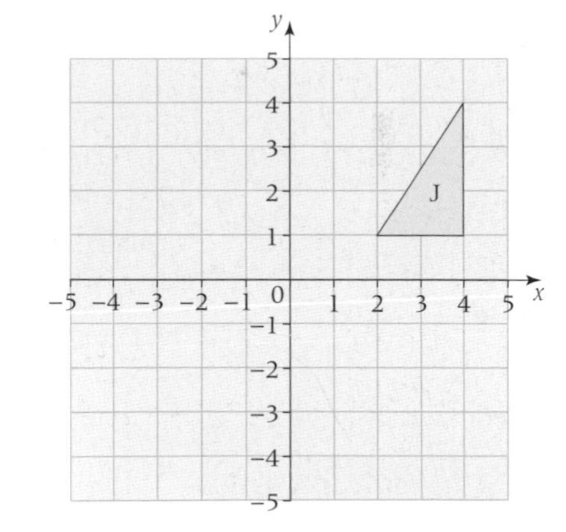

a Reflect triangle J in the x-axis. Label it K.

b Rotate triangle K 180° about centre (0, 0). Label it L.

c Describe fully the single transformation that takes triangle L to triangle J.

4 Copy this diagram.

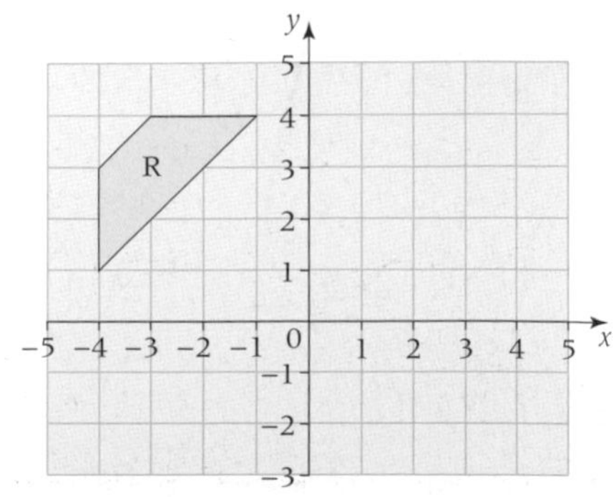

a Translate trapezium R by the vector $\begin{pmatrix} 5 \\ -3 \end{pmatrix}$. Label the image S.

b Translate trapezium S by the vector $\begin{pmatrix} -1 \\ 2 \end{pmatrix}$. Label the image T.

c Describe fully the single transformation that takes trapezium T to trapezium R.

G2.3 Congruence

This spread will show you how to:
- Understand congruence in the context of transforming 2-D shapes
- Recognise the properties of congruent triangles

Keywords
Congruent
Hypotenuse

- In **congruent** shapes:
  - corresponding lengths are equal
  - corresponding angles are equal.

Congruent shapes are rotations, reflections or translations of each other.

- You can prove that two triangles are congruent by showing they satisfy one of four sets of conditions.

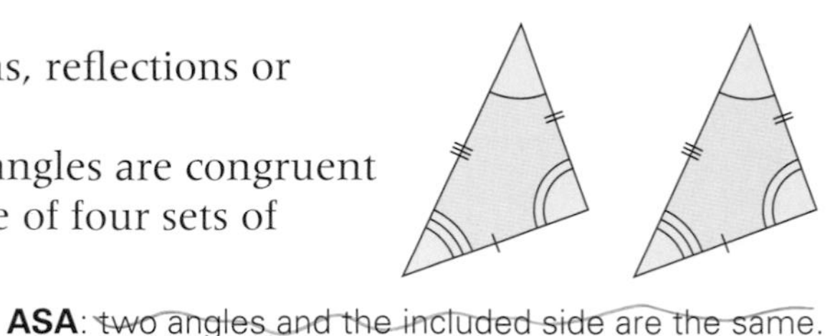

Rotation, reflection and translation preserve lengths and angles.

SSS: three sides are the same.

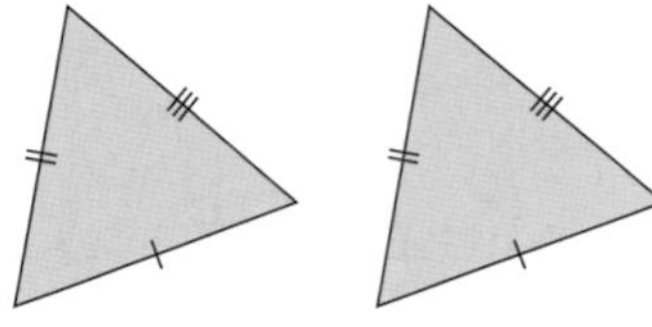

ASA: ~~two angles and the included side are the same.~~

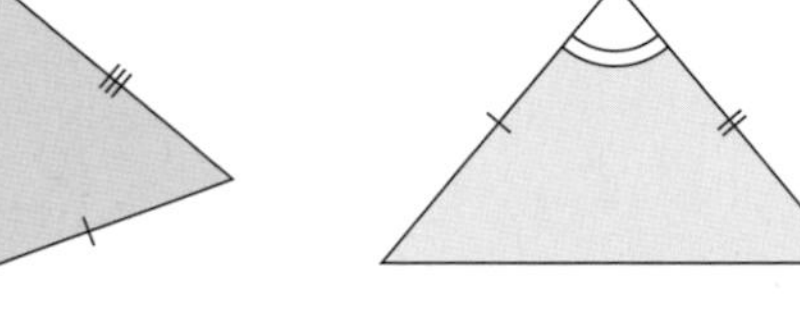

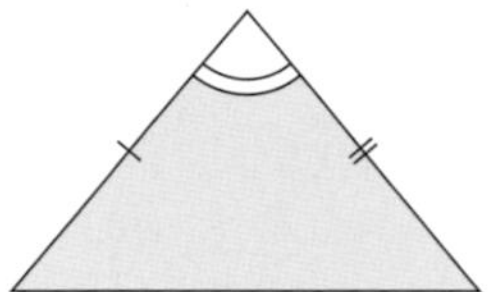

ASA: two angles and the included side are the same.

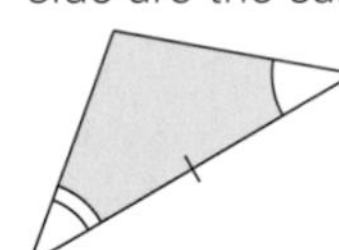

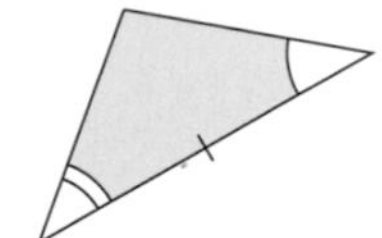

RHS: Right-angled triangles with **hypotenuse** and one other side the same.

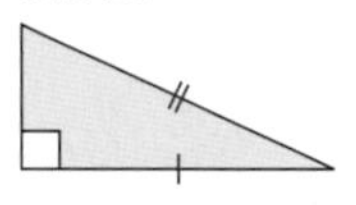

Example

Explain why the following pairs of triangles are congruent.

a 7cm, 3cm; 3cm, 7cm

b 32°, 6 cm; 74°, 6 cm

a The triangles are both right-angled.
The hypotenuses are both 7 cm.
One other side is the same, 3 cm.
The triangles are congruent by RHS.

b The triangles are isosceles.
The angles in the triangles are:

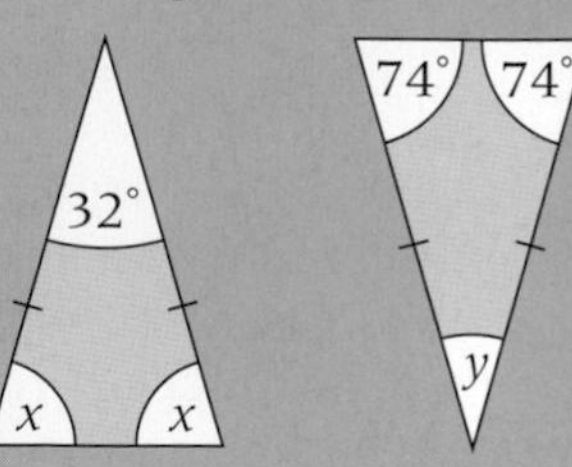

ASA (32°, 6 cm, 74°) so the triangles are congruent.

Base angles in an isosceles triangle are equal.
$2x + 32° = 180°$
$x = \frac{148}{2} = 74°$
$y + 2 \times 74° = 180°$
$y = 32°$

You could also use: SAS (6 cm, 32°, 6 cm)

Exercise G2.3 **Grade B/A**

1 Explain whether or not these pairs of triangles are congruent.

a

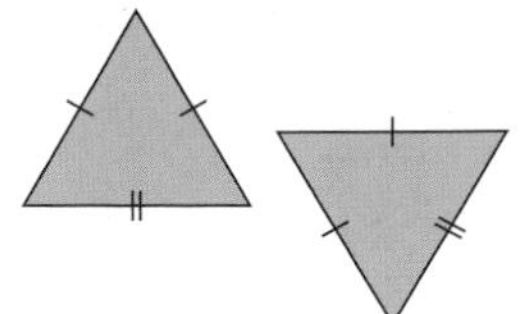

b

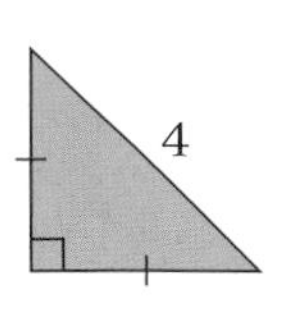

c

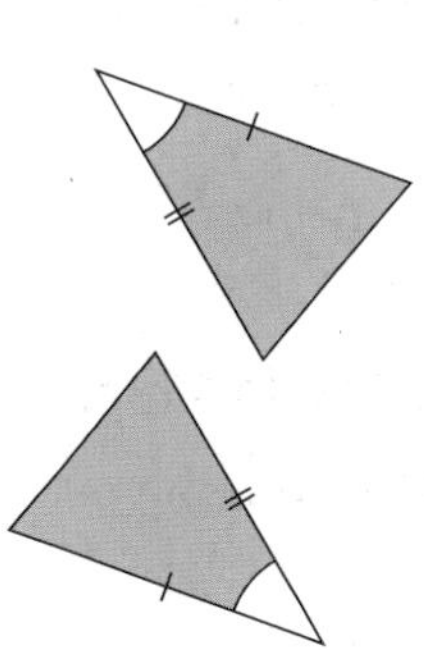

2 ABCD is a rectangle.

Prove that triangles ABD and CDB are congruent.

3 Explain why triangles PQR and WXT are congruent.

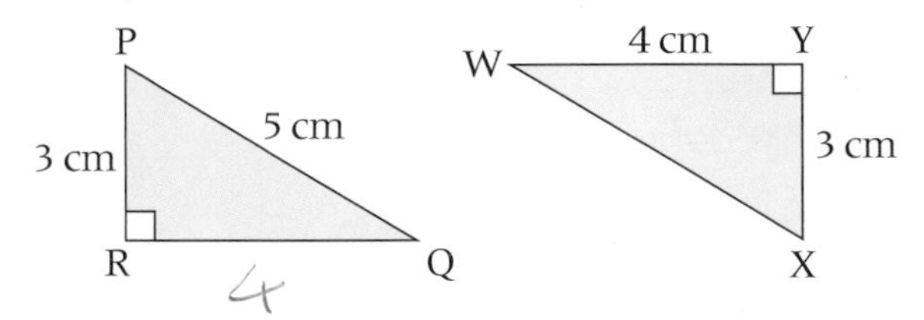

4 KLMN is a kite.

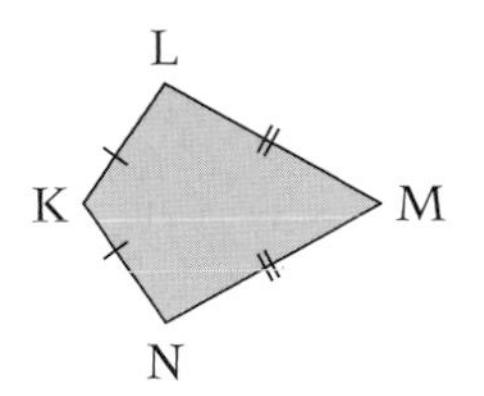

a Explain why triangles KLN and MLN are not congruent.
b Explain why triangles KLM and KNM are congruent.

5 EFGH is a parallelogram.

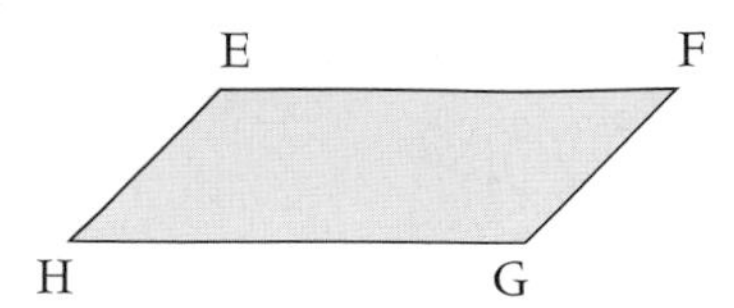

a Find two different pairs of congruent triangles.
b Explain how you know they are congruent.

G2.4 Congruence and proof

This spread will show you how to:

- Use congruent triangles to prove geometrical results and constructions

Keywords
Congruent
Construction

You can use **congruent** triangles to prove geometrical results.

Proof

Prove that two tangents drawn from a point to a circle are equal in length.

Tangents from P touch the circle with centre O at X and Y.

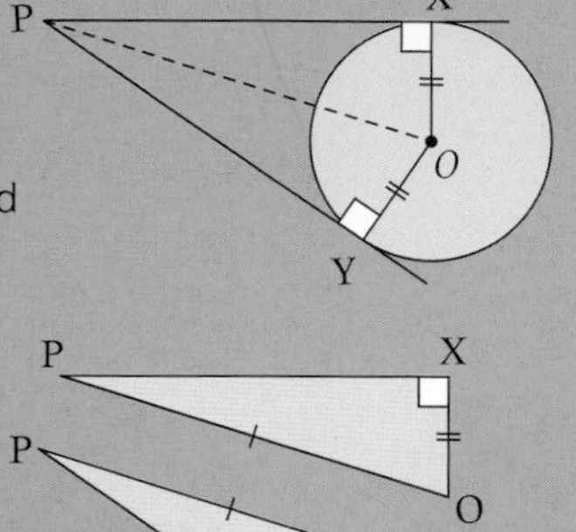

Start with a sketch and mark on all the facts you know.

OX = OY — Radii of circle.

$\angle OXP = \angle OYP = 90°$ — Angle between tangent and radius is a right angle.

p.156

Split the kite PXOY into two triangles.
The hypotenuse of each triangle is OP.
The triangles are congruent by RHS.
So PX = PY.

- You can use congruent triangles to prove constructions.

Example

The diagram shows the **construction** of the bisector of $\angle PQR$.

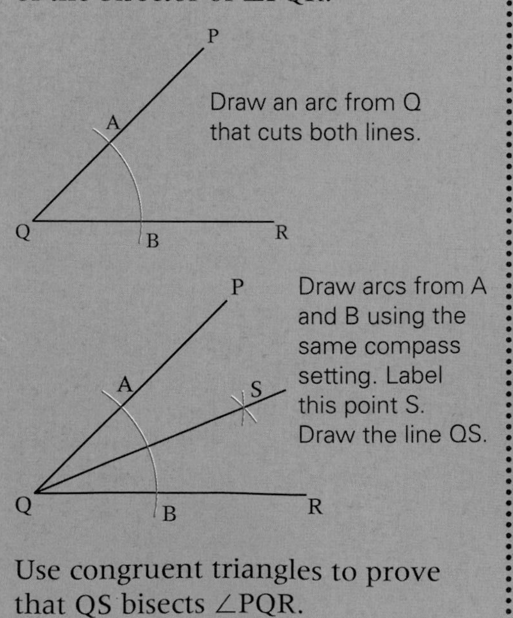

Use congruent triangles to prove that QS bisects $\angle PQR$.

Draw in the lines AS and BS.

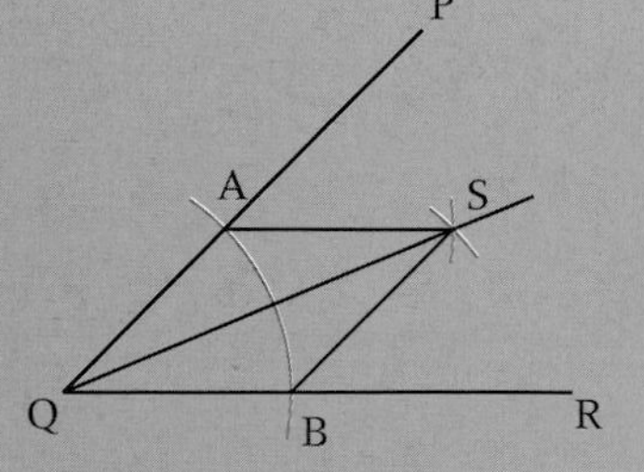

This is the standard method for constructing an angle bisector.

Bisect means 'cut in half'.

Compare the two triangles QAS and QBS.

QA = QB — Radii of circle.

AS = BS — Radii of circle.

QS = QS — Side in common.

Triangles QAS and QBS are congruent by SSS.

So $\angle AQS = \angle BQS$, which means that QS bisects $\angle PQR$.

Exercise G2.4

Grade A*

1 EFGH is a parallelogram.
The diagonals EG and FH meet at the point M.

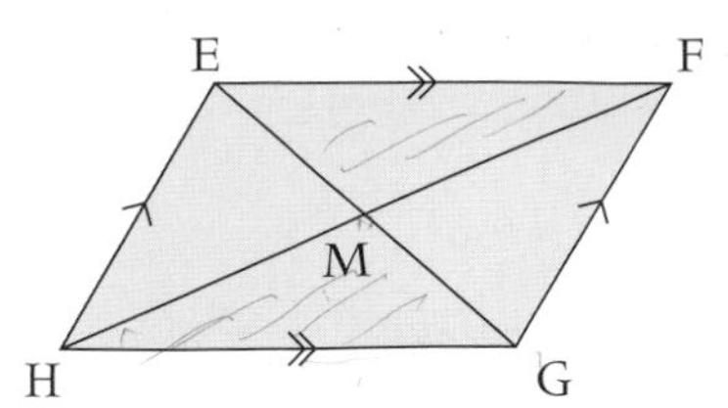

a Prove that triangles MEF and MGH are congruent.
b Hence prove that M is the midpoint of EG.

Use alternate angles.

2 RSTU is a rhombus.
O is the point where the diagonals cross.
a Prove that the triangles ORS, OTU, ORU and OTS are all congruent.
b Hence prove that diagonals RT and SU cross at right angles.

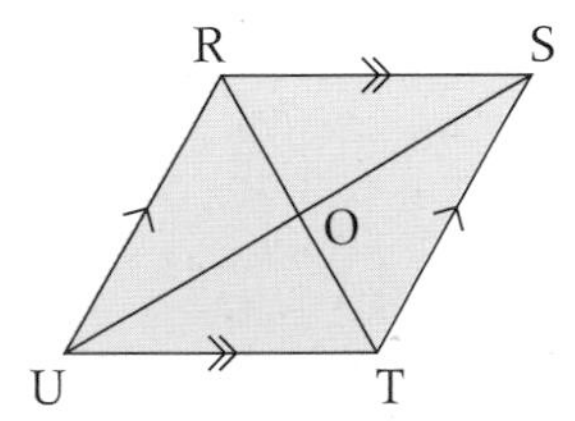

3 The diagram shows the construction of the line XY using arcs of equal radius centred on A and on B.

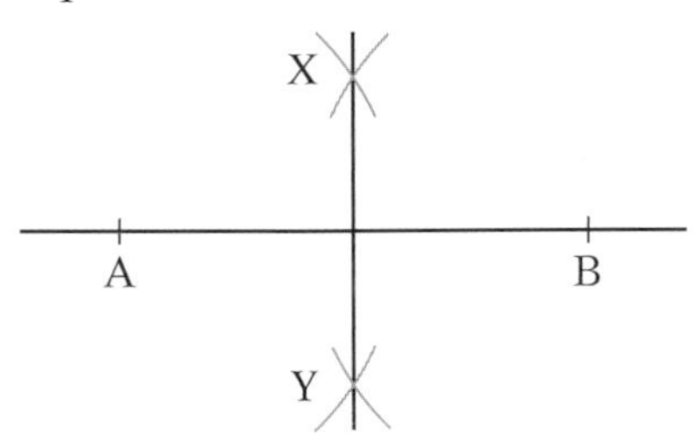

Use congruent triangles to explain why the line XY is the perpendicular bisector of AB.

4 XS and XT are equal chords of a circle with centre O.
Using congruent triangles, prove that OX bisects angle SXT.

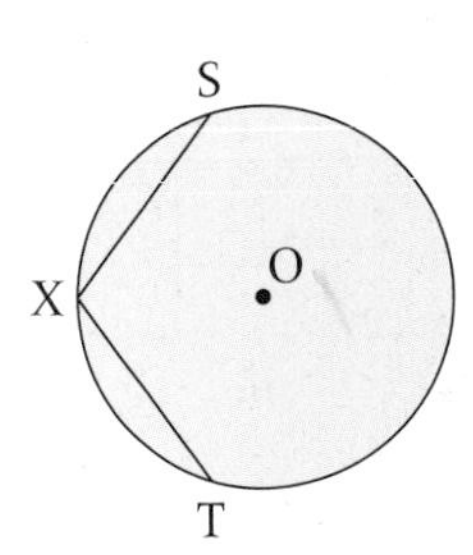

AO3 Problem

5 Prove that the perpendicular line from the centre of a circle to a chord bisects the chord.

The diagram for this proof is given in G 2.3

G2.5 Enlargement

This spread will show you how to:

- Recognise and construct enlargements
- Understand the significance of the scale factor of an enlargement

Keywords
Enlargement
Ratio

- In an **enlargement**:
  - corresponding angles are the same
  - corresponding lengths are in the same **ratio**.

Enlargement of scale factor 2.

14 cm

10 cm

28 cm

20 cm

Enlargement with a fractional scale factor reduces the size of the shape.

object

image

Enlargement scale factor $\frac{1}{2}$, centre (0, 0)

Centre of enlargement

Scale factor $\frac{1}{2}$: all lengths on the image are half the corresponding lengths on the object.

Enlargement with a negative scale factor produces a shape upside down on the opposite side of the centre.

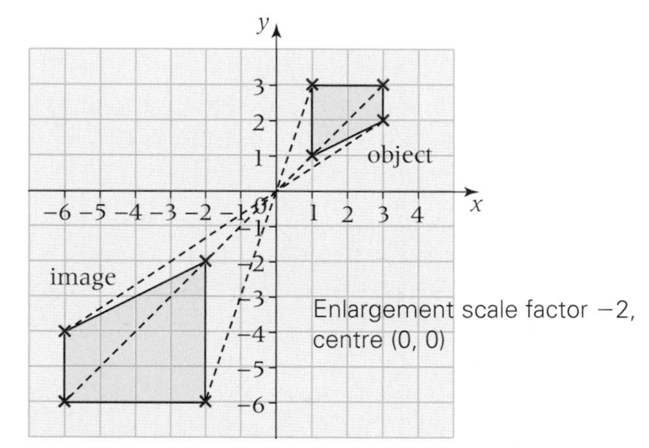

Scale factor −2 : all lengths on the image are twice the corresponding lengths on the object; the image is inverted.

- To describe an enlargement, you give the scale factor and the centre of enlargement.

Example

Describe fully the enlargement that maps ABCD on to A′B′C′D′.

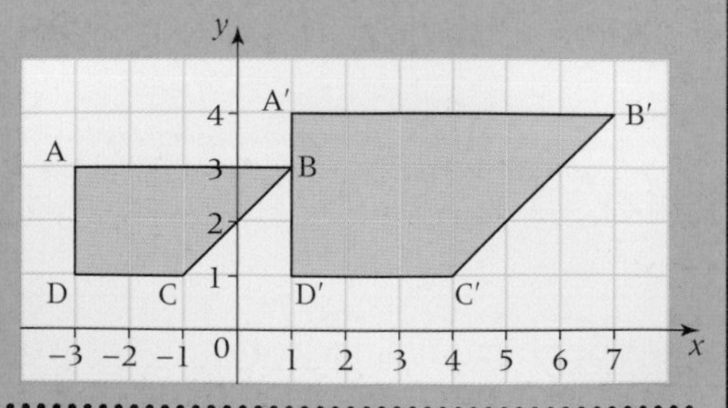

p.264

The ratio of corresponding sides is $= \frac{A'D'}{AD} = \frac{3}{2}$

The scale factor of the enlargement is $\frac{3}{2}$.

Join corresponding points on the image and object. They cross at the centre of enlargement, (−11, 1).

The enlargement has scale factor $\frac{3}{2}$, centre (−11, 1).

Check: all pairs of corresponding sides should give the same ratio.

Exercise G2.5

1 **a** Copy this diagram.

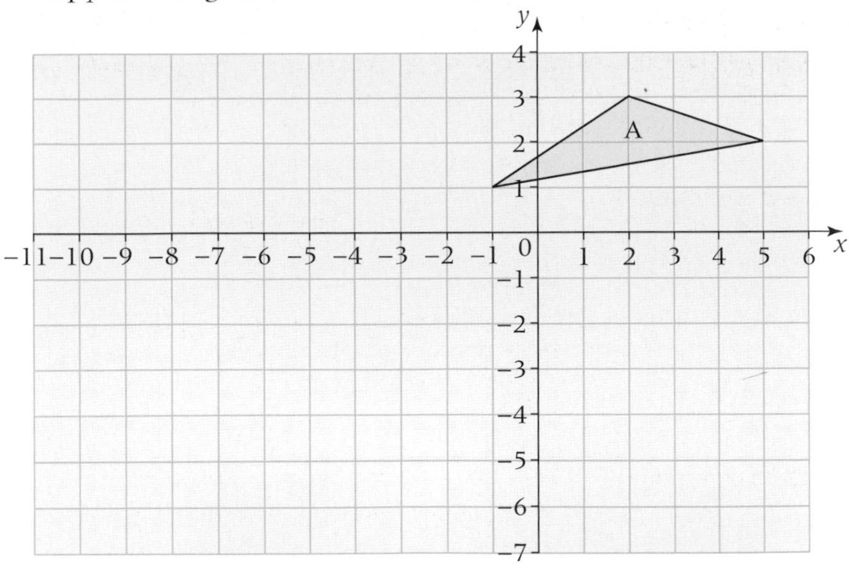

b Enlarge triangle A by scale factor −1, centre (0, 0). Label the image B.
c Enlarge triangle A by scale factor −2, centre (0, 0). Label the image C.
d What can you say about triangles A and B?

2 **a** Draw an x-axis from −13 to 5 and a y-axis from −7 to 3.
Draw triangle D with vertices at (−1, 1), (2, 2) and (4, 0).
b Enlarge triangle D by scale factor −3, centre (0, 0). Label the image E.
c Enlarge triangle D by scale factor $-\frac{1}{2}$, centre (0, 0). Label the image F.
d How many times longer are the side lengths of triangle E than triangle D?

3 **a** Draw an x-axis from −7 to 6 and a y-axis from −9 to 8.
Draw triangle J with vertices at (−2, 4), (−2, 7) and (4, 4).
b Enlarge triangle J by scale factor $-1\frac{1}{2}$, centre (1, 1). Label the image K.
c Enlarge triangle J by scale factor $-\frac{1}{3}$, centre (1, 1). Label the image L.
d How many times smaller are the side lengths of triangle L than triangle J?

4 **a** Draw x- and y-axis from −7 to 5.
Draw triangle X with vertices at (−1, 1), (0, 3) and (3, 2).
b Enlarge triangle X by scale factor −2, centre (0, 0). Label the image Y.
c Enlarge triangle Y by scale factor $-1\frac{1}{2}$, centre (0, 0). Label the image Z.
d Describe the enlargement that will transform triangle X to triangle Z.
e Describe the enlargement that will transform triangle Z to triangle X.
f Comment on your answers to **d** and **e**.

G2.6 Similar shapes

This spread will show you how to:

- Understand similarity of 2-D shapes, using this to find missing lengths and angles

Keywords
Ratio
Similar

In an enlargement, the object and image are **similar**.

p.254

- In similar shapes:
  - corresponding angles are equal
  - corresponding sides are in the same **ratio**.

You can use ratio to find missing lengths in similar shapes.

Example

These two pentagons are similar.
Work out the side lengths *a* and *b*.

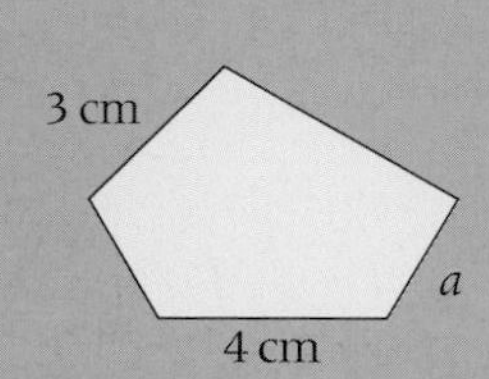

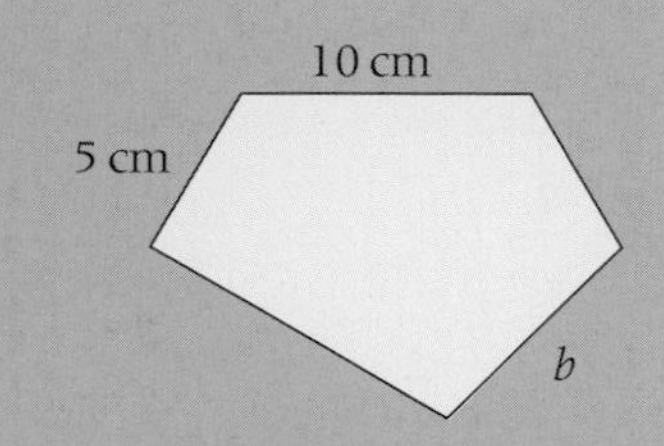

Side 4 cm corresponds with side 10 cm.
The ratio of the sides is 4 : 10 or $\frac{4}{10}$. Side *a* cm corresponds to side 5 cm. So $\frac{a}{5} = \frac{4}{10}$

$$a = \frac{5 \times 4}{10} = 2\text{ cm}$$

Side *b* cm corresponds to side 3 cm.

So $\frac{3}{b} = \frac{4}{10}$ Invert both sides of the equation.

$\frac{b}{3} = \frac{10}{4}$

$b = \frac{10 \times 3}{4} = 7.5\text{ cm}$

Sketch the pentagons the same way up, to help you identify corresponding sides and angles.

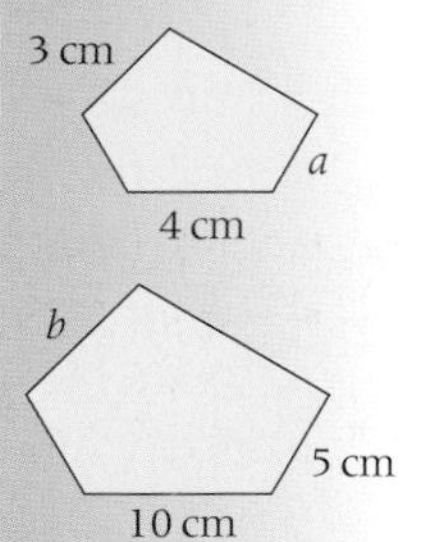

Write all the ratios in the correct order. These are all smaller : larger.

Example

Find the length of QS.

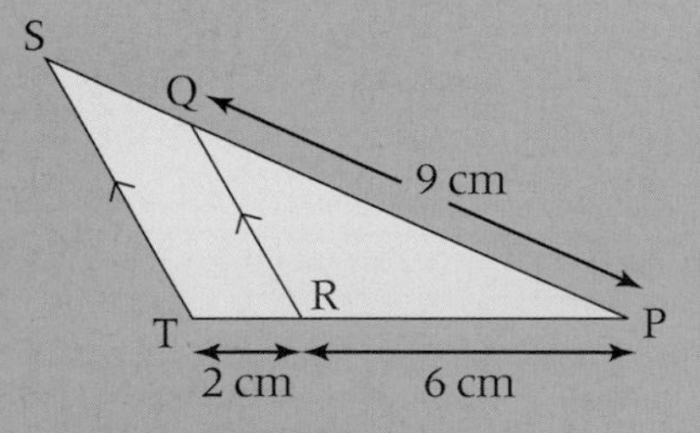

$\angle QPR = \angle SPT$

$\angle QRP = \angle STP$ Corresponding angles TS || RQ.

$\angle PQR = \angle PST$ Corresponding angles TS || RQ.

So triangle PQR is similar to triangle PST.

PT corresponds to PR.

PT : PR $= \frac{PT}{PR} = \frac{8}{6}$

So $\frac{PS}{PQ} = \frac{8}{6} = \frac{PS}{9}$

$PS = \frac{9 \times 8}{6} = 12$

QS = PS − PQ = 12 − 9 = 3 cm

PST is an enlargement of PQR, centre P.

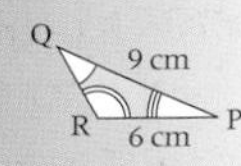

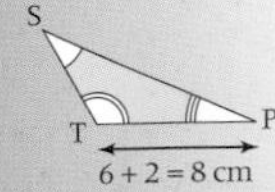

The symbol || means ‘parallel to’.

Exercise G2.6 **Grade C/B**

1 The two trapezia are similar.
Find the lengths a and b.

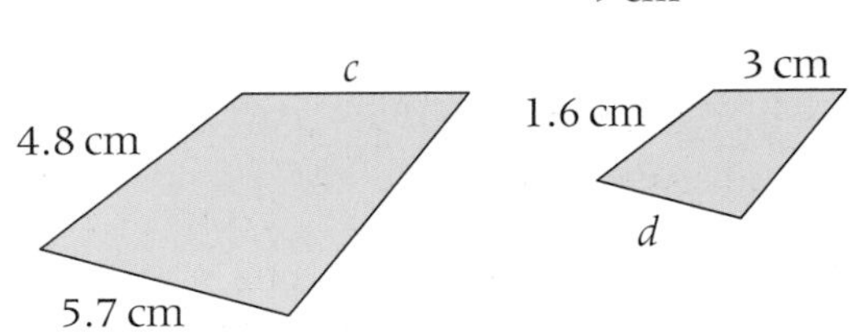

2 The two quadrilaterals are similar.
Find the lengths c and d.

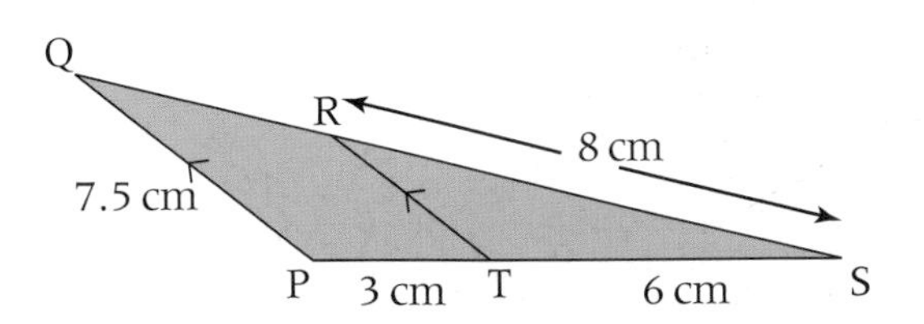

3 In the diagram PQ is parallel to TR.
Work out the lengths RT, QR and QS.

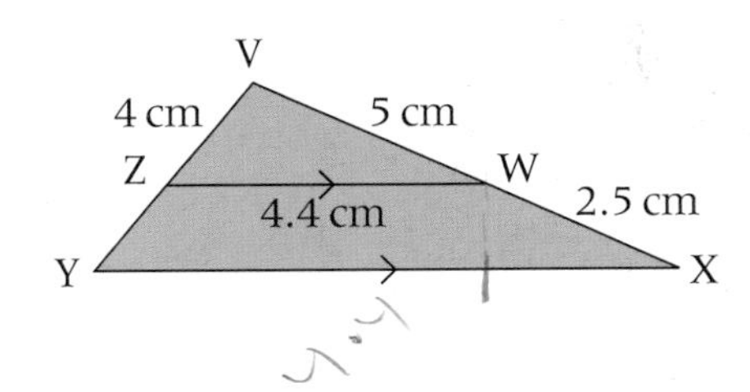

4 In the diagram WZ is parallel to XY.
a Work out the lengths XY and VY.
b Work out the perimeter of the trapezium WXYZ.

5 KN is parallel to LM.

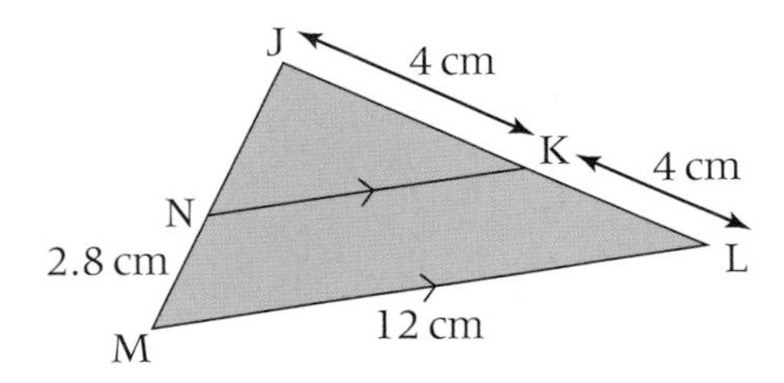

Work out the perimeters of triangle JKN and trapezium KLMN.

6 A and B are similar circles.

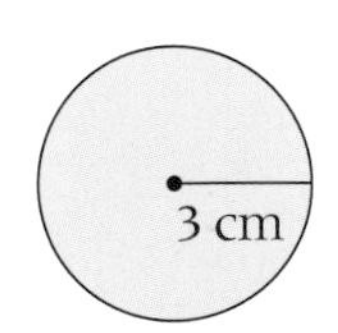

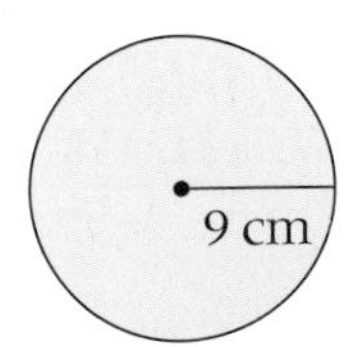

a What is the ratio of their radii?
b What is the ratio of their circumferences?
c Explain why all circles are similar.

Unit 3

G2 Summary

Check out

You should now be able to:

- Describe and transform 2-D shapes using reflections, rotations, translations and a combination of these transformations
- Understand congruence triangles and prove geometrical results and constructions
- Use congruence to show that reflections, rotations and translations preserve length and angle
- Enlarge 2-D shapes using a centre of enlargement and a positive, fractional or negative scale factor
- Understand similarity of 2-D shapes, using this to find missing lengths and angles

Worked exam question

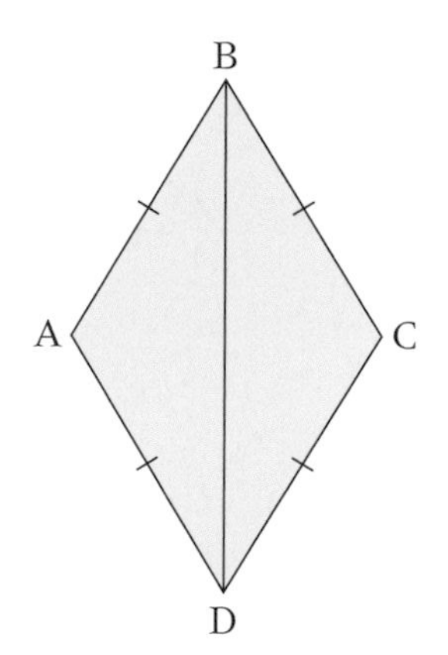

Diagram NOT accurately drawn.

In the diagram $AB = BC = CD = DA$
Prove that triangle ADB is congruent to triangle CDB.

(Edexcel Limited 2008)

$AD = CD$
$AB = CB$
BD is common

Triangle ADB is congruent to triangle CDB by SSS

There must be 3 statements.

The reason SSS must be stated.

Exam questions

1

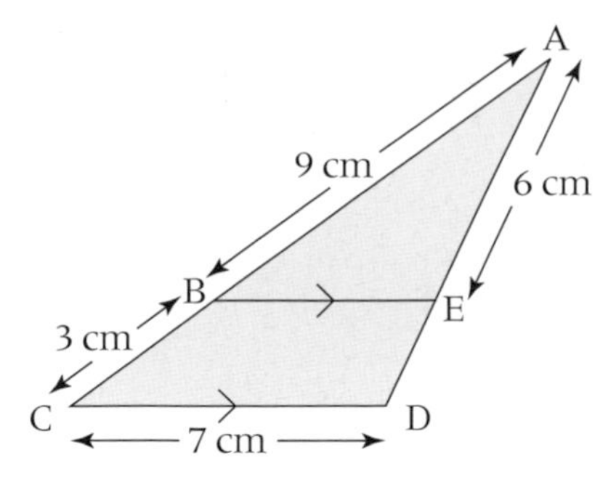

Diagram NOT accurately drawn.

BE is parallel to *CD*.

$AB = 9\,\text{cm}$, $BC = 3\,\text{cm}$, $CD = 7\,\text{cm}$, $AE = 6\,\text{cm}$.

a Calculate the length of *ED*. (2)

b Calculate the length of *BE*. (2)

(Edexcel Limited 2005)

2

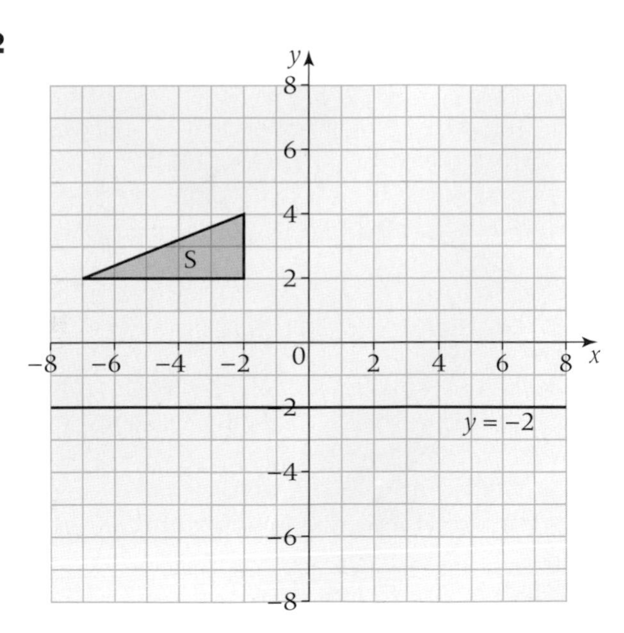

Triangle **S** is reflected in the *y*-axis to give triangle **T**.
Triangle **T** is reflected in the $y = -2$ to give triangle **U**.

Describe the single transformation that takes triangle **S** to triangle **U**. (3)

Quadratic equations

When a traffic policeman arrives at the scene of a road accident, they measure the length of the skid marks and assess the road conditions. They can then use a quadratic equation to calculate the speed of the vehicles and hence reconstruct exactly what happened.

What's the point?

Quadratic equations have many applications in science and engineering and have played a central role within the development of mathematics itself.

Check in

You should be able to

- **factorise quadratic expressions**

1 Factorise each of these expressions.

a $x^2 - 5x$ **b** $x^2 + 10x + 21$

c $x^2 - 25$ **d** $y^2 - 6y + 9$

e $p^2 - 100$ **f** $3ab - 9a^2$

g $16m^2 - 49$ **h** $3x^2 + 7x + 2$

- **use quadratic formulae**

2 Given that $s = ut + \frac{1}{2}at^2$, find:

a s when $u = 8$, $t = 5$ and $a = -4$

b u when $s = 100$, $t = 4$ and $a = -2$

Orientation

What I need to know	What I will learn	What this leads to
A1 Factorise quadratic expressions	■ Solve quadratic equations by factorisation and completing the square ■ Sketch quadratic graphs	**A6** Solve and use quadratic equations **A7** Drawing parabolas and solving simultaneous equations
A3 Solve linear equations		**A-level** Maths, Engineering, Science,

Rich task

A man wants to enclose part of his garden into a rectangle. He uses a wall in his garden as one side of the rectangle, and uses 40 m of fencing to enclose the other three sides.

What is the maximum area of rectangle he can enclose?

What is the maximum area of rectangle he can enclose for L metres of fencing?

A4.1 Solving quadratic equations

This spread will show you how to:

- Solve quadratic equations by factorisation

Keywords
DOTS
Factorise
Quadratic

- **Quadratic** equations contain a squared term as the highest power, for example $2x^2 + 7x - 9 = 0$.

- Quadratic equations can have 0, 1 or 2 solutions.

For example	$x^2 = 100$	$x^2 = 0$	$x^2 = -25$
Solution(s)	$x = 10$ or -10	$x = 0$	Impossible
Number of solutions	2	1	0

- Many quadratic equations can be solved by:
  - rearranging so that one side equals zero
  - **factorising**.

p.122

Example

Solve $x^2 = 6x$.

$x^2 - 6x = 0$ — Rearrange so that one side = zero.

$x(x - 6) = 0$ — Factorise.

Either $x = 0$ or $(x - 6) = 0$

Either $x = 0$ or $x = 6$

Spot the x^2 term, the equation is quadratic.

Example

Solve **a** $x^2 + 5x + 6 = 0$ **b** $4x^2 = 81$.

a $x^2 + 5x + 6 = 0$

$(x + 3)(x + 2) = 0$

Either $x + 3 = 0$ so $x = -3$ — This equation has two solutions.

or $x + 2 = 0$ so $x = -2$

p.126

b $4x^2 = 81$

$4x^2 - 81 = 0$ — Factorise – **DOTS**.

$(2x - 9)(2x + 9) = 0$

Either $2x - 9 = 0$ so $2x = 9$ so $x = 4\frac{1}{2}$

or $2x + 9 = 0$ so $2x = -9$ so $x = -4\frac{1}{2}$

If two expressions multiply to give zero, one of them must be zero.

Example

Solve $5x^2 + 14x - 3 = 0$.

$5x^2 + 14x - 3 = 0$

$(5x - 1)(x + 3) = 0$

Either $5x - 1 = 0$ or $x + 3 = 0$

$x = \frac{1}{5}$ $\quad$ $x = -3$

Exercise A4.1 Grade A/A*

1 First factorise these quadratic equations, by using the common factor, then solve them.

a $x^2 - 3x = 0$ **b** $x^2 + 8x = 0$
c $2x^2 - 9x = 0$ **d** $3x^2 - 9x = 0$
e $x^2 = 5x$ **f** $x^2 = 7x$
g $12x = x^2$ **h** $4x = 2x^2$
i $6x - x^2 = 0$ **j** $9y - 3y^2 = 0$
k $0 = 7w - w^2$

2 Solve these quadratic equations by factorising.

a $x^2 + 7x + 12 = 0$ **b** $x^2 + 8x + 12 = 0$
c $x^2 + 10x + 25 = 0$ **d** $x^2 + 2x - 15 = 0$
e $x^2 + 5x - 14 = 0$ **f** $x^2 - 4x - 5 = 0$
g $x^2 - 5x + 6 = 0$ **h** $x^2 - 12x + 36 = 0$
i $2x^2 + 7x + 3 = 0$ **j** $3x^2 + 7x + 2 = 0$
k $2x^2 + 5x + 2 = 0$ **l** $6y^2 + 7y + 2 = 0$
m $x^2 = 8x - 12$ **n** $2x^2 + 7x = 15$
o $0 = 5x - 6 - x^2$ **p** $x(x + 10) = -21$

3 Solve these quadratic equations by factorising using the difference of two squares.

a $x^2 - 16 = 0$ **b** $x^2 - 64 = 0$
c $y^2 - 25 = 0$ **d** $9x^2 - 4 = 0$
e $4y^2 - 1 = 0$ **f** $x^2 = 169$
g $4x^2 = 25$ **h** $36 = 9y^2$

4 Solve these quadratic equations.

a $3x^2 - x = 0$ **b** $x^2 - 2x - 15 = 0$
c $3x^2 - 11x + 6 = 0$ **d** $9y^2 - 16 = 0$
e $25 = 16x^2$ **f** $x^2 = x$
g $20x^2 = 7x + 3$ **h** $8x = 12 + x^2$

5 Solve these equations.

a $5 = x + 6x(x + 1)$ **b** $(x + 1)^2 = 2x(x - 2) + 10$
c $10x = 1 + \frac{3}{x}$ **d** $\frac{2}{x - 2} + \frac{4}{x + 1} = 0$
e $x^4 - 13x^2 + 36 = 0$

AO3 Problem

6 A rectangle has a length that is 7 cm more than its width, w. The area of the rectangle is 60 cm^2.

a Write an algebraic expression for the area of the rectangle.
b Show that $w^2 + 7w - 60 = 0$.
c Find the dimensions of the rectangle.

w

Unit 3

Completing the square

This spread will show you how to:
- Solve quadratic equations by completing the square

Keywords
Coefficient

- Completing the square means writing a quadratic as a squared bracket plus an extra term.
- Some quadratics factorise into squared brackets without an extra term.

For example,

$$x^2 + 2x + 1 = (x + 1)(x + 1) = (x + 1)^2$$
$$x^2 + 4x + 4 = (x + 2)(x + 2) = (x + 2)^2$$
$$x^2 + 6x + 9 = (x + 3)(x + 3) = (x + 3)^2$$
$$x^2 + 8x + 16 = (x + 4)(x + 4) = (x + 4)^2$$

Notice that the number in the squared bracket is half the **coefficient** of the x term in the original quadratic expression.

- Some quadratics factorise into the form $(x + p)^2 + q$.

For example,

$x^2 + 10x + 15 \longrightarrow (x + 5)^2 + ? \longrightarrow (x + 5)^2 - 10$

Coefficient of $x = 10$
$\frac{1}{2} \times 10 = 5 \rightarrow (x + 5)(x + 5)$

Expanding $(x + 5)(x + 5)$ gives constant term 25. You only want 15, so subtract 10.

$x^2 - 6x + 20 \longrightarrow (x - 3)^2 + ? \longrightarrow (x - 3)^2 + 11$

$\frac{1}{2} \times -6 = -3 \rightarrow (x - 3)^2$

$(x - 3)^2 \rightarrow$ constant 9.
You want 20, so add 11.

Example

Complete the square on $x^2 - 8x - 12$.

$$x^2 - 8x - 12 = (x - 4)^2 + q$$
$$= (x - 4)^2 - 28$$

$(-4)^2$ is 16, so subtract 28 to give the -12 required.

Example

Write these equations in the form $(x + p)^2 + q$.

a $x^2 + 20x + 1$ **b** $x^2 + 5x + 8$ **c** $2x^2 + 8x + 6$

This means 'complete the square'.

a $x^2 + 20x + 1 = (x + 10)^2 + q$
$= (x + 10)^2 + 1 - 100$
$= (x + 10)^2 - 99$

You can write the constant you want (1) and subtract 100 (10^2).

b $x^2 + 5x + 8 = \left(x + \frac{5}{2}\right)^2 + q$
$= \left(x + \frac{5}{2}\right)^2 + 8 - \left(\frac{5}{2}\right)^2$
$= \left(x + \frac{5}{2}\right)^2 + \frac{7}{4}$

With odd coefficients, it may be best to work with fractions.
$8 - \frac{25}{4} = \frac{32 - 25}{4} = \frac{7}{4}$

c $2x^2 + 8x + 6 = 2(x^2 + 4x + 3)$
$= 2[(x + 2)^2 - 1]$
$= 2(x + 2)^2 - 2$

Factorise first.

Exercise A4.2 — Grade A/A*

1 Complete the square on these quadratic expressions.

a $x^2 + 4x + 6$ **b** $x^2 + 8x + 15$
c $x^2 + 10x + 26$ **d** $x^2 + 4x$
e $x^2 + 12x + 10$ **f** $x^2 + 14x + 25$
g $x^2 + 4x - 10$ **h** $x^2 + 8x - 3$
i $x^2 + 16x - 1$ **j** $x^2 + 3x + 4$
k $x^2 + 5x + 6$ **l** $x^2 + 7x + 10$
m $x^2 + 9x$ **n** $x^2 + 5x - 2$
o $x^2 + 11x - 4$

2 Write these expressions in the form $(x + p)^2 + q$ where p and q are positive or negative integers.

a $x^2 + 30x + 90$ **b** $x^2 - 2 + 16x$
c $7x + x^2$ **d** $x^2 + 17x - 2$
e $x^2 + \frac{1}{2}x + 1$ **f** $x^2 - 9x - 11$

3 Complete the square on these expressions.

a $2x^2 + 6x + 4$ **b** $3x^2 + 6x + 9$
c $-x^2 + 6x - 2$ **d** $5x^2 + 10x + 15$
e $6x - 8 - x^2$ **f** $2x^2 + 7x - 3$

4 **a** Explain why it is difficult to make x the subject of $x^2 + 2bx = c$.
b Complete the square on $x^2 + 2bx$.
c Using your answers from parts **a** and **b**, show that it is possible to make x the subject of $x^2 + 2bx = c$ and that
$x = \pm\sqrt{c + b^2} - b$

5 Make x the subject of these equations by completing the square first.

a $x^2 + 4cx = k$ **b** $x^2 + 6x = t^3$
c $x^2 + m - 6gx = 0$ **d** $2x^2 + 4cx = p$

AO3 Problem

6 By completing the square on $ax^2 + bx + c = 0$, prove the quadratic equation formula.

$$x = \frac{-b \pm \sqrt{b^2 - 4ac}}{2a}$$

Unit 3

Solving quadratics by completing the square

This spread will show you how to:

- Solve quadratic equations by completing the square
- Use completing the square to find the minimum value

Keywords
Minimum
Solve

- You can use completing the square to **solve** quadratic equations.

For example:

$$x^2 + 8x + 7 = 0$$
$$(x + 4)^2 - 9 = 0$$
$$(x + 4)^2 = 9$$
$$x + 4 = 3 \quad \text{or } x + 4 = -3$$
$$x = -1 \qquad x = -7$$

A positive integer has two square roots.

- You can use completing the square to find the **minimum** value that a quadratic expression can have.

$$x^2 + 6x + 14 = (x + 3)^2 + 5$$

This part of the expression is a square. The smallest it can be is zero, when $x = -3$.

This part of the expression is constant, it is always equal to 5.

The minimum value of $x^2 + 6x + 14$ is $0 + 5 = 5$. This occurs when $x = -3$.

When $x = -3$, $y = x^2 + 6x + 14 = 5$.
On a graph of this quadratic expression, the lowest (minimum) point would be at the point $(-3, 5)$.

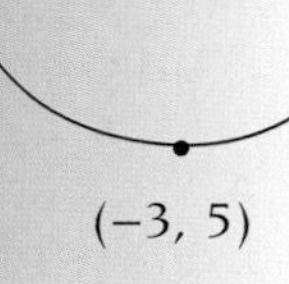

Example

Solve $y^2 + 10y + 21 = 0$ by completing the square.

$$y^2 + 10y + 21 = 0$$
$$(y + 5)^2 - 25 + 21 = 0$$
$$(y + 5)^2 - 4 = 0$$
$$(y + 5)^2 = 4$$

Either $y + 5 = 2$ or $y + 5 = -2$

Hence $y = -3$ or $y = -7$

Example

Show that, if $y = x^2 + 8x - 10$, then $y \geqslant -26$ for all values of x.

$y = x^2 + 8x - 10$
$y = (x + 4)^2 - 10 - 16$
$y = (x + 4)^2 - 26$
Since $(x + 4)^2 \geqslant 0$, $y \geqslant 0 - 26 \quad \Rightarrow \quad y \geqslant -26$

Any number squared is $\geqslant 0$.

Example

Find the co-ordinates of the minimum point on the graph of $y = x^2 - 4x - 5$.

$y = x^2 - 4x - 5$
$\quad = (x - 2)^2 - 9$
Minimum value of $y = -9$, where $x = 2$. Minimum point is $(2, -9)$.

Exercise A4.3 — Grade A*

1 Solve these quadratic equations by completing the square.

Don't forget the two square roots.

a $x^2 - 12x + 20 = 0$ **b** $x^2 + 2x - 15 = 0$
c $x^2 - 4x - 5 = 0$ **d** $x^2 + 2x + 1 = 0$
e $x^2 + 2x - 63 = 0$ **f** $x^2 - 14x + 49 = 0$
g $x^2 - 8x = 0$ **h** $y^2 = 1 - 12y$
i $p^2 = 3p + 2$

2 Use completing the square to explain why $x^2 + 8x + 17 = 0$ has no solutions.

3 Solve these equations by completing the square.
a $y(y + 3) = 88$ **b** $x(x + 2) = 143$
c $x(x + 7) = -21$ **d** $2x^2 + 6x + 3 = 0$
e $(w + 1)^2 - 2w(w - 2) = 10$

4 The area of Painting A is twice the area of Painting B.

Painting A

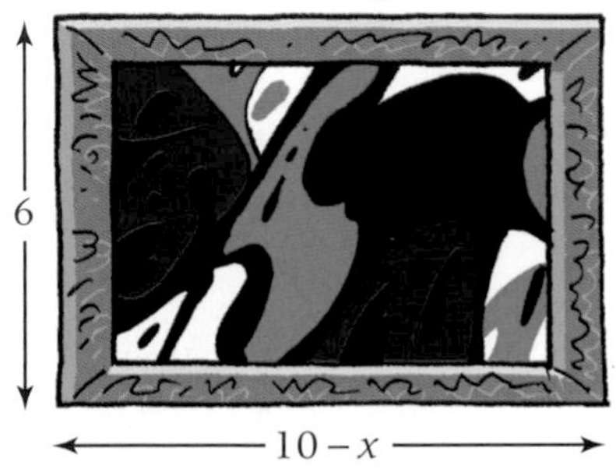

Painting B

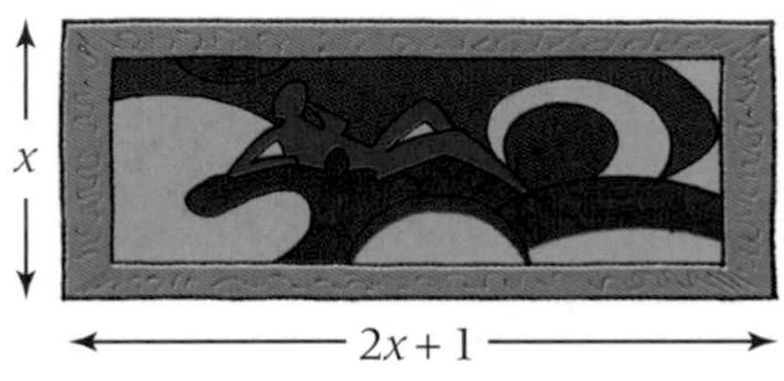

a Show that $x^2 + 2x - 15 = 0$.
b By completing the square, find the value of x and the dimensions of each painting.

5 Use completing the square to explain why $x^2 + 4x + 10$ is never less than 6.

6 Find the coordinates of the minimum point of each of these functions.
a $y = x^2 + 8x + 12$ **b** $y = x^2 + 10x - 5$
c $y = x^2 - 12x - 4$ **d** $y = x^2 + 3x + 1$
e $y = \dfrac{1}{x^2 + 6x + 3}$

AO3 Problem

7 Prove that $p^2 + 6p + 9$ can never be negative.

Sketching quadratic graphs

This spread will show you how to:
- Recognise the shape of a graph of a quadratic function
- Generate points and plot graphs of general quadratic functions

Keywords
Function
Intercept
Maximum
Minimum
Parabola

- You can write a quadratic expression as a **function**, for example $f(x) = x^2 + 2$.
- The graph of a quadratic function is always a **parabola**.

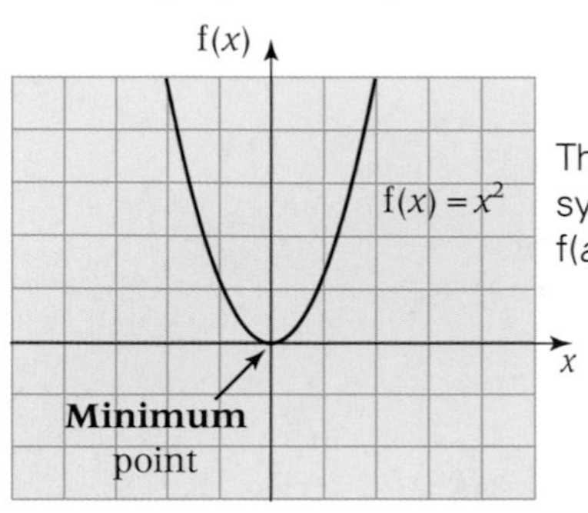

The graphs are symmetrical.
$f(a) = a^2 = f(-a)$

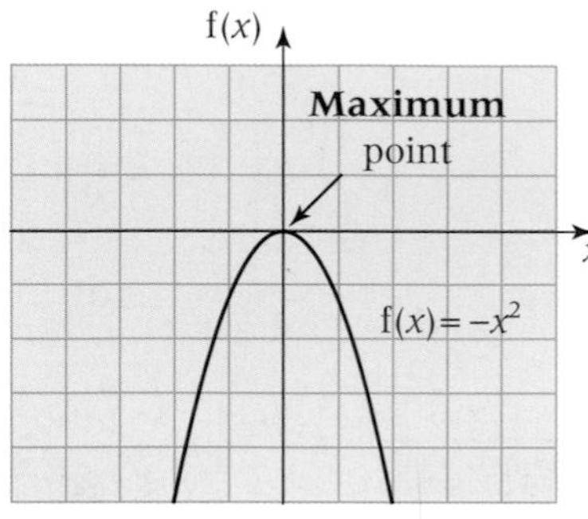

You can plot a quadratic function by tabulating (x, y) values.

- To sketch a quadratic graph, you should show
  - where it intersects the axes
  - the coordinates of its turning point (minimum or maximum point).

You can work these out from the quadratic function.
For example, for $y = f(x) = x^2 - 6x + 8$

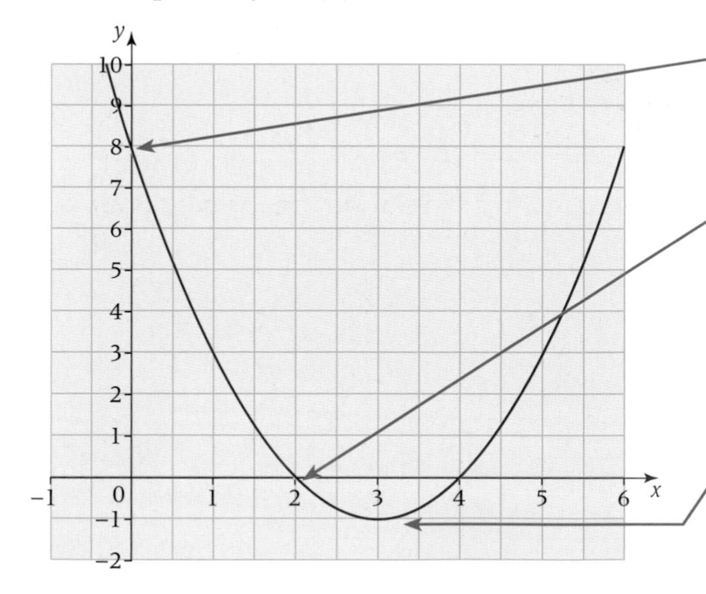

***y*-axis intercept**
On the y-axis, $x = 0$.
Substitute $x = 0$ into the function:
$f(0) = 0^2 - 6 \times 0 + 8 = 8$
y-intercept is $(0, 8)$.

***x*-axis intercept(s)**
On the x-axis, $y = 0$.
Substitute $y = f(x) = 0$ and solve the quadratic:
$0 = x^2 - 6x + 8$
$0 = (x - 2)(x - 4)$
Either $x - 2 = 0$ so $x = 2$ or $x - 4 = 0$ so $x = 4$
The intercepts are $(2, 0)$ and $(4, 0)$.

Turning point
Find the minimum value by completing the square.
$f(x) = x^2 - 6x + 8$
$f(x) = (x - 3)^2 - 1$
For the minimum value $x - 3 = 0$, $x = 3$ and $f(3) = -1$.
The minimum value or turning point is $(3, -1)$.

Example

Which quadratic functions cross the x-axis at $(-3, 0)$ and $(5, 0)$?

$x = -3$ and $x = 5$ are the solutions to $0 = (x + 3)(x - 5)$. So the functions are of the form: $f(x) = a(x + 3)(x - 5) = a(x^2 - 2x - 15)$. where a is any non-zero constant.

Exercise A4.4 Grade A/A*

1 Find the y-intercept of each of these quadratic functions.

a $y = x^2 + 8x + 12$ **b** $f(x) = x^2 - 8x + 15$

c $y = x^2 + 6x + 5$ **d** $f(x) = x^2 + 5x + 6$

e $y = x^2 + 10x + 25$

2 Find the coordinates where each of the quadratic graphs in question **1** intercept the x-axis.

3 Find the coordinates of the minimum point of each of the quadratic graphs in question **1**.

4 Explain why $y = x^2 + 8x + 20$ does not intersect the x-axis.

5 Sketch a graph of the quadratic function $y = x^2 + 7x + 12$, indicating the coordinates of any points of intersection with the axes and of the minimum point.

6 Match these sketch graphs with the equations given. Sketch the remaining function, labelling its points of intersection with the axes.

a

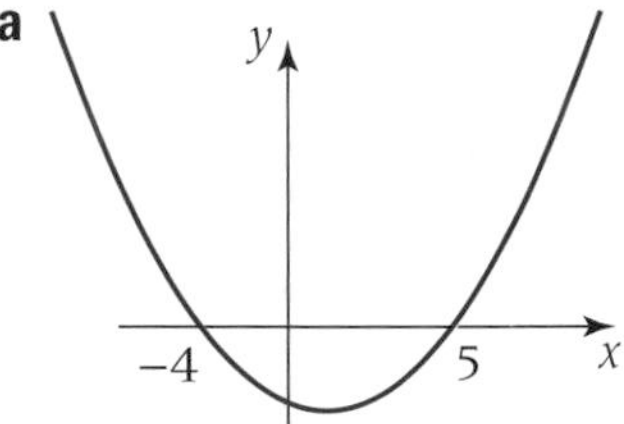

b

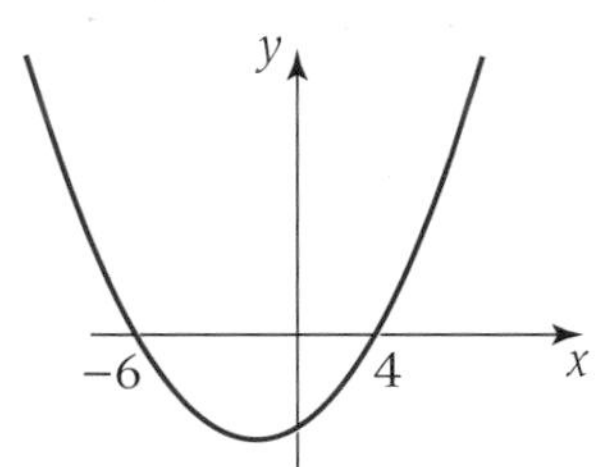

c

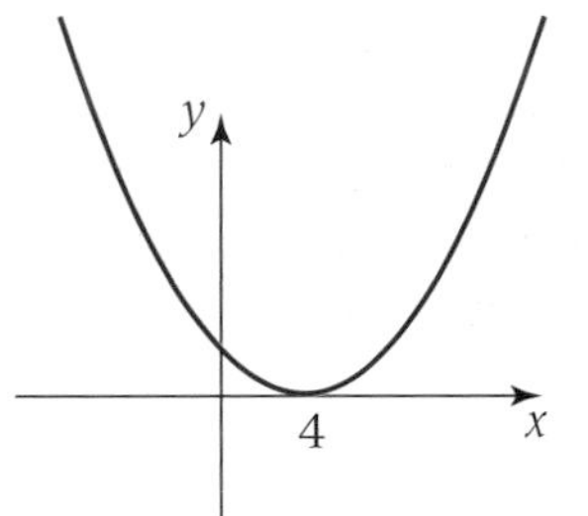

i $y = (x - 4)^2$ **ii** $y = x^2 - x - 20$ **iii** $y = (x - 6)(x + 4)$ **iv** $y = x^2 + 2x - 24$

7 State the equation of these quadratic functions, using the information given in their sketch graphs.

a

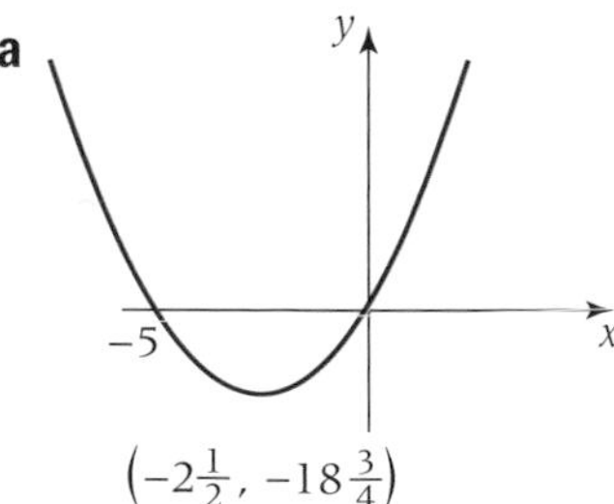

b

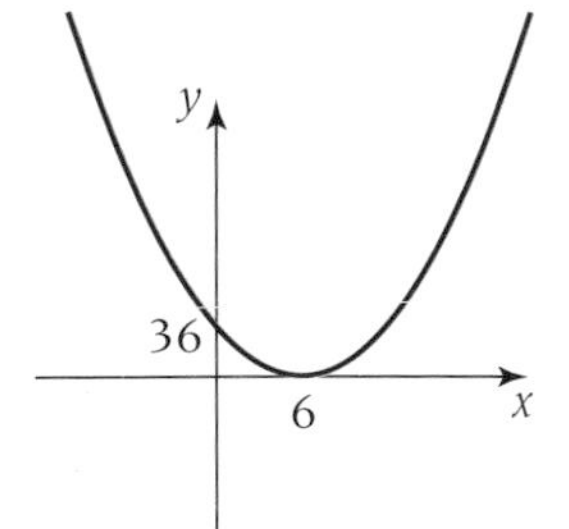

c

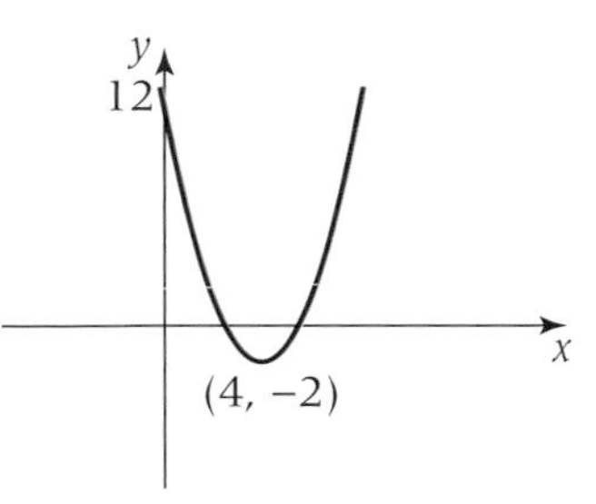

8 Give the equation of a quadratic function that

a intersects the y-axis at (0, 5) but that does not intersect the x-axis

b intersects the x-axis just once, at (−12, 0)

c has a maximum value of (4, 9).

AO3 Problem

9 a Explain why $f(x) = (x - 2)(x + 3)(x + 4)$ is not a parabola. What shape is it?

b Sketch the graph of $f(x) = (x - 2)(x + 3)(x + 4)$.

Where does f(x) intersect the x-axis?

Unit 3

A4 Summary

Check out

You should now be able to:

- Solve simple quadratic equations by factorisation and completing the square
- Sketch graphs of quadratic functions

Worked exam question

a Expand and simplify $(2x + 5)(3x - 2)$ (3)

b Given that $x^2 + 6x - 5 = (x + p)^2 + q$ for all values of x, find the value of

i p,

ii q. (3)

(Edexcel Limited 2006)

---

a

$$(2x + 5)(3x - 2) = 6x^2 - 4x + 15x - 10$$
$$= 6x^2 + 11x - 10$$

There should be 4 terms.

b

$$x^2 + 6x - 5 = (x + 3)^2 + q$$
$$= (x + 3)^2 - 5 - 9$$
$$= (x + 3)^2 - 14$$

$q = -14$

$p = 3$

State the values of p and q.

OR

b

$$(x + p)^2 + q = (x + p)(x + p) + q$$
$$= x^2 + px + px + p^2 + q$$
$$= x^2 + 2px + p^2 + q$$

$2p = 6$ and $p^2 + q = -5$

$p = 3$ and $3^2 + q = -5$

$9 + q = -5$

$q = -14$

$p = 3$

State the values of p and q.

Exam questions

1 **a** Factorise $2x^2 + 7x - 15$ (2)

b Solve $2x^2 + 7x - 15 = 0$

2 For all values of x, $x^2 - 6x + 15 = (x - p)^2 + q$

a Find the value of p and the value of q. (2)

b On a copy of the axes, draw a sketch of the graph $y = x^2 - 6x + 15$

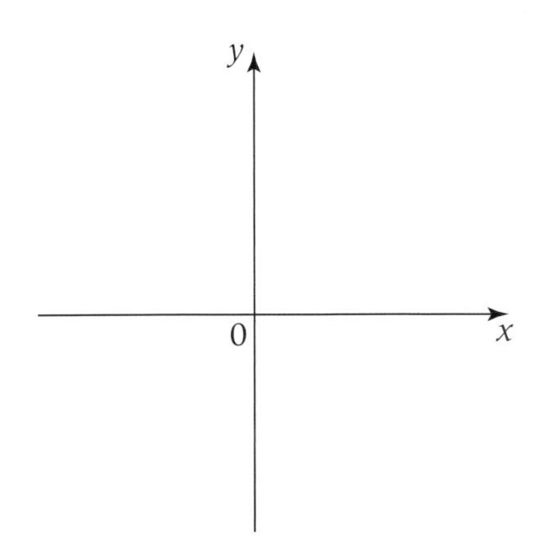

(2)

(Edexcel Limited 2007)

AO3

3 $y = x^2 + 4x - 5$

Give the coordinates of the minimum point on this graph. (4)

Functional Maths 5: Holiday

Mathematics can help you to plan and budget for a holiday, as well as to understand currency, temperature and other units of measure at your destination.

holiday

Paris

LOUISE'S family are planning to go on holiday. Her parents will pay for the trip, but she must raise her own spending money.

1 How much money could Louise save in the three months from March to May if she hired a DVD once a week instead of going to the cinema?

Screen 13
29/11/09
17:50
A GOOD NIGHT £4.50

Hire charge £1.99 per film!

SIGN UP NOW!!

2 A neighbour offers to pay Louise £10 per week if she takes her dog for a 30-minute walk every weekday before school. What hourly rate of pay does this represent? How much would Louise earn if she walked the dog every weekday throughout March, April and May?

3 Louise's brother sold 20 of his CDs and 11 of his DVDs to raise his holiday money. In the first month he sold 9 CDs and 4 DVDs for a total of £36.50. In the second month he sold 5 CDs and 5 DVDs for a total of £30. How much money does he raise from selling all of the CDs and DVDs?

How could you raise money towards a holiday fund or to buy a new item? How long would it take you to reach your target amount?

If you are going on holiday outside of the UK, then you will need to convert your money from £ Sterling to the local currency of your destination. Many European countries now use the Euro, €.

Suppose that you are charged £70 (with no commission) to buy 91.7EUR.

What is the exchange rate? Give your answer as a ratio £ Sterling : Euro.

Some companies charge a commission fee to exchange currency. With an added charge of 1%, how many Euros would you now receive (at the same exchange rate) for £70?

In 2000 Tim went on a holiday to France and Germany. To prepare for his trip he bought 700 French Francs (FRF) and 100 German Deutsche Marks (DEM) for £99.86. He was then given 200 FRF and 40 DEM as a present from his parents, who paid £32.23 for this currency. Calculate the unit per £ Sterling (GBP) rates for the FRF and the GEM (assuming that the rates were the same for both Tim and his parents).

Use the Internet to research the conversion rates used to convert the national currencies to the Euro in 2002. How did this affect the strength of the currencies of the Eurozone compared to the GBP? Compare this to today's unit per GBP exchange rate.

Deciding on your method of transport is an important part of planning a holiday.

Some travel options between Oxford and Paris are shown.

WHICH travel option would you choose? Explain your response with reference to the travel times and costs. All times given are local. Paris is in the time zone GMT + 1 hour.

The foreign travel legs of the same journey options can be paid for in Euros for the following prices:

Return flight BHX to Paris CDG 151.49€; return Eurostar journey 130€, return flight London Heathrow to Paris CDG 162.82€.

How does each of the prices in Euros compare with the corresponding price in GBP?

Explore travel options from your hometown to different destinations. Be careful, there are some times hidden costs such as additional taxes and fees.

■ Different countries often use different units of measure for quantities such as temperature.

An internet site states that the maximum and minimum temperatures in Rome on a particular day are 34°C (93°F) and 22°C (63°F) respectively. The formula used to convert temperatures is of the form °C = aF° + b where a and b are constants.

Use the information to set up two simultaneous equations involving a and b.

Hence find the values of a and b and derive the formula used by the website.

What are the maximum and minimum temperatures in your home town today? Use the formula in the example to convert the temperatures you have found from °C to °F.

G3

Length, area and volume

In modern manufacturing and architecture designers uses computer aided design (CAD) packages to create virtual, 3-dimensional models of their products. They build up complex shapes and surfaces by putting together simpler shapes like triangles and cuboids. After developing their design the computer can analyse the structure and produce detailed plans.

What's the point?

Since complex shapes can be broken down into simpler shapes, understanding the properties of a few basic shapes allows a mathematician to deduce the properties of any shape.

Length, area and volume

Check in

You should be able to

- **calculate the circumference and area of a circle**

1 Work out the
 i area
 ii circumference
 of each circle.

a

b

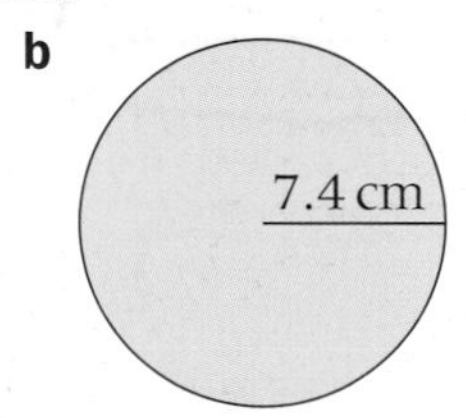

- **calculate the surface area and volume of prisms**

2 Find the volume of these prisms.
State the units of your answers.

a

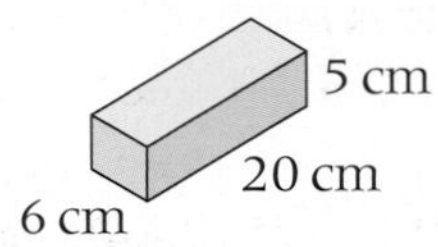

b

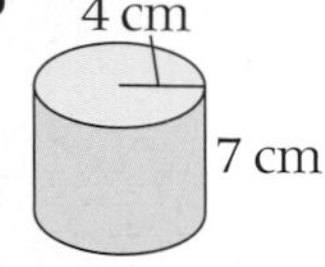

c

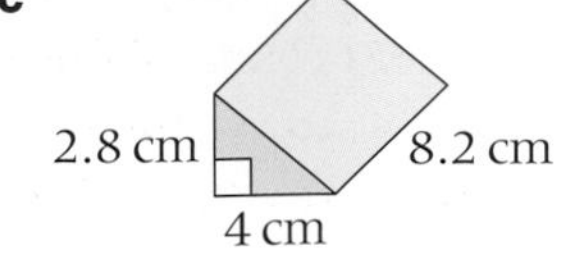

3 Work out the surface area of the prisms in question **2**.

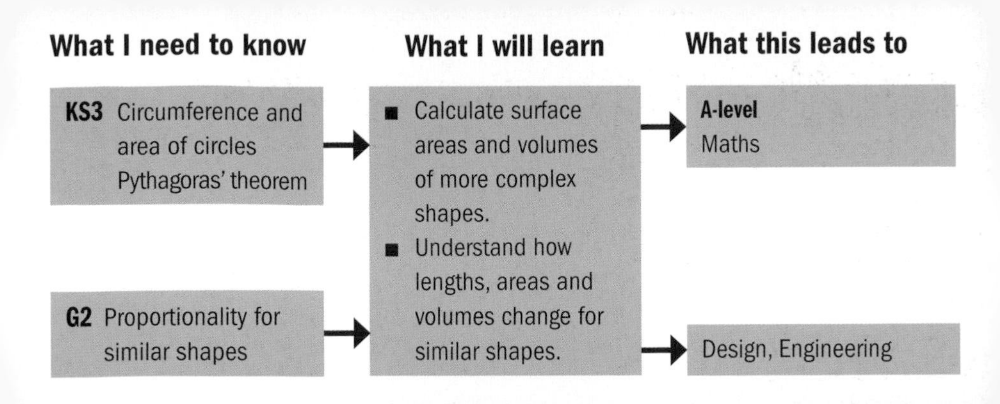

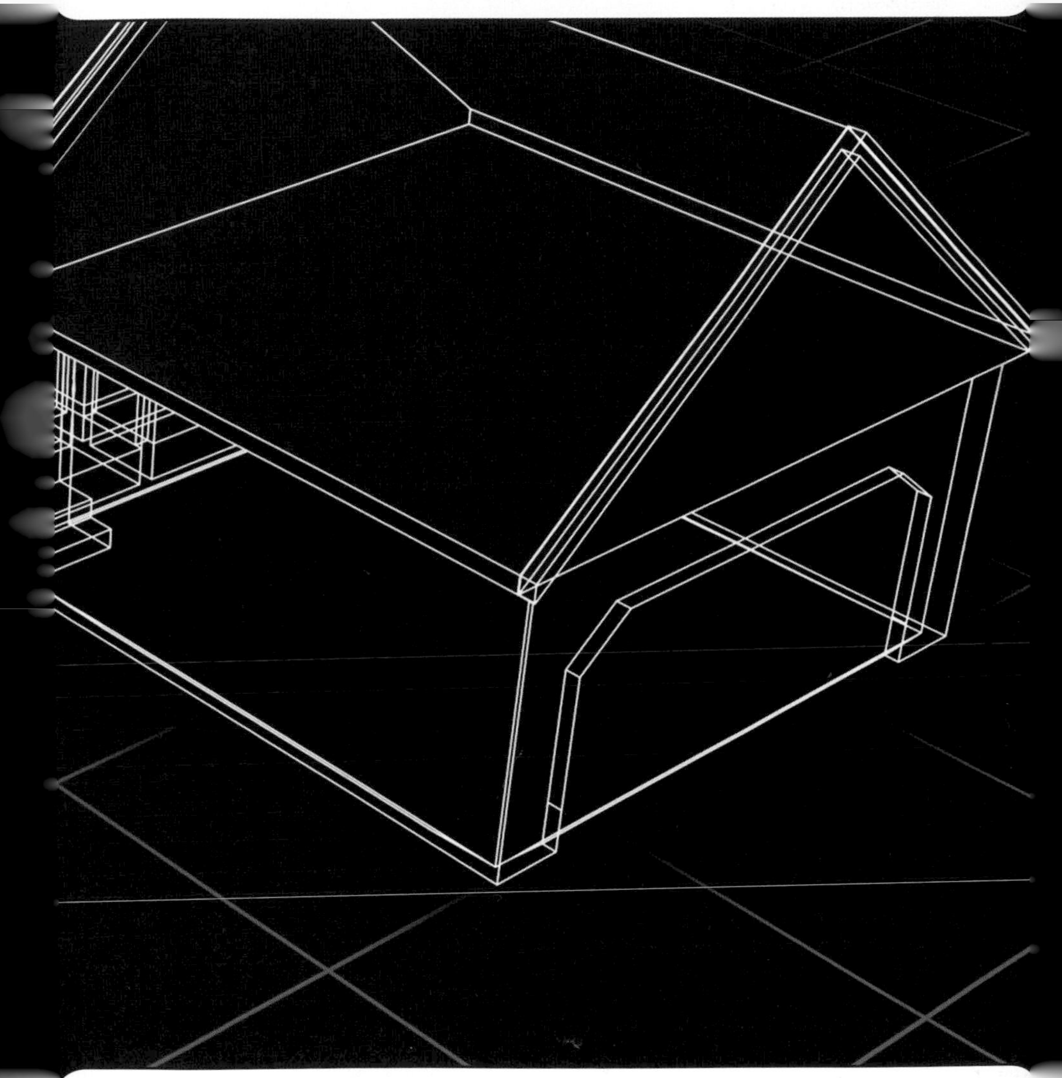

A drinks company needs to design a container to hold exactly 360 ml. They are aware of environmental issues and want to minimise the surface area of the container.

a Design a container to hold exactly 360 ml which has the minimum surface area.

b Refine your design in light of any practical considerations.

G3.1 Arc length and sector area

This spread will show you how to:

- Calculate the area and arc length of a sector of a circle

Keywords
Arc
Circle
Diameter
Perimeter
Radius
Sector

The circumference of a **circle** is in proportion to its **diameter**.

$C \propto d$ $\qquad$ $C \approx 3 \times d$

The constant of proportionality is an irrational number. You can represent it with the symbol π (pi).

$\pi = 3.14159\ldots$

- $C = \pi \times d = \pi d$ or $C = 2\pi r$

$d = 2 \times r$

The area of a circle is in proportion to the square of its **radius**.

- $A = \pi \times r^2 = \pi r^2$

An **arc** is a fraction of the circumference of a circle.

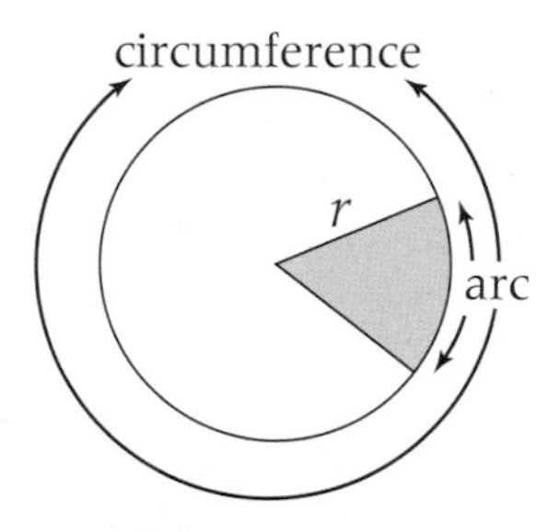

A **sector** is a fraction of a circle, shaped like a slice of pie.

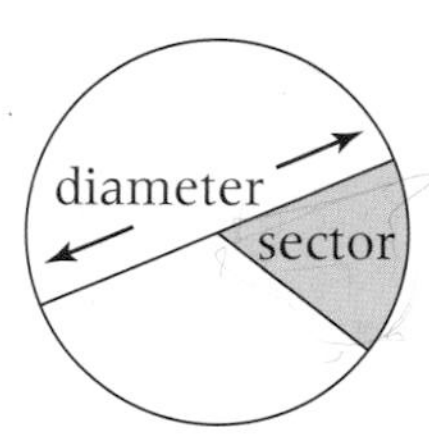

To calculate the area and arc length of a sector, you use the angle at the centre of the sector.

- Arc length $= \frac{\theta}{360} \times$ circumference of whole circle
  $= \frac{\theta}{360} \times 2\pi r$
- Sector area $= \frac{\theta}{360} \times$ area of whole circle
  $= \frac{\theta}{360} \times \pi r^2$

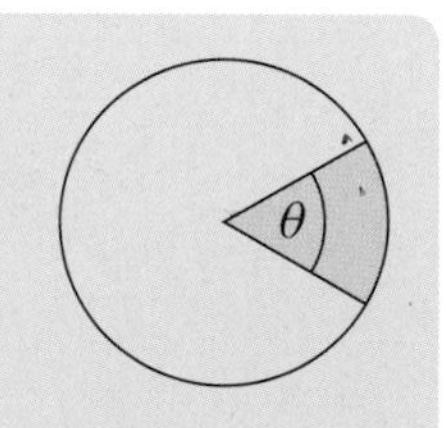

θ is the Greek letter 'theta'.

p.336

Example

This fan is a sector. Find

a the arc length

b the area

c the perimeter of the fan.

Give your answers to 1 decimal place.

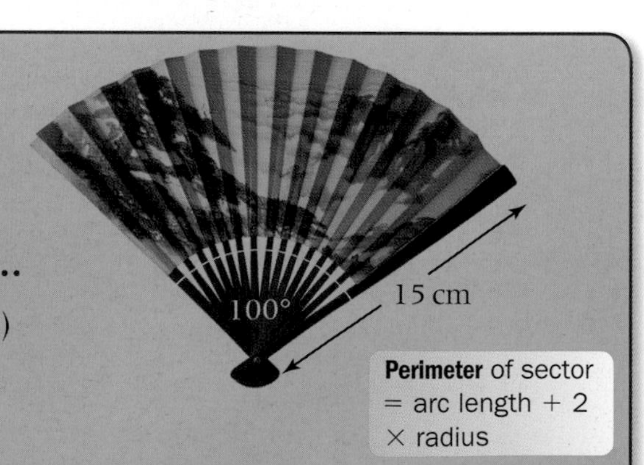

a arc length $= \frac{100}{360} \times 2 \times \pi \times 15 = 26.2$ cm (1 dp)

b sector area $= \frac{100}{360} \times \pi \times 15^2 = 196.3\text{ cm}^2$ (1 dp)

c perimeter = arc length + 2 × 15
= 26.2 + 30
= 56.2 cm

Perimeter of sector = arc length + 2 × radius

Exercise G3.1 — Grade A

In this exercise give your answers to 1 decimal place where appropriate.

1 Find the area of each sector.

a

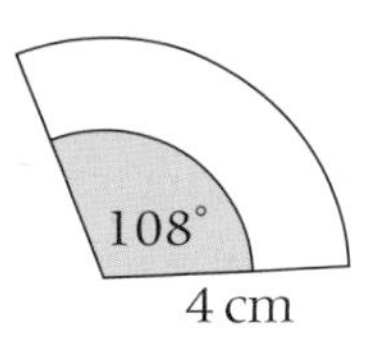

b

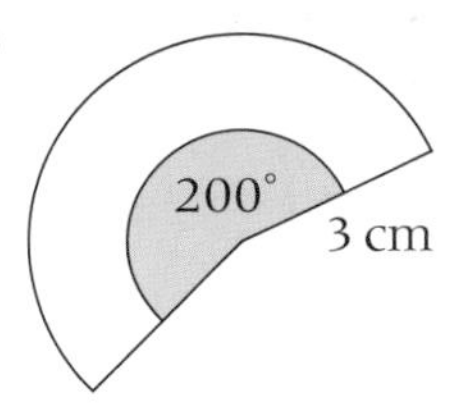

c

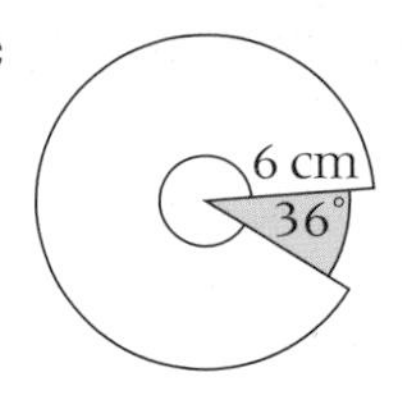

d

72°
5 cm

e

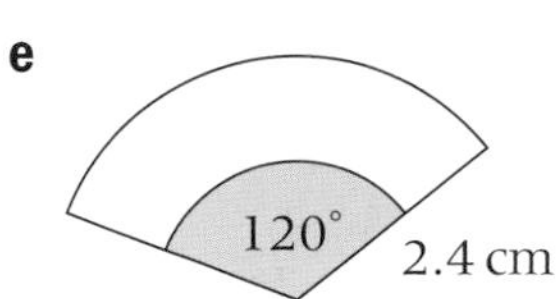

f

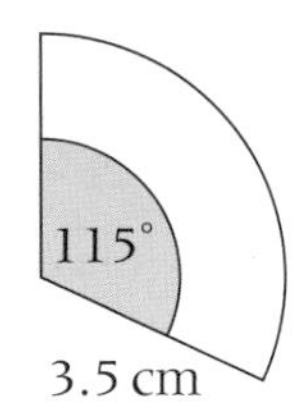

2 Find the arc length of each sector in question **1**.

3 Find the perimeter of each sector.

a

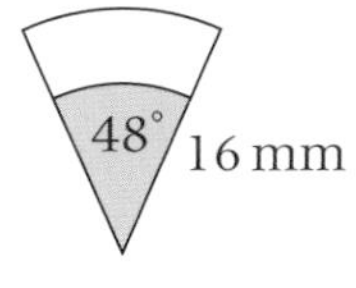

b

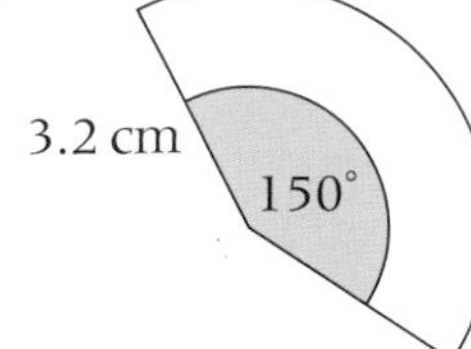

c

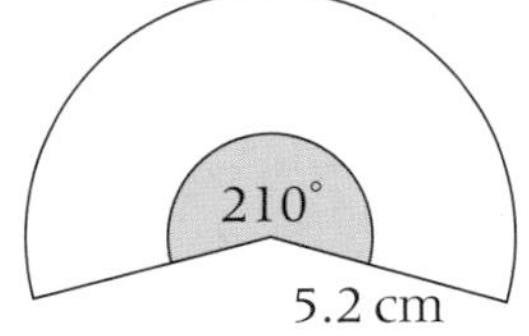

4 Find the area of this incomplete annulus.

An annulus is the region between two concentric circles, shown here as the purple ring.

a

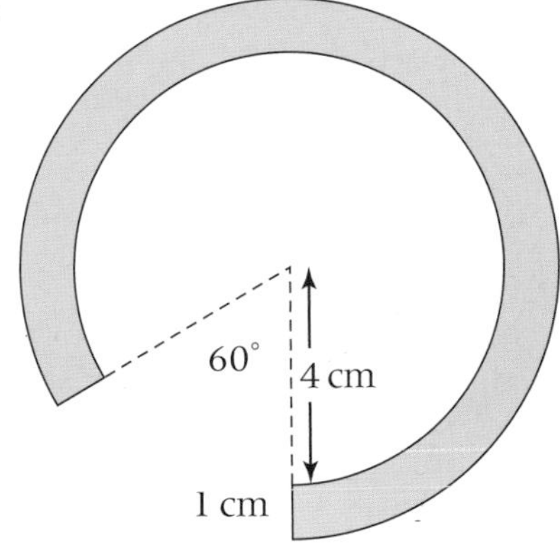

Functional Maths AO2

5 Jan wants to lay a patio in a corner of her garden. The patio is shaped like the sector of a circle.

a What is the area of the patio?

She wants to lay a brick path around the curved edge of the patio. Each brick is 30 cm long.

b Calculate the number of bricks she needs.

5.7 m
Bricks
Patio
39°

Unit 3

G3.2 Volume of a pyramid and a cone

This spread will show you how to:

- Calculate the volume of pyramids and cones

Keywords
Cone
Pyramid
Tetrahedron
Volume

Pyramids and **cones** are 3-D shapes with sides that taper to a point.

Regular tetrahedron	**Square-based pyramid**	**Cone**	**Irregular pyramid**
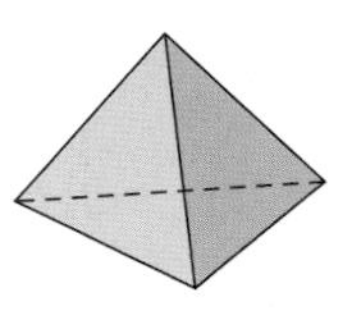	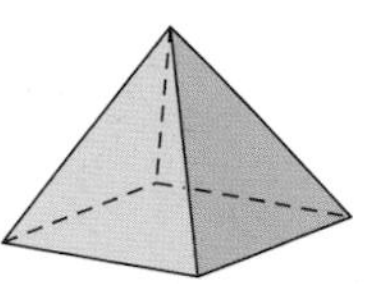	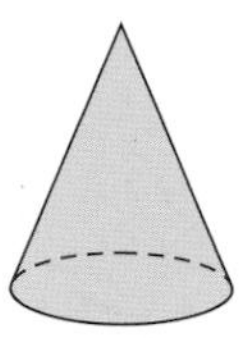	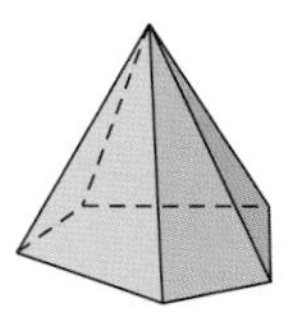
All four faces are identical equilateral triangles.	Base is a square, all four sides are identical isosceles triangles.	Base is a circle.	Base is an irregular polygon.

- Volume of a pyramid = $\frac{1}{3}$ of base area × vertical height

Volume is the amount of space taken up by an object.

This formula also holds true for a cone.

Example

Find the volume of each solid.

a

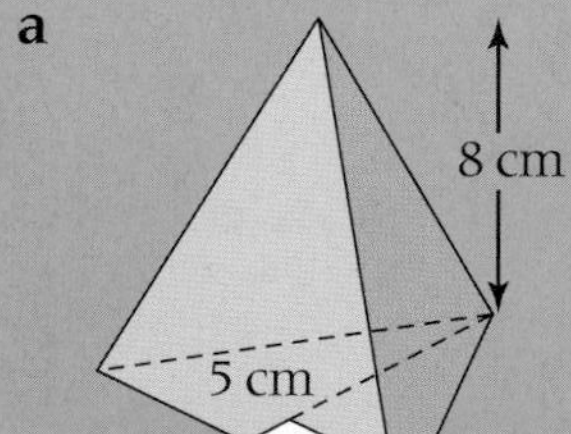

a Volume of pyramid

$= \frac{1}{3}$ base area × height

$= \frac{1}{3} \times (\frac{1}{2} \times 6 \times 5) \times 8$

$= 40\,\text{cm}^3$

Area of triangle $= \frac{1}{2} \times$ base $\times$ height $= \frac{1}{2}bh$

b

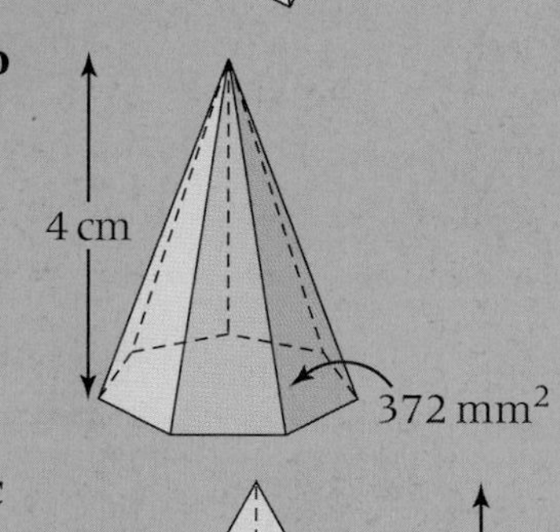

b Volume of pyramid

$= \frac{1}{3}$ base area × height

$= \frac{1}{3} \times 372 \times 40$

$= 4960\,\text{mm}^3$

Dimensions need to be in the same units.

c

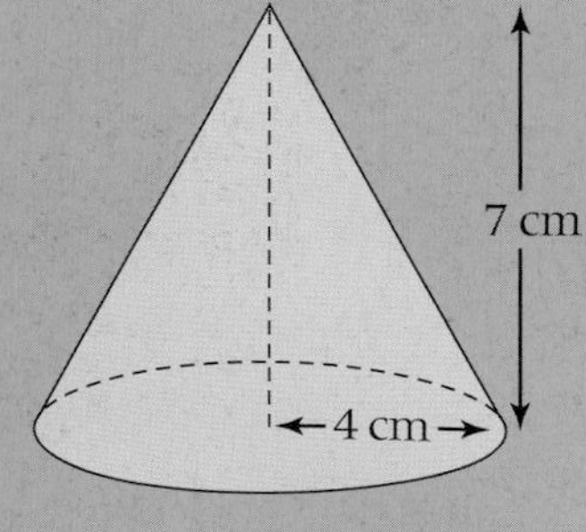

c Volume of cone

$= \frac{1}{3}$ base area × height

$= \frac{1}{3} \times \pi \times 4^2 \times 7$

$= 117.3\,\text{cm}^3$ (1 dp)

Area of base $= \pi r^2$
Volume of cone $= \frac{1}{3}\pi r^2 h$

Round your answer to a sensible degree of accuracy.

Exercise G3.2 Grade B/A

1 Find the volume of each solid.

a

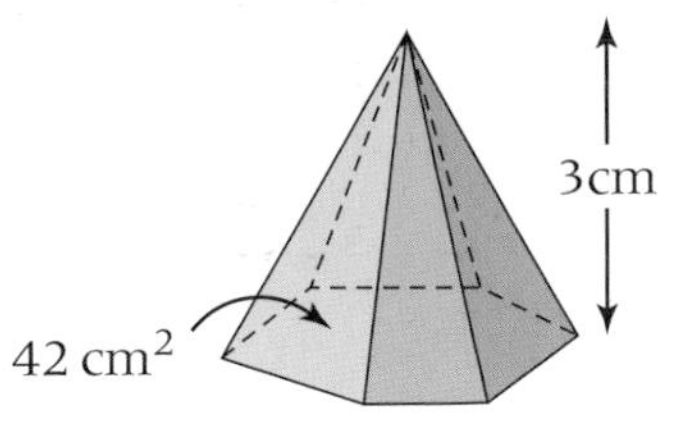

b

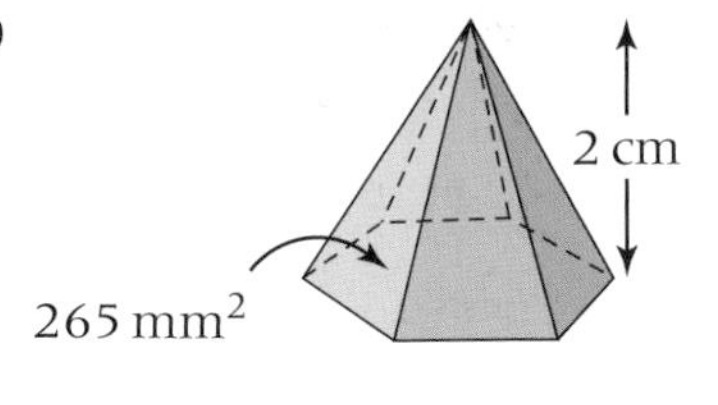

c

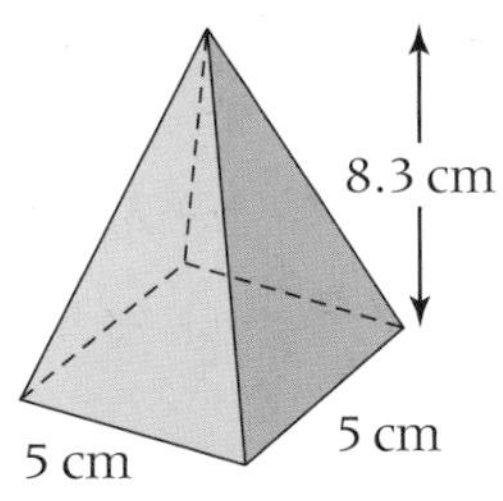

d

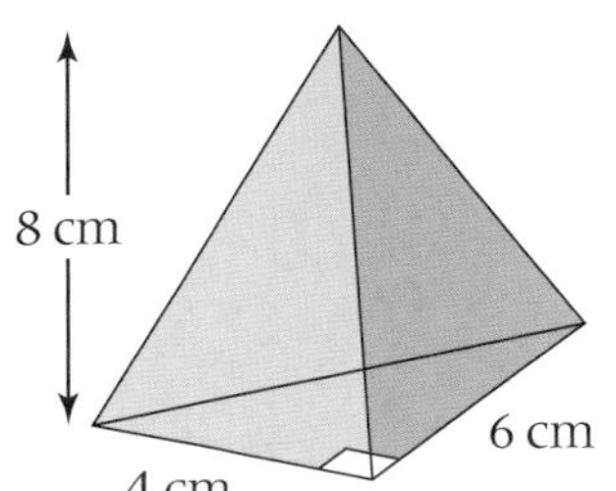

e

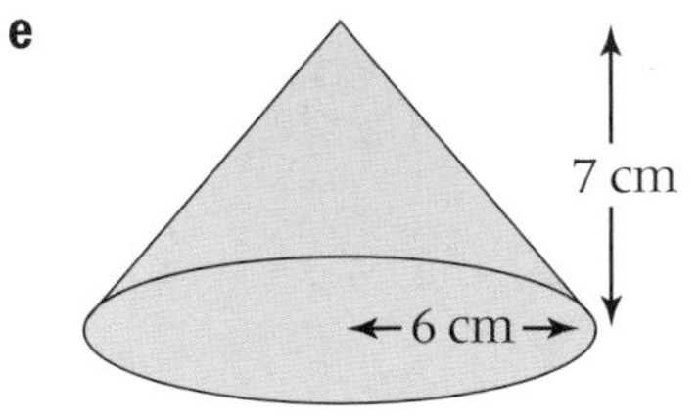

f

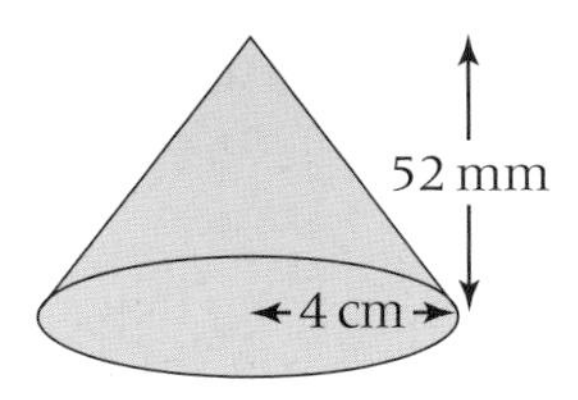

2 James is going to fill a paper cone with sweets.
He can choose between cone X and cone Y.
Cone X has top radius 4 cm and height 8 cm.
Cone Y has top radius 8 cm and height 4 cm.

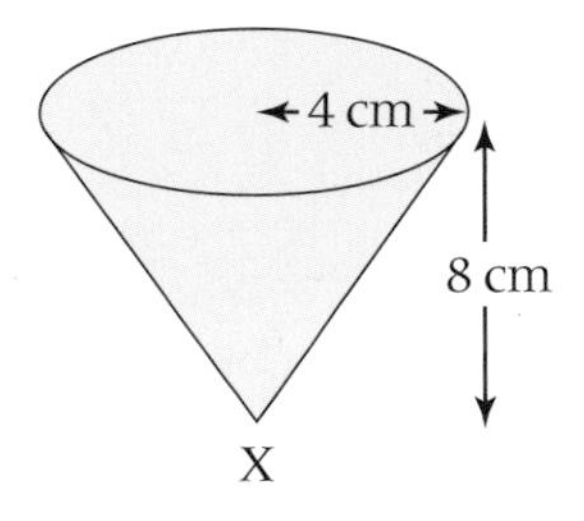

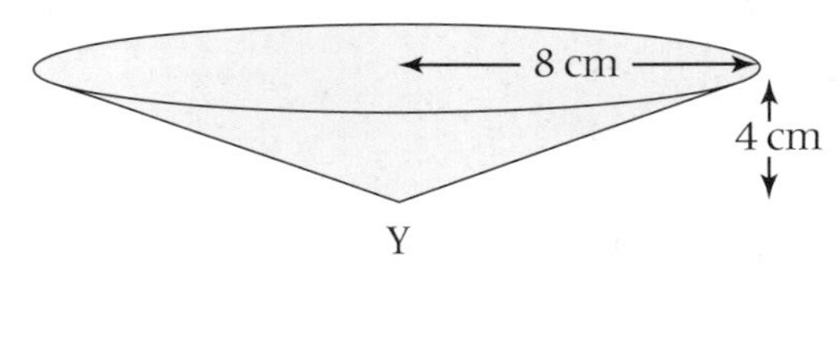

a Which cone has the greater volume?
b What is the difference in volume between cone X and cone Y?

3 A square-based pyramid has base side length x cm and vertical height $2x$ cm.
The volume of the pyramid is 18 cm^3.
Work out the value of x.

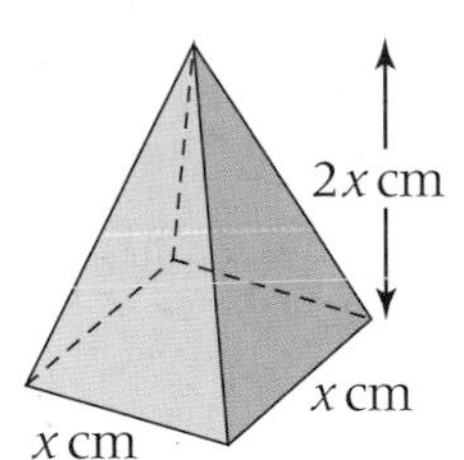

4 Laura made a model rocket from a cylinder with height 6 cm and a cone with vertical height 2.4 cm.
The radius of the cylinder and the cone is 1.8 cm.

Find the volume of the rocket.

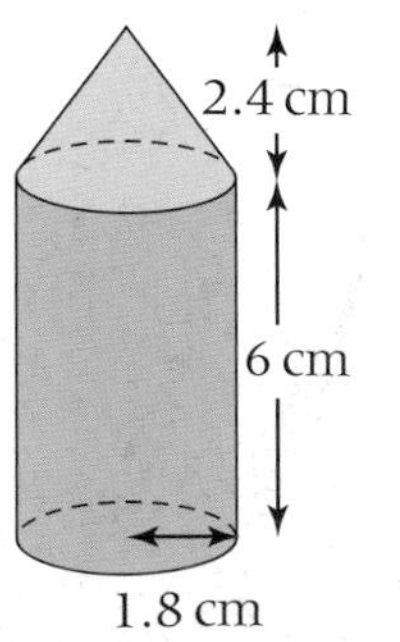

Volume of cylinder = area of base × height

G3.3 Surface area of a pyramid

This spread will show you how to:

- Calculate the surface area of pyramids

Keywords
Pythagoras' theorem
Surface area

Surface area is the total area of all the surfaces of a 3-D solid.

- For 3-D solids with flat surfaces, you work out the area of each surface and add them together.

p.294

You use **Pythagoras' theorem** to find missing lengths in right-angled triangles.

- $a^2 + b^2 = c^2$

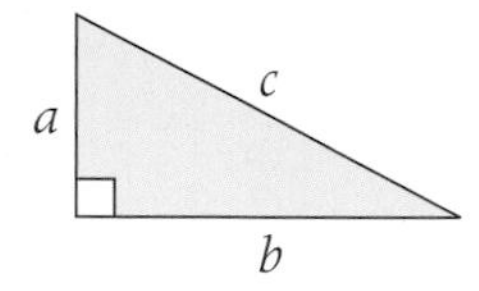

Example

Find the surface areas of these 3-D solids. Give your answer to 1 dp.

a

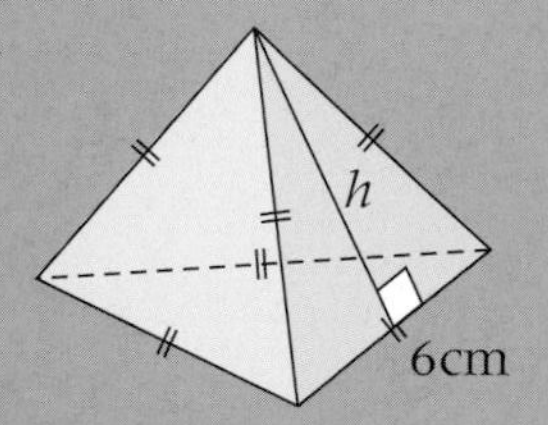

b

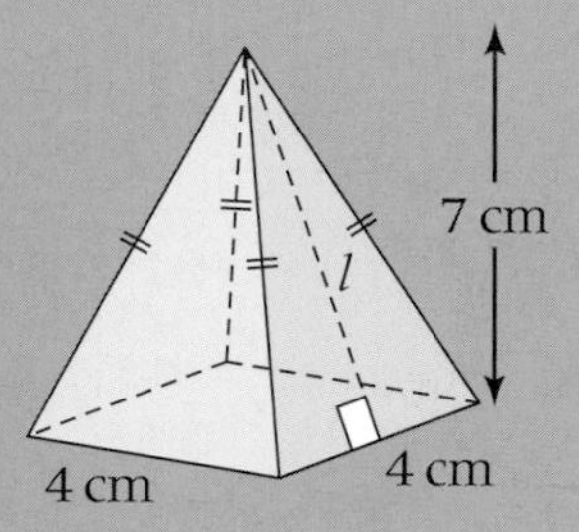

a Each face is an equilateral triangle.

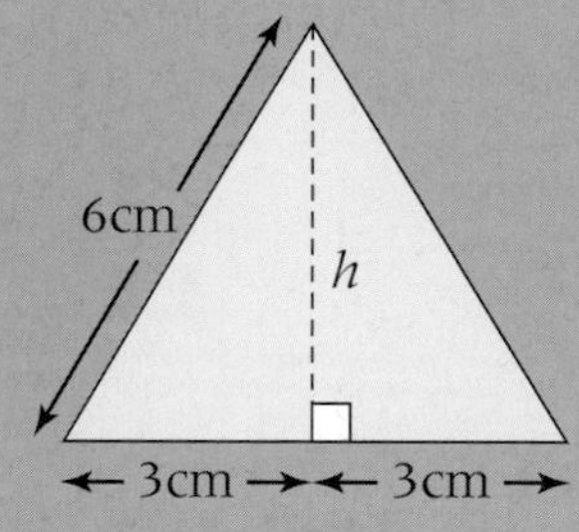

Using Pythagoras: $6^2 = 3^2 + h^2$

$h = \sqrt{6^2 - 3^2} = 5.196\,\text{cm}$

Area of one face

$= \frac{1}{2} \times 6 \times 5.196 = 15.588\,\text{cm}^2$

Surface area

$= 4 \times 15.588 = 62.4\,\text{cm}^2$ (1 dp)

b Base area $4 \times 4 = 16\,\text{cm}^2$

Each face is an isosceles triangle.

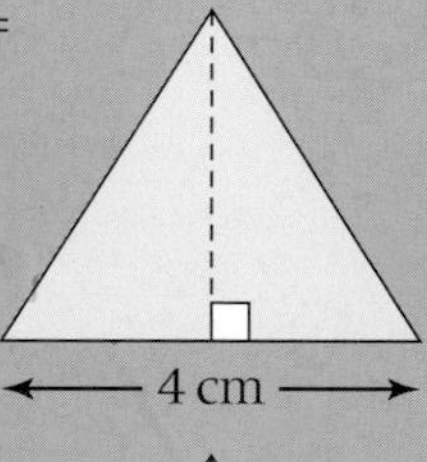

The sloping side of this triangle is l, the height of the triangular face of the pyramid.

You are only given the base measurement. To find the height of each side triangle, imagine a vertical slice through the pyramid:

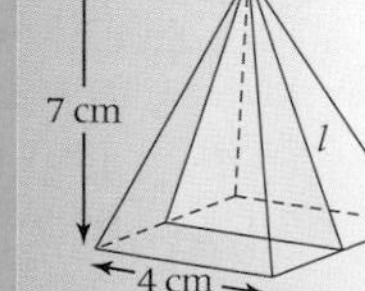

Using Pythagoras: $7^2 + 2^2 = l^2$

$l = \sqrt{7^2 + 2^2} = \sqrt{53} = 7.28\,\text{cm}$

Area of each triangular face

$= \frac{1}{2} \times 4 \times 7.28 = 14.56\,\text{cm}^2$

Surface area

= base area + area of 4 triangular faces

$= 16 + (4 \times 14.56) = 74.2\,\text{cm}^2$ (1 dp)

Exercise G3.3

Grade A*

1 Find the surface area of each 3-D solid.

a

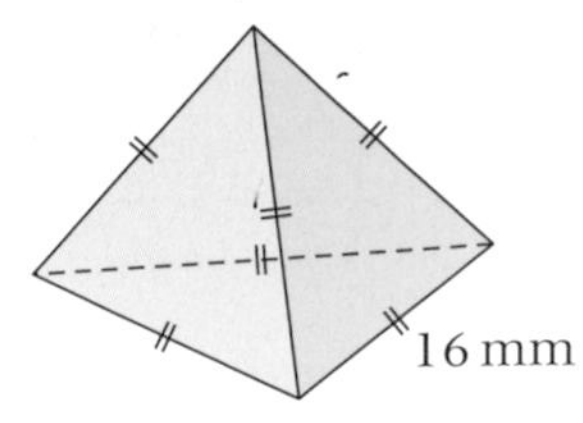

b

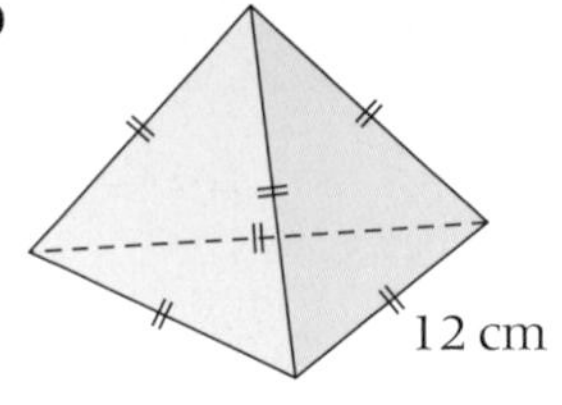

c

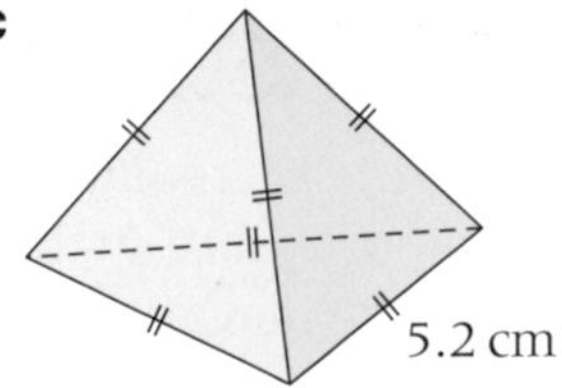

d

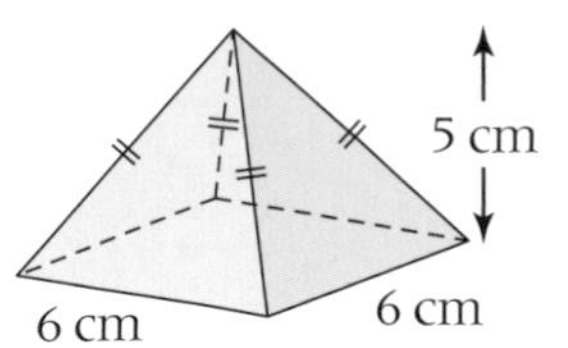

e

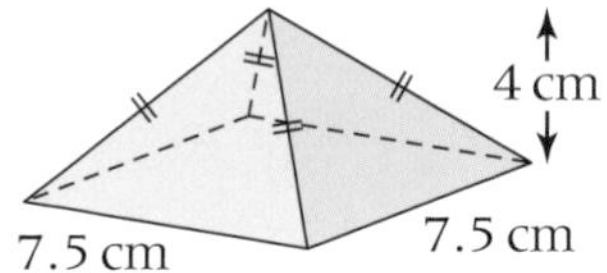

f

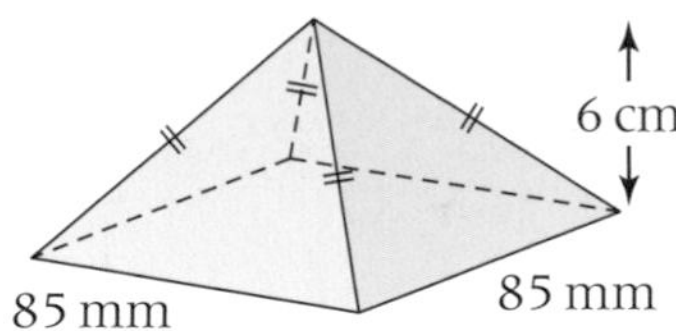

g

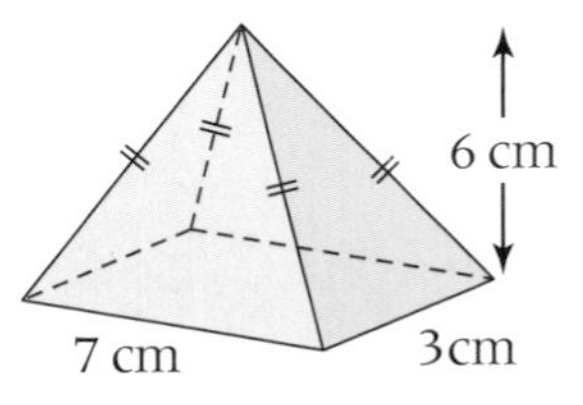

h

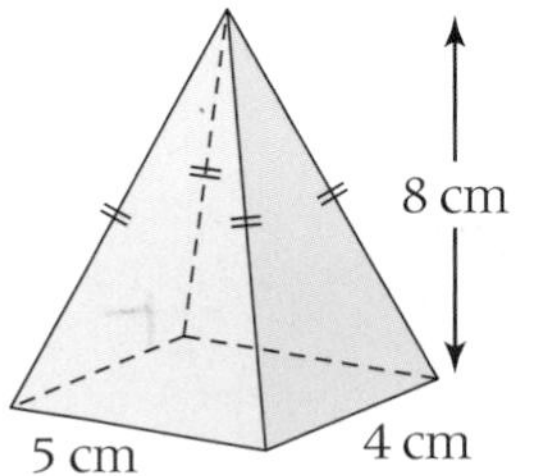

i

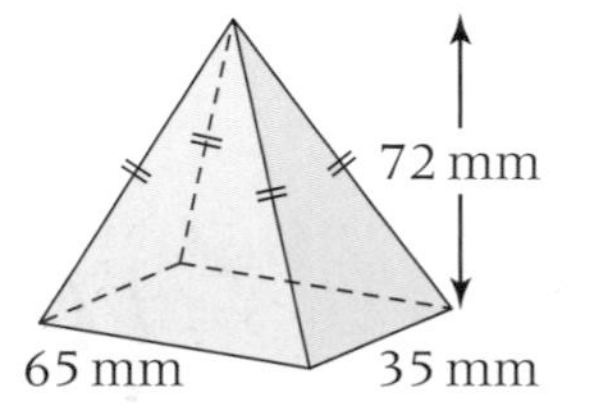

Unit 3

Problem

AO3

2 A pencil is in the shape of a regular hexagonal prism.
Each side of the hexagon is 6 mm.

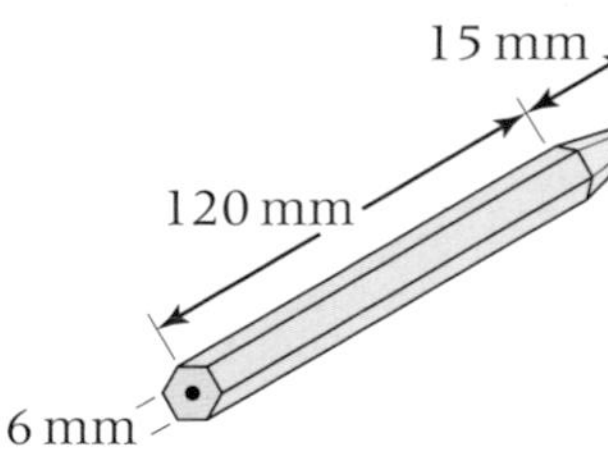

6 mm
6 mm
6 mm

a Find the area of the base of the prism.
The pencil is sharpened at one end to form a pyramid.
The pyramid has height 15 mm and sides 6 mm.
The prism that makes the remainder of the pencil is 120 mm long.

b Find the total surface area of the pencil.

G3.4 Curved surface area of a cone

This spread will show you how to:

- Solve problems involving surface areas of cones

Keywords
Circumference
Cone
Radius
Surface area

A **cone** is a solid with a circular base.
You form the curved surface by folding a sector of a circle.

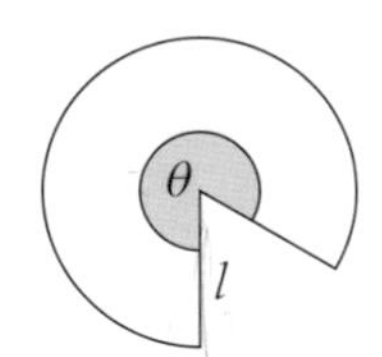

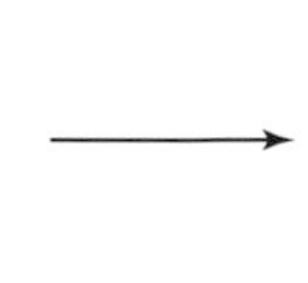

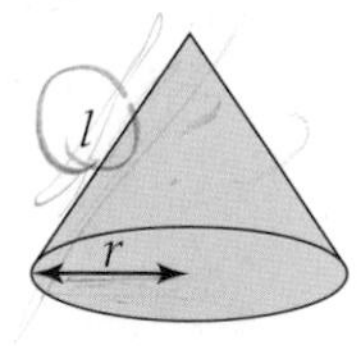

The **radius** of the sector, l, is the sloping side of the cone.

Arc length of the sector = **circumference** of the base of the cone

so $\frac{\theta}{360} \times 2\pi l = 2\pi r$

- Radius of cone $r = \frac{\theta}{360} \times l$

Remember this formula.
$r = \frac{\theta}{360} \times l$

Area of sector = curved **surface area** of cone

$= \frac{\theta}{360} \times \pi l^2$

$= \frac{\theta}{360} \times l \times \pi \times l$

$= r \times \pi \times l$

- Curved surface area of cone = πrl

- Surface area of a cone = curved surface area + area of base
  $= \pi rl + \pi r^2$

Example

This sector is folded to form a cone.
Find, giving your answers to 1 dp:

a the radius of the cone
b the curved surface area of the cone
c the total surface area of the cone
d the vertical height of the cone.

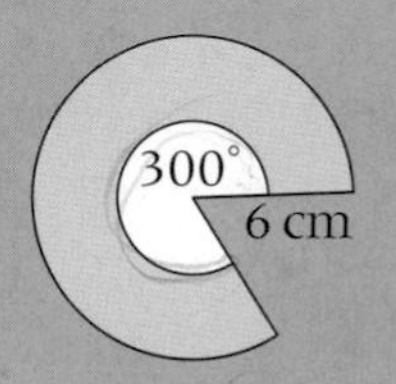

a $\theta = 300°$, $l = 6$ cm

$r = \frac{\theta}{360} \times l = \frac{300}{360} \times 6 = 5$ so $r = 5$ cm

b Curved surface area $= \pi rl = \pi \times 5 \times 6 = 94.2\text{ cm}^2$

c Base area $= \pi r^2 = \pi \times 5^2 = 78.5\text{ cm}^2$

Total surface area $94.24 + 78.53 = 172.8\text{ cm}^2$

d Vertical height

$h^2 = 6^2 - 5^2$

$h = \sqrt{6^2 - 5^2} = 3.3$ cm

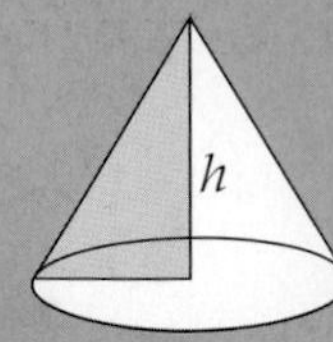

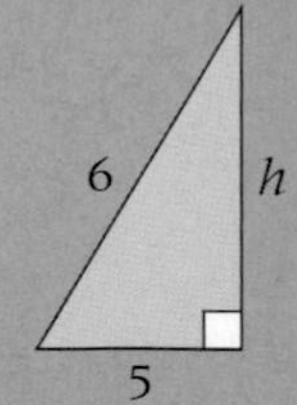

Base radius, vertical height and slant side make a right-angled triangle, so use Pythagoras.

Exercise G3.4

Grade A/A*

> In this exercise, give your answers to 1 dp where appropriate.

1 Find the curved surface areas of cones with

a base radius 5 cm, slant height 8.2 cm
b base radius 3 cm, slant height 6 cm
c base radius 45 mm, slant height 30 mm
d base radius 2.5 cm, slant height 30 mm
e base radius 6.7 cm, slant height 10.5 cm
f base radius 135 mm, slant height 18.5 cm

2 Find the curved surface area of each cone.

> Remember to find the slant height first.

a

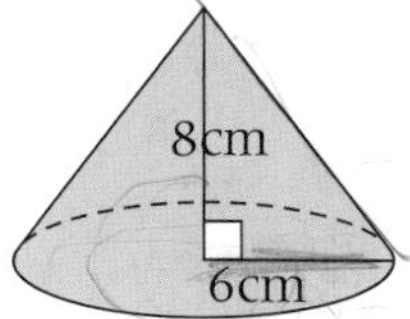

b

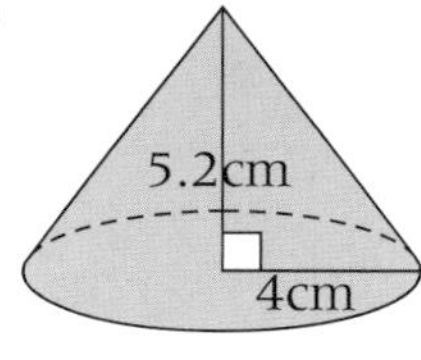

c

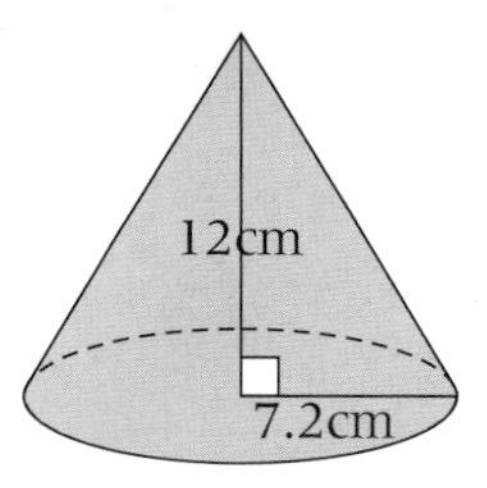

3 These sectors are folded to form cones.
Find the curved surface area of each cone.

a

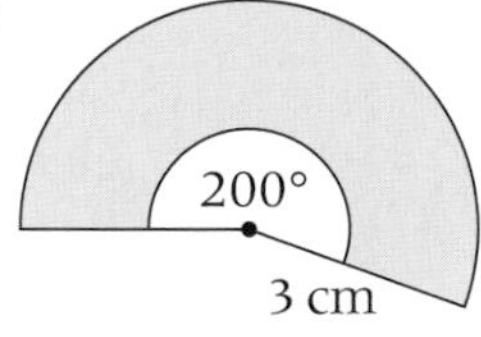

b

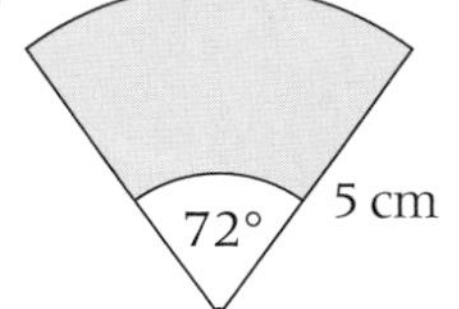

c

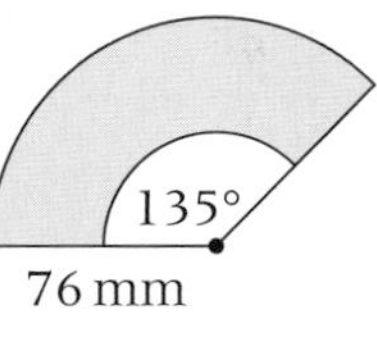

d

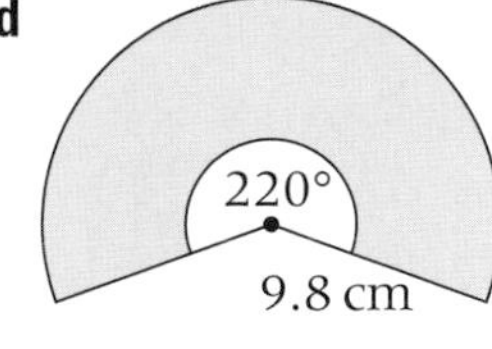

e

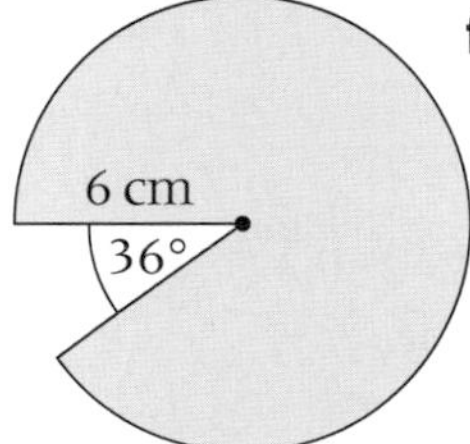

f

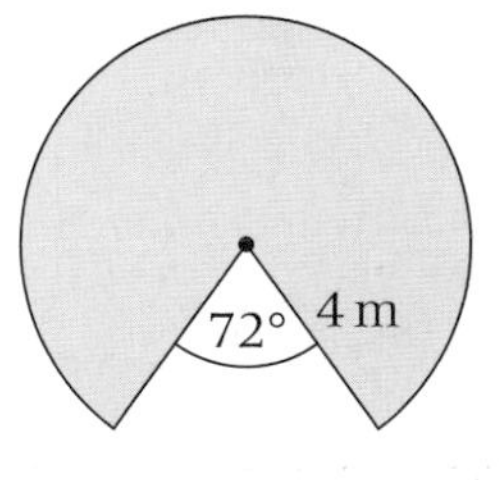

A03 Problem

4 Charlie made a model rocket from a cylinder with height 12 cm and a cone with vertical height 3.2 cm. The radius of the cylinder and the cone is 1.7 cm.

Find the total surface area of the rocket.

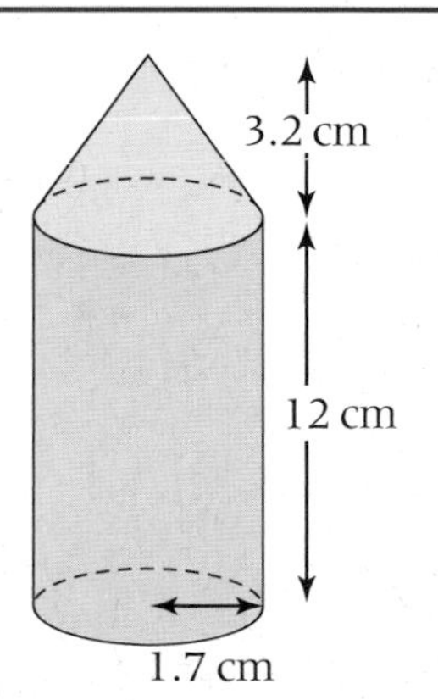

Volume and surface area of a sphere

This spread will show you how to:

- Solve problems involving volumes and surface areas of spheres

Keywords
Sphere
Surface area
Volume

A **sphere** is like a ball. It has one curved surface, no edges and no vertices.

For a sphere with radius = r cm

- **Volume** $V = \frac{4}{3}\pi r^3$ cm^3

- **Surface area** $SA = 4\pi r^2$ cm^2

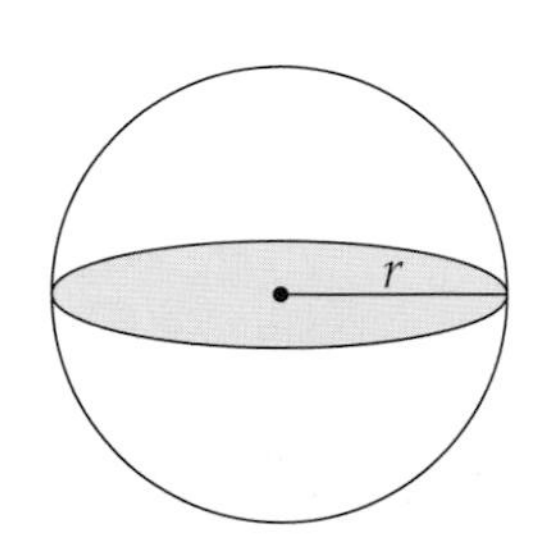

Example

A tennis ball has radius 4.2 cm.
Find, giving your answers to 1 dp
a its volume
b its surface area.

a Volume of tennis ball $= \frac{4}{3}\pi r^3$

$= \frac{4}{3}\pi \times 4.2^3 = 310.3$ cm^3

b Surface area $= 4\pi r^2$

$= 4\pi \times 4.2^2 = 221.7$ cm^2

The earth is split into two hemispheres.

- A hemisphere is half a sphere.

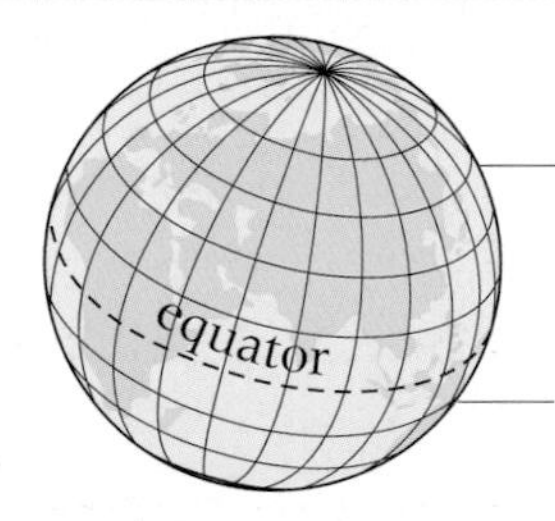

Example

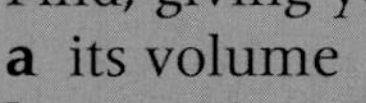

A paperweight is in the shape of a hemisphere.
It has radius 25 mm.
Find, giving your answer to the nearest whole unit
a its volume
b its surface area.

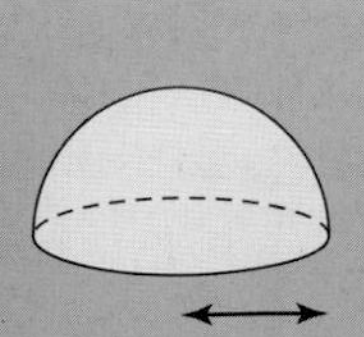
25 mm

a Volume of paperweight $= \frac{1}{2} \times \frac{4}{3}\pi \times 25^3 = 32\,725$ mm^3

b Curved surface area $= \frac{1}{2} \times 4\pi \times 25^2$
$= 3926.99$ mm^2

Area of base $= \pi \times 25^2$
$= 1963.49$ mm^2

Total surface area $= 3926.99 + 1963.49$
$= 5890$ mm^2 (to the nearest mm^2)

$V = \frac{1}{2} \times \frac{4}{3}\pi r^3$ for a hemisphere.

Curved SA $= \frac{1}{2} \times 4\pi r^2$

Exercise G3.5

Grade A/A*

1 Find the volumes of these spheres.

a

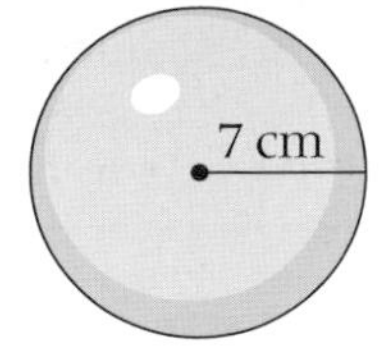

b

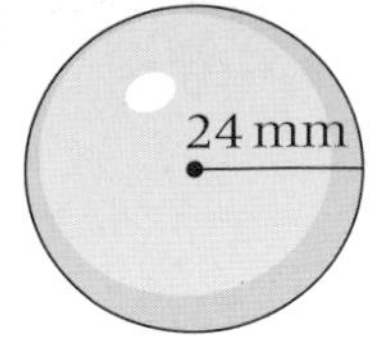

c

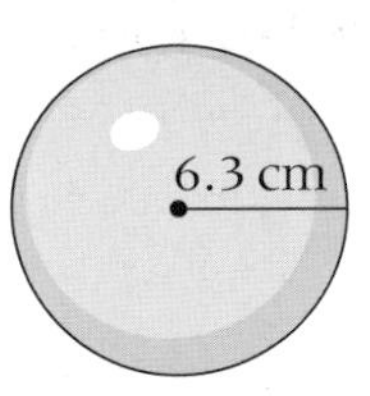

d

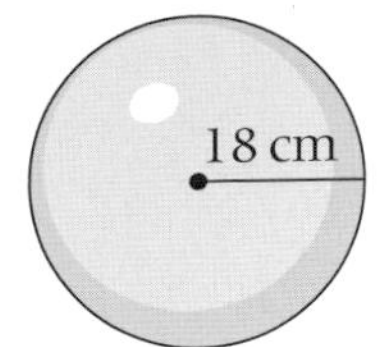

e

f

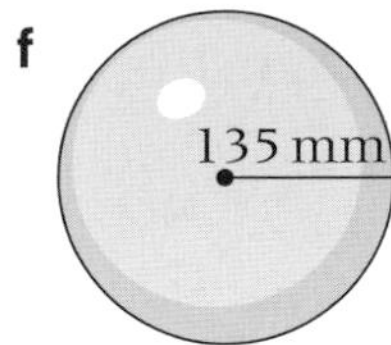

2 Find the surface area of each sphere in question **1**.

3 The surface area of a sphere is $616\,\text{cm}^2$.
Work out the radius of the sphere.

4 Find
a the volume
b the curved surface area
of a hemisphere with radius 4 cm.

5 Rachel bought a cylindrical tube containing 3 power balls.
Each power ball is a sphere of radius 5 cm.
The power balls touch the sides of the tube.
The balls touch the top and bottom of the tube.

Work out the volume of empty space in the tube.
Give your answer both as a numerical value correct to 1 dp and as an exact expression in terms of π.

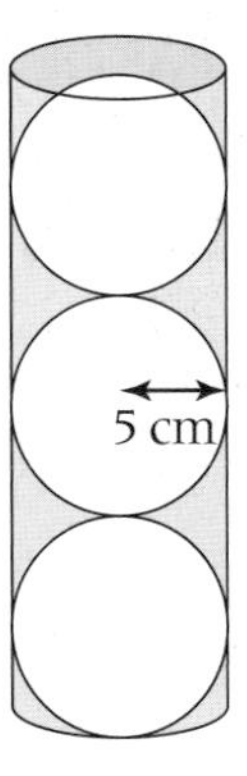

6 **a** Which has the greater volume, a sphere with diameter 3 cm or a cube with side length 3 cm?
b Which has the greater surface area, a sphere with diameter 3 cm or a cube with side length 3 cm?

AO3 Problem

7 A sphere with radius 6 cm fits exactly inside a cylinder.
a Write **i** the radius
ii the height of the cylinder.
b Work out the surface area of the sphere.
c Work out the curved surface area of the cylinder.
d For a sphere radius r that fits exactly inside a cylinder, find an expression for:
i the surface area of the sphere
ii the curved surface area of the cylinder.
e Explain why your answers to **b** and **c** must be the same. You may want to use your answer to part **d** to help you.

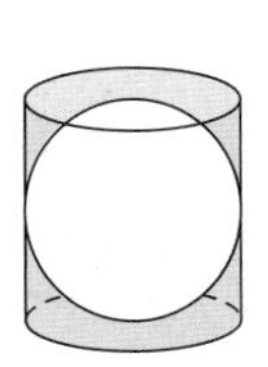

Unit 3

G3.6 Measures and dimensions

This spread will show you how to:

- Identify dimensions by looking at units of measurement
- Convert between length measures, area measures and volume measures

Keywords
Area
Dimension
Perimeter
Volume

You can identify **dimensions** by looking at units of measurement.

- **Perimeter** is the distance drawn around a shape.
  It is a length.
  Perimeter is measured in mm, cm and m.
- **Area** is the space covered by a two-dimensional shape.
  It is the product of length × length.
  Area is measured in square units: mm^2, cm^2 and m^2.
- **Volume** is the space inside a three-dimensional object.
  It is the product of length × length × length.
  Volume is measured in cubic units: mm^3, cm^3 and m^3.

1 m, 1 m — $1\,m^2 = 100\,cm \times 100\,cm$

1 m, 1 m, 1 m — $1\,m^3 = 100\,cm \times 100\,cm \times 100\,cm$

Example

Change **a** 5 400 000 cm to m **b** 5 400 000 cm^2 to m^2 **c** 5 400 000 cm^3 to m^3.

a $5\,400\,000 \div 100 = 54\,000\,m$ **b** $5\,400\,000 \div (100 \times 100) = 540\,m^2$

c $5\,400\,000 \div (100 \times 100 \times 100) = 5.4\,m^3$

Larger unit means smaller number → divide.

Example

Change **a** 0.026 m to mm **b** 0.041 m^2 to cm^2 **c** 8 cm^3 to mm^3.

a $0.026 \times 1000 = 26\,mm$ **b** $0.041 \times (100 \times 100) = 410\,cm^2$

c $8 \times (10 \times 10 \times 10) = 8000\,mm^3$

Smaller unit means larger number → multiply.

Example

Change 0.042 m^3 to **a** cm^3 **b** litres.

a $0.042 \times (100 \times 100 \times 100) = 42\,000\,cm^3$

b $42\,000 \div 1000 = 4.2$ litres

1 litre = 1000 cm^3

You must be careful to use consistent dimensions in any expression.
Constants, such as π and numbers, have no dimensions.

Example

A cylinder has radius r mm and height h cm, write expressions for its

a volume

b total surface area.

h cm = $10h$ mm

a Volume $= \pi r^2 \times 10h = 10\pi r^2 h\ mm^3$

b Surface area $= 2 \times \pi r^2 + 2\pi r \times 10h$

$= 2\pi r^2 + 20\pi rh\ mm^2$

Exercise G3.6 Grade B/A

1 Change each amount to the unit given.

a $320\,000\,\text{cm}^2$ to m^2 **b** $0.004\,\text{m}$ to mm
c $0.02\,\text{m}^3$ to cm^3 **d** $1900\,\text{cm}^3$ to litres
e 5100 m to km **f** 580 mm to m
g 6 300 000 mm to km **h** $24\,\text{cm}^3$ to mm^3
i $630\,000\,000\,\text{mm}^3$ to cm^3 **j** $900\,000\,000\,\text{cm}^2$ to m^2
k $0.0007\,\text{km}^2$ to m^2 **l** $10\,\text{cm}^2$ to mm^2

2 m, n and l are all lengths.
Explain why the expression
a $m + n + l$ represents a length
b mnl can represent a volume.

3 The diagram shows a pool that is rectangular with semicircular ends.
Write an expression for
a the perimeter **b** the area.

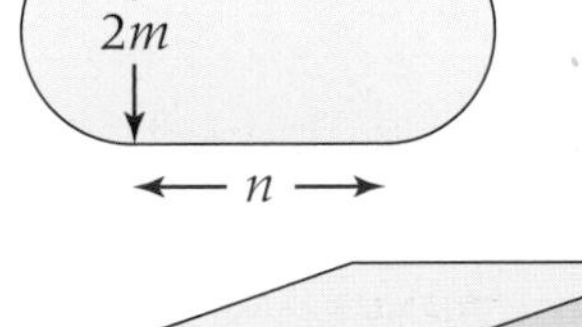

4 The diagram shows a prism with a cross-section that is a parallelogram.
Write an expression for
a the surface area **b** the volume.

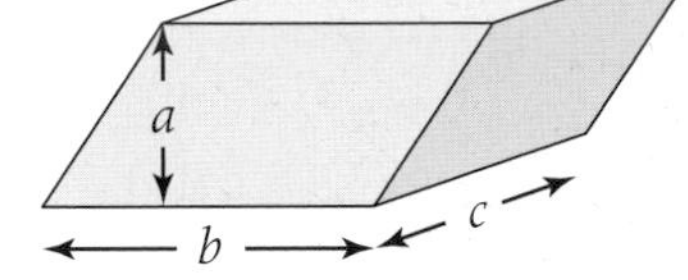

5 An unusually shaped window has area given by the expression

$4ab + b^2\,\text{cm}^2$ where a and b are lengths in cm.

The glass fitted in the window has thickness 6 mm.
Write an expression for the volume of the glass.

6 The cross-sectional area of a prism is given by the expression

wl where w and l are lengths in cm.

The height of the prism is h where h is in mm.
Write an expression for the volume of the prism, stating the correct unit.

AO3 Problem

7 Jade said the volume of this shape is

$\frac{5}{8}\pi r l^3$

Explain why this expression connot be correct

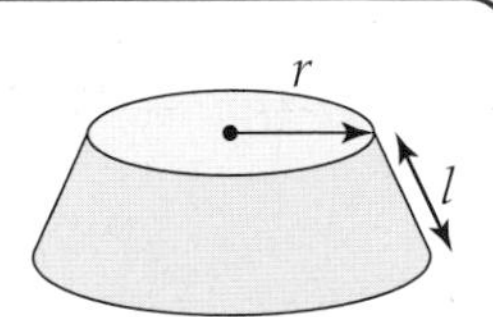

Unit 3

G3.7 Similar shapes – area and volume

This spread will show you how to:

- Understand and use the effect of enlargement on areas and volumes of 3-D shapes and solids

Keywords
Area
Length
Ratio
Similar
Volume

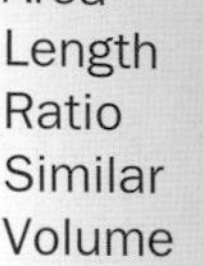

p.222

In **similar** shapes, corresponding sides are in the same **ratio**.
You can use the ratio to work out **areas** and **volumes** in similar shapes and solids.

p.264

For similar shapes with length ratio $1:n$

- the ratio of the areas is $1:n^2$
- the ratio of the volumes is $1:n^3$

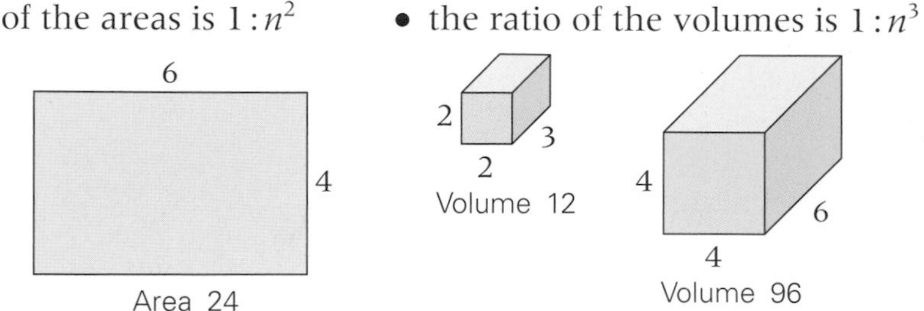

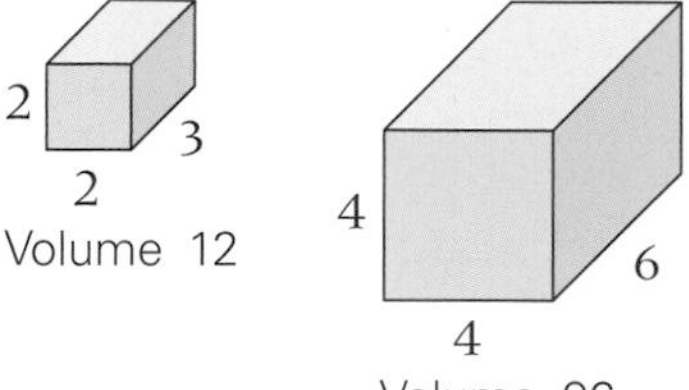

Length ratio is 1 : 2
Area ratio is $6:24 = 1:4 = 1:2^2$

Length ratio is 1 : 2
Volume ratio is $12:96 = 1:8 = 1:2^3$

Area = length × length (square units)
Volume = length × length × length (cubic units)

Example

Two similar Russian dolls are on display.

a The surface area of the smaller doll is $7.2\,cm^2$.
Work out the surface area of the larger doll.

b The volume of the larger doll is $145.8\,cm^3$.
Work out the volume of the smaller doll.

Ratio of lengths, smaller : larger = $1.2:3.6 = 1:3$

a Area ratio is $1:3^2 = 1:9$
Surface area of larger doll = $9 \times 7.2 = 64.8\,cm^2$

b Volume ratio is $1:3^3 = 1:27$
Volume of smaller doll = $145.8 \div 27 = 5.4\,cm^3$

Example

P and Q are similar shapes.
Calculate the surface area of shape Q.

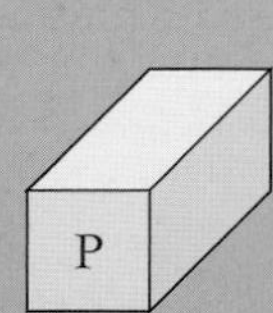

Volume $260\,cm^3$
Surface area $144\,cm^2$

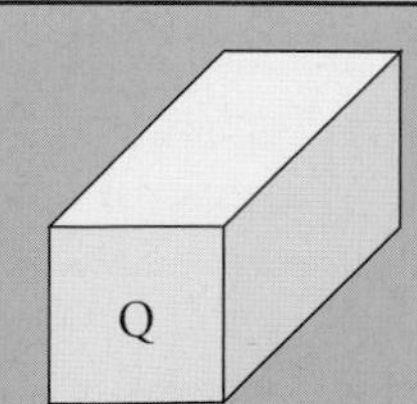

Volume $877.5\,cm^3$
Surface area = ?

Volume ratio

$$\frac{V_Q}{V_P} = \frac{877.5}{260} = \frac{27}{8}$$

Length ratio

$$\frac{L_Q}{L_P} = \frac{\sqrt[3]{27}}{\sqrt[3]{8}} = \frac{3}{2}$$

Area ratio

$$\frac{A_Q}{A_P} = \frac{3^2}{2^2} = \frac{9}{4}$$

Surface area of Q = $144 \times \frac{9}{4} = 324\,cm^2$

Work out the length, area and volume ratios first.

Rearrange the area ratio equation and substitute in the area of P.

Exercise G3.7 — Grade A

1 P and Q are two similar cuboids.

a The surface area of cuboid P is $17.2\,\text{cm}^2$.
Work out the surface area of cuboid Q.

b The volume of cuboid P is $12.4\,\text{cm}^3$.
Work out the volume of cuboid Q.

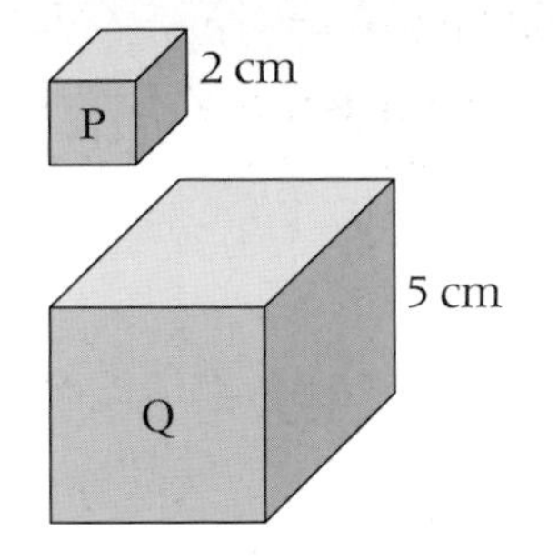

2 J and K are two similar boxes.
The volume of J is $702\,\text{cm}^3$.
The volume of K is $208\,\text{cm}^3$.
The surface area of J is $549\,\text{cm}^2$.
Calculate the surface area of K.

3 X and Y are two similar solids.
The total surface area of X is $150\,\text{cm}^2$.
The total surface area of Y is $216\,\text{cm}^2$.
The volume of Y is $216\,\text{cm}^3$.
Calculate the volume of X.

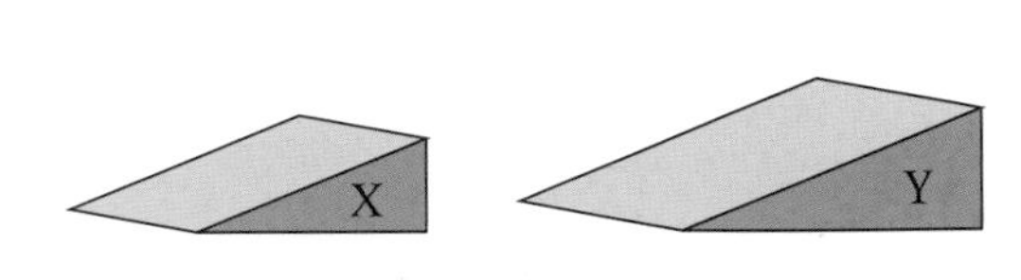

4 C and D are two cubes.

a Explain why any two cubes must be similar.

b Explain why any two cuboids are not necessarily similar.

The ratio of the side lengths of cube C and cube D is 1 : 7.

c Write down the ratio of **i** their surface areas **ii** their volumes.

5 Two model cars made to different scales are mathematically similar.
The overall widths of the cars are 3.2 cm and 4.8 cm respectively.

a What is the ratio of the radii of the cars' wheels?

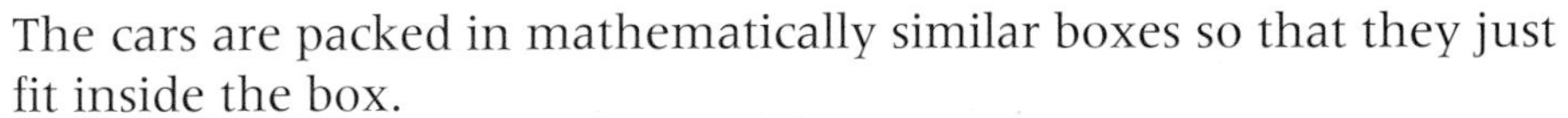

The cars are packed in mathematically similar boxes so that they just fit inside the box.

b The surface area of the larger box is $76.5\,\text{cm}^2$.
Work out the surface area of the smaller box.

c The volume of the smaller box is $24\,\text{cm}^3$.
Work out the volume of the larger box.

A02 Functional Maths

6 At the local pizzeria Gavin is offered two deals for £9.99.

> Deal A: One large round pizza with radius 18 cm
> Deal B: Two smaller round pizzas each with radius 9 cm

Which deal gives the most pizza?

Frustums

This spread will show you how to:

- Use the formulae for surface area and volume of a cone
- Solve problems involving more complex shapes, including frustums of cones

Keywords
Cone
Frustum

When you cut the top off a **cone** with a cut parallel to the base, the part left is called the **frustum**.

The cone removed is similar to the whole cone.

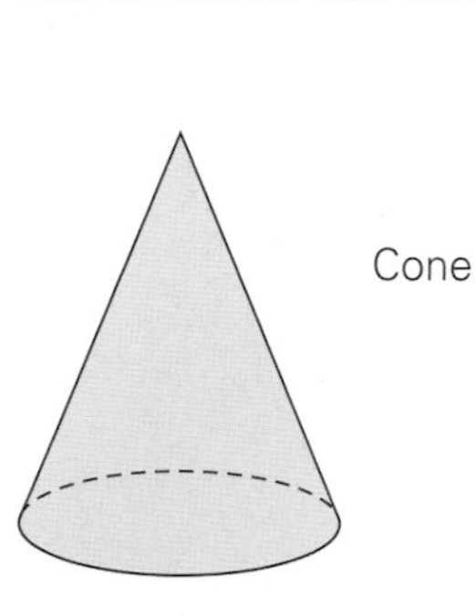

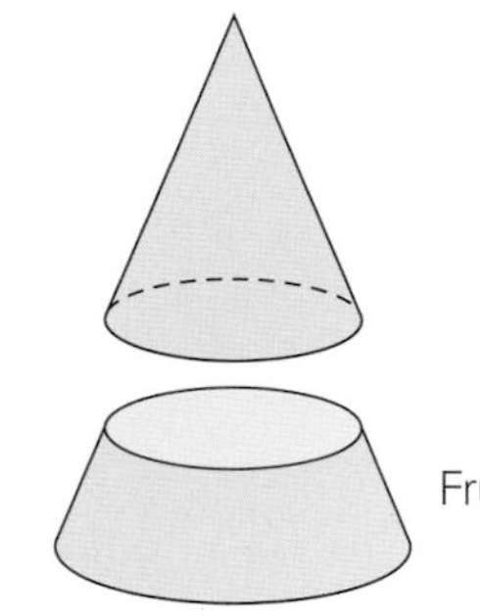

- Volume of a frustum = volume of the whole cone − volume of the smaller cone

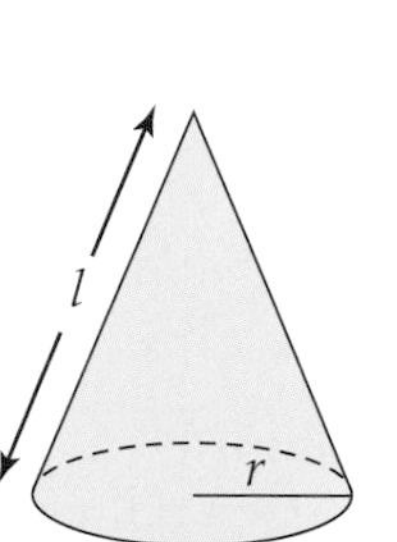

- Volume of a cone $= \frac{1}{3}\pi r^2 h$
- Curved surface area of a cone $= \pi r l$

Example

Find **a** the volume
b the surface area of this frustum.

8 cm
5 cm
12 cm

Use similar triangles to find the height of the whole cone.

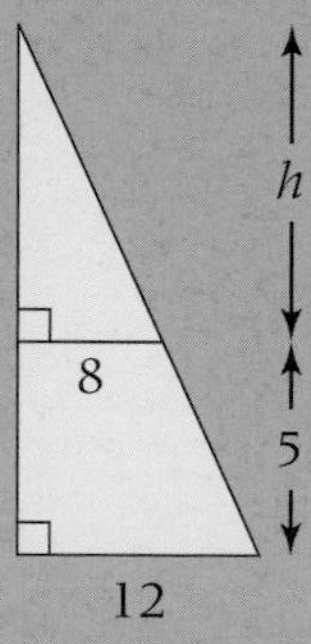

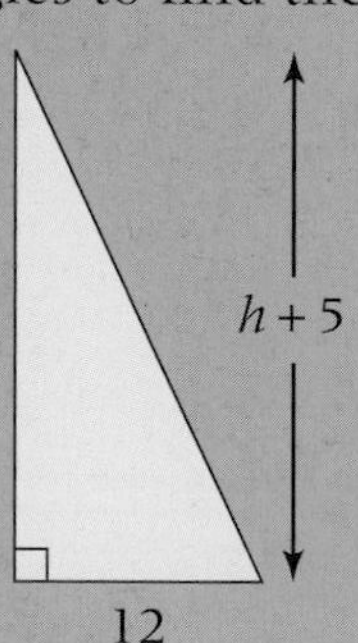

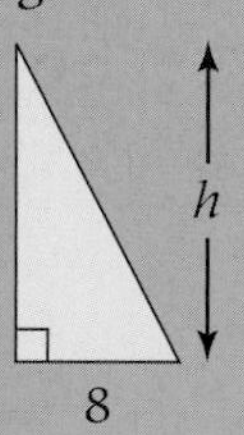

$$\frac{h}{8} = \frac{h+5}{12}$$
$$12h = 8(h+5)$$
$$4h = 40$$
$$h = 10$$

The triangles are similar, so the ratio of the sides is constant.

You can leave π in your working to avoid rounding. It is more accurate.

a Volume of whole cone: $\frac{1}{3}\pi \times 12^2 \times 15 = 720\pi$

Volume of small cone: $\frac{1}{3}\pi \times 8^2 \times 10 = 213\frac{1}{3}\pi$

Volume of frustum: $720\pi - 213\frac{1}{3}\pi = 1592\,\text{cm}^3$

b Curved surface area of whole cone
$\pi \times 12 \times \sqrt{369} = 12\pi\sqrt{369}$
Curved surface area of small cone
$\pi \times 8 \times \sqrt{164} = 8\pi\sqrt{164}$

Surface area of frustum
= section of curved surface + area top circle + area bottom circle
$= (12\pi\sqrt{369} - 8\pi\sqrt{164}) + \pi \times 8^2 + \pi \times 12^2 = 1056\,\text{cm}^2$

Use Pythagoras to find the sloping length l for the whole cone:
$l^2 = 12^2 + (h+5)^2$
$l^2 = 12^2 + 15^2$
$l = \sqrt{369}$
For the small cone:
$l^2 = 8^2 + h^2$
$l^2 = 8^2 + 10^2$
$l = \sqrt{164}$

Exercise G3.8 **Grade A***

1 Work out
i the volume
ii the surface area of each of these frustums.

a
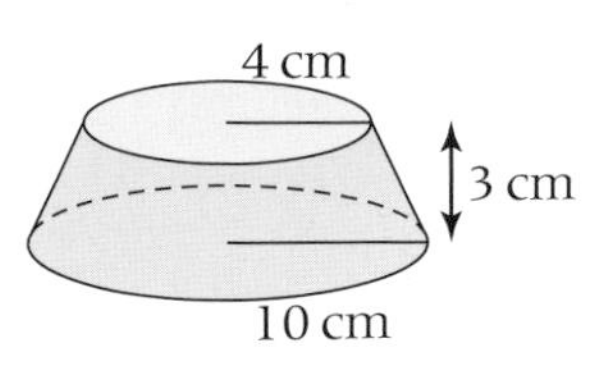

b
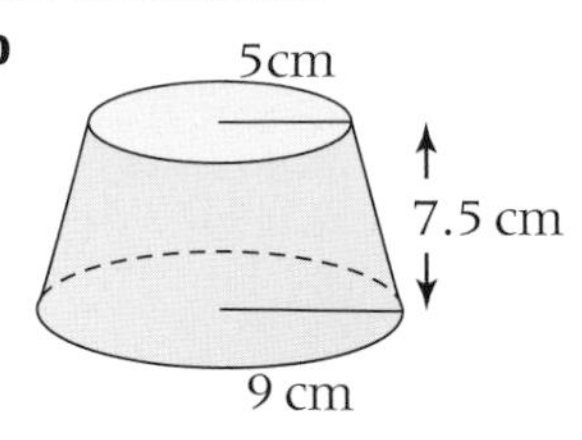

c
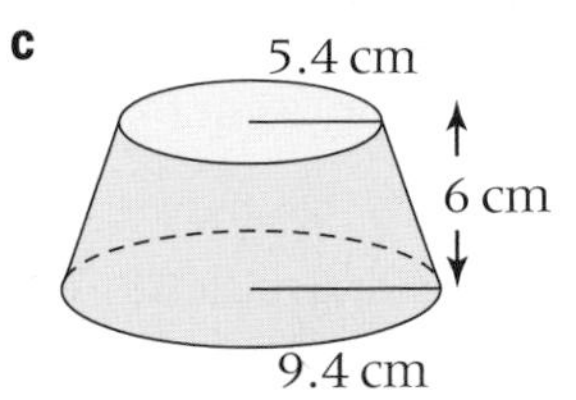

d
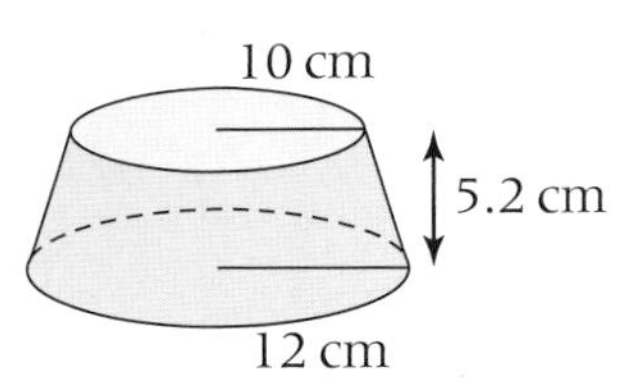

e
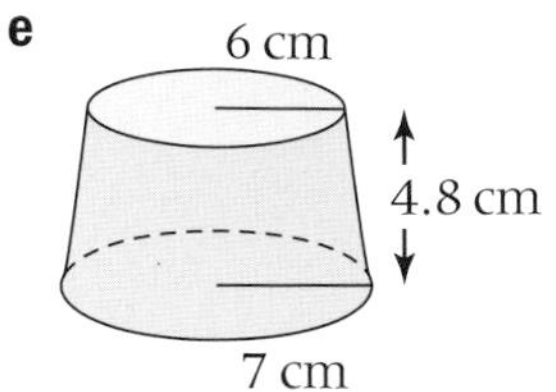

f
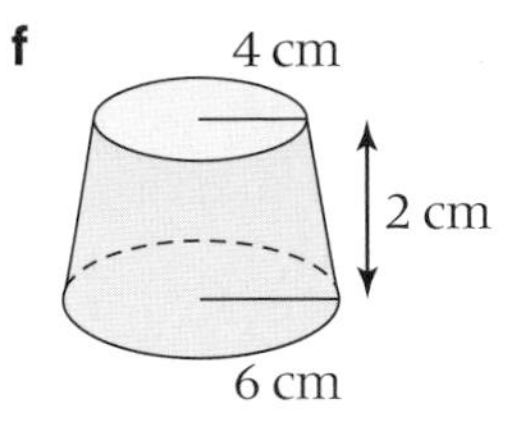

g
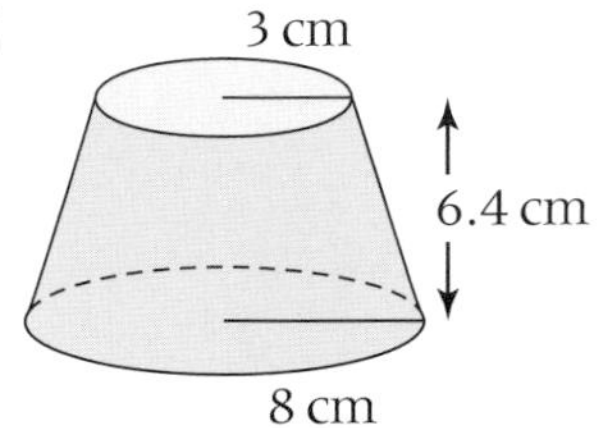

h
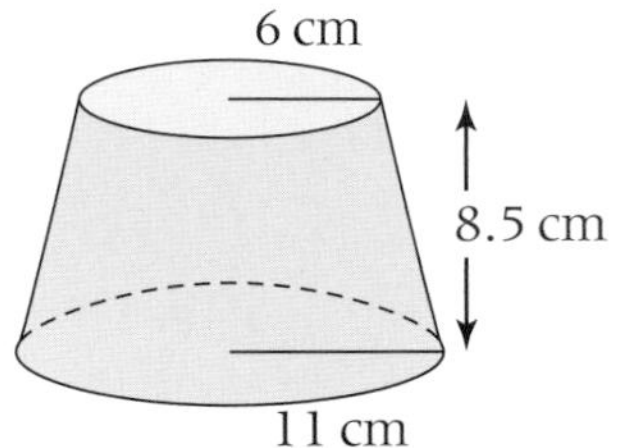

i
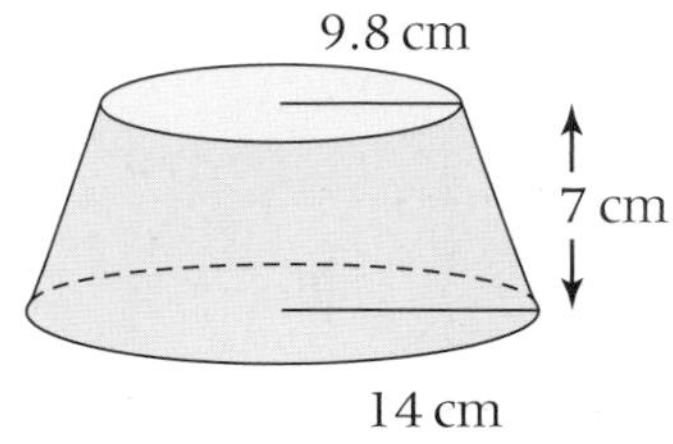

j
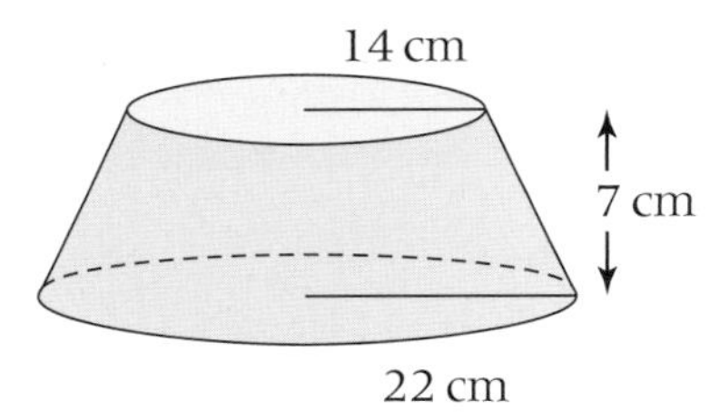

k
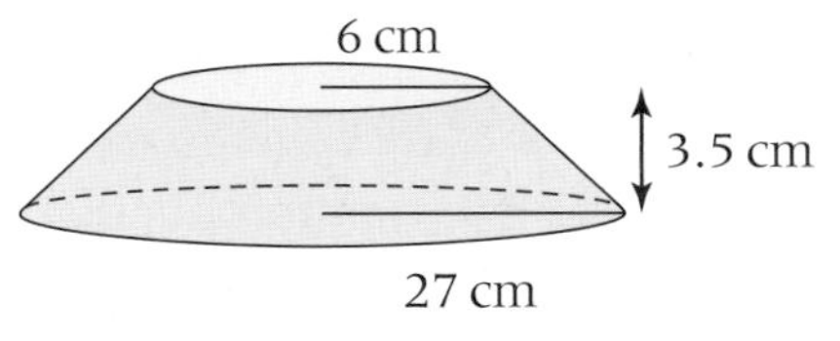

l
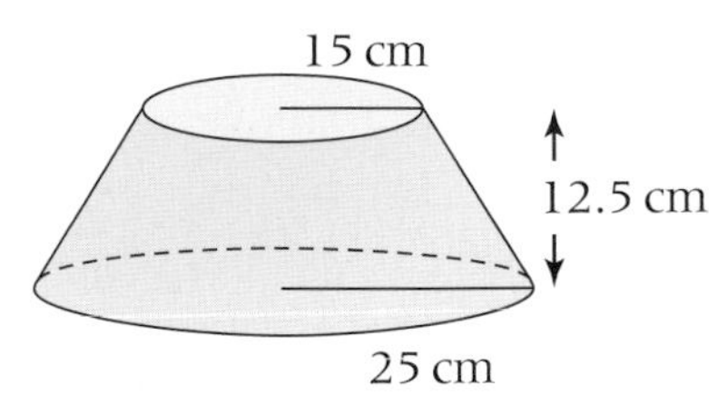

Unit 3

Problem

AO3

2 The diagram shows a frustum.
The base radius, $2r$ cm, is twice the radius of the top of the frustum, r cm.
The height of the frustum is h cm.
Write down an expression for
a the volume
b the surface area of this frustum.

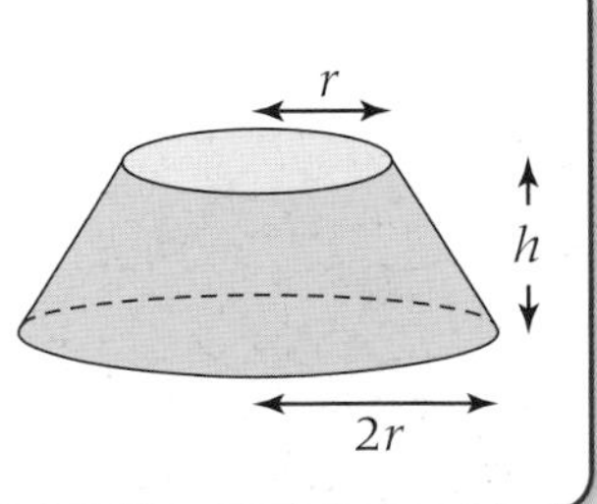

Summary

Check out

You should now be able to:

- Calculate the length of arcs and area of sectors of a circle
- Calculate the surface area and volume of pyramids, cones, spheres, cylinders and frustum of cones
- Convert between length measures, area measures and volume measures
- Understand and use the effect of enlargement for perimeter, area and volume of 2-D and 3-D shapes

Worked exam question

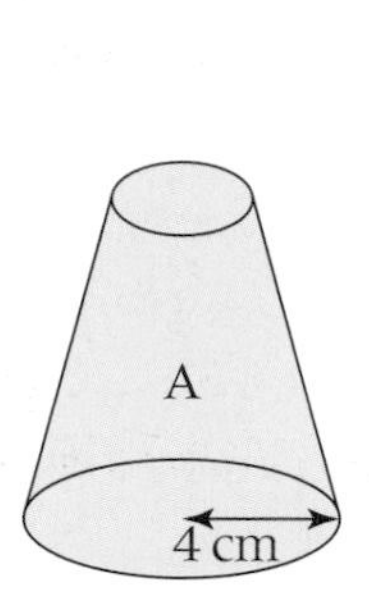

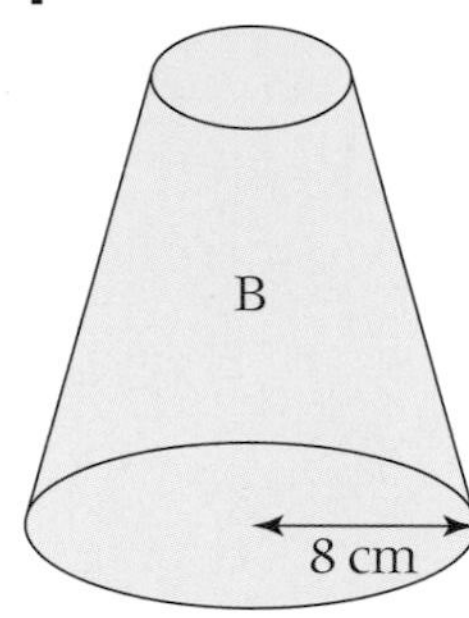

Diagrams NOT accurately drawn.

Two solid shapes, **A** and **B**, are mathematically similar.
The base of shape **A** is a circle with radius 4 cm.
The base of shape **B** is a circle with radius 8 cm.

The surface area of shape **A** is 80 cm^2.

a Work out the surface area of shape **B**. (2)

The volume of shape **B** is 600 cm^3.

b Work out the volume of shape **A**. (2)

(Edexcel Limited 2008)

---

a

Ratio of the lengths is $4 : 8 = 1 : 2$
Ratio of surface areas is $1^2 : 2^2 = 1 : 4$
$80 \times 4 = 320\text{ cm}^2$

The ratio of the surface areas could be given as $4^2 : 8^2$

b

Ratio of volumes is $1^3 : 2^3 = 1 : 8$
$600 \div 8 = 75\text{ cm}^3$

This calculation could also be shown as $\left(\frac{1}{2}\right)^3 \times 600$ or $\left(\frac{4}{8}\right)^3 \times 600$

Exam questions

1

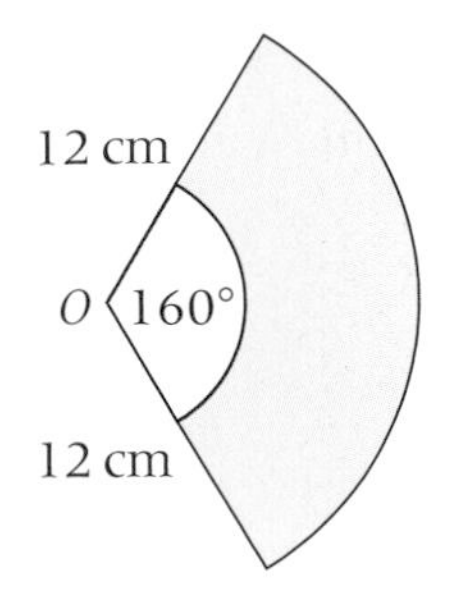

Diagram NOT accurately drawn.

The diagram shows a sector of a circle, centre O.
The radius of the circle is 12 cm.
The angle of the sector is 160°.
Calculate the area of the sector.
Give your answer correct to 3 significant figures. (2)

(Edexcel Limited 2008)

2 Jim makes a model of his school.
He uses a scale of 1 : 50
The area of the door on his model is 8 cm^2.
Work out the area of the door on the real school. (2)

(Edexcel Limited 2006)

3 The diagram shows a storage tank.

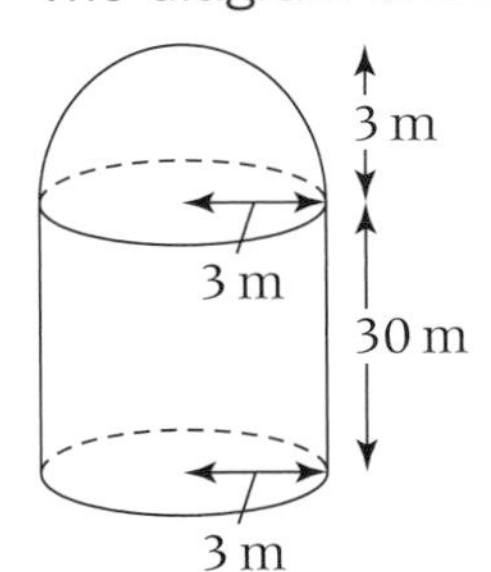

Diagram NOT accurately drawn.

The storage tank consists of a hemisphere on top of a cylinder.

The height of the cylinder is 30 metres.
The radius of the cylinder is 3 metres.
The radius of the hemisphere is 3 metres.

a Calculate the total volume of the storage tank.
Give your answer correct to 3 significant figures. (3)

A sphere has a volume of 500 m^3.

b Calculate the radius of the sphere.
Give your answer correct to 3 significant figures. (3)

(Edexcel Limited 2008)

Proportionality and accurate calculation

Ships are used to transport much of the world's goods. To do this efficiently it is important to optimise the shape of the ship's hull. Rather than experiment with costly, full sized ships, naval architects use scale models and water tanks to perfect their designs.

What's the point?

To work out the actual forces on a full sized ship requires an understanding of proportion. First to make an accurate scale model then to allow for effects such as using fresh rather than salt water or the speed at which the model is towed.

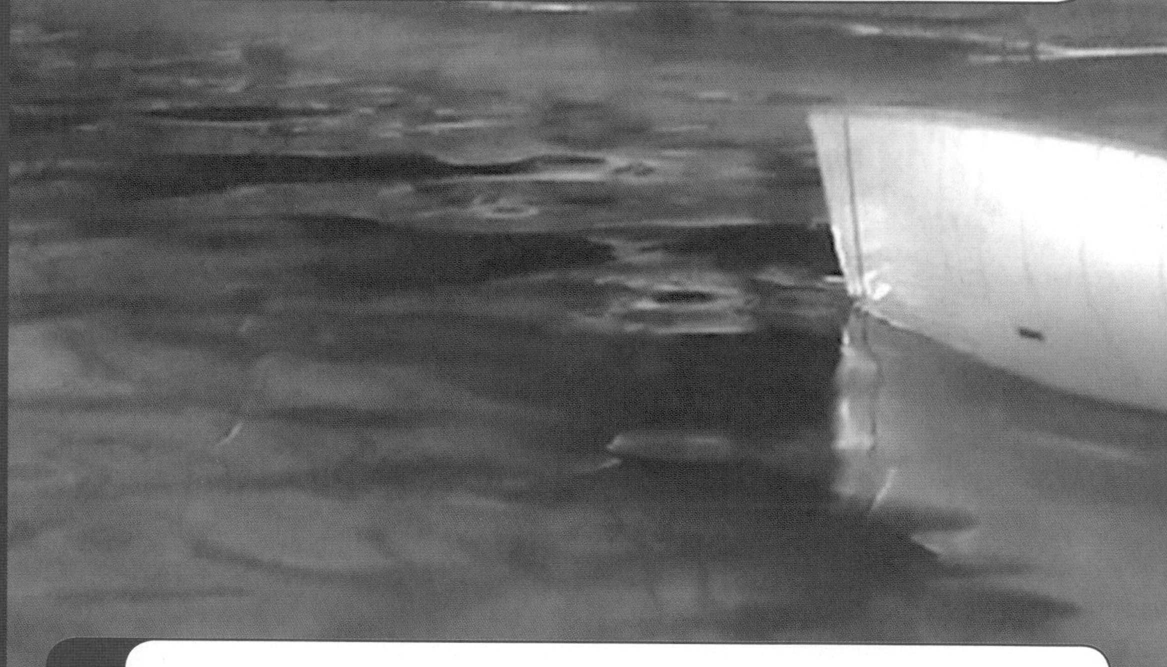

Check in

You should be able to

- **use proportional reasoning**

1 If three identical books weigh 288 g altogether, how much would seven of the same books weigh?

2 A car completes a particular journey in 4 hours, at an average speed of 30 miles per hour.
How long would it take to complete the same journey at an average speed of 40 miles per hour?

- **work with simple formulae**

3 **a** Given that $y = 4x$, find **i** y when $x = 2.5$ **ii** x when $y = 22$.
b Given that $y = 3x^2$, find **i** y when $x = 4$ **ii** x when $y = 75$.
c Given that $y = \frac{12}{x}$, find **i** y when $x = 8$ **ii** x when $y = 10$.

Orientation

What I need to know	What I will learn	What this leads to
N1 Calculate with fractions Round numbers **N2** Proportion	■ Solve ratio and proportion problems ■ Use direct and inverse proportion	**A-level** Maths, Sciences
G3 Similar shapes		Engineering, Science

Rich task

The fraction $\frac{2}{13} = 0.\dot{1}5384\dot{6}$ is a recurring decimal. To convert this back into a fraction can be difficult. Here is a method using reciprocals. Use the $\frac{1}{x}$ button on your calculator as follows.

$\frac{1}{0.153846} \approx 6.5\ldots$ which suggests that the fraction is $\frac{1}{6.5} = \frac{2}{13}$.

Investigate converting some other recurring decimals into fractions using this method.

What about some irrational numbers like π, $\sqrt{2}$, *etc*.

N5.1 Direct proportion

This spread will show you how to:

- Understand direct proportion
- Solve problems involving direct proportion

Keywords
Constant
Direct proportion
Ratio
Variable

- Quantities which are allowed to change are called **variables**.

Amounts of red and white paint in a mixture of pink paint are variables.

- Two variables are in **direct proportion** if the **ratio** between them stays the same as the actual values vary.

One litre of a shade of pink paint is made by mixing 200 ml of red paint and 800 ml of white paint. The ratio of red paint to white paint is 1 : 4. To make the same shade of pink paint, this ratio should stay the same even if the quantities change.

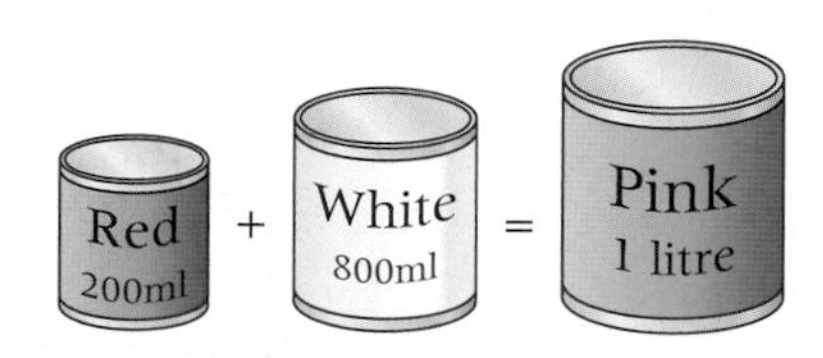

- When you multiply (or divide) one of the variables by a certain number, you have to multiply or divide the other variable by the same number.

250 ml of red paint has to be mixed with $4 \times 250 = 1000$ ml of white paint.

- You write 'y is proportional to x' as $y \propto x$. This can also be written as $y = kx$, where k is the **constant** of proportionality.

You can always find the amount of white paint by multiplying the amount of red paint by a fixed number, the constant of proportionality, in this case 4.

Example

If 35 metres of steel cable weigh 87.5 kilograms, how much do 25 metres of the same cable weigh?

If x = length of the cable (metres)

w = weight (kilograms),

then $w = kx$, where the constant of proportionality k represents the weight of one metre of cable.

$87.5 = k \times 35 \Rightarrow k = 87.5 \div 35 = 2.5$

so $w = 2.5x$

Substitute $x = 25$ into the formula

$w = 2.5 \times 25$

$= 62.5$ kg

Exercise N5.1 — Grade A

1 A 750 ml can of paint covers 5 m^2. What area will a 2.5 litre can of the same paint cover? Show your working.

2 A 4 metre length piece of pipe weighs 28.7 kg. How much does a 3.6 m length of the same pipe weigh?

3 The table shows corresponding values of the variables w, x, y and z.

w	3	6	9	15
x	8	14	20	32
y	5	10	15	25
z	4	7	10	16

Which of these statements could be true?

a $w \propto x$ **b** $z \propto x$ **c** $z \propto w$ **d** $y \propto w$

e $w = ky$, where k is a constant

4 Using the values from the table in question **3**, plot graphs of

a w against x **b** w against y **c** x against y **d** x against z.

Use your results to describe the key features of a graph showing the relationship between two variables that are in direct proportion.

AO2 Functional Maths

5 The weight, w, of a piece of wooden shelving is directly proportional to its length, l.

a Write the statement 'w is directly proportional to l' using algebra.

b Rewrite your answer to part **a** as a formula including a constant of proportionality, k.

c Given that 2.5 m of the shelving weighs 6.2 kg, find the value of the constant k.

d Use your previous answers to calculate the weight of a 2.9 m length of the shelving.

6 450 ml of olive oil costs €2.45. Calculate the cost of 750 ml of the same oil.

7 Anna and Betty buy gravel from a garden centre. Anna buys 4.7 kg of gravel for £12.98. Betty buys 6.2 kg of the same gravel. How much does she pay? Show your working.

Did you know?

It takes more than 5 kilograms of olives to produce 1 litre of olive oil.

AO3 Problem

8 A store sells the same type of ribbon in two different packs.

Regular	**Super**
5 metres	12 metres
Cost £13.25	Cost £31.50

Which of these two packs gives the best value for money? You must show all your working.

N5.2 Further direct proportion

This spread will show you how to:

- Solve problems involving direct proportion

Keywords
Proportional to ($\propto$)
Proportional
Varies

- One quantity can be **proportional** to the square of another quantity.

For example, consider a cuboid of length 5 m and square cross-section which **varies** in size:

p.254

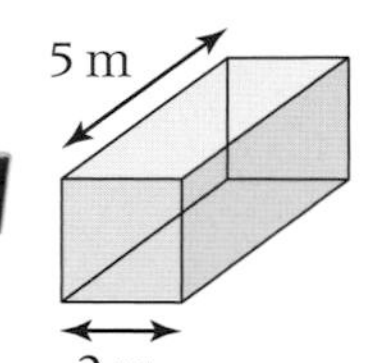

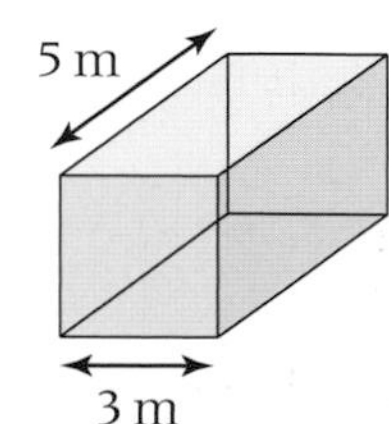

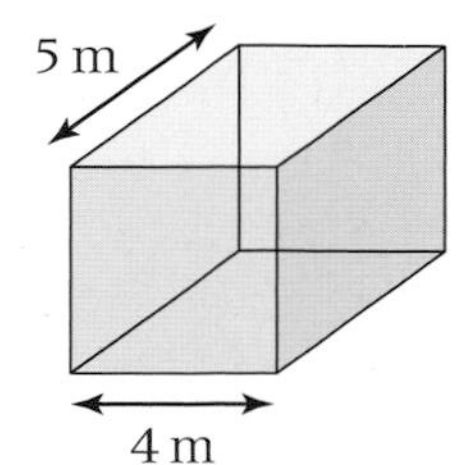

Volume of a cuboid = Area of cross-section × Length
So the volume is proportional to the area of the cross-section.

$V \propto A$ or $V = kA$ ... in this case, $V = 5A$

Also, the area of the cross-section is the square of its side length.

$V \propto l^2$ or $V = kl^2$ ... in this case, $V = 5l^2$

- Similarly, a quantity can be proportional to the cube of another quantity, $y \propto x^3$, or to the square root of another quantity, $y \propto \sqrt{x}$.

Example

Given that y varies directly with the square of x and that y is 75 when x is 5, find y when x is 15.

$y \propto x^2$ so $y = kx^2$
When $y = 75$, $x = 5$ $\quad 75 = k \times 5^2$
$75 = 25k$
$k = 3$
Hence, $y = 3x^2$
When x is 15, $y = 3 \times 15^2 = 675$

Use the values given to find the constant of proportionality, k.

Example

A wedding cake is going to be made from two round tiers.
The baking time, T minutes, is directly proportional to the square of the individual cake's radius, R mm.
When $R = 150$, $T = 50$.
Find T when $R = 180$.

$T \propto R^2$ so $T = kR^2$
When $R = 150$, $T = 50$, so $50 = k \times 150^2$
$k = 0.002222\,2...$
Hence, $T = 0.00222... \times 180^2$
$T = 72$

Exercise N5.2 Grade A

1 Norman thinks of a number, squares it and then multiplies the answer by 4.

a Copy and complete the table to show Norman's starting number, x and his final answer, y.

x	1	2	3	4	5	6
x^2						
y						

b Explain what $y \propto x^2$ means and why it is true for Norman's numbers.

c Write an algebraic relationship between y and x and use it to find Norman's starting number when his final answer is 784.

d Which graph best represents the relationship between y and x?

i

ii

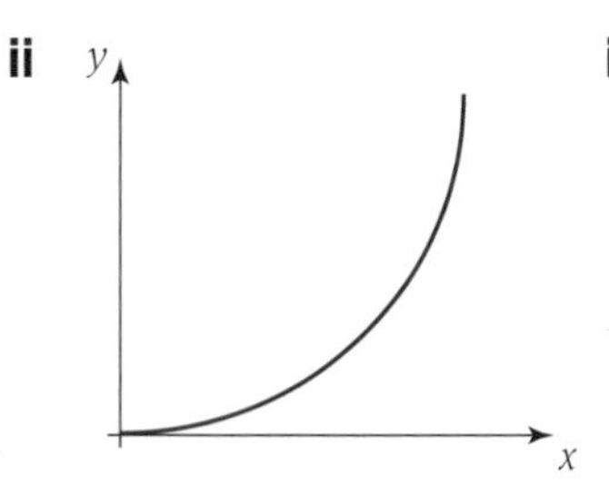

iii

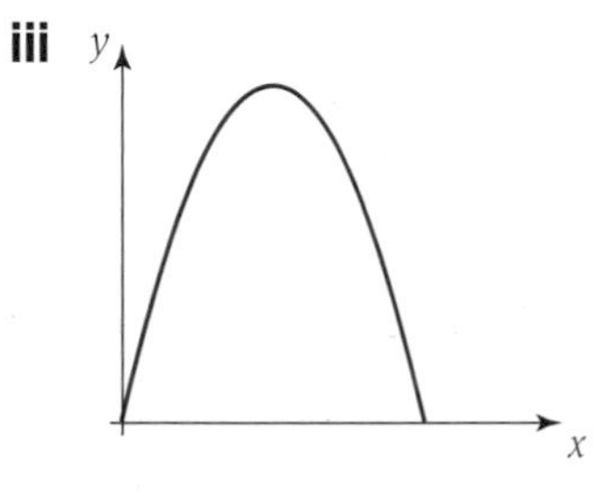

2 R varies with the square of s. If R is 144 when s is 1.2, find

a a formula for R in terms of s

b the value of R when s is 0.8

c the value of s when R is 200.

3 For a circle, explain how you know that the area is directly proportional to the square of its radius.
State the value of the constant of proportionality in this case.
What other formulae do you know that show direct proportion?

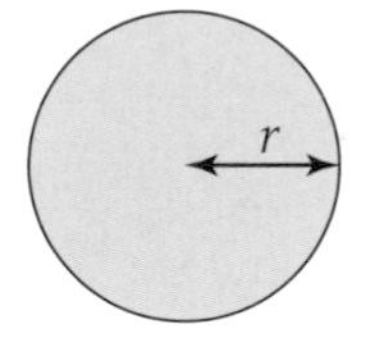

4 The surface area of a sphere varies with the square of its radius. If the surface area of a sphere with radius 3 cm is 113 cm^2, find the surface area of a sphere with radius 7 cm.

5 Given that y varies with the cube of x and that y is 16 when x is 2, find

a a formula connecting y and x

b the value of x when y is 128.

6 Given that P is directly proportional to the square root of k and that P is 20 when k is 16, find the value of P when k is 100.

N5.3 Inverse proportion

This spread will show you how to:

- Understand inverse proportion
- Solve problems involving inverse proportion

Keywords
Constant
Inverse proportion
Product

When two variables are in direct proportion, as the value of one of the variables increases, the other one increases; as the value of one of the variables decreases, the other one decreases.

- When variables are in **inverse proportion**, one of the variables increases as the other one decreases, and vice-versa.
- 'y is inversely proportional to x' can be written as $y \propto \frac{1}{x}$, or $y = \frac{k}{x}$, where k is the **constant** of proportionality.
- If two variables are in inverse proportion, the **product** of their values will stay the same.

For example, the number of bricklayers building a wall and the time taken are in inverse proportion.

5 bricklayers	$\times$ 4 hours	= 20	5 people take 4 hours.
10 bricklayers	$\times$ 2 hours	= 20	10 people take 2 hours.
1 bricklayer	$\times$ 20 hours	= 20	1 person takes 20 hours.

The amount of labour needed is often called 'man-hours'.

Example

A pot of paint, which will cover an area of $12\,\text{m}^2$, is used to paint a garage floor.

a Show that the width of the floor that can be painted is inversely proportional to its length.

b Find the width of the floor when the length is

i 1 m **ii** 3 m **iii** 6.5 m

a Use l for the length of the floor and w for the width

$$lw = 12 \Rightarrow w = \frac{12}{l}$$

This equation shows that w is inversely proportional to l.

b Substitute the given values into the formula.

i $w = \frac{12}{1} = 12\,\text{m}$ **ii** $w = \frac{12}{3} = 4\,\text{m}$ **iii** $w = \frac{12}{6.5} = 1.85\,\text{m}$

Example

A trolley travels down a 10 m long test track at a constant speed.

a Show that the time required for the journey is inversely proportional to the speed of the trolley.

b Find the speed required to make the trolley complete the journey in exactly 4 seconds.

a Use v to represent the speed (in metres per second), and t to represent the time (in seconds). The relationship 'Speed $\times$ Time = Distance' gives:

$$vt = 10 \Rightarrow t = \frac{10}{v}$$

This equation shows that t is inversely proportional to v.

b The equation from part **a** can be rearranged to give $v = \frac{10}{t}$.

Substituting $t = 4$ gives $v = \frac{10}{4} = 2.5$ metres per second.

Exercise N5.3 Grade A

1 You are told that the variable y is directly proportional to the variable x. Explain what will happen to the value of y when the value of x is

a doubled **b** halved **c** multiplied by 6
d divided by 10 **e** multiplied by a factor of 0.7.

2 You are told that the variable w is inversely proportional to the variable z. Explain what will happen to the value of w when the value of z is

a doubled **b** halved **c** multiplied by 6
d divided by 10 **e** multiplied by a factor of 0.7.

3 You are told that y is inversely proportional to x, and that when $x = 4$, $y = 4$. Find the value of y when x is equal to

a 8 **b** 2 **c** 40 **d** 1 **e** 100.

4 Given that $y \propto \frac{1}{w}$, and that $y = 10$ when $w = 50$, write an equation connecting y and w.

5 You are told that $y = \frac{k}{x}$, and that $y = 20$ when $x = 40$. Find the value of the constant k.

AO2 Functional Maths

6 The number of hours, t, required to dig a hole is inversely proportional to n, the number of men digging the hole. It would take 2 men 5 hours to dig the hole.

a Write a formula to give t in terms of n.

b Find the time required to dig the hole when the number of men is

i 1 **ii** 4 **iii** 8 **iv** 5 **v** 7

c Do you think that the assumption $t \propto \frac{1}{n}$, is realistic? Explain your answer.

7 The electric current, I amps, flowing through a component is inversely proportional to the resistance, R ohms, of the component. When $R = 240$, $I = 1$. Find the value of I when R is equal to

a 120 **b** 60 **c** 40 **d** 30
e 24 **f** 360 **g** 480 **h** 100.

8 Use your answers from question **7** to plot a graph of R against I. Comment on the main features of your graph.

9 The variables u and v are in inverse proportion to one another. When $u = 6$, $v = 8$. Find the value of u when $v = 12$.

AO3 Problem

10 Prove that

a if x is directly proportional to y, then y is directly proportional to x

b if y is inversely proportional to x, then x is inversely proportional to y.

N5.4 Further inverse proportion

This spread will show you how to:

- Solve problems involving inverse proportion

Keywords
Inverse proportion
Proportional
Square
Varies

- One quantity can be **inversely proportional** to the **square** of another quantity.

For example, the number of tiles needed to cover a wall is related to the area of the tile.
As the size of the tile gets larger, the number required gets smaller.
The number of tiles is inversely proportional to the area of each tile.
Also, the area of a tile is **proportional** to the square of its length, l^2.

So the number of tiles, N is inversely proportional to l^2: $N \propto \frac{1}{l^2}$ or $N = \frac{k}{l^2}$.

- Similarly, a quantity can be inversely proportional to the cube of another quantity $y \propto \frac{1}{x^3}$, or to the square root of another quantity, $y \propto \frac{1}{\sqrt{x}}$.

Example

Given that y is inversely proportional to the square of x and that y is 4 when x is 5, write a formula for y in terms of x.

$y \propto \frac{1}{x^2}$ so $y = \frac{k}{x^2}$

When x is 5, y is 4 $\quad 4 = \frac{k}{5^2}$

$k = 4 \times 5^2 = 100$

Hence, $y = \frac{100}{x^2}$

Example

The force of attraction, F, between two magnets is inversely proportional to the square of the distance, d, between them.
Two magnets are 0.5 cm apart and the force of attraction between them is 50 newtons.
When the magnets are 3 cm apart what will be the force of attraction between them?

$F \propto \frac{1}{d^2}$, so $F = \frac{k}{d^2}$

When F is 50, d is 0.5 $\quad 50 = \frac{k}{0.5^2}$

$k = 50 \times 0.5^2$
$= 12.5$

So $F = \frac{12.5}{d^2}$

When $d = 3 \quad F = \frac{12.5}{3^2}$
$= 1.4$ newtons

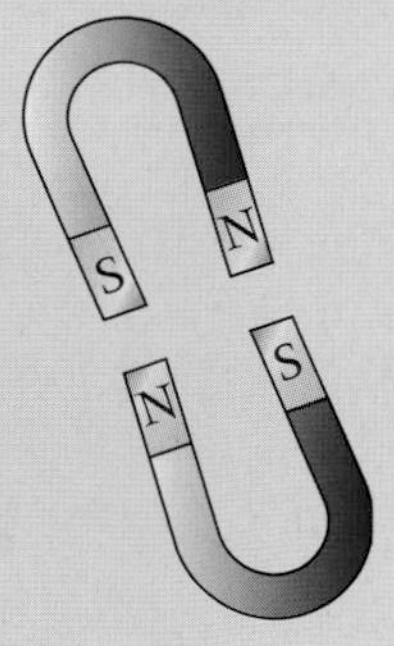

Exercise N5.4 Grade A*

1 Imogen carries out a scientific experiment. She sets two magnets at varying distances apart and notices that the force between them decreases as the square of the distance increases. She performs the experiment three times. These are her results.

Distance d (cm)	1	2	3	4	5
d^2	1	4	9	16	25
Force F (newtons)	100	25	?	?	4

a Study Imogen's results. Write a formula for F in terms of d and use it to complete the rest of the table.

b Plot a graph of F against d and join the points to form a smooth curve.

2 Match the four sketch graphs with the four proportion statements.

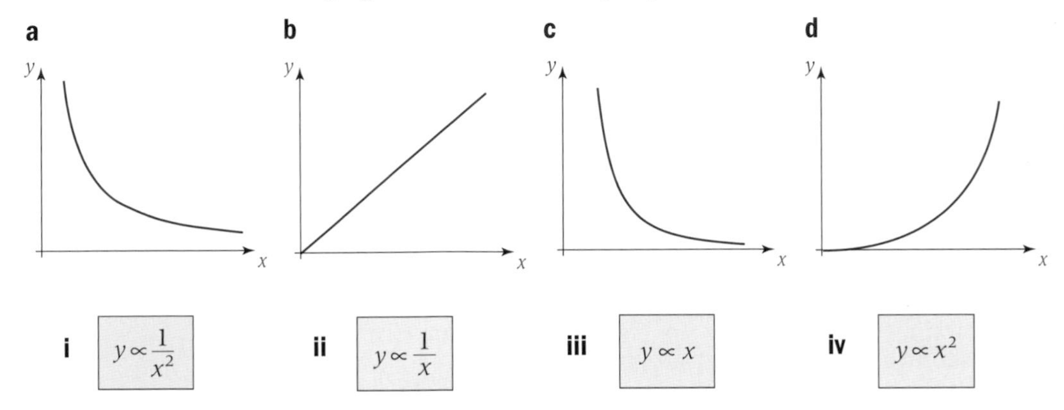

i $y \propto \frac{1}{x^2}$ **ii** $y \propto \frac{1}{x}$ **iii** $y \propto x$ **iv** $y \propto x^2$

Functional Maths AO2

3 A carpet layer is laying square floor tiles in two identical apartments. The time t (mins) that it takes to lay the tiles is inversely proportional to the square of the length of each tile. Given that tiles with a side length of 20 cm take 4 hours and 10 minutes to lay, how long will the second apartment take using tiles with side length 40 cm?

4 Given that y is inversely proportional to the square root of x and that y is 5 when x is 4, find

a a formula for y in terms of x

b the value of

i y when x is 100 **ii** x when y is 12.

Problem AO3

5 Given that P varies inversely with the cube of t and that P is 4 when t is 2, find

a a formula for P in terms of t

b without a calculator

i P when t is 3 **ii** t when P is $\frac{1}{2}$.

Unit 3

N5.5 Exact calculations

This spread will show you how to:

- Use π and surds in exact calculations

Keywords
Irrational
π (pi)
Recurring
Surds
Terminating

- A fraction will have a **terminating** decimal equivalent if the only prime factors of the denominator are 2 or 5. $\frac{3}{20} = \frac{3}{(2 \times 2 \times 5)}$
- All other fractions give **recurring** decimals. $\frac{1}{18} = \frac{1}{(2 \times 3 \times 3)} = 0.05555\ldots = 0.0\dot{5}$

When you calculate with fractions, you should work with the numbers in fraction form as far as possible.

p.28

Example

Calculate

a $\frac{1}{2} + \frac{2}{3}$ **b** $\frac{3}{4} \times \frac{2}{9}$ **c** $\frac{4}{5} \div \frac{3}{10}$

a $\frac{1}{2} + \frac{2}{3} = \frac{3}{6} + \frac{4}{6} = \frac{7}{6} = 1\frac{1}{6}$ **b** $\frac{{}^{1}\cancel{3}}{{}_{2}\cancel{4}} \times \frac{\cancel{2}^{1}}{\cancel{9}_{3}} = \frac{1}{2} \times \frac{1}{3} = \frac{1}{6}$

c $\frac{4}{5} \div \frac{3}{10} = \frac{4}{\cancel{5}_{1}} \times \frac{\cancel{10}^{2}}{3} = \frac{4}{1} \times \frac{2}{3} = \frac{8}{3} = 2\frac{2}{3}$

- Some numbers cannot be written as fractions. These **irrational** numbers have no repeating decimal patterns.
- Irrational numbers such as $\sqrt{2}$ are called **surds**.

$\sqrt{2} = 1.414213562\ldots$

You can work with combinations of rational numbers and surds.

$$\frac{1 + \sqrt{5}}{2} \quad \text{or} \quad \frac{1}{2\pi}\sqrt{\frac{7}{9.8}}$$

Again, work with the numbers in surd form as far as possible.

p.194

Example

Calculate

a $\frac{7 + \sqrt{20}}{4} - \frac{3 + \sqrt{5}}{5}$ **b** $(1 + \sqrt{5}) \div (2 + \sqrt{3})$

a $\frac{7 + \sqrt{20}}{4} - \frac{3 + \sqrt{5}}{5} = \frac{35 + 5\sqrt{20} - 12 - 4\sqrt{5}}{20}$

$= \frac{35 + 5\sqrt{20} - 12 - 4\sqrt{5}}{20}$

$= \frac{23 + 10\sqrt{5} - 4\sqrt{5}}{20}$

$= \frac{23 + 6\sqrt{5}}{20}$

b $(1 + \sqrt{5}) \div (2 + \sqrt{3})$

$= \frac{1 + \sqrt{5}}{2 + \sqrt{3}}$

$= \frac{1 + \sqrt{5}}{2 + \sqrt{3}} \times \frac{2 - \sqrt{3}}{2 - \sqrt{3}}$

$= 2 - \sqrt{3} + 2\sqrt{5} - \sqrt{15}$

Exercise N5.5 Grade A*

1 Evaluate exactly

a $\frac{1}{3}+\frac{1}{5}$ **b** $\frac{2}{5}+\frac{3}{7}$ **c** $\frac{3}{8}-\frac{2}{7}$ **d** $\frac{8}{15}+\frac{4}{9}$

e $6\frac{1}{2}-1\frac{5}{8}$ **f** $\frac{2}{5}+\frac{1}{3}+\frac{1}{4}$ **g** $\frac{3}{8}+\frac{1}{2}-\frac{2}{5}$ **h** $7\frac{2}{9}+2\frac{1}{4}$

2 Evaluate exactly

a $\frac{2}{3}\times\frac{3}{4}$ **b** $\frac{5}{9}\div\frac{1}{3}$ **c** $2\frac{1}{2}\times\frac{5}{8}$ **d** $\frac{8}{9}\div 1\frac{2}{3}$

e $3\frac{1}{2}\div 2\frac{1}{4}$ **f** $5\frac{1}{5}\times 2\frac{3}{4}$ **g** $8\frac{2}{5}\div 3\frac{1}{7}$ **h** $3\frac{1}{2}\times 7\frac{5}{9}$

3 For each answer from questions **1** and **2**, either give an exact decimal equivalent of the answer, or explain why it is not possible to do so.

4 Evaluate exactly

a $\frac{2}{5}\times\left(\frac{3}{4}+\frac{2}{3}\right)$ **b** $\left(\frac{2}{3}+\frac{4}{5}\right)\div\left(\frac{2}{5}+\frac{3}{7}\right)$ **c** $\frac{5}{6}\div\frac{3^2+4^2}{10}$

d $\left(\frac{3}{7}\right)^2\times\left(\frac{4}{5}-\frac{1}{7}\right)$ **e** $\left(\frac{1}{2}+\frac{5}{9}\right)^2+\left(\frac{2}{3}\right)^3$ **f** $\left(\frac{5}{6}+\frac{1}{2}\right)-\left(\frac{4}{7}\times 3\right)$

5 Simplify these expressions

a $\sqrt{20}+3\sqrt{5}$ **b** $(5+3\sqrt{3})-(2+4\sqrt{3})$

c $\frac{6+\sqrt{27}}{3}-\frac{2+\sqrt{3}}{4}$ **d** $\frac{20+3\sqrt{7}}{3}-(5+\sqrt{28})$

e $\frac{6+\sqrt{8}}{2}+\frac{3+\sqrt{2}}{9}$ **f** $\frac{8+\sqrt{45}}{3}-\frac{4+\sqrt{20}}{9}$

6 Evaluate, giving your answers in surd form.

a $\sqrt{5}(1+\sqrt{5})$ **b** $\sqrt{3}(5-\sqrt{3})$

c $(2+\sqrt{3})(3+\sqrt{3})$ **d** $(3-\sqrt{2})^2$

e $(5-\sqrt{7})(5+\sqrt{7})$ **f** $(4+3\sqrt{5})(6-\sqrt{5})$

7 Write each expression without irrational numbers in the denominator.

a $\frac{1}{\sqrt{2}}$ **b** $\frac{1+\sqrt{3}}{\sqrt{3}}$ **c** $\frac{5-\sqrt{20}}{\sqrt{5}}$

d $\frac{3}{1+\sqrt{7}}$ **e** $\frac{8}{5-2\sqrt{3}}$ **f** $\frac{3-\sqrt{3}}{4+\sqrt{5}}$

8 Simplify these expressions, writing your solutions without irrational numbers in the denominators.

a $(1+\sqrt{5})\div(1+\sqrt{3})$ **b** $(2+\sqrt{7})\div(2+\sqrt{3})$

c $(5-\sqrt{11})\div(2+\sqrt{11})$ **d** $(1+2\sqrt{3})\div(2-\sqrt{3})$

e $(5+2\sqrt{3})\div(3-2\sqrt{3})$ **f** $(7-2\sqrt{5})\div(8-2\sqrt{5})$

N5.6 Limits of accuracy

This spread will show you how to:

- Recognise limitations on the accuracy of data and measurements and calculate their upper and lower bounds
- Give answers to an appropriate degree of accuracy

Keywords
Implied accuracy
Lower bound
Upper bound

In calculations, give your answers to an appropriate degree of accuracy.

A measurement of 2.3 cm has an **implied accuracy** of 1 decimal place.

p.22

A measurement given as 3.50 m has an implied accuracy of 2 decimal places.

If all measurements are to 1 dp, give your answers to 1 or 2 dp.

$2.25 \leqslant x < 2.35$

2.2 2.25 2.3 2.35 2.4

All values in the range $2.25 \le x < 2.35$ round to 2.3 to 1 dp.
The upper bound is 2.35 cm and the lower bound is 2.25 cm.

Example

A pile of 16 sheets of card is 7.3 mm thick.
Calculate the thickness of one sheet.

$7.3 \text{ mm} \div 16 = 0.456\,25 \text{ mm} = 0.46 \text{ mm (2 sf)}$

0.456 25 mm implies that you could measure the thickness of a piece of card to the nearest 0.000 01 mm – unlikely!

Use upper and lower bounds to calculate precise limits of accuracy.

- The **upper bound** is the smallest value that is greater than or equal to any possible value of the measurement.
- The **lower bound** is the greatest value that is smaller than or equal to any possible value of the measurement.

For continuous data, the upper bound is *not* a possible value of the data.

Example

In a science experiment, a trolley travelled 8.4 m in 3.6 seconds.
Calculate the upper and lower bound of the average speed of the trolley in metres per second (ms^{-1}), writing down all the digits on your calculator display.

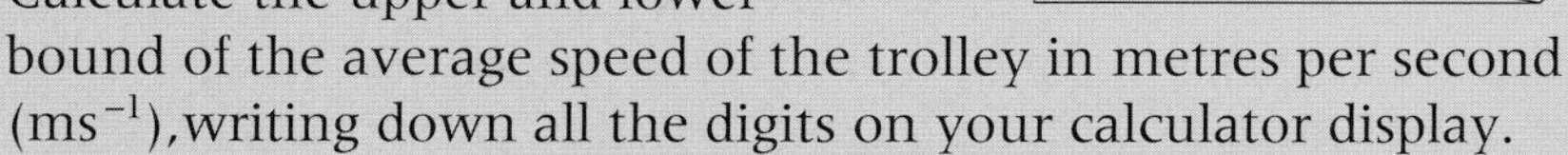

Speed = distance ÷ time

$\text{Speed}_{\text{Upper}} = \text{Distance}_{\text{Upper}} \div \text{Time}_{\text{Lower}} = 8.45 \div 3.55 = 2.380\,281\,7 \text{ m s}^{-1}$

$\text{Speed}_{\text{Lower}} = \text{Distance}_{\text{Lower}} \div \text{Time}_{\text{Upper}} = 8.35 \div 3.65 = 2.287\,671\,2 \text{ m s}^{-1}$

Distance is in the range $8.35 \le d < 8.45$ m.
Time is in the range $3.55 \le t < 3.56$ seconds.

Example

A measurement is given as 3.8 cm, correct to the nearest 0.1 cm.
Pritesh writes:

Lower bound = 3.75 cm
Upper bound is a bit less than 3.85 cm,
say 3.84999

Is he correct?

The lower bound is correct; 3.75 is the largest value that is less than or equal to any possible value of the measurement.
The upper bound is incorrect. It should be 3.85.

Exercise N5.6 — Grade A/A*

Unit 3

1 The length of a rod is 0.45 m, correct to the nearest centimetre. Write the upper and lower bounds for the length of the rod.

2 Use the implied accuracy of these measurements to write down the upper and lower bound for each one.

a 6.4 mm **b** 4.72 m **c** 18 s **d** 0.388 kg **e** 6.5 volts

3 A car travels a distance of 32 m (to the nearest metre) in a time of 1.6 seconds (to the nearest 0.1 s). Find the upper and lower bounds of the average speed of the car.

4 A rectangle has a length of 4.3 m and a width of 3.1 m, both measured to the nearest 0.1 m.
Find

a the lower bound of the perimeter of the rectangle
b the upper bound of the area of the rectangle.

5 Find the upper and lower bounds on

i the surface area **ii** volume of a cube of side length

a 100 cm **b** 1000 mm.

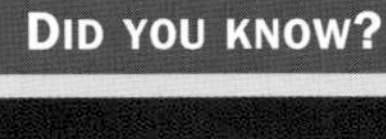
Did you know?

The largest tower crane in the world can lift a maximum load of 120 tonnes.

Functional Maths — AO2

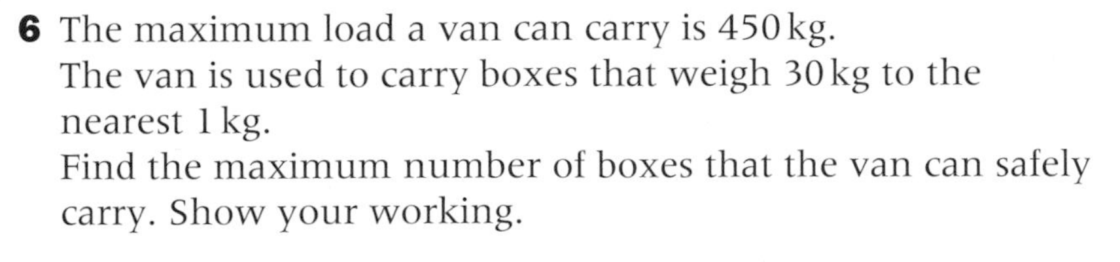

6 The maximum load a van can carry is 450 kg.
The van is used to carry boxes that weigh 30 kg to the nearest 1 kg.
Find the maximum number of boxes that the van can safely carry. Show your working.

7 A crane has a maximum working load of 670 kg to 2 sf.
It is used to lift crates that weigh 85 kg, to the nearest 5 kg.
What is the greatest number of crates that the crane can safely lift at one time?

8 A lift can carry a maximum of five people, and the total load must not exceed 440 kg. Five members of a judo team enter the lift. Each person weighs 87 kg to the nearest kilogram.
Is it possible that the total weight of the group exceeds the maximum load of the lift? Show your working.

9 This trapezium has area 450 cm^2 to 2 sf. It has parallel sides 18 cm and 22 cm, each to the nearest centimetre.

Calculate the lower bound of the height, h, of the trapezium.

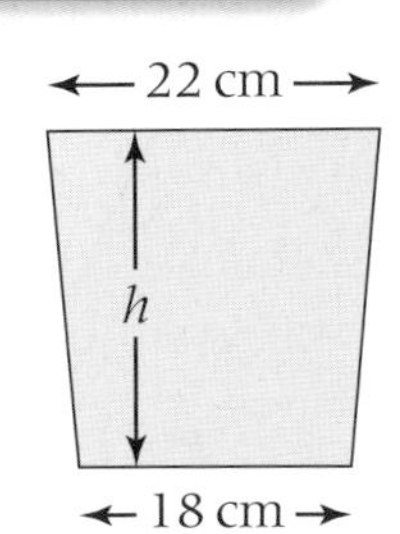

10 The length of Eva's stride is 86 cm, to the nearest centimetre.

a Write the upper and lower bounds of Eva's stride.
b The length of a path is 28 m, to the nearest metre.
Starting at the beginning of the path, Eva takes 32 strides in a straight line along the path. Explain, showing all your working, why Eva may not reach the end of the path.

N5.7 Efficient use of a calculator

This spread will show you how to:

- Use calculators effectively and efficiently for complex calculations
- Use calculators to calculate in standard form

Keywords
Function
Mode
Standard form

You need a calculator to find the results of complex calculations.

In such calculations you have to follow the order of operations. Make sure you know how to enter complex calculations in your calculator.

- You need to know how to use the **function** keys on your calculator.
- You need to know how to enter and interpret numbers in **standard form**.

p.182

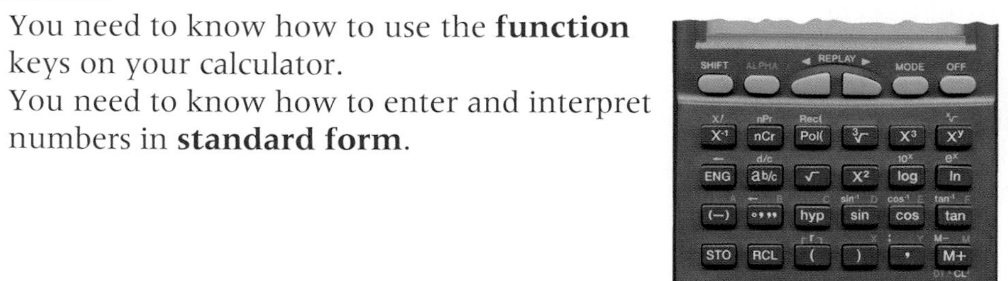

Not all calculators are the same!
For example, to calculate $\sqrt{74}$:
- on some you press [√] then [7] [4]
- on some you press [7] [4] then [√].

Example

Use your calculator to work out $\sqrt{\dfrac{42\,389}{31.6^2}}$.
Give your answer to 3 significant figures.

`6.515376289`

$\sqrt{\dfrac{42389}{31.6^2}} = 6.52$ (3 sf)

Example

Use your calculator to work out $(6.43 \times 10^6) \div (4.21 \times 10^{-2})$.

`1.527315914 08`

$(6.43 \times 10^6) \div (4.21 \times 10^{-2}) = 1.53 \times 10^8$ (to 3 sf)

Example

Calculate $\sqrt{7.9^2 - 8.8 \cos 35°}$.
Rashima calculated the answer 8.39.
Is she correct?

No. The correct answer is 7.43 to 3 st.

p.298

You can estimate the correct answer

$$7.9^2 - 8.8 \cos 35° \simeq 8^2 - 10 = 54$$

$$\sqrt{54} \simeq 7\tfrac{1}{2} < 8.$$

Rashima had her calculator in the wrong **mode** for the angle 35°. When calculating with angles, make sure your calculator is in degree mode.

Exercise N5.7 — Grade B/A

1 Use a calculator to find the value of these in standard form.

a $(6.4 \times 10^{-4}) + (7.1 \times 10^{-3})$ **b** $(9.9 \times 10^{5}) - (2.7 \times 10^{4})$
c $(4.8 \times 10^{-6}) + (3.9 \times 10^{-5})$ **d** $(3.3 \times 10^{2}) - (7.5 \times 10^{1})$
e $(9.8 \times 10^{5}) - (6.4 \times 10^{5})$ **f** $(3.5 \times 10^{-2}) + (9.7 \times 10^{-3})$

2 Use a calculator to find the value of these in standard form.

a $(5.3 \times 10^{-4}) \times (4.1 \times 10^{-7})$ **b** $(5.4 \times 10^{-5}) \div (3.1 \times 10^{2})$
c $(8.9 \times 10^{-6}) \div (6.5 \times 10^{-4})$ **d** $(4.7 \times 10^{-2}) \times (9.2 \times 10^{-8})$
e $(3.8 \times 10^{3}) \times (1.7 \times 10^{5})$ **f** $(2.4 \times 10^{-2}) \div (3.8 \times 10^{-6})$

3 Use your calculator to evaluate these.

a $6.34 + \sqrt{\frac{8.79}{0.35}}$ (Give your answer correct to 2 significant figures.)
b $11\frac{3}{4} - 2\frac{4}{7}$ (Give your answer as a mixed number.)
c $(4.78 \times 10^{-4}) \times (6.1 \times 10^{6})$ (Give your answer in standard form.)

4 Evaluate these.

a $3 \times 4.7^3 + 2 \times 4.7 - 3$ **b** $\left(1 + \frac{4.5}{0.65}\right)^3$ **c** $\left(\frac{1.54 - 0.79}{0.03}\right)^3$
d $105\left(1 + \frac{5.6}{100}\right)^8$ **e** $\left(\frac{4.9^2 \times 3.2 - 0.75}{2.8 - 0.75}\right)^2$ **f** $\frac{3.2^2}{1.7} \times \frac{2.9}{1.6^2}$

5 Evaluate these, giving your results to 3 significant figures.

a $\sqrt{31.9^2 - 8.77^2}$ **b** $6.75 + \sqrt{\frac{3.92}{4.15}}$ **c** $\sqrt{\frac{5.68^4}{4.75^2 - 2.59^2}}$
d $\sqrt[3]{3.87 \times 4.36^2}$ **e** $\frac{1}{2\pi}\sqrt{\frac{1.55}{9.81}}$ **f** $\left(\frac{1 + \sqrt{5}}{2}\right)^{-\frac{3}{2}}$

6 Use the fraction facility on your calculator to evaluate these.

a $\frac{2}{4} + \frac{4}{5}$ **b** $\frac{3}{4} + \frac{5}{6}$ **c** $\frac{7}{9} - \frac{2}{3}$ **d** $\frac{3}{8} \times \frac{2}{9}$
e $\frac{4}{5} \div \frac{2}{3}$ **f** $2\frac{1}{2} \times \frac{3}{4}$ **g** $4\frac{3}{4} + 3\frac{1}{3}$ **h** $4\frac{1}{2} \div 1\frac{2}{3}$
i $1\frac{5}{8} - \frac{11}{16}$ **j** $5\frac{1}{4} \times 2\frac{3}{5}$

7 Evaluate these expressions, giving your answers to 3 significant figures.

a $\frac{2.5 \sin 60°}{4.6^2 - 3.8^2}$ **b** $\sqrt{19.7^2 + 14.5^2 - 2 \times 19.7 \times 14.5 \times \cos 63°}$
c $\frac{\left(\frac{4.7}{9.8}\right)^2 \times \tan 84°}{\sqrt{4.7^2 + 5.2^2}}$ **d** $\frac{\sqrt{3.65 \times 1.93^2 - 8.75 \cos 45°}}{8.75^2 + 3.65^2}$

Functional Maths

8 Amy earns £182 per week and Beth earns £1950 per month. The first £7745 of their annual income is untaxed, the remainder is taxed at 20%. What percentage of their incomes are paid in tax?

N5 Summary

Check out

You should now be able to:

- Understand and solve problems involving direct proportion and inverse proportion
- Recognise the difference between rational and irrational numbers
- Use surds and π in exact calculations
- Rationalise a denominator, including when it involves a surd form
- Use algebraic techniques to expand and simplify brackets containing surds

Worked exam question

a Write down the value of $8^{\frac{1}{3}}$ (1)

$8\sqrt{8}$ can be written in the form 8^k

b Find the value of k. (1)

$8\sqrt{8}$ can also be expressed in the form $m\sqrt{2}$ where m is a positive integer.

c Express $8\sqrt{8}$ in the form $m\sqrt{2}$ (2)

d Rationalise the denominator of $\frac{1}{8\sqrt{8}}$

Give your answer in the form $\frac{\sqrt{2}}{p}$ where p is a positive integer. (2)

(Edexcel Limited 2006)

a

$8^{\frac{1}{3}} = \sqrt[3]{8} = 2$

b

$8\sqrt{8} = 8^1 \times 8^{\frac{1}{2}} = 8^{\frac{3}{2}}$

So $k = \frac{3}{2}$

State the value of k.

c

$8\sqrt{8} = 8\sqrt{4 \times 2}$

$= 8\sqrt{4} \times \sqrt{2}$

$= 8 \times 2 \times \sqrt{2} = 16\sqrt{2}$

Show that $\sqrt{8}$ is the same as $2\sqrt{2}$

d

$\frac{1}{8\sqrt{8}} = \frac{1}{8\sqrt{8}} \times \frac{8\sqrt{8}}{8\sqrt{8}}$

$= \frac{8\sqrt{8}}{8 \times 8 \times 8} = \frac{16\sqrt{2}}{64 \times 8}$

$= \frac{\sqrt{2}}{32}$

Show the multiplication by $\frac{8\sqrt{8}}{8\sqrt{8}}$

Exam questions

A02

1 A ball falls vertically after being dropped.
The ball falls a distance d metres in a time of t seconds.
d is directly proportional to the square of t.

The ball falls 20 metres in a time of 2 seconds.

a Find a formula for d in terms of t. (3)

b Calculate the distance the ball falls in 3 seconds. (1)

c Calculate the time the ball takes to fall 605 m. (3)

(Edexcel Limited 2006)

2 f is inversely proportional to d.

When $d = 50$, $f = 256$

Find the value of f when $d = 80$ (3)

(Edexcel Limited 2007)

A03

3 A ball is thrown vertically upwards with speed V metres per second.
The height, H metres, to which it rises is given by

$$H = \frac{V^2}{2g}$$

where g m/s^2 is the acceleration due to gravity.

$V = 24.4$ correct to 3 significant figures.
$g = 9.8$ correct to 2 significant figures.

Calculate the upper bound of H.
Give your answer correct to 3 significant figures. (3)

(Edexcel Limited 2008)

A5 Inequalities and simultaneous equations

Inequalities and simultaneous equations

In the business world, people try to increase profits by maximising their productivity and minimising their costs. However, there are many factors (called constraints) such as the number of workers, the capacity of their factories, cost of materials, *etc*., which they must take into account.

What's the point?

Mathematicians solve such problems by representing the different constraints as straight line graphs and using a process called linear programming to find the optimal solution.

Check in

You should be able to

- **plot straight line graphs**

1 On axes labelled from −8 to + 8, plot these lines.

a $y = 6$ **b** $x = -2$ **c** $y = 2x + 1$ **d** $x + y = 8$

- **recognise and write inequalities**

2 Insert the symbol $<$, $>$ or $=$ into the $\square$ to make them true.

a 10% of 360 $\square$ 15% of 250 **b** $(-2)^2 \square (-2)^3$

c $-\frac{3}{8} \square -\frac{1}{3}$ **d** $(-2) \times 15 \square 60 \div (-2)$

- **solve simultaneous equations by inspection**

3 Use inspection to find

a two numbers that differ by three and have a product of 40

b two numbers that have sum 9 and difference 5

c two values which add to 7 and whose squares add to 25.

- **manipulate algebraic expressions**

4 Given that $2x + 3y = 9$, find each value.

a $4x + 6y$ **b** $x + 1.5y$ **c** $20x + 30y$ **d** $14x + 21y$

Orientation

What I need to know	What I will learn	What this leads to
A2 Plot and describe straight line graphs	■ Solve inequalities in one and two variables ■ Solve simultaneous linear equations by elimination	**A7** Simultaneous linear, quadratic and circle equations
A3 Solve linear equations		Economics

Rich task

In a store there are a range of different mobile phone packages available.

Package 1 Pay as you go 10p per min.

Package 2 £5 per month and then all calls at 5p per min.

Package 3 £12 per month, with a 100 mins of free calls, and then all calls at 3p per min.

Package 4 £25 per month, with 600 mins of free calls, and then all calls at 2p per min.

Which package is the best value for money?

A5.1 Solving inequalities

This spread will show you how to:

- Solve simple linear inequalities in one variable, and represent the solution set on a number line

Keywords
Inequality
Integer

- An **inequality** is a mathematical statement including one of these symbols:

$<$	$>$	$\leqslant$	$\geqslant$
less than	more than	less than or equal to	more than or equal to

For example $3 < 5$.

- You can solve an inequality by rearranging and using inverse operations, in a similar way to solving an equation.

p.202

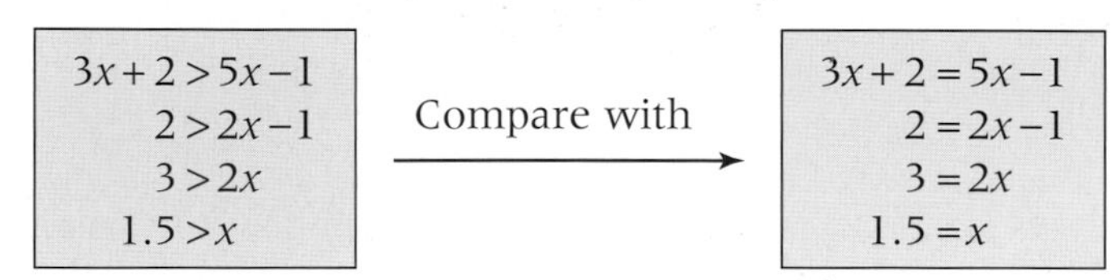

- If you multiply or divide an inequality by a positive number, the inequality remains true.
- If you multiply or divide an inequality by a negative number you need to reverse the inequality sign to keep it true.
- The solution to an inequality can be a range of values, which you can show on a number line:

$4 < 6$ and $8 < 12$ and $2 < 3$

$4 < 6$ but $-2 > -3$
$5 > 2$ but $-15 < -6$

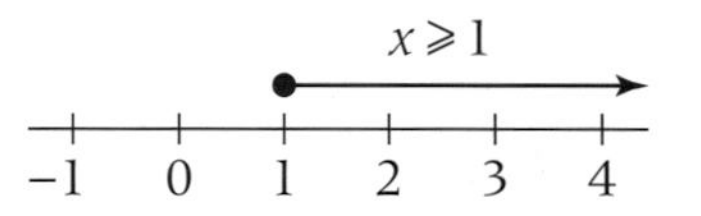

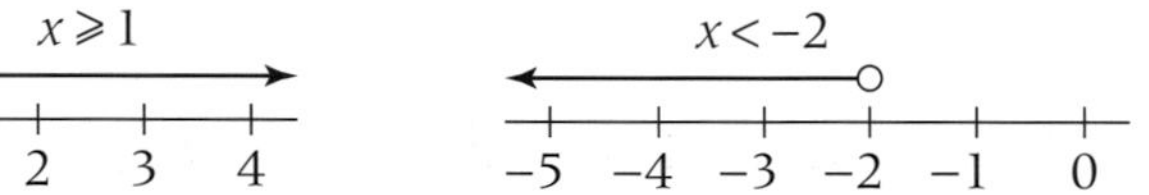

Use an 'empty' circle for $<$ and $>$. Use a 'filled' circle for $\leqslant$ and $\geqslant$.

Example

a Find the range of values of x that satisfy both $3x \geqslant 2(x - 1)$ and $10 - 3x > 6$.
Represent the solution set on a number line.

b List the **integer** values of x that satisfy both inequalities.

Integers: positive and negative whole numbers and zero.

a $3x \geqslant 2(x - 1)$ $\quad$ $10 - 3x > 6$
$3x \geqslant 2x - 2$ $\quad$ $10 > 6 + 3x$
$x \geqslant -2$ $\quad$ $4 > 3x$
$1\frac{1}{3} > x$

So $-2 \leqslant x < 1\frac{1}{3}$.

Combine the two inequalities. $1\frac{1}{3} > x$ is the same as $x < 1\frac{1}{3}$.

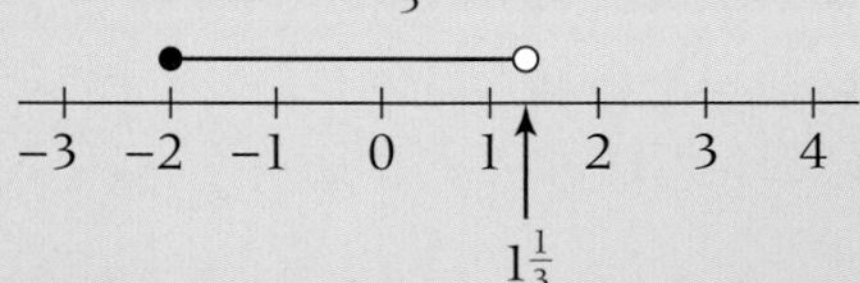

b The integer values of x that satisfy both inequalities are −2, −1, 0 and 1.

Exercise A5.1

Grade B/A

1 **a** What is the smallest prime number p such that $p > 30$?
b What is the smallest value of x such that $2^x \geqslant 1024$?

Use trial and improvement.

2 True or false: if x is a number, then $x^2 > x$?
If false, write the range of values of x for which it is false, as an inequality.

3 Solve these inequalities, representing each solution on a number line.

a $3x - 5 > 18$ **b** $\frac{p}{4} + 6 \leqslant -2$ **c** $6x + 3 \leqslant 2x - 8$

d $-3y > 12$ **e** $\frac{q}{-5} \geqslant -2$ **f** $4z - 3 \leqslant 3(z - 2)$

g $3(y - 2) < 8(y + 6)$

4 **a** The area of this rectangle exceeds its perimeter. Write an inequality and solve it to find the range of values of x.
b Given that x is an integer, find the smallest possible value that x can take.

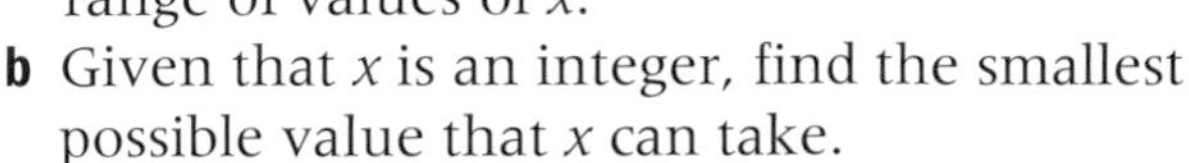

5 Find the range of values of x that satisfy *both* inequalities.
a $3x + 6 < 18$ and $-2x < 2$
b $10 > 5 - x$ and $3(x - 9) < 27$

6 Explain why it is not possible to find a value of y such that $3y \leqslant 18$ and $2y + 3 > 15$.

7 Solve, by treating the two inequalities separately.
a $y \leqslant 3y + 2 \leqslant 8 + 2y$
b $z - 8 < 2(z - 3) < z$
c $4p + 1 < 7p < 5(p + 2)$

In part **a**, solve $y \leqslant 3y + 2$ and $3y + 2 \leqslant 8 + 2y$.

Unit 3

AO3 Problem

8 **a** Explain why the solution to $x^2 \leqslant 25$ is not simply $x \leqslant 5$.
b Hence, solve:
i $3x^2 + 1 > 49$ **ii** $10 - p^2 < 6$

9 By drawing a graph or otherwise, and using a calculator, find the range of values between 0° and 360° such that
a $\sin x > 0.5$ **b** $2 \cos x > \sqrt{3}$

A5.2 Regions

This spread will show you how to:

- Represent inequalities in two variables

Keywords
Inequality
Region

- An **inequality** is a mathematical statement using one of these signs:
  $<$ less than $\leqslant$ less than or equal to
  $>$ greater than $\geqslant$ greater than or equal to

5 > 3 and 3 < 5 are inequalities.

- You can represent inequalities as **regions** on a graph.

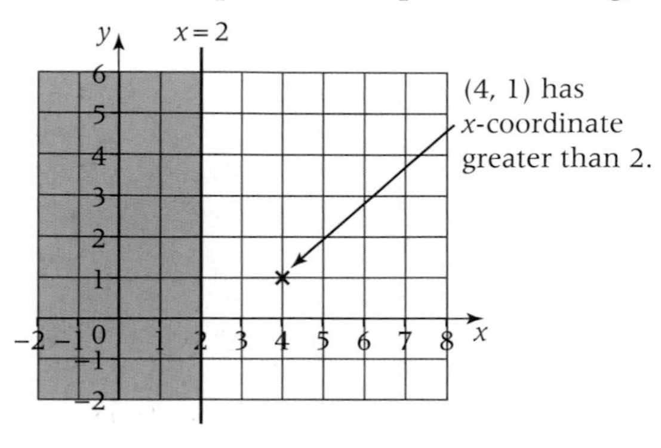

The unshaded region shows all points with x-coordinate greater than 2, or the inequality $x \geqslant 2$.

The line $x = 2$ is solid, to show that points on this line are included in the region required.

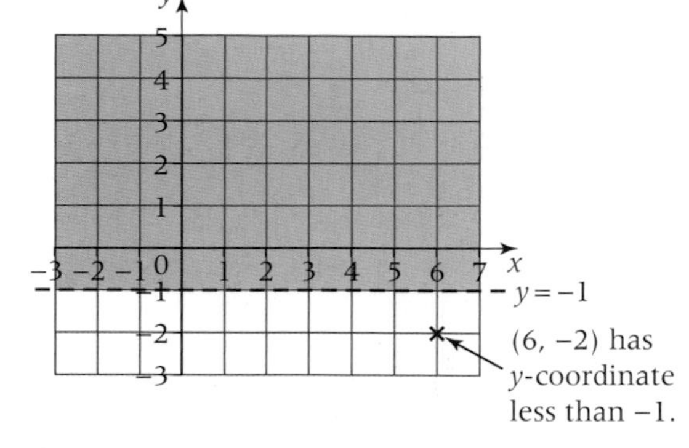

The *unshaded* region shows all points with y-coordinate less than -1 or the inequality $y < -1$.

The line $y = -1$ is dashed, to show that points on this line are *not* included in the region required.

Example

On a pair of axes, construct an *unshaded* region to represent the inequalities $x > 1$ and $y \geqslant 2$.

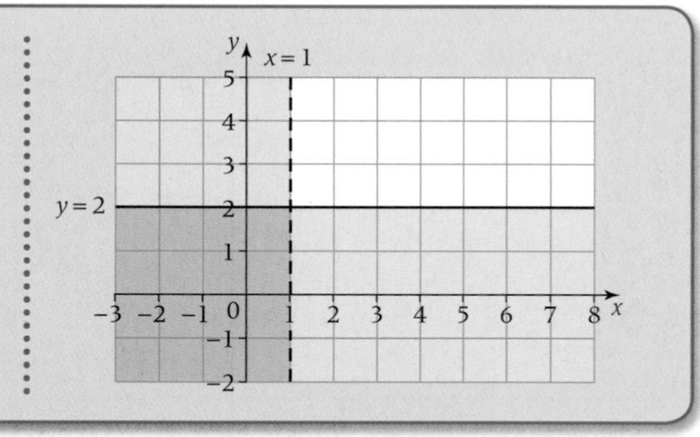

First draw the lines $x = 1$ and $y = 2$.

Use a solid line for $\leqslant$ or $\geqslant$. Use a dashed line for $<$ or $>$.

Example

Write the inequality represented by the *shaded* region in this diagram.

$-1 < x \leqslant 4$

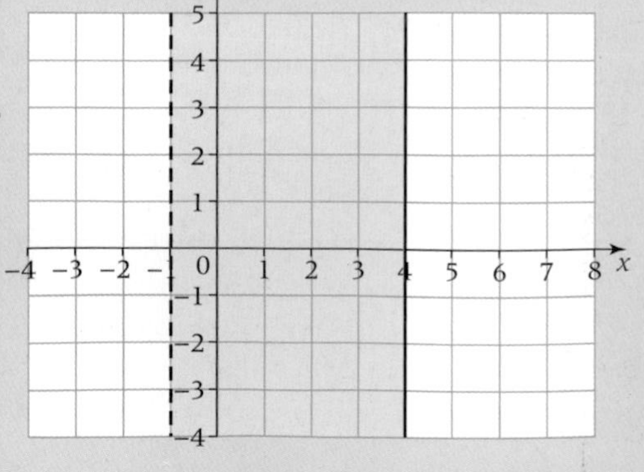

Examiner's tip
Some questions ask you to shade the region required. Some ask you to leave it unshaded. Read the question carefully.

Exercise A5.2

Grade C/B

1 Draw suitable diagrams to show these inequalities. You should leave the required region *unshaded* and label it **R**.

a $x \geqslant 4$ **b** $y \leqslant 3$ **c** $x > -2$

d $y < -3$ **e** $-1 \leqslant x \leqslant 6$ **f** $1 < y < 8$

g $x > 8$ or $x \leqslant -2$ **h** $y < 2$ or $y > 9$ **i** $2x \geqslant 5$

j $8 > y$ **k** $-2.5 < x < 7$ and $0 \leqslant y \leqslant 5\frac{1}{4}$

2 For each diagram, write the inequalities shown by the *shaded* region.

a

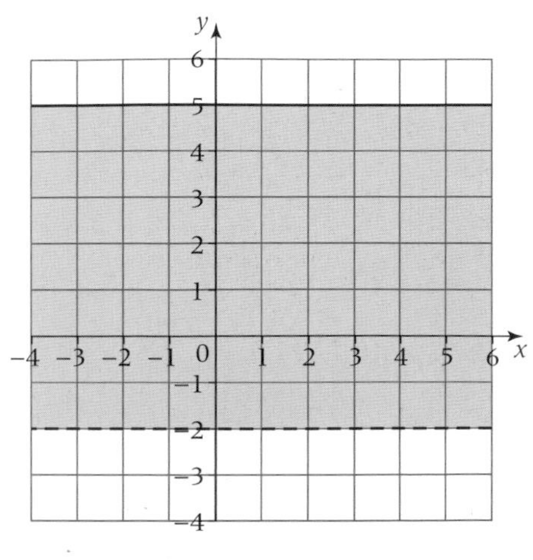

b

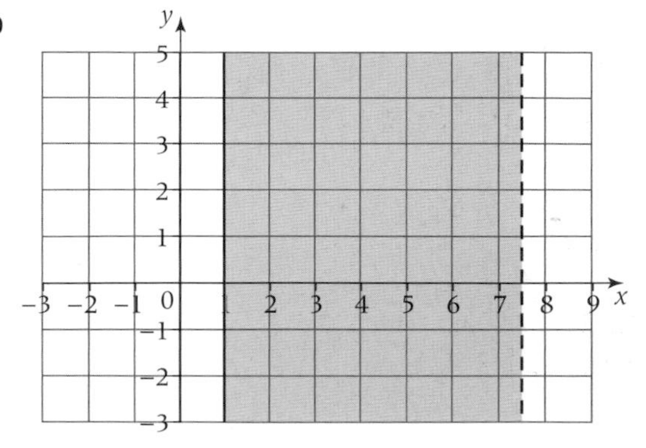

3 List all the points with integer coordinates that satisfy both the inequalities $-2 < x \leqslant 1$ and $2 > y \geqslant 0$.

4 Draw diagrams and write inequalities that when shaded would give

a a shaded rectangle

b an unshaded square

c a shaded rectangle with length double its width.

A03 Problem

5 Here are two graphs of quadratic equations, with horizontal lines drawn as shown. Write the inequalities represented by the *shaded* regions.

a

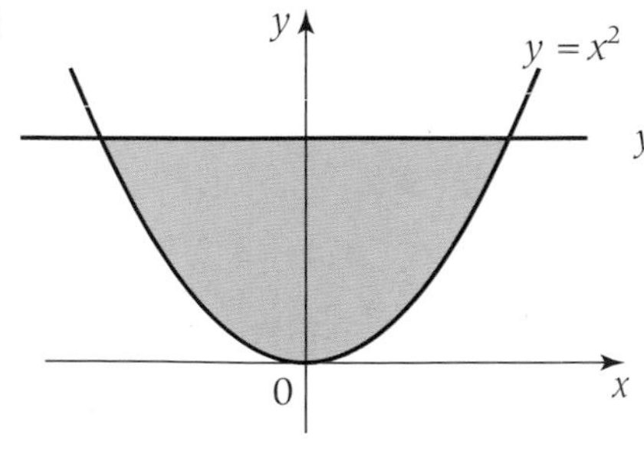

b

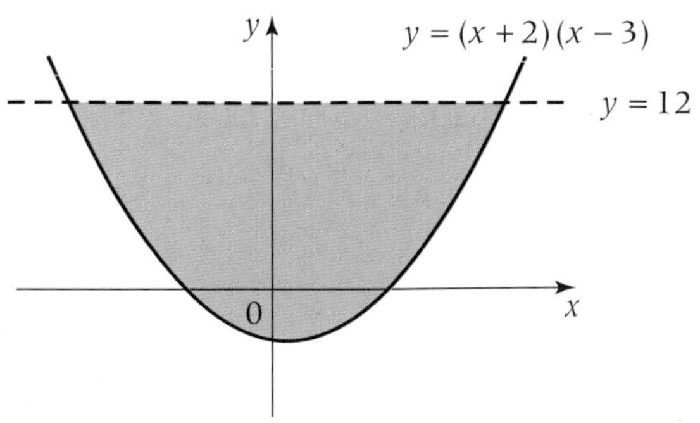

A5.3 Inequalities in two variables

This spread will show you how to:

- Solve simple inequalities in one or two variables

Keywords
Inequality
Point test
Region

- You can represent **inequalities** in two variables on a graph.

For example, to draw $y > 2x - 1$, start by drawing the line $y = 2x - 1$ (dashed for $>$).

- You can use a **point test** to decide which side of the line is the required **region**.

Choose one point and see if it fits the inequality. If it does, this is the region you need.

For $y > 2x - 1$ choose (0, 0):

Is $0 > 2 \times 0 - 1$?

Yes, so the point (0, 0) is in the region required, shown unshaded on the graph.

Example

Show the region of points with coordinates that satisfy the four inequalities $y > 0$, $x > 0$, $2x < 5$ and $3y + 2x < 9$.

Draw dashed lines at $y = 0$ (x-axis), $x = 0$ (y-axis) and $x = \frac{5}{2}$.

$2x = 5 \rightarrow x = \frac{5}{2}$

Plot the graph of $3y + 2x = 9$.

x	−3	0	3
y	5	3	1

Find the y-values for three values of x and plot the points. Draw the line dashed.

Choose (1, 1) for the point test for all the inequalities.

$y > 0$: Is $1 > 0$? Yes.
$x > 0$: Is $1 > 0$? Yes.
$2x < 5$: Is $2 \times 1 < 5$? Yes.
$3y + 2x < 9$: Is $3 \times 1 + 2 \times 1 < 9$? Yes.

Choose a point that does not lie on any of the graphs.

(1, 1) satisfies all the inequalities, so it is in the required region.

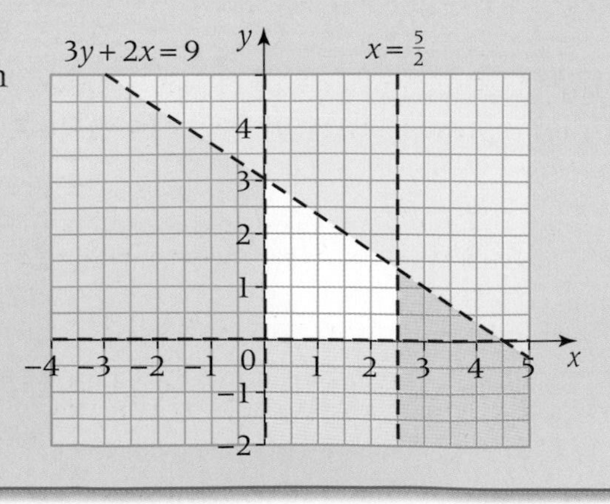

Required region left unshaded.

Exercise A5.3 Grade B

1 Shade the regions satisfied by these inequalities.

a $y \leqslant x + 5$ **b** $y > 2x + 1$ **c** $y \geqslant 1 - 3x$ **d** $y < \frac{1}{4}x - 2$

e $2y \geqslant 3x + 4$ **f** $3y < 9x - 7$ **g** $x + y < 9$ **h** $2x + 4y > 7$

2 What inequalities are shown by the shaded region in each of these diagrams?

a

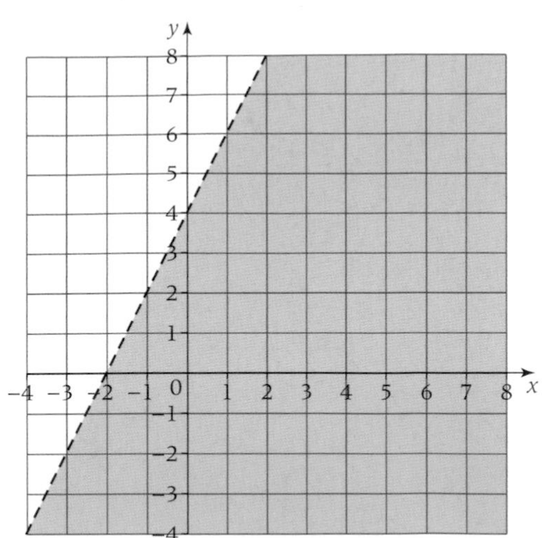

b

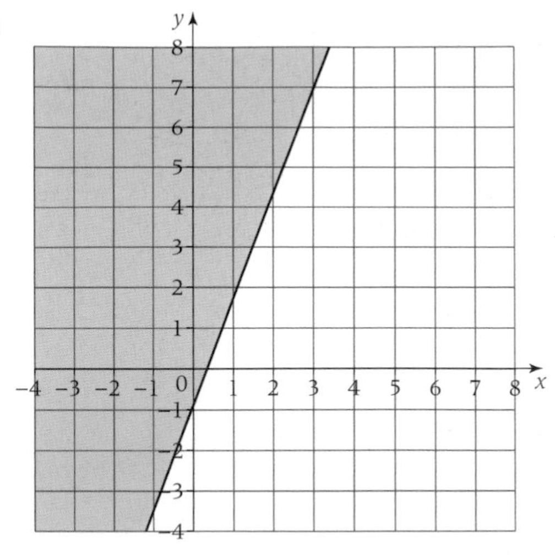

3 On one diagram, show the region satisfied by the three inequalities. List all the integer coordinates that satisfy all three inequalities.

a $x \geqslant -1$ $y < 2$ $y \geqslant 3x - 1$

b $2x + 3y < 10$ $x > 2$ $2y + 1 > 0$

4 Draw the region satisfied by all three inequalities

$y \geqslant x^2 - 4,\quad y < 2,\quad y + 1 < x.$

AO2 Functional Maths

5 300 students are going on a school trip and 16 adults will accompany them. The head teacher needs to hire coaches to take them on the trip. She can hire small coaches that seat 20 people or large coaches that seat 48 people. There must be at least 2 adults on each coach to supervise the students.
If x is the number of small coaches hired and y is the number of large coaches, explain why these inequalities model the situation.

$x \geqslant 0,\ y \geqslant 0,\ 5x + 12y \geqslant 79,\ x + y \leqslant 8$

AO3 Problem

6 Draw diagrams and write inequalities that when shaded would give

a a shaded, right-angled triangle with right angle at the origin

b an unshaded trapezium.

A5.4 Solving simultaneous linear equations

This spread will show you how to:

- Solve simultaneous equations in two unknowns by eliminating a variable

Keywords
Eliminate
Simultaneous
Variable

- **Simultaneous** equations are true at the same time.
  They share a solution.
  For example $x + y = 10$
  $x - y = 4$

  x and y add to make 10 and their difference is 4.
  The shared solution must be $x = 7$ and $y = 3$.

- You can solve simultaneous equations by **eliminating** one of the two **variables**.
  - Multiply one or both equations to get equal numbers of one variable in both equations.
  - Add or subtract the equations to eliminate the variable.

$3x + 2y = 16$
$x - 6y = 2$

→ Obtain equal numbers of x's by multiplying the second equation by 3.

$3x + 2y = 16$ (1)
$3x - 18y = 6$ (2)

→ Subtract (1) − (2) to eliminate x.

$20y = 10$
$y = \frac{1}{2}$

→ Substitute for y in (1) to find x.

$3x + 1 = 16$
$x = 5$

Example

Solve the simultaneous equations $2x - 4y = 8$
$3x + 3y = -15$

$2x - 4y = 8$ (1) multiply by 3 $6x - 12y = 24$ (3)
$3x + 3y = -15$ (2) multiply by 4 $12x + 12y = -60$ (4)
Add (3) + (4): $18x = -36$
$x = -2$
Substituting in (2): $-6 + 3y = -15$ so $3y = -9$
$y = -3$

Check the solution by substituting in one of the original equations.

Example

The perimeter of this isosceles triangle is 50 cm. Find the length of its base.

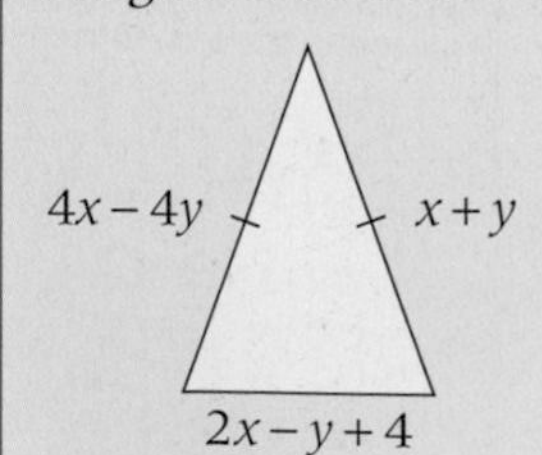

Isosceles: $4x - 4y = x + y \rightarrow \quad 3x - 5y = 0$ (1)
Perimeter: $7x - 4y + 4 = 50 \rightarrow 7x - 4y = 46$ (2)
$7 \times (1)$ $\quad 21x - 35y = 0$ (3)
$3 \times (2)$ $\quad 21x - 12y = 138$ (4)
(3) − (4) $\quad 23y = 138$
$y = 6$
Substituting in (1): $3x - 30 = 0$
$3x = 30$
$x = 10$
Hence, the base is $(2 \times 10) - 6 + 4 = 18$ cm.

There are two unknowns, so you need to set up two equations to find them.

Equal terms have the **S**ame **S**igns so **S**ubtract.

Finish by answering the question.

Exercise A5.4

Grade B/A

1 Solve these simultaneous equations.

a $2x + y = 18$
$x - 2y = -1$

b $5x + 2y = -30$
$3x + 4y = -32$

c $14(c + d) = 14$
$5d - 3c = -11$

d $5x = 7 + 6y$
$8y = x + 2$

e $24a + 12b + 7 = 0$
$6a + 12b - 5 = 0$

f $4p - 1\frac{1}{2}q = 5\frac{1}{2}$
$6p - 2q = 21$

2 For each question, set up a pair of simultaneous equations and solve them to find the required information.

a Two numbers have a sum of 23 and a difference of 5. What numbers are they?

b Two numbers have a difference of 6. Twice the larger plus the smaller number also equals 6. What numbers are they?

3 Use simultaneous equations to find the value of each symbol in the puzzle.

☾	✵	☾	✵	92
✵	☾	✵	✵	104

4 A straight-line graph passes through the points (3, 7) and (21, 09). Apply $y = mx + c$ and use simultaneous equations to find the equation of the line.

Functional Maths AO2

5 Tickets for a theatre production cost £3.50 per child and £5.25 per adult. 94 tickets were sold for a total of £365.75. How many children attended the production?

Problem AO3

6 The equation of a parabola is $y = ax^2 + bx + c$.
A parabola crosses the y-axis at (0, 7) and passes through (2, 5) and (5, 42).

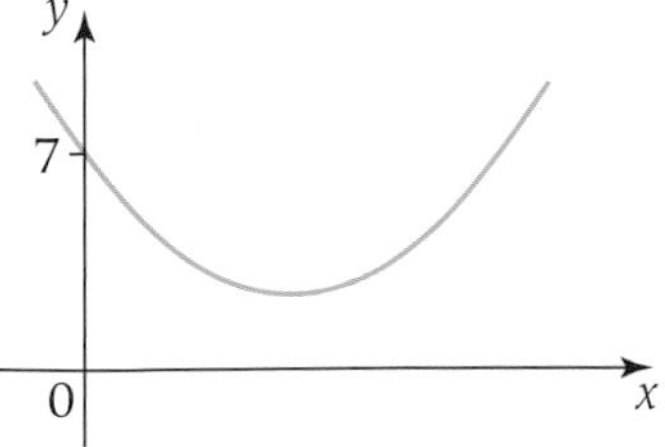

Find the values of a, b and c.

You need two points to determine a straight line $y = mx + c$ and three points to determine a parabola $y = ax^2 + bx + c$.

Summary

Check out

You should now be able to:

- Solve simple linear inequalities in one variable and represent the solution set on a number line
- Solve simple linear inequalities in two variables and represent the solution set on a coordinate grid
- Solve linear simultaneous equations in two unknowns by eliminating a variable

Worked exam question

Solve the simultaneous equations

$$x + 6y = 7$$
$$4x - 2y = -11$$

(3)

$x + 6y = 7$	(1)	multiply by 1	$x + 6y = 7$	(3)
$4x - 2y = -11$	(2)	multiply by 3	$12x - 6y = -33$	(4)

Add (3) + (4): $13x = -26$

$x = -2$

This eliminates y

Substituting in (1): $-2 + 6y = 7$

$6y = 9$

$y = 1.5$

Checking in (2): $4 \times -2 - 2 \times 1.5 = -11$

$-8 - 3 = -11$

Correct

$x = -2$ and $y = 1.5$

OR

$x + 6y = 7$	(1)	multiply by 4	$4x + 24y = 28$	(3)
$4x - 2y = -11$	(2)	multiply by 1	$4x - 2y = -11$	(4)

Subtract (3) – (4): $24y - -2y = 28 - -11$

$26y = 39$

$y = 1.5$

This eliminates x

Substituting in (1): $x + 9 = 7$

$x = -2$

Checking in (2): $4 \times -2 - 2 \times 1.5 = -11$

$-8 - 3 = -11$

Correct

$x = -2$ and $y = 1.5$

Exam questions

1 The region **R** satisfies the inequalities $x \geq 2, y \geq 1, x + y \leq 6$

On a copy of the grid below, draw straight lines and use shading to show the region **R**.

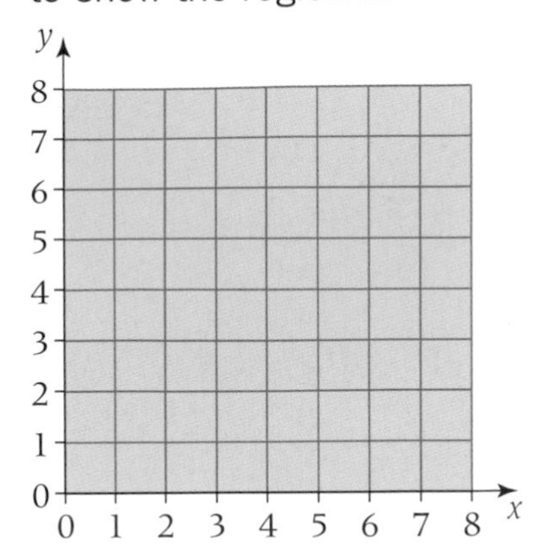

(3)

(Edexcel Limited 2008)

2 Solve the simultaneous equations

$$9x + 3y = 18$$
$$3x - 5y = 12$$

(3)

3 The graphs of the straight lines with equations $3y + 2x = 12$ and $y = x - 1$ have been drawn on the grid.

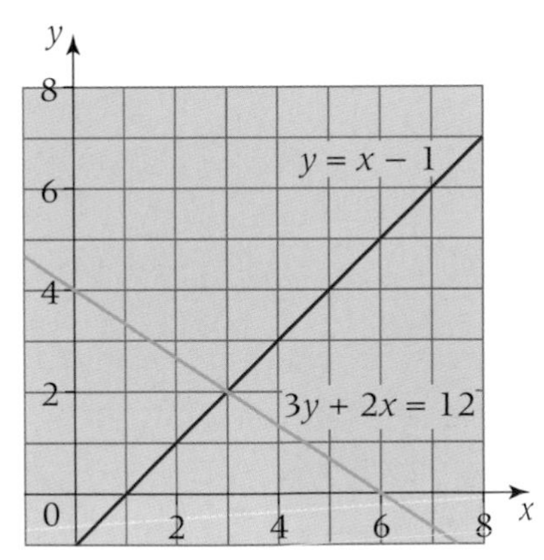

a Use the graphs to solve the simultaneous equations

$3y + 2x = 12$

$y = x - 1$

(1)

b $3y + 2x > 12$ $\quad\quad$ $y < x - 1$ $\quad\quad$ $x < 6$

x and y are integers.

On a copy of the grid, mark with a cross (×) each of the four points which satisfy all these 3 inequalities.

(3)

(Edexcel Limited 2006)

Functional Maths 6: Business

One out of every two small businesses goes bust within its first two years of trading. Mathematics can be applied to reduce the risk of failure for a business as well as to maximise its profits.

A manager needs to know how much cash is coming into and going out of the business. Accountants must set a suitable budget that includes realistic performance targets, and limits expenditure to what the business can afford.

Example

Annie sells hand made cards at a monthly craft fair. She has two ranges of cards; standard and deluxe. The production costs and selling prices per card are:

	Materials used	Time to make	Wages paid	Selling price	Profit
Standard	£0.30	15 minutes	£1.00	£2.55	£1.25
Deluxe	£0.20	30 minutes	£2.00	£3.60	£1.40

This is Annie's cash flow budget for her first three craft fairs (some of the information is missing):

	January (£)	February (£)	March (£)
Standard card sales	45.90	40.80	
Deluxe card sales	43.20		32.40
TOTAL INCOME	89.10	91.20	93.60
Materials used	7.80	7.60	9.00
Wages			
Craft fair fees	10.00	10.00	10.00
Advertising	5.00	5.00	5.00
TOTAL EXPENDITURE	64.80		
NET CASH SURPLUS/DEFICIT	24.30		
CASH BALANCE BROUGHT FORWARD	-	24.30	
CASH BALANCE TO CARRY FORWARD	24.30		

How many of each type of card did Annie sell in each of the three months?

Calculate her spend on materials and wages for each month.

The net surplus (profit) or net deficit (loss) is calculated using the formula Balance = Income − Expenditure

Copy the table and complete the missing values.

On separate copies of the table template, show how the cash flow could change if

* the craft fair fees were increased to £15
* the cost of the materials used to make each type of card increased by 40%
* Annie sold the cards at a discount price of 20% off each type of card.

Investigate how other changes to costs/income might affect Annie's cash flow.

Managers can use mathematical models to make decisions about their business. These techniques are widely used in production planning to obtain the maximum profit or to incur the minimum cost in a given situation.

Annie wants to know how much of each type of card she should produce in order to maximise her profits in a particular month. She has £9.00 cash available for materials and a maximum of £50 to spend on wages for that month.

Use s to represent the number of standard cards, and d to represent the number of deluxe cards.

For the material costs, you have $0.3s + 0.2d \leq 9$

For the wage costs, you have $s + 2d \leq 50$

The aim is to maximise the profit, £P, where

$P = 1.25s + 1.40d - 15$

The '- 15' is for the fixed costs.

The graph shows the feasible region (solution set) for the inequalities.

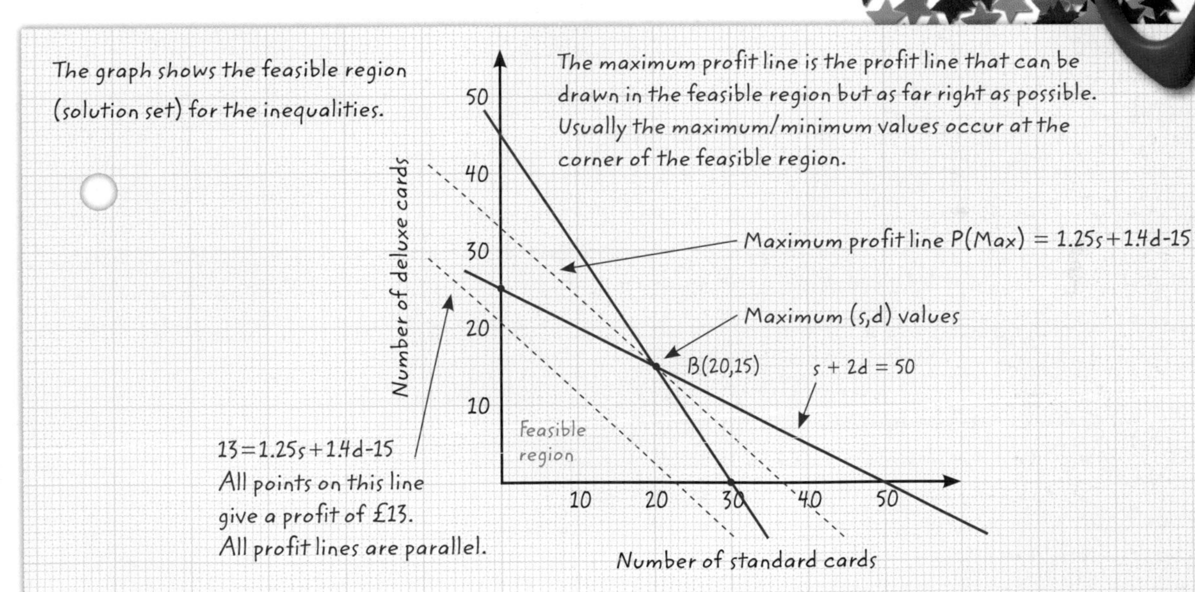

Annie should make 20 standard and 15 deluxe cards, which would give her a profit of £31.00.

Given these constraints, investigate how changes to the production costs and selling prices might affect the maximum possible profit.

For another month, Annie has only £8.00 cash available for materials and a maximum of £40 to spend on wages. Using the method shown, calculate the maximum possible profit and the number of standard and deluxe cards Annie should make to achieve this value.

How do you think you would need to adapt the method to find the minimum amount Annie would need to spend on materials to guarantee a specified minimum profit and wage for a given month?

G4

Pythagoras and trigonometry

Pilots have to learn how to navigate without the use of GPS. They use the stars as landmarks (celestial navigation) and calculate the angles two separate stars make with the horizon. They look up the position of the stars on their charts and then they can use trigonometry to calculate their position. This technique has been used for centuries as sailors were able to navigate across the widest oceans.

What's the point?

Once a right-angled triangle can be seen in a particular situation or problem, a mathematician only needs two pieces of information from which any other lengths or angles can then be calculated.

Check in

You should be able to

- **be familiar with square numbers**

1 Work out each of these.

a 7^2 **b** $4^2 + 6^2$ **c** $3^2 + 5^2$

d $8^2 - 4^2$ **e** $7^2 - 2^2$ **f** $\sqrt{17^2 - 15^2}$

- **rearrange and solve simple equations involving fractions**

2 Rearrange these equations to make x the subject.

a $y = \frac{x}{6}$ **b** $y = \frac{x}{5}$ **c** $y = \frac{x}{10}$

d $y = \frac{2}{x}$ **e** $y = \frac{5}{x}$ **f** $y = \frac{8}{x}$

3 a Solve $5x + 7 = 32$. **b** Find y where $\frac{y}{7} = 11$.

4 a Solve $3x - 2.8 = 12$. **b** Find y where $3y + 1 = 9$.

5 a Solve $\frac{x}{4} = 5.7$. **b** Find y where $2 + 5y = 6$.

Orientation

What I need to know	What I will learn	What this leads to
KS3 Apply Pythagoras' theorem	■ Apply and use Pythagoras' theorem ■ Apply and use trigonometric ratios	**G5** Sine and cosine rules Pythagoras in 3D **G7** Graphs of trigonometric functions
A3 Solve simple equations involving fractions		**A-level** Maths, Physics, Engineering

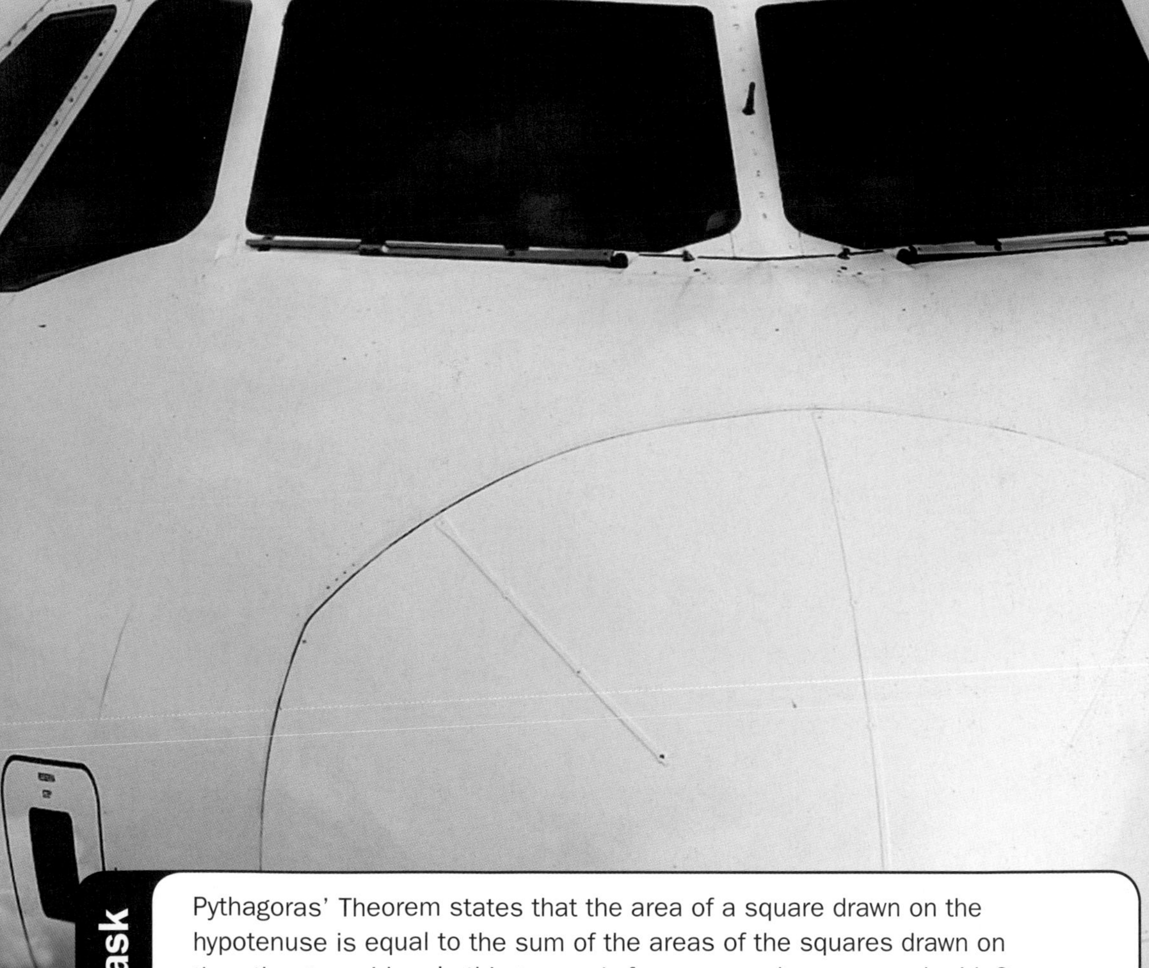

Rich task

Pythagoras' Theorem states that the area of a square drawn on the hypotenuse is equal to the sum of the areas of the squares drawn on the other two sides. Is this true only for squares drawn on each side? Is Pythagoras' theorem true for equilateral triangles drawn on each side of a right-angled triangle? Investigate

G4.1 Pythagoras' theorem and coordinates

This spread will show you how to:

- Understand, recall and use Pythagoras' theorem in 2-D problems
- Given the coordinates of points A and B, calculate the length of the line segment AB

Keywords
Hypotenuse
Midpoint
Pythagoras' theorem
Right-angled triangle

Pythagoras' theorem for **right-angled triangles** states:

- In a right-angled triangle, the length of the **hypotenuse** squared equals the sum of the squares of the lengths of the other two sides.

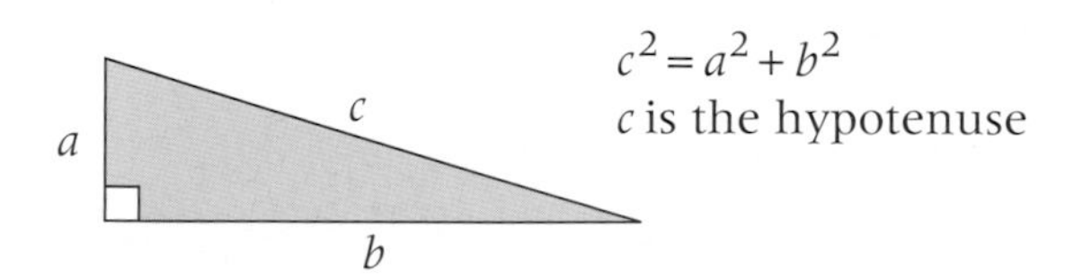

$c^2 = a^2 + b^2$
c is the hypotenuse

The hypotenuse is always opposite the right angle.

You can use Pythagoras' theorem to find the length of a line joining two points on a grid.

- sketch a diagram and label the right angle
- label the unknown side
- round your answer to a suitable degree of accuracy.

Unless the question tells you otherwise, round to 2 dp.

Example

Work out the length of the line joining A(2, 3) and C(−2, 1).

Draw a right-angled triangle and label the lengths of the two shorter sides.

$AC^2 = 2^2 + 4^2$
$AC^2 = 20$
$AC = 4.47$ (2 dp)

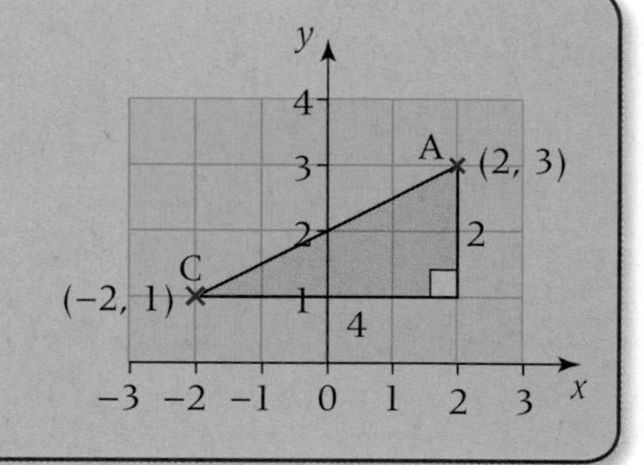

- The length of the line segment joining (x_a, y_a) and (x_b, y_b) is $\sqrt{(x_a - x_b)^2 + (y_a - y_b)^2}$

Example

For the points A(−2, 4) and B(8, −1)

a Find their midpoint M

b show that AM = MB

The midpoint of (x_a, y_a) and (x_b, y_b) is $\left(\frac{x_a + x_b}{2}, \frac{y_a + y_b}{2}\right)$.

a $M = \left(\frac{-2+8}{2}, \frac{4-1}{2}\right) = \left(3, 1\frac{1}{2}\right)$

b $AM^2 = (-2-3)^2 + \left(4 - 1\frac{1}{2}\right)^2$
$= 5^2 + \left(2\frac{1}{2}\right)^2$

$MB^2 = (8-3)^2 + \left(-1 - 1\frac{1}{2}\right)^2$
$= 5^2 + \left(2\frac{1}{2}\right)^2$

So $AM^2 = MB^2 = \frac{125}{4}$ or $AM = MB = \frac{5\sqrt{5}}{2} = 5.59$ (2dp)

Exercise G4.1 Grade B/A

1 The sketch shows the points (3, 1) and (8, 4).
Work out the length of the line joining these two points.

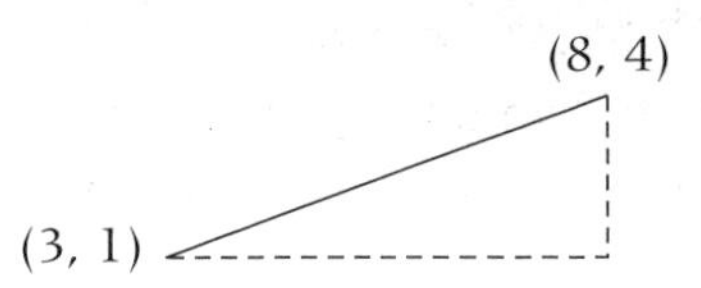

2 Use Pythagoras' theorem to work out the length of the line joining
a (0, 6) and (1, −2) **b** (−4, 3) and (5, −2).

Sketch the line first.

3 Find the midpoints, M, of
a A(6, 4) and B(−4, 5) **b** A(−2, 5) and B(7, −1)
and show that AM = MB.

4 Find the length of the diagonal of a square with side 10 cm.

5 **a** Find the length of the side of a square with diagonal 10 cm.
b Write the area of a square with a diagonal 10 cm.

6 PQR and PRS are right-angled triangles.
Work out the length PS.

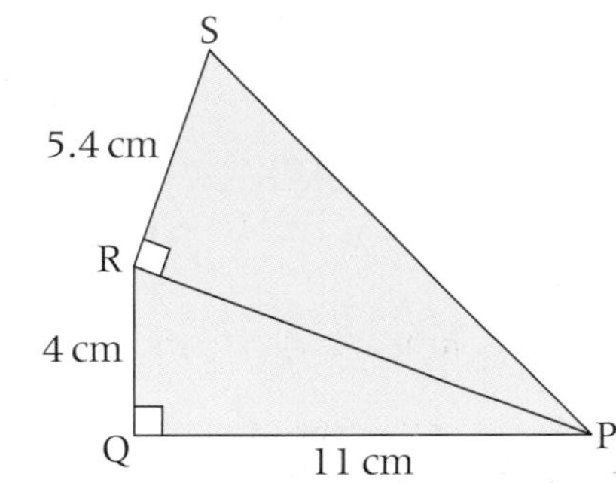

Work out PR first.

7 ABC and ACD are right-angled triangles.
Work out the length AB.

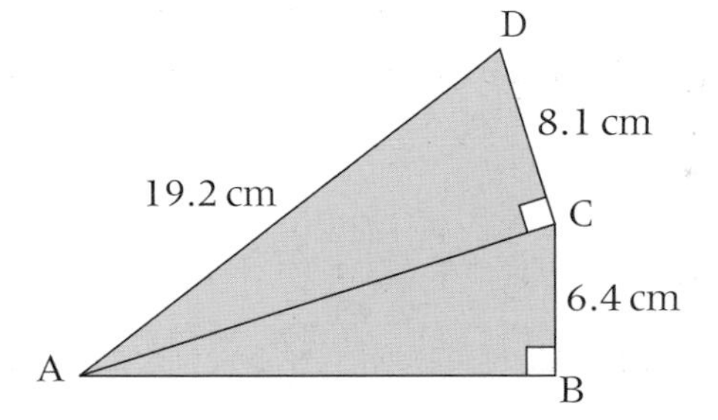

AO3 Problem

8 A cylinder has base diameter 7.5 cm and height 18 cm.
A thin rod just fits diagonally inside the cylinder.
What is the longest length that the rod could be so that it fits exactly inside the cylinder?

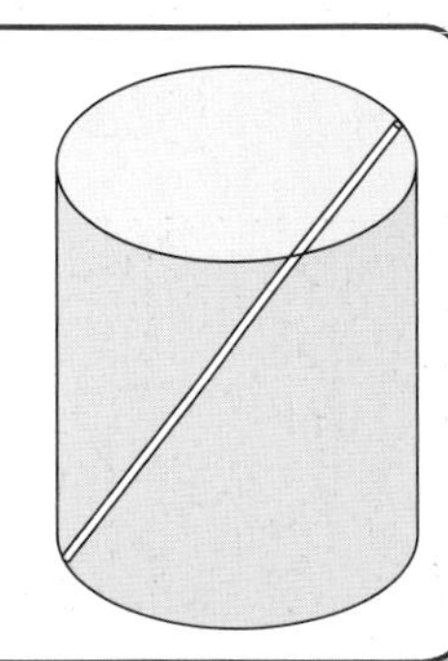

Unit 3

G4.2 Tangent ratio

This spread will show you how to:

- Understand, recall and use trigonometry in right-angled triangles

Keywords
Gradient
Right-angled triangle
Tangent

A road sign gives the **gradient** of a hill as a percentage.

Hills with the same percentage slope climb at the same angle.

The gradient is the ratio of the vertical climb to the horizontal distance.

You can draw a **right-angled triangle** to represent the hill.

- $\tan \theta = \dfrac{\text{opposite side}}{\text{adjacent side}}$

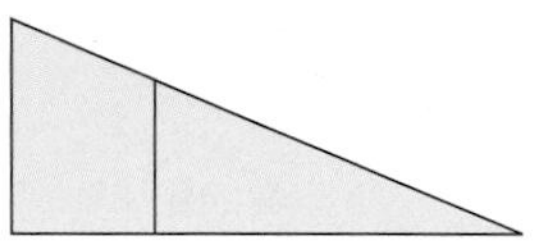

is the same, however large or small you draw the triangle.

This ratio is called the **tangent** ratio.

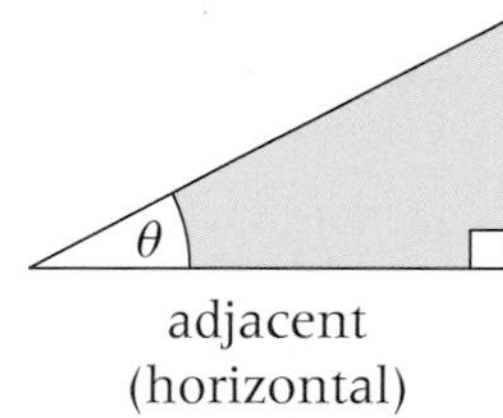

Adjacent means 'next to'.

θ is used to represent an angle.

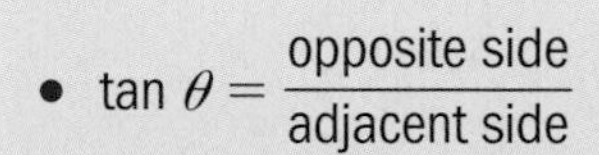

- $\tan \theta = \dfrac{\text{opposite side}}{\text{adjacent side}}$

Example

Find the missing sides in these triangles.

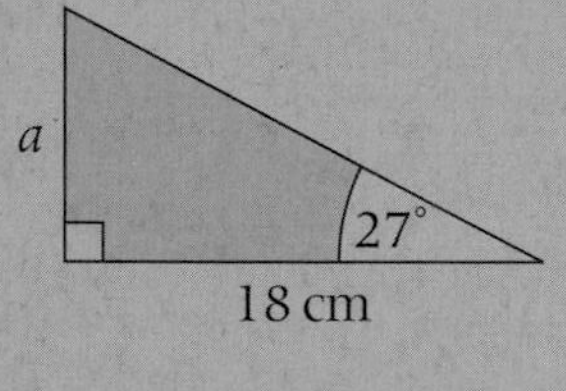

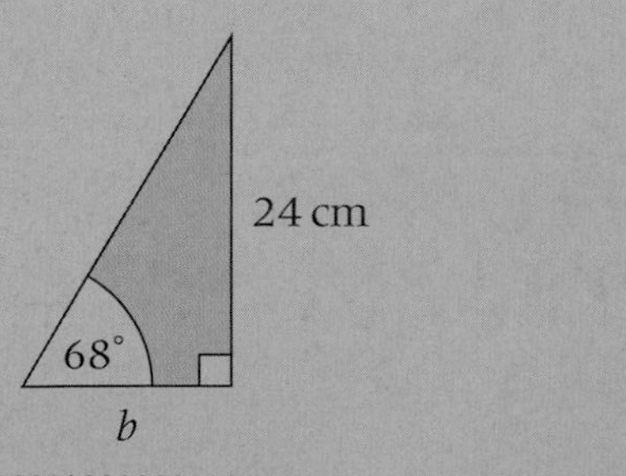

$$\tan 27° = \frac{a}{18}$$

$$18 \times \tan 27° = a$$

$$a = 9.17\text{ cm}$$

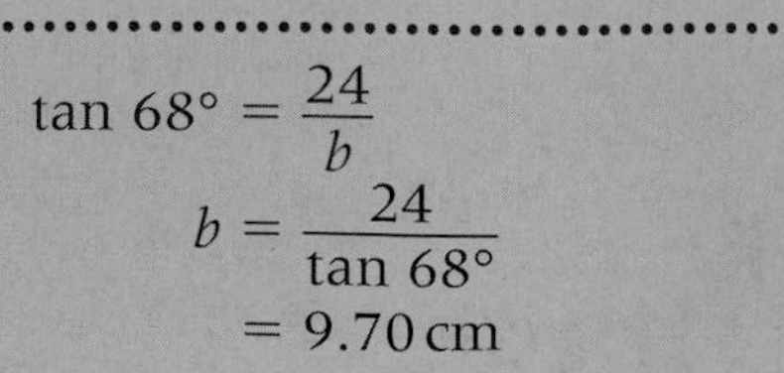

$$\tan 68° = \frac{24}{b}$$

$$b = \frac{24}{\tan 68°}$$

$$= 9.70\text{ cm}$$

Remember to put your calculator into 'degree' mode.

You could use the other acute angle in this triangle:
$90° - 68° = 22°$
$\tan 22° = \frac{b}{24}$
$24 \times \tan 22° = b$
$b = 9.70\text{ cm}$

Exercise G4.2 Grade B

You can use the tangent ratio in any right-angled triangle.

1 Find the missing side in each of these right-angled triangles.
Give your answers to 3 significant figures.

a

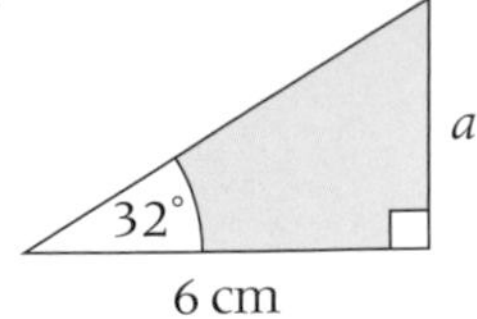

b

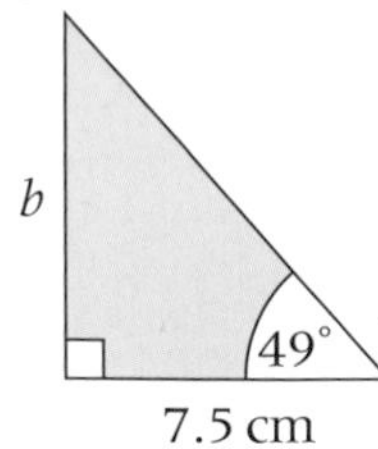

c

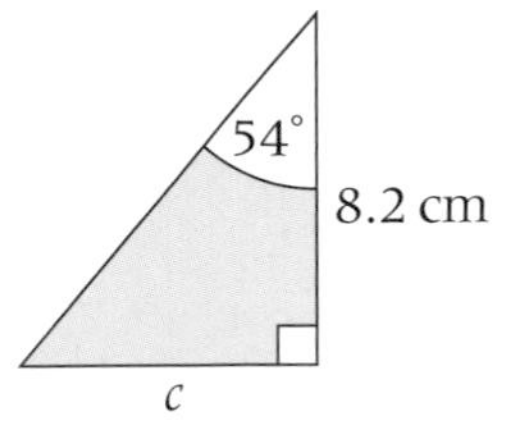

d

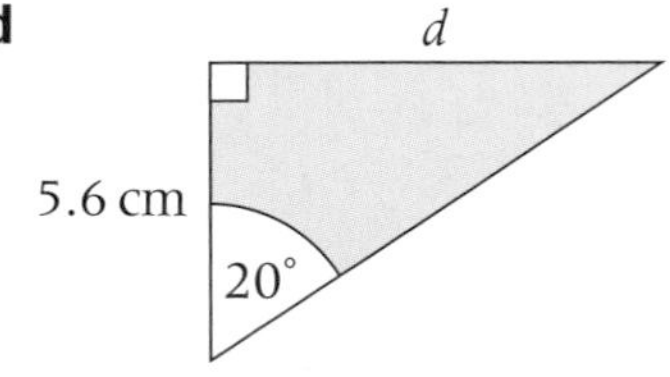

e

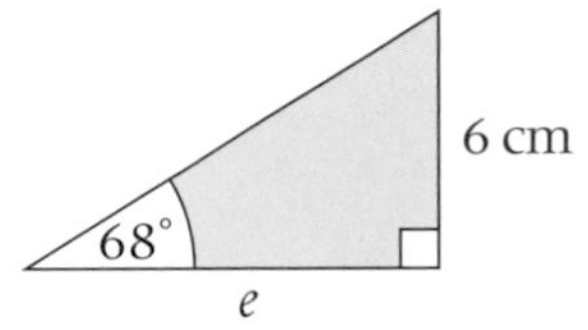

f

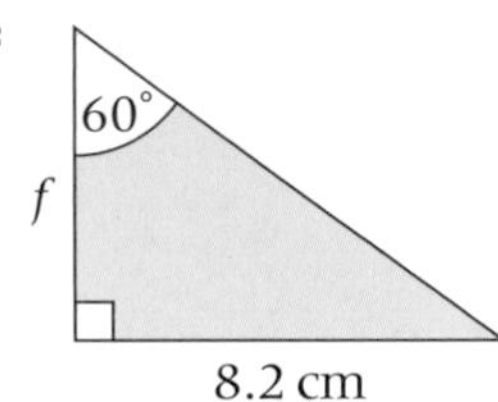

g

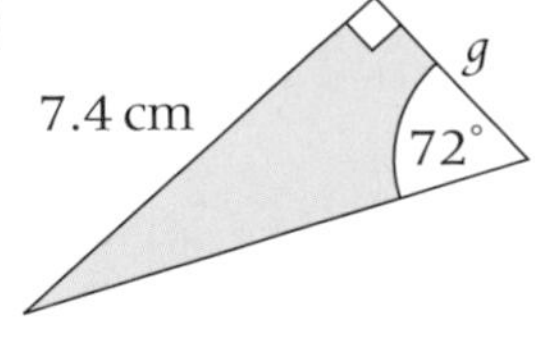

h

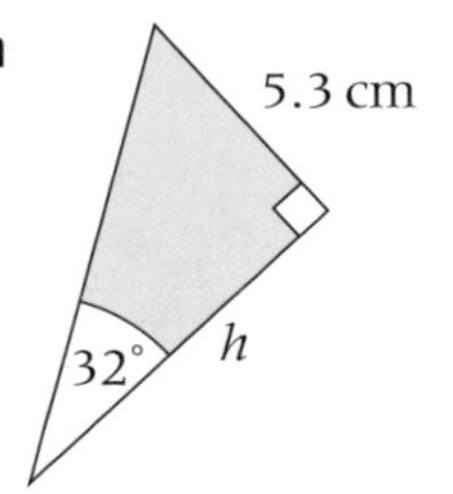

i

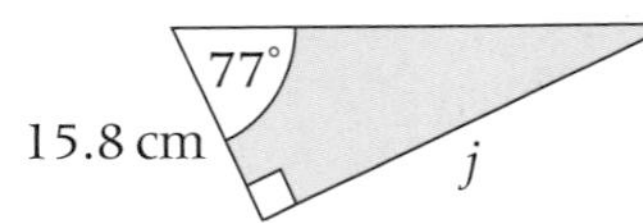

j

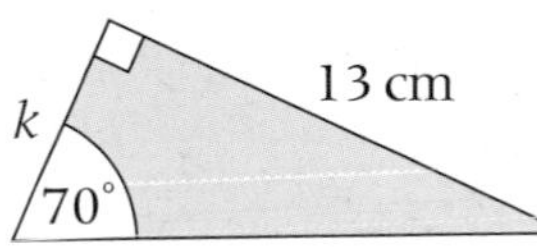

k

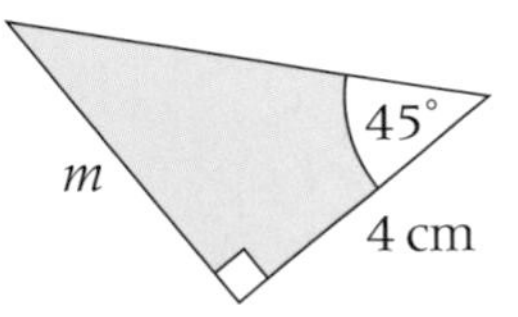

l

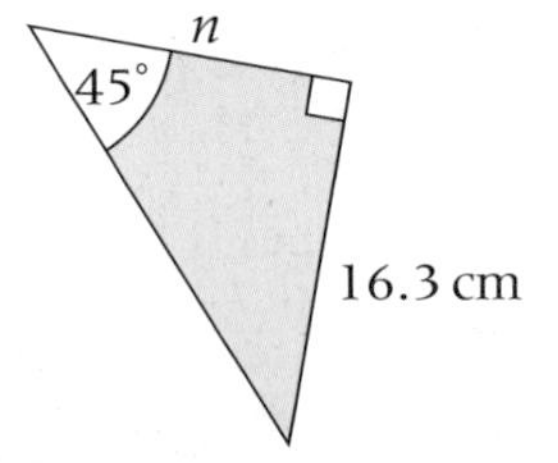

2 What type of triangles are question **1** parts **k** and **l**?
How could you find the missing sides m and n without using the tangent ratio?

G4.3 Sine and cosine ratios

This spread will show you how to:
- Understand, recall and use trigonometry in right-angled triangles

Keywords
Cosine
Hypotenuse
Right-angled triangle
Sine

The tangent ratio relates opposite and adjacent sides in a **right-angled triangle**.

Two other ratios you can use are **sine** and **cosine**.

- $\sin\theta = \dfrac{\text{opposite side}}{\text{hypotenuse}}$
- $\cos\theta = \dfrac{\text{adjacent side}}{\text{hypotenuse}}$

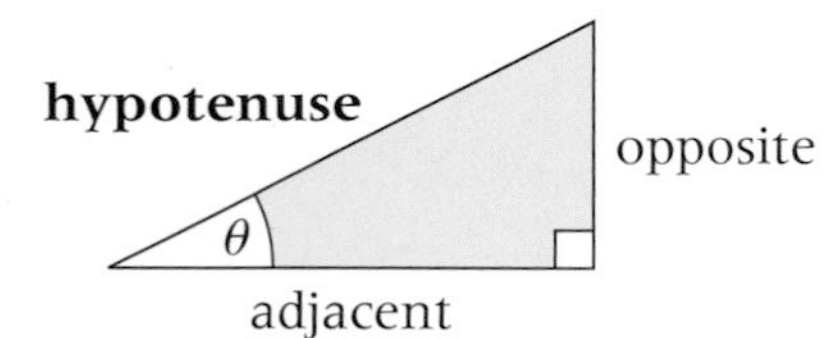

You can use the sine ratio and cosine ratio in any right-angled triangle.

Find the missing sides in these triangles.

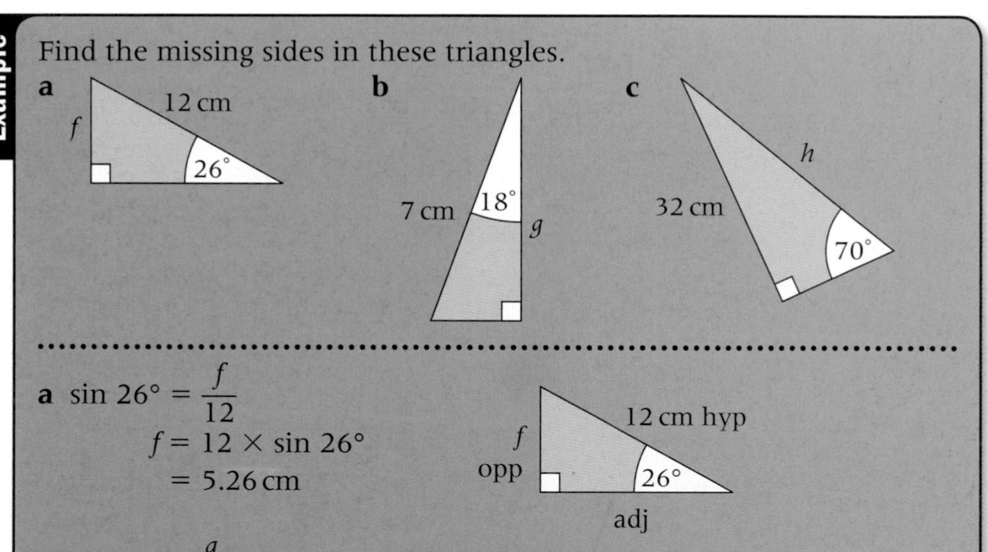

a $\sin 26° = \dfrac{f}{12}$

$f = 12 \times \sin 26°$

$= 5.26\text{ cm}$

12 cm hyp, f opp, 26°, adj

b $\cos 18° = \dfrac{g}{7}$

$g = 7 \times \cos 18$

$= 6.66\text{ cm}$

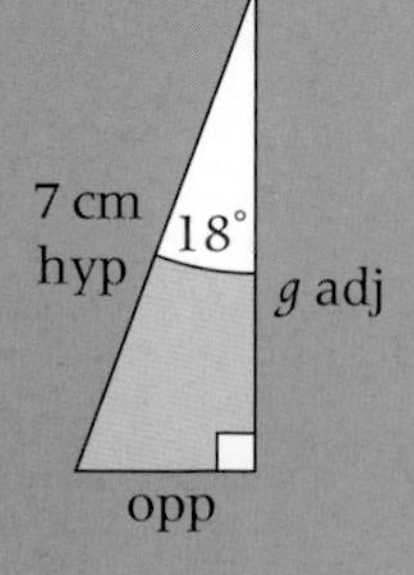

c $\sin 70° = \dfrac{32}{h}$

$h \times \sin 70° = 32$

$h = \dfrac{32}{\sin 70°}$

$= 34.05\text{ cm}$

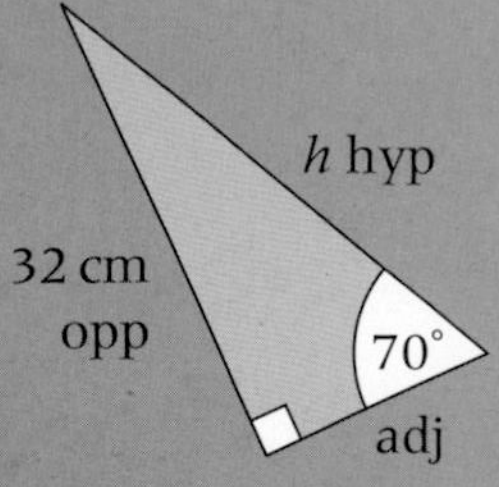

Sketch the triangle and label the sides in relation to the angle given. Label the hypotenuse first.

You always need to divide by either $\sin\theta$ or $\cos\theta$ to find the hypotenuse.

Exercise G4.3 Grade B

Find the missing side in each of these right-angled triangles.
Give your answers to 3 significant figures.

1

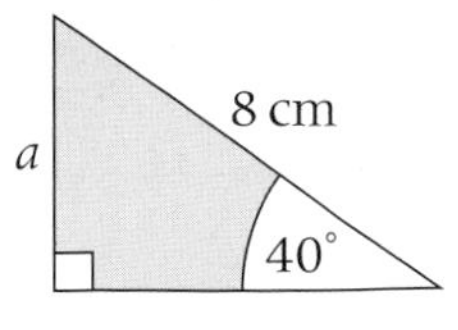

2

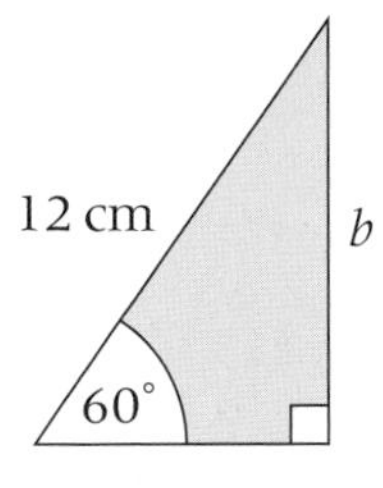

3

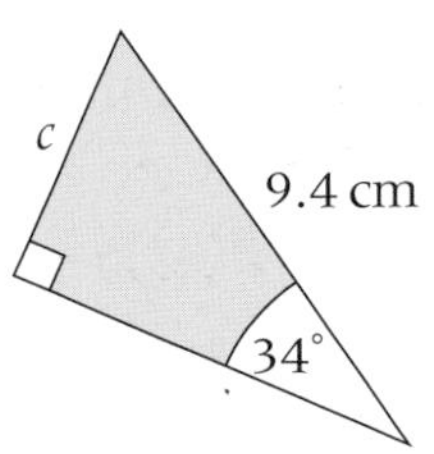

4

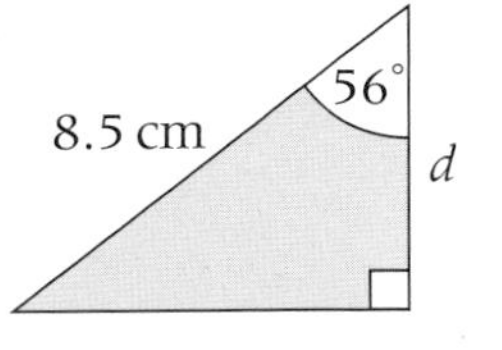

5

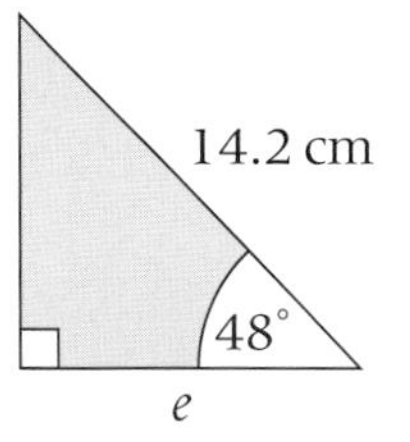

6

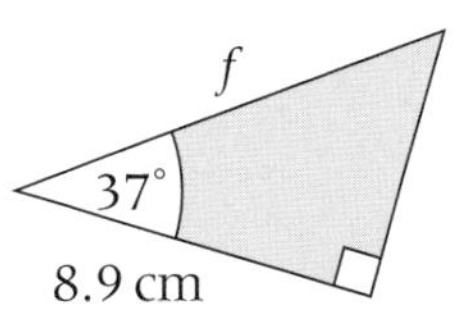

7

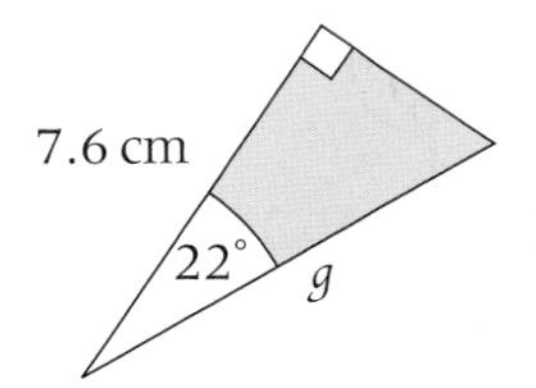

8

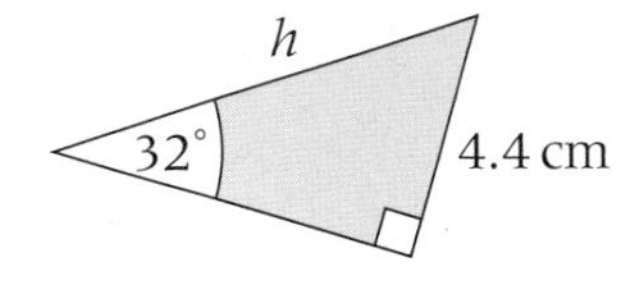

9

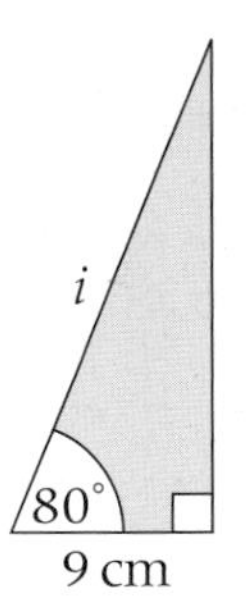

10

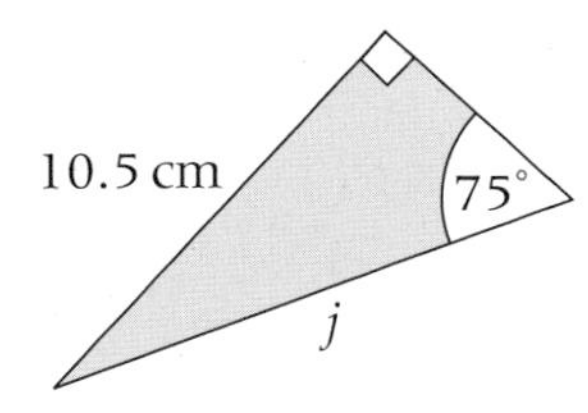

11

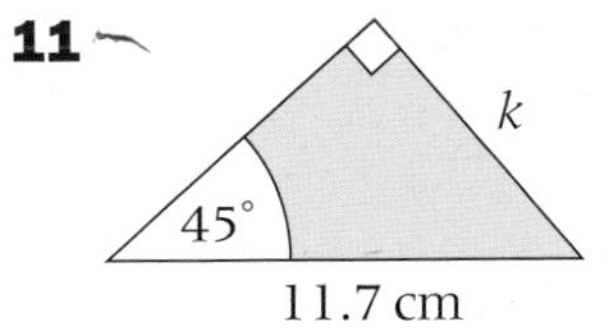

12

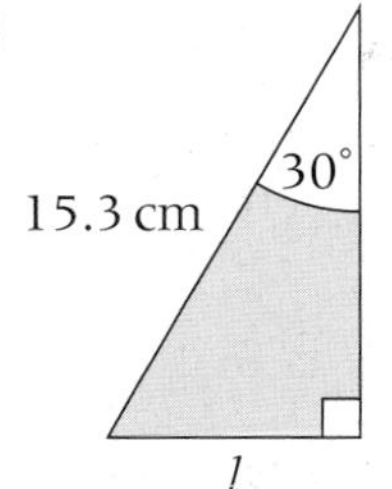

13

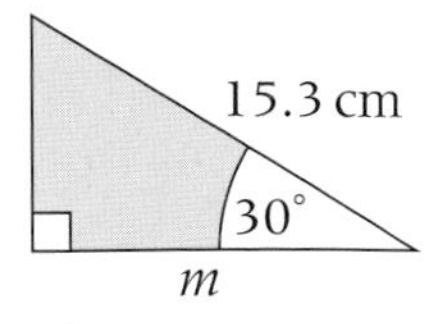

14

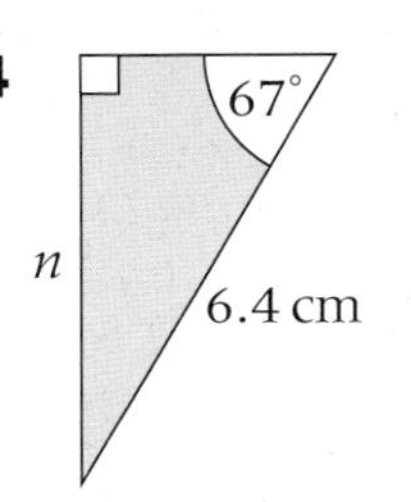

15

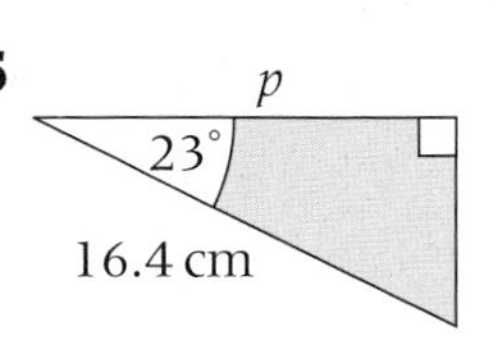

Unit 3

G4.4 Finding angles in right-angled triangles

This spread will show you how to:

- Understand, recall and use trigonometry in right-angled triangles

Keywords
Cosine
Right-angled triangle
Sine
Tangent

You can use the **sine**, **cosine** and **tangent** ratios to find a missing angle in a **right-angled triangle**.

Choose the appropriate ratio for the angle you want to find.

- $\sin\theta = \dfrac{\text{opp}}{\text{hyp}}$ $\quad\cos\theta = \dfrac{\text{adj}}{\text{hyp}}$ $\quad\tan\theta = \dfrac{\text{opp}}{\text{adj}}$

Use the inverse operations $\sin^{-1}$, $\cos^{-1}$ and $\tan^{-1}$ to find the angle.

You can also use the terms arcsin, arcos and arctan for $\sin^{-1}$, $\cos^{-1}$ and $\tan^{-1}$ respectively.

Example

Find the missing angles in these triangles.
Give your answers to 1 dp.

a

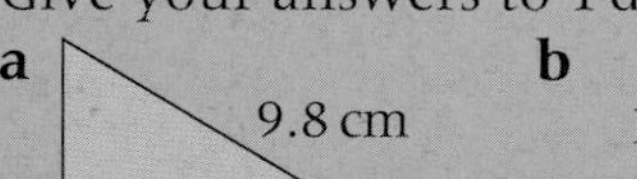

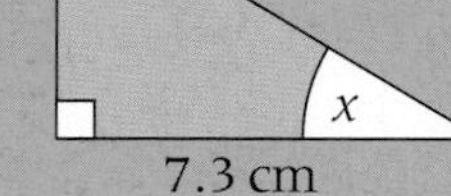

b

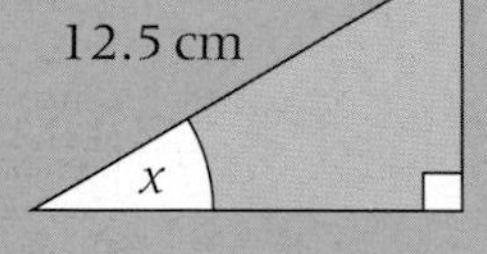

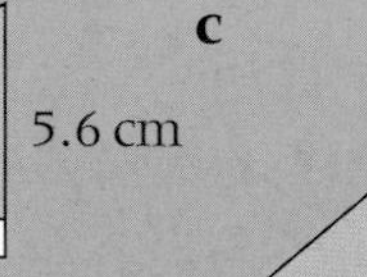

c

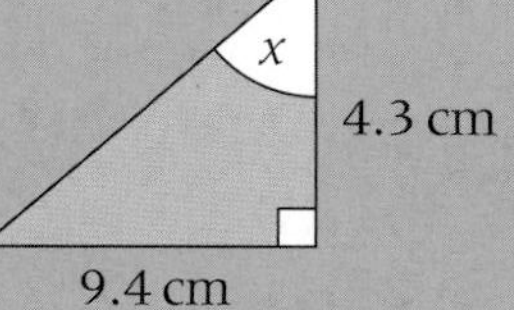

a opp and adj so use cos

$$\cos x = \frac{7.3}{9.8}$$
$$x = \cos^{-1}\left(\frac{7.3}{9.8}\right)$$
$$= 41.9°$$

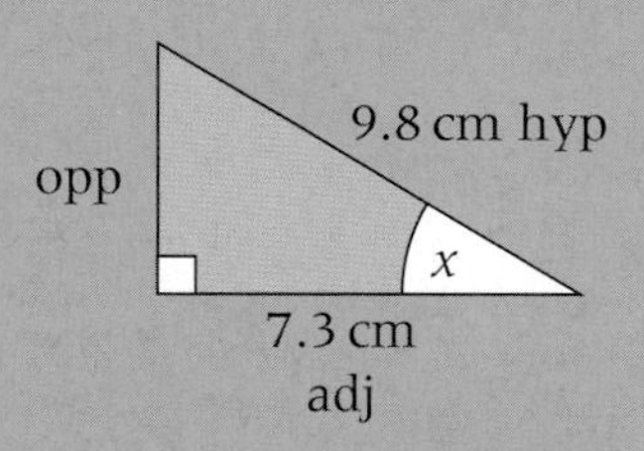

Sketch the triangle and label the sides in relation to the missing angle. Label the hypotenuse first.

b opp and hyp so use sin

$$\sin x = \frac{5.6}{12.5}$$
$$x = \sin^{-1}\left(\frac{5.6}{12.5}\right)$$
$$= 26.6°$$

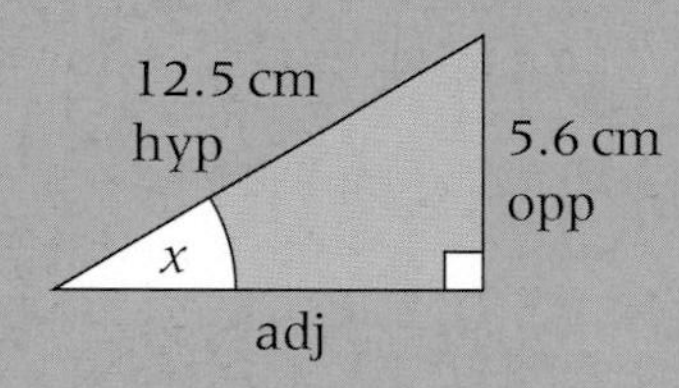

Sines and cosines are always proper fractions.

c opp and adj so use tan

$$\tan x = \frac{9.4}{4.3}$$
$$x = \tan^{-1}\left(\frac{9.4}{4.3}\right)$$
$$= 65.4°$$

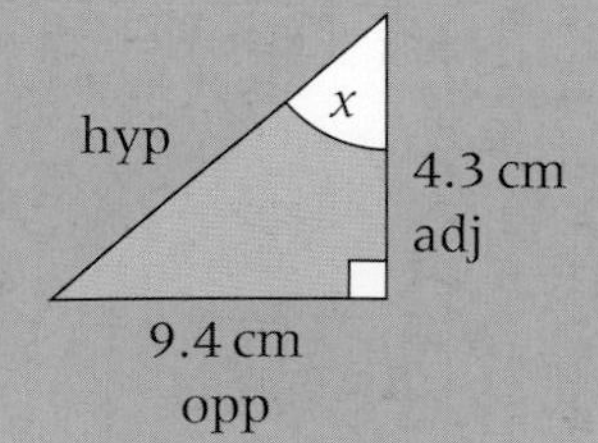

Tan can be an improper fraction.

Exercise G4.4 Grade B

Find the missing angle in each of these right-angled triangles.
Give your answers to 3 significant figures.

1

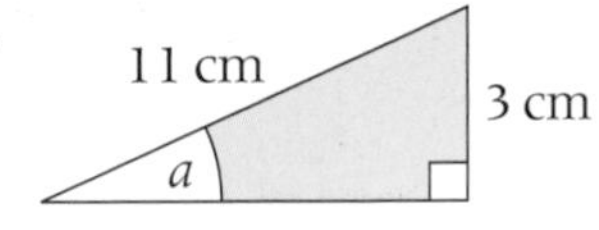

2

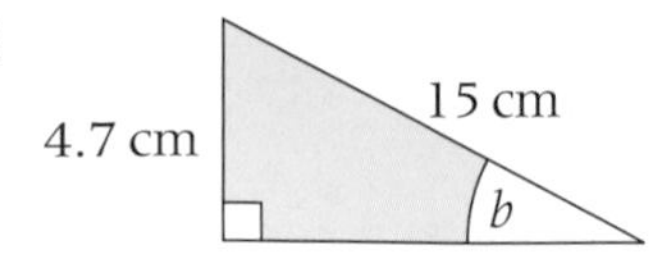

3

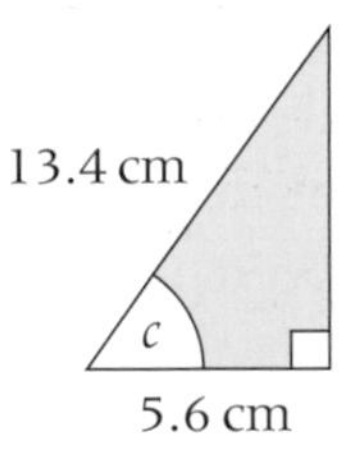

4

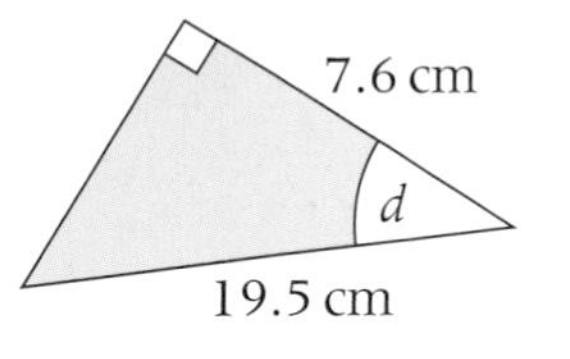

5

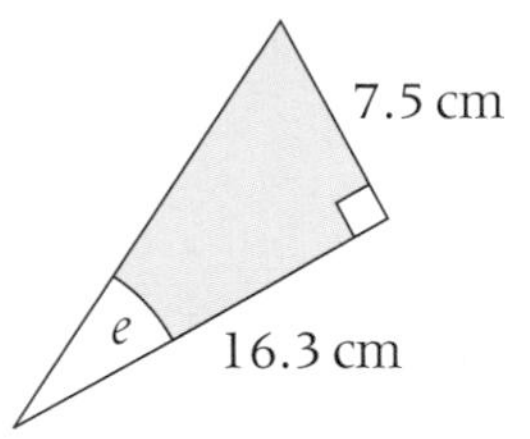

6

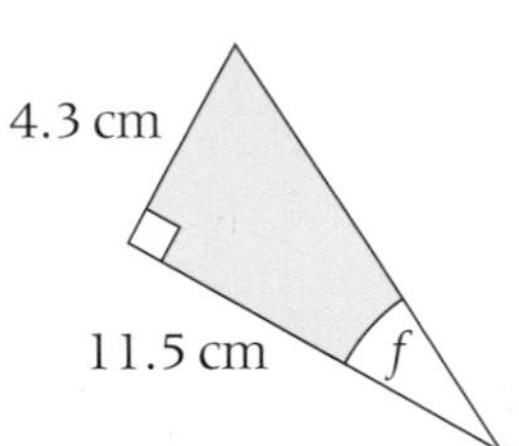

7

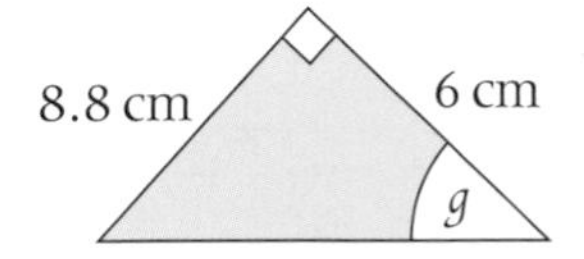

8

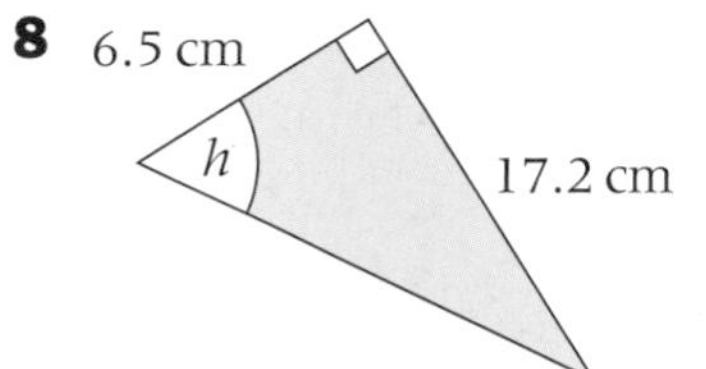

9

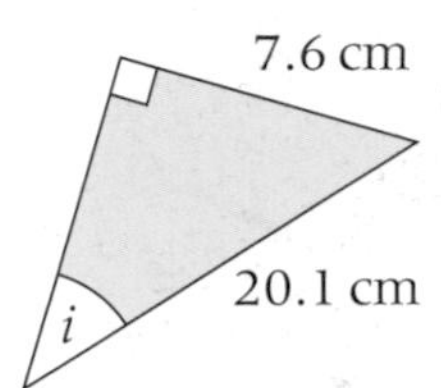

10

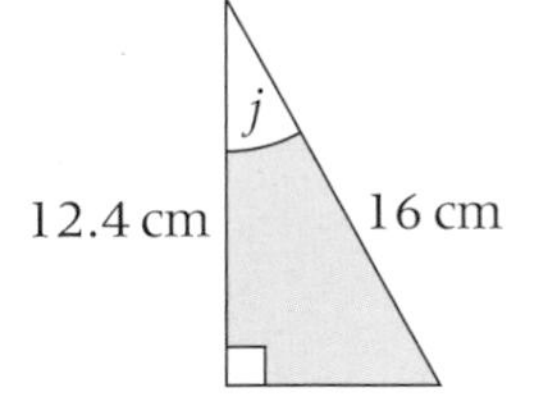

11

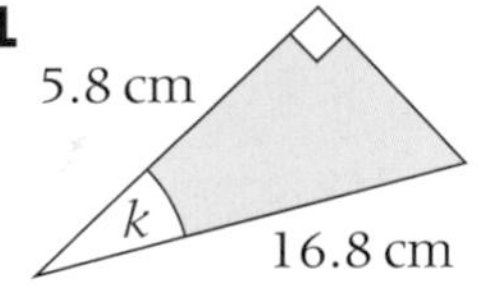

12

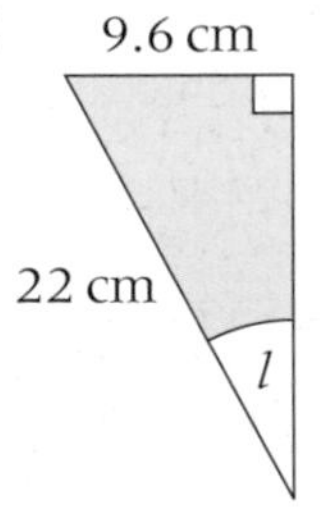

13

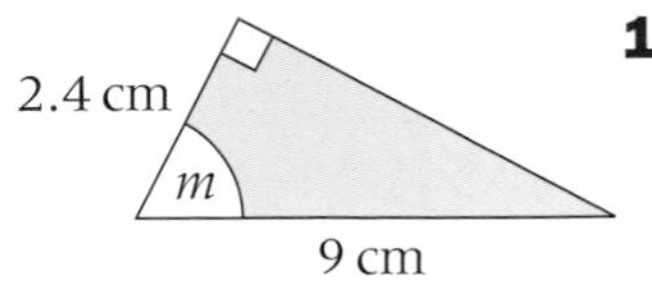

14

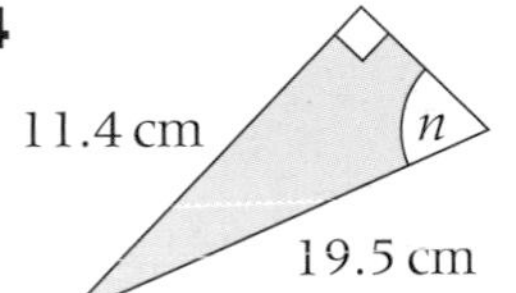

15

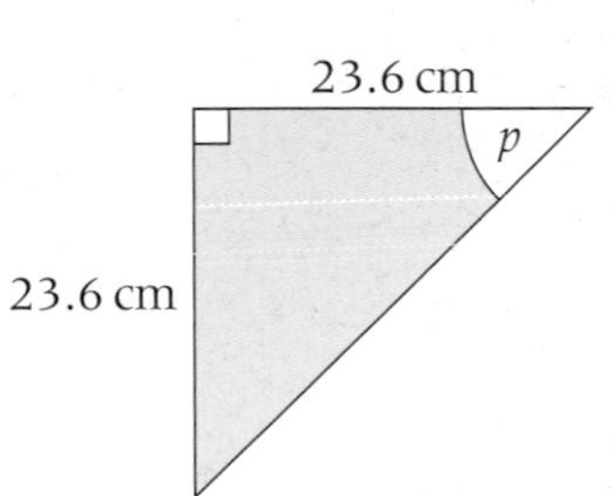

16

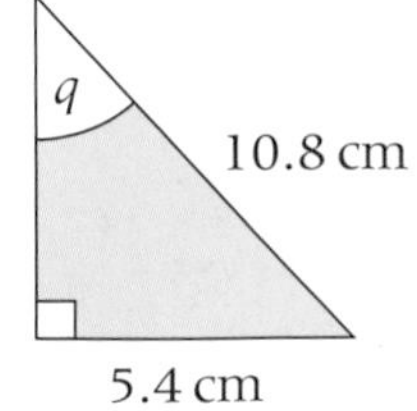

17

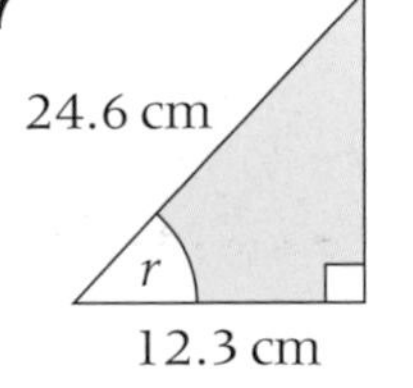

Unit 3

G4.5 Problem solving using trigonometry

This spread will show you how to:

- Use trigonometry in right-angled triangles to solve problems

Keywords

Right-angled triangle

You can use trigonometric ratios to solve problems involving **right-angled triangles**.

You should always sketch a diagram to represent the situation.
You may need to work out extra information to solve the problem.

Example

ABCD is a parallelogram.
AB = 8.2 cm, BC = 6.6 cm and angle ABC = 53°.
Work out the area of the parallelogram.

Sketch a diagram.
Area of a parallelogram = $b \times h$
You need to find the perpendicular height, h.

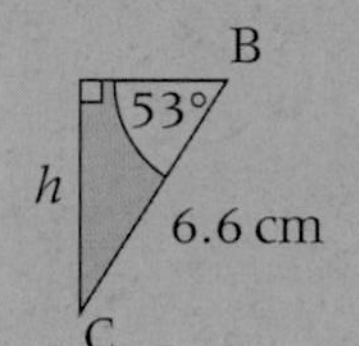

$$\sin 53° = \frac{h}{6.6}$$

$$h = 6.6 \times \sin 53°$$
$$h = 5.27...$$

$$\text{Area} = b \times h$$
$$= 8.2 \times 5.27...$$
$$= 43.2\text{ cm}^2$$

The height h must be at right angles to the base b.

Finding h is an intermediate step. Do not round any values until the end of the calculation.

Example

A vertical flagpole FP has P at the top of the pole. It is held by two wires, PX and PY, fixed on horizontal ground at X and Y. Angle FXP = 36° and angle FYP = 49°. FX is 20.5 m. Find the length of PY.

Sketch a diagram.
FP is vertical, so it is perpendicular to the horizontal ground.
It forms two right-angled triangles: PFX and PFY.
First find the height of PF.

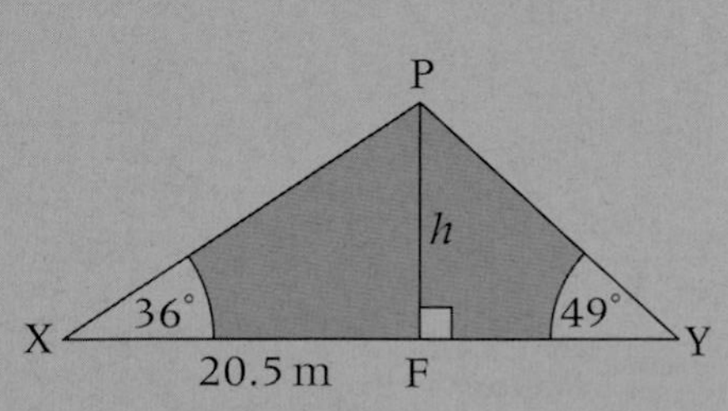

$$\tan 36° = \frac{h}{20.5}$$

$$h = 20.5 \times \tan 36° = 14.894...$$

Use trigonometry to find the length PY.

$$\sin 49° = \frac{h}{PY}$$

$$PY = \frac{h}{\sin 49°} = \frac{14.894...}{\sin 49°} = 19.734...$$

So PY = 19.7 m

Exercise G4.5

Grade A

1 Find the missing lengths.

a

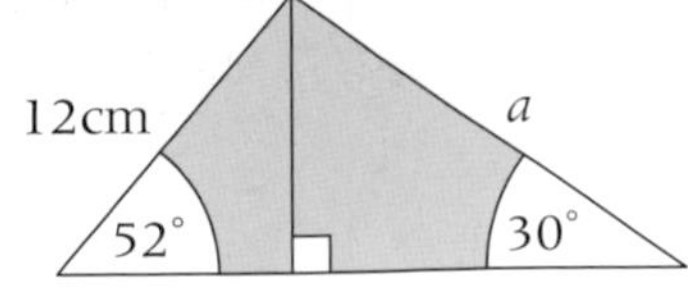

b

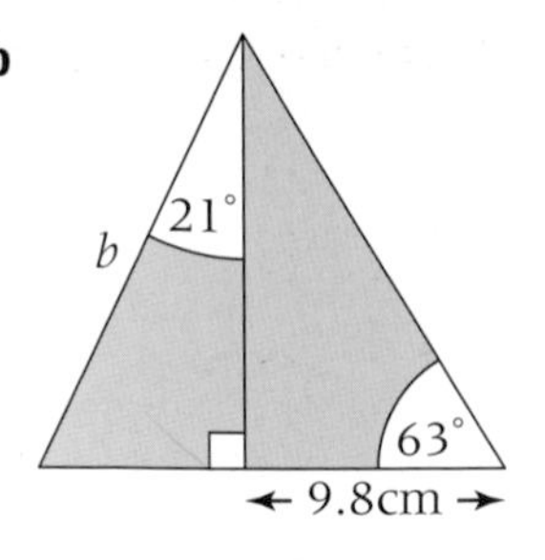

c

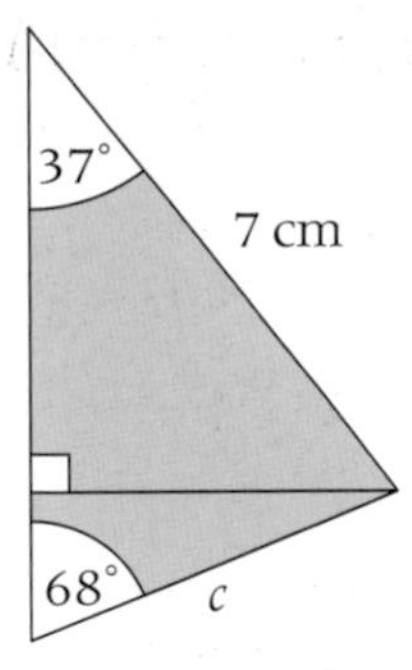

2 Find the missing angles.

a

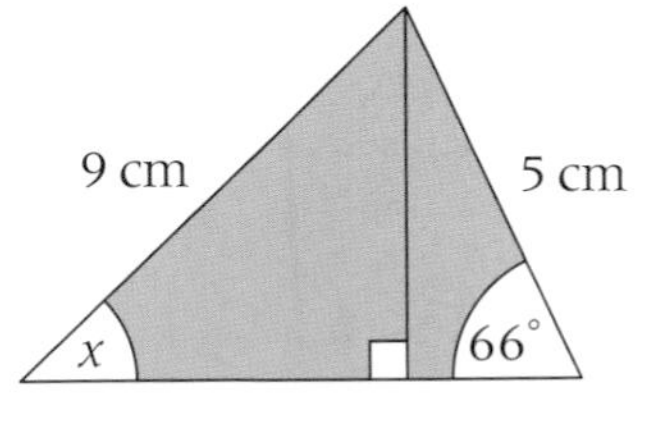

b

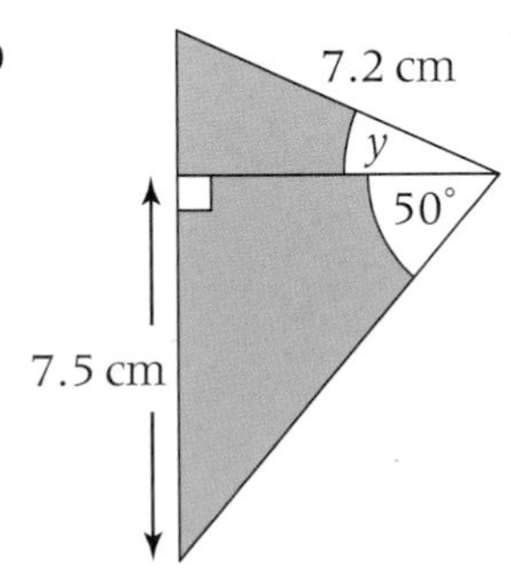

c

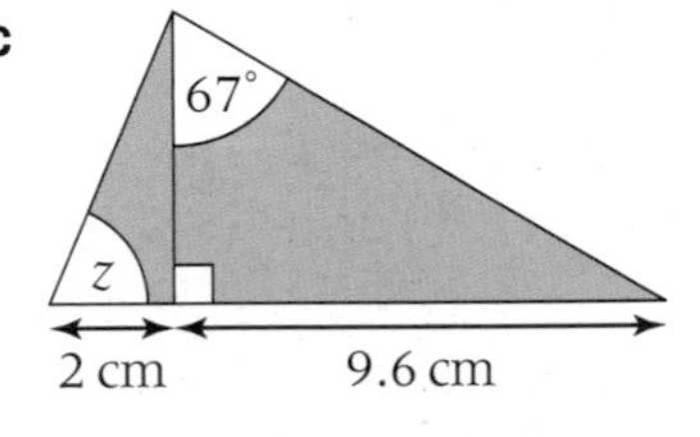

3 A parallelogram has sides of length 5 cm and 9 cm.
The smaller angles are both 45°.
Find the area of the parallelogram.

4 A rhombus has side length 8 cm and smaller angle 35°.
Find the area of the rhombus.

5 A chord AB, of length 12 cm, is drawn inside a circle with centre O and radius 8.5 cm.
Find the angle AOB.

6 A chord PQ is drawn inside a circle with centre O and radius 6 cm, such that angle POQ = 104°.
Find the length PQ.

7 An isosceles triangle has sides of length 7 cm, 7 cm and 9 cm.
Find the interior angles of the triangle.

AO3 Problem

8 Edina and Patsy are estimating the height of the same tree.
Edina stands 20 m from the tree and measures the angle of elevation of the treetop as 52°.
Patsy stands 28 m from the tree and measures the angle of elevation of the treetop as 44°.
Can they both be correct? Explain your reasoning.

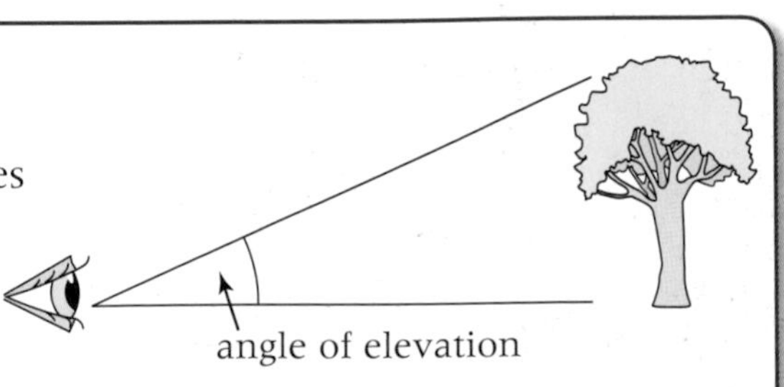

G4.6 Pythagoras' theorem and trigonometry

This spread will show you how to:

- Understand, recall and use Pythagoras' theorem in 2-D problems
- Understand, recall and use trigonometrical relationships in right-angled triangles

Keywords
Cosine
Pythagoras' theorem
Sine
Tangent

You can use **Pythagoras' theorem** to find a side in a right-angled triangle if you know the other two sides.

- Pythagoras' theorem for right-angled triangles: $c^2 = a^2 + b^2$

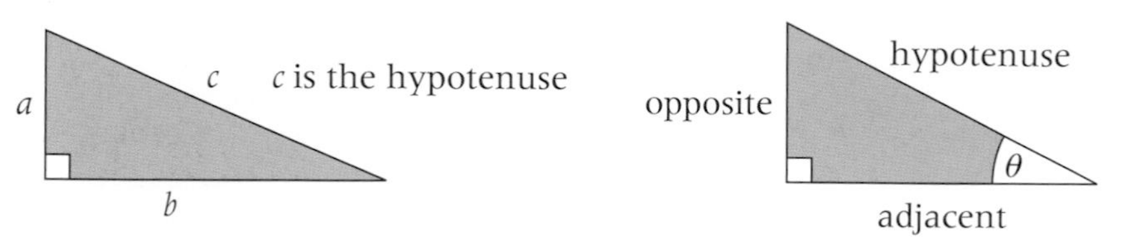

The hypotenuse is always opposite the right-angle.

In a right angle triangle you can use trigonometry to find
an unknown side if you know one side and an angle
an unknown angle if you know two sides

- $\sin\theta = \dfrac{\text{opposite side}}{\text{hypotenuse}}$ $\cos\theta = \dfrac{\text{adjacent side}}{\text{hypotenuse}}$ $\tan\theta = \dfrac{\text{opposite side}}{\text{adjacent side}}$

Example

Find the missing sides and angles in these triangles.

a

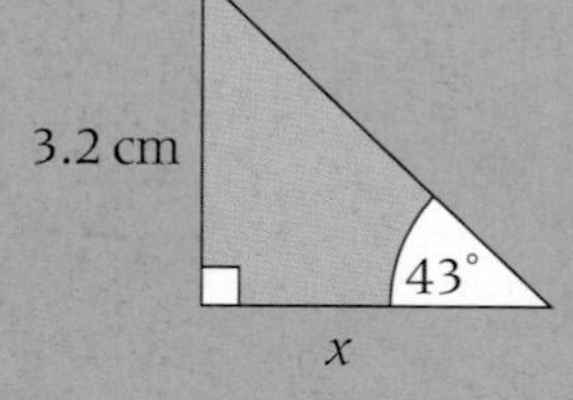

b

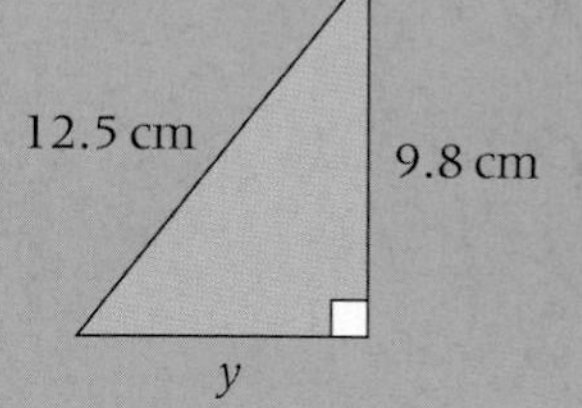

Give your answer to the same degree of accuracy as the measurements given in the question.

a $\tan 43° = \dfrac{3.2}{x}$

$x = \dfrac{3.2}{\tan 43°}$

$x = 3.4\text{ cm}$ (to 1 dp)

b $9.8^2 + y^2 = 12.5^2$

$y = \sqrt{12.5^2 - 9.8^2}$

$y = 7.8\text{ cm}$

Example

Find the missing angle.
Give your answer to the nearest degree.

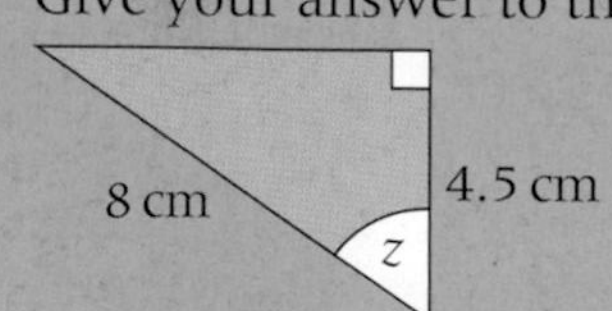

$\cos z = \dfrac{4.5}{8}$

$z = \cos^{-1}\left(\dfrac{4.5}{8}\right)$

$z = 56°$ to nearest degree

You use the inverse trig functions $\sin^{-1}$, $\cos^{-1}$ and $\tan^{-1}$.

Make sure your calculator is in degree mode.

Exercise G4.6 — Grade B/A

1 Find the missing sides in these triangles.

a

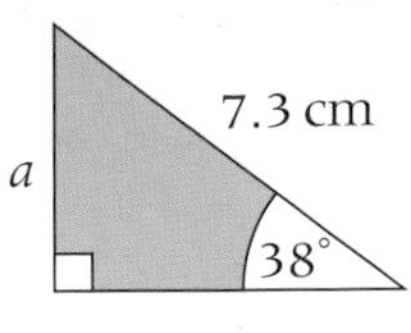

b

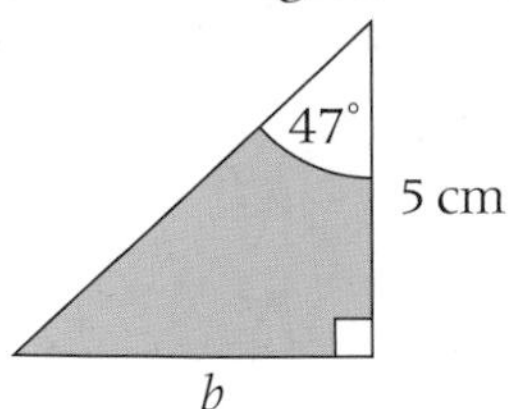

c

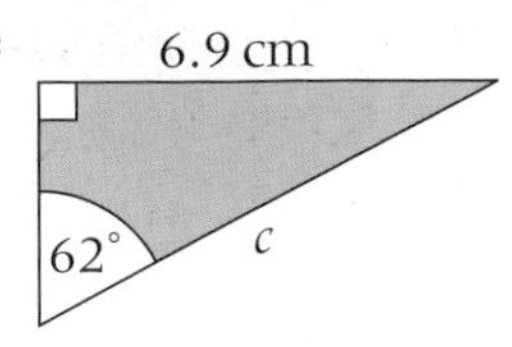

d

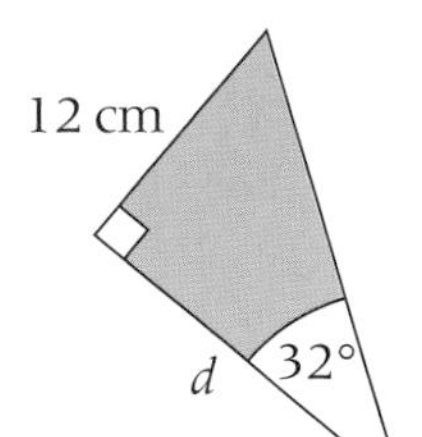

e

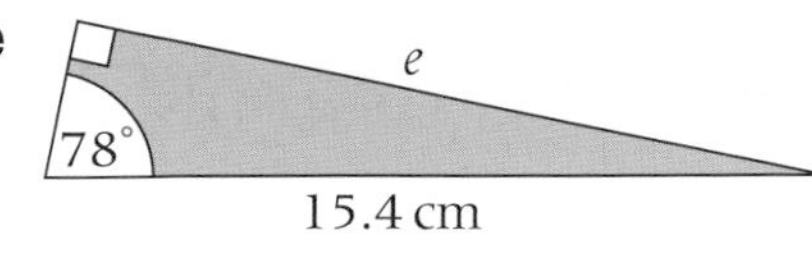

f

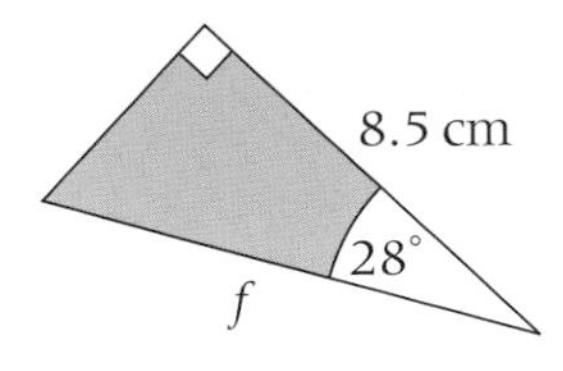

2 Work out the missing angles.

a

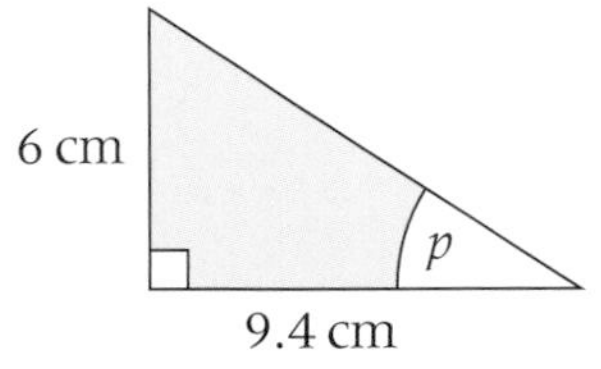

b

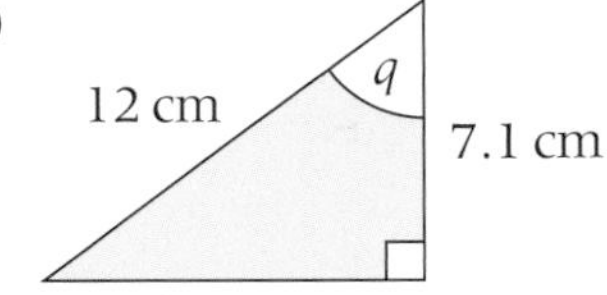

c

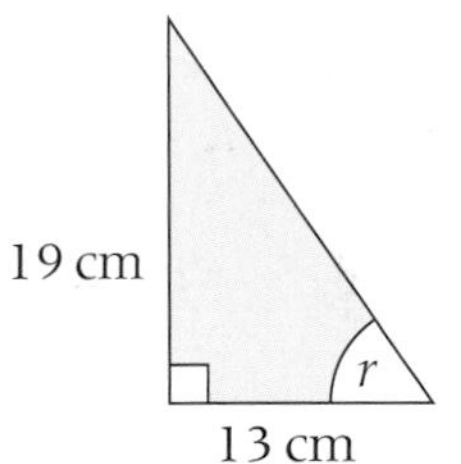

d

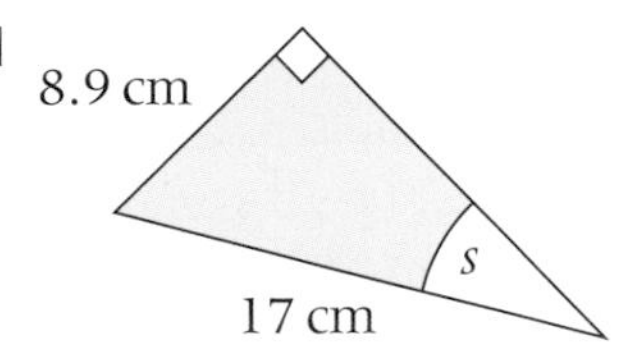

e

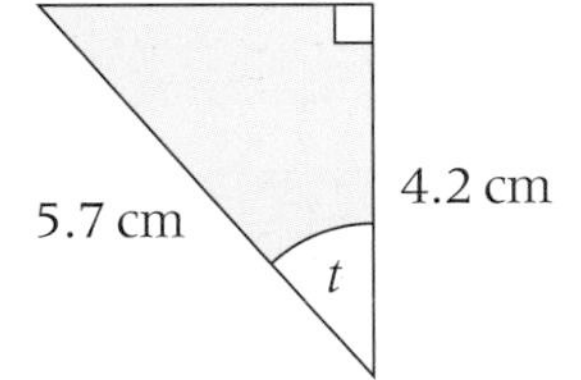

3 ABC and ACD are right-angled triangles.
AB = 12 cm
BC = 3 cm
CD = 4 cm

a Show that AD = 13 cm.

b Work out AD if the lengths are changed so that AB = 16 cm, BC = 5 cm and CD = 7 cm.

Find AC first.

Problem AO3

4 a Find PQ.

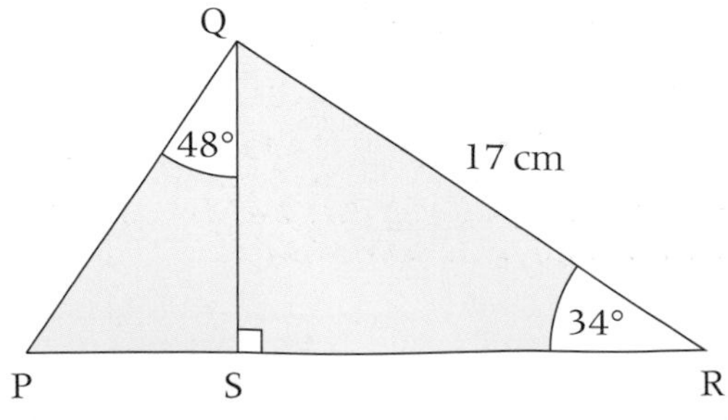

b Find AB.

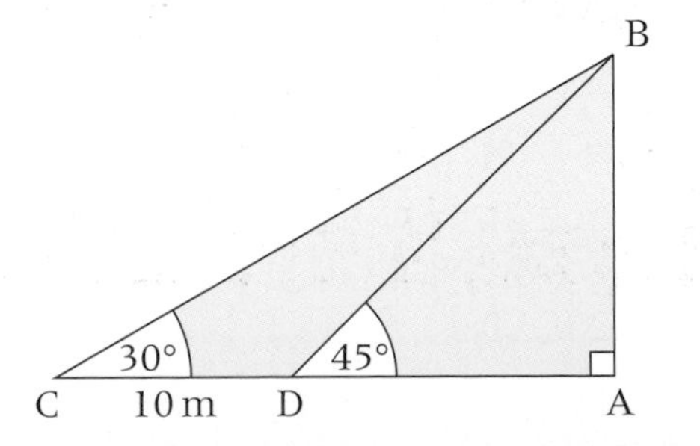

Unit 3

G4

Summary

Check out

You should now be able to:

- Understand, recall and use Pythagoras' theorem in 2-D problems
- Understand, recall and use trigonometric relationships in right-angled triangles to solve 2-D problems
- Use the trigonometric functions of a scientific calculator

Worked exam question

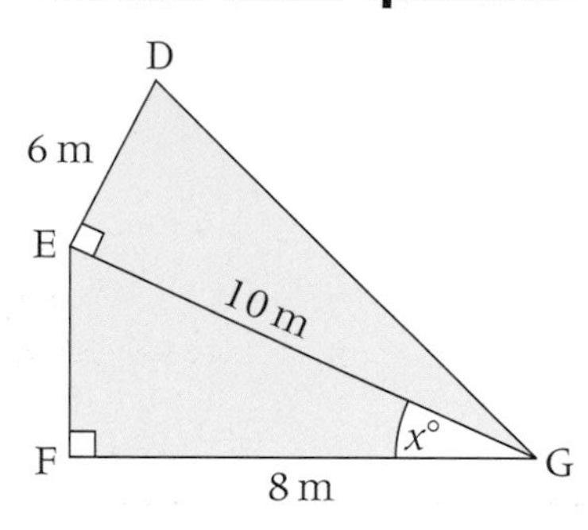

Diagram NOT accurately drawn.

$DE = 6$ m.
$EG = 10$ m.
$FG = 8$ m.
Angle $DEG = 90°$. Angle $EFG = 90°$.

a Calculate the length of DG.
Give your answer correct to 3 significant figures. (3)

b Calculate the size of the angle marked $x°$.
Give your answer correct to one decimal place. (3)

(Edexcel Limited 2004)

a

$$10^2 + 6^2 = 100 + 36$$
$$= 136$$
$$DG = \sqrt{136}$$
$$= 11.7 \text{ m}$$

Show $10^2 = 6^2$ and $\sqrt{136}$ in the working.

b

$$\cos x = \frac{8}{10}$$
$$x = \cos^{-1}\left(\frac{8}{10}\right)$$
$$x = 36.9°$$

Write out an equation with the trig ratio.

The angle must be rounded to one decimal place.

Exam questions

1 **a** Calculate the size of angle a in this right-angled triangle.
Give your answer correct to 3 significant figures.

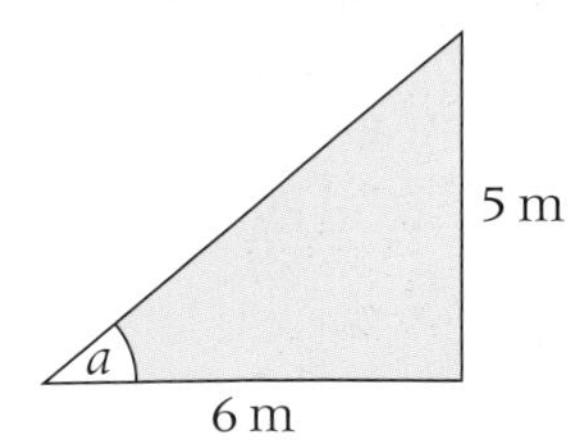

Diagram NOT accurately drawn.

(3)

b Calculate the length of the side x in this right-angled triangle.
Give your answer correct to 3 significant figures.

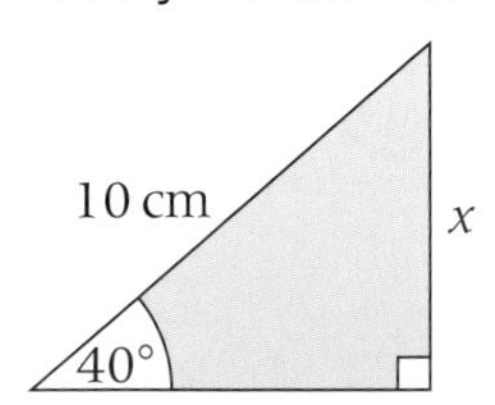

Diagram NOT accurately drawn.

(3)
(Edexcel Limited 2006)

A03

2

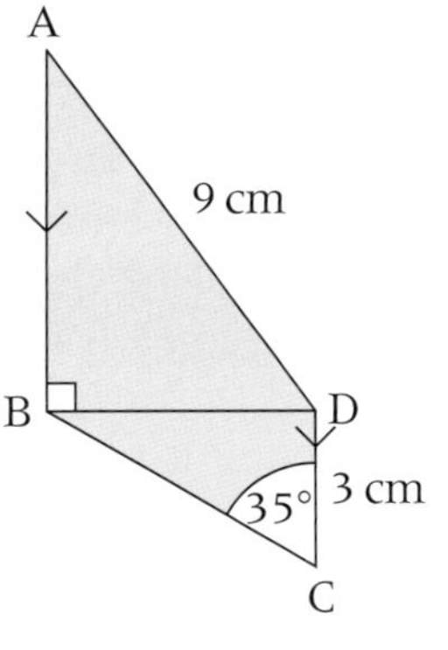

Diagram NOT accurately drawn.

AB is parallel to DC.
$AD = 9\text{cm}$, $DC = 3\text{cm}$.
Angle $BCD = 35°$.
Angle $ABD = 90°$.

Calculate the size of angle BAD.
Give your answer correct to one decimal place.

(4)
(Edexcel Limited 2007)

A6

Formulae and quadratic equations

Sports scientists analyse the performance of elite sportsmen and women. They look for small improvements in technique which will lead to corresponding improvements in performance. For example, the use of video and computer technology allows them to analyse the flight of objects thrown by the athletes and help the athletes improve their accuracy.

What's the point?

The path taken by the object is described using a quadratic equation. By having a greater range of techniques for solving quadratic equations it is possible to apply mathematics to an increasing number of real life applications.

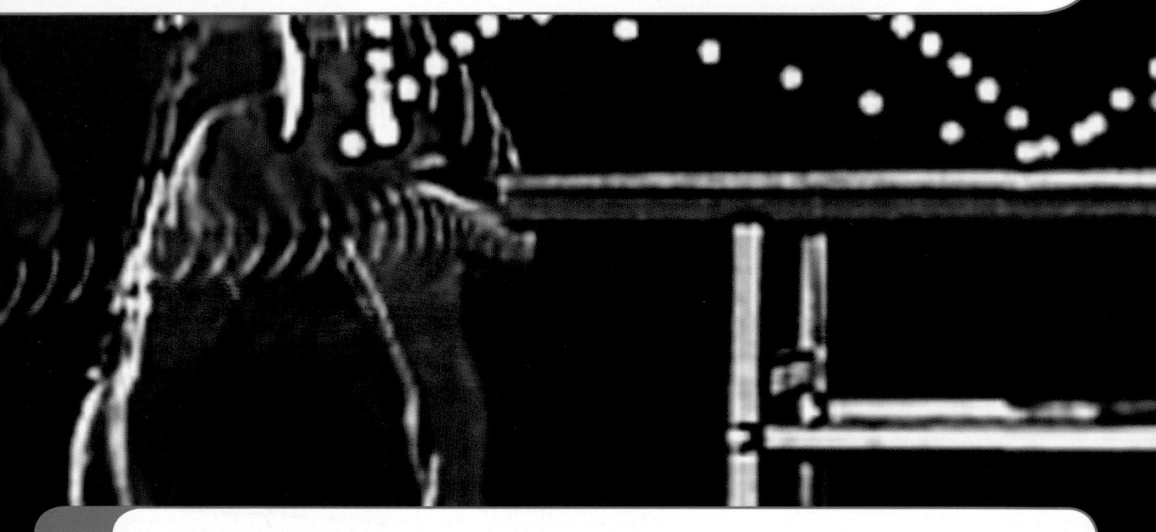

Formulae and quadratic equations

Check in

You should be able to

- **use simple formulae**

1 The cost, in pounds, C of a mobile phone bill is found using the formula

$c = 0.5m + 12$

where m is the number of minutes used. Use the formula to find:

a The cost of using 39 minutes

b The number of minutes used if the cost is £39.

- **rearrange and solve equations**

2 Solve these equations.

a $2x - 7 = 23$ **b** $10 - 2x = 15$

c $\frac{5}{x} - 2 = 15$ **d** $3x + 2 = 5 - 4x$

- **solve quadratic equations using factorisation**

3 **a** Given that $y = x^2 + 5x + 6$, find the value of y when x is zero and the values of x when y is zero.

b Repeat part **a** for $y = 2x^2 - 7x + 6$.

Orientation

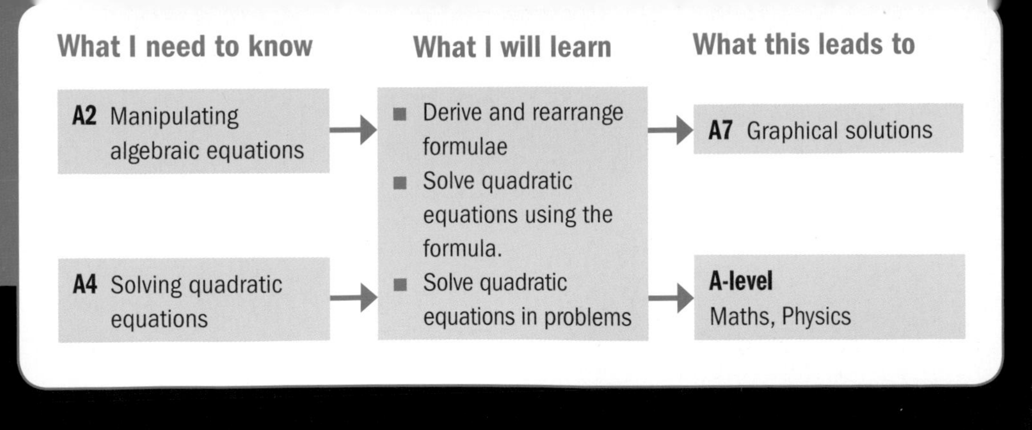

Rich task

The table shows the distance from the sun (million kilometres) and the time to orbit the sun (years) for the planets in our solar system.

Planet	Mercury	Venus	Earth	Mars	Jupiter	Saturn	Uranus	Neptune	Pluto
Distance	57.91	108.2	149.6	227.9	778.5	1433	2877	4503	5906
Time	0.2408	0.6152	1	1.881	11.86	29.46	84.32	164.8	248.1

Draw a graph of distance from the sun against time to orbit the sun.
Write down what you notice.
Can you find a mathematical function which approximates your graph?
(Hint: you may need to use a spreadsheet to hep you here)

A6.1 Rearranging formulae

This spread will show you how to:

- Change the subject of a formula, including cases where a power of the subject appears

Keywords
Rearrange
Subject

- The **subject** of a formula is the variable on its own, on one side of the equals sign.

For example, in the formula $A = \pi r^2$, A is the subject.

- You can **rearrange** a formula to make another variable the subject.

You use inverse operations to do this, as when rearranging equations.

Example

Make h the subject of $A = \frac{1}{2}(a + b)h$.

$$A = \frac{1}{2}(a + b)h$$

$$2A = (a + b)h \quad \text{Multiply both sides by 2.}$$

$$\frac{2A}{a + b} = h$$

h is multiplied by $(a + b)$. The inverse operation is 'divide by $(a + b)$'. $\frac{2A}{a+b} = \frac{(a+b)h}{a+b}$

Example

Make x the subject of $V = p(w - ax^2y)$.

$$V = p(w - ax^2y) \quad \text{Divide both sides by } p.$$

$$\frac{V}{p} = w - ax^2y \quad \text{Add } ax^2y \text{ to both sides.}$$

$$\frac{V}{p} + ax^2y = w \quad \text{Subtract } \frac{V}{p} \text{ from both sides.}$$

$$ax^2y = w - \frac{V}{p} \quad \text{Divide both sides by } ay \text{ (or by } a \text{ and then by } y\text{).}$$

$$x^2 = \frac{w - \frac{V}{p}}{ay} \quad \text{Take square roots of both sides.}$$

Now the term including x is on its own.

$$x = \sqrt{\frac{w - \frac{V}{p}}{ay}} \quad \text{To simplify further, multiply each term in the fraction by } p.$$

Hence $x = \sqrt{\dfrac{pw - V}{apy}}$

Example

Make l the subject of $T = 2\pi\sqrt{\frac{l}{g}}$.

$$T = 2\pi\sqrt{\frac{l}{g}} \quad \text{Divide both sides by } 2\pi.$$

$$\frac{T}{2\pi} = \sqrt{\frac{l}{g}} \quad \text{Square both sides.}$$

$$\frac{T^2}{4\pi^2} = \frac{l}{g} \quad \text{Multiply both sides by } g.$$

Exercise A6.1 — Grade B/A

1 Make m the subject of each formula.

a $y = mx + c$ **b** $t = \frac{m - k}{w}$ **c** $m^2 + kt = p$ **d** $\sqrt[3]{m} - k = l$

2 Make x the subject of each formula.

a $y = \frac{1}{2}x + kw$ **b** $m = \frac{1}{4}(ax - t^2)$ **c** $2y = \sqrt{x}$ **d** $y = \sqrt{k - lx}$

e $k = \frac{t - a\sqrt{x}}{h}$ **f** $m = \frac{p}{x} - t$ **g** $w - \frac{p}{x} = c$ **h** $\frac{y}{ax} - b = j$

3 Clare and Isla have each tried to rearrange a formula to make x the subject. Each has made a mistake – where have they gone wrong?

Clare's attempt:

$C = p - ax$

$C - p = ax$

$\frac{C - p}{a} = x$

Isla's attempt:

$W = p - \frac{k}{x}$

$xW = p - k$

$x = \frac{p - k}{W}$

4 These boxes give the lines of the rearrangement of a formula but they have been scrambled up. Write them in the correct order to show the full rearrangement.

$\frac{c}{df(b-a)} = e$	$a + \frac{c}{def} = b$	$\frac{c}{def} = b - a$	$a = b - \frac{c}{def}$	$c = def(b - a)$

5 Show that each formula can be rearranged into the given form.

a $\frac{p(c - qt)}{m - x^2} = wr$ into $x = \sqrt{\frac{mwr + pqt - pc}{wr}}$

b $\frac{1}{a} + \frac{1}{b} = \frac{1}{c}$ into $b = \frac{ac}{a - c}$

c $\sqrt[4]{t - qx} = 2p$ into $x = \frac{t - 16p^4}{q}$

6 These three formulae are equivalent – true or false? Explain your answer.

$x = \frac{t}{p} - wx$ and $x = \frac{t - wpx}{p}$ and $\frac{w - px}{-p} = x$

A02 Functional Maths

7 Rearrange the lens formula $\frac{1}{f} = \frac{1}{u} + \frac{1}{v}$

a to make f the subject

b to make v the subject.

Unit 3

A6.2 Further rearranging formulae

This spread will show you how to:

- Change the subject of a formula, including cases where the subject occurs twice

Keywords
Factorise
Rearrange
Subject

- You can **rearrange** formulae using inverse operations.
- When the **subject** appears twice, you can rearrange the formula by collecting like terms and then factorising.
  For example, to make x the subject of $ax + b = cx + d$

p.122

$$ax + b = cx + d \longrightarrow ax - cx = d - b \longrightarrow x(a - c) = d - b \longrightarrow x = \frac{d-b}{a-c}$$

Collect x-terms on one side → **Factorise** → Divide by $a - c$

Example

Rearrange these formulae to make x the subject.

a $k(p - x) = q(x + w)$ **b** $\dfrac{bx + c}{x} = 8$

a $k(p - x) = q(x + w)$ Expand first.
$kp - kx = qx + qw$ Collect x terms.
$kp - qw = qx + kx$ Factorise
$kp - qw = x(q + k)$
$\dfrac{kp - qw}{q + k} = x$

b $\dfrac{bx + c}{x} = 8$ Eliminate the fraction.
$bx + c = 8x$
$c = 8x - bx$
$c = x(8 - b)$
$x = \dfrac{c}{8 - b}$

Take the x terms to the side that avoids negatives: $qx + kx$ is easier to work with than $-kx - qx$

p.206

Example

Rearrange this formula to make R the subject.

a $\dfrac{2p + 10}{50} = \dfrac{R}{15 + R}$

a $(2p + 10)(15 + R) = 50R$
$30p + 150 + 2pR + 10R = 50R$
$30p + 150 + 2pR = 40R$
$30p + 150 = 40R - 2pR$
$30p + 150 = R(40 - 2p)$
$R = \dfrac{30p + 150}{40 - 2p} \Rightarrow R = \dfrac{15p + 75}{20 - p}$

Exercise A6.2 — Grade A/A*

1 Rearrange each formula to make x the subject.

a $y = ax^2 + b$ **b** $t - \frac{1}{2}(x - c)$ **c** $k = c - bx$

d $p = \frac{b}{x} - q$ **e** $\sqrt{axz - b} = 3t$ **f** $\frac{k - x^3}{t} = t$

2 Explain why you cannot make x the subject of $x^2 + 5x + 6 = 0$ using the method of collecting the x-terms on one side and then factorising.

3 Rearrange each formula to make w the subject.

a $pw + t = qw - r$ **b** $a - cw = k - lw$ **c** $p(w - y) = q(t - w)$

d $r - w = t(w - 1)$ **e** $w + g = \frac{w + c}{r}$ **f** $\frac{wx - t}{r} = 5 - w$

g $\frac{w + t}{w - 5} = k$ **h** $\frac{w + p}{q - w} = \frac{3}{4}$ **i** $\sqrt{\frac{w - t}{w + q}} = 5$

Unit 3

4 **a** The diagram shows a cylinder with radius r and height h. Write a formula for the total volume of the cylinder.

b Rearrange your formula to make r the subject.

5 Repeat question **4** for a cone of height h and base radius r.

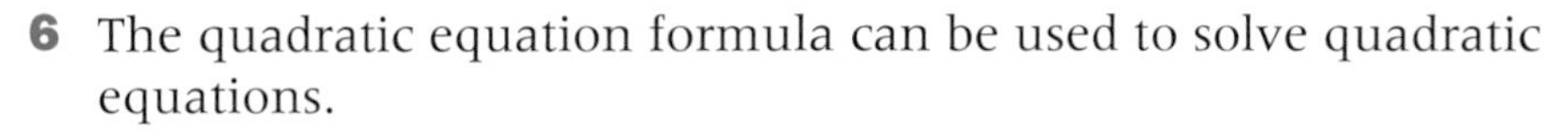

6 The quadratic equation formula can be used to solve quadratic equations.

$$x = \frac{-b \pm \sqrt{b^2 - 4ac}}{2a}$$

a Use the formula to solve $3x^2 - 2x - 9 = 0$.

b Rearrange the formula to make c the subject.

c Rearrange the formula to make b the subject.

d Why is it too difficult to make a the subject of this formula?

7 The double angle formula in trigonometry is:

$$\sin(A + B) = \sin A \cos B + \cos A \sin B$$

Explain why you cannot rearrange this to give

$$\sin A = \frac{\cos A \sin B - \sin B}{1 - \cos B}.$$

Hint: Does sin(30° + 15°) equal sin 30° + sin 15°?

A6.3 The quadratic formula

This spread will show you how to:

- Solve quadratic equations using the quadratic formula

Keywords
Coefficient
Constant
Quadratic

Some **quadratic** equations do not factorise.

p.228

- You can solve all quadratic equations of the form $ax^2 + bx + c = 0$ using the quadratic equation formula:

$$x = \frac{-b \pm \sqrt{b^2 - 4ac}}{2a}$$

In the formula
– a is the **coefficient** of the x^2-term
– b is the coefficient of the x-term
– c is the **constant**.

$x^2 - 7x + 11 = 0$ does not factorise. You cannot find two numbers that multiply to give 11 and add to give -7.

The constant is the number on its own.

Example

Solve the quadratic equation $3x^2 - 5x = 1$.

$$3x^2 - 5x = 1$$
$$3x^2 - 5x - 1 = 0$$

In the formula: $a = 3$, $b = -5$ and $c = -1$

$$x = \frac{-b \pm \sqrt{b^2 - 4ac}}{2a}$$

$$x = \frac{5 \pm \sqrt{(-5)^2 - 4 \times 3 \times (-1)}}{2 \times 3}$$

$$x = \frac{5 \pm \sqrt{25 + 12}}{6}$$

Either $x = \frac{5 + \sqrt{37}}{6} = 1.847\ 127\ 088... = 1.85$ (to 3 sf)

Or $x = \frac{5 - \sqrt{37}}{6} = -0.180\ 460\ 421... = -0.180$ (to 3 sf)

First rearrange into the form . $ax^2 + bx + c = 0$

Write the values of a, b and c.

b is -5, so $-b$ is $+5$.

Example

Given that the side of a rectangle measures $(x + 4)$ and its area satisfies the equation $x^2 + 3x - 14 = 0$, find the length to 2 dp.

$x^2 + 3x - 14 = 0$
$a = 1$, $b = 3$ and $c = -14$, so

$$x = \frac{-3 \pm \sqrt{9 - 4 \times 1 \times (-14)}}{2 \times 1}$$

Hence, $x = \frac{-3 + \sqrt{65}}{2} = 2.531...$ or $x = \frac{-3 - \sqrt{65}}{2} = -5.531...$

Since the length is $x + 4$, $x = 2.531$ gives a length of 6.53 (to 2 dp).

The formula gives two answers, but a negative value of x does not make sense in the context of the question.

Exercise A6.3 — Grade A/A*

1 Solve these quadratic equations using the quadratic formula. Where necessary, give your answers to 3 significant figures.

a $3x^2 + 10x + 6 = 0$
b $5x^2 - 6x + 1 = 0$
c $2x^2 - 7x^2 - 15 = 0$
d $x^2 + 4x + 1 = 0$
e $2x^2 + 6x - 1 = 0$
f $6y^2 - 11y - 5 = 0$
g $3x^2 + 3 = 10x$
h $6x + 2x^2 - 1 = 0$
i $20 - 7x - 3x^2 = 0$

2 Solve these equations using the quadratic formula. Give your answers to 3 sf.

Rearrange the equation in the form $ax^2 + bx + c = 0$.

a $x(x + 4) = 9$
b $2x(x + 1) - x(x + 4) = 11$
c $(3x)^2 = 8x + 3$
d $y + 2 = \dfrac{14}{y}$
e $\dfrac{3}{x + 1} + \dfrac{4}{2x - 1} = 2$

3 Here is a rectangle.

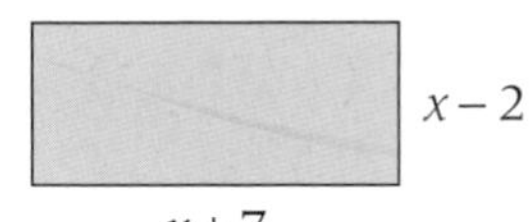

a Write an expression for the area of this rectangle.
b Given that the area of the rectangle is 20 cm², show that

$$x^2 + 5x - 34 = 0.$$

c Solve the equation to find the dimensions of the rectangle to 2 dp.

AO3 Problem

4 a Solve both $2x^2 + 11x + 12 = 0$ and $2x^2 + 11x + 10 = 0$, using the quadratic equation formula.
b Now try and solve the equations by factorisation.
c Compare your answers to parts **a** and **b**. Which part of the formula helps you to decide whether a quadratic will factorise or not?
d Use your suggestion to decide if these equations will factorise.
i $3x^2 + 19x - 12$
ii $12x^2 + 14x - 3$

5 Use the quadratic formula to solve $2x^2 + 7x + 9 = 0$.
What do you notice?
Why does this happen?

A6.4 Using quadratic equations

This spread will show you how to:

- Solve quadratic equations by factorisation and using the quadratic formula
- Set up simple equations

Keywords
Factorise
Formula
Quadratic

- A **quadratic** equation has a squared term as its highest power.
- To solve a quadratic equation, you:

Make the equation equal to zero. → **Factorise** using common factors, double brackets or DOTS. / Use the **formula**: $x = \dfrac{-b \pm \sqrt{b^2 - 4ac}}{2a}$ → Obtain up to two solutions.

Always try to factorise first.

To solve a problem, you may be able to set up an equation and solve it.

For quadratics you may need to reject one solution, depending on the context.

For example, if the solutions are $x = -3$ or $x = 5$ and x is a length, then reject $x = -3$.

Example

Two numbers have a product of 105 and a difference of 8.
If the larger number is x:

a show that $x^2 - 8x - 105 = 0$

b solve this equation to find the two numbers.

a The two numbers are x and $x - 8$.
So $x(x - 8) = 105$
$x^2 - 8x - 105 = 0$ (as required)

p.228

b $x^2 - 8x - 105 = 0$
$(x + 7)(x - 15) = 0$
Either $x + 7 = 0$ or $x - 15 = 0$
So $x = -7$ and $x - 8 = -15$ or $x = 15$ and $x - 8 = 7$.
The two numbers are -7 and -15 or 7 and 15.

Two answers for x lead to two answers for $x - 8$.

Example

If the area of the trapezium is $400\,\text{cm}^2$, show that $x^2 + 10x = 400$ and find the value of x correct to 3 dp.

Area of trapezium $= \dfrac{(a + b)}{2}h$

$$A = \tfrac{1}{2}(x + 10) \times 2x = x(x + 10)$$
$$400 = x(x + 10)$$
$$x^2 + 10x = 400 \text{ (as required)}$$
$$x^2 + 10x - 400 = 0$$
$$x = \frac{-10 \pm \sqrt{10^2 - 4 \times 1 \times -400}}{2}$$
$$= 15.61552... \text{ or } -25.61552...$$

Since x is a length, it must be positive, so $x = 15.616$ to 3 dp.

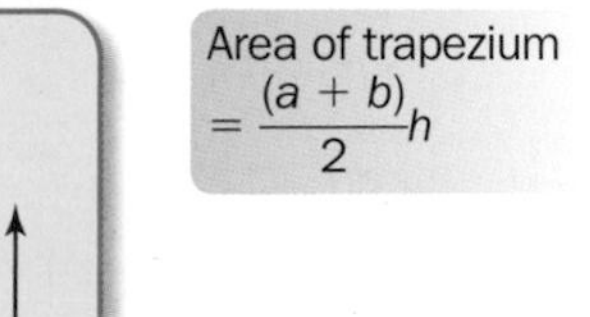

Exercise A6.4

Grade A*

1 Solve these quadratic equations by factorisation.

a $x^2 - 9x + 18 = 0$ **b** $x^2 - 100 = 0$ **c** $7x = x^2$

2 Solve these quadratic equations using the formula, giving your answers to 3 significant figures.

a $x^2 - 8x - 2 = 0$ **b** $6x^2 - 5x - 5 = 0$ **c** $7x^2 = 10 - 3x$

3 In each question

i write a quadratic equation to represent the information given

ii solve the equation to find the necessary information.

Give answers to 2 decimal places where appropriate.

a Two numbers which differ by 4 have a product of 117. Find the numbers.

b The length of a rectangle exceeds its width by 4 cm. The area of the rectangle is 357 cm^2. Find the dimensions of the rectangle.

c The square of three less than a number is ten. What is the number?

d Three times the reciprocal of a number is one less than ten times the number. What is the number?

e The diagonal of a rectangle is 17 mm. The length is 7 mm more than the width. Find the dimensions of the rectangle.

f If this square and rectangle have equal areas, find the side length of the square.

$3x - 2$

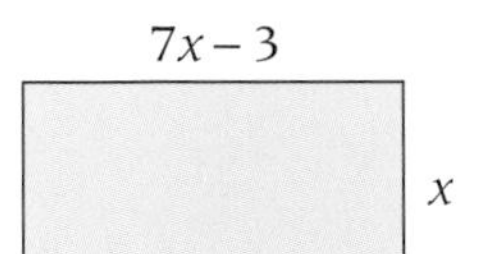

Unit 3

AO3 Problem

4 a The height of a closed cylinder is 5 cm and its surface area is 100 cm^2.

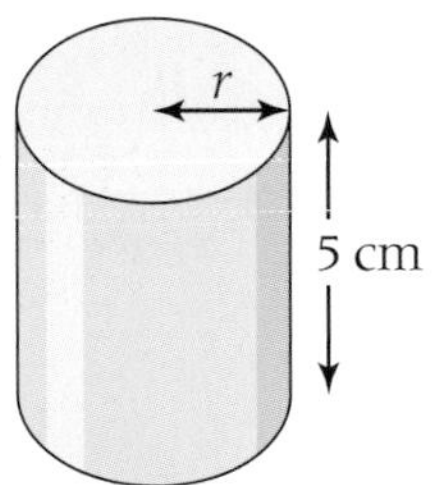

Given that the radius is r, show that

$\pi r^2 + 5\pi r - 50 = 0$

b Hence, find the diameter of the base of the cylinder.

A6.5 Further quadratic equations

This spread will show you how to:

- Solve quadratic equations by factorisation and using the quadratic formula

Keywords
Discriminant
Quadratic

- You can solve some **quadratic** equations by making them equal to zero, then either factorising or using the quadratic equation formula.
- Quadratic equations can have one, two or no solutions.

Quadratic	Does it factorise?	Solution	Comment
$x^2 + 8x + 12 = 0$	Yes: $(x + 6)(x + 2)$	$x = -2$ or -6	• Factorises • Two solutions • Integer solutions
$x^2 + 8x + 11 = 0$	No	$x = \dfrac{-8 \pm \sqrt{8^2 - 4 \times 1 \times 11}}{2 \times 1}$ $x = \dfrac{-8 \pm \sqrt{20}}{2}$ $x = -6.236...$ or $-1.7639...$	• Does not factorise • Two solutions • Non-integer solutions
$x^2 + 8x + 16 = 0$	Yes: $(x + 4)(x + 4)$ or $(x + 4)^2$	$x = -4$ twice	• Factorises • One repeated solution • Integer solution
$x^2 + 8x + 20 = 0$	No	$x = \dfrac{-8 \pm \sqrt{8^2 - 4 \times 1 \times 20}}{2 \times 1}$ $x = \dfrac{-8 \pm \sqrt{-16}}{2}$	• Does not factorise • No solutions You cannot find the square root of -16

- The value in the square root of the formula $b^2 - 4ac$ is called the **discriminant**. It tells you how to solve the equation and how many solutions there will be.

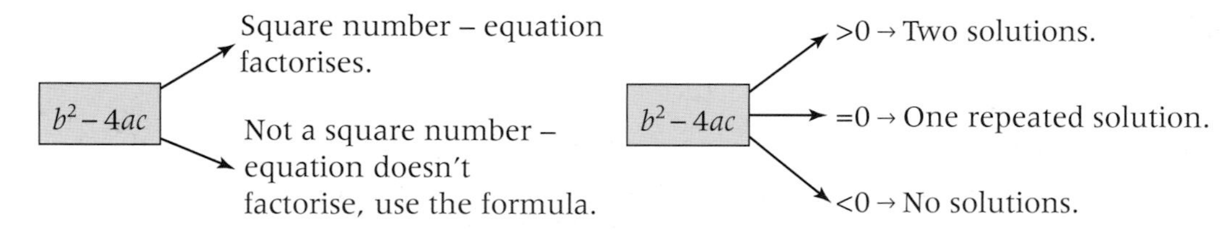

Example

By calculating the discriminant, decide how many solutions $2x^2 - 3x - 7 = 0$ has and whether to solve the equation by factorising or by formula.

$a = 2$, $b = -3$ and $c = -7$ so $b^2 - 4ac = (-3)^2 - 4 \times 2 \times (-7) = 65$

Discriminant > 0, so there are two solutions. Discriminant is not a square, so the equation will not factorise – use the formula.

$$x = \frac{+3 \pm \sqrt{65}}{4}$$

$x = 2.77$ or -1.27 (both to 3 sf.)

Exercise A6.5 Grade A/A*

1 Solve these quadratic equations by factorisation.

a $x^2 + 7x + 10 = 0$ **b** $x^2 + 4x - 12 = 0$ **c** $x^2 - 49 = 0$

d $x^2 - 8x = 0$ **e** $(x + 2)^2 = 16$ **f** $3y^2 - 7y + 2 = 0$

2 Solve these quadratic equations using the formula.

a $x^2 + 8x + 6 = 0$ **b** $7x^2 + 6x + 1 = 0$ **c** $x^2 - 2x - 1 = 0$

d $10y^2 - 2y - 3 = 0$ **e** $4x^2 + 3x = 2$ **f** $(2x - 3)^2 = 2x$

3 For each equation, use the discriminant test to decide

i if it factorises

ii how many solutions it has.

a $x^2 + 2x - 15 = 0$ **b** $x^2 - 6x + 9 = 0$ **c** $2x^2 - 3x - 4 = 0$

d $3y^2 + 11x + 6 = 0$ **e** $10x^2 - x - 3 = 0$ **f** $6x^2 - 11x + 3 = 0$

4 Solve these quadratics by the most efficient method.

a $y(y + 3) = 88$ **b** $(2x)^2 = 25$

c $(7 + 2x)^2 + 4x^2 = 37$ **d** $(x + 2)^2 + 45 = 2x + 1$

e $(y + 1)^2 = 2y(y - 2) + 10$ **f** $\frac{15}{p} = p + 22$

g $\frac{3}{x} + \frac{3}{x + 1} = 7$

If you are not sure a quadratic will factorise, use the discriminant test.

5 True or false? $x^2 - 5x + 10 = 0$ can be solved by factorisation, producing two solutions. Explain your answer.

6 The perimeter of a rectangle is 46 cm and its diagonal is 17 cm.

You should not need to use the formula.

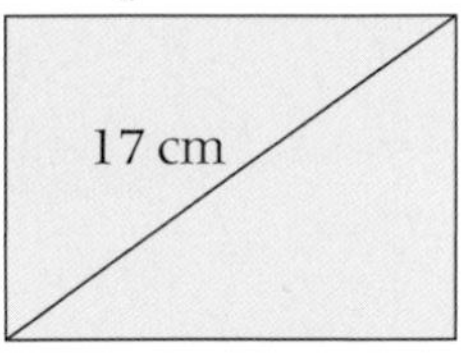

Set up a quadratic equation and solve it to find the dimensions of the rectangle.

Problem **AO3**

7 You can solve the equation $x^2 - 5x = 0$ by factorisation, $x(x - 5) = 0$, and then letting either $x = 0$ or $x - 5 = 0$. Extend this method to find three solutions for each of these equations.

a $x^3 - 5x^2 + 6x = 0$

b $2x^3 + 5x^2 - x = 0$

c $x^3 = 6x - x^2$

Unit 3

A6.6 Solving problems involving quadratics

This spread will show you how to:

- Solve quadratic equations by factorisation, completing the square and using the quadratic formula.
- Select and use appropriate and efficient techniques and strategies to solve problems.

Keywords
Quadratic
Solve

- When a problem involves a **quadratic** equation:

Spot a quadratic by its x^2 term.

Make the equation equal to zero, keeping the x^2 term positive.
Solve by:

Factorisation
Always, if possible, factorise, that is if $b^2 - 4ac$ is a square number.

The formula
Use any time, but especially when $b^2 - 4ac$, is not a square.

Completing the square
Use any time, but especially when $b^2 - 4ac$ is not a square.

The solutions
$b^2 - 4ac > 0 \rightarrow 2$ solutions
$b^2 - 4ac = 0 \rightarrow 1$ solution
$b^2 - 4ac < 0 \rightarrow$ no solutions

Examiner's tip
You could also use a graph to find the number of solutions to a quadratic equation.

Example

p.154

Find the radius of this circle.

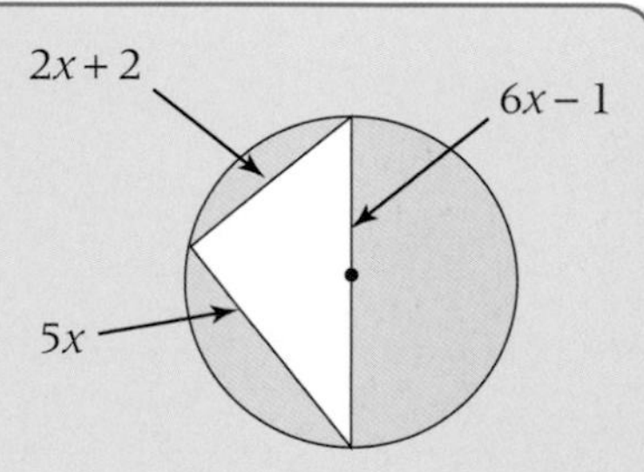

Using Pythagoras' theorem:

Angle in a semicircle is 90°.

The diameter is the hypotenuse of the right-angled triangle.

$$a^2 + b^2 = c^2$$
$$(5x)^2 + (2x + 2)^2 = (6x - 1)^2$$
$$25x^2 + (2x + 2)(2x + 2) = (6x - 1)(6x - 1)$$
$$25x^2 + 4x^2 + 4x + 4x + 4 = 36x^2 - 6x - 6x + 1$$
$$29x^2 + 8x + 4 = 36x^2 - 12x + 1$$
$$0 = 7x^2 - 20x - 3$$
$$0 = 7x^2 - 21x + x - 3$$
$$0 = 7x(x - 3) + 1(x - 3)$$
$$0 = (7x + 1)(x - 3)$$

Keep the x^2 term positive.

$b^2 - 4ac = (-20)^2 - 4 \times 7 \times (-3) = 484 = 22$ a square number, so it does factorise.

Either $7x + 1 = 0$ or $x - 3 = 0$
$x = -\frac{1}{7}$ $\quad x = 3$

In this context, x cannot be negative, hence, $x = 3$.
Diameter $= 6 \times 3 - 1 = 17$, so radius $= 8.5$.

Exercise A6.6 — Grade A*

1 Solve these, by writing them as quadratic equations.

a $2x^2 = 5x - 1$ **b** $(y - 5)^2 = 20$ **c** $p(p + 10) + 21 = 0$

d $10x + 7 = \frac{3}{x}$ **e** $\frac{x^2 + 3}{4} + \frac{2x - 1}{5} = 1$

2 In parts **a–d**, form an equation and solve it to answer the problem.

a The diagonal of this rectangle is 13 cm.
What is the perimeter of the rectangle?

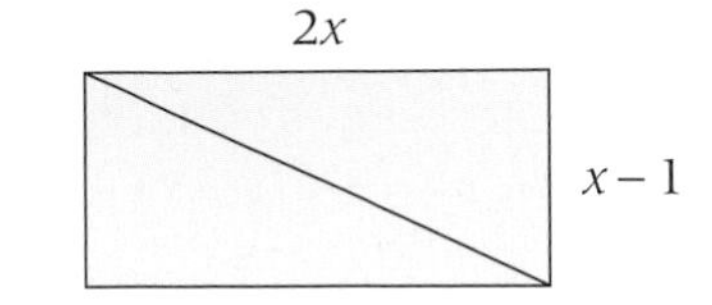

b The area of this hexagon is $25\,m^2$.
Find the perimeter of the hexagon.

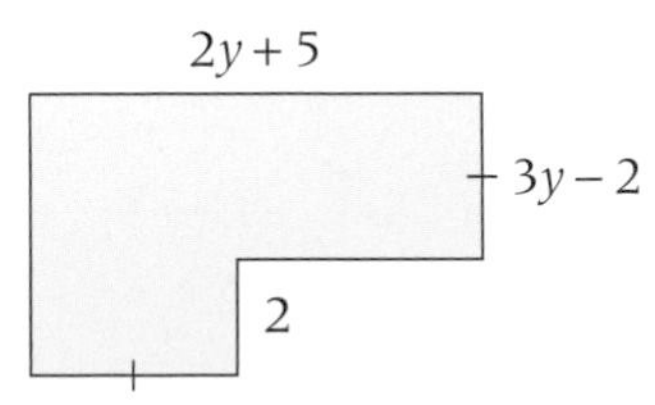

c A circle with radius 5 and centre (0, 0) intersects the line $y = 2(x - 1)$ in two places. Where do they intersect?

d The surface area of the sphere is equal to the curved surface area of the cylinder. Which has the largest volume?

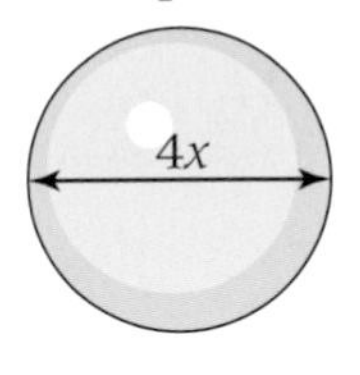

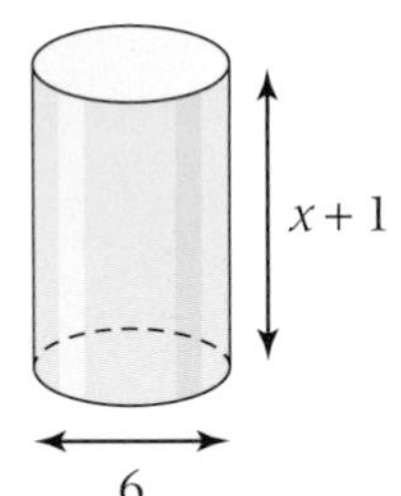

3 **a** Sketch the graphs $y = 3 - 2x$ and $y = x^2 - 4x + 3$ on the same axes.

b Hence write down one point of intersection of these graphs.

c Use algebra to find their second point of intersection.

A03 Problem

4 For what value of x is the *total* surface area of the cylinder equal to three times the area of the *curved* surface of the cone?

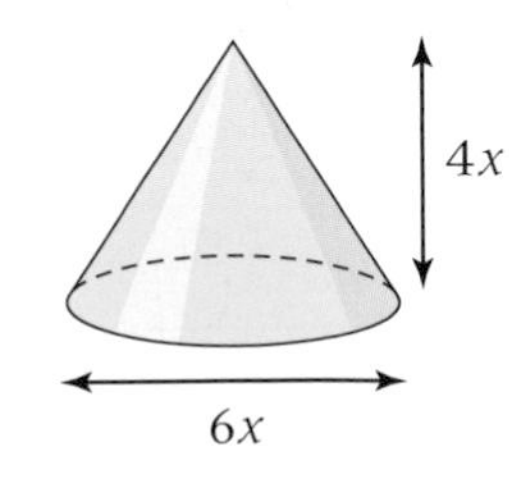

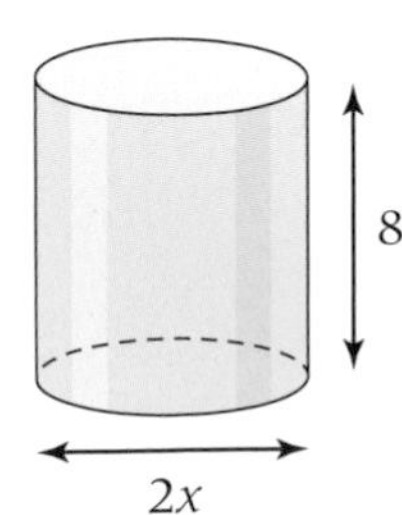

Unit 3

A6 Summary

Check out

You should now be able to:

- Change the subject of a formula
- Solve quadratic equations by using the quadratic formula
- Set up and solve simple quadratic equations by factorisation and completing the square

Worked exam question

$2x + 5$

$x - 3$ | $x - 3$

$2x + 5$

A rectangle has sides of length $x - 3$, $x - 3$, $2x + 5$ and $2x + 5$.
All the measurements are in centimetres.
The area of the rectangle is $90\,\text{cm}^2$.

a Show that $2x^2 - x - 105 = 0$ (3)

b Hence calculate the perimeter of the rectangle. (4)

a

Area of the rectangle $= (2x + 5)(x - 3)$
$= 2x^2 - 6x + 5x - 15$
$= 2x^2 - x - 15$

$2x^2 - x - 15 = 90$
$2x^2 - x - 105 = 0$

Write down this equation.

b

$2x^2 - x - 105 = 0$
$(2x - 15)(x + 7) = 0$
$2x - 15 = 0$ or $x + 7 = 0$
$x = 7.5$ or $x = -7$

$x = -7$ is impossible as $x - 3 = -10$ which is a negative length.
$x = 7.5$

State that one solution is impossible.

$x - 3 = 7.5 - 3 = 4.5\,\text{cm}$
$2x + 5 = 2 \times 7.5 + 5 = 20\,\text{cm}$

Substitute $x = 7.5$ to find the length and width of the rectangle.

Perimeter $= 2 \times (4.5 + 20)$
$= 49\,\text{cm}$

Exam questions

1 Make b the subject of the formula

$$a = \frac{2 - 7b}{b - 5}$$

(4)
(Edexcel Limited 2008)

2 $\frac{1}{u} + \frac{1}{v} = \frac{1}{f}$

$u = 2\frac{1}{2}, \quad v = 3\frac{1}{3}$

a Find the value of f (3)

b Rearrange $\frac{1}{u} + \frac{1}{v} = \frac{1}{f}$

to make u the subject of the formula.
Give your answer in its simplest form. (2)
(Edexcel Limited 2009)

A03

3

AT is a tangent at T to a circle, centre O.
$OT = x$ cm, $\quad AT = (x + 5)$ cm, $\quad OA = (x + 8)$ cm.

Find the radius of the circle.

Give your answer correct to 3 significant figures. (7)
(Edexcel Limited 2004)

Sine and cosine rules

Historically it is more accurate to measure angles than measure distances directly. Surveyors take this into account when measuring land to make maps by measuring angles in a triangle of known base and then using trigonometry to calculate the position of the third vertex. Using an expanding network of triangles surveyors can accurately map whole countries, including the heights of mountains.

What's the point?

Many objects can be described in terms of component triangles. Knowing how to calculate missing angles and sides in such triangles allows you to accurately describe them and the world around you.

Check in

You should be able to

- **apply Pythagoras' theorem**

1 Use Pythagoras' theorem to find the missing side in these triangles:

a

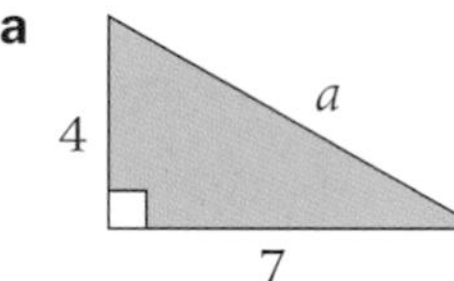

b

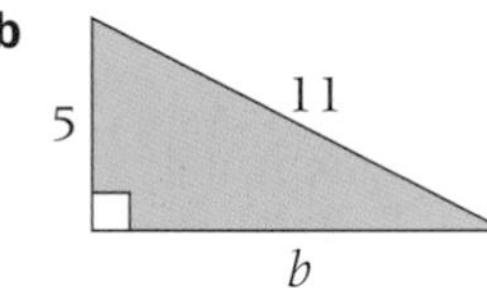

- **solve equations involving fractions**

2 Solve these equations.

a $\frac{x}{4} = \frac{5}{7}$ **b** $\frac{2}{11} = \frac{x}{9}$ **c** $\frac{5}{4} = \frac{7}{x}$

- **apply the rules of precedence in arithmetic**

3 Evaluate these expressions.

a $4^2 + 3 - 2 \times 7$ **b** $5 - 7^2 + 2 \times 3$ **c** $\frac{8^2 + 3^2 - 5^2}{2 \times 4}$

- **apply trigonometry**

4 For the triangles drawn, **i** Use trigonometry to find the height.
ii Use the height to find the area.

a

b

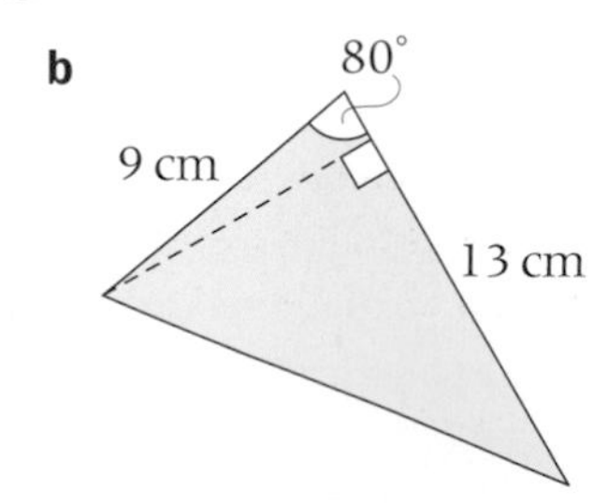

Orientation

What I need to know	What I will learn	What this leads to
N5 Work with surds **G3** Find areas and volumes of shapes and solids	■ Use the sine, cosine and area formulae for scalene triangles ■ Use Pythagoras' theorem in 3D	**A-level** Maths
G4 Use Pythagoras' theorem and trigonometric functions		Design, Engineering, Surveying

Rich task

a Three circles are bound together with a band as shown in the diagram.
Each circle has a radius of 20 cm.
Calculate the length of the band.

b Draw three circles with different radii and calculate the perimeter of the band that binds them together.

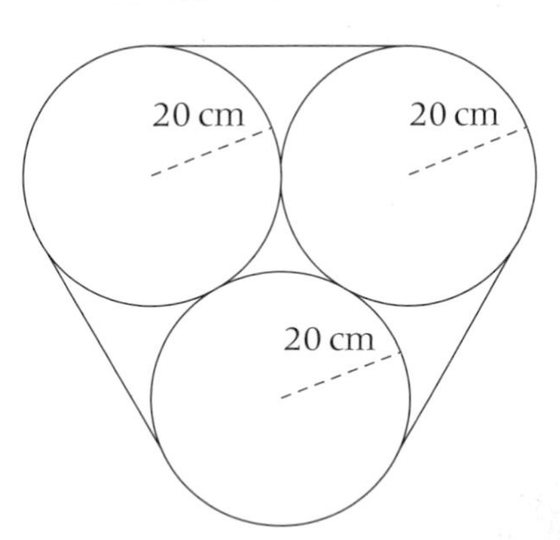

G5.1 The sine rule

This spread will show you how to:

- Understand, recall and use trigonometrical relationships in triangles that are not right-angled

Keywords
Sine rule

You can split any triangle into two right-angled triangles.

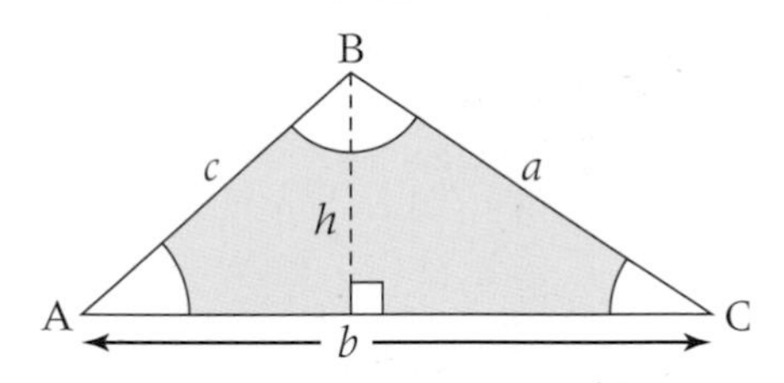

The angles are A, B and C.
Label the side opposite angle A as *a*, etc.

p.298

In the left-hand triangle:

$\sin A = \frac{h}{c}$

so $h = c \times \sin A$

In the right-hand triangle:

$\sin C = \frac{h}{a}$

so $h = a \times \sin C$

$$c \times \sin A = a \times \sin C$$

$$\text{so} \quad \frac{\sin A}{a} = \frac{\sin C}{c} \quad \text{or} \quad \frac{a}{\sin A} = \frac{c}{\sin C}$$

You can extend the rule to include the third side:

- This is the **sine rule**: $\frac{a}{\sin A} = \frac{b}{\sin B} = \frac{c}{\sin C}$

You use the sine rule when a problem involves two sides and an angle or two angles and a side.

The sine rule works in triangles whether they are right-angled or not.

Example

a Find the lengths PQ and PR.

b Find the angles X and Y.

P, 112°, 36°, R, 14.5 cm, Q

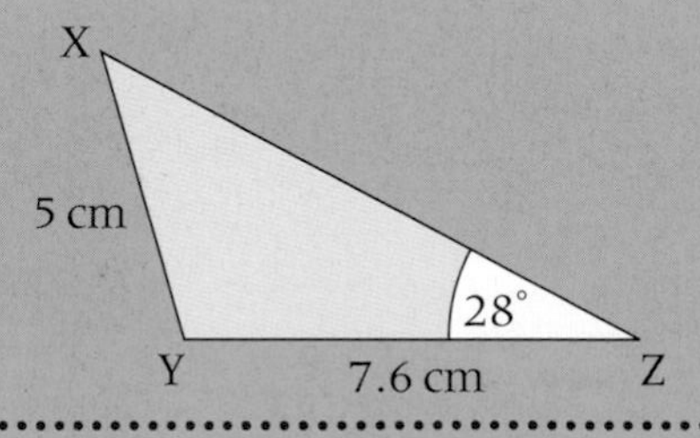

a $\frac{PQ}{\sin 36°} = \frac{14.5}{\sin 112°}$

$PQ = \sin 36° \times \frac{14.5}{\sin 112°}$

$PQ = 9.2\text{ cm}$

$Q = 180° - (112° + 36°) = 32°$

$\frac{PR}{\sin 32°} = \frac{14.5}{\sin 112°}$

$PR = \sin 32° \times \frac{14.5}{\sin 112°}$

$PR = 8.3\text{ cm}$

b $\frac{\sin X}{7.6} = \frac{\sin 28°}{5}$

$\sin X = 7.6 \times \frac{\sin 28°}{5}$

$X = \sin^{-1}\left(7.6 \times \frac{\sin 28°}{5}\right)$

$X = 46°$

$Y = 180° - (28° + 46°)$

$= 106°$

To find a side, use $\frac{a}{\sin A} = \frac{c}{\sin C}$

To find an angle, use $\frac{\sin A}{a} = \frac{\sin C}{c}$

Exercise G5.1 — Grade A

1 Find the sides marked x.

Use $\dfrac{a}{\sin A} = \dfrac{b}{\sin B} = \dfrac{c}{\sin C}$

a

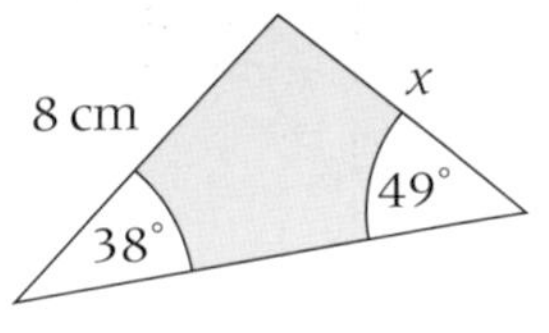

b

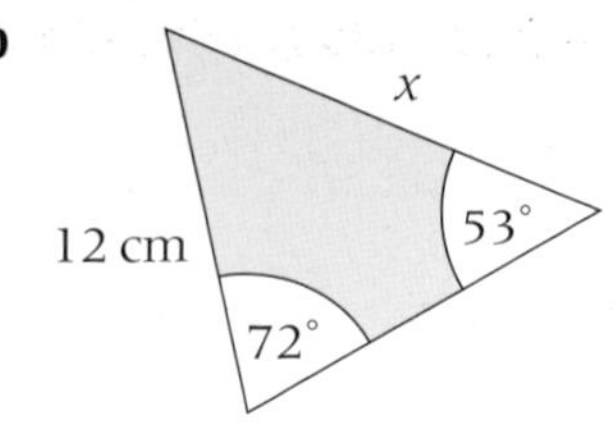

c

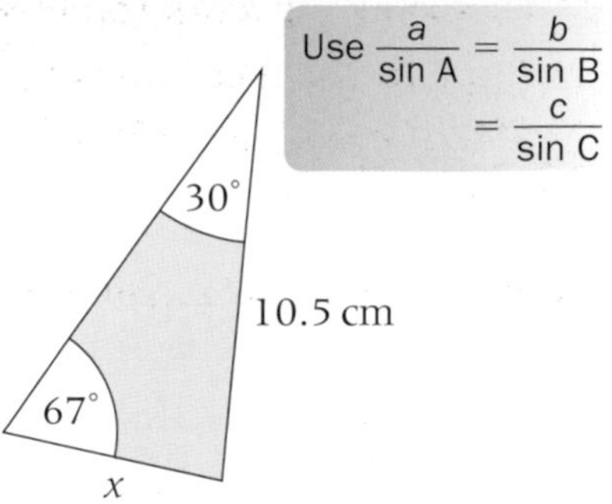

d

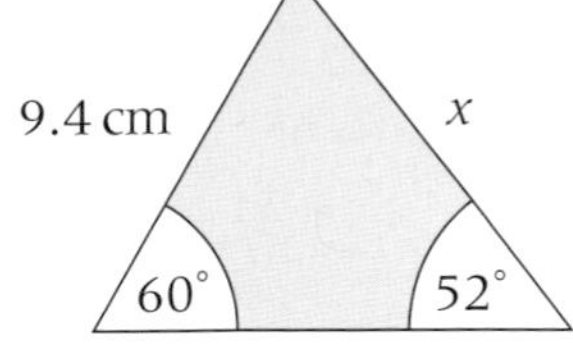

e

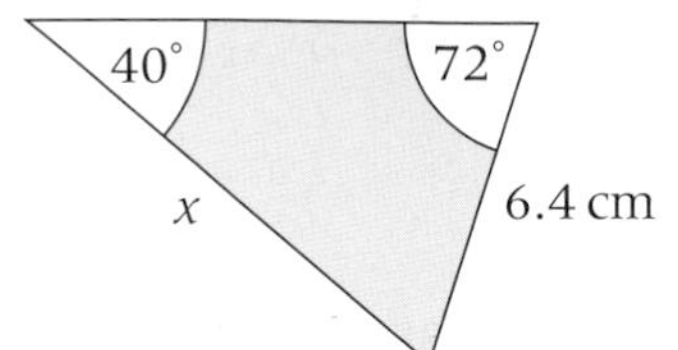

f

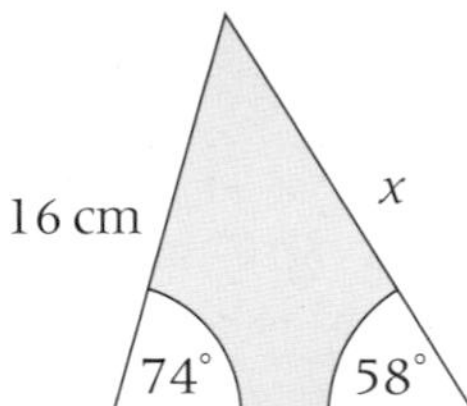

g

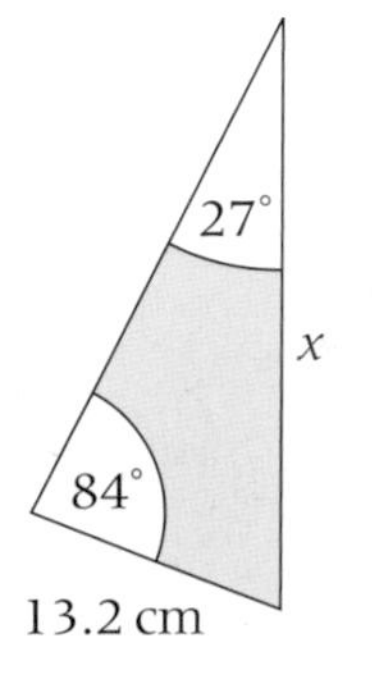

h

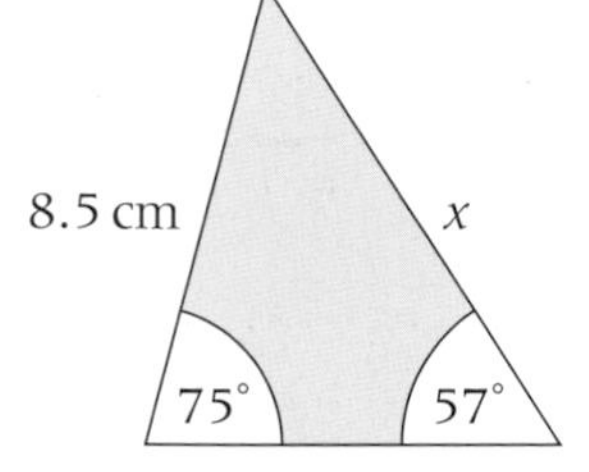

Use $\dfrac{\sin A}{a} = \dfrac{\sin B}{b} = \dfrac{\sin C}{c}$

2 Find the angles marked θ.

a

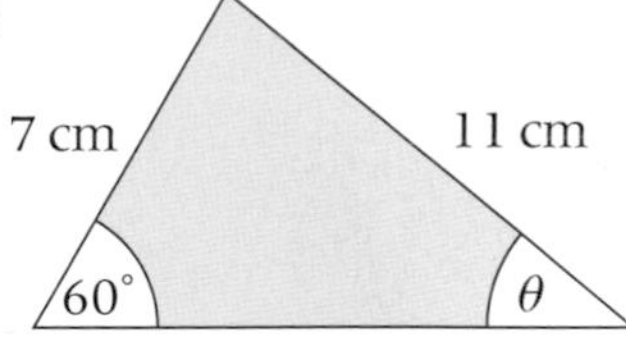

b

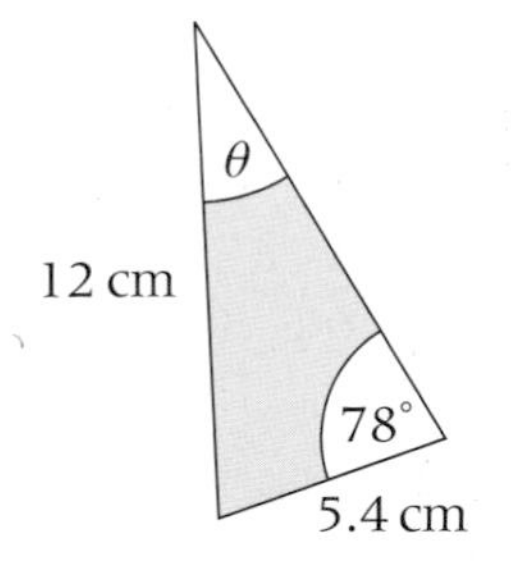

c

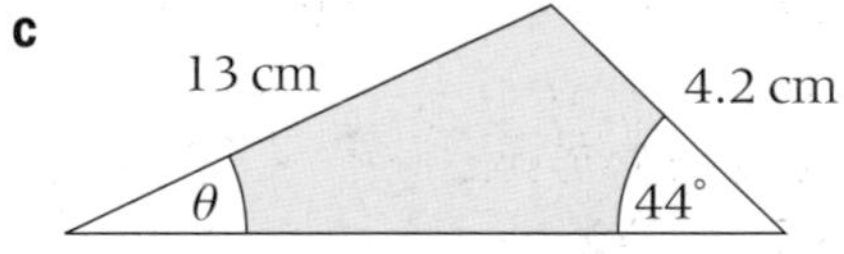

d

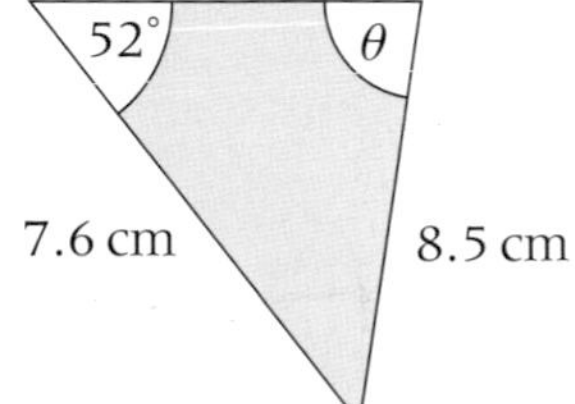

e

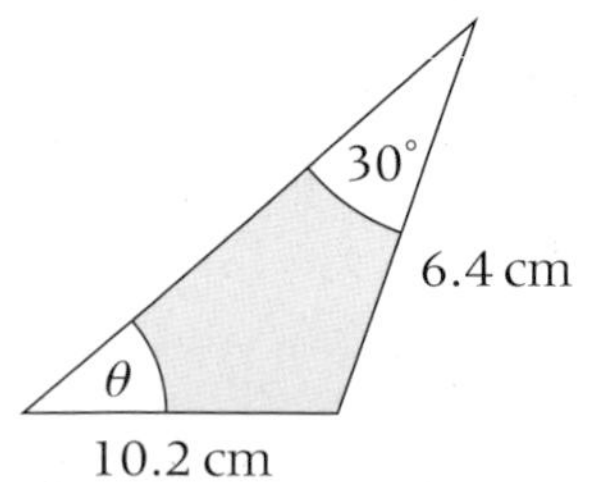

f

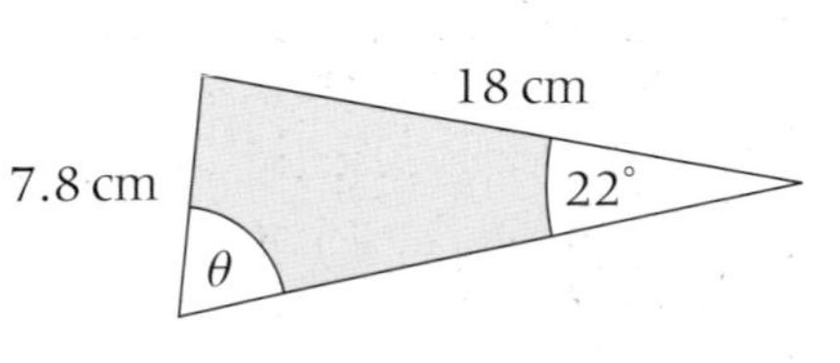

g

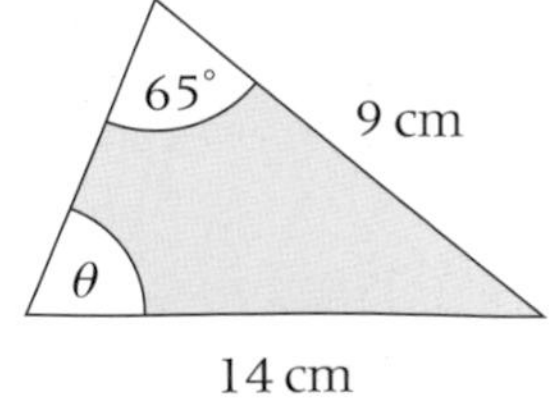

h

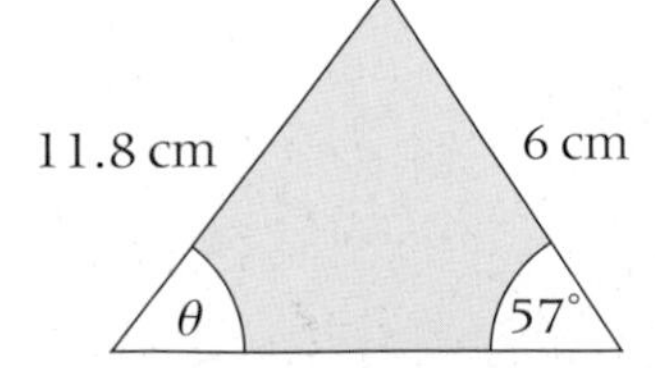

G5.2 The cosine rule

This spread will show you how to:

- Understand, recall and use trigonometrical relationships in triangles that are not right-angled

Keywords
Cosine rule
Pythagoras' theorem

You need the **cosine rule** if a problem involves all three sides and one angle of a triangle.

Split side b as shown into lengths x and $b - x$.

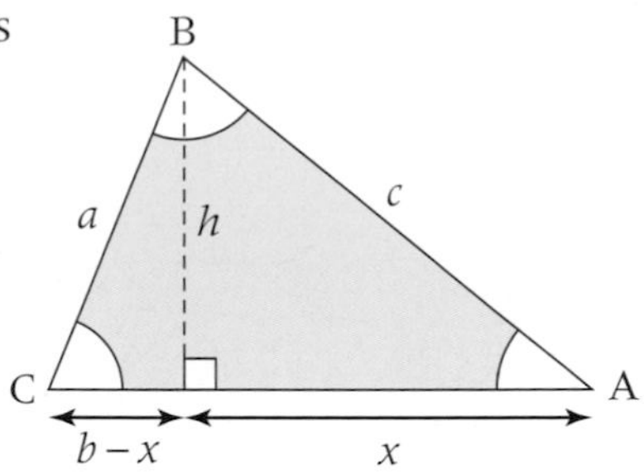

Use **Pythagoras' theorem**

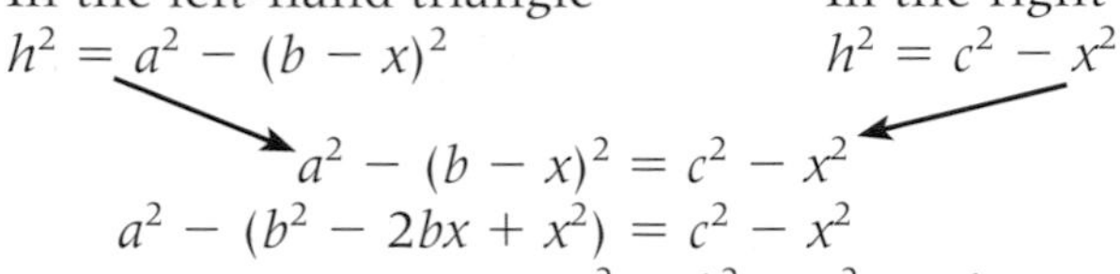

In the left-hand triangle $h^2 = a^2 - (b - x)^2$

In the right-hand triangle $h^2 = c^2 - x^2$

$$a^2 - (b - x)^2 = c^2 - x^2$$ Eliminate h.

$$a^2 - (b^2 - 2bx + x^2) = c^2 - x^2$$

$$a^2 = b^2 + c^2 + 2bx \quad (1)$$

In the right-hand triangle

$\cos A = \frac{x}{c}$ so $x = c \cos A$

Substitute this expression for x in equation (1)

$$a^2 = b^2 + c^2 - 2bc \cos A$$

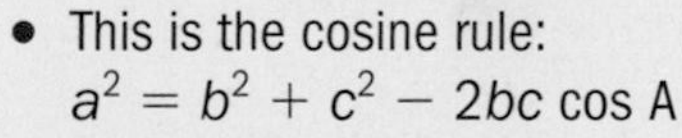

- This is the cosine rule:
  $a^2 = b^2 + c^2 - 2bc \cos A$
- It can be rearranged to give:
  $\cos A = \dfrac{b^2 + c^2 - a^2}{2bc}$

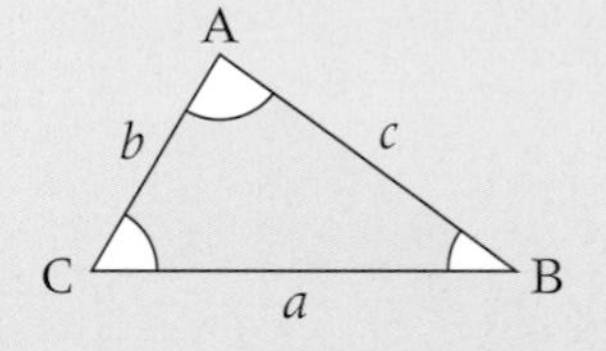

Example

a Work out the length EF.

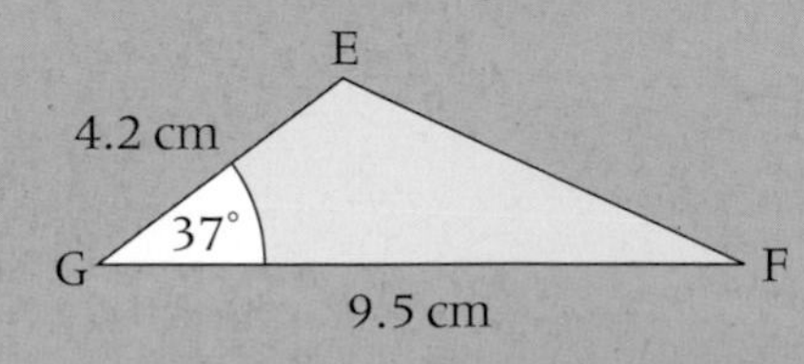

b Work out the angle P.

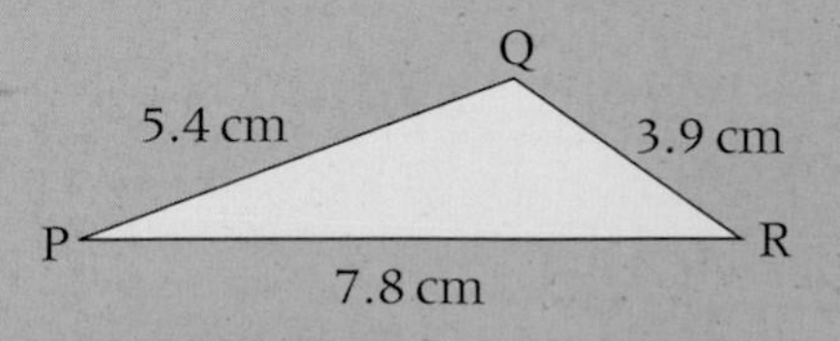

The cosine of an obtuse angle is negative. You will solve problems involving the cosine of the negative angles in G7.2.

a Use the cosine rule:

$$a^2 = b^2 + c^2 - 2bc \cos A$$

$$EF^2 = 9.5^2 + 4.2^2 - 2 \times 9.5 \times 4.2 \times \cos 37°$$

$$= 90.25 + 17.64 - 63.73...$$

$$EF = 6.6\,\text{cm (1dp)}$$

b Use the rearranged cosine rule:

$$\cos A = \frac{b^2 + c^2 - a^2}{2bc}$$

$$\cos P = \frac{7.8^2 + 5.4^2 - 3.9^2}{2 \times 7.8 \times 5.4}$$

$$P = \cos^{-1}\left(\frac{7.8^2 + 5.4^2 - 3.9^2}{2 \times 7.8 \times 5.4}\right)$$

$$= 27.4° \text{ (nearest degree)}$$

Exercise G5.2

Grade A

1 Find the sides marked x.

Use $a^2 = b^2 + c^2 - 2bc \cos A$.

a

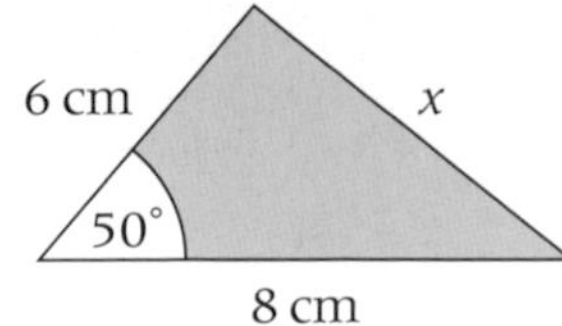

b

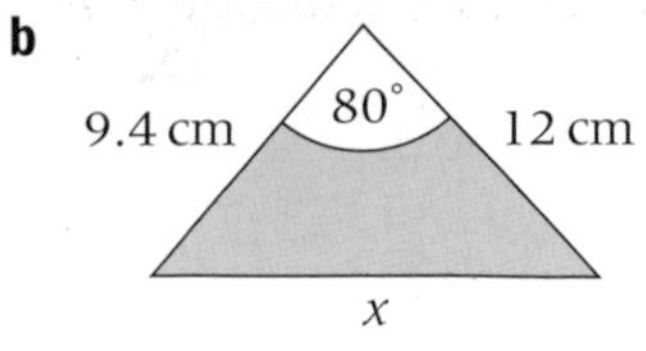

c

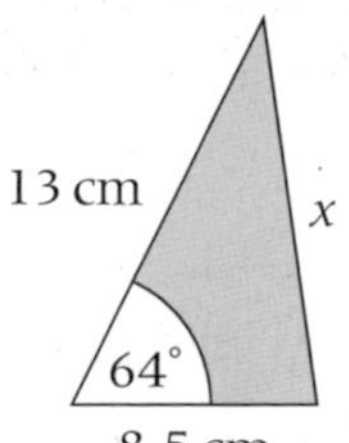

d

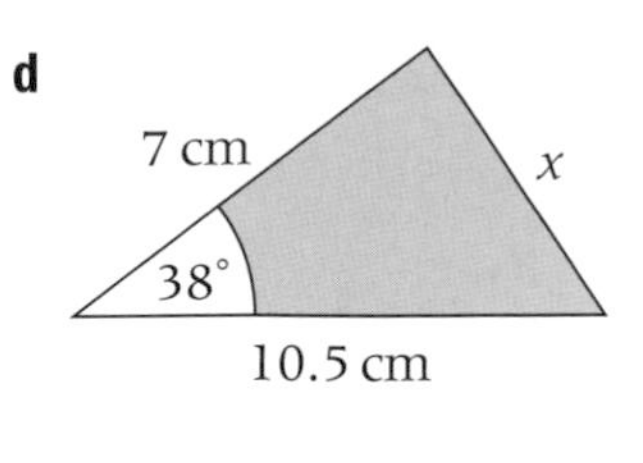

e

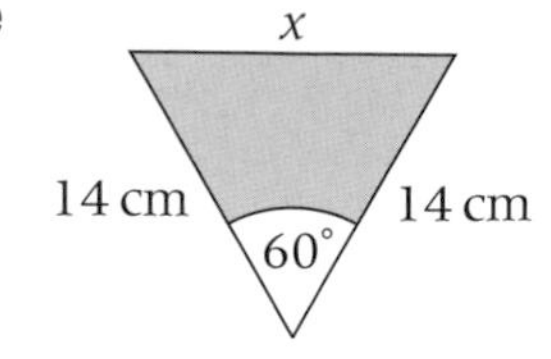

f

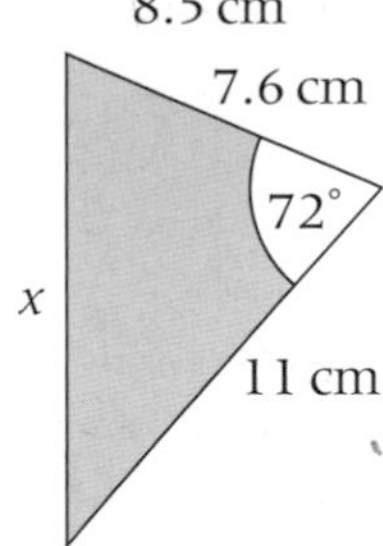

g

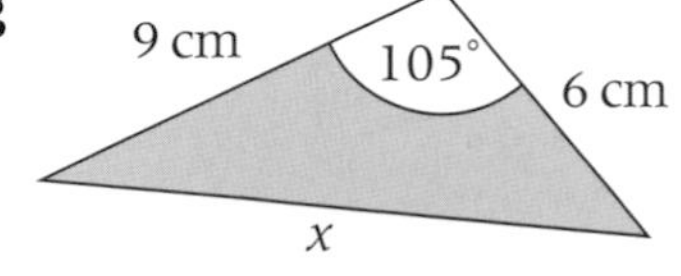

h

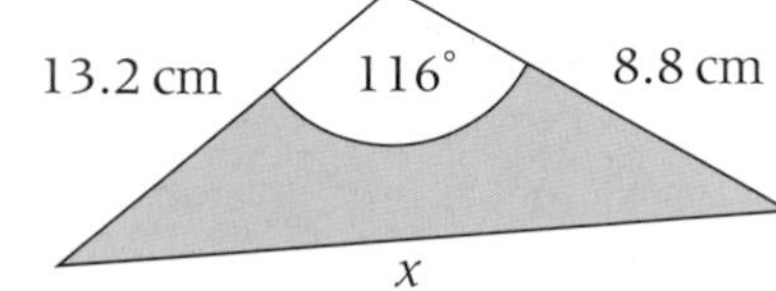

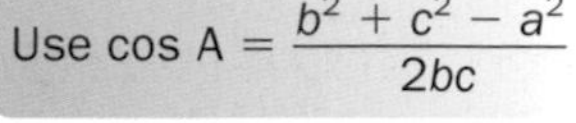

Use $\cos A = \dfrac{b^2 + c^2 - a^2}{2bc}$

2 Find the angles marked θ.

a

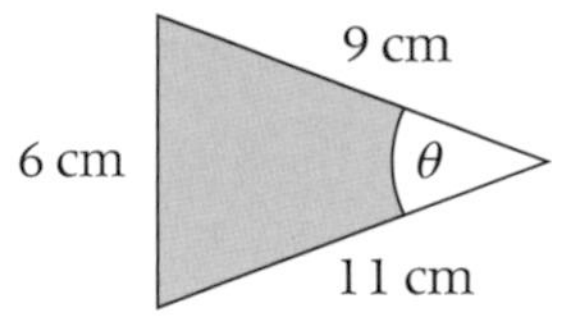

b

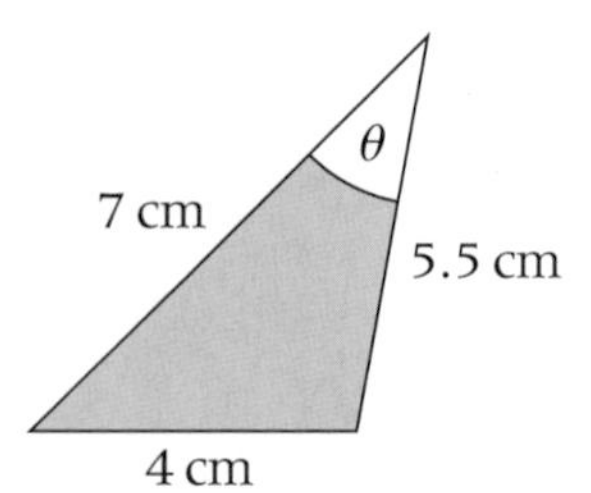

c

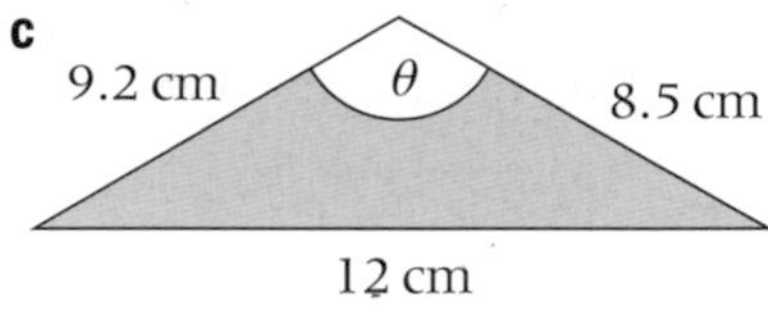

d

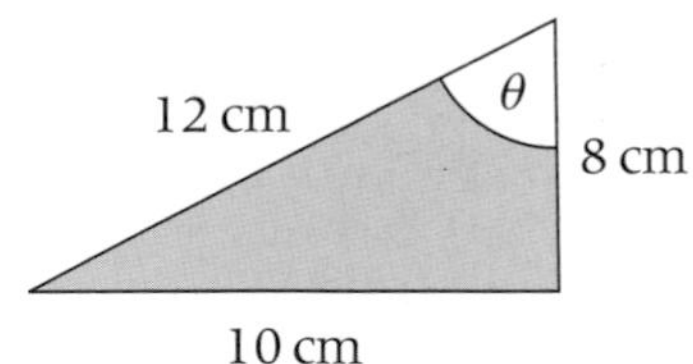

e

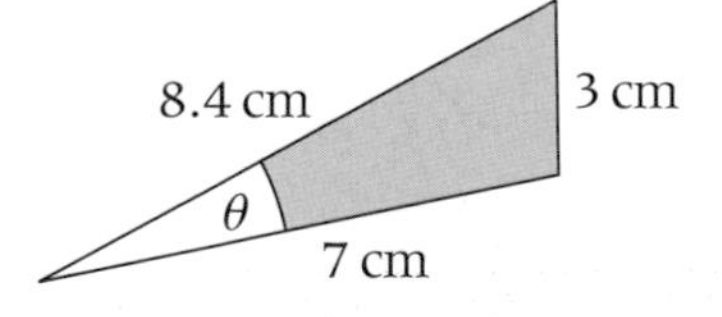

f

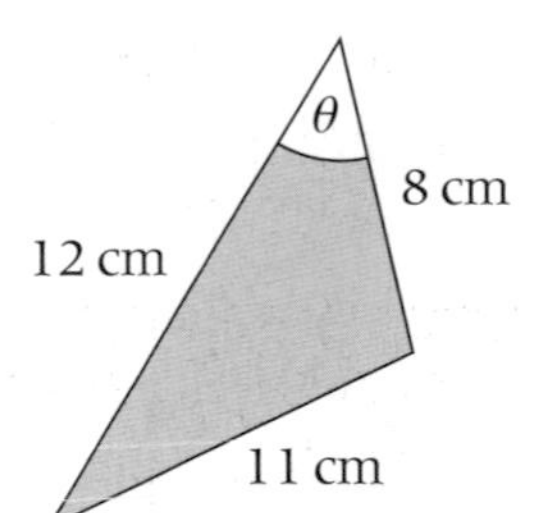

g

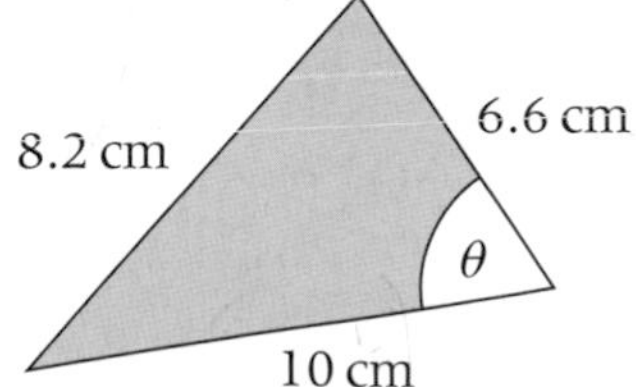

h

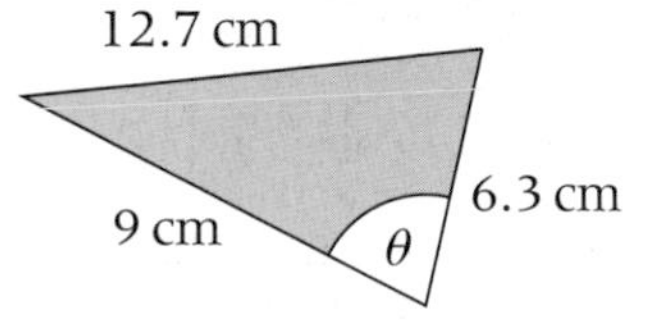

AO3 Problem

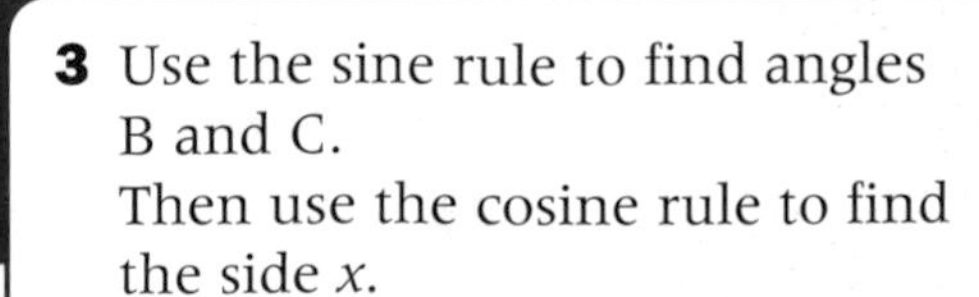

3 Use the sine rule to find angles B and C.
Then use the cosine rule to find the side x.

A, 80°, 8 cm, x, B, C, 12 cm

Unit 3

G5.3 Solving problems using the sine and cosine rules

This spread will show you how to:

- Understand, recall and use trigonometrical relationships in triangles that are not right-angled

Keywords
Bearing
Cosine rule
Sine rule

You use the **sine rule** and the **cosine rule** to solve problems in triangles that are not right-angled.

You may need to use a combination of rules to solve a problem.

- To use the sine rule you need
  - an angle and the side opposite to it
  - one other angle or side.
- To use the cosine rule you need
  - all three sides
  - or two sides and the angle between them.

$$\frac{a}{\sin A} = \frac{b}{\sin B} = \frac{c}{\sin C}$$

$$a^2 = b^2 + c^2 - 2bc \cos A$$

Always start by sketching a diagram.

Example

Peter walks 5 km from S, on a **bearing** of 063°.
At C he changes direction and walks a further 3.2 km on a bearing of 138°, to F.
Find the distance, SF, from where he began.

Make a sketch.
Shaded angle = 180° − 63° = 117° Angles in parallel lines.
∠SCF = 360° − 117° − 138° = 105° Angles at a point.
Use the cosine rule.
$SF^2 = 5^2 + 3.2^2 - 2 \times 5 \times 3.2 \cos 105°$
SF = 6.6 km

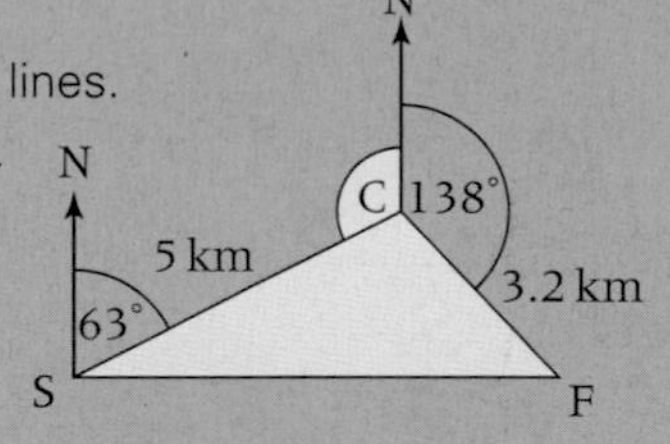

You know two sides and the included angle.

Example

The diagram shows a bicycle frame.
PQ is parallel to SR.
Work out the length QR.

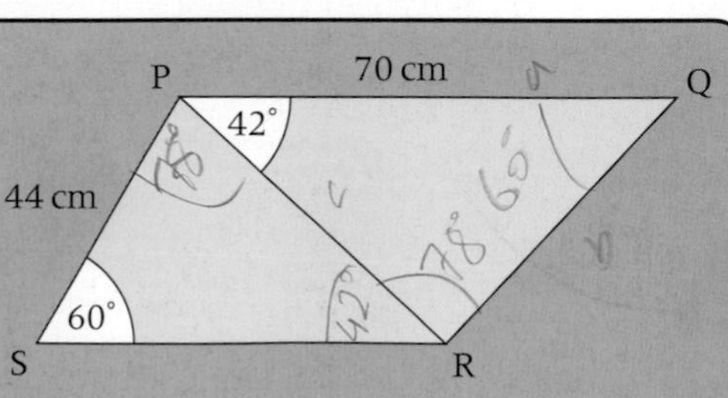

Use the sine rule to find the length PR.
PQ is parallel to RS so ∠SRP = ∠QPR = 42°.

$$\frac{PR}{\sin 60°} = \frac{44}{\sin 42°} \qquad PR = \frac{44 \times \sin 60°}{\sin 42°} \qquad PR = 56.947...$$

In triangle PRS you know an angle, the side opposite it, one other angle and one other side.

Use the cosine rule to find QR.
$QR^2 = 56.947...^2 + 70^2 - 2 \times 56.947... \times 70 \cos 42°$
QR = 47 cm

In triangle PQR you know two sides and the included angle.

Exercise G5.3 — Grade A*

Functional Maths — AO2

1 Debbie runs 8 km on a bearing of 310°.
She stops, changes direction and continues running for 10 km on a bearing of 055°.
Find the distance and bearing on which Debbie should run to return to her starting point.

2 Clare cycles 12 km on a bearing of 050°.
She stops, changes direction and continues cycling 10 km on a bearing of 120°.
Find the distance and bearing on which Clare should cycle to return to her starting point.

3 AB is parallel to DC.
Work out the length BC.

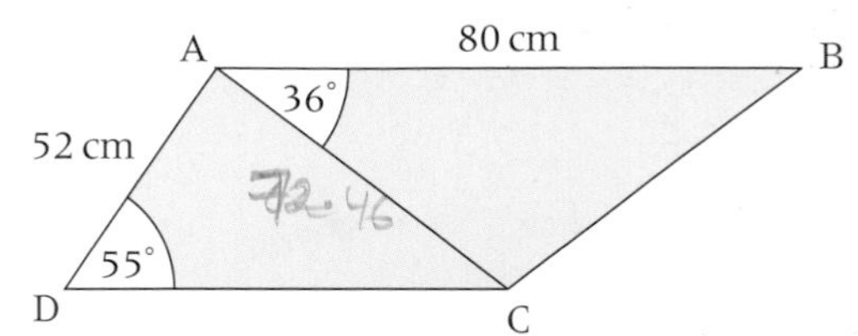

4 JK is parallel to ML.
Work out the length JM.

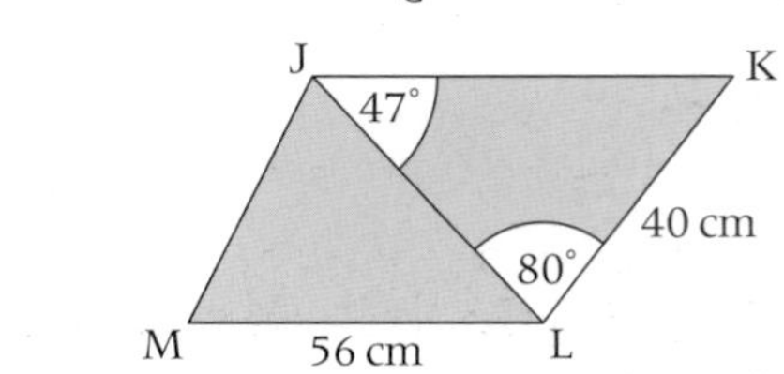

5 PQ is parallel to SR.
Work out the length PS.

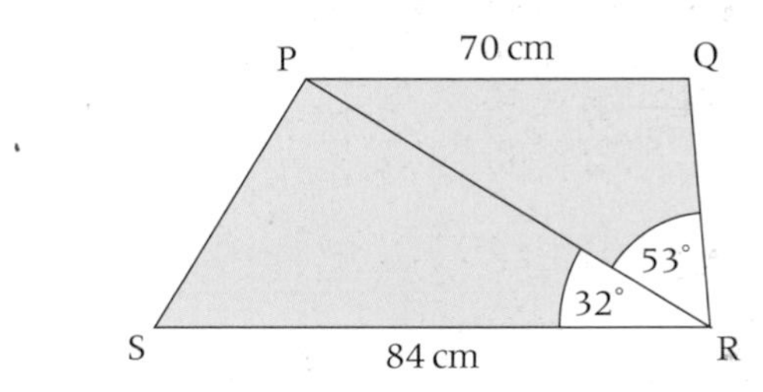

6 **a** Find x.

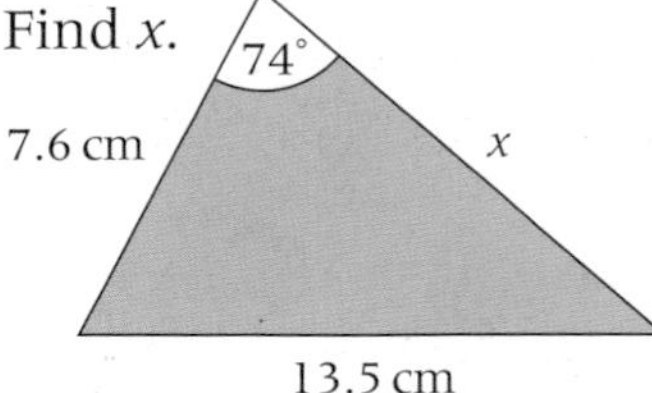

b Find y.

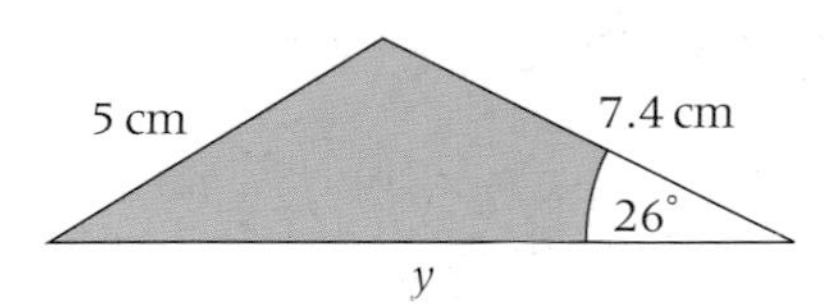

7 Two sides of a triangle are 15.4 cm and 12 cm.
The angle between them is 72°.
Work out the perimeter of the triangle.

8 Adjacent sides of a parallelogram are 6 cm and 8.3 cm.
The shorter diagonal is 7 cm.
Work out the length of the other diagonal.

Problem — AO3

9 Adjacent sides of a parallelogram are 8 cm and 11 cm.
The longer diagonal is 15.2 cm.
Work out the length of the other diagonal.

G5.4 Pythagoras' theorem and trigonometry in 3-D

This spread will show you how to:

- Use coordinates to identify a point on a 3-D grid
- Use Pythagoras' theorem and the rules of trigonometry to solve 2-D and 3-D problems

Keywords
Pythagoras' theorem

In 3-D, a point has *x*-, *y*- and *z*-coordinates.

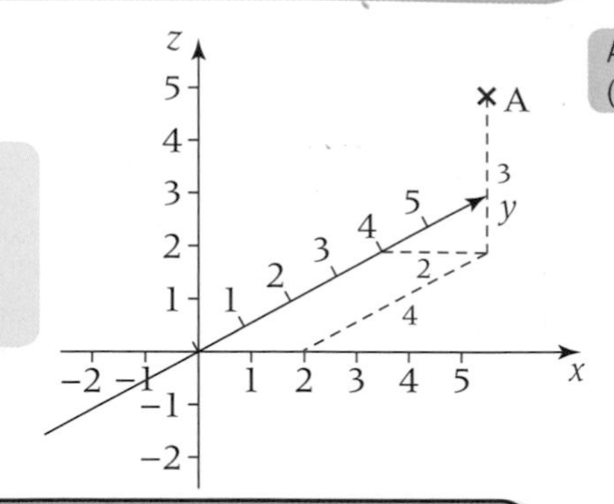

A is the point (2, 4, 3)

p.294

- You can use **Pythagoras' theorem** to find the distance d between two points in 3-D.
  $d^2 = x^2 + y^2 + z^2$

Example

Find the distance from A (3, 4, 1) to B (1, 3, 5).

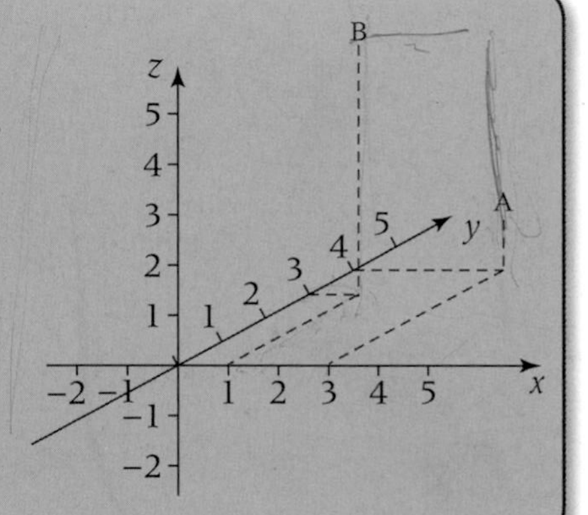

$AB^2 = (3 - 1)^2 + (4 - 3)^2 + (1 - 5)^2$

$= 2^2 + 1^2 + (-4)^2 = 21$

$AB = \sqrt{21} = 4.6$

Example

Work out the angle θ between the diagonal and the base in this cuboid.

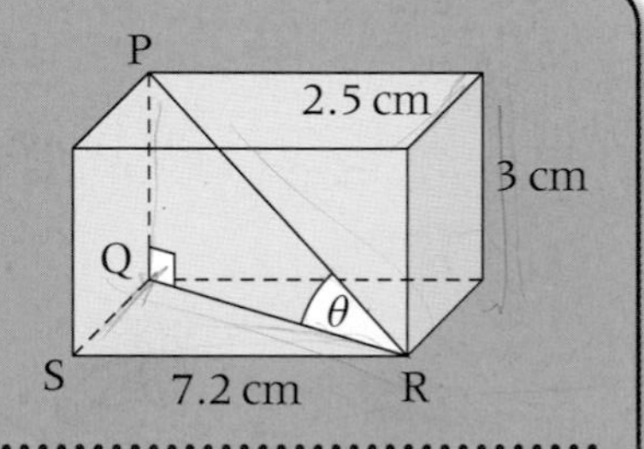

SRQ and PQR are right-angled triangles:

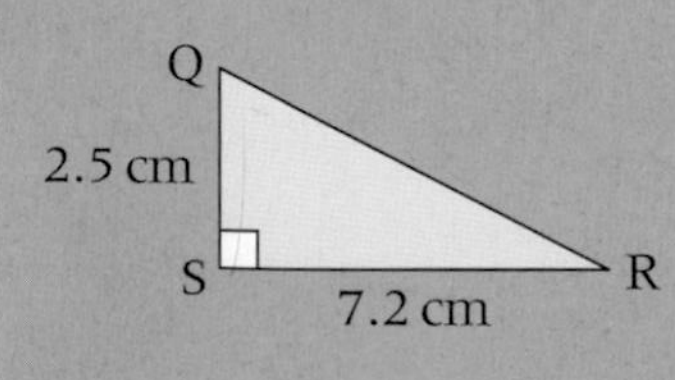

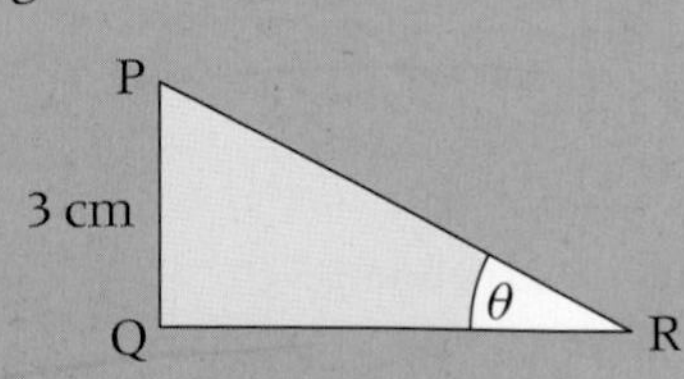

$QR^2 = 7.2^2 + 2.5^2$

$QR = \sqrt{58.09}$

$= 7.62\text{ cm}$

Use Pythagoras in SRQ to find QR.

$\tan\theta = \frac{3}{7.62}$

$\theta = 21.49 \approx 21^\circ$

Use trigonometry to find θ in triangle PQR.

Exercise G5.4 — Grade A*

1 Find the distances between these points.

a (2, 4, 5) and (3, 7, 10) **b** (1, 5, 3) and (6, 2, 6)

c (−1, 9, 2) and (−3, 0, 5) **d** (4, −2, 3) and (2, −4, 7)

2 Work out the length of the diagonal d in these cuboids.

a

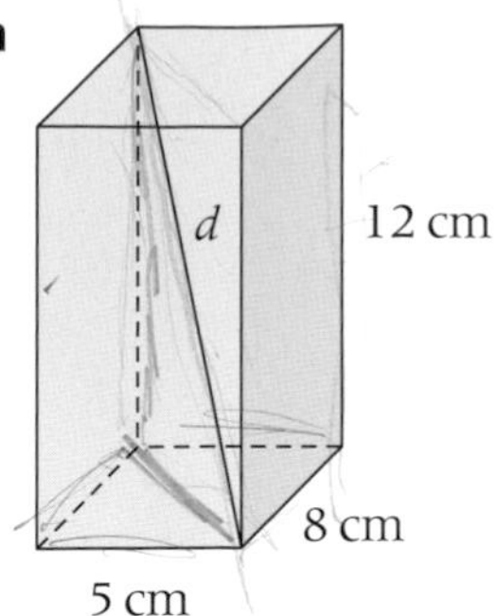

b

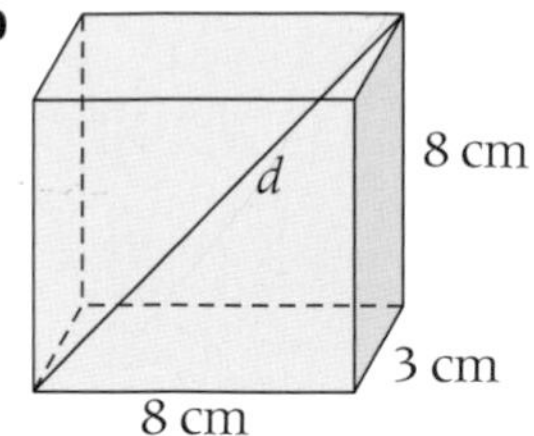

c

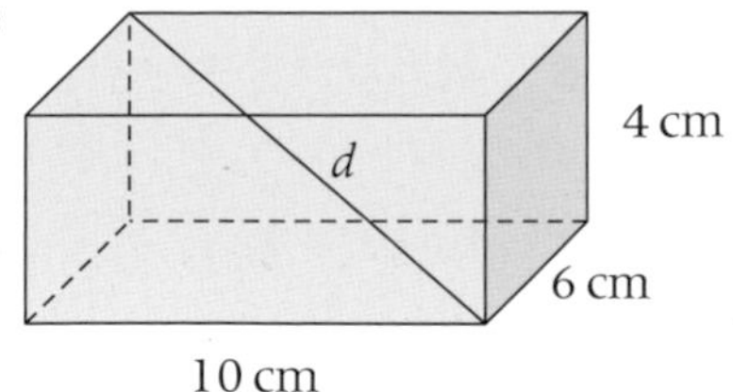

3 For each cuboid in question **2**, work out the angle between the base and the diagonal d.

4 **a** Find the angle between the base and the diagonal d in this cube.

b Repeat part **a** for a cube with the side length 10 cm.

4 cm

4 cm

4 cm

d

AO2 Functional Maths

5 A vertical pole is placed at the corner of a horizontal, rectangular garden.

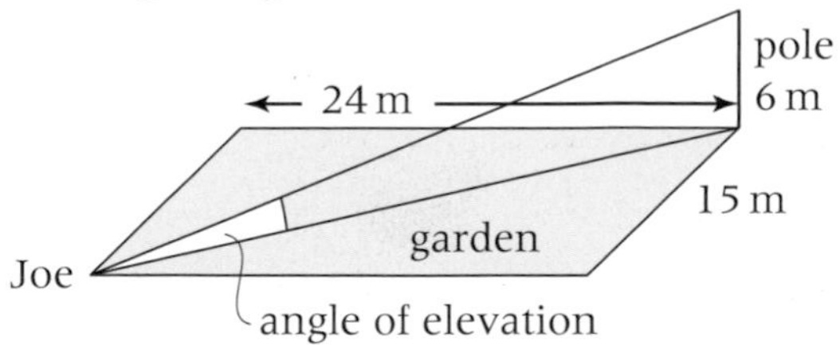

The garden is 15 m by 24 m.
The pole is 6 m high.
Joe is at a corner of the garden diagonally opposite the pole.
Work out the angle of elevation from Joe to the top of the pole.

6 The diagram shows a wedge.
The rectangular base ABCD is perpendicular to the side CDEF.
AB = 8 cm, BC = 6 cm and CF = 3 cm.
Work out

a AF **b** angle FAC.

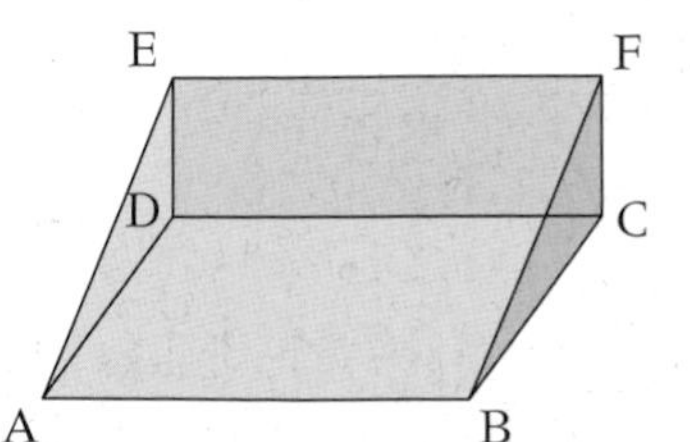

Unit 3

G5.5 The area of a triangle

This spread will show you how to:

- Calculate the area of a triangle using $\frac{1}{2}ab \sin C$

Keywords
Sine

You can use the **sine** ratio to find a formula for the area of any triangle.

Divide the triangle into two right-angled triangles.

h is perpendicular to side b.

Area of triangle ABC $= \frac{1}{2} \times b \times h$

In the right-hand triangle, $\sin C = \frac{h}{a}$ so $h = a \sin C$.

Area of triangle ABC $= \frac{1}{2} \times b \times a \sin C$

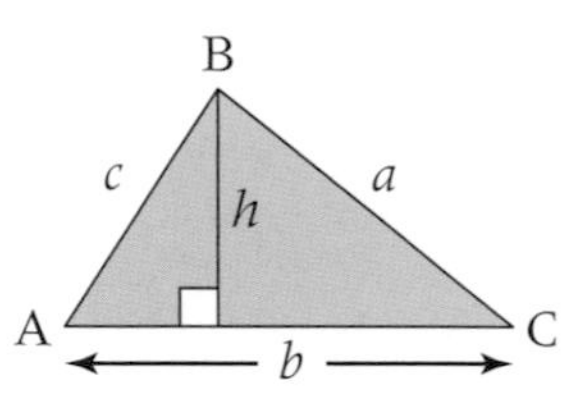

Label the angles A, B and C. Label the sides opposite these angles a, b and c.

- The area of a triangle $= \frac{1}{2}ab \sin C$

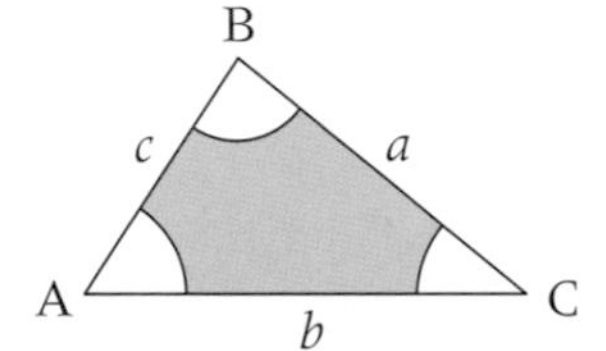

The formula can also be written as:
Area $= \frac{1}{2}bc \sin A$
Area $= \frac{1}{2}ac \sin B$

You can use this formula when you know two sides and the angle between them.

Example

Work out the areas of these triangles.

a

b

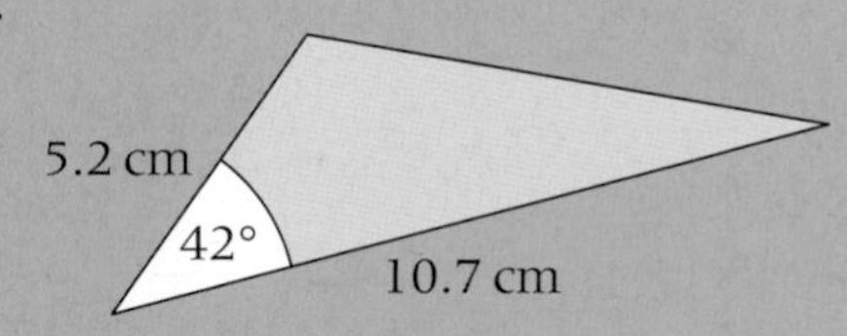

a Area $= \frac{1}{2} \times 3.6 \times 4.9 \times \sin 108° = 8.4\,\text{cm}^2$

b Area $= \frac{1}{2} \times 5.2 \times 10.7 \times \sin 42° = 18.6\,\text{cm}^2$

You can split other shapes into triangles to work out their areas.

Example

Work out the area of the parallelogram.

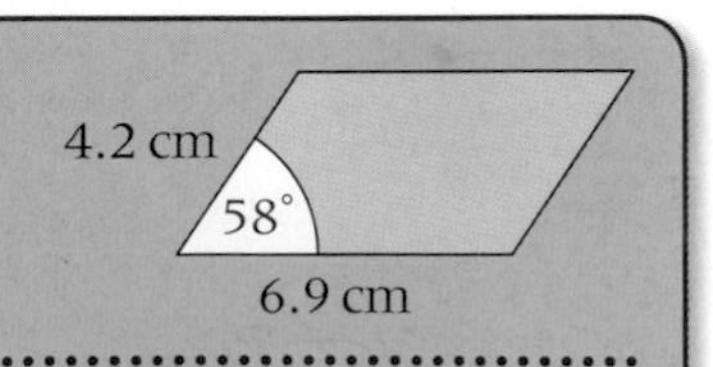

Draw a diagonal to divide the parallelogram into two congruent triangles.

Area of one triangle:

$\frac{1}{2} \times 6.9 \times 4.2 \times \sin 58° = 12.288...$

Area of parallelogram:

$2 \times 12.288... = 24.6\,\text{cm}^2$

Congruent triangles are identical.

Don't round until the end.

Exercise G5.5 — Grade A/A*

1 Work out the area of each triangle.
Give your answers correct to 3 significant figures.

a
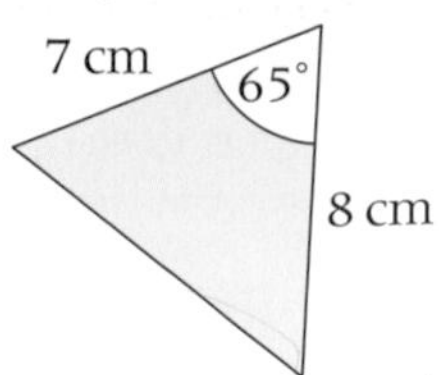

b
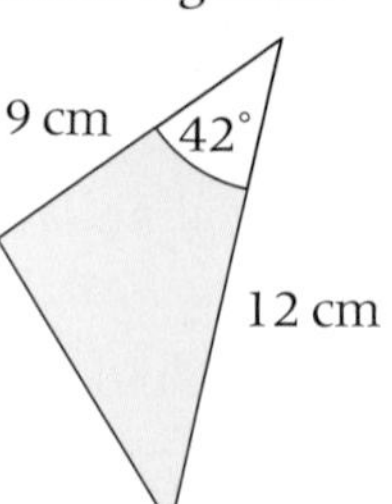

c
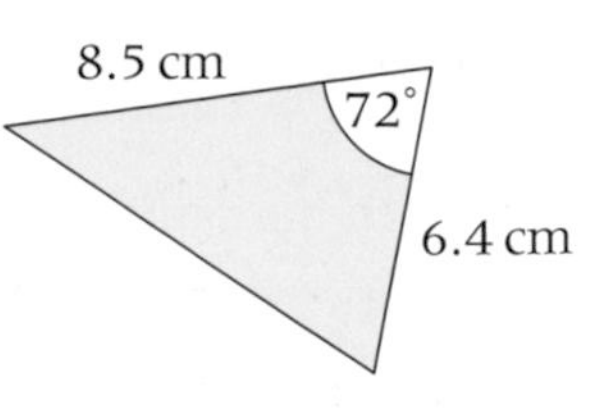

d
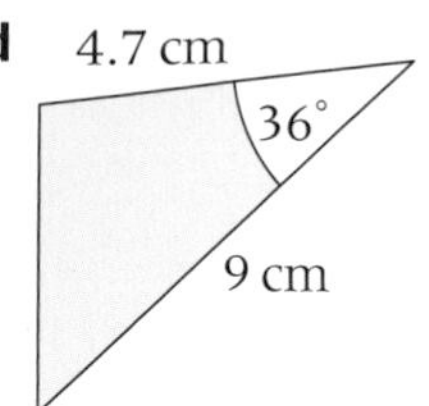

e
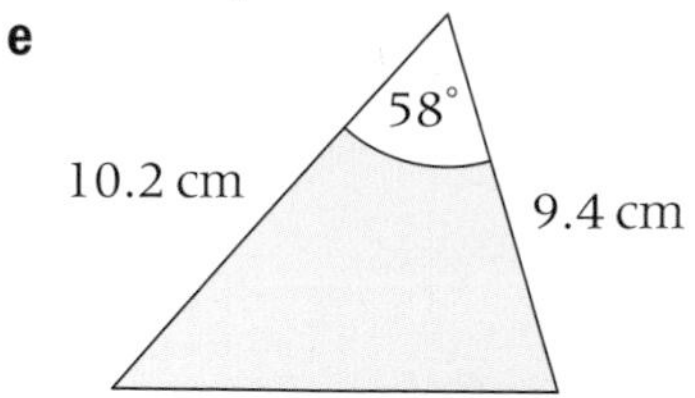

f
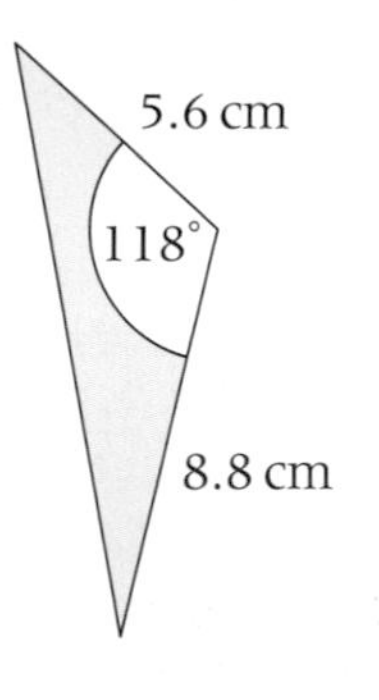

g
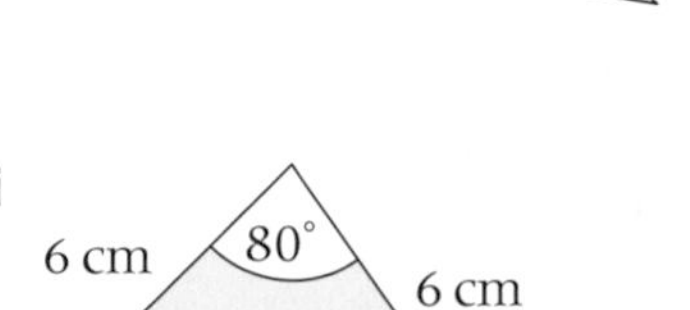

h
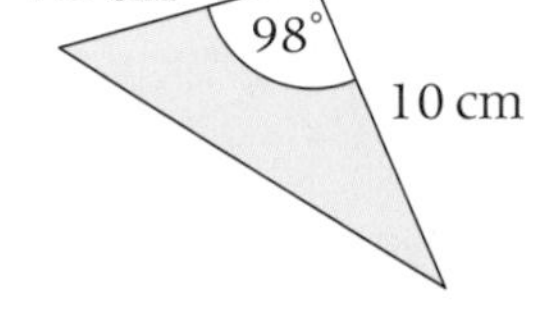

i

6 cm 80° 6 cm

j
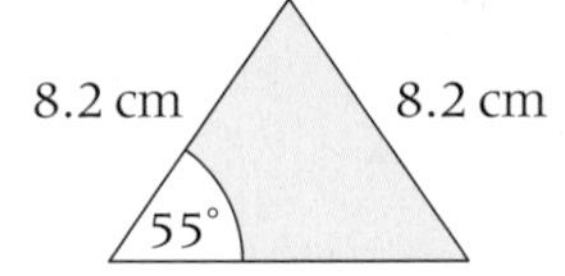

2 Work out the area of each parallelogram or rhombus.
Give your answers correct to 3 significant figures.

a
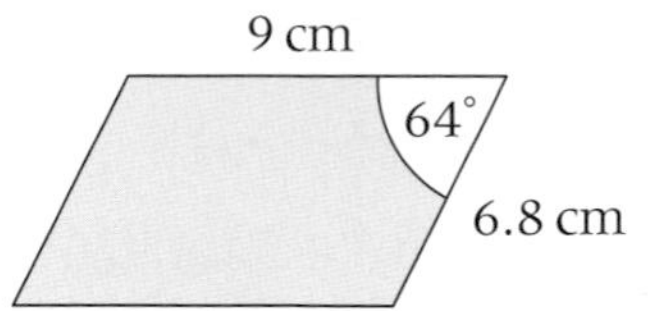

b
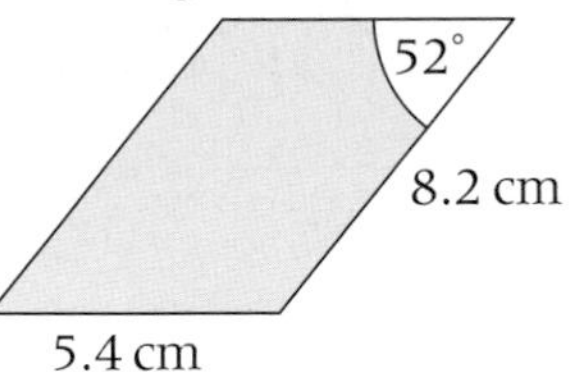

c
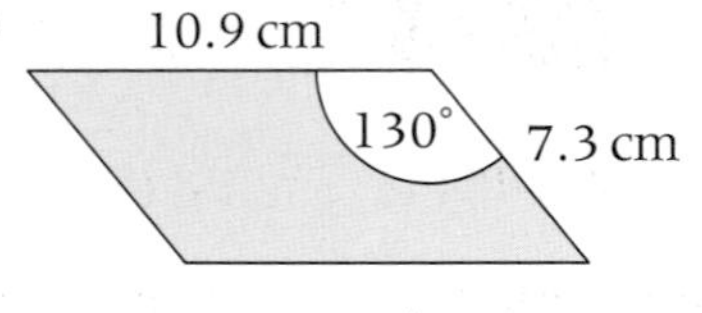

d
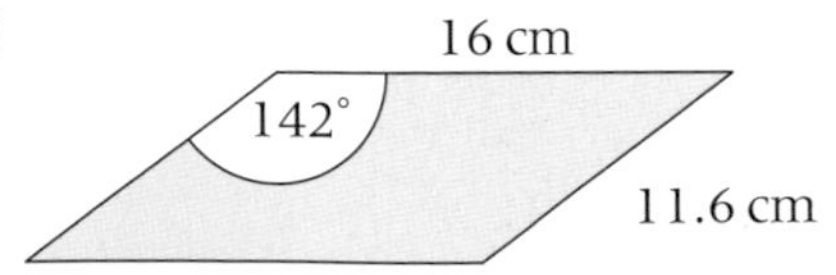

e
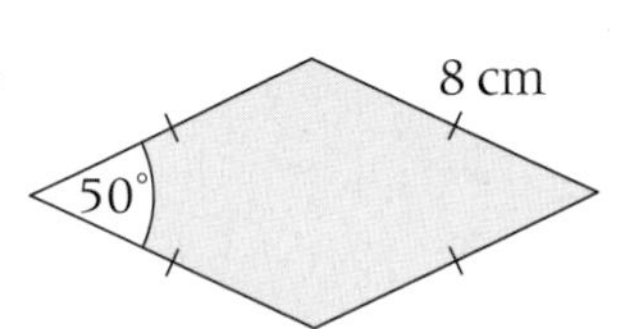

f
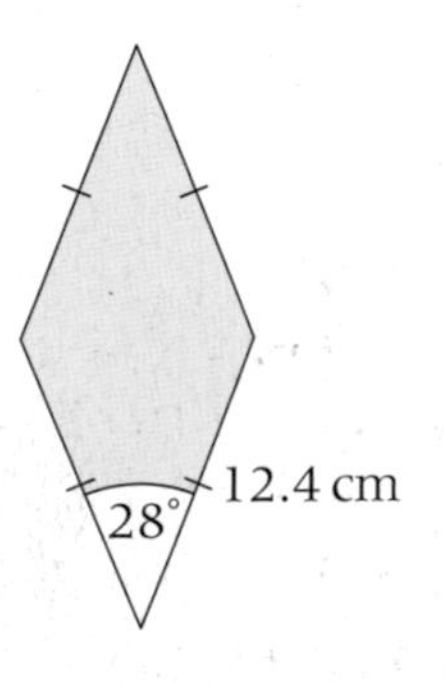

g

9.4 cm 115°

h
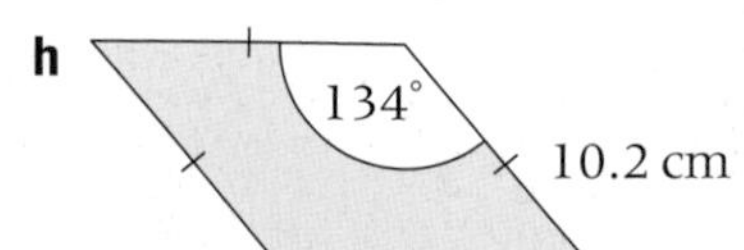

G5.6 Solving problems using trigonometry

This spread will show you how to:
- Use the sine and cosine rules to solve 2-D problems
- Solve problems involving more complex shapes, including segments of circles
- Improve the accuracy of solutions to multi-step problems

Keywords
Area of triangle
Cosine rule
Segment
Sine rule

In multi-stage problems you often need to calculate extra information.

Your answers will be more accurate if you do not round numbers at intermediate steps.

For any triangle
- **Cosine rule**: $a^2 = b^2 + c^2 - 2bc \cos A$
- **Sine rule**: $\frac{a}{\sin A} = \frac{b}{\sin B} = \frac{c}{\sin C}$
- **Area of triangle** $= \frac{1}{2}ab \sin C$

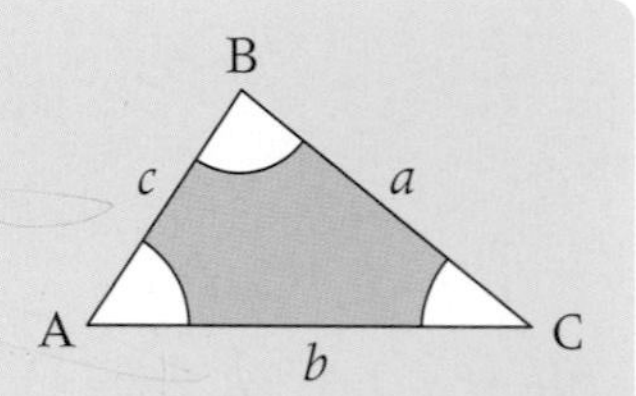

Example

p.242

Work out the area of the shaded **segment**.

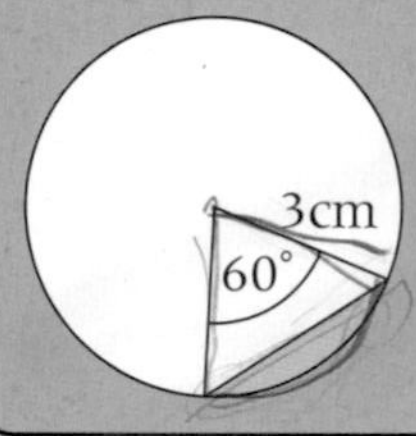

Area of sector

$\frac{60}{360}\pi \times 3^2 = 4.712...$

Area of triangle

$\frac{1}{2} \times 3 \times 3 \sin 60 = 3.897...$

Area of segment

$4.712... - 3.897...$

$= 0.815\,\text{cm}^2$

360° at centre of circle, so area of sector $= \frac{60}{360}$ area of circle.

Area of segment = area of sector − area of triangle

Example

The diagram shows a four-sided field.
Find the area of the field.

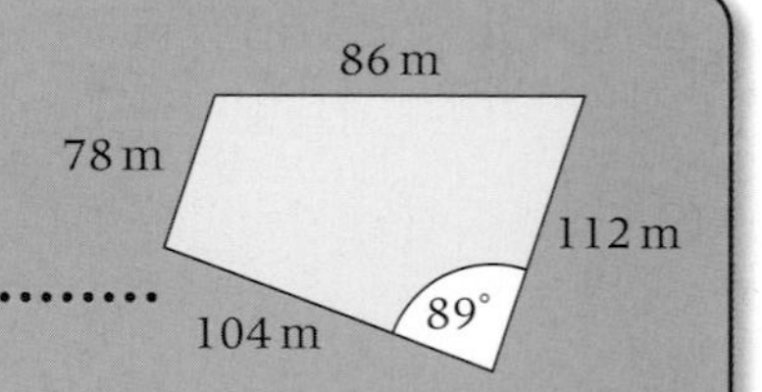

Divide the shape into two triangles by drawing in the diagonal DB.

$DB^2 = 112^2 + 104^2 - 2 \times 112 \times 104 \times \cos 89°$

$DB^2 = 22953.42...$

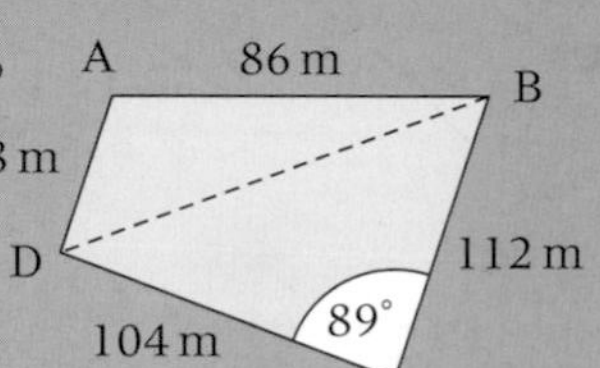

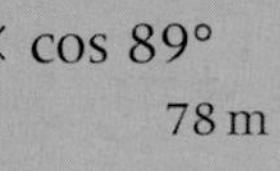

$\cos DAB = \frac{78^2 + 86^2 - 22953.42...}{2 \times 78 \times 86}$

$\angle DAB = \cos^{-1}(-0.7061292...)$

$= 134.920...°$

Area of triangle ADB $= \frac{1}{2} \times 78 \times 86 \times \sin 137.405°...$

$= 2374.9106...\,\text{m}^2$

Area of triangle BCD $= \frac{1}{2} \times 104 \times 112 \times \sin 89°$

$= 5823.112...\,\text{m}^2$

Area of field $= 2374.9106... + 5823.112...$

$= 8198\,\text{m}^2$

Use the cosine rule in triangle DCB to find DB.

Use the cosine rule in triangle ADB to work out angle DAB.

Use $\frac{1}{2}ab \sin C$.

Exercise G5.6 — Grade A*

1 Work out the area of the shaded segment of each circle.

a

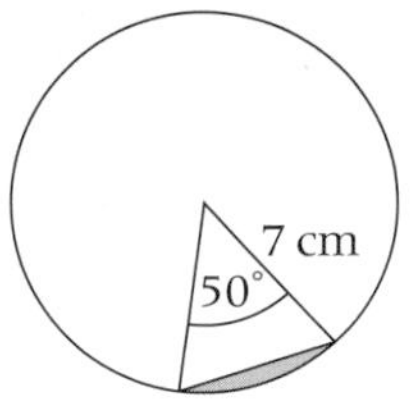

b

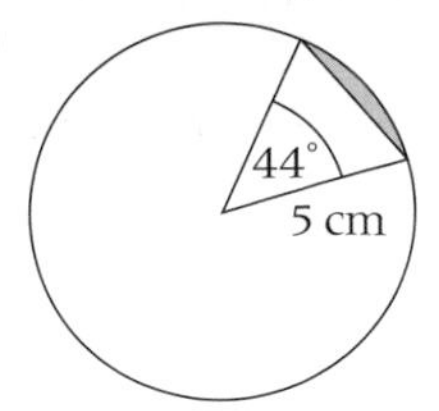

c

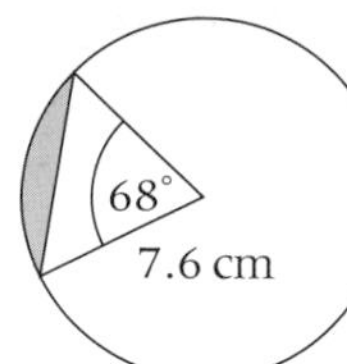

d

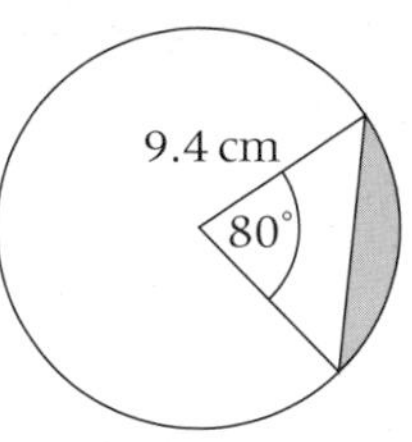

e

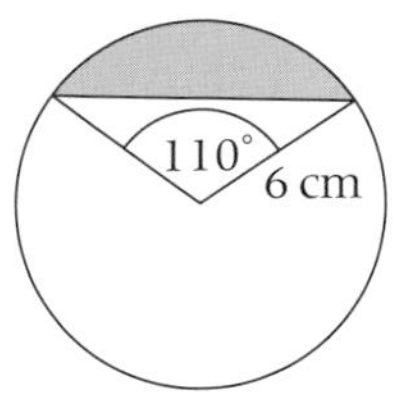

f

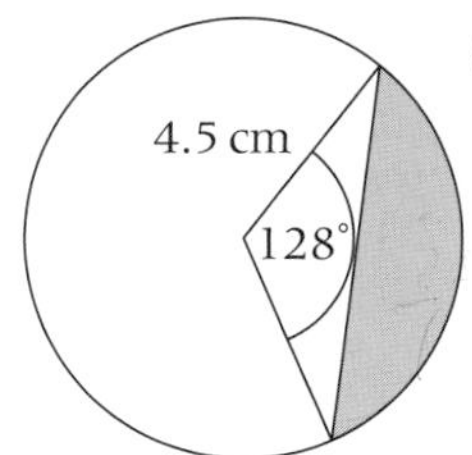

g

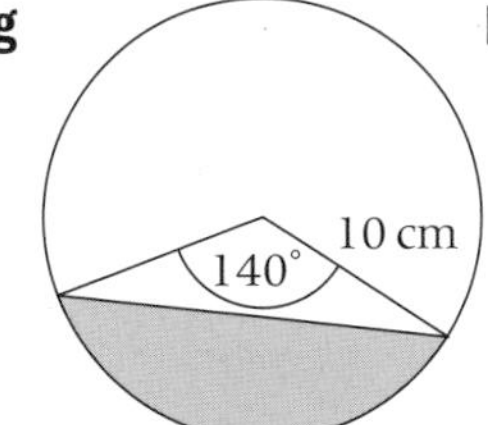

h

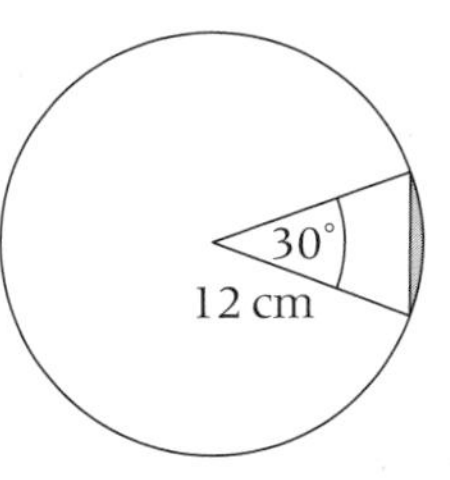

2 **a** Use the sine rule to find angle Q.
b What is angle R?
c Work out the area of triangle PQR.

P
80°
5 cm
R
Q
7.8 cm

3 Work out the area of each of these triangles.

a

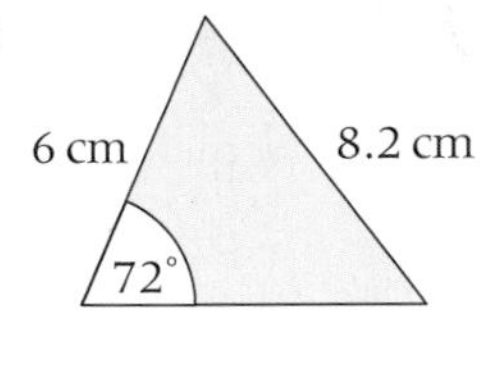

b

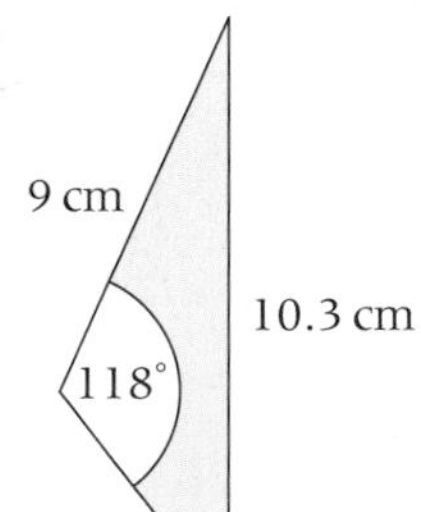

c

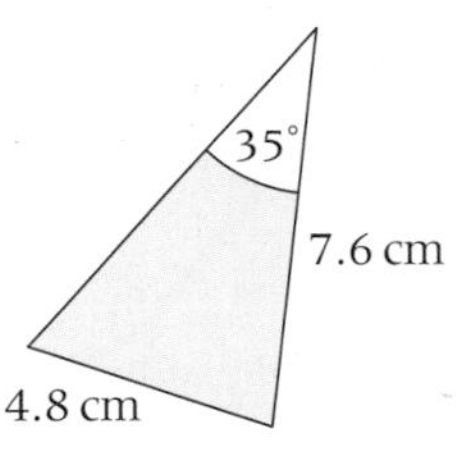

4 **a** Use the cosine rule to find angle B.
b Work out the area of triangle ABC.

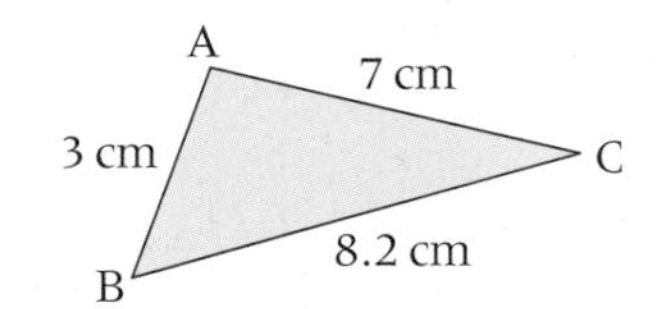

5 Work out the area of these triangles.

a

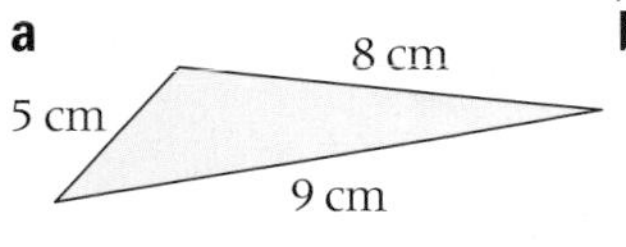

b

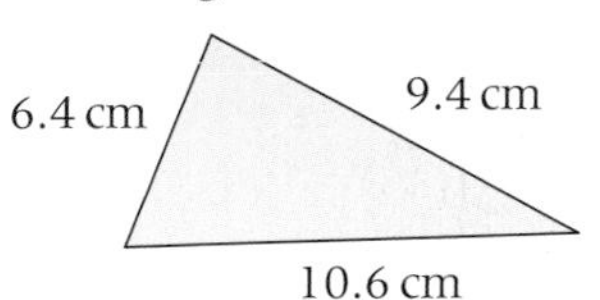

c

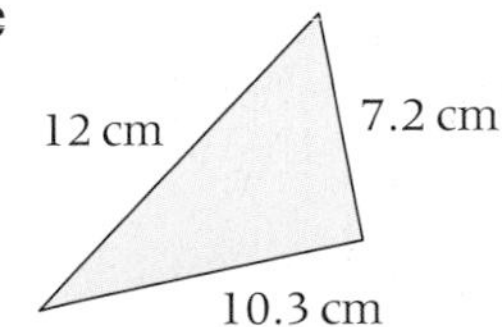

AO3 Problem

6 Work out the area of quadrilateral PQRS.

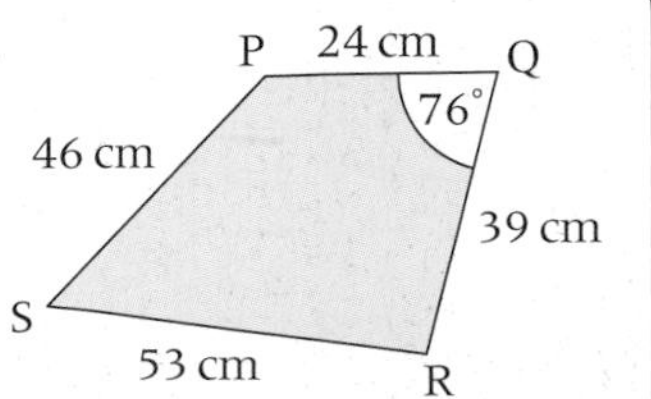

Unit 3

G5.7 Solving 3-D problems

This spread will show you how to:
- Use the formulae for surface area and volume of a cone
- Solve 3-D problems involving surface areas and volumes of cones and spheres

Keywords
Constant
Net
Sector
Surd

You can make answers to multi-stage problems exact by
- leaving **constants** such as π in your answer
- giving your answer as a fraction
- giving your answer in **surd** form.

p.194

Example

This is a **net** for a cone.
Show that the volume of the cone is $36\pi\sqrt{5}$.

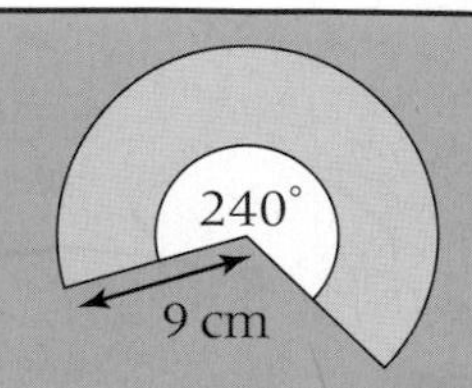

p.244

The made-up cone will look like this:

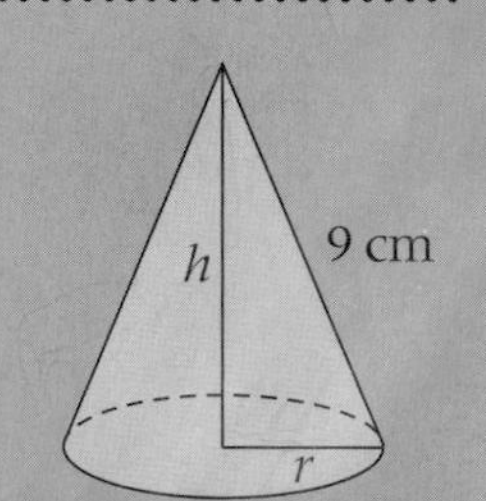

Volume of a cone $= \frac{1}{3}\pi r^2 h$

The arc length of the **sector** is the circumference of the base of the cone.

Circumference of base of cone $= 2\pi r$

So $\frac{240}{360} \times 2\pi \times 9 = 2\pi r$

$12\pi = 2\pi r$

$r = 6$

$h^2 = 9^2 - 6^2$

$h = \sqrt{45} = 3\sqrt{5}$

Volume of cone: $\frac{1}{3}\pi r^2 h = \frac{1}{3}\pi \times 6^2 \times 3\sqrt{5} = 36\pi\sqrt{5}$

You need to find:
- r, the radius of the base of the cone
- h, the height of the cone.

Arc length $= \frac{\theta}{360} \times 2\pi r$

Use Pythagoras to find h.

- Volume of a sphere $= \frac{4}{3}\pi r^3$
- Surface area of a sphere $= 4\pi r^2$

Example

A sphere of radius $2r$ has the same volume as a cylinder with base radius r and height 8 cm.
Work out the surface area of the sphere. Give your answer in terms of π.

p.250

Volume of sphere

$= \frac{4}{3}\pi(2r)^3 = \frac{32}{3}\pi r^3$

Volume of cylinder

$= 8\pi r^2$

So $\frac{32}{3}\pi r^3 = 8\pi r^2$

$r = \frac{3}{4}$

Surface area of sphere $= 4\pi r^2 = 4\pi\left(\frac{3}{4}\right)^2 = \frac{9\pi}{4}\text{ cm}^2$

Exercise G5.7 Grade A*

1 This is a net for a cone.
Show that the volume of the cone is $\frac{200}{3}\pi\sqrt{11}$.

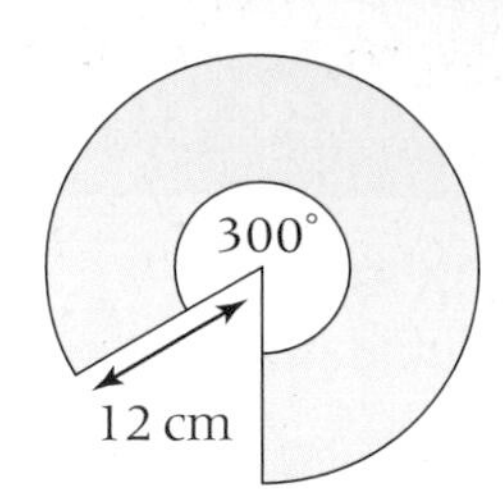

2 This is a net for a cone.
Show that the volume of the cone is $\frac{250}{3}\pi\sqrt{2}$.

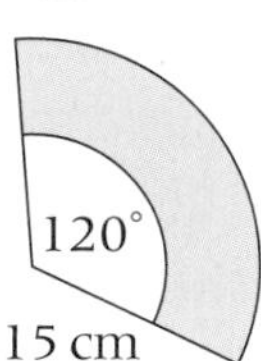

3 The surface area of a sphere of radius r is the same as the curved surface area of a cylinder with base radius 3 cm and height 2 cm.

Show that the volume of the sphere is $4\pi\sqrt{3}$.

4 The radius of a sphere is 5 cm.
The radius of the base of a cone is 10 cm.
The volume of the sphere is twice the volume of the cone.

Work out the curved surface area of the cone, give your answer as a multiple of π.

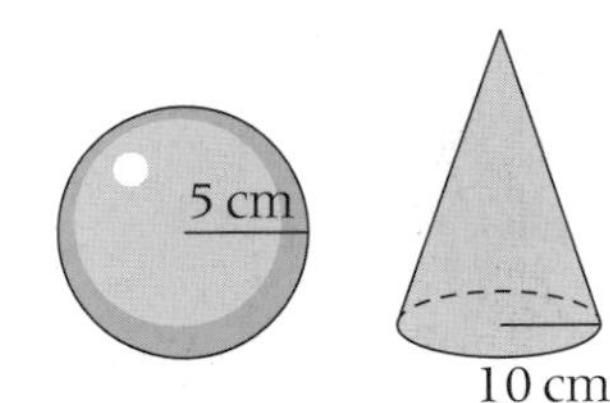

5 The radius of a sphere is three times the radius of the base of a cone.
The base radius of the cone is 2 cm.
The volume of the sphere is nine times the volume of the cone.

Work out the curved surface area of the cone.

6 The curved surface area of a cylinder is equal to the curved surface area of a cone.
The base radius of the cylinder is twice the base radius of the cone.

Show that the height of the cylinder is a quarter of the slant height of the cone.

7 A solid metal sphere with radius 3 cm is melted down and re-formed as a cylinder.
The base radius of the cylinder is 3 cm.

Work out the height of the cylinder.

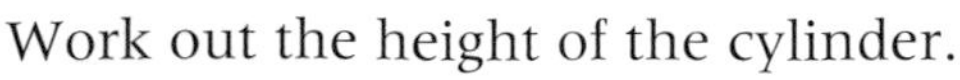

G5 Summary

Check out

You should now be able to:

- Use Pythagoras' theorem and the trigonometric ratios to solve 2-D and 3-D problems
- Calculate the area of any triangle using $\frac{1}{2}ab\sin C$
- Use the sine and cosine rules to solve 2-D and 3-D problems
- Find the area of a segment of a circle
- Solve 3-D problems involving surface areas and volumes of cylinders, cones and spheres
- Use surds and π in exact calculations

Worked exam question

The diagram shows an equilateral triangle.

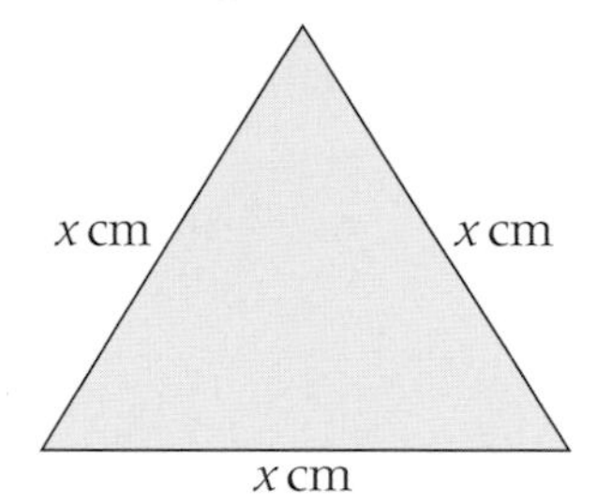

Diagram NOT accurately drawn.

The area of the equilateral triangle is 36 cm^2.

Find the value of x.
Give your answer correct to 3 significant figures. (3)

(Edexcel Limited 2007)

$$\frac{1}{2} \times x \times x \times \sin 60° = 36$$
$$x^2 \times \sin 60° = 72$$
$$x^2 = \frac{72}{\sin 60°}$$
$$x^2 = 83.138439$$
$$x = \sqrt{83.138439}$$
$$x = 9.12$$

Write out an equation for the area.

Rearrange to find x^2.

x must be rounded to 3 significant figures.

Exam questions

1 A cuboid has length 3 cm, width 4 cm and height 12 cm.

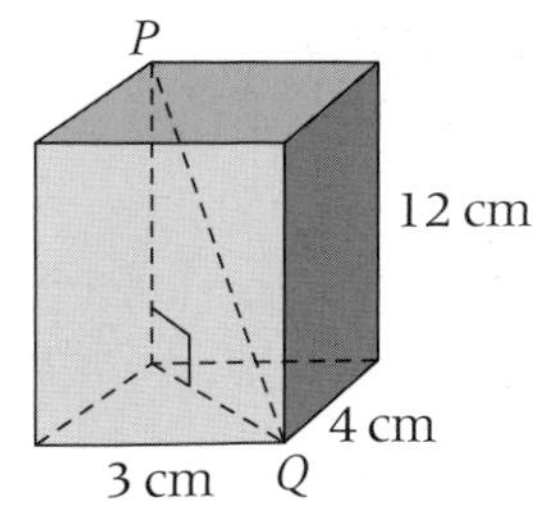

Diagram NOT accurately drawn.

Work out the length of *PQ*. (3)

(Edexcel Limited 2007)

2

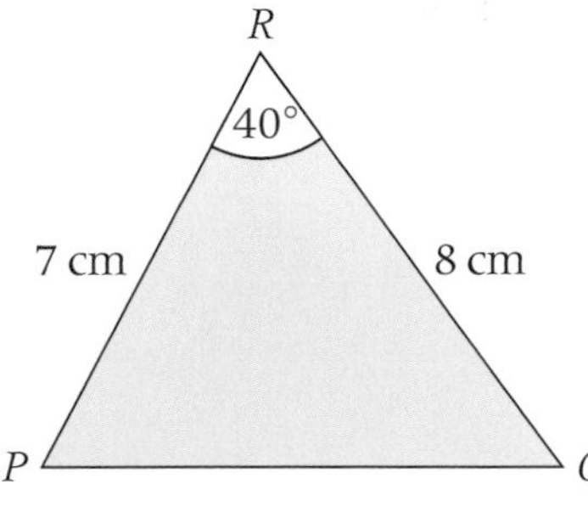

Diagram NOT accurately drawn.

PQR is a triangle.
PR = 7 cm.
QR = 8 cm.
Angle *PRQ* = 40°.
Calculate the length *PQ*.
Give your answer correct to 3 significant figures. (3)

3

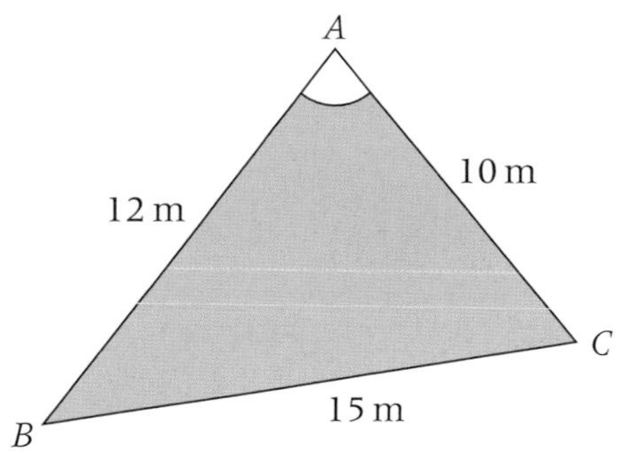

Diagram NOT accurately drawn.

ABC is a triangle.
AB = 12 m.
AC = 10 m.
BC = 15 m.
Calculate the size of angle *BAC*.
Give your answer correct to one decimal place. (3)

(Edexcel Limited 2008)

Functional Maths 7: Art

Graffiti artists often sketch their designs before projecting them onto the surface. They sometimes use grids or parts of their body as measuring tools to help them copy the proportions accurately.

1. A graffiti artist projects an image from a sketchpad of length 20cm and width 14.8cm onto a wall of length 6m.
   a. What scale factor is being used?
   b. What is the width of the graffiti wall?

   The artist's hand-span is 150mm.
   c. What are the dimensions of the wall in terms of hands?
   d. If a 1cm square grid was used in the sketch, what size would the grid squares be on the graffiti wall?
   e. What effect does the enlargement have on the area of the graffiti image?

Graffiti designs often involve arcs and spirals.
You can plot these on a polar grid using polar coordinates.

Polar coordinates describe the position of a point, **P**, in terms of **r**, the distance P is from the origin and ϑ, the angle that the line **OP** makes with the horizontal axis. **P** has polar coordinates (**r**, ϑ).

2. a. Match points **A**, **B** and **C** with their polar coordinates:
   i. (2, 60) ii. (1, 210) iii. (3, 300)
   b. Write down the polar coordinates of point **D**.

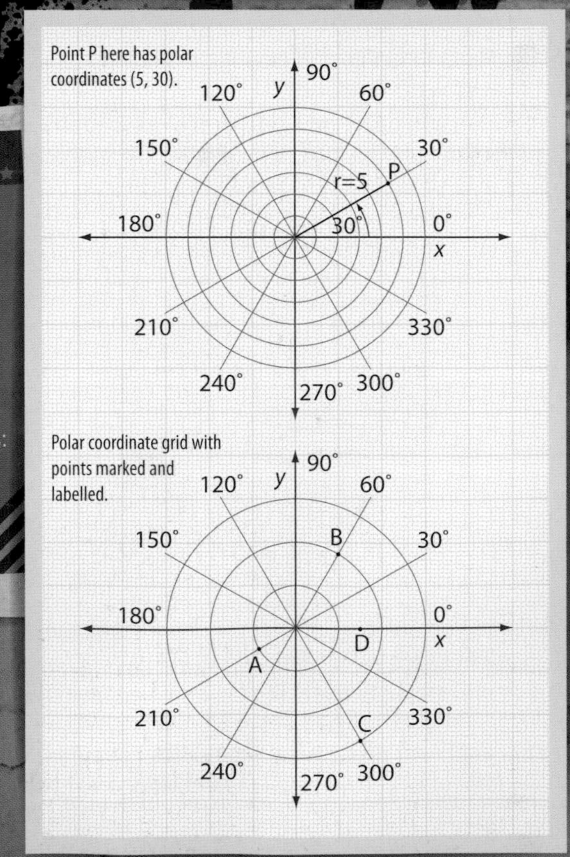

Plot and label some other points on a polar coordinate grid.

Crop circles are geometric patterns that are displayed in crop fields. They are often based on circles and spirals.

A circle has a fixed radius.

The radius of a spiral changes as the angle changes.

3. This diagram shows a circle (red) and a spiral (blue) through 360°.

a. What is the radius of the circle?
The circle has an equation of the form
$r = a$ where a is a constant.

b. What is the value of a?
The radius of the spiral increases by 0.025 units every degree.

c. How many units does the radius increase by in total?
The spiral has an equation of the form
$r = k\vartheta$ where k is a constant.

d. What is the value of k?

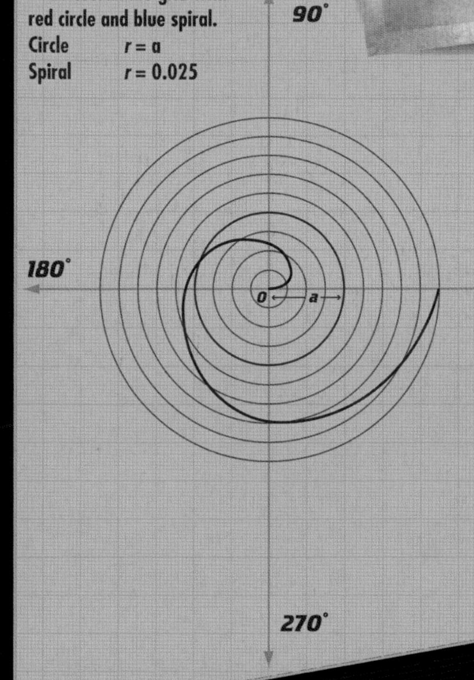

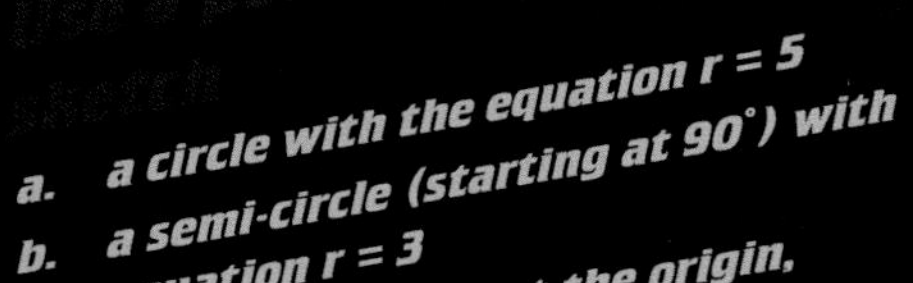

4. Use a polar coordinate grid to sketch

a. a circle with the equation $r = 5$

b. a semi-circle (starting at 90°) with equation $r = 3$

c. a spiral, starting at the origin, with equation $r = 0.01\vartheta$

This crop circle was found in Wiltshire in 2006.

5. Which parts of the design are based on
a. a circle
b. a spiral?

Explain your answers, referring to the radii.

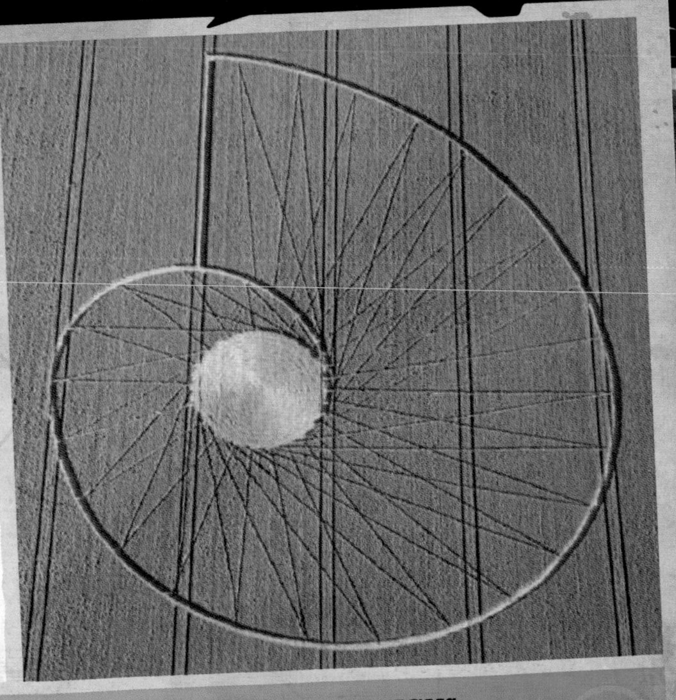

Study some photos of crop circles.

What geometric shapes do they contain?

CREATE YOUR OWN CROP CIRCLE DESIGNS WITH SPIRALS AND ARCS USING A POLAR GRID.

A7 Sketching graphs

Sketching graphs

The world is currently trying to respond to the effects on the environment of global warming. These effects are being predicted by use of complex mathematical models. The models use mathematical functions along with current data on a wide range of variables.

What's the point?

The graphs of mathematical functions show how one quantity changes in relation to another. A mathematical model uses this idea to represent the real world by using a number of mathematical functions to represent a range of changing quantities and factors.

Check in

You should be able to

- **solve quadratic equations**

1 Solve these using factorisation or the quadratic formula.

a $x^2 - 7x + 12 = 0$ **b** $y^2 - 8y = 0$ **c** $x^2 - 3x - 2 = 0$

d $2x^2 + 7x + 3 = 0$ **e** $y^2 = 11y + 24$ **f** $3x^2 - 2x - 1 = 0$

- **evaluate functions**

2 Evaluate each expression for the given value of x.

a $3x^2 - 2x \quad x = 4$ **b** $x^2 - x \quad x = -2$ **c** $x^2(2x + 3) \quad x = -4$

- **sketch real-life graphs**

3 Sketch a graph to show each situation.

a Your height as you age from a baby to a twenty year old. (x-axis time, y-axis height.)

b The temperature of a cup of tea as it is left to stand for half an hour. (x-axis time, y-axis temperature.)

Orientation

What I need to know	What I will learn	What this leads to
A4, A6 Solve quadratic equations and complete the square	■ Sketch quadratic, cubic, reciprocal and exponential graphs ■ Solve simultaneous linear, quadratic and circle equations	**A-Level** Maths, Science
A5 Solve simultaneous linear equations		Economic, Engineering, Medicine

Rich task

Investigate the gradient of the curve $y = x^2$ at different points along the curve.

A tangent is a line that touches the curve at one point. The tangent to the curve at any given point allows the gradient of the curve to be calculated at that point. Here the tangent has been drawn at the point (2, 4).

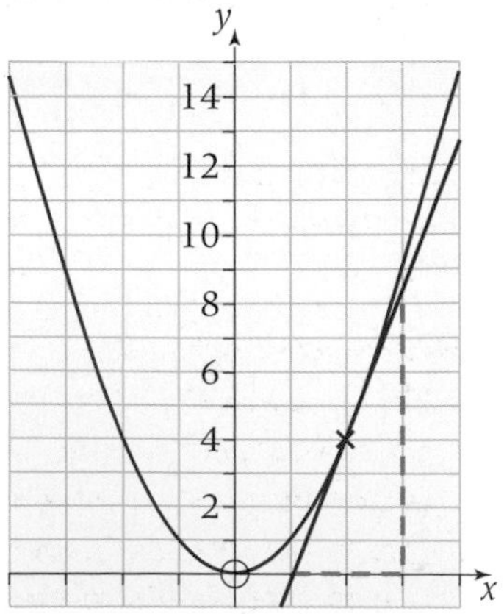

A7.1 Simultaneous linear and quadratic equations

This spread will show you how to:

- Use substitution to solve simultaneous equations where one equation is linear and one is quadratic

Keywords
Linear
Quadratic
Simultaneous

- You can use substitution to solve **simultaneous** equations where one is **linear** and one **quadratic**.
- You rearrange the linear equation, if necessary, to make one unknown the subject. Then substitute this expression into the quadratic equation and solve.

A linear equation contains no square or higher terms.
A quadratic equation contains a square term, but no higher powers.

Example

Solve the simultaneous equations $x + y = 7$
$x^2 + y = 13$

$x + y = 7 \quad (1)$
$x^2 + y = 13 \quad (2)$
Rearranging (1): $y = 7 - x$
Substitute in (2): $x^2 + (7 - x) = 13$
$x^2 - x - 6 = 0$
$(x + 2)(x - 3) = 0$
Either $x = -2$ and $y = 7 - (-2) = 9$
or $x = 3$ and $y = 7 - 3 = 4$

p.228

Equation (1) is linear. Rearrange it to make y the subject.
Equation (2) is quadratic.

Example

The line $x + 3y = 5$ crosses the parabola $y = x^2 - 25$ at two points. Find the coordinates of these points.

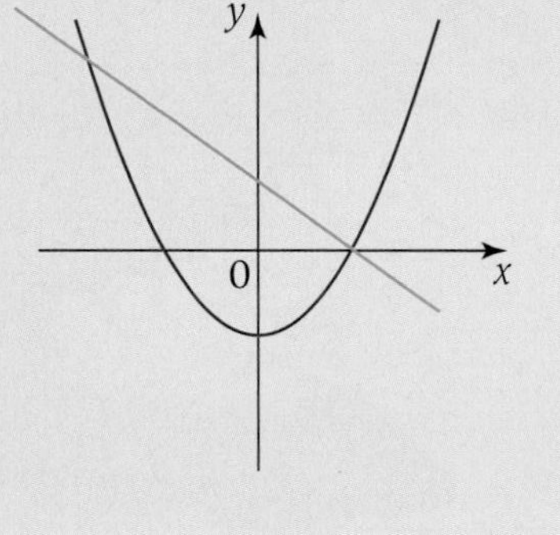

$y = x^2 - 25 \quad (1)$
$x + 3y = 5 \quad (2)$
Rearranging (2): $y = \frac{5 - x}{3}$
Substitute in (1): $\frac{5 - x}{3} = x^2 - 25$
$5 - x = 3x^2 - 75$
$3x^2 + x - 80 = 0$
$(3x + 16)(x - 5) = 0$
Either $x = -\frac{16}{3}$ or $x = 5$
When $x = -\frac{16}{3}$, $y = \frac{31}{9}$ and when $x = 5$, $y = 0$.
The line crosses the parabola at $\left(-\frac{16}{3}, \frac{31}{9}\right)$ and $(5, 0)$.

Check that you have correctly paired the x- and y-values.

Exercise A7.1

Grade A*

1 Solve these simultaneous equations.

a $x^2 + y = 55$
$y = 6$

b $x + y^2 = 32$
$x = 7$

c $x^2 - 3y = 73$
$y = 9$

2 Solve these simultaneous equations.

a $y = x^2 - 2x$
$y = x + 4$

b $y = x^2 - 1$
$y = 2x - 2$

c $x = 2y^2$
$x = 9y - 4$

3 Solve these simultaneous equations.

a $y = x^2 - 3x + 7$
$y - 5x + 8 = 0$

b $p^2 + 3pq = 10$
$p = 2q$

The equation of a parabola is $y = ax^2 + bx + c$.

4 The graph shows a parabola and a line.

a The parabola crosses the x-axis at (2, 0) and (−3, 0) and the y-axis at (0,−6). Find its equation.

b The line intersects the y-axis at (0, 10) and the x-axis at (−10, 0). Find its equation.

c Use the graph to find the coordinates of the points where the graphs intersect. Check this solutions using simultaneous equations.

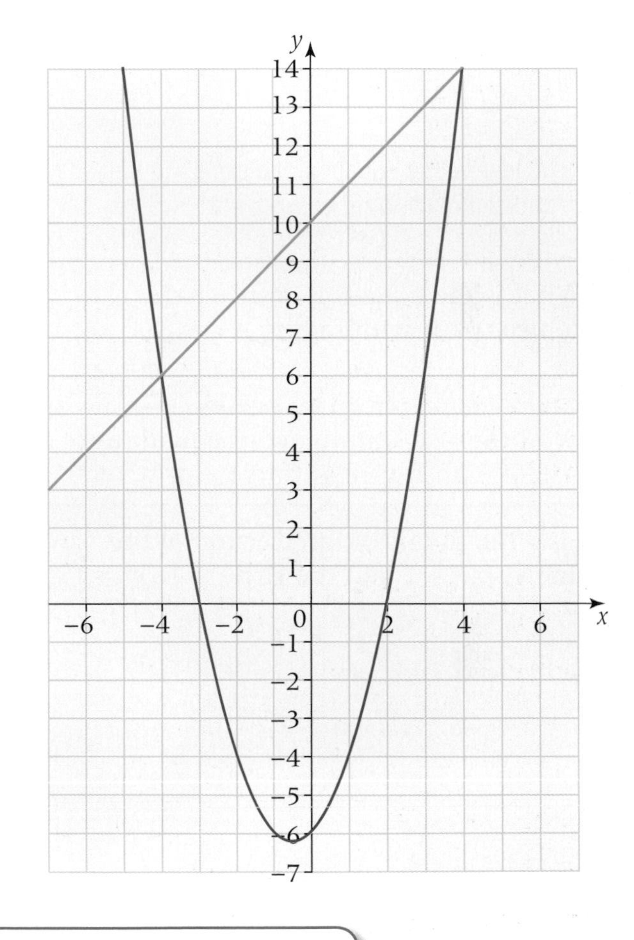

Unit 3

A03 Problem

5 Solve simultaneously $xy^5 = -96$
$2xy^3 = -48$

6 Solve simultaneously $2^{p+q} = 32$
$\frac{3^q}{3^{2p}} = 6561$

Write 32 as a power of 2 and 6561 as a power of 3

A7.2 The equation of a circle

This spread will show you how to:

- Use Pythagoras' theorem to find the equation of a circle centred at the origin
- Use simultaneous equations to find where a line intersects a circle centred at the origin

Keywords
Circle
Origin
Radius

- You can use Pythagoras' theorem to find the equation of a **circle** of **radius** 4 and centre the **origin**.
  Choose a point (x, y) on the circumference of the circle.
  Draw in a right-angled triangle.

p.294

$$x^2 + y^2 = 4^2$$
$$x^2 + y^2 = 16$$

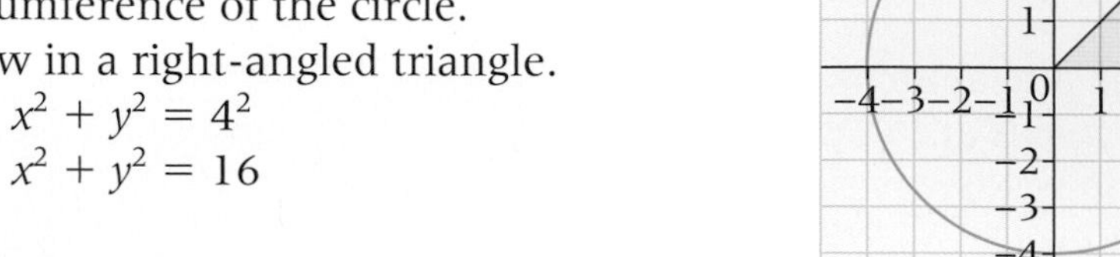

- In general the equation of a circle with radius r and centre $(0, 0)$ is $x^2 + y^2 = r^2$

Example

What is the equation of the circle, centre $(0, 0)$ with radius $\frac{3}{4}$?

$$x^2 + y^2 = \left(\frac{3}{4}\right)^2$$
$$x^2 + y^2 = \frac{9}{16}$$

- You can use simultaneous equations to find where a line intersects a circle with centre at the origin.

Example

Find the points of intersection of the line $y = x + 1$ and the circle $x^2 + y^2 = 25$.

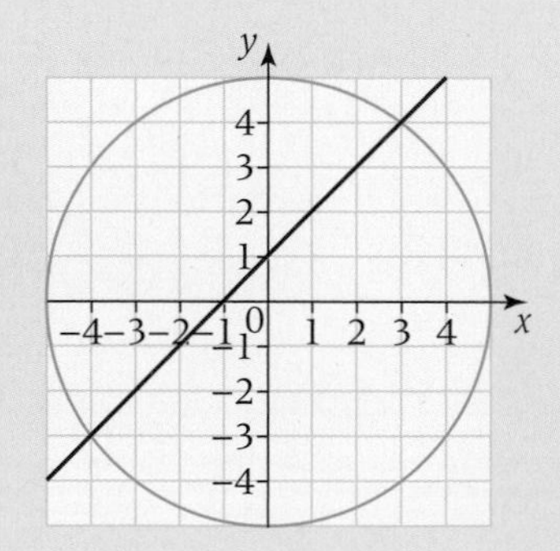

Substituting $y = x + 1$ into $x^2 + y^2 = 25$ gives:

$$x^2 + (x + 1)^2 = 25$$
$$2x^2 + 2x + 1 = 25$$
$$2x^2 + 2x - 24 = 0$$
$$x^2 + x - 12 = 0$$

$(x + 4)(x - 3) = 0$ so either $x = -4$ or $x = 3$

When $x = -4$, $y = (-4) + 1 = -3$

When $x = 3$, $y = 3 + 1 = 4$

Hence, the circle and line intersect at $(-4, -3)$ and $(3, 4)$.

Simplify by dividing each term by 2.
Solve the quadratic by factorising.

Exercise A7.2 — Grade A/A*

1 Write the equations of these circles.

a centre origin, radius 6 **b** centre origin, radius $\frac{1}{2}$

c centre origin, radius 0.4 **d** centre origin, radius $\sqrt{5}$

2 Find the equations of these circles.

a

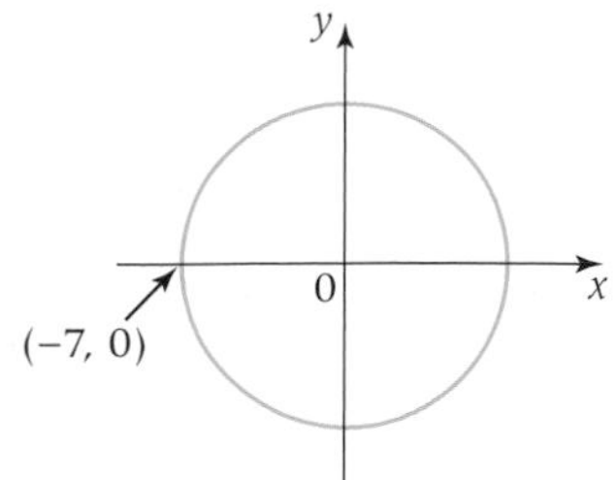

b

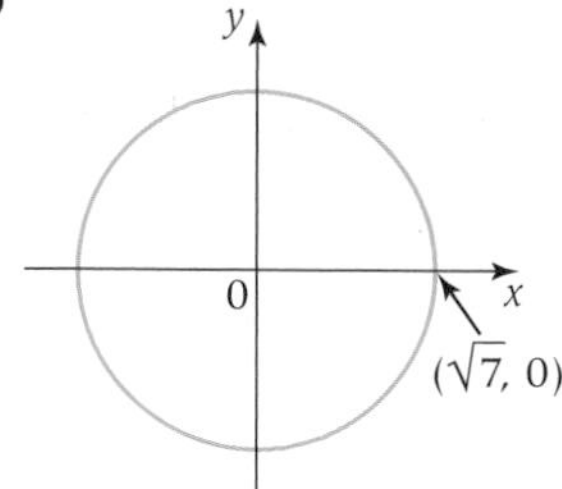

3 The diagram shows the circle, $x^2 + y^2 = 169$ and the line $y = 2x + 2$.
Find the coordinates of the two points of intersection of the line and the circle.

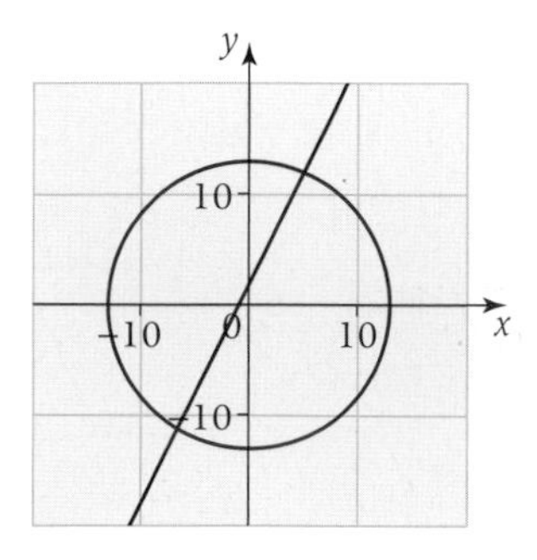

4 Where do the circles $x^2 + y^2 = 25$ and $x^2 + y^2 = 49$ intersect?

Imagine/draw a diagram.

5 **a** How many times do these lines intersect with the circle $x^2 + y^2 = 25$?

i $y = -5$ **ii** $y = 3$ **iii** $4y + 3x = 25$

b Hence, what word could you use to describe the line in **a** part **i**?

Imagine/draw a diagram if necessary.

6 Find, using algebraic methods, the points of intersection of these lines and circles.

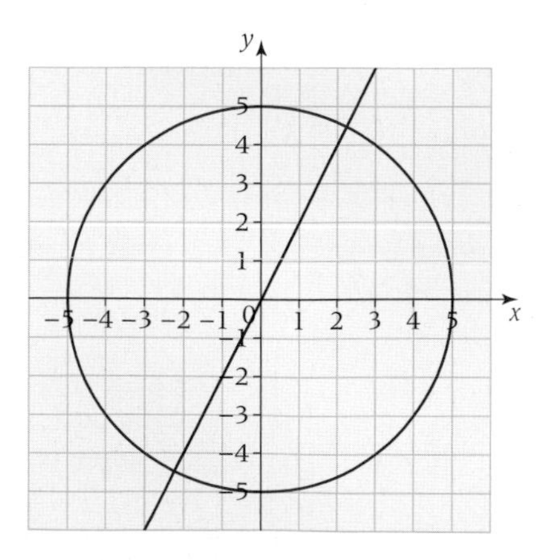

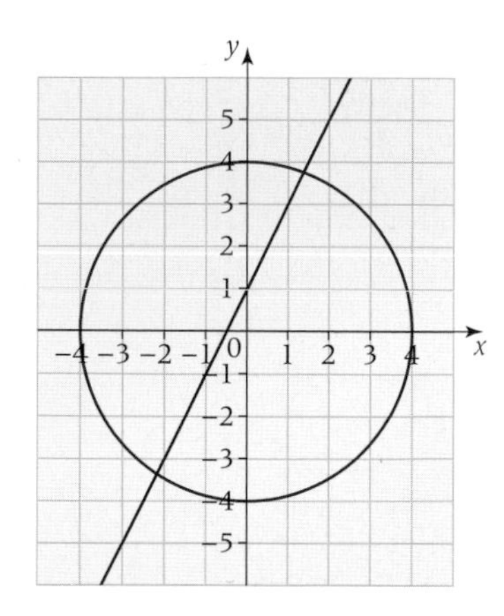

AO3 Problem

7 **a** Use Pythagoras' theorem to find the equation of a circle, centre (3, 5) and radius 6.

b By imagining a diagram, find where the circles $x^2 + y^2 = 36$ and $(x - 12)^2 + y^2 = 36$ intersect.

Unit 3

A7.3 Further equations of circles

This spread will show you how to:

- Recognise the equation of a circle
- Solve two simultaneous equations representing a line and a circle graphically

Keywords
Intersect
Origin
Radius
Simultaneous

- A circle with centre (0, 0) and **radius** r has equation $x^2 + y^2 = r^2$
- You can solve two **simultaneous** equations representing a line and a circle graphically.

Example

Solve $x^2 + y^2 = 16$ and $y = 2x - 1$ graphically.

$x^2 + y^2 = 16$ is a circle, centre the **origin**, radius 4.
$y = 2x - 1$ is a straight line with gradient 2 and y-intercept $(0, -1)$.

Compare with $x^2 + y^2 = r^2$

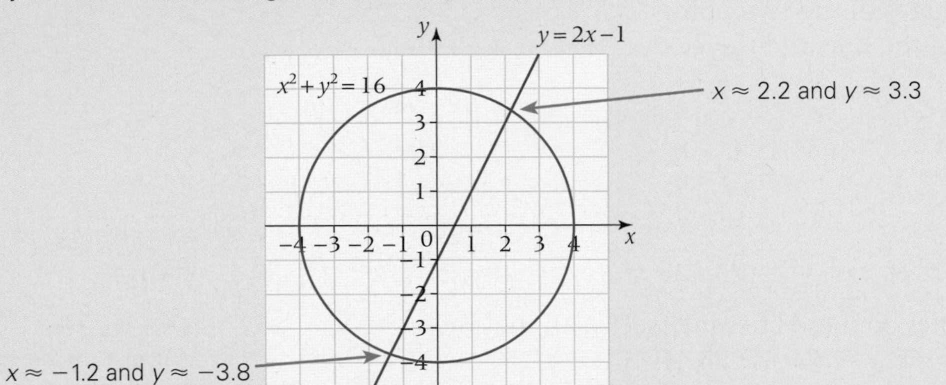

Read off the approximate x- and y-values where the graphs **intersect**. Give them in pairs as they include x and y.

Example

Solve $x^2 + y^2 = 25$ and $y = x^2 - 2$ **a** graphically **b** algebraically.

a The solutions are $x \approx 2.5$, $y \approx 4.3$ and $x \approx -2.5$, $y \approx 4.3$.

b $y = x^2 - 2$, so $x^2 = y + 2$

Substitute $x^2 = y + 2$ into the equation of the circle

$y + 2 + y^2 = 25 \quad \Rightarrow \quad y^2 + y - 23 = 0$

p.314

$$y = \frac{-1 \pm \sqrt{1^2 + 4 \times 1 \times 23}}{2 \times 1} = \frac{-1 \pm \sqrt{93}}{2}$$

Solve the quadratic using the formula.

$y = 4.321\,825\,381\ldots$ or $-5.321\,825\,381\ldots$
From the diagram, $y = -5.321\,825\,381$ is impossible in this case.
Hence, $y = 4.3$ (to 1 dp).
Since $x^2 = y + 2$, then $x^2 = 6.3218\ldots$,
so $x = \pm 2.5$ (to 1 dp).

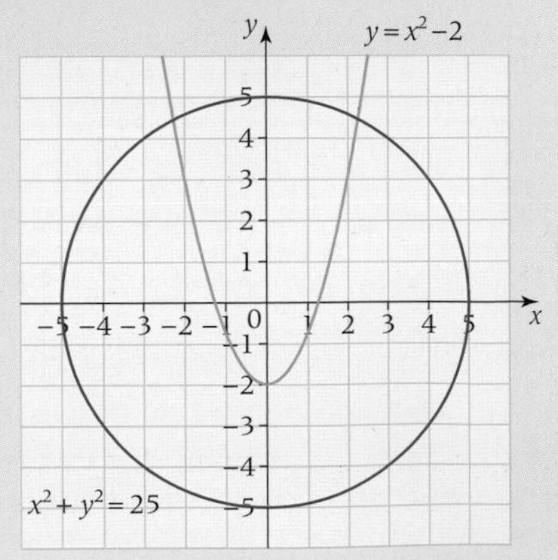

Exercise A7.3 Grade A*

1 The diagram shows the circle $x^2 + y^2 = 9$ and the line $y = x + 1$.
Use the diagram to find the approximate solution of the simultaneous equations
$x^2 + y^2 = 9$
$y = x + 1$

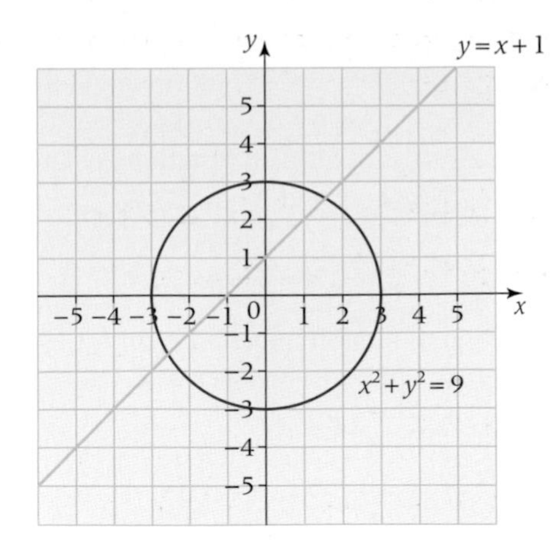

2 Determine the number of solutions to the simultaneous equations by imagining an extra graph on the diagram that currently shows $x^2 + y^2 = 16$.

a $x^2 + y^2 = 16$
$y = 2$

b $x^2 + y^2 = 16$
$y = x$

c $x^2 + y^2 = 16$
$x = 4$

d $x^2 + y^2 = 16$
$y = x^2$

e $x^2 + y^2 = 16$
$y = 5$

f $x^2 + y^2 = 16$
$y = x^3$

g $x^2 + y^2 = 16$
$x + y = 4$

h $x^2 + y^2 = 16$
$y = 4 - x^2$

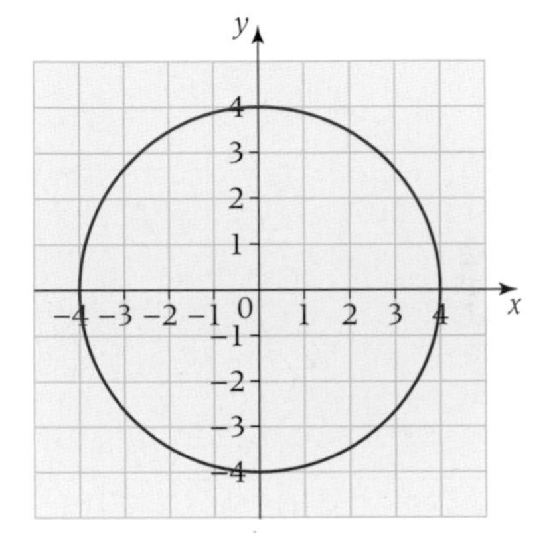

3 By drawing a suitable graph in each case, find approximate solutions to these simultaneous equations.

a $x^2 + y^2 = 25$
$y = 3$

b $x^2 + y^2 = 4$
$y = x$

c $x^2 + y^2 = 36$
$y = 3x - 1$

d $x^2 + y^2 = 1$
$y = x^2$

4 For each pair of simultaneous equations, find their solutions graphically and then confirm them algebraically.

a $x^2 + y^2 = 16$
$y = 2x$

b $x^2 + y^2 = 25$
$y = x$

c $x^2 + y^2 = 9$
$y = 3x - 1$

d $x^2 + y^2 = 49$
$y = x^2$

AO3 Problem

5 Is it possible to have three intersections between a circle and another function? Give an example to support your findings.

Unit 3

Graphs of quadratic and cubic functions

This spread will show you how to:

- Recognise the graphs of quadratic and cubic functions
- Draw graphs of quadratic and cubic functions by plotting points

Keywords
Cubic
Function
Parabola
Quadratic
S-shaped

- The graph of a **quadratic function** is a U-shaped curve called a parabola.
  $f(x) = x^2$ or $y = x^2$

- The graph of a **cubic** function is an **'S'-shaped** curve.
  $f(x) = x^3$ or $y = x^3$

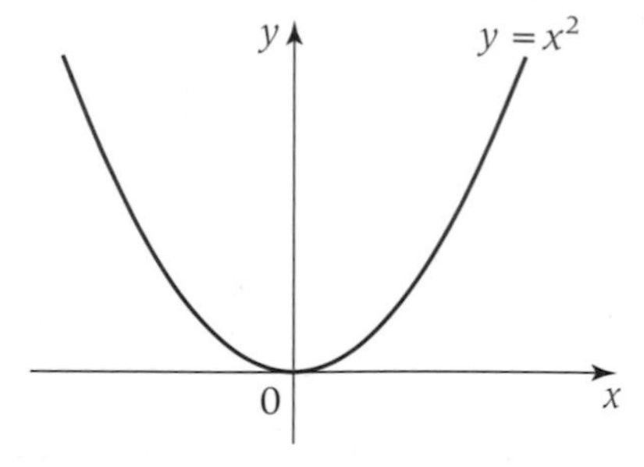

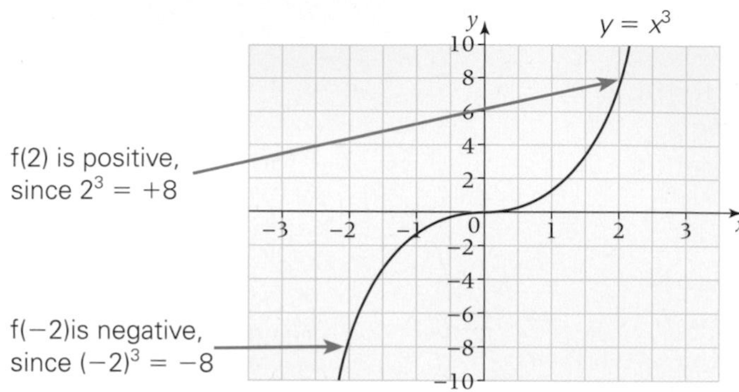

- You draw graphs of quadratic and cubic functions by plotting points.

Example

Plot the graph of $f(x) = x^2 - 3x + 5$ for $-3 \leqslant x \leqslant 3$.
From your graph, find the coordinates of the minimum point and confirm this using an algebraic technique.

x	−3	−2	−1	0	1	2	3
x^2	9	4	1	0	1	4	9
$-3x$	9	6	3	0	−3	−6	−9
$+5$	5	5	5	5	5	5	5
$f(x)$	23	15	9	5	3	3	5

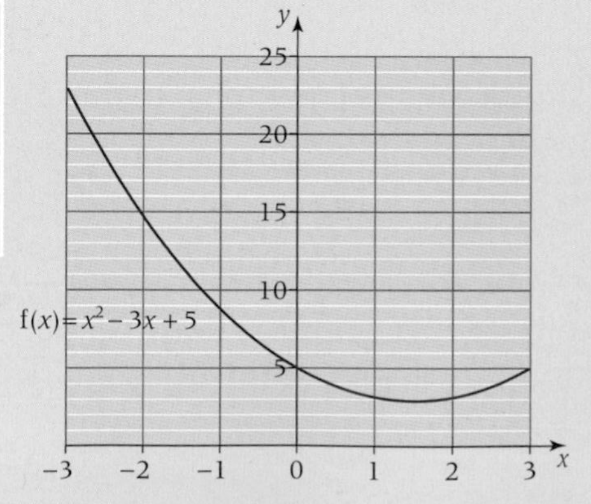

From the graph, the minimum point is at approximately $\left(1\frac{1}{2}, 2\frac{3}{4}\right)$.
Complete the square on $x^2 - 3x + 5$.

$$x^2 - 3x + 5 = \left(x - \frac{3}{2}\right)^2 + 5 - \frac{9}{4}$$
$$= \left(x - 1\frac{1}{2}\right)^2 + 2\frac{3}{4}$$

So the minimum point is $\left(1\frac{1}{2}, 2\frac{3}{4}\right)$.

$f(x) = x^2 - 3x + 5$ is the same as $y = x^2 - 3x + 5$.

Draw up a table of values. Keep signs with the terms. Add down a column for each value of $f(x)$.

Plot the points $(-3, 23)$, $(-2, 15)$, $(-1, 9)$, ..., $(3, 5)$, joining them with a smooth curve. Make sure the bottom of the curve is 'rounded'.

For the minimum value of $f(x)$, $x - 1\frac{1}{2} = 0$.

Exercise A7.4 Grade B

1 Match the graphs with the equations.

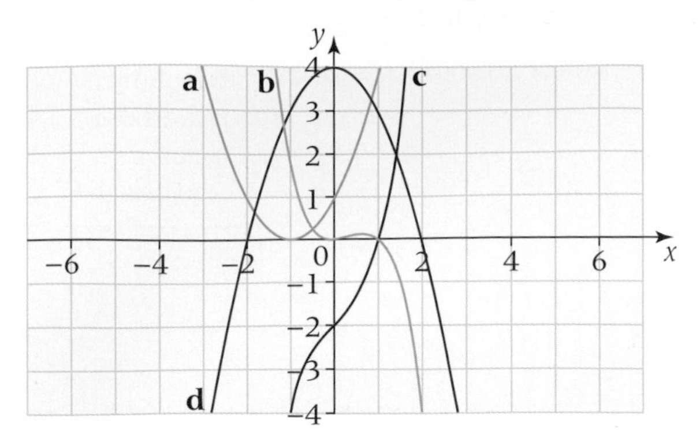

i $y = x^2 + 2x + 1$

ii $f(x) = x^3 + x - 2$

iii $y = 4 - x^2$

iv $g(x) = x^2 - x^3$

2 **a** Copy and complete this table of values for the graph of $y = 2x^2 + 3x - 6$.

x	−4	−3	−2	−1	0	1	2	3	4
$2x^2$		18							
$3x$		−9							
−6	−6	−6	−6	−6	−6	−6	−6	−6	−6
y		3							

b Draw suitable axes and plot the graph of $y = 2x^2 + 3x - 6$, joining the points with a smooth curve.

c From your graph, write down the coordinates of the minimum point of $y = 2x^2 + 3x - 6$.

3 Draw graphs of these functions for the range of x-values given.

a $y = x^2 + 3x$, for $-3 \leqslant x \leqslant 3$ **b** $y = x^2 + x - 2$, for $-3 \leqslant x \leqslant 3$

c $y = 2x^2 - 3x$, for $-2 \leqslant x \leqslant 5$ **d** $f(x) = 3 - x^2$, for $-3 \leqslant x \leqslant 3$

e $y = x^3 + x - 4$, for $-2 \leqslant x \leqslant 3$ **f** $f(x) = x^3 - x^2 + 3x$, for $-3 \leqslant x \leqslant 3$

4 **a** Plot the graph $f(x) = x^3 - 2x^2 + x + 4$ for $-3 \leqslant x \leqslant 3$.

b Use your graph to find

i the value of x when $y = -20$ **ii** the value of y when $x = 1.7$.

Functional Maths AO2

5 A farmer has 100 metres of fencing with which to make a chicken pen. He wants to build the pen against the side of his barn, as shown.

BARN

CHICKEN PEN

The farmer wants to enclose the smaximum possible area in the pen.

a Let the width of the pen be w. Write a formula in terms of w for **i** the length **ii** the area of the pen.

b Plot a graph of area against width and use it to find the dimensions the farmer should use.

A7.5 Graphs of exponential and reciprocal functions

This spread will show you how to:

- Recognise the shapes of graphs of quadratic and cubic functions
- Draw graphs of quadratic and cubic functions by plotting points

Keywords
Asymptote
Exponential
Function
Hyperbola
Index
Reciprocal

- Graphs of **functions** involving **reciprocals** have a characteristic shape, called a **hyperbola**.

For example, $f(x) = \frac{1}{x}$.
As x gets larger, y gets smaller and vice versa.
If x is 1000, y is $\frac{1}{1000}$.
If x is $\frac{1}{1\,000\,000}$, y is 1 000 000.
x cannot take the value zero, since you cannot evaluate $\frac{1}{0}$. The same is true for y. So the x- and y-axes are **asymptotes** – the graphs will never touch them but can get as close as you like.

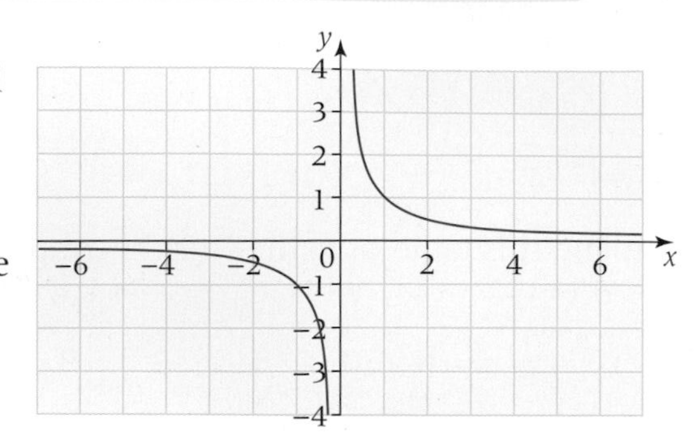

- **Exponential** functions include a term with a variable **index**, for example $f(x) = 2^x$. Graphs of exponential functions have a characteristic shape.

As x gets larger, y gets larger and vice versa.
If x is 10, y is 2^{10} or 1024.
If x is -4, y is 2^{-4} or $\frac{1}{16}$.
Since $2^x (= y)$ can never be zero for any value of x, the x-axis is an asymptote, the curve gets near but never actually touches it.

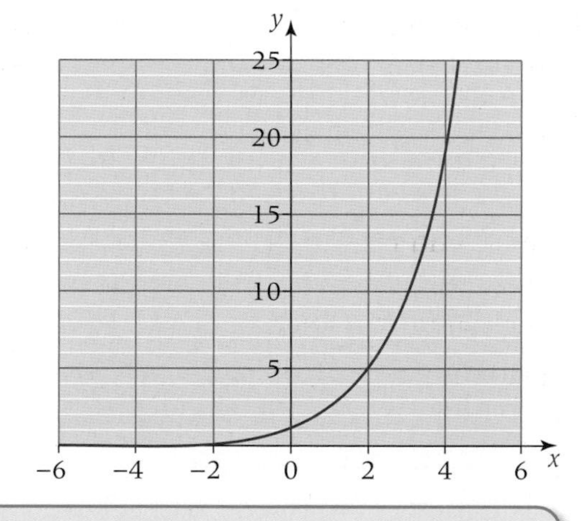

Example

A piece of paper is 1 unit thick. It is folded in half, then in half again, then again repeatedly.
If x is the number of folds and y is the thickness of the paper, form an equation connecting x and y and draw the graph of this equation.

x	0	1	2	3	4	5
y	1	2	4	8	16	32

The thickness doubles each time. After 2 folds, it is 2^2 thick, after 5 folds it is 2^5. Hence the equation is $y = 2^x$.

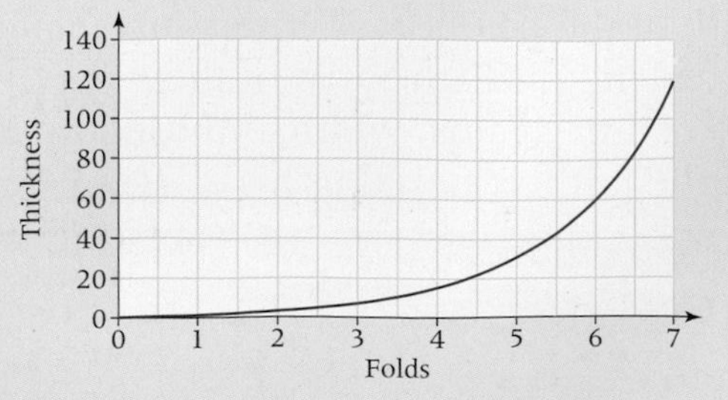

Write the information in a table and spot the pattern.

Exercise A7.5 Grade A

1 Match each graph with its equation.

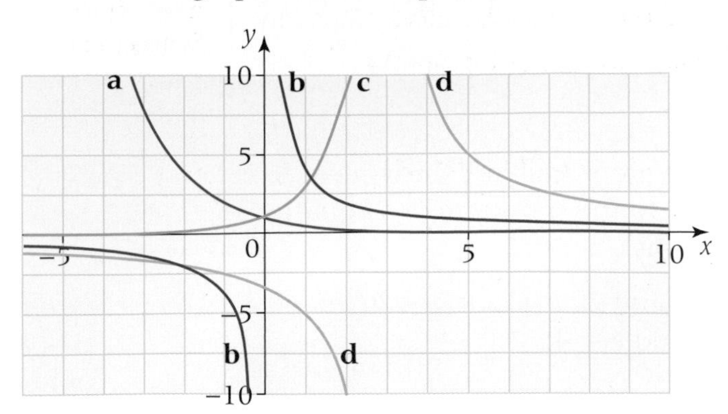

i $y = \frac{4}{x}$

ii $y = 3^x$

iii $y = \frac{10}{x - 3}$

iv $y = 2^{-x}$

2 a Copy and complete this table of values for the function $f(x) = \frac{12}{x}$

x	−6	−5	−4	−3	−2	−1	0	1	2	3	4	5	6
f(x)		−2.4					Asymptote				3		

b Draw suitable axes and plot the graph of $f(x) = \frac{12}{x}$, joining your points with a smooth curve.

c What shape is your graph and where are its asymptotes?

d Use your graph to estimate the value of f(2.5).

3 a Without using a calculator, copy and complete this table of values for the function $g(x) = 2^{x+1}$

x	−4	−3	−2	−1	0	1	2	3	4
g(x)		$\frac{1}{4}$					8		

b Draw suitable axes and plot the graph of $g(x) = 2^{x+1}$, joining your points with a smooth curve.

c Approximate the value of x for which $g(x) = 25$.

4 Copy and complete the table of values below and use it to plot the graph of $y = \frac{20}{x} + x - 5$.

x	−4	−3	−2	−1	0	1	2	3	4	5
$\frac{20}{x}$										
−5										
y										

5 Plot these functions for the range of x-values given.

a $y = \frac{12}{x - 2}$ for $-2 \leqslant x \leqslant 6$

b $f(x) = 4^{x-2}$, for $-2 \leqslant x \leqslant 6$

c $y = \frac{x}{x + 4}$ for $-4 \leqslant x \leqslant 4$

d $f(x) = \frac{6}{x + x} - 2$ for $-3 \leqslant x \leqslant 3$

e $y = 3^{-x} - 1$ for $-4 \leqslant x \leqslant 4$

A7.6 Solving equations using graphs

This spread will show you how to:

- Find the intersection points of graphs of a linear and a quadratic function
- Solve simultaneous equations graphically by drawing a graph to represent each equation

Keywords
Intersection
Simultaneous

- You can solve **simultaneous** equations graphically by drawing a graph to represent each one. A solution to the simultaneous equations is at a point of **intersection** of the graphs.

For example, for the equations $3x - y = 2$ and $2x + y = 8$, the lines intersect at (2, 4) so the solution is $x = 2$ and $y = 4$.

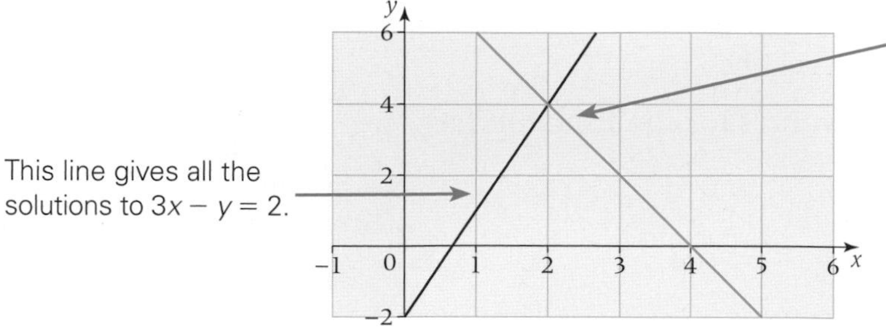

- You can solve some complex equations graphically by splitting them into two equations and treating these as simultaneous equations.

For example, $x^2 - 5x + 6 = 3$ splits into the two equations $y = x^2 - 5x + 6$ and $y = 3$.

Example

Solve $x^2 - 5x + 6 = 3$ graphically.

Plot the graphs of $y = x^2 - 5x + 6$ and $y = 3$ on the same axes.

Draw up a table of values to plot $y = x^2 - 5x + 6$.

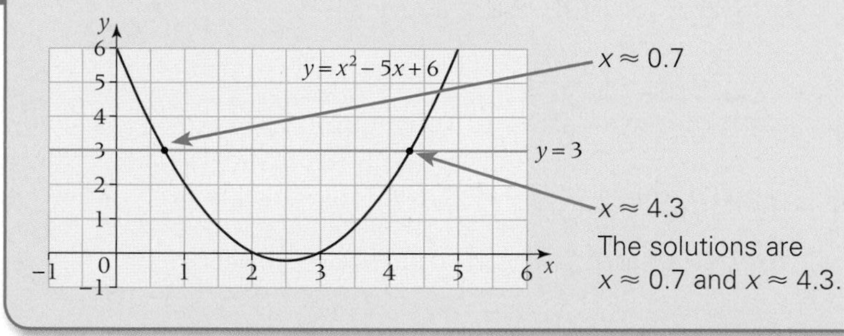

The solutions are $x \approx 0.7$ and $x \approx 4.3$.

To solve the equation you need to find the values of x. Read these off the graph.

Example

Solve the equation $5x - x^2 = 2x - 1$ graphically.

Plot the graphs of $y = 5x - x^2$ and $y = 2x - 1$.

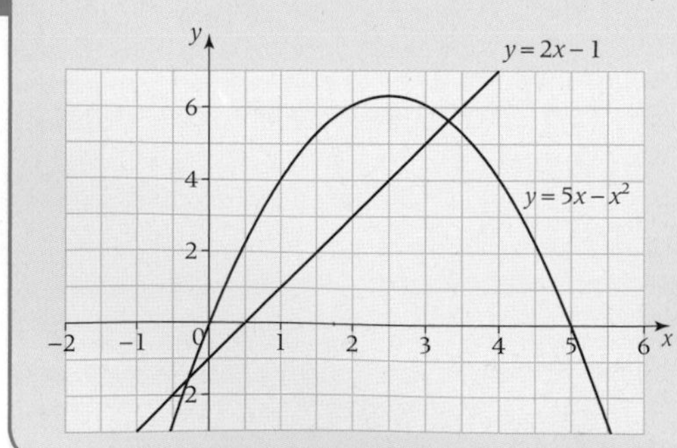

The solutions are $x \approx -0.3$ and $x \approx 3.3$.

The answers are approximate, since you read them off the graph, rather than finding them by a direct algebraic approach.

Exercise A7.6

Grade A/A*

1 Solve these simultaneous equations graphically.

a $y = 2x + 1$
$x + y = 10$

b $y = 3x - 2$
$x + y = 2$

c $2x + y = 5$
$x - y = 4$

Confirm your solution to each pair algebraically.

2 **a** Use these graphs to find the approximate solutions of these equations.

i $x^2 - x - 2 = 2$ **ii** $x^2 - x - 2 = -1$ **iii** $x^2 - x - 2 = x + 1$

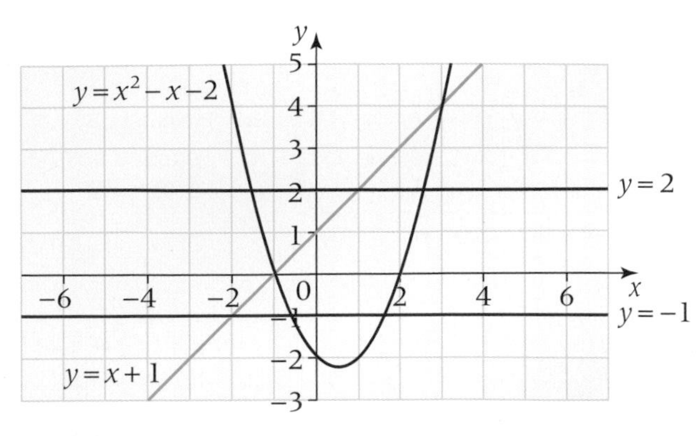

b Where would you find the solutions to $x^2 - x - 2 = 0$? Approximately what are these? Confirm your answer algebraically.

c Which graph would you need to add to the diagram to solve $2 - x = x^2 - x - 2$?

3 Draw graphs to find the approximate solutions of each of these equations. Draw all the graphs on one pair of axes for the range $-3 \leq x \leq 4$.

a $x^2 - 2x - 2 = 0$

b $x^2 - 2x - 2 = 2$

c $x^2 - 2x - 2 = x + 1$

d $x^2 - 2x - 2 = 3 - x$

e $x^2 - 2x - 2 = \frac{1}{2}x + 1$

f $y - x = 1$ and $x + y = 3$

4 Draw appropriate graphs to solve each of these equations. For each part of the question draw a new pair of axes for the range $-3 \leq x \leq 4$.

a $\frac{12}{x} = 2.5$

b $2^x = 5$

c $3^x = 3x - 2$

d $\frac{1}{x - 3} = 5$

AO3 Problem

5 Using *sketch* graphs only, decide how many solutions each of these equations will have.

a $2^x = x - 1$

b $2^{-x} = -3$

c $\frac{2}{x} = 4$

d $\frac{1}{x} = 2^x$

Unit 3

A7.7 Further graphical solutions

This spread will show you how to:

- Find the intersection points of graphs of a linear and quadratic function
- Solve complex equations graphically by splitting them into two equations

Keywords
Function
Intersection
Transform

You can solve $x^2 - 5x + 7 = 2$ by drawing the graphs of the **functions** $y = x^2 - 5x + 7$ and $y = 2$ and finding the x-value(s) at their point(s) of **intersection**.

You can use the same graphs to solve other equations, such as $x^2 - 8x + 1 = 0$.

To **transform** $x^2 - 8x + 1$ to $x^2 - 5x + 7$, you need to add $3x + 6$. Adding $3x + 6$ to both sides of $x^2 - 8x + 1 = 0$ gives $x^2 - 5x + 7 = 3x + 6$.

Plot the graph of $y = 3x + 6$ and see where the two graphs intersect.

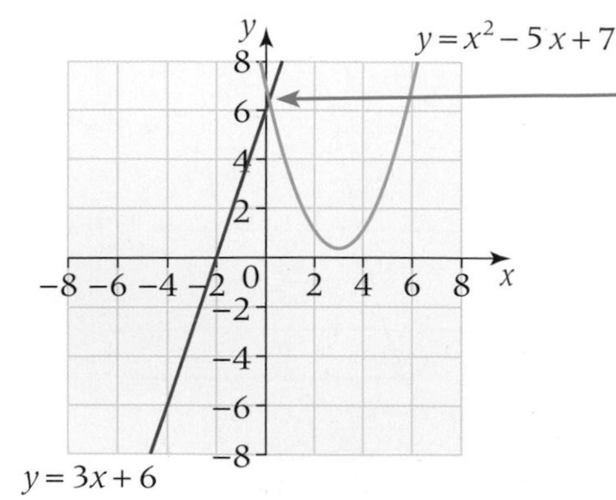

Example

Given the graph of $y = 3x^2 + 2x - 4$, what graph do you need to draw to solve

a $3x^2 + 2x - 4 = 3x - 1$ **b** $3x^2 + 6x = 0$?

By drawing the graphs, solve the equations.

---

a Draw the graph of $y = 3x - 1$.

b To transform $3x^2 + 6x$ to $3x^2 + 2x - 4$ you subtract $4x$ and subtract 4.

Doing the same to both sides: $3x^2 + 6x - 4x - 4 = 0 - 4x - 4$

$$3x^2 + 2x - 4 = -4x - 4$$

Draw the graph of $y = -4x - 4$.

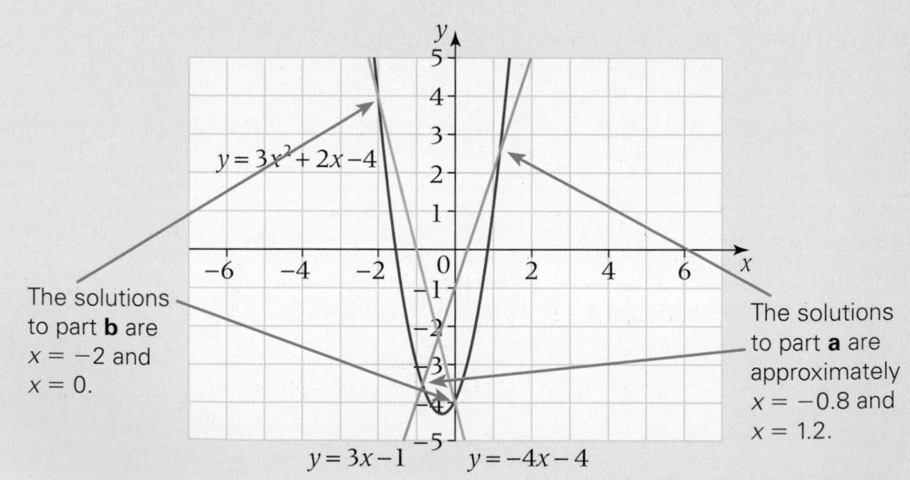

Exercise A7.7 Grade A*

1 Use the graphs to find approximate solutions of the these equations.

a $x^2 + 2x - 3 = 0$ **b** $x^2 + 2x - 3 = x + 1$

c $x^2 + 2x - 5 = 0$ **d** $x^2 + x = 0$

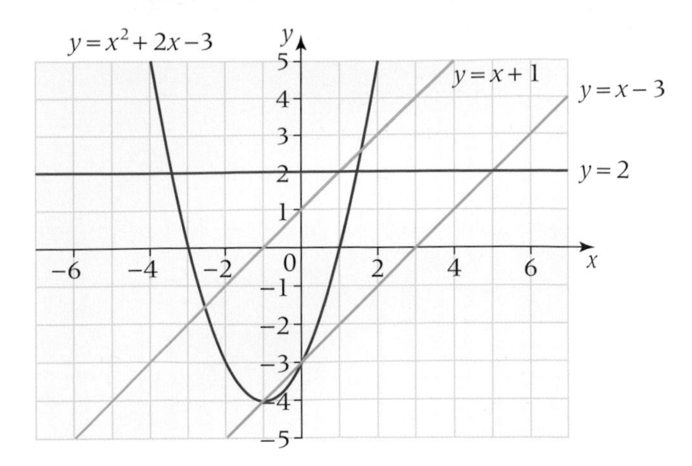

2 Which graph, if any, would you need to add to the grid in question **1** in order to solve each of these equations?

a $x^2 + 2x = 3$ **b** $x^2 + x = 5$ **c** $2x - 3 = 0$

3 If you have the graph of $y = x^2 + 4x - 2$, which one line would you need to draw in order to solve each of these equations?

You do not need to drawn them.

a $x^2 + 4x - 2 = 3$ **b** $x^2 + 4x - 2 = 0$ **c** $x^2 + 4x - 2 = 2x + 1$

d $x^2 + 4x = 6$ **e** $x^2 + 5x = x + 4$ **f** $x^2 + 2x - 3 = 6x$

4 If you have the graph of $y = x^3 + 2x^2 + 5x - 1$, which graph would you need to draw in order to solve each of these equations?

a $x^3 + 2x^2 + 5x - 1 = 2x - 1$ **b** $x^3 + 2x^2 + 2 = 0$

c $x^3 + 2x^2 + 3x = 4$ **d** $x^3 + 2x^2 + 5x - 1 = 0$

e $x^3 + x^2 + 5x = 0$ **f** $x(x + 5) = 0$

5 **a** Draw the graph $f(x) = 2^x$ for $-4 \leq x \leq 4$.

b Draw suitable graphs to solve

i $2^x = 3$ **ii** $2^x + 4x = 2$ **iii** $2^x - x^2 = 0$

AO3 Problem

6 **a** Draw the graph $f(x) = \frac{24}{x} + 1$ for $-4 \leq x \leq 4$.

b Draw suitable graphs to find the approximate solutions of

i $\frac{24}{x} = 3$ **ii** $\frac{24}{x} = x^2$

c What graph could you draw in order to find the approximate value of $\sqrt{24}$?

Unit 3

A7.8 Modelling real situations

This spread will show you how to:

- Recognise the characteristic shape of quadratic, cubic, exponential and reciprocal functions
- Discuss and interpret graphs modelling real situations

Keywords
Cubic
Exponential
Quadratic
Reciprocal

You can recognise some functions from the shapes of their graphs.

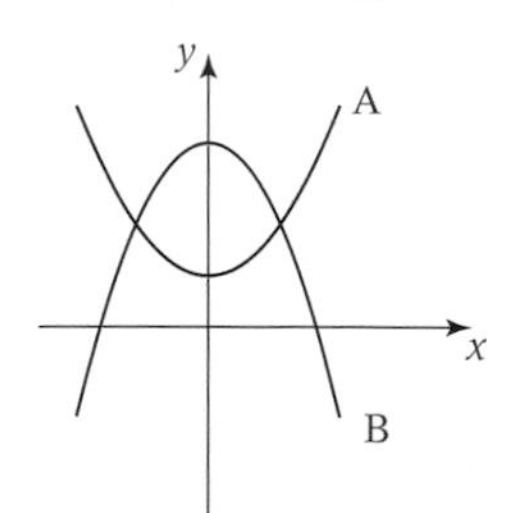

Quadratic functions- parabola or U shape

A $x^2 > 0$

B $x^2 < 0$

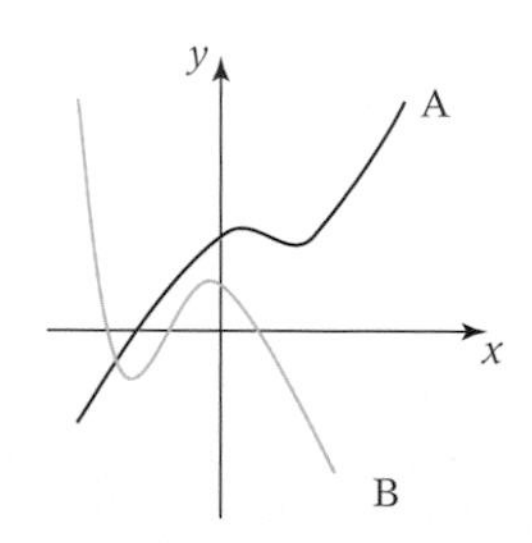

Cubic functions- S-shape

A $x^3 > 0$

B $x^3 < 0$

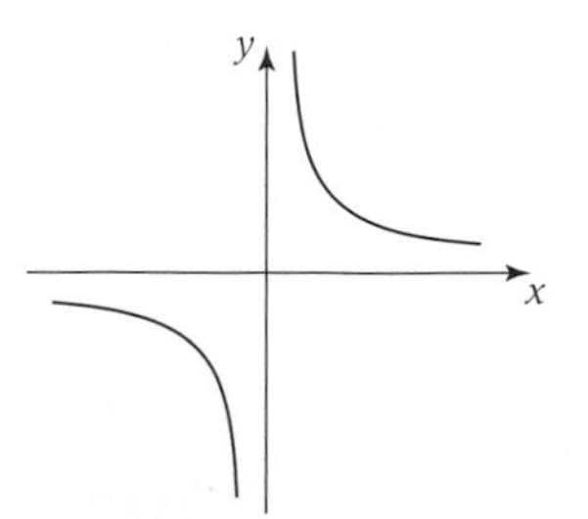

Reciprocal functions - hyperbola
For $y = \frac{1}{x}$, the axes are asymptotes.

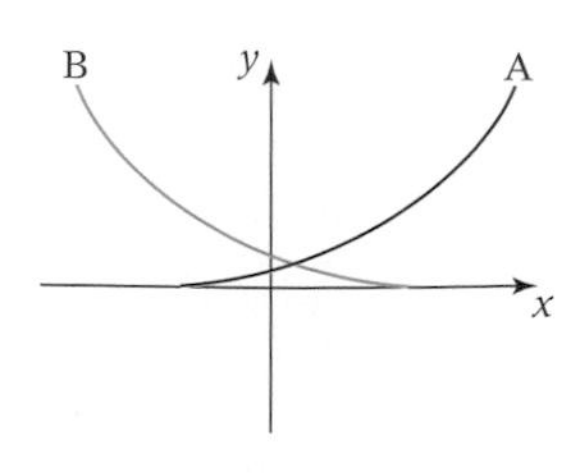

Exponential functions
$f(x) = a^x$
The x-axis is an asymptote.

A $f(x) = a^x$

B $f(x) = a^{-x}$

- You can use curves to model some real-life situations.

For example, an exponential curve is often a suitable model for population growth or the half-life of a radioactive isotope as it decays.

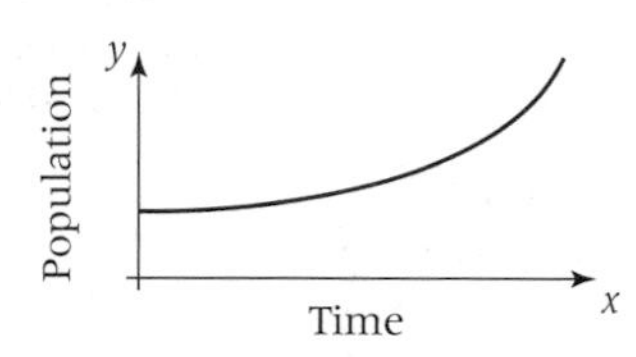

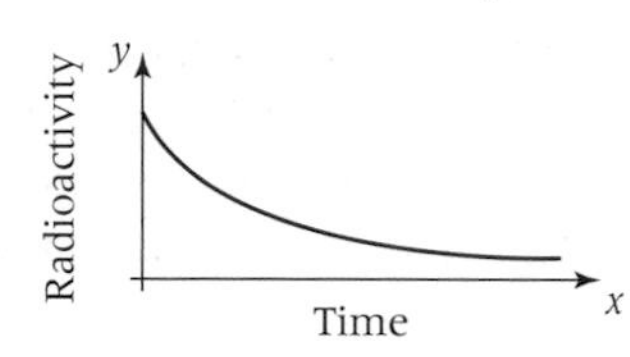

Example

The graph shows the population P of a village as it grows over time t. After a year, the population of the village is 120 people and after 2 years it is 144 people. Given that $P = ab^t$, find the values of a and b and the time when the village's population will exceed 500.

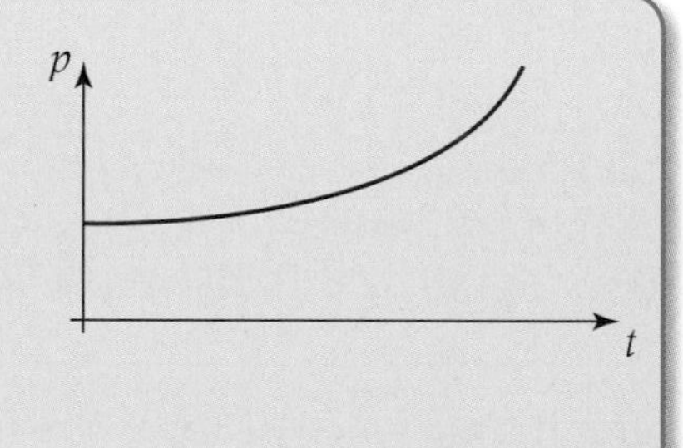

$P = ab^t$

At (1, 120) $120 = a \times b^1$ or $ab = 120$ (1)

When $t = 1$, $P = 120$.

At (2, 144) $144 = a \times b^2$ or $ab^2 = 144$ (2)

When $t = 2$, $P = 144$.

From (1), $a = \dfrac{120}{b}$. Substituting in (2): $\dfrac{120}{b} \times b^2 = 144$

$120b = 144$

$b = 1.2$

Hence $a = 120 \div 1.2 = 100$, so $P = 100 \times 1.2^t$.

When $t = 9$, $P = 100 \times 1.2^9 = 515.978$.

The population will exceed 500 after 9 years.

Use trial and improvement to find the value of t such that $P > 500$.

Exercise A7.8 Grade A*

1 Match each of the functions with its graph.

i $y = \frac{1}{x-2}$ ii $f(x) = x(1 - x^2)$ iii $y = 4^x$ iv $f(x) = x^2 + x - 2$ v $y = x^3 + x^2 - 6x$

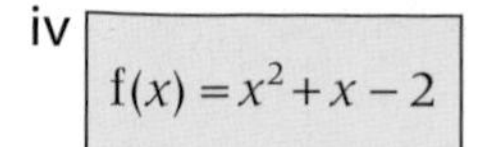

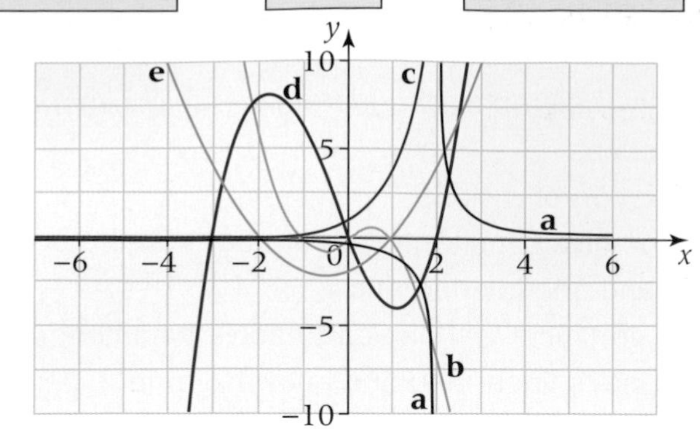

Unit 3

A02 Functional Maths

2 Bacteria reproduce by splitting in two. If you begin with one bacterium and this splits, after 1 minute, to make two bacteria and these split, after a further minute, to make four bacteria and so on, sketch a graph of the number of bacteria (N) against time (t).
Sketch a graph for a bacterium that splits into 3 to reproduce.

DID YOU KNOW?

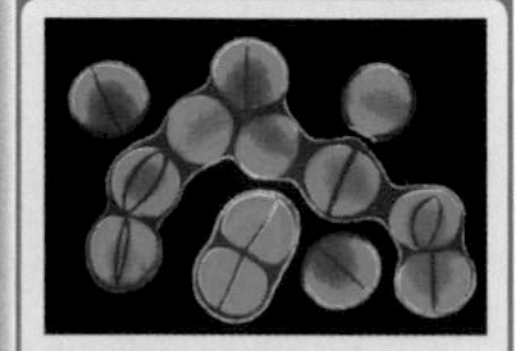

Bacteria with an average generation (doubling) time of 20 minutes can produce 1 billion new cells in just 10 hours.

3 The population P, over time t, of a herd of elephants living in a Thai jungle is modelled using the formula $P = 5 \times 1.5^t$

a What is the population of the herd initially (when $t = 0$)?
b What is the population of the herd after **i** 1 year **ii** 2 years?
c Find, by trial and error, the time at which the population will exceed 20 elephants.
d Sketch a graph of P against t.

4 The graph shows the profits P of a company over time t.
The profit is modelled using the equation $P = mn^t$.

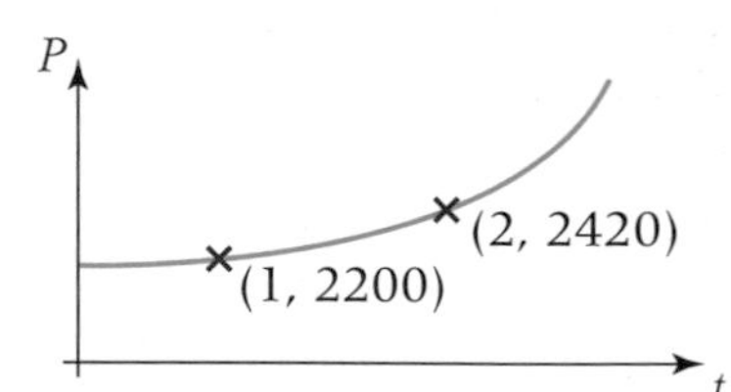

a Use the information in the graph to find the values of m and n.
b When will the profits reach £5000?

5 Here is a function: $y = 2^{-x}$

a Draw a table of values for $-2 \leqslant x \leqslant 2$ and sketch a graph of $y = 2^{-x}$.
b Use your calculator to explore what happens to y for very large positive and negative values of x.
c Describe what would happen to your graph if it were extended from $x = -100$ to $x = 100$.
d How would your results differ if the original function was $y = 2^x$?

Use the power key on your calculator. This is often x^y or ^.

A7 Summary

Check out

You should now be able to:

- Recognise and draw the graphs of quadratics, cubic, exponential and reciprocal functions
- Recognise the equation of a circle
- Use substitution to solve simultaneous equations where one equation is linear and one is quadratic
- Use algebraic and graphical techniques to solve simultaneous equations, where one is linear and one is of the form $x^2 + y^2 = r^2$
- Interpret a pair of simultaneous equations as a pair of straight lines and their solution as the point of intersection
- Find approximate solutions of equations from their graphs, including one linear and one quadratic
- Use graphical techniques to solve simultaneous equations, where one is linear and one is quadratic
- Discuss and interpret graphs modelling real-life situations

Worked exam question

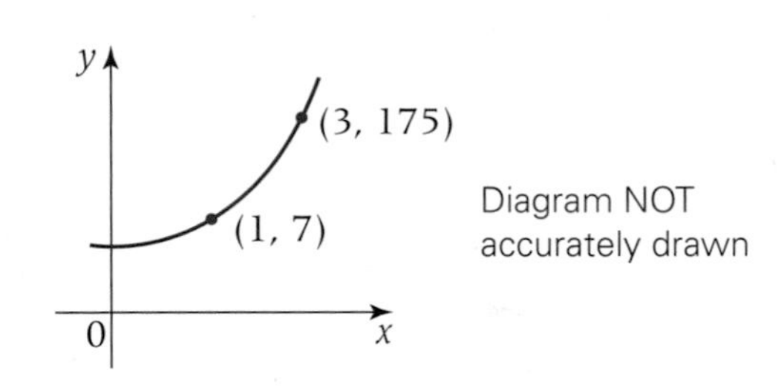

The sketch shows a curve with equation $y = ka^x$
where k and x are constants, and $a > 0$
The curve passes through the points (1, 7) and (3, 175).
Calculate the value of k and the value of a. (3)

(Edexcel Limited 2008)

$y = ka^x$

At (1, 7), $x = 1$ and $y = 7$ and so $\quad 7 = ka \quad (1)$

At (3, 175), $x = 3$ and $y = 175$ and so $\quad 175 = ka^3 \quad (2)$

From (1), $k = \frac{7}{a}$ Substitute in (2) $\quad 175 = \frac{7}{a}a^3$

$25 = a^2$

$a = 5$ only as $a > 0$

Substitute in (1) $\quad k = \frac{7}{5} = 1.4$

Show your method.

Another method would be to cube (1) to give $343 = k^3a^3$ and to substitute for a^3.

Exam questions

1 **A**

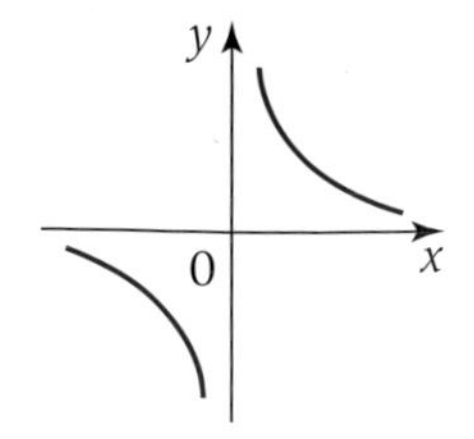

B

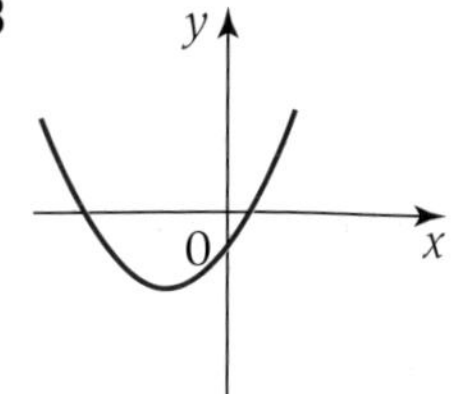

C

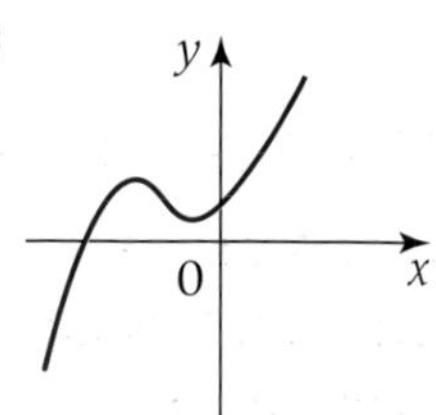

D

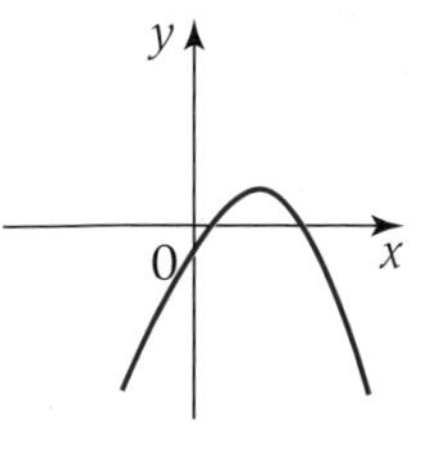

E

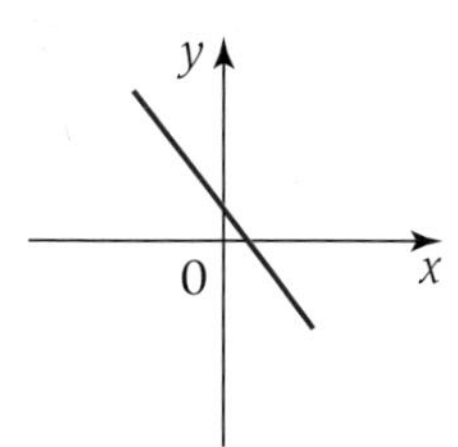

F

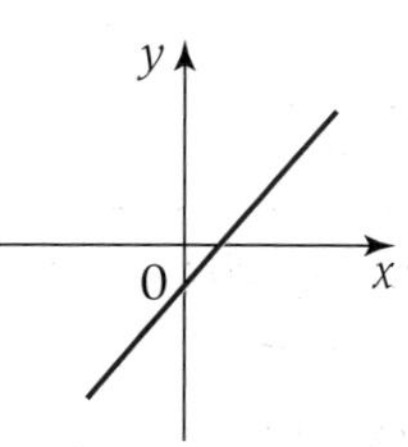

G

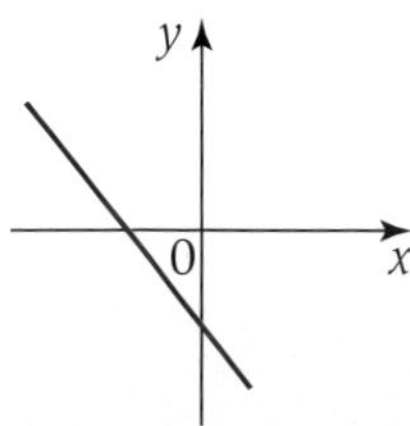

H

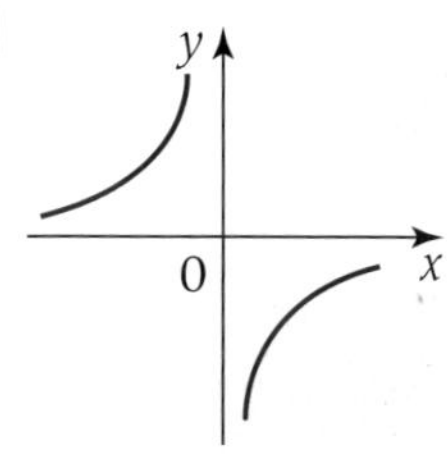

I

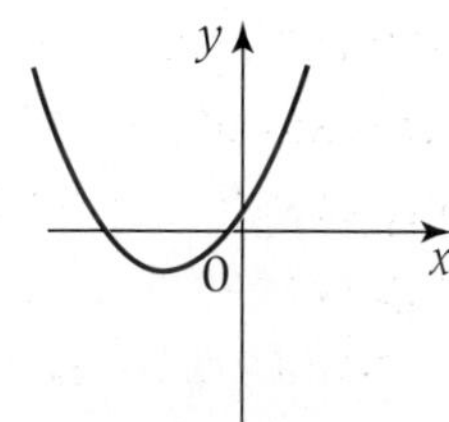

Write down the letter of the graph which could have the equation

i $y = 1 - 3x$ **ii** $y = \frac{1}{x}$ **iii** $y = 2x^2 + 7x + 3$ (3)

(Edexcel Limited 2006)

AO3

2 Solve the simultaneous equations

$x^2 + y^2 = 5$ $y = 3x + 1$ (6)

(Edexcel Limited 2008)

3 The graph of $y = x^2 - x - 2$ is shown on the grid.

By drawing suitable graphs on the grid, solve the equations

a $x^2 - x - 2 = 0$ (1)

b $x^2 - x - 2 = 4$ (1)

c $x^2 - x - 2 = x - 2$ (2)

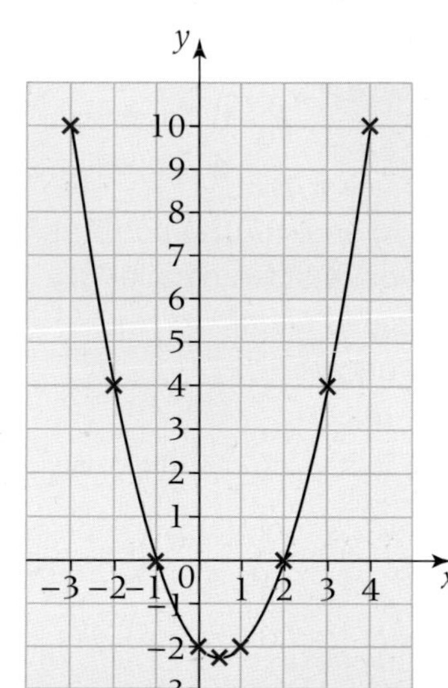

G6

Vectors

Vectors

A satellite in orbit will feel a gravitational attraction towards both the Earth and the Sun, which is inversely proportional to the distance from each, together with a tendency to fly outwards due to its circular orbit. At certain 'Lagrange' points the magnitudes and directions of these forces all add up to zero so that the satellite will appear stationary from either the Earth or Sun and be permanently visible.

What's the point?

Something with both a magnitude and a direction is called a vector. Examples include, position, velocity, acceleration, all forces, wind, *etc*. Knowing how to calculate with vectors is an essential skill for scientists and engineers.

Check in

You should be able to

- **recall facts about basic geometric shapes**

1 For the following quadrilaterals

a Square **b** Rhombus **c** Rectangle
d Parallelogram **e** Trapezium **f** Kite

draw a table to show if they possess any of the following properties

i 1 pair opposite sides parallel **ii** 2 pairs opposite sides parallel
iii Opposite sides equal **iv** All sides equal
v All angles equal **vi** Opposite angles equal
vii Diagonals equal **viii** Diagonals perpendicular
ix Diagonals bisect each other **x** Diagonals bisect the angle

2 Copy these regular polygons and draw all lines of symmetry.

a **b** **c**

Orientation

What I need to know	What I will learn	What this leads to
KS3 Recall geometric properties of basic shapes	■ Use vector notation ■ Add and subtract vectors and recognise parallel vectors ■ Use vectors in geometric proofs	**A-level** Maths, Physics
G2 Understand translations		Engineering

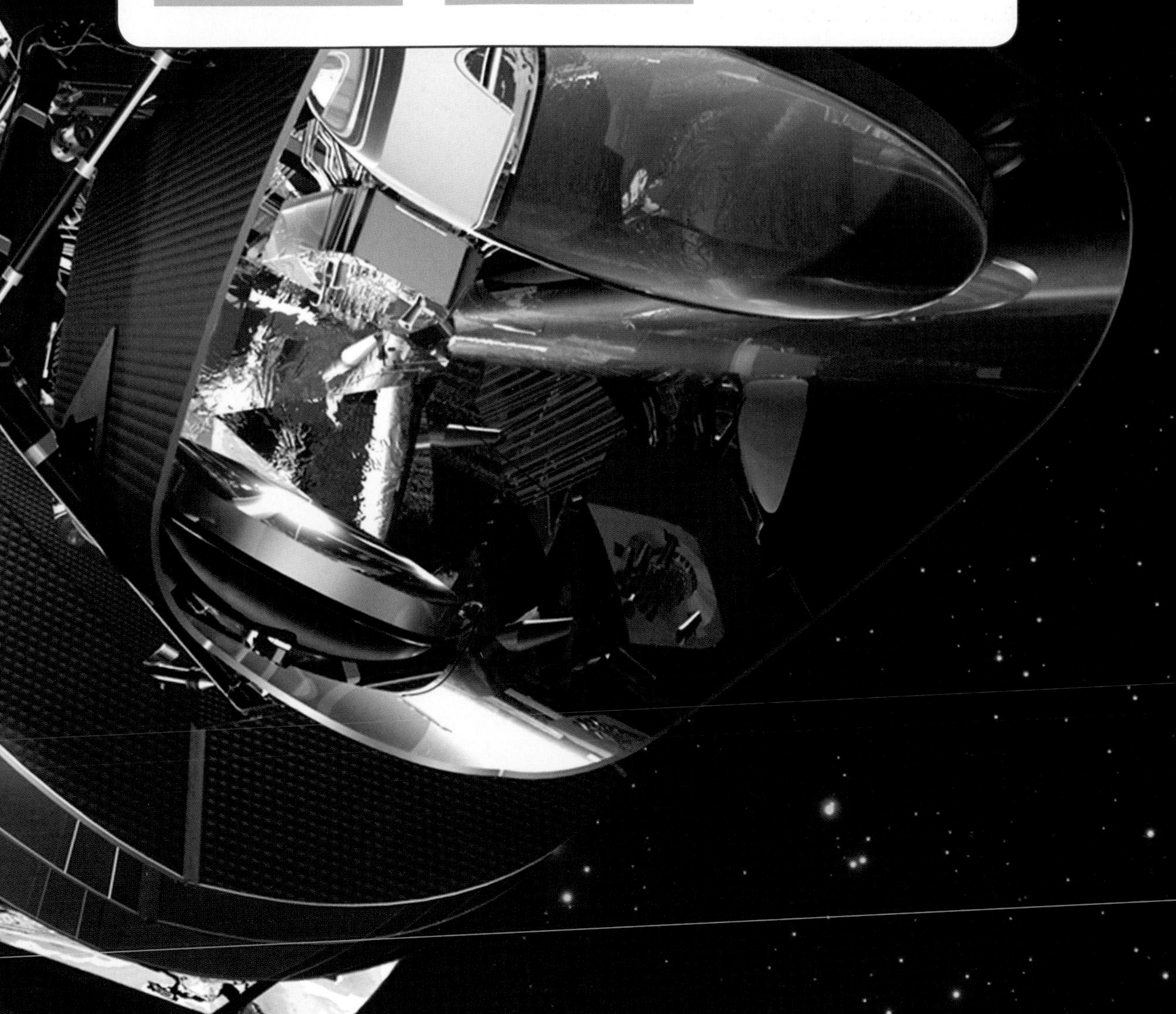

Rich task

In a triangle ABC, let P, Q and R be the midpoints of the sides AB, BC and CA respectively. Show that the lines joining AQ, BR and CP all meet at a point which is two-thirds of the distance along each line from the vertex.

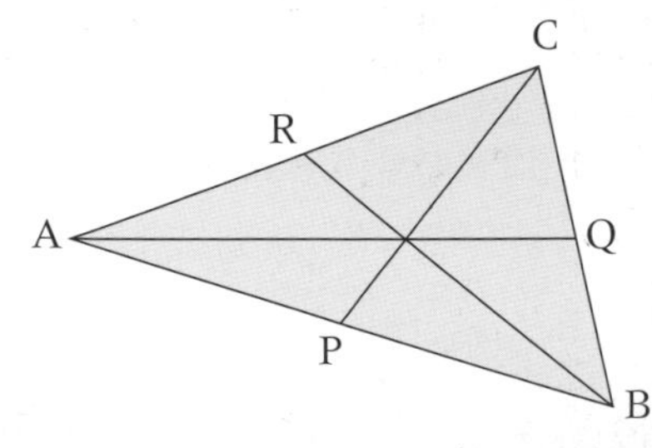

G6.1 Vector notation

This spread will show you how to:

- Understand and use vector notation and the associated vocabulary
- Describe a translation by a vector

Keywords
Displacement
Parallel
Pythagoras' theorem
Vector

You describe a translation by a **vector**. You specify the distance moved left or right, then up or down.

p.212

The vector $\begin{pmatrix} 4 \\ 3 \end{pmatrix}$ takes ABC to A′B′C′.

The vector $\begin{pmatrix} -4 \\ -3 \end{pmatrix}$ takes A′B′C′ to ABC.

You can represent the translation $\begin{pmatrix} 4 \\ 3 \end{pmatrix}$ by an arrowed line parallel to AA′

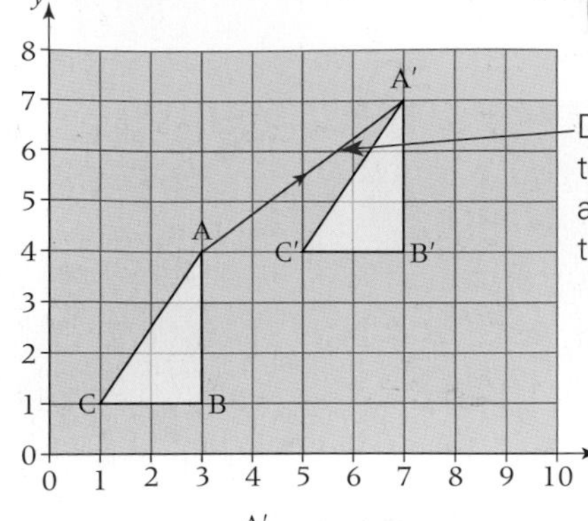

Draw an arrowed line to show the direction and distance of the translation.

Lines joining B and B' or C and C', are **parallel** to AA' and the same length.

p.294

You can find the length of the vector taking A′ to A′ using Pythagoras' theorem.
length AA′ $= \sqrt{3^2 + 4^2} = 5$

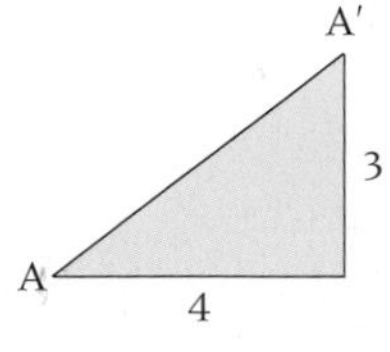

- A vector is a **displacement** that has a fixed length in a fixed direction.

You can draw a vector as an arrowed line. Its orientation gives the direction of movement, its length gives the distance.

These lines are all parallel and the same length.

They all represent the same vector **a**.

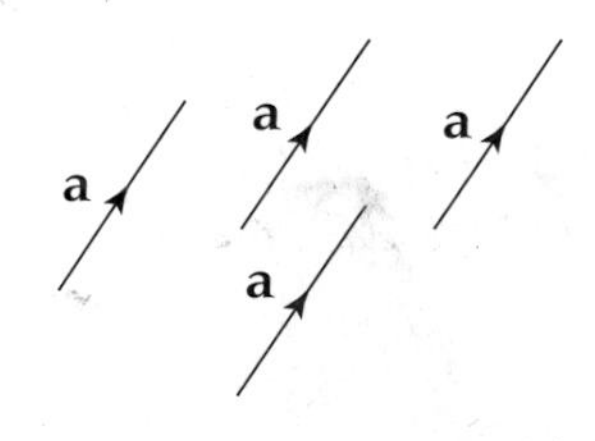

You can tie a vector to a starting point.

The line PQ represents the vector $\overrightarrow{PQ} = \mathbf{p}$.

P —p→ Q

Note that $\overrightarrow{QP} = -\mathbf{P}$

A vector does not need to be tied to a fixed starting point.

The line ST represents the vector $\overrightarrow{ST} = \mathbf{s}$.

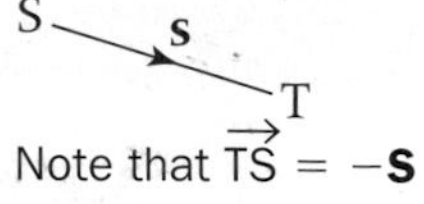

Note that $\overrightarrow{TS} = -\mathbf{S}$

The vector $-\mathbf{a}$ is parallel, the same length, in-the opposite direction to **a**.

- Vectors can be described using the notation $\overrightarrow{AB}$ or bold type, **a**.

The arrow shows the direction is from A to B.

In handwriting, vectors can be shown with an underline, $\underline{a}$.

Exercise G6.1 — Grade B

1 Draw these vectors on squared paper.

a $\begin{pmatrix}4\\3\end{pmatrix}$ **b** $\begin{pmatrix}2\\5\end{pmatrix}$ **c** $\begin{pmatrix}-1\\4\end{pmatrix}$ **d** $\begin{pmatrix}-3\\-3\end{pmatrix}$ **e** $\begin{pmatrix}0\\2\end{pmatrix}$ **f** $\begin{pmatrix}-4\\0\end{pmatrix}$

2 **a** On squared paper draw the vectors $\begin{pmatrix}1\\-2\end{pmatrix}$ and $\begin{pmatrix}-1\\2\end{pmatrix}$.

b Write what you notice about these two vectors.

3 Use $\overrightarrow{AB}$ notation to identify equal vectors in this diagram.

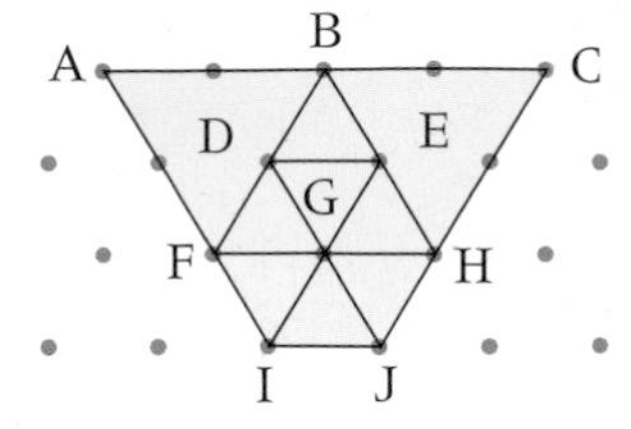

For example, $\overrightarrow{FG} = \overrightarrow{IJ}$ because they are parallel, in the same direction (left to right) and the same length.

4 ABCDEF is a regular hexagon.
X is the centre of the hexagon.
$\overrightarrow{XA} = \mathbf{a}$ and $\overrightarrow{AB} = \mathbf{b}$.

a **i** Write all the vectors that are equal to **a**.

ii Write all the vectors that are equal to **b**.

b **i** Write all the vectors that are equal to $-\mathbf{a}$.

ii Write all the vectors that are equal to $-\mathbf{b}$.

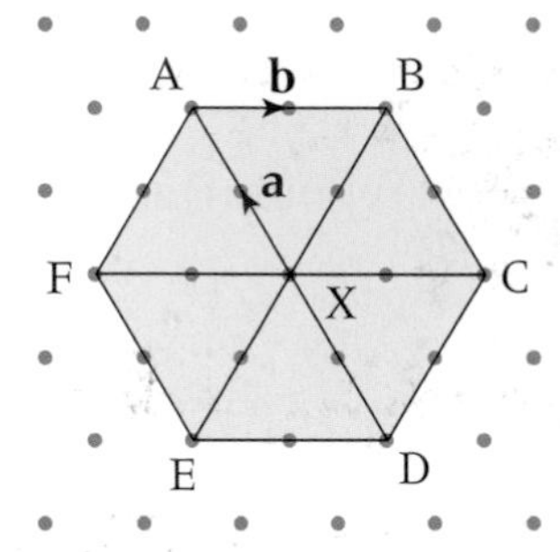

5 JKLMNOPQ is a regular octagon.
$\overrightarrow{OJ} = \mathbf{j}$ $\overrightarrow{OM} = \mathbf{m}$ $\overrightarrow{OM} = \mathbf{p}$.

a Write all the vectors that are

i equal to **j**

ii equal to **m**

iii equal to **p**.

b Write all the vectors that are

i equal to $-\mathbf{j}$

ii equal to $-\mathbf{m}$

iii equal to $-\mathbf{p}$.

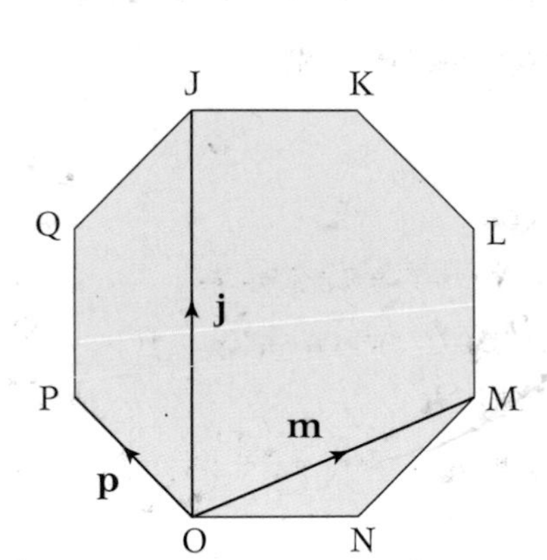

G6.2 Combining vectors

This spread will show you how to:

- Represent, add and subtract vectors graphically

Keywords
Resultant vector
Vector

You can add or subtract **vectors** graphically by arranging them 'nose to tail'.

This is the same as combining two translations.

- The **resultant vector** completes the triangle of vectors.

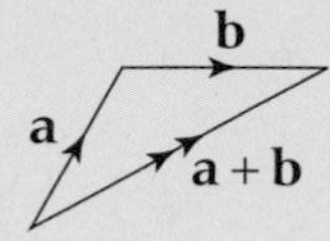

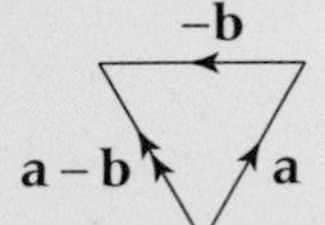

You show resultant vectors with a double arrow.

A resultant vector can be tied to a starting point.

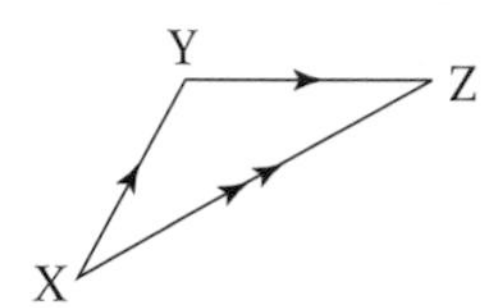

$$\overrightarrow{XY} + \overrightarrow{YZ} = \overrightarrow{XZ}$$

Note the sequence of the letters.

You can add or subtract any number of vectors in this way.

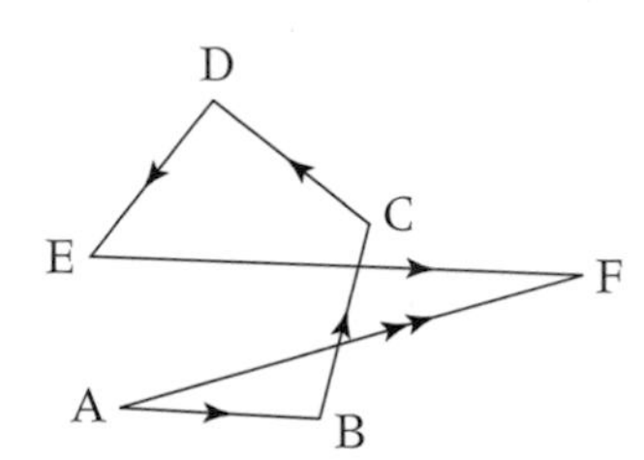

$$\overrightarrow{AB} + \overrightarrow{BC} + \overrightarrow{CD} + \overrightarrow{DE} + \overrightarrow{EF} = \overrightarrow{AF}$$

The pairs of letters in the sequence fit together.

Example

OABC is a parallelogram.

$\overrightarrow{OA} = \mathbf{a}$ $\quad \overrightarrow{OC} = \mathbf{c}$

Write the vector that represents the diagonal

a $\overrightarrow{OB}$ **b** $\overrightarrow{CA}$.

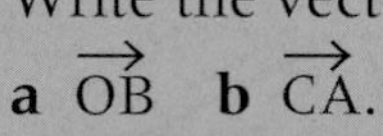

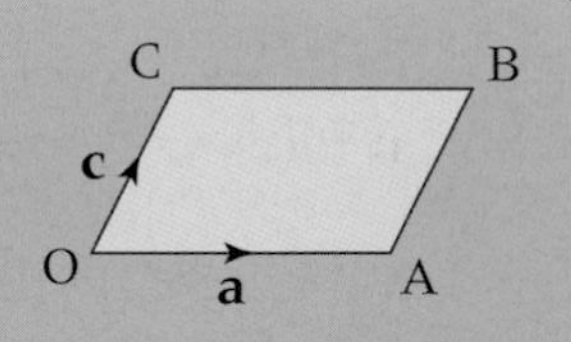

a $\overrightarrow{OB} = \overrightarrow{OA} + \overrightarrow{AB}$
$= \mathbf{a} + \mathbf{c}$

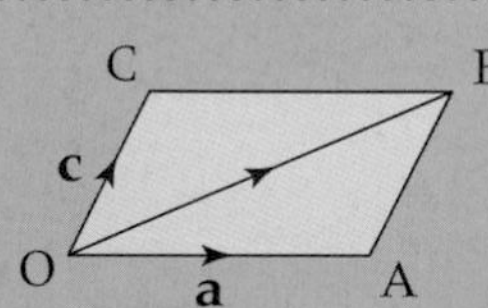

$\overrightarrow{AB} = \mathbf{c}$ because OC is equal and parallel to AB.

b $\overrightarrow{CA} = \overrightarrow{CO} + \overrightarrow{OA}$
$= -\overrightarrow{OC} + \overrightarrow{OA}$
$= -\mathbf{c} + \mathbf{a}$

Exercise G6.2 **Grade A**

1 The diagram shows vectors **s** and **t**.
On squared paper draw the vectors that represent

a $\mathbf{s} + \mathbf{s}$ **b** $\mathbf{s} + \mathbf{t}$ **c** $\mathbf{t} + \mathbf{s}$
d $\mathbf{s} - \mathbf{t}$ **e** $\mathbf{t} - \mathbf{s}$ **f** $\mathbf{t} + \mathbf{t} - \mathbf{s}$

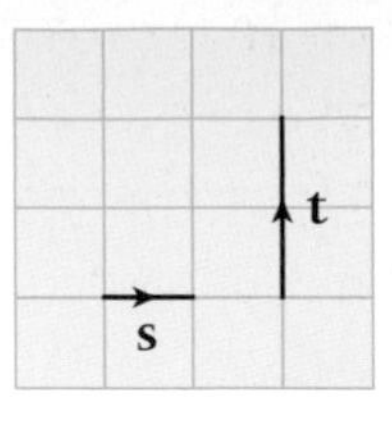

2 The diagram shows vectors **g** and **h**.
On isometric paper draw the vectors that-represent

a $\mathbf{g} + \mathbf{g}$ **b** $\mathbf{g} + \mathbf{h}$
c $\mathbf{h} - \mathbf{g}$ **d** $-\mathbf{g} + \mathbf{h}$
e $\mathbf{g} - \mathbf{h} - \mathbf{h}$

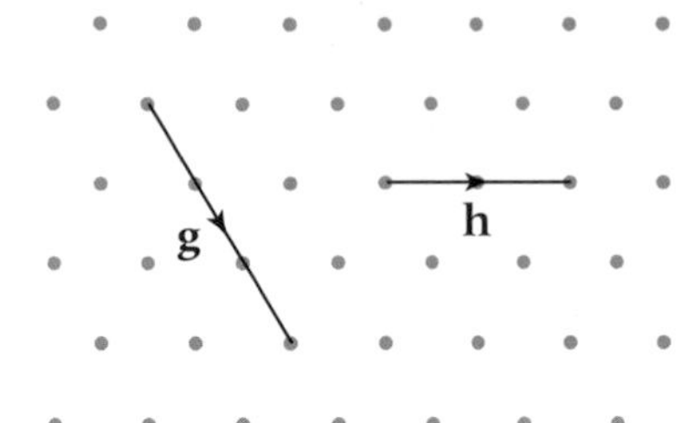

3 OPQR is a rectangle
$\overrightarrow{OP} = \mathbf{p}$ and $\overrightarrow{OR} = \mathbf{r}$.

Work out the vector, in terms of **p** and **r**, that represents

a $\overrightarrow{PQ}$ **b** $\overrightarrow{OQ}$
c $\overrightarrow{QO}$ **d** $\overrightarrow{RP}$

O P
R Q

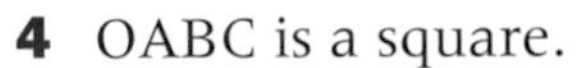

4 OABC is a square.
$\overrightarrow{OA} = \mathbf{a}$ and $\overrightarrow{OB} = \mathbf{b}$.

Work out the vector, in terms of **a** and **b**, that-represents

a $\overrightarrow{BA}$ **b** $\overrightarrow{AB}$
c $\overrightarrow{BC}$ **d** $\overrightarrow{OC}$

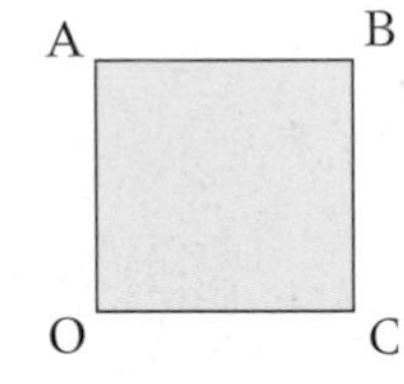

5 OJKL is a rhombus.
$\overrightarrow{OJ} = \mathbf{j}$ and $\overrightarrow{OL} = \mathbf{l}$.

Work out the vector, in terms of **j** and **l**, that-represents

a $\overrightarrow{JK}$ **b** $\overrightarrow{JL}$
c $\overrightarrow{KO}$ **d** $\overrightarrow{KL}$

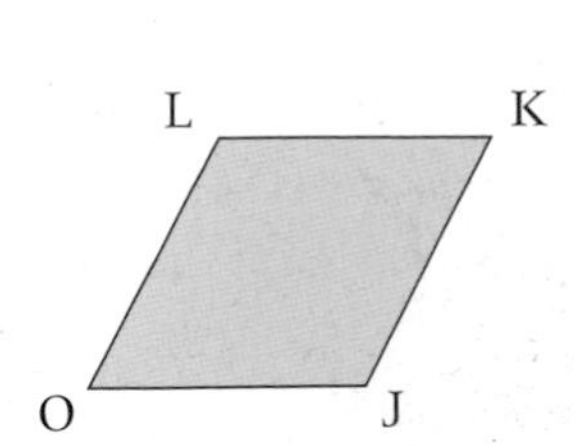

G6.3 Parallel vectors

This spread will show you how to:

- Understand and use the commutative properties of vector addition
- Calculate a scalar multiple of a vector and represent it graphically
- Understand that vectors represented by parallel lines are multiples of each other

Keywords
Commutative
Vector
Multiple
Parallel
Scalar

You can add **vectors** in any order.

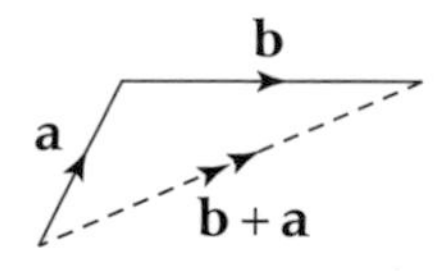

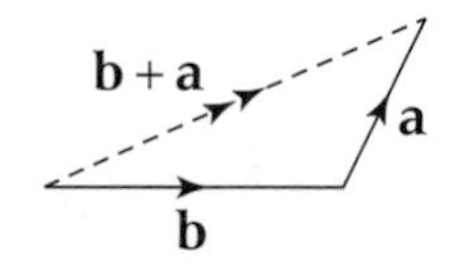

- Vector addition is **commutative:** $\mathbf{a} + \mathbf{b} = \mathbf{b} + \mathbf{a}$

You can extend addition to more than two vectors.

You can write this vector as 4**b**.

$\mathbf{b} + \mathbf{b} + \mathbf{b} + \mathbf{b} = 4\mathbf{b}$

The vector 4**b** is **parallel** to the vector **b**, and four times as long as **b**.

In the vector 4**b**, 4 is a **scalar**.
A scalar has magnitude (size) but no direction.

Speed is a scalar; velocity is a vector.

- You can multiply a vector by a scalar.

$-1 \times \mathbf{p} = -\mathbf{p}$

The vector 2**p** is **parallel** to the vector **p** and twice the length.
The vector 3**p** is parallel to the vector **p** and three times the length.
The vector $-$**p** is parallel to the vector **p** and the same length, but in the opposite direction.

- Vectors represented by parallel lines are **multiples** of each other.

Lines representing multiples of the same vector are parallel.

Example

OABC is a trapezium.
The parallel sides CB and OA are such that CB = 3OA.
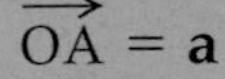
$\overrightarrow{OA} = \mathbf{a}$
Write, in terms of **a**, the vector

a $\overrightarrow{CB}$ **b** $\overrightarrow{BC}$

a $\overrightarrow{CB} = 3\mathbf{a}$ **b** $\overrightarrow{BC} = -3\mathbf{a}$

Exercise G6.3 Grade A

1 $\overrightarrow{OP} = \mathbf{p}$ $\quad \overrightarrow{OQ} = \mathbf{q}$

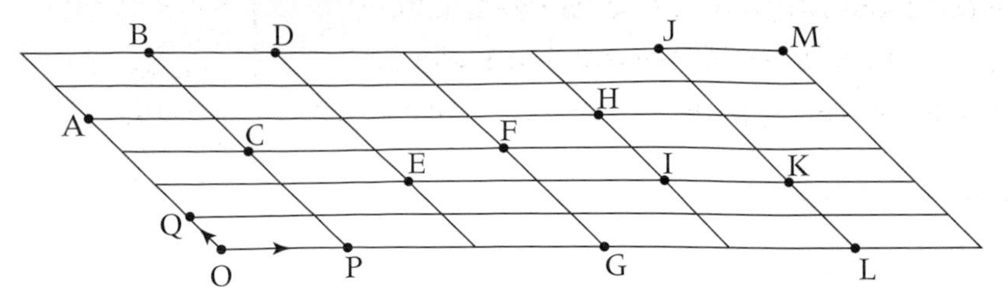

Write, and simplify, the vectors

a $\overrightarrow{OG}$ **b** $\overrightarrow{OL}$ **c** $\overrightarrow{OK}$ **d** $\overrightarrow{OJ}$

e $\overrightarrow{OA}$ **f** $\overrightarrow{OC}$ **g** $\overrightarrow{OB}$ **h** $\overrightarrow{OF}$

i $\overrightarrow{PE}$ **j** $\overrightarrow{PD}$ **k** $\overrightarrow{EF}$ **l** $\overrightarrow{CA}$

m $\overrightarrow{JK}$ **n** $\overrightarrow{JI}$ **o** $\overrightarrow{JE}$ **p** $\overrightarrow{FD}$

q $\overrightarrow{DF}$ **r** $\overrightarrow{DC}$ **s** $\overrightarrow{ME}$ **t** $\overrightarrow{HG}$

u $\overrightarrow{KD}$

2 The diagram shows vectors **x** and **y**.

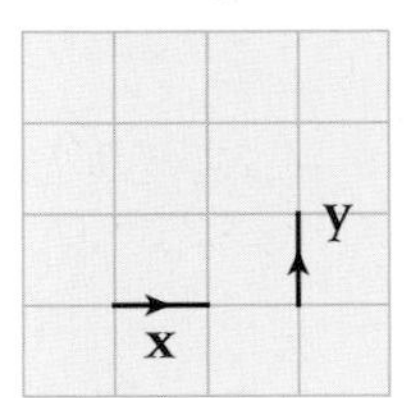

On squared paper draw the vectors

a $2\mathbf{x}$ **b** $3\mathbf{y}$ **c** $2\mathbf{x} + 3\mathbf{y}$

d $3\mathbf{x} - \mathbf{y}$ **e** $\mathbf{y} - 2\mathbf{x}$ **f** $1\frac{1}{2}\mathbf{x} + 1\frac{1}{2}\mathbf{y}$

g $2(\mathbf{x} + \mathbf{y})$ **h** $\frac{1}{2}(2\mathbf{x} - 3\mathbf{y})$ **i** $3\mathbf{x} + 4\mathbf{y}$

3 On squared paper draw a trapezium ABCD with
- parallel sides AB and DC and AB = 4DC
- angle CDA = 90°
- angle DAB = 90°.

If $\overrightarrow{AD} = \mathbf{a}$ and $\overrightarrow{DC} = \mathbf{d}$, write in terms of **a** and **d**, the vectors

a $\overrightarrow{AC}$ **b** $\overrightarrow{DB}$

4 OJKL is a trapezium.
The parallel sides OJ and LK are such that $\frac{1}{4}$ OJ.
OJ = 6**j**
Write, in terms of **j**, the vectors that represent

a $\overrightarrow{LK}$ **b** $\overrightarrow{KL}$

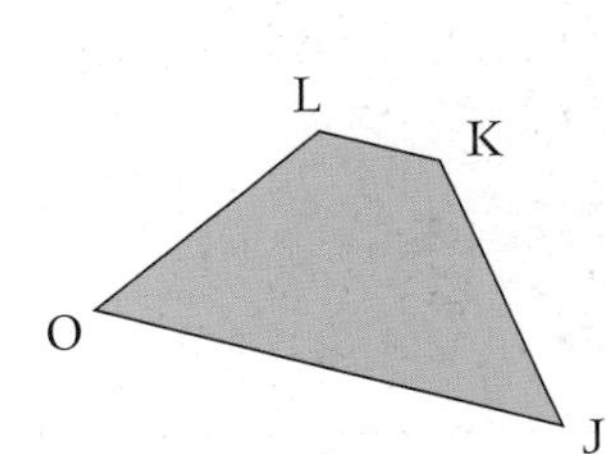

G6.4 Using vectors in geometry

This spread will show you how to:
- Understand how the sign of a vector relates to its direction
- Solve simple geometric problems in 2-D using vector methods

Keywords
Parallel
Resultant vector
Vector

- **Vectors** represented by **parallel** lines are multiples of each other.
- An equal vector in the opposite direction is negative.

Example

OPQR is a square. X is the midpoint of the side PQ.

$\overrightarrow{OP} = \mathbf{p}$ $\quad \overrightarrow{OR} = \mathbf{r}$

Find these vectors in terms of **p** and **r**.

a $\overrightarrow{OQ}$ **b** $\overrightarrow{PX}$ **c** $\overrightarrow{OX}$

$\mathbf{a}$, $2\mathbf{a}$, $-3\mathbf{a}$, $\frac{1}{2}\mathbf{a}$ are parallel vectors. $-3\mathbf{a}$ is in the opposite direction.

a $\overrightarrow{OQ} = \overrightarrow{OP} + \overrightarrow{PQ} = \mathbf{p} + \mathbf{r}$

or $\overrightarrow{OQ} = \overrightarrow{OR} + \overrightarrow{RQ} = \mathbf{r} + \mathbf{p}$

b $\overrightarrow{PX} = \frac{1}{2}\overrightarrow{PQ} = \frac{1}{2}\mathbf{r}$

c $\overrightarrow{OX} = \overrightarrow{OP} + \overrightarrow{PX} = \mathbf{p} + \frac{1}{2}\mathbf{r}$

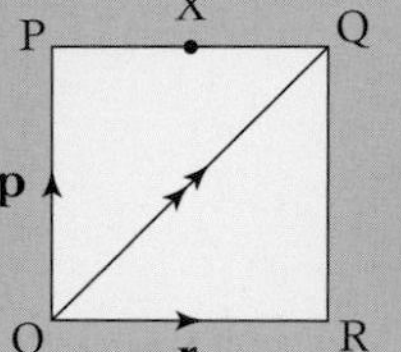

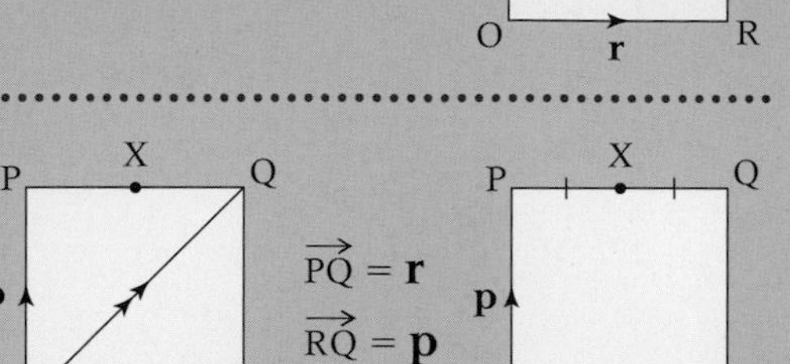

$\overrightarrow{PQ} = \mathbf{r}$

$\overrightarrow{RQ} = \mathbf{p}$

Vector addition is commutative, so $\mathbf{r} + \mathbf{p} = \mathbf{p} + \mathbf{r}$.

$\overrightarrow{PX}$ is parallel to $\overrightarrow{PQ}$ and half the length.

You can use a polygon's geometric properties to help you write vectors.

Example

OABCDE is a regular hexagon.
P is a point on AD such that the ratio AP : PD = 1 : 2.
$\overrightarrow{OA} = \mathbf{a}$ and $\overrightarrow{OE} = \mathbf{e}$.

Find these vectors in terms of **a** and **e**.

a $\overrightarrow{AE}$ **b** $\overrightarrow{AD}$ **c** $\overrightarrow{AP}$ **d** $\overrightarrow{AB}$ **e** $\overrightarrow{OB}$

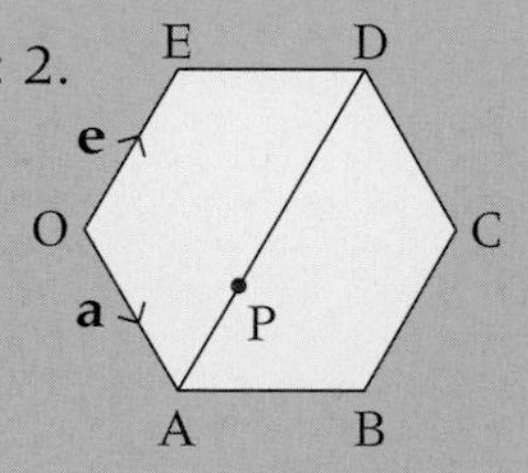

a $\overrightarrow{AE} = \overrightarrow{AO} + \overrightarrow{OE}$
$= -\mathbf{a} + \mathbf{e}$
$= \mathbf{e} - \mathbf{a}$

$\overrightarrow{AO} = -\overrightarrow{OA}$

b $\overrightarrow{AD} = 2\overrightarrow{OE}$
$= 2\mathbf{e}$

c $\overrightarrow{AP} = \frac{1}{3}\overrightarrow{AD}$
$= \frac{1}{3} \times 2\mathbf{e} = \frac{2}{3}\mathbf{e}$

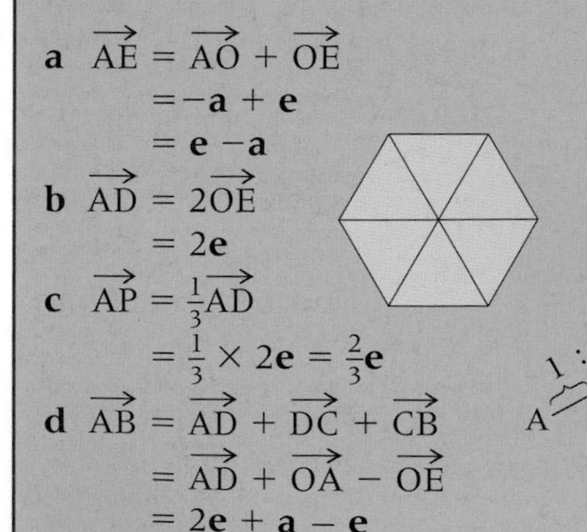

Ratio 1 : 2 means that P is $\frac{1}{3}$ of the way along AD.

d $\overrightarrow{AB} = \overrightarrow{AD} + \overrightarrow{DC} + \overrightarrow{CB}$
$= \overrightarrow{AD} + \overrightarrow{OA} - \overrightarrow{OE}$
$= 2\mathbf{e} + \mathbf{a} - \mathbf{e}$
$= \mathbf{a} + \mathbf{e}$

e $\overrightarrow{OB} = \overrightarrow{OA} + \overrightarrow{AB}$
$= \mathbf{a} + \mathbf{a} + \mathbf{e}$
$= 2\mathbf{a} + \mathbf{e}$

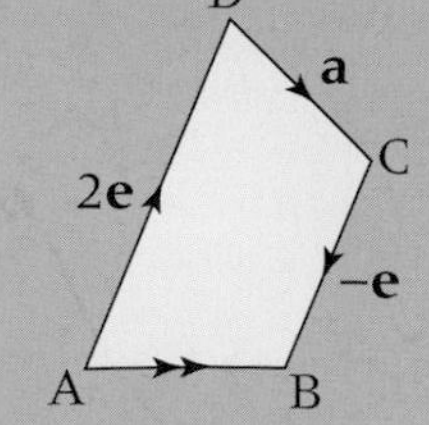

Using symmetry
$\overrightarrow{DC} = \overrightarrow{OA}$
$\overrightarrow{CB} = \overrightarrow{EO} = -\overrightarrow{OE}$

Exercise G6.4 Grade A*

1 WXYZ is a square.
The diagonals WY and XZ intersect at M.
$\overrightarrow{WX} = \mathbf{w}$ and $\overrightarrow{WZ} = \mathbf{z}$.
Write these vectors in terms of **w** and **z**.

a $\overrightarrow{XY}$ **b** $\overrightarrow{WY}$ **c** $\overrightarrow{WM}$
d $\overrightarrow{MY}$ **e** $\overrightarrow{MX}$ **f** $\overrightarrow{XM}$

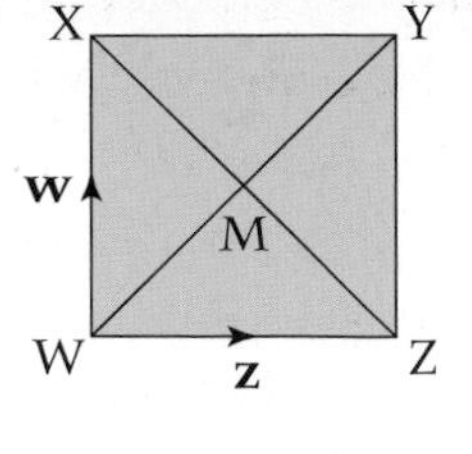

2 ABCDEFGH is a regular octagon.
$\overrightarrow{AB} = \mathbf{a}$ and $\overrightarrow{DE} = \mathbf{d}$.
Work out the vector, in terms of **a** and **d**, that represents

a $\overrightarrow{AH}$ **b** $\overrightarrow{FE}$
c $\overrightarrow{FD}$ **d** $\overrightarrow{HB}$

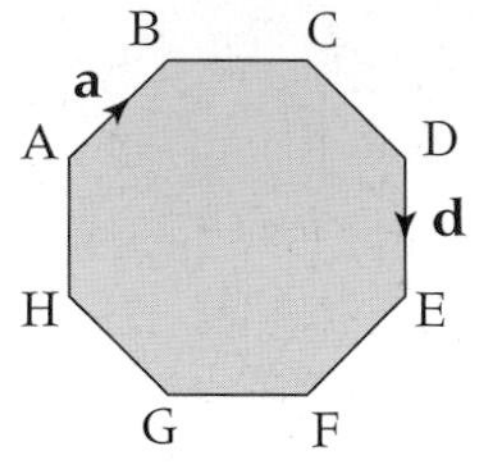

3 ABCDEF is a regular hexagon.
O is the centre of the hexagon.
$\overrightarrow{OB} = \mathbf{b}$ and $\overrightarrow{OC} = \mathbf{c}$.
Work out the vector, in terms of **b** and **c**, that represents

a $\overrightarrow{FC}$ **b** $\overrightarrow{BE}$
c $\overrightarrow{BC}$ **d** $\overrightarrow{AD}$

4 ORST is a rhombus.
$\overrightarrow{OR} = \mathbf{r}$ and $\overrightarrow{OT} = \mathbf{t}$.
P lies on OS such that OP : PS = 3 : 1.
Work out the vector, in terms of **r** and **t**, that represents

a $\overrightarrow{RS}$ **b** $\overrightarrow{OS}$ **c** $\overrightarrow{OP}$
d $\overrightarrow{SP}$ **e** $\overrightarrow{TP}$

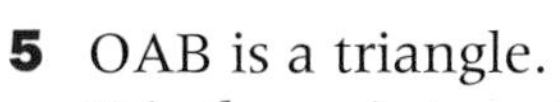

5 OAB is a triangle.
X is the point on AB for which AX : XB = 1 : 4.
$\overrightarrow{OA} = \mathbf{a}$ and $\overrightarrow{OB} = \mathbf{b}$.

a Write, in terms of **a** and **b**, an expression for $\overrightarrow{AB}$.

b Express $\overrightarrow{OX}$ in terms of **a** and **b**.
Give your answer in its simplest form.

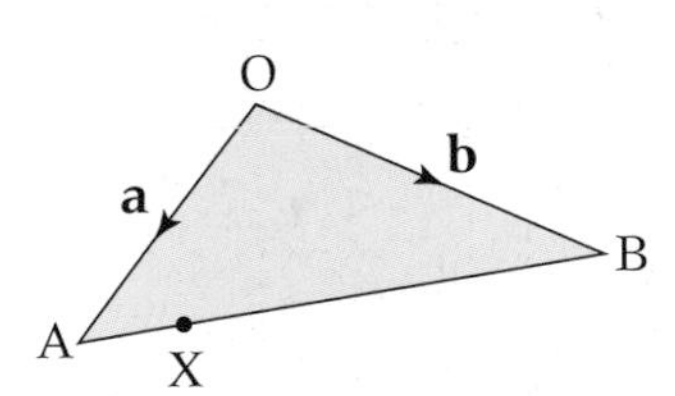

Proof using vectors

This spread will show you how to:
- Solve simple geometric problems in 2-D using vector methods

Keywords
Collinear
Parallel
Vector

You can use **vectors** in geometric proofs.
- To prove that lines are **parallel**, you show that the vectors they represent are multiples of each other.

Proof

OAB is a triangle.
M is the midpoint of OA.
N is the midpoint of OB.
$\overrightarrow{OA} = \mathbf{a}$ and $\overrightarrow{OB} = \mathbf{b}$.
Show that MN is parallel to AB.
So MN is parallel to AB.

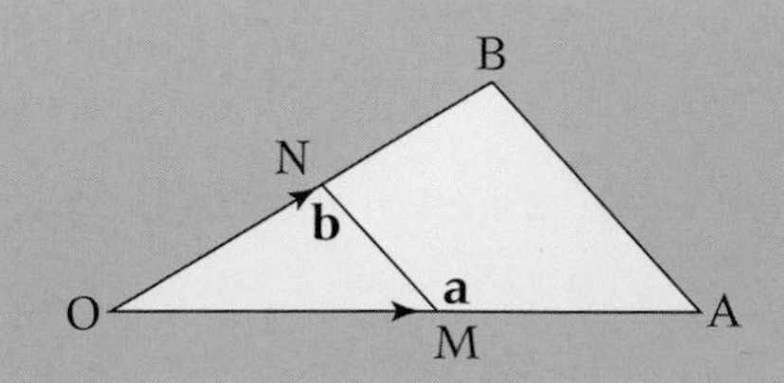

$\overrightarrow{AB} = \overrightarrow{AO} + \overrightarrow{OB}$
$= -\mathbf{a} + \mathbf{b}$
$\overrightarrow{MN} = \overrightarrow{MO} + \overrightarrow{ON}$
$= -\frac{1}{2}\mathbf{a} + \frac{1}{2}\mathbf{b}$
$= \frac{1}{2}(-\mathbf{a} + \mathbf{b})$
$-\mathbf{a} + \mathbf{b} = 2 \times \frac{1}{2}(-\mathbf{a} + \mathbf{b})$
So MN is parallel to AB.

Find the vectors $\overrightarrow{AB}$ and $\overrightarrow{MN}$.

$\overrightarrow{AB}$ is a multiple of $\overrightarrow{MN}$.

Always end your proof with a clear statement.

- To prove that points are **collinear**, you show that
  - the vectors joining pairs of the points are parallel
  - the vectors share a common point.

'Collinear' means 'lie on the same straight line'.

Proof

PQRS is an isosceles trapezium.
PQ and SR are parallel sides with PQ = 2 × SR.
$\overrightarrow{PQ} = 2\mathbf{p}$ $\quad \overrightarrow{QR} = \mathbf{q}$
X lies on QR such that QX : XR = 1 : 2.
Y lies on PQ extended such that PQ : QY = 1 : 1.
Prove that S, X and Y are collinear.

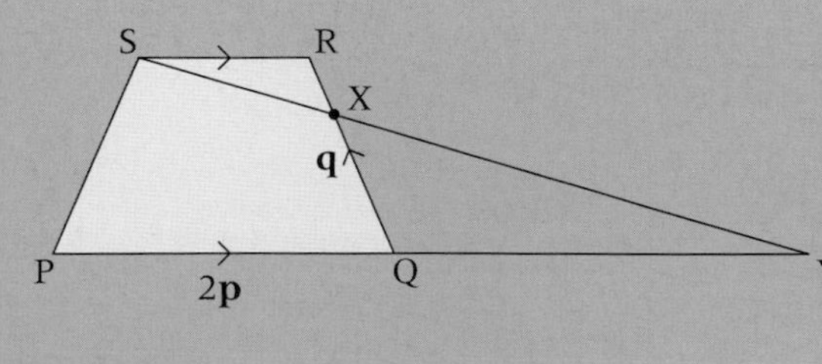

PQ = 2SR, so $\overrightarrow{SR} = \mathbf{p}$
$\overrightarrow{RX} = \frac{1}{3}\overrightarrow{RQ} = -\frac{1}{3}\mathbf{q}$

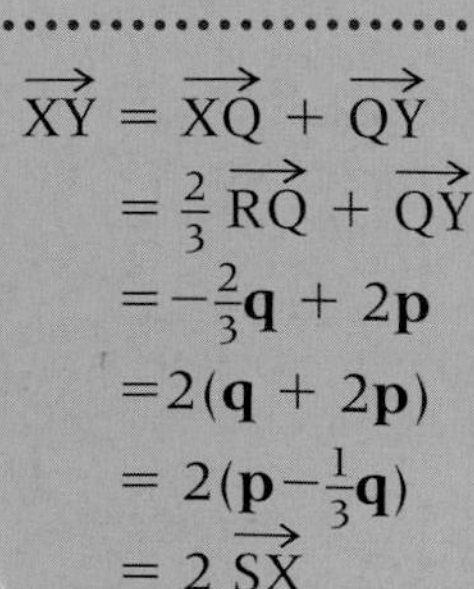

$\overrightarrow{XY} = \overrightarrow{XQ} + \overrightarrow{QY}$
$= \frac{2}{3}\overrightarrow{RQ} + \overrightarrow{QY}$
$= -\frac{2}{3}\mathbf{q} + 2\mathbf{p}$
$= 2(\mathbf{q} + 2\mathbf{p})$
$= 2(\mathbf{p} - \frac{1}{3}\mathbf{q})$
$= 2\,\overrightarrow{SX}$

$\overrightarrow{SX} = \overrightarrow{SR} + \overrightarrow{RX}$
$= \mathbf{p} - \frac{1}{3}\mathbf{q}$
So $\overrightarrow{SX}$ and $\overrightarrow{XY}$ are parallel.
SX and XY have the point X in common.

First show that SX and XY are parallel.

QY is PQ extended so it is parallel to PQ. PQ : QY = 1 : 1, so PQ are QY are the same length. Therefore $\overrightarrow{QY} = \overrightarrow{PQ} = 2\mathbf{p}$.

Exercise G6.5 Grade A*

So S, X and Y are collinear.

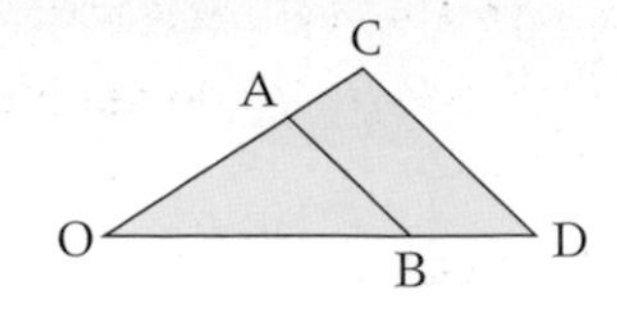

1 In the diagram, $\overrightarrow{OA} = 2\mathbf{a}$, $\overrightarrow{OB} = 2\mathbf{b}$, $\overrightarrow{OC} = 3\mathbf{a}$ and $\overrightarrow{BD} = \mathbf{b}$.
Prove that AB is parallel to CD.

2 RSTU is a rectangle.
M is the midpoint of the side RS.
N is the midpoint of the side ST.

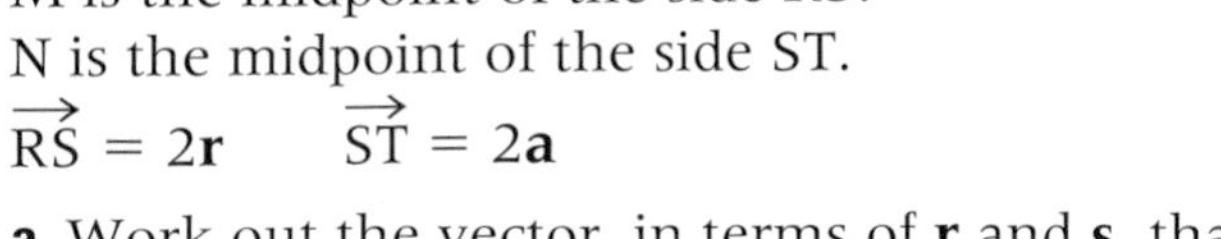

$\overrightarrow{RS} = 2\mathbf{r}$ $\quad\overrightarrow{ST} = 2\mathbf{a}$

a Work out the vector, in terms of **r** and **s**, that represents
i $\overrightarrow{RT}$ **ii** $\overrightarrow{RM}$ **iii** $\overrightarrow{SN}$ **iv** $\overrightarrow{MN}$.

b Show that MN is parallel to RT.

3 OPQR is a trapezium.
OP is parallel to RQ and OP = $\frac{1}{3}$RQ.
$\overrightarrow{OP} = \mathbf{p}$ and $\overrightarrow{OR} = \mathbf{r}$.

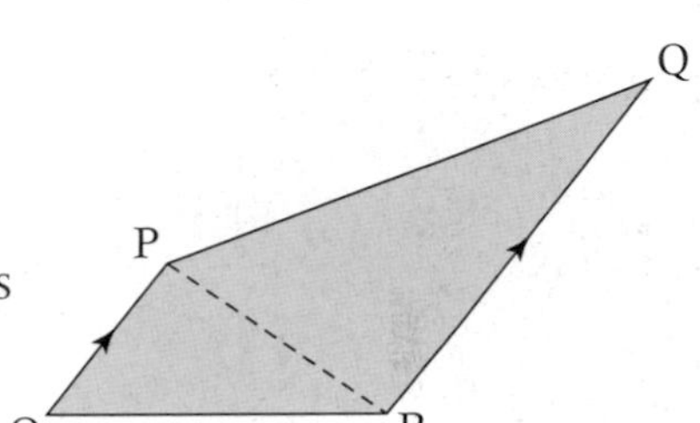

a Work out the vector, in terms of **p** and **r**, that represents
i $\overrightarrow{RQ}$ **ii** $\overrightarrow{OQ}$ **iii** $\overrightarrow{RP}$ **iv** $\overrightarrow{PQ}$

b The point X lies on PR such that PX : XR = 1 : 3.
Show that O, X and Q lie on the same straight line.

4 OJKL is a parallelogram.
$\overrightarrow{OJ} = \mathbf{j}$ $\quad\overrightarrow{JK} = \mathbf{k}$

a Express, in terms of **j** and **k** these vectors.
i $\overrightarrow{OK}$ **ii** $\overrightarrow{OL}$ **iii** $\overrightarrow{JL}$

b M is the point on KL extended such that KL : LM = 1 : 2.
X is a point on OL such that OX = $\frac{1}{3}$OL.

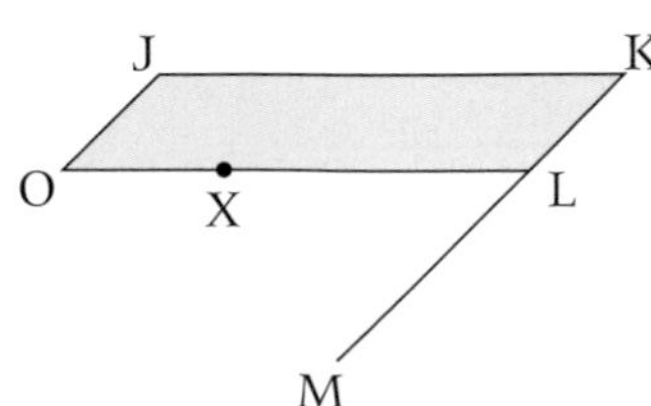

Show that J, X and M lie on the same straight line.

G6

Summary

Check out

You should now be able to:

- Understand and use vector notation for translations
- Calculate and represent graphically
  - **i** the sum of two vectors
  - **ii** the difference of two vectors
  - **iii** a scalar multiple of a vector
- Calculate the resultant of two vectors
- Understand and use the commutative and associative properties of vector addition
- Solve simple geometrical problems in 2-D using vector methods

Worked exam question

OABC is a parallelogram.
M is the midpoint of *CB*.
N is the midpoint of *AB*.
$\overrightarrow{OA} = \mathbf{a}$
$\overrightarrow{OC} = \mathbf{c}$

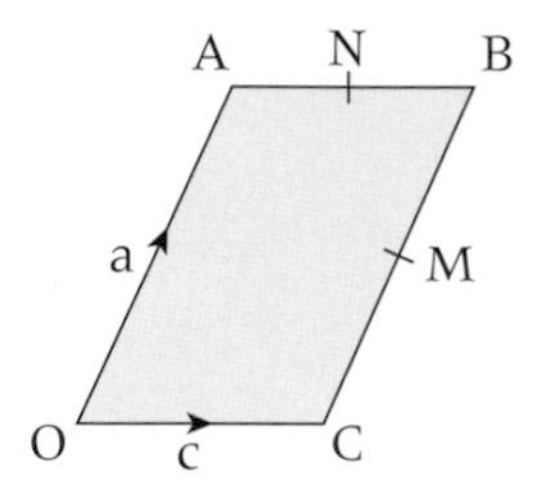

Diagram NOT accurately drawn

a Find, in terms of **a** and/or **c**, the vectors

i $\overrightarrow{MB}$,

ii $\overrightarrow{MN}$. (2)

b Show that *CA* is parallel to *MN*. (2)

(Edexcel Limited 2008)

a **i** $\overrightarrow{MB} = \frac{1}{2}\mathbf{a}$

ii $\overrightarrow{MN} = \frac{1}{2}\mathbf{a} - \frac{1}{2}\mathbf{c}$

The direction of the vectors is important.

b $\overrightarrow{CA} = \mathbf{a} - \mathbf{c}$

$\overrightarrow{MN} = \frac{1}{2}\mathbf{a} - \frac{1}{2}\mathbf{c}$

$= \frac{1}{2}(\mathbf{a} - \mathbf{c})$

$= \frac{1}{2}\overrightarrow{CA}$ So *CA* is parallel to *MN*

$\overrightarrow{CA} = 2\overrightarrow{MN}$

Exam questions

1

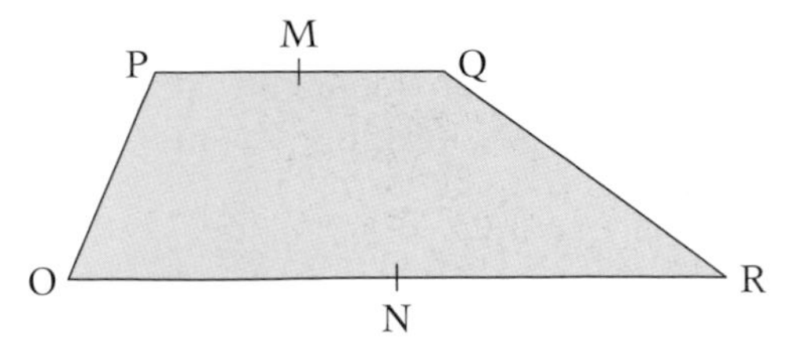

Diagram **NOT** accurately drawn

OPQR is a trapezium with PQ parallel to OR.

$\overrightarrow{OP} = 2\mathbf{b}$ $\overrightarrow{PQ} = 2\mathbf{a}$ $\overrightarrow{OR} = 6\mathbf{a}$

M is the midpoint of PQ and N the midpoint of OR.

a Find the vector $\overrightarrow{MN}$ in terms of **a** and **b**. (2)

X is the midpoint of MN and Y is the midpoint of QR

b Prove that XY is parallel to OR. (2)

(Edexcel limited 2005)

2

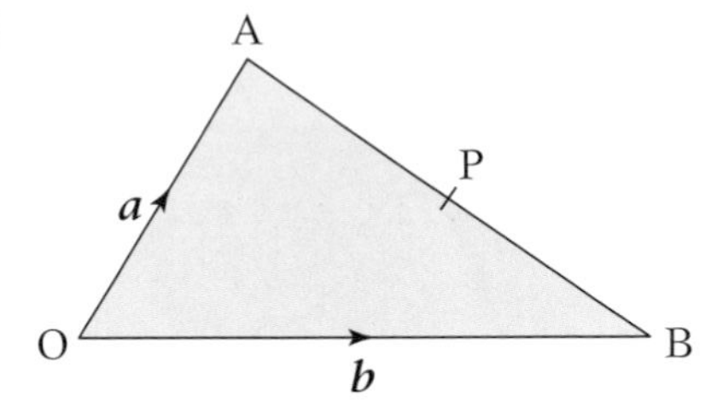

Diagram **NOT** accurately drawn

OAB is a triangle.

$\overrightarrow{OA} = \mathbf{a}$ $\overrightarrow{OB} = \mathbf{b}$

a Find the vector $\overrightarrow{AB}$ in terms of **a** and **b**. (1)

P is the point on AB such that AP : PB = 3 : 2

b Show that OP $= \frac{1}{5}(2\mathbf{a} + 3\mathbf{b})$ (3)

(Edexcel limited 2009)

A8 Transforming graphs and proof

When you transform graphs and shapes, being careful not to 'tear' or 'glue', certain properties remain the same. For example, if you stretch a graph horizontally then its number of maxima and minima stay the same, though their positions may change. Topology is the study of properties which are invariant under continuous transformations.

What's the point?

Ideas from topology have found applications to all sorts of networks, road, electricity, telephone, internet, *etc.* where they are used to design and optimise layouts.

Check in

You should be able to

- **recognise the characteristic shapes of graphs**

1 What shape would each graph have?

a $y = x^2 - 4x + 2$ **b** $y = 3x - 4$ **c** $y = x^3 - x$ **d** $y = \frac{2}{x}$

- **find the key features of graphs**

2 Where would each graph intersect

i the y-axis **ii** the x-axis?

a $y = x^2 + 6x + 8$ **b** $y = x^2 - 10x + 24$ **c** $y = x^2 + 5x + 6$

3 For the graphs in question **2**, what would be the coordinate of the minimum point?

- **identify and describe translations**

4 On what column vector would you need to move in order to move between these pairs of coordinates?

a (3, 5) and (5, 9) **b** (−2, 7) and (3, −5) **c** (−2, −5) and (−5, −8)

Orientation

What I need to know	What I will learn	What this leads to
G2 Apply translations	■ Translate and stretch graphs	**A-level** Maths, Physics
A7 Sketch graphs and recognise their characteristic shapes	■ Prove mathematical statements	Computer graphics, Reasoning

Rich task

The graph of $y = x^2$ is reflected in the x-axis.
To find the new coordinates of each vertex of the shape (image), simply multiply the y-coordinates by -1.
The new graph is the graph $y = -x^2$
Investigate rules for finding the image coordinates for this and other transformations of graphs. Write the equation of each new graph after it has been transformed.
You may wish to use quadratic and cubic graphs in your work.

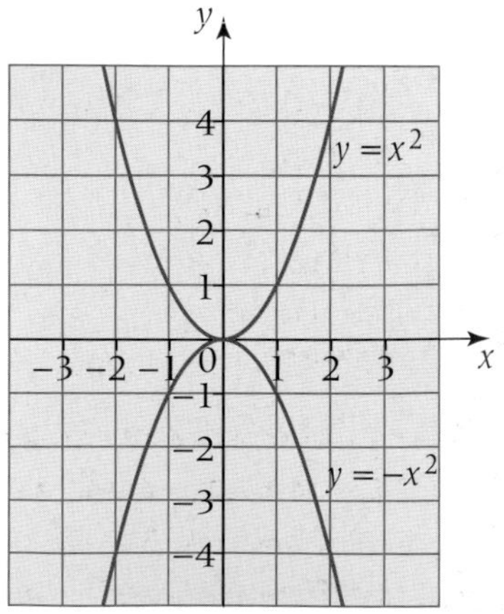

A8.1 Translating graphs vertically

This spread will show you how to:

- Transform a function f(x) to f(x) + a where a is a constant

Keywords
Column vector
Transform
Translation

- You can **transform** a function $f(x)$ to $f(x) + a$, by adding a constant.

For example, if $f(x) = x^2$ $\quad f(x) + 2 = x^2 + 2$
$\quad f(x) - 3 = x^2 - 3$

The diagram shows the graphs of these functions.

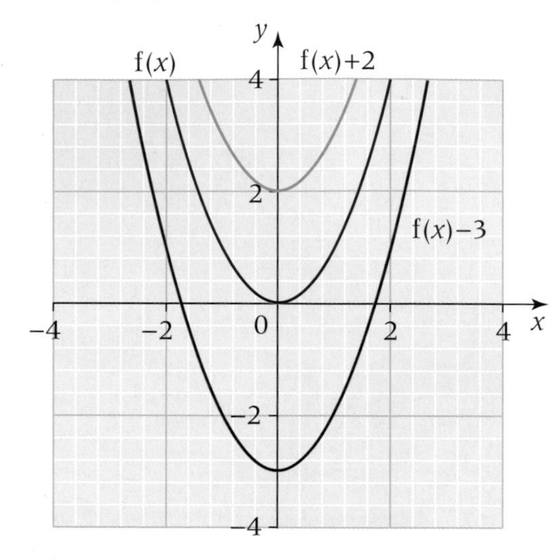

The graph of $y = f(x) + 2$ is the graph of $y = f(x)$ translated 2 units up, that is, translated by $\begin{pmatrix} 0 \\ 2 \end{pmatrix}$.

p.212

The graph of $y = f(x) - 3$ is the graph of $y = f(x)$ translated 3 units down, that is, translated by $\begin{pmatrix} 0 \\ -3 \end{pmatrix}$.

$\begin{pmatrix} 0 \\ 2 \end{pmatrix}$ is the **column vector** for the translation:
0 units in the x-direction
2 units in the y-direction.

- In general, $y = f(x) + a$ is a translation of $\begin{pmatrix} 0 \\ a \end{pmatrix}$ on $y = f(x)$.

Translation a units parallel to the y-axis.

Example

The graph shows the function $f(x)$, passing through points A (0, 8) and B (2, 4).
The graph is transformed to $y = f(x) - 5$.
What are the coordinates of the new points A and B?

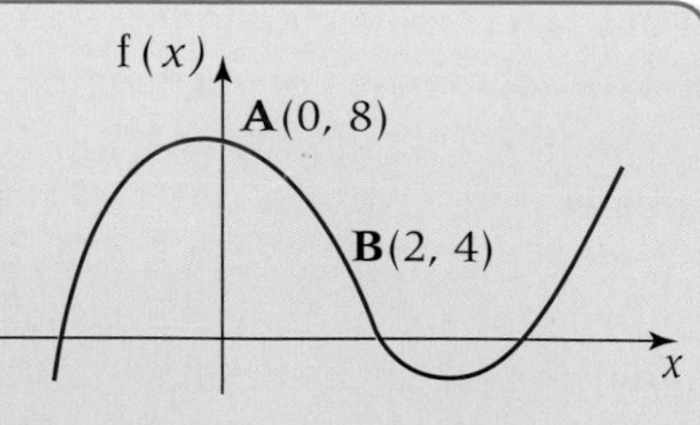

Each point will be translated by $\begin{pmatrix} 0 \\ -5 \end{pmatrix}$, that is, 5 units down.

A (0, 8) will move down to A (0, 3).
B (2, 4) will move down to (2, −1).

The translation is parallel to the y-axis, so the x-coordinate does not change.

Exercise A8.1 Grade B/A

1 The graph of f(x) is transformed. Match each of the transformed functions **a–d** with one of the graphs A–D.

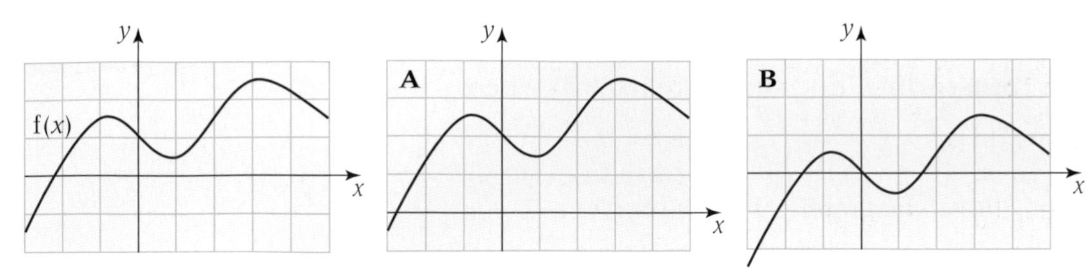

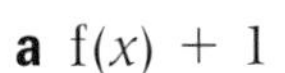
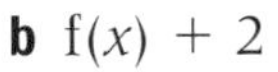

a f(x) + 1
b f(x) + 2
c f(x) − 2
d f(x) − 1

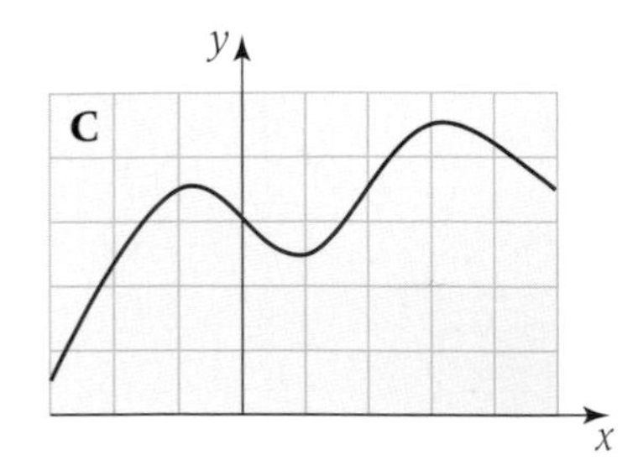

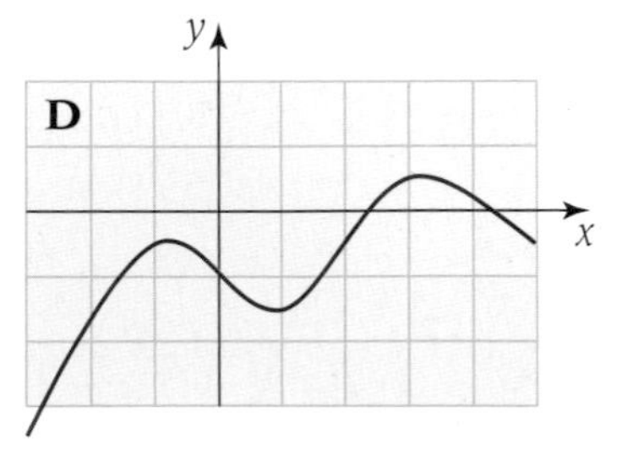

2 For each of parts **a–d**, sketch the graph of f(x) and the transformation on the same axes.

a
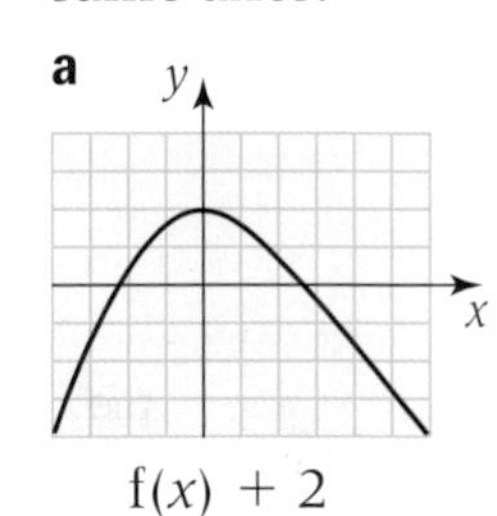

f(x) + 2

b
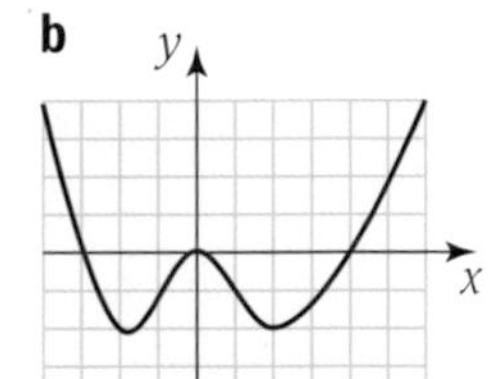

f(x) − 1

c
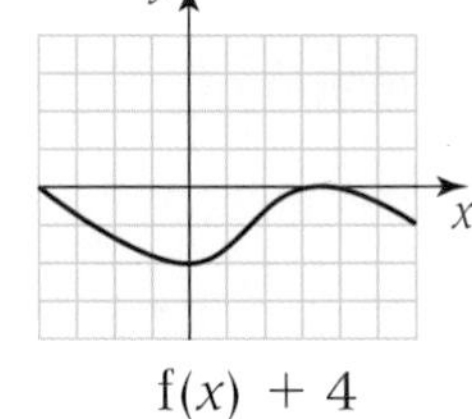

f(x) + 4

d
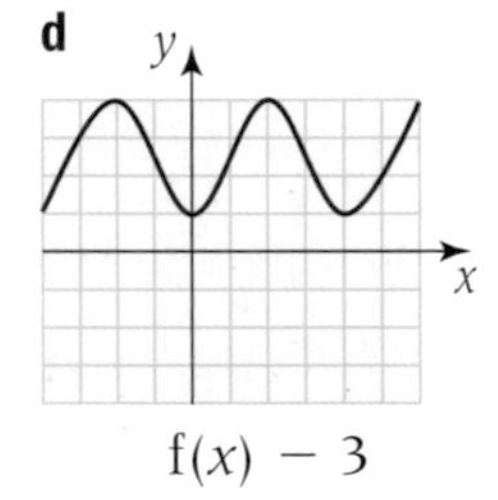

f(x) − 3

3 The graph shows the function f(x). Where would the points A, B and C be translated to under these transformations?

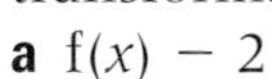

a f(x) − 2
b f(x) + 3
c f(x) − 5
d f(x) + 10
e 2 + f(x)

C(5,7)
B(0, 2)
A(−3, 0)

4 For each of the functions
i sketch the graph y = f(x)
ii sketch the transformed graph, labelling any y-intercepts with their coordinates.

a $f(x) = x^3$ transformed to $y = x^3 - 2$
b $f(x) = x^2$ transformed to $y = x^2 - 6.5$
c $f(x) = \frac{1}{x}$ transformed to $y = 3 + \frac{1}{x}$
d $f(x) = 2^x$ transformed to $y = 2^x - 2$

A8.2 Translating graphs horizontally

This spread will show you how to:

- Transform a function f(x) to f(x + a) where a is a constant

Keywords
Column vector
Transform
Translation

- You can **transform** a function $f(x)$ to $f(x + a)$ where a is a constant.

For example, if $f(x) = x^2$ $\quad f(x + 2) = (x + 2)^2$
$\quad f(x - 1) = (x - 1)^2$

The diagram shows the graphs of these functions.

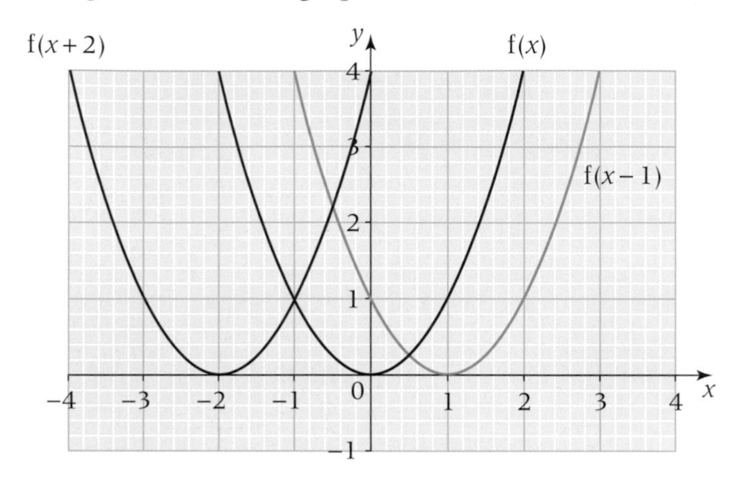

The graph of $y = f(x + 2)$ is the graph of $y = f(x)$ translated 2 units left, that is, translated by $\begin{pmatrix} -2 \\ 0 \end{pmatrix}$.

The graph of $y = f(x - 1)$ is the graph of $y = f(x)$ translated 1 unit right, that is, translated by $\begin{pmatrix} 1 \\ 0 \end{pmatrix}$.

$\begin{pmatrix} -2 \\ 0 \end{pmatrix}$ is the **column vector** for the translation: −2 units in the x-direction 0 units in the y-direction.

- In general, $y = f(x + a)$ is a translation of $\begin{pmatrix} -a \\ 0 \end{pmatrix}$ on $y = f(x)$.

Translation −a units parallel to the x-axis.

Example

The graph shows the function $y = g(x)$.
What will be the position of the point (2, 5) on the transformed graph $y = g(x + 3)$?
Sketch the transformed graph.

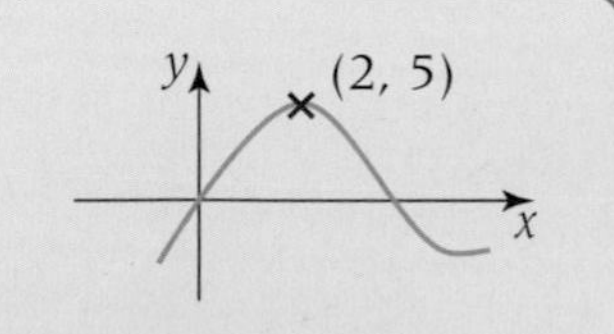

The transformation $g(x + 3)$ translates the graph of $g(x)$ − 3 units parallel to the x-axis, $\begin{pmatrix} -3 \\ 0 \end{pmatrix}$.

So the point (2, 5) will be translated to (−1, 5).

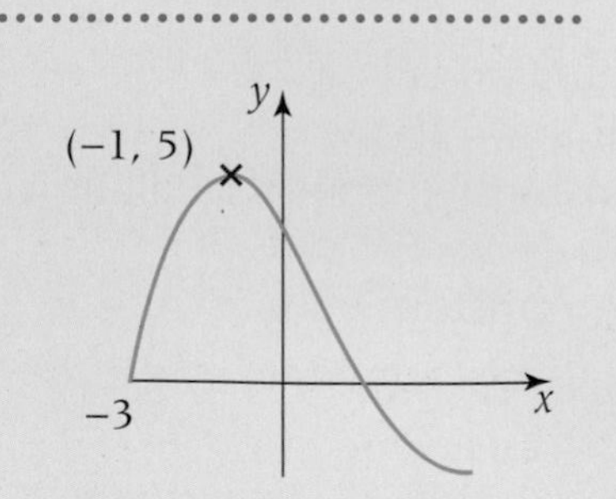

The y-coordinate remains unchanged as the function is only moved left.

Exercise A8.2 — Grade A/A*

1 The graph of f(x) is shown, with the graphs of different transformations of f(x). Match the graphs **a–c** with their equations.

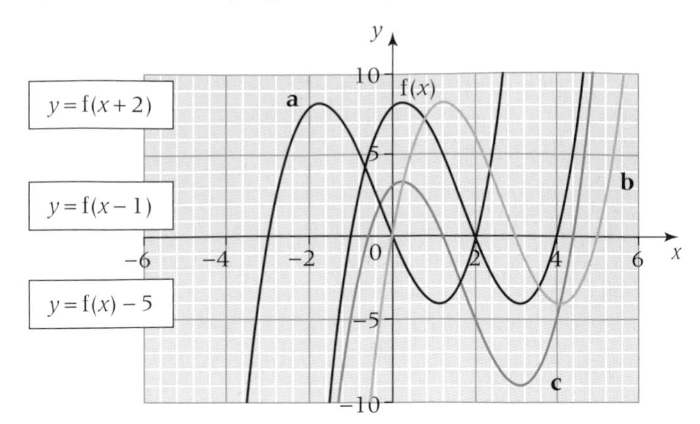

2 For each of parts **a–d**, sketch the graph of f(x) and the transformation on the same axes.

a

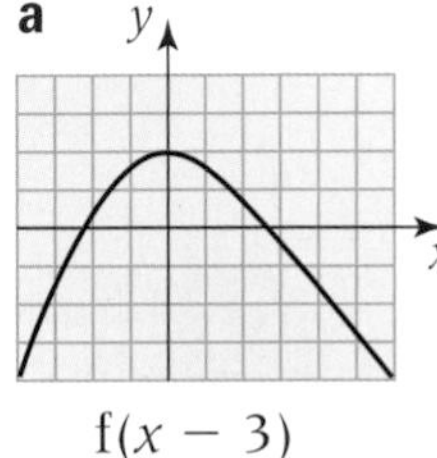

f(x − 3)

b

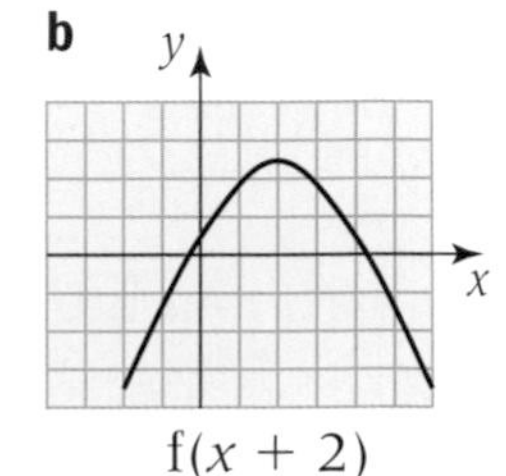

f(x + 2)

c

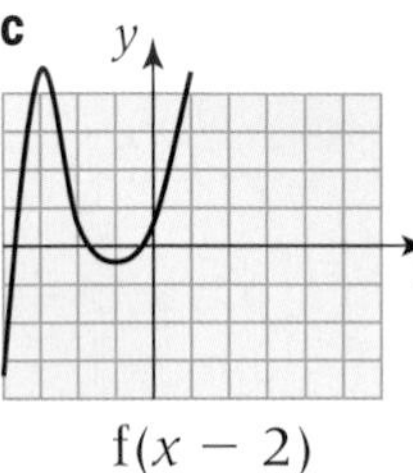

f(x − 2)

d

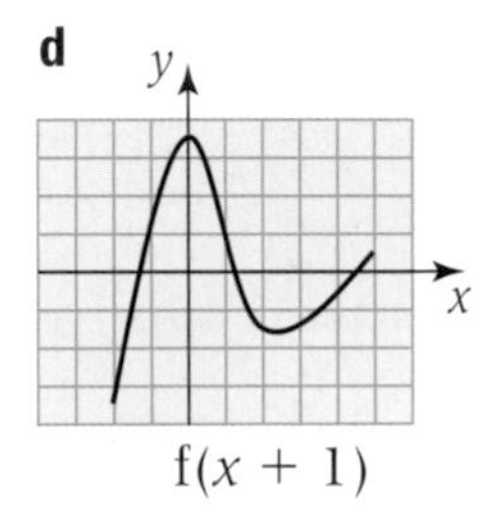

f(x + 1)

3 The graph shows the function f(x). Where would the points A, B and C be translated to under these transformations?

- **a** f(x − 2)
- **b** f(x + 3)
- **c** f(4 + x)
- **d** f(x) − 2
- **e** f(x) + 5

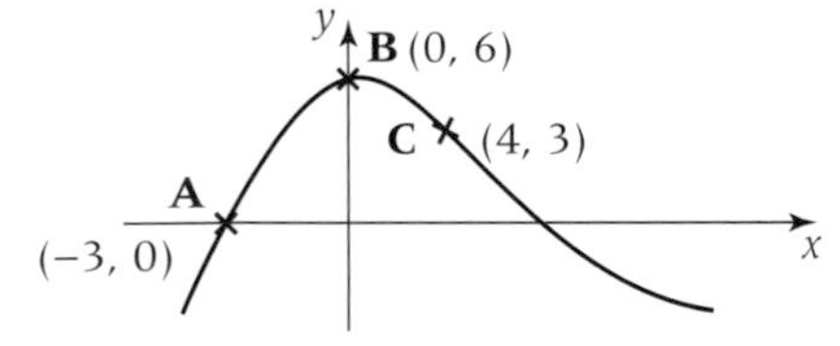

4 Write the equations of graphs **a–d** using function notation.

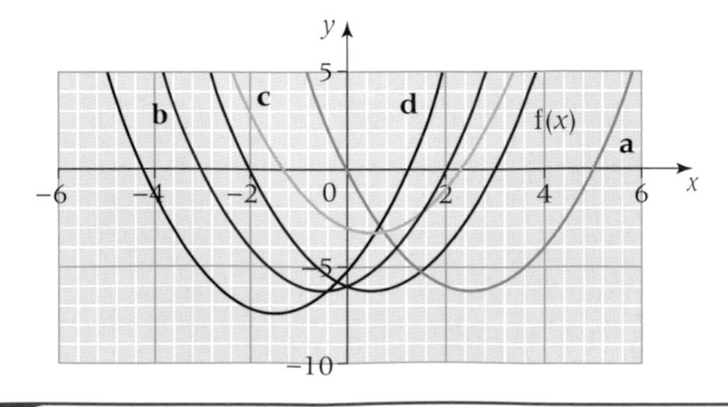

d requires two transformations.

AO3 Problem

5 The graph $y = x^2$ is transformed to $y = x^2 + 8x + 14$. By completing the square, find the column vector that achieves this transformation.

Stretching graphs

This spread will show you how to:

- Transform a function f(x) to $af(x)$ where a is a constant

Keywords
Scale factor
Stretch
Transform

- You can **transform** a function $f(x)$ to $af(x)$, by multiplying by a constant.

For example, if $f(x) = x^3$

$$2f(x) = 2x^3$$
$$10f(x) = 10x^3$$
$$\frac{1}{100}f(x) = \frac{1}{100}x^3$$

The diagram shows the graphs of these functions.

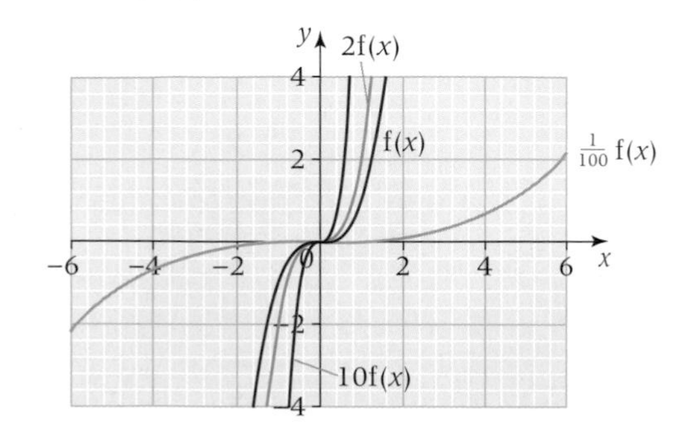

$y = 2f(x)$ is $f(x)$ **stretched** twice as far parallel to the y-axis, that is by a **scale factor** of 2.

$y = 10f(x)$ is $f(x)$ stretched by a scale factor of 10 parallel to the y-axis.

$y = \frac{1}{100}f(x)$ is $f(x)$ stretched by a scale factor of $\frac{1}{100}$ parallel to the y-axis.

(2, 8) becomes (2, 16).

(2, 8) becomes (2, 80).

(2, 8) becomes (2, 0.08). The graph is flatter than the original.

- In general, $y = af(x)$ is a stretch parallel to the y-axis on $f(x)$ by a scale factor of a.

Example

The graph of $f(x)$ is shown. Curves A, B and C are the graphs of transformations of $f(x)$. Express the functions for A, B and C in terms of $f(x)$.

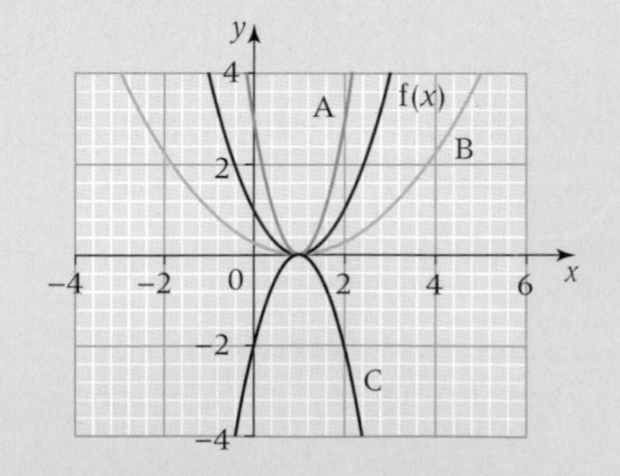

A: (2, 1) becomes (2, 3) so $f(x)$ is stretched by a scale factor of 3.
$y = 3f(x)$

B: (2, 1) becomes $\left(2, \frac{1}{4}\right)$ so $f(x)$ is stretched by a scale factor of $\frac{1}{4}$.
$y = \frac{1}{4}f(x)$

C: (2, 1) becomes (2, −2) so $f(x)$ is stretched by a scale factor of −2.
$y = -2f(x)$

$y = -f(x)$ is a reflection in the x-axis.

Under a stretch parallel to the y-axis, points on the x-axis remain unchanged since $a \times 0 = 0$.

Stretched by scale factor 2 in the opposite direction.

Exercise A8.3

Grade B/A

1 The graph of $f(x)$ is shown in the diagram, with the graphs of different transformations of $f(x)$. Match the graphs **a–c** with their equations.

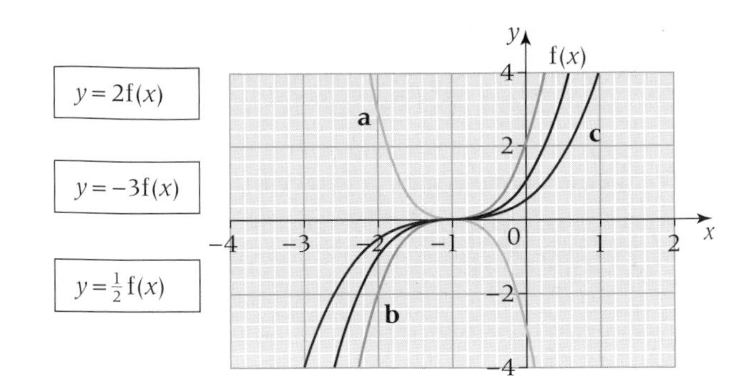

2 For each of **a–d**, sketch the graph of $f(x)$ and the transformation on the same axes.

a

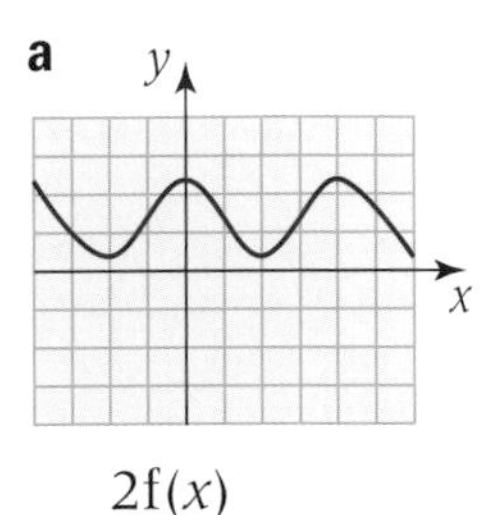

$2f(x)$

b

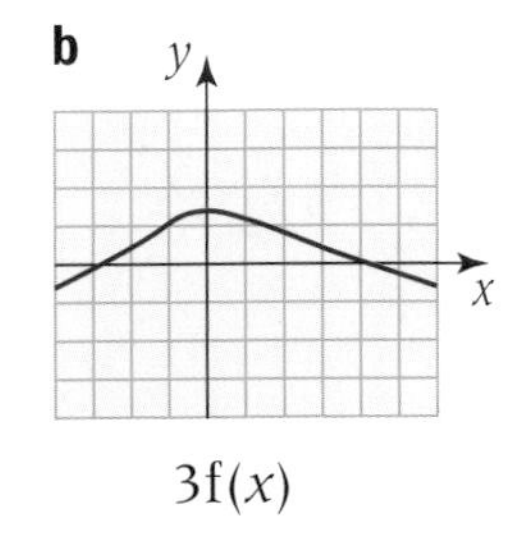

$3f(x)$

c

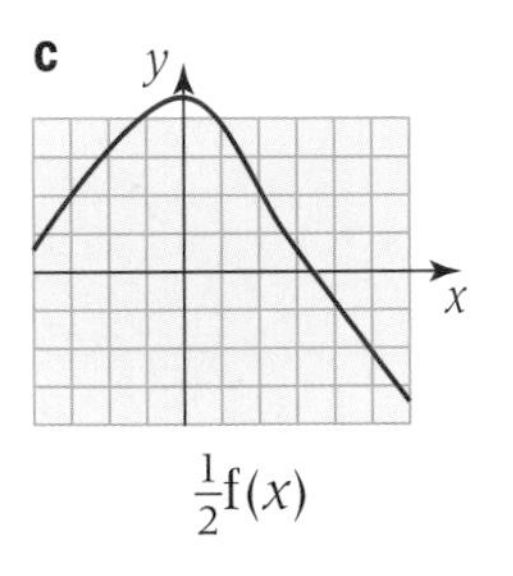

$\frac{1}{2}f(x)$

d

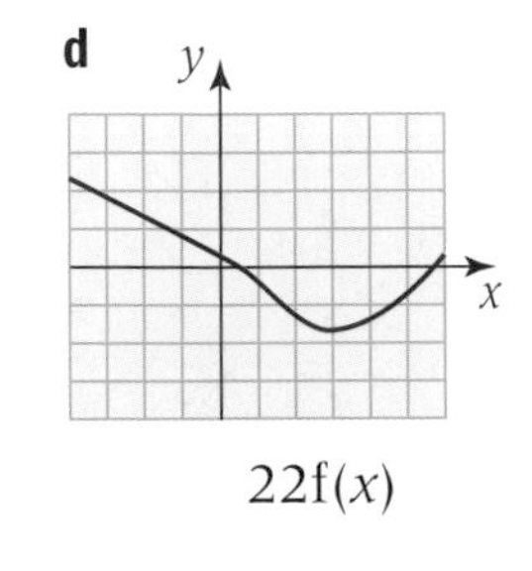

$22f(x)$

3 Copy and complete these function boxes for the functions that have been transformed in some way.

Original function:	New function:
$2 \rightarrow$ $f(x)$ $\rightarrow 6$	$2 \rightarrow$ $3f(x)$ $\rightarrow$?
$-4 \rightarrow$ $g(x)$ $\rightarrow -2$	$-4 \rightarrow$ $\frac{1}{2}g(x)$ $\rightarrow$?

4 The graph shows the function $f(x)$. Where would the points A, B and C be translated to under these transformations?

a $y = 2f(x)$ **b** $y = 10f(x)$
c $y = -3f(x)$ **d** $y = \frac{1}{2}f(x)$
e $y = 2f(x) + 1$ **f** $y = 4f(x - 2)$

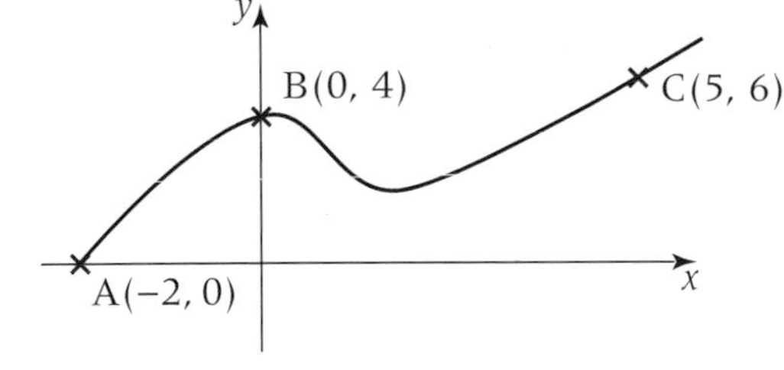

5 Describe the transformation $y = -f(x)$.

6 **a** Sketch the graph of $f(x) = x^2 + 8x + 12$, showing the y-axis intercept, the x-axis intercepts and the coordinate of the minimum point.

b On the same axes, sketch $y = 3f(x)$, showing how the coordinates of the points you had labelled now change.

c Repeat for $y = x^2 - 4x + 1$ and the transformation $y = -f(x)$.

A8.4 Further stretching graphs

This spread will show you how to:

- Transform a function f(x) to f(ax) where a is a constant

Keywords
Scale factor
Stretch
Transform

- You can **transform** a function $f(x)$ to $f(ax)$ where a is a constant.

For example, if $f(x) = x^2$ $\quad f\left(\frac{1}{2}x\right) = \frac{1}{4}x^2$

$f(2x) = 4x^2$

The diagram shows the graphs of these functions.

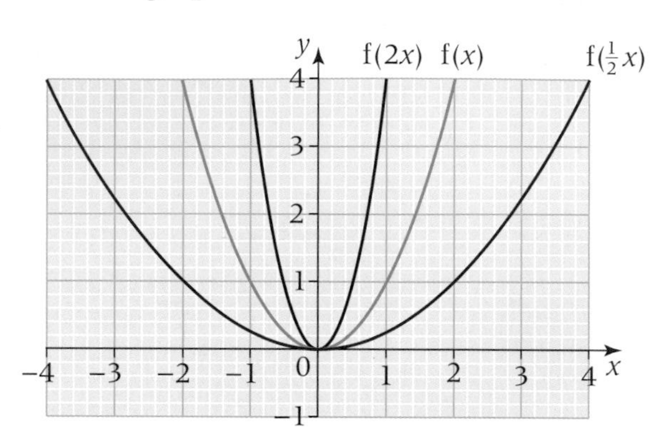

You call these a stretch, even though the graph is squashed.

$y = f\left(\frac{1}{2}x\right)$ is $f(x)$ **stretched** twice as far parallel to the x-axis, that is by a **scale factor** of 2.

$y = f(2x)$ is $f(x)$ stretched by a scale factor of $\frac{1}{2}$ parallel to the x-axis.

(2, 4) becomes (4, 4).

(2, 4) becomes (1, 4).

- In general, $y = f(ax)$ is a stretch parallel to the x-axis by a scale factor of $\frac{1}{a}$.

$y = f(-x)$ is a reflection in the y-axis.

Example

The graph shows the function $f(x)$. What will be the position of each point shown on the transformed graph $y = f(2x)$? Sketch the transformed graph.

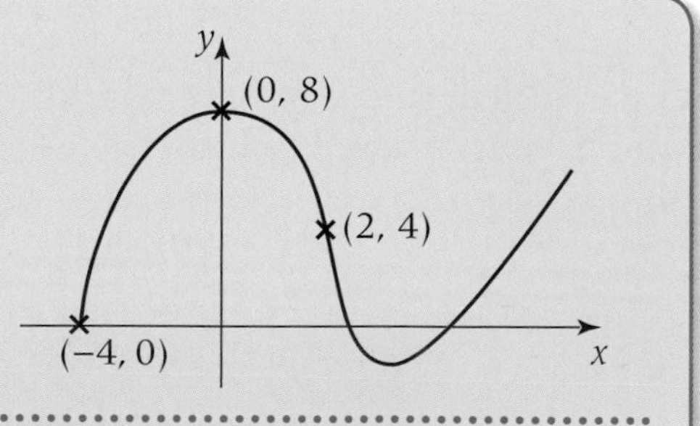

Each x-coordinate will be stretched by a scale factor of $\frac{1}{2}$ parallel to the x-axis. Hence,

(−4, 0) becomes (−2, 0)
(0, 8) remains (0, 8)
(2, 4) becomes (1, 4).

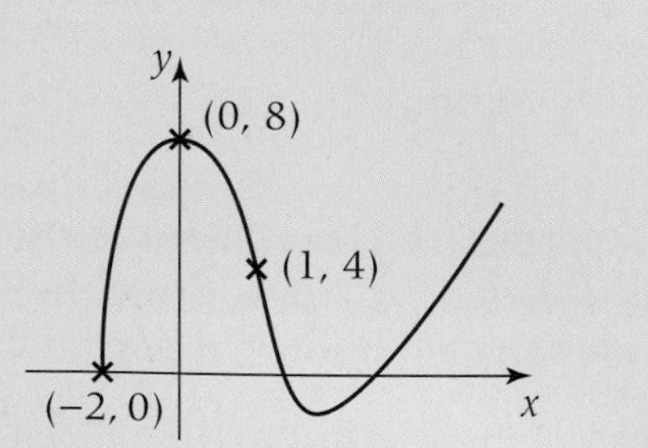

The y-coordinate remains unchanged as the graph only moves parallel to the x-axis.

Exercise A8.4 — Grade A/A*

1 The graph of the function $f(x)$ is shown. Match the transformed graphs **a–c** with their equations.

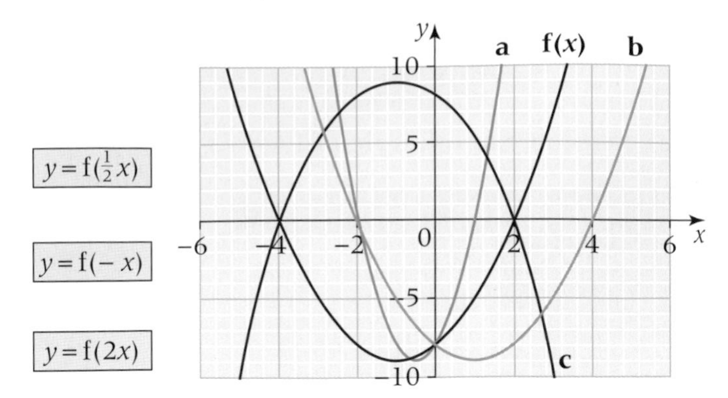

2 For each of **a–d**, sketch the graph of $f(x)$ and the transformed function on the same axes.

a $f(2x)$

b $f(\frac{1}{4}x)$

c $f(-x)$

d

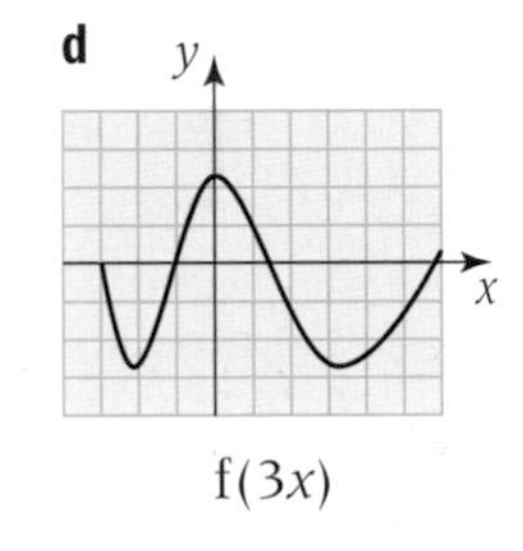

$f(3x)$

3 Copy and complete these function boxes for the functions that have been transformed in some way.

Original function:	New function:
2 → $f(x)$ → 6	? → $f(2x)$ → 6
−4 → $g(x)$ → −2	? → $g(\frac{1}{4}x)$ → −2

4 The graph shows the function $f(x)$. Where will the points A, B and C be translated to under these transformations?

a $f(2x)$
b $f(\frac{1}{2}x)$
c $f(4x)$
d $f(-x)$
e $f(2x) + 5$

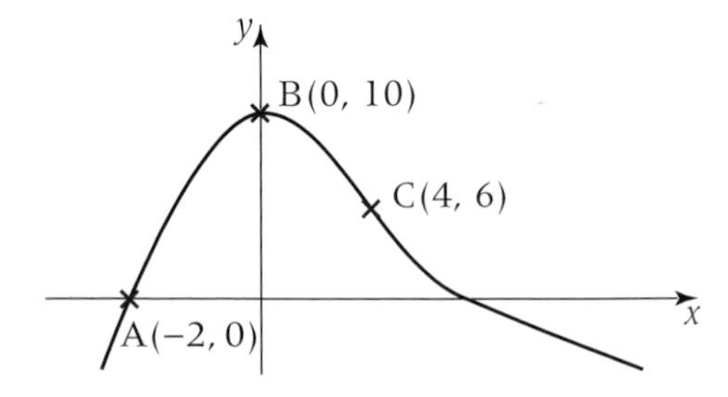

5 Given that $f(x) = 2^x$, sketch $f(x)$ and $f(-2x)$ on the same axes. Give the equation of the second sketch in terms of x.

A8.5 Proof

This spread will show you how to:

- Understand the difference between a practical demonstration and a proof
- Derive proofs using short chains of deductive reasoning
- Use counter-examples to show that a statement is false

Keywords
Counter-example
Demonstrate
Prove

- To demonstrate that a statement is true, you find examples that fit the statement.
- To **prove** that a statement is true you can use algebra to generalise it to all possible examples.
- To show that a statement is false, you can use a **counter-example**.

Some useful algebraic generalisations:	
Even numbers	$2n$
Odd numbers	$2n + 1$
Consecutive even	$2n$, $2n + 2$
Consecutive odd	$2n + 1$, $2n + 3$

To disprove the statement 'All cube numbers are even', you could use the counter-example 125 ($=5^3$) is odd.

Example

'The square of an odd number is always odd.'

a Demonstrate this statement with an example.

b Use algebra to prove this statement.

a 3 is odd and $3^2 = 9$ is odd.

b Any odd number is one more than an even number, so you can represent any odd number as $2n + 1$, where n is an integer.

$(2n + 1)^2 = 4n^2 + 4n + 1$

$= 2(2n^2 + 2n) + 1$

For any integer n, $(2n^2 + 2n)$ is an integer, so

$(2n + 1)^2 = 2 \times \textit{integer} + 1$, which is an odd number.

Find an expression for the square of an odd number.

Example

Prove that, for any three consecutive integers, the difference between the product of the first two and the product of the last two is always twice the middle number.

For example, for 6, 7, 8: $(8 \times 7) - (7 \times 6) = 56 - 42 = 14 = 2 \times 7$

Writing three consecutive integers as $n - 1$, n, $n + 1$:

$n(n + 1) - n(n - 1) = n^2 + n - n^2 + n$

$= 2n$

$=$ twice the middle number

It may help to demonstrate the statement to yourself using an example.

Example

$y = 2x^2 + 11$ The value of y is prime when $x = 0, 1, 2$ or 3.

The following statement is *not* true:

'$y = 2x^2 + 11$ is *always* a prime number when x is an integer'.

Show that the statement is not true.

For $x = 11$, $y = 2x^2 + 11 = 2 \times 11^2 + 11 = 253$

253 is not prime, since $253 = 23 \times 11$.

Therefore the statement is not true.

Try different values of x until you find a counter-example.

Exercise A8.5 — Grade A*

1 'For any three consecutive integers, the difference between the sum of the first two and the sum of the last two is always two.'
  a Demonstrate this statement with an example.
  b Letting the three consecutive integers be n, $n + 1$ and $n + 2$, write:
    i an expression for the sum of the first two
    ii an expression for the sum of the last two.
  c Using your expressions from part **b**, prove that the statement given is always true.

2 Find a counter-example to disprove each of these statements.
  a All square numbers are even.
  b The difference between two cube numbers is never odd.
  c Squaring a number will always give you a value greater than the number you started with.
  d $(a + b)^2 \neq a^2 + b^2$ for any values of a and b.
  e $\sin(A + B) \neq \sin A + \sin B$ for any values of A or B.
  f The value of $x^2 - 11x + 121$ is never a square number for any value of x.

$\neq$ means 'not equal to'.

3 Use algebraic generalisations to prove these statements.
  a For any three consecutive integers, the square of the middle integer is always one more than the product of the other two.
  b The product of two consecutive odd numbers is always one less than a multiple of four.
  c For any three consecutive integers, the difference between the square of the middle integer and the product of the other two is always equal to 1.

Unit 3

A03 Problem

4 Prove that, when two ordinary dice are rolled, the sum of these four products is always equal to 49.
- the product of the numbers on the top faces
- the product of the numbers on the bottom faces
- the product of the number on the top of dice A and the number on the bottom of dice B
- the product of the number on the bottom of dice A and the number on the top of dice B.

A B

Investigate the relationship between numbers on opposite faces of a dice.

5 Prove that
'The square of the mean of five consecutive integers differs from the mean of the squares of the same five consecutive integers by two.'

A8.6 Further proof

This spread will show you how to:

- Derive proofs using short chains of deductive reasoning

Keywords
Prove

- You can **prove** results from any area of mathematics by generalising using algebra.

Example

Show that these triangles are similar *only* when x is equal to 3.

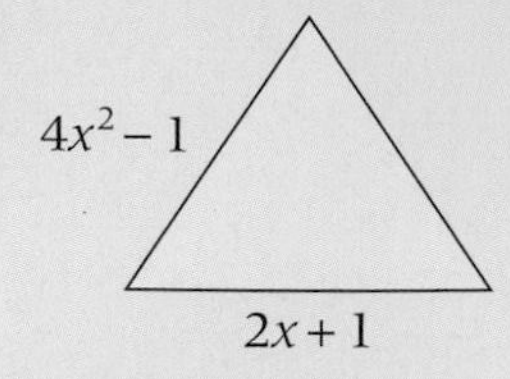

$x^2 + 5x + 6$

$x + 3$

p222

For the triangles to be similar, pairs of sides must be in the same ratio, that is

$$\frac{4x^2 - 1}{2x + 1} = \frac{x^2 + 5x + 6}{x + 3}$$

$$\frac{(2x - 1)(2x + 1)}{2x + 1} = \frac{(x + 2)(x + 3)}{x + 3}$$

$$2x - 1 = x + 2$$

$$x = 3$$

Consider how you would tackle a numerical example, for example

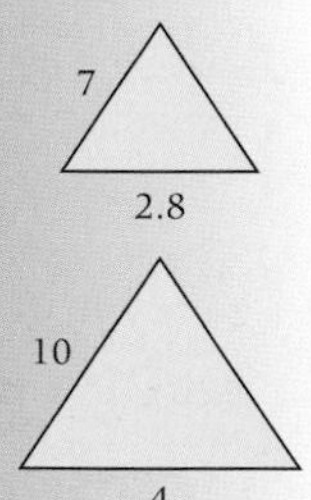

10

4

You would show that $\frac{10}{4} = \frac{7}{2.8}$

Example

The distance of the point P from the origin(0, 0) is the same as the distance of the point P from the line $y = -2$.

Show that $y = \frac{1}{4}x^2 - 1$.

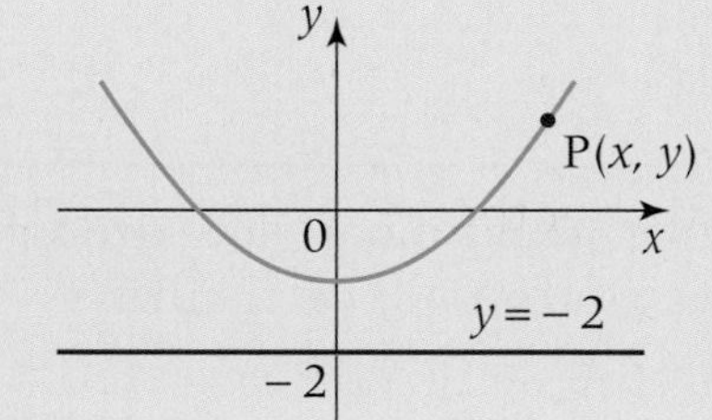

p.294

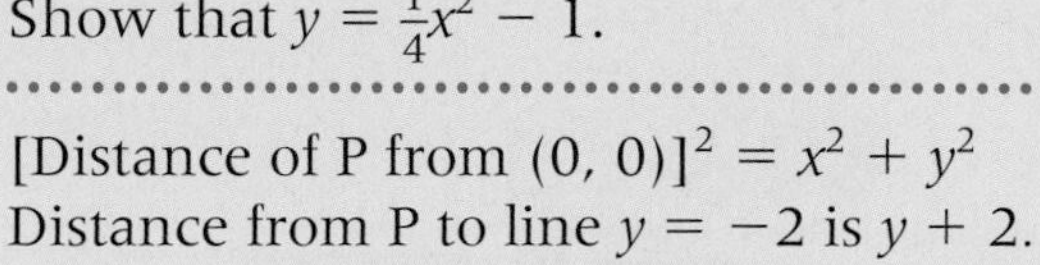

[Distance of P from (0, 0)]$^2 = x^2 + y^2$

Distance from P to line $y = -2$ is $y + 2$.

Since two distances are equal

$x^2 + y^2 = (y + 2)^2$

$x^2 + y^2 = y^2 + 4y + 4$

$x^2 = 4y + 4$

So, $y = \frac{1}{4}x^2 - 1$ as required

Find the distance from P to (0, 0) using Pythagoras:

P(x, y,)
y
(0, 0) x

Exercise A8.6

Grade A*

1 Given that this triangle is right-angled, prove that $x^2 - 6x - 39 = 0$.

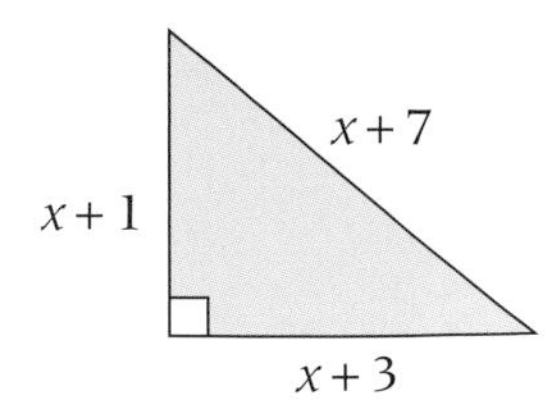

2 Given that n is a positive integer, prove that the perimeter of this triangle will always be even.

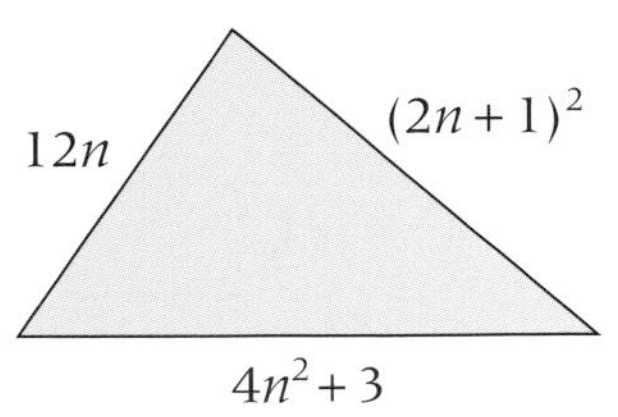

3 The formula $d = \sqrt{(x_2-x_1)^2 + (y_2-y_1)^2}$ can be used to find the distance between two points (x_1, y_1) and (x_2, y_2).
Show that the triangle joining A(2, 3) to B(4, 10) to C(7, 5) is scalene.

4 **a** Write a formula for the area of this trapezium.

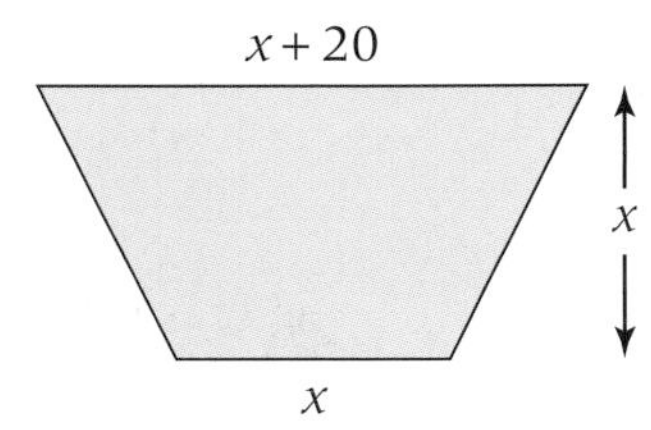

b Show that the area of the trapezium is 24 units^2 *only* when $x = 2$.

A03 Problem

5 **a** A bucket contains m red balls and 6 blue balls.
Write an expression for the probability that when I pick a ball from the bucket it will be red.

b I pick a second ball without replacing the first one.
Write an expression for the probability that the second ball I pick is blue given that the first was red.

c Given that the probability that the two balls I picked in parts **a** and **b** were both red is $\frac{2}{11}$, show that $3m^2 - 11m = 20$.

d Hence, find the number of red balls in the bucket to start with.

6 Show that it is not possible for this triangle to be right-angled.

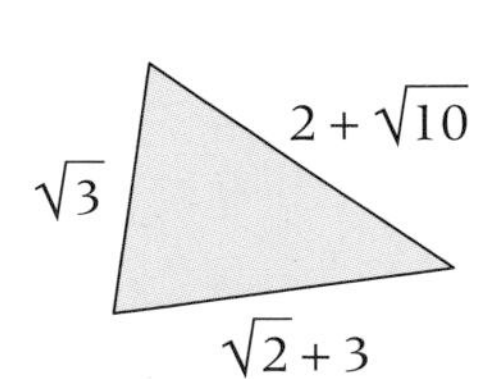

Summary

Check out

You should now be able to:

- Transform the function f(x) to **a** $f(x) + a$
  b $f(x + a)$
  c $af(x)$
  d $f(ax)$, where a is a constant
  for linear, quadratic, sine and cosine functions
- Understand the difference between a practical demonstration and a proof
- Use a counter-example to show a statement is false

Worked exam question

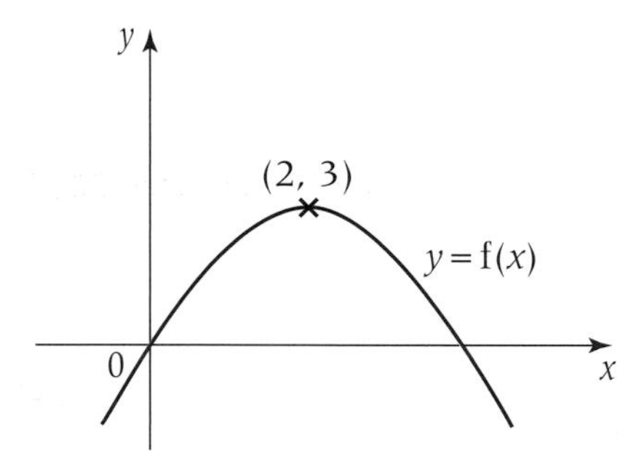

The diagram shows part of the curve with equation $y = f(x)$.
The coordinates of the maximum point of this curve are (2, 3).

Write down the coordinates of the maximum point of the curve with equation

a $y = f(x - 2)$ (1)

b $y = 2f(x)$ (1)

(Edexcel Limited 2008)

a (4, 3)

$y = f(x - 2)$ is a translation of $\begin{pmatrix} 2 \\ 0 \end{pmatrix}$ on $y = f(x)$

b (2, 6)

$y = 2f(x)$ is a stretch parallel to the y-axis on $y = f(x)$ by scale factor 2.

Exam questions

1 The graph of $y = f(x)$ is shown on the grid.
On an accurate copy of this diagram

a sketch the graph of $y = f(x) + 2$ (2)

b sketch the graph of $y = -f(x)$ (2)

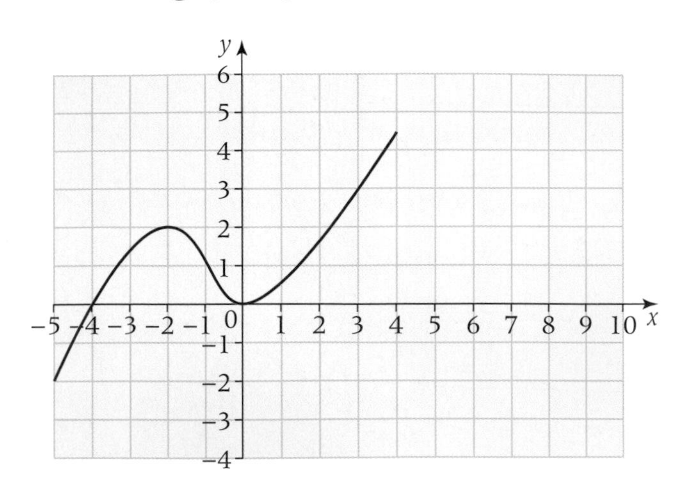

(Edexcel Limited 2007)

2 The nth even number is $2n$.
The next even number after $2n$ is $2n + 2$.

a Explain why (1)

b Write down an expression, in terms of n, for the next even number after $2n + 2$. (1)

c Show algebraically that the sum of any 3 consecutive even numbers is always a multiple of 6. (3)

(Edexcel Limited 2008)

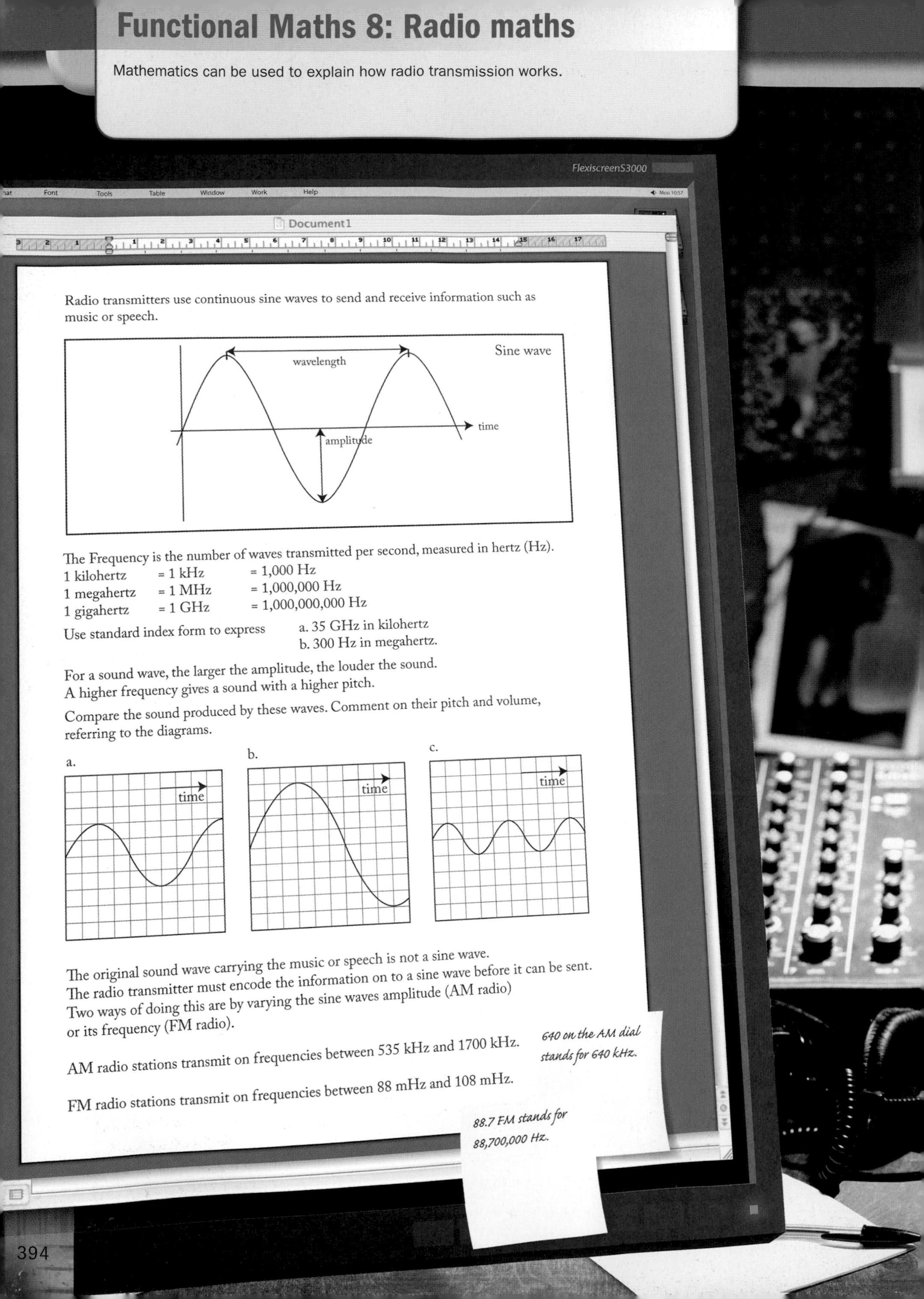

Functional Maths 8: Radio maths

Mathematics can be used to explain how radio transmission works.

Radio transmitters use continuous sine waves to send and receive information such as music or speech.

The Frequency is the number of waves transmitted per second, measured in hertz (Hz).

1 kilohertz = 1 kHz = 1,000 Hz
1 megahertz = 1 MHz = 1,000,000 Hz
1 gigahertz = 1 GHz = 1,000,000,000 Hz

Use standard index form to express
a. 35 GHz in kilohertz
b. 300 Hz in megahertz.

For a sound wave, the larger the amplitude, the louder the sound.
A higher frequency gives a sound with a higher pitch.

Compare the sound produced by these waves. Comment on their pitch and volume, referring to the diagrams.

The original sound wave carrying the music or speech is not a sine wave.
The radio transmitter must encode the information on to a sine wave before it can be sent.
Two ways of doing this are by varying the sine waves amplitude (AM radio) or its frequency (FM radio).

AM radio stations transmit on frequencies between 535 kHz and 1700 kHz.

640 on the AM dial stands for 640 kHz.

FM radio stations transmit on frequencies between 88 mHz and 108 mHz.

88.7 FM stands for 88,700,000 Hz.

Wave speed (m/s) = frequency (Hz) × wavelength (m)

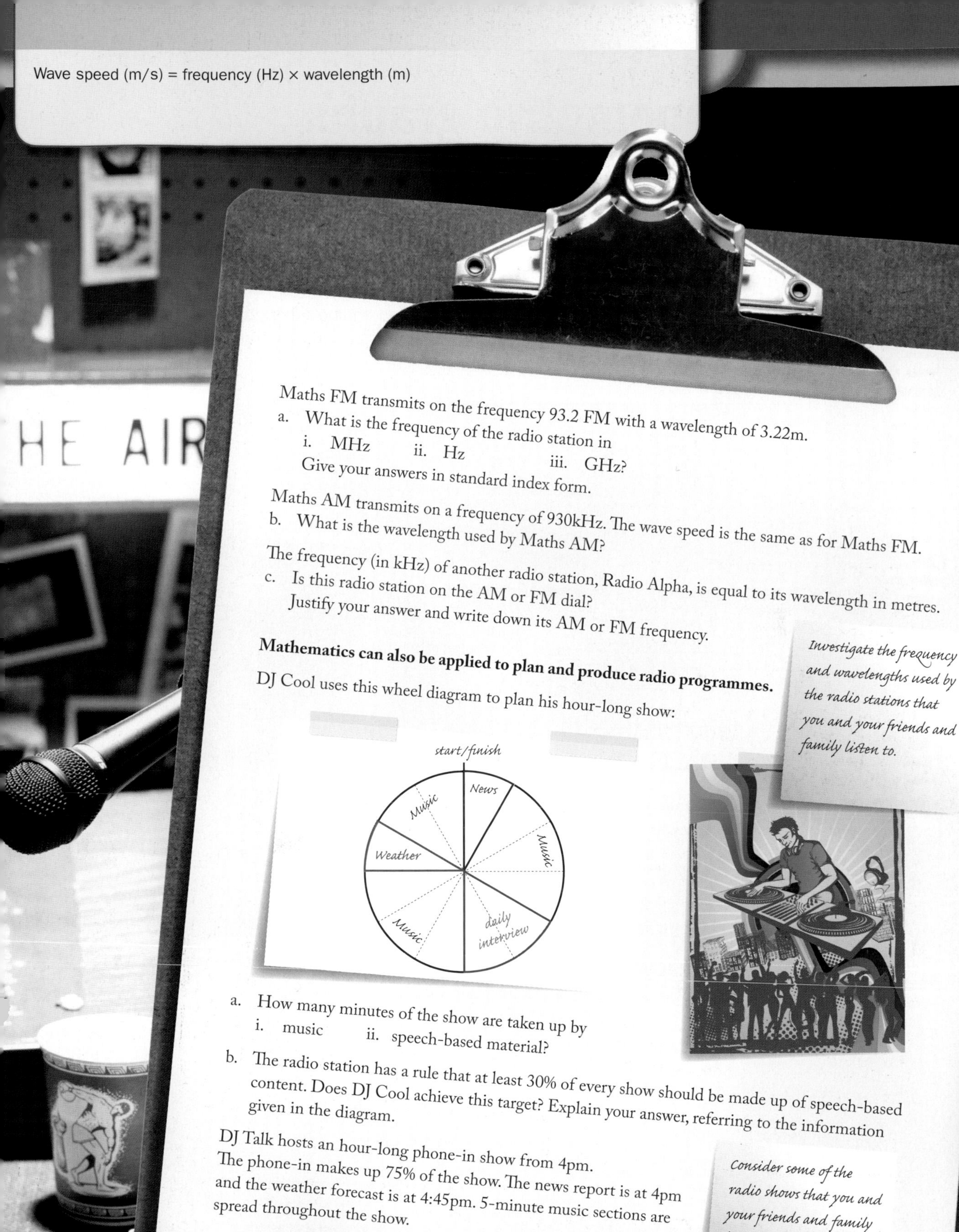

Maths FM transmits on the frequency 93.2 FM with a wavelength of 3.22m.

a. What is the frequency of the radio station in
   i. MHz   ii. Hz   iii. GHz?
   Give your answers in standard index form.

Maths AM transmits on a frequency of 930kHz. The wave speed is the same as for Maths FM.

b. What is the wavelength used by Maths AM?

The frequency (in kHz) of another radio station, Radio Alpha, is equal to its wavelength in metres.

c. Is this radio station on the AM or FM dial?
   Justify your answer and write down its AM or FM frequency.

Investigate the frequency and wavelengths used by the radio stations that you and your friends and family listen to.

Mathematics can also be applied to plan and produce radio programmes.

DJ Cool uses this wheel diagram to plan his hour-long show:

a. How many minutes of the show are taken up by
   i. music   ii. speech-based material?

b. The radio station has a rule that at least 30% of every show should be made up of speech-based content. Does DJ Cool achieve this target? Explain your answer, referring to the information given in the diagram.

DJ Talk hosts an hour-long phone-in show from 4pm.
The phone-in makes up 75% of the show. The news report is at 4pm and the weather forecast is at 4:45pm. 5-minute music sections are spread throughout the show.

c. Draw a wheel diagram to show how DJ Talk's show might look.

Consider some of the radio shows that you and your friends and family listen to. Do they use a format that could be shown on a wheel?

G7

Trigonometric graphs

Knowing the amount of daylight in a day or predicting tides is vital to many people. The number of hours of daylight rises in the summer and falls in the winter and then repeats. Tides rise and fall and then repeat. This cyclical behaviour means that hours of daylight and tides can be predicted using sine and cosine functions.

What's the point?

By adding together combinations of basic sine and cosine functions, differing only by stretches from sin x and cos x, mathematicians can describe any periodic function. Using data to fix the specific combination they can then predict future behaviour.

Check in

You should be able to

- **evaluate trigonometric functions**

1 **a** Use a calculator to find **i** sin 30 **ii** cos 60.

b Use a calculator to find **i** sin 40 **ii** cos 50.

c Draw a right-angled triangle.
Use this to explore other trig ratios with the same value.

- **translate and stretch functions**

2 The graph shows a sketch of $y = x^2$.

Copy the graph and add graphs for each of these.

a $y = x^2 + 2$

b $y = 2x^2$

c $y = (x - 2)^2$

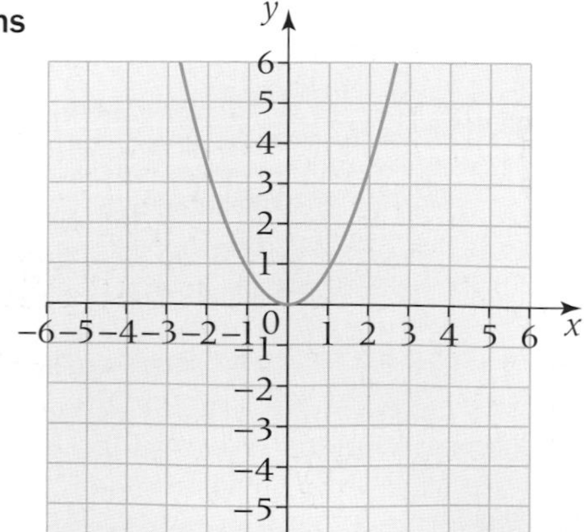

Orientation

What I need to know	What I will learn	What this leads to
G4 Use trigonometric functions	■ Plot and transform graphs of trigonometric functions	**A-level** Maths
A8 Apply translations and stretches to functions	■ Find multiple solutions to trigonometric equations	Electronics

Rich task

Use your calculator to construct the following graphs from $0^\circ \le \theta \le 360^\circ$.

a $y = \sin\theta + \cos\theta$

b $y = \sin\theta \div \cos\theta$

c $y = (\sin\theta)^2 + (\cos\theta)^2$

Investigate some combinations of your own.

G7.1 Sine graph

This spread will show you how to:

- Draw, sketch and describe the graph of the sine function for angles of any size
- Use the graph of the sine function to solve equations

Keywords
Acute
Obtuse
Sine

In right-angled triangles, two angles are always **acute**.

In obtuse-angled triangles, one angle is always **obtuse**.

p.298

Your calculator will give a value for the **sine** of any angle.

You can draw a graph of the sine ratio from this table of values.

x	0	45°	90°	135°	180°	225°	270°	315°	360°
$\sin x$	0	0.707	1	0.707	0	−0.707	−1	−0.707	0

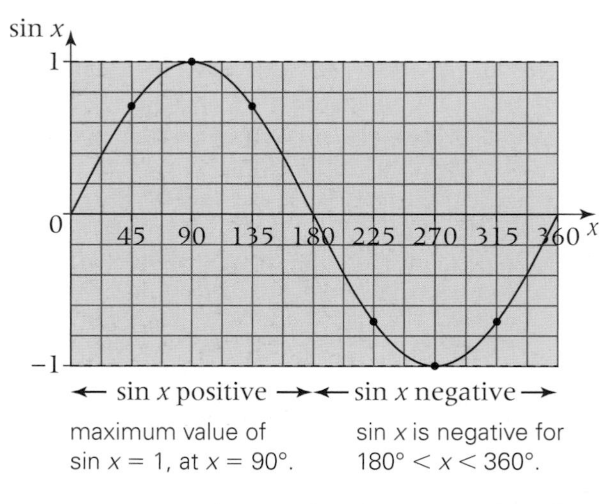

These four triangles are congruent:

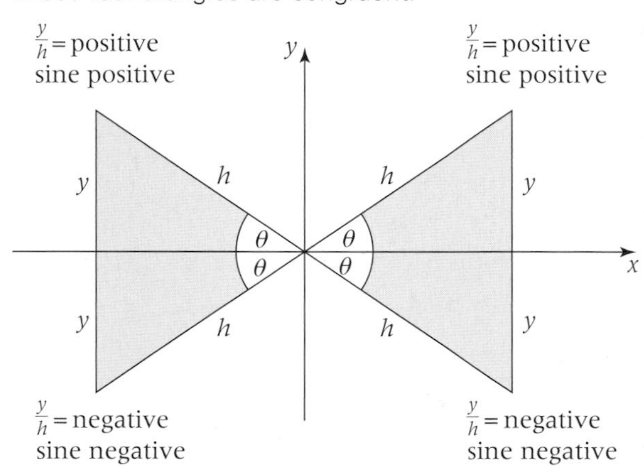

- You can use the graph of $y = \sin x$ to solve equations for angles between 0° and 360°. The graph is symmetrical.

Most values of $\sin x$ give two solutions.

Example

Solve these equations for angles between 0° and 360°.

a $\sin x = 0.6$ **b** $\sin x = -0.4$ **c** $\sin x = 1$

a $x = 37°$ and $143°$

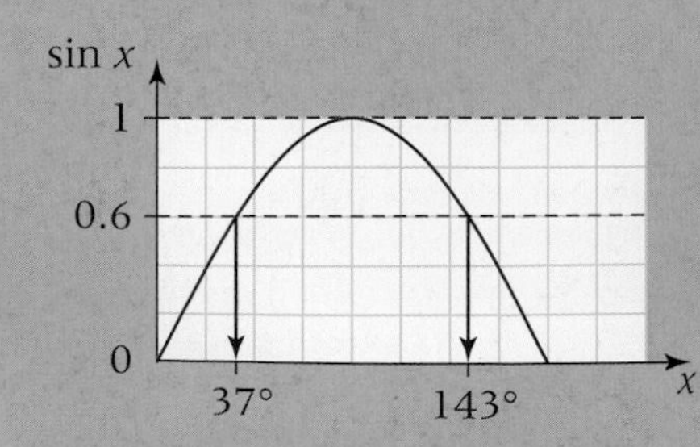

b $x = 204°$ and $336°$

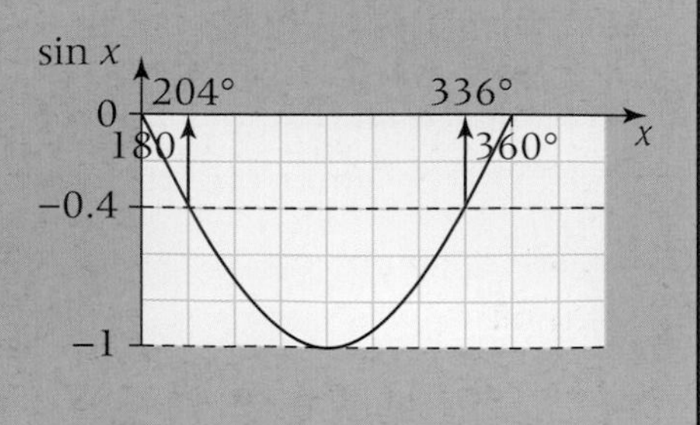

c $\sin x = 1$ is the maximum point of the graph, $x = 90°$.

The symmetry of the graph shows that the two answers should add to give a multiple of 180°.

Exercise G7.1

Grade A*

1 Use the graph of $y = \sin x$ to solve these equations for angles between 0° and 360°.
Use a calculator to check the accuracy of your answers.

a $\sin x = 0.7$ **b** $\sin x = 0.8$ **c** $\sin x = -0.8$
d $\sin x = -0.5$ **e** $\sin x = -0.2$ **f** $\sin x = 0.4$
g $\sin x = -0.6$ **h** $\sin x = 0.3$ **i** $\sin x = -0.3$
j $\sin x = -1$ **k** $\sin x = 0$ **l** $\sin x = 0.5$

2 Use the graph of $y = \sin x$ to find the sine of these angles.
Use a calculator to check the accuracy of your answers.

a sin 30° **b** sin 150° **c** sin 45°
d sin 225° **e** sin 315° **f** sin 90°
g sin 270° **h** sin 180° **i** sin 75°
j sin 240° **k** sin 100° **l** sin 300°

3 Use the graph of $y = \sin x$ to find the other angle between 0° and 360° with sine the same value as

a sin 50° **b** sin 68° **c** sin 140°
d sin 217° **e** sin 262° **f** sin 98°
g sin 330° **h** sin 243° **i** sin 12°
j sin 309° **k** sin 172° **l** sin 287°

4 **i** Use the sine rule to find the angle marked θ in each triangle.
ii Use the graph of $y = \sin x$ to work out a second value for θ, if a second value is possible.

a

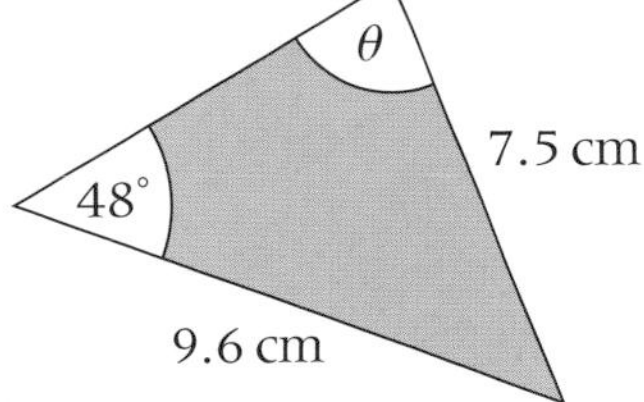

b

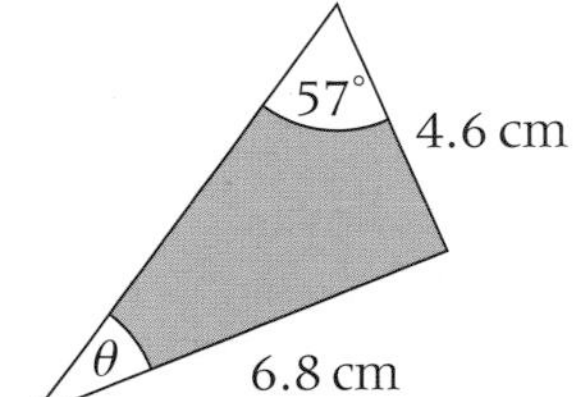

c

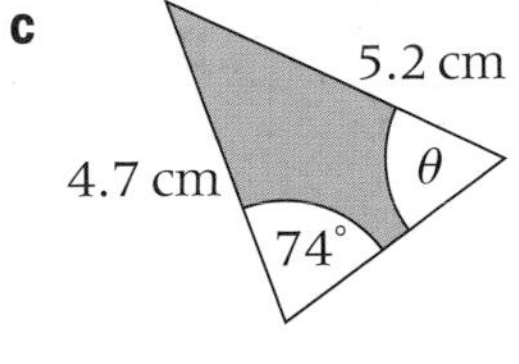

5 Plot the graph of $y = \sin x$ for angles between −360° and 0°.

Problem AO3

6 JKL is an equilateral triangle of side 2 cm.

a Show that $\tan 30° = \frac{\sqrt{3}}{3}$.

b Work out tan 60° from the triangle. Leave your answer in surd form.

c XYZ is an isosceles right-angled triangle. ZY = XZ = 2 cm and angle XZY = 90°.

i Work out tan 45°.

ii Work out sin 45°, leaving your answer in surd form.

iii Write the value of cos 45°.

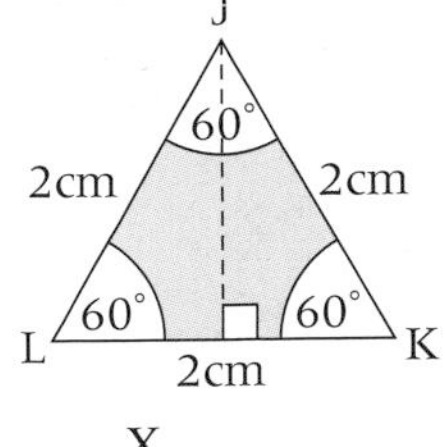

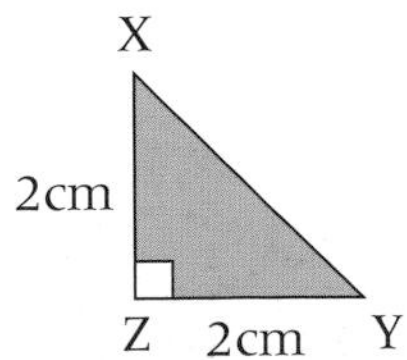

Use the relationship between cos and sin.

G7.2 Cosine graph

This spread will show you how to:

- Draw, sketch and describe the graph of the cosine function for angles of any size
- Use the graph of the cosine function to solve equations

Keywords
Cosine

Your calculator will give a value for the **cosine** of any angle.

You can draw a graph of the cosine ratio from this table of values.

x	0	45°	90°	135°	180°	225°	270°	315°	360°
$\cos x$	1	0.707	0	−0.707	−1	−0.707	0	0.707	1

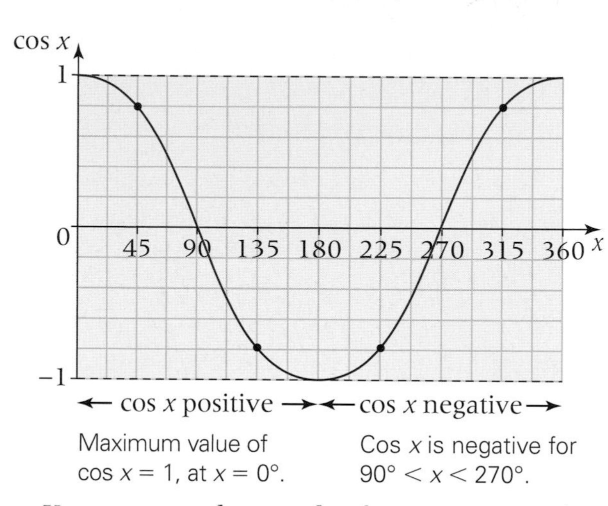

Maximum value of $\cos x = 1$, at $x = 0°$.

Cos x is negative for $90° < x < 270°$.

These four triangles are congruent:

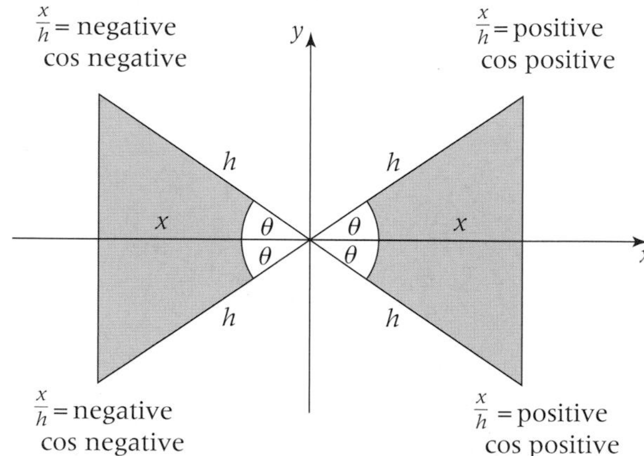

- You can use the graph of $y = \cos x$ to solve equations for angles between 0° and 360°. The graph is symmetrical.

Most values of cos x give two solutions.

Example

Solve these equations for angles between 0° and 360°.

a $\cos x = 0.6$
b $\cos x = -0.4$
c $\cos x = -1$

a $x = 53°$ and $307°$
b $x = 114°$ and $246°$
c This is the minimum value of the graph, $x = 180°$.

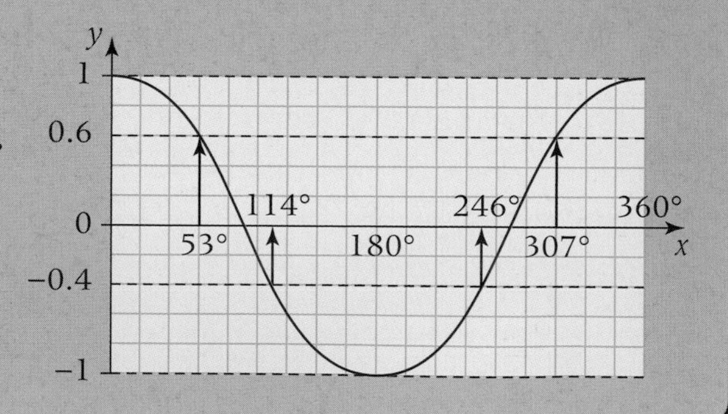

The symmetry of the graph shows that the two answers should add to 360°.

The graph of $y = \cos x$ is the same as the graph of $y = \sin x$ translated 90° along the x-axis.

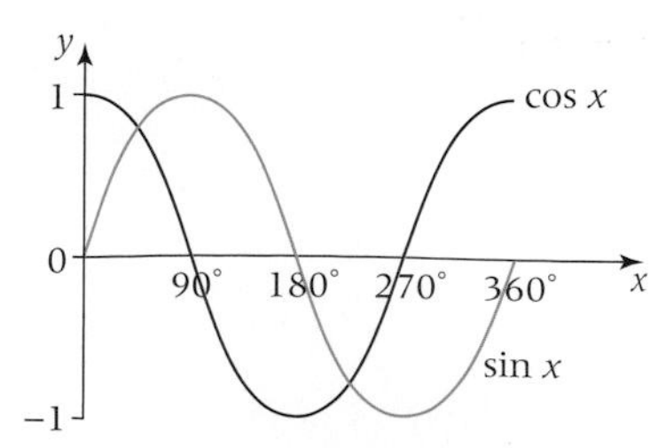

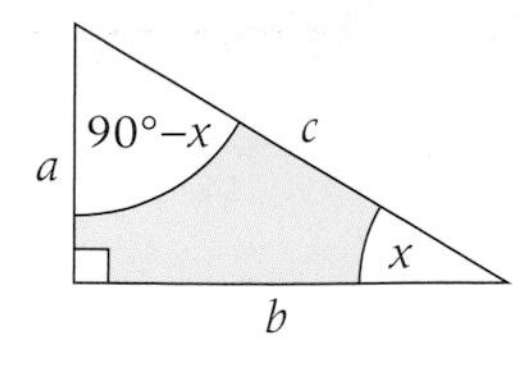

$$\sin x = \frac{a}{c} = \cos(90° - x)$$

$$\sin(90° - x) = \frac{b}{c} = \cos x$$

Exercise G7.2 Grade A*

1 Use the graph of $y = \cos x$ to solve these equations for angles between 0° and 360°.
Use a calculator to check the accuracy of your answers.

a $\cos x = 0.7$
b $\cos x = 0.8$
c $\cos x = -0.8$
d $\cos x = -0.5$
e $\cos x = -0.2$
f $\cos x = 0.4$
g $\cos x = -0.6$
h $\cos x = 0.3$
i $\cos x = -0.3$
j $\cos x = 1$
k $\cos x = 0$
l $\cos x = 0.5$

2 Use the graph of $y = \cos x$ to work out the cosine of each angle.
Use a calculator to check the accuracy of your answers.

a cos 60°
b cos 140°
c cos 45°
d cos 225°
e cos 315°
f cos 90°
g cos 270°
h cos 180°
i cos 55°
j cos 250°
k cos 80°
l cos 330°

3 Use the graph of $y = \cos x$ to find the other angle between 0° and 360° with cosine the same value as

a cos 30°
b cos 66°
c cos 150°
d cos 237°
e cos 164°
f cos 78°
g cos 125°
h cos 234°
i cos 18°
j cos 199°
k cos 302°
l cos 268°

4 Plot the graph of $y = \cos x$ for angles between −360° and 0°.

AO3 Problem

5 **a** Use your calculator to check that
cos 60° = sin 30°, cos 52° = sin 38° and cos 17° = sin 73°.

b Use the right-angled triangle, ABC, to explain the equalities in part **a**.

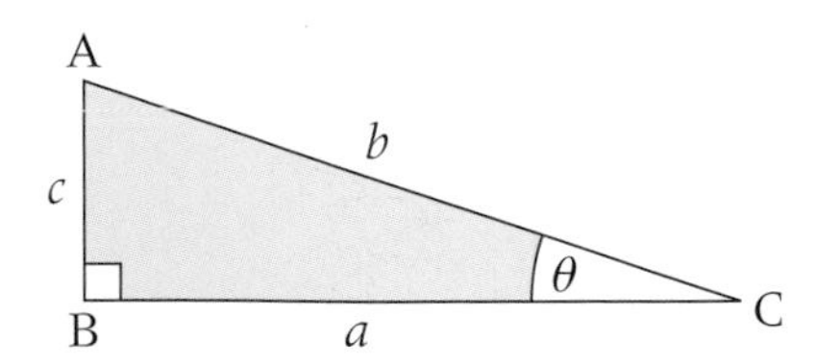

6 Plot the graph of $y = \tan x$ for angles between 0° and 360°.
Comment on what you observe.

Unit 3

G7.3 Solving trigonometric equations

This spread will show you how to:

- Use a calculator to find one solution to equations involving trigonometric functions
- Use the symmetry of the graphs of trigonometric functions to find further solutions to equations

Keywords
Cosine
Sine

You can use your calculator to solve equations involving trigonometric ratios. It will only give one solution.

You use the symmetry of the trigonometric graphs to find other solutions.

Sine graph

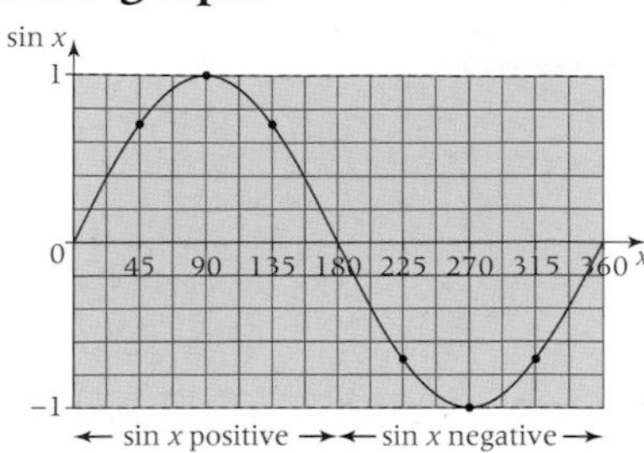

The two solutions add to a multiple of 180°.

You use x and $180° - x$.

Cosine graph

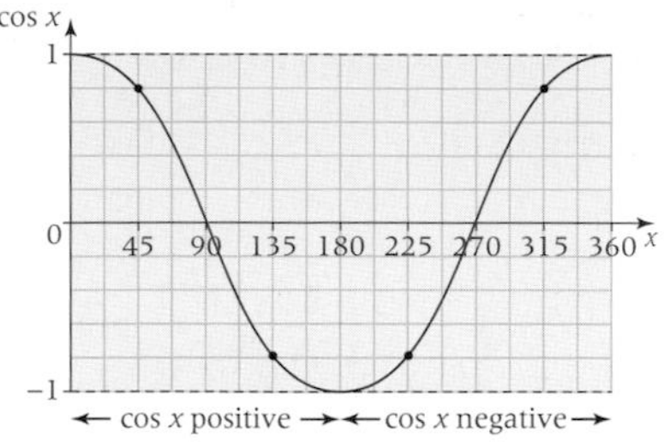

The two solutions add to 360°.

You use x and $360° - x$.

Example

Use your calculator to solve these equations for angles 0° to 360°.
Give your answers to 1 decimal place.

a $\sin x = 0.904$ **b** $\cos x = -0.849$ **c** $\sin x = -0.176$

a $x = 64.7°$ From calculator
or $x = 180° - 64.7° = 115.3°$ From graph

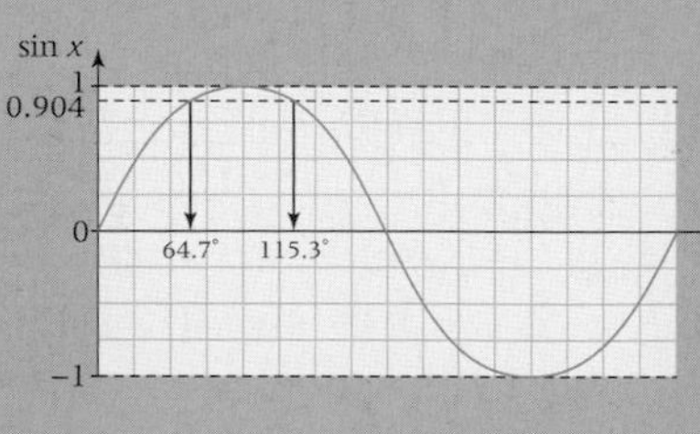

b $x = 148.1°$ From calculator
or $x = 360° - 148.1° = 211.9°$ From graph

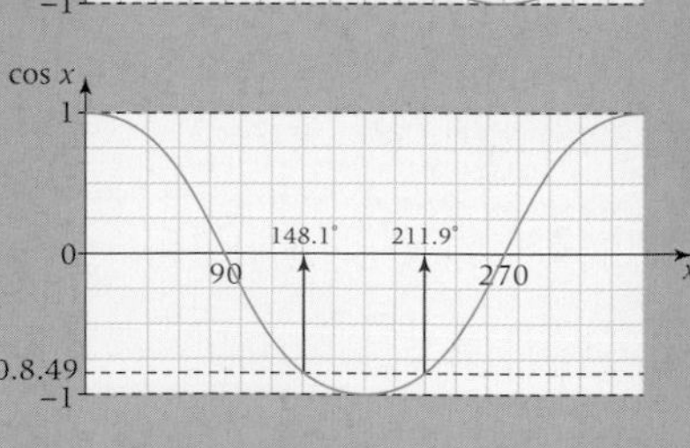

c $x = -10.1°$ From calculator

The positive solutions are:
$x = 180° + 10.1° = 190.1°$ and
$360° - 10.1° = 349.9°$
in the range 0° to 360°

Extend the graph to −90°, using symmetry.

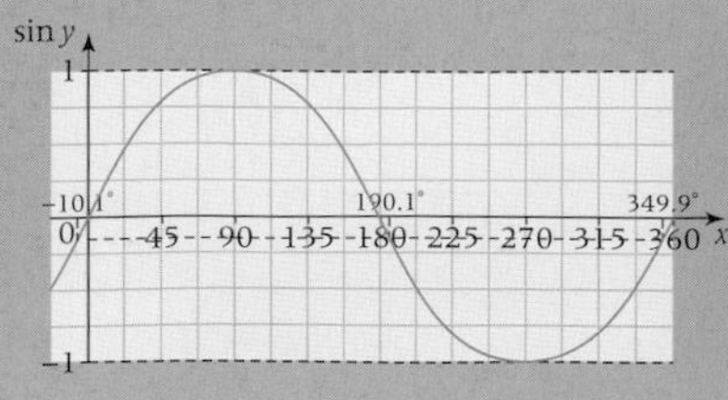

Exercise G7.3

Grade A*

Use your calculator to solve these equations for angles 0° to 360°.

Use the graphs of $y = \sin x$ and $y = \cos x$ to find other solutions in the range 0° to 360°.

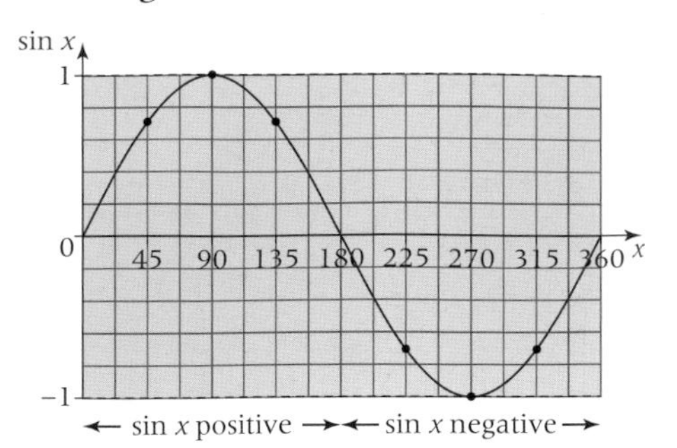

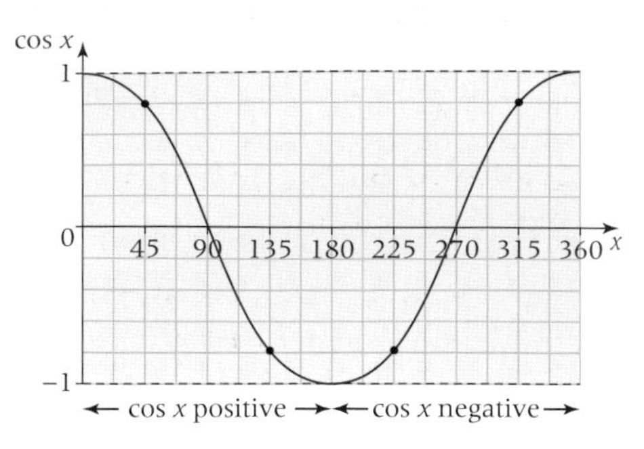

Give all your answers to 1 decimal place.

1 $\sin x = 0.534$ **2** $\sin x = 0.934$ **3** $\sin x = -0.997$

4 $\cos x = 0.521$ **5** $\cos x = 0.124$ **6** $\sin x = 0.124$

7 $\sin x = 0.62$ **8** $\cos x = -0.508$ **9** $\sin x = 0.866$

10 $\cos x = -0.866$ **11** $\sin x = 0.732$ **12** $\cos x = -0.430$

13 $\cos x = -0.725$ **14** $\sin x = -0.467$ **15** $\sin x = 0.321$

16 $\cos x = -0.5$ **17** $\sin x = 0.5$

18 $\cos x = 0.402$ **19** $\cos x = -0.695$

Use symmetry to help you sketch the graph for negative values of x to −90°.

AO3 Problem

Using the graph of $y = \tan x$ and your calculator find all the solutions in the range 0° to 360°.

20 $\tan x = 1.25$ **21** $\tan x = 2.61$

22 $\tan x = -1.67$ **23** $\tan x = 1.889$

24 $\tan x = -1.889$ **25** $\tan x = 4.5$

26 $\tan x = -4.5$ **27** $\tan x = 0.807$

28 $\tan x = 1$ **29** $\tan x = 8.68$

30 $\tan x = 5.923$

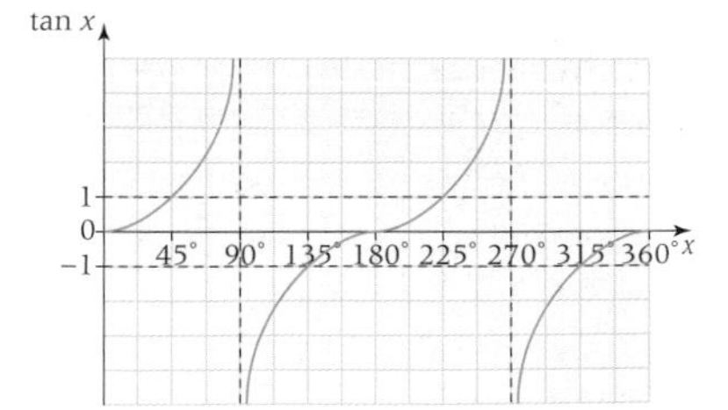

G7.4 Transformations of trigonometric graphs

This spread will show you how to:

- Transform graphs of trigonometric functions

Keywords
Cosine
Sine
Tangent
Transformation

You can transform graphs of trigonometric functions in the same way as graphs of other functions.

p.380-386

- $f(x + a)$ means a shift along the x-axis. — a positive → shift to left; a negative → shift to right
- $f(x) + a$ means a shift along the y-axis. — a positive → shift up; a negative → shift down
- $af(x)$ means a stretch parallel to the y-axis with scale factor a.
- $f(ax)$ means a stretch parallel to the x-axis with scale factor $\frac{1}{a}$.

On a sketch you indicate where the graph crosses the axes and where the graph turns.

Example

Sketch these graphs for angles 0° to 360°.

a $y = 3 + \sin x$ **b** $y = \cos 2x$ **c** $y = \sin(x + 30)$

a $y = 3 + \sin x$

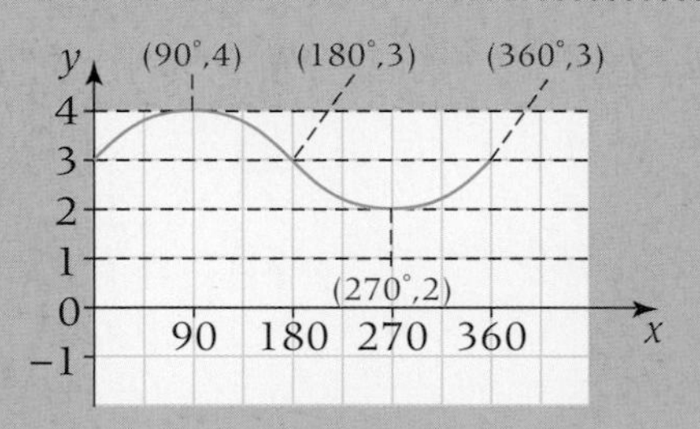

This is $y = f(x) + a$.

The graph has the same shape, but moves 3 units up on the y-axis.

b $y = \cos 2x$

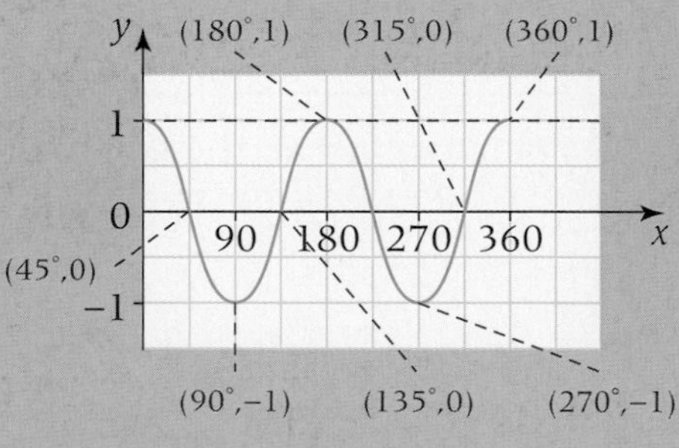

This is $y = f(ax)$.

The graph has the same shape, but is squashed in the direction of the x-axis.

c $y = \sin(x + 30)$

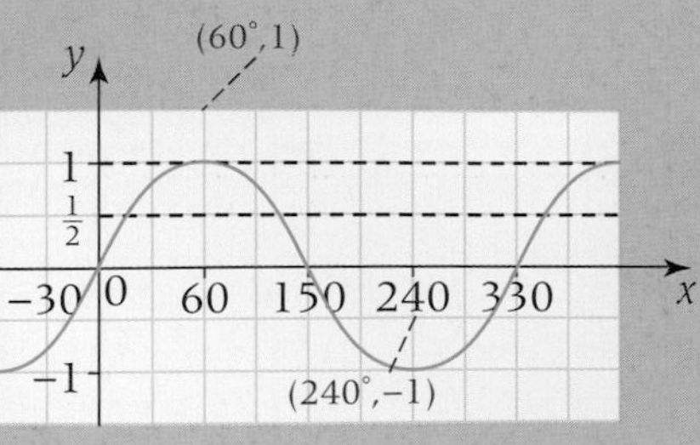

This is $y = f(x + a.)$.

The graph has the same shape, but moves 30° to the left.

Exercise G7.4 Grade A*

1 This is a sketch of the graph $y = \sin x$.
On separate diagrams sketch the graphs of
a $y = \sin 2x$ for angles 0° to 360°
b $y = 2 \sin x$ for angles 0° to 360°
c $y = -2 + \sin x$ for angles 0° to 360°

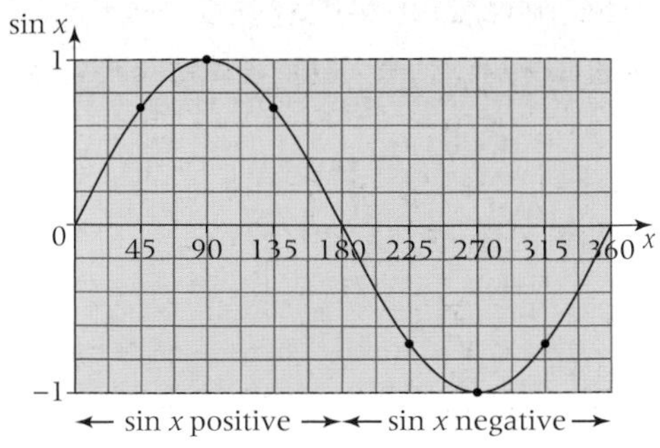

2 This is a sketch of the graph $y = \cos x$.
On separate diagrams sketch the graphs of
a $y = \cos \frac{1}{2}x$ for angles 0° to 360°
b $y = \frac{1}{2} \cos x$ for angles 0° and 360°
c $y = \cos (x - 60°)$ for angles 0° to 360°

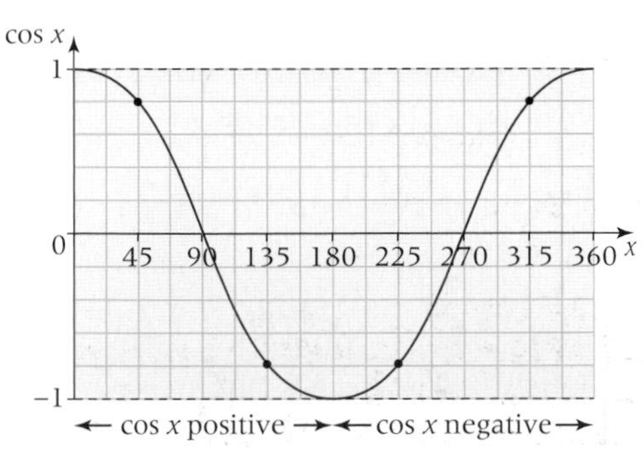

3 This is a sketch of the graph $y = a + b \sin x$ for angles 0° to 360°.
Use the graph to find estimates for the values of a and b.

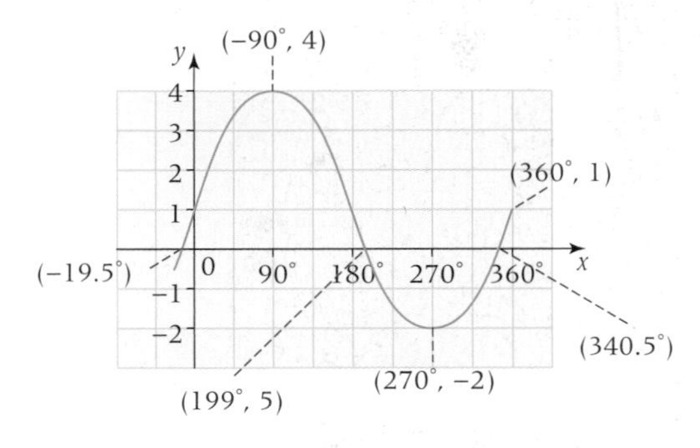

4 This is a sketch of the graph $y = p \cos qx$ for angles 0° to 360°.
Use the graph to find estimates for the values of p and q.

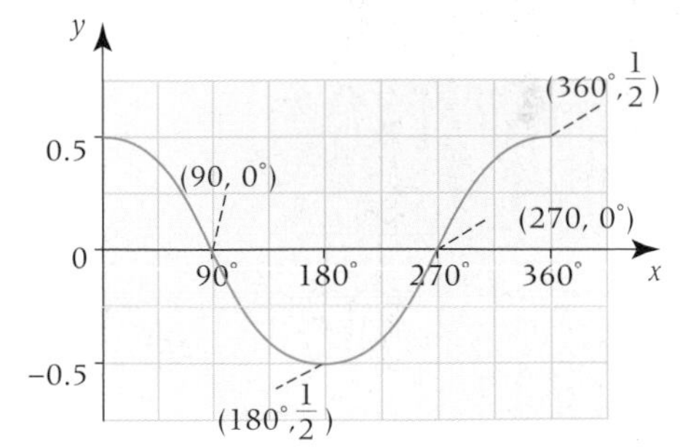

Problem AO3

5 This is a sketch of the graph $y = \tan x$.
On separate diagrams sketch the graphs of
a $y = 1 + \tan x$ for angles 0° to 360°
b $y = \tan (x + 30°)$ for angles 0° to 360°
c $y = 1 + \tan (x + 30°)$ for angles 0° to 360°

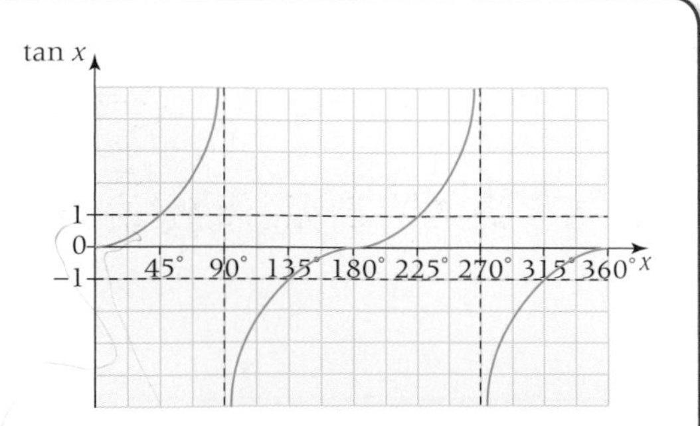

Summary

Check out

You should now be able to:

- Draw and describe the graphs of the trigonometric functions $y = \sin x$ and $y = \cos x$ for $-360° \leq x \leq 360°$
- Transform graphs of trigonometric functions

Worked exam question

Here is the graph of $y = \cos x°$, where $0 \leq x \leq 360$

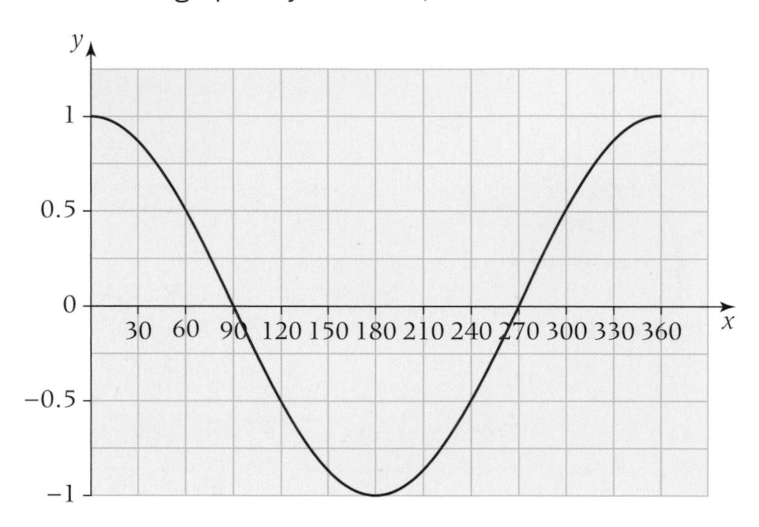

Use the graph to solve $\cos x° = 0.75$ for $0 \leq x \leq 360$ (2)

(Edexcel Limited 2007)

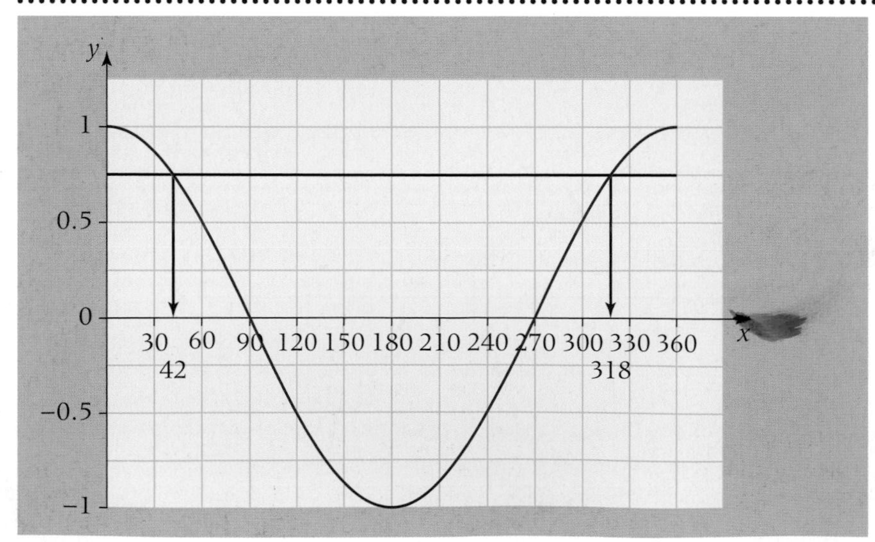

There are two values of x.

42° and 318°

Exam question

1 Here is the graph of $y = \sin x$, where $0° \leq x \leq 360°$

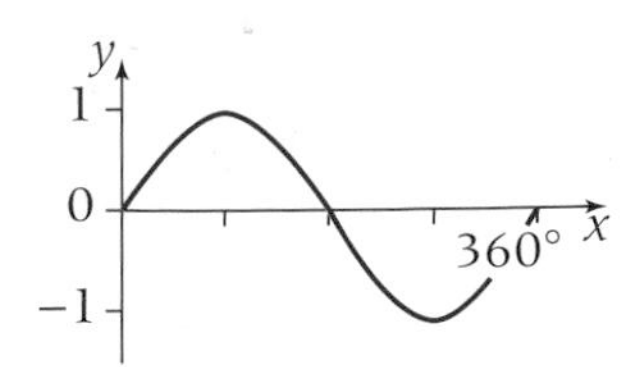

Graph A

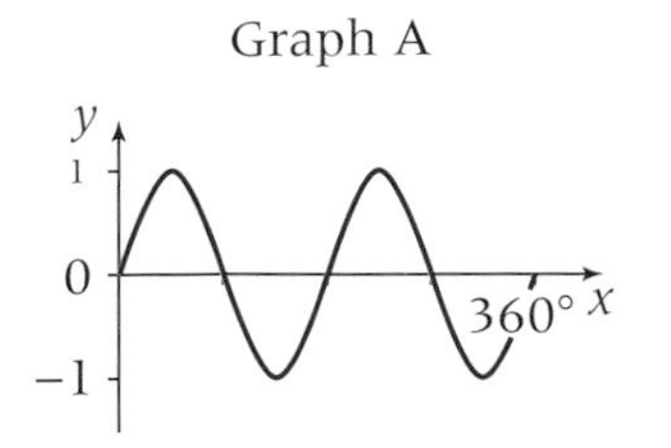

Graph B

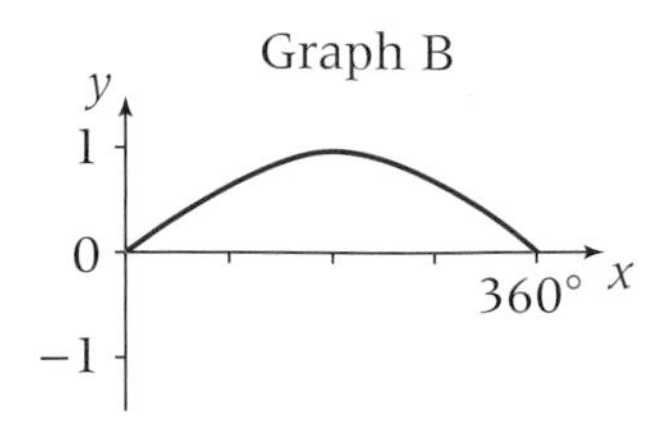

Graph C

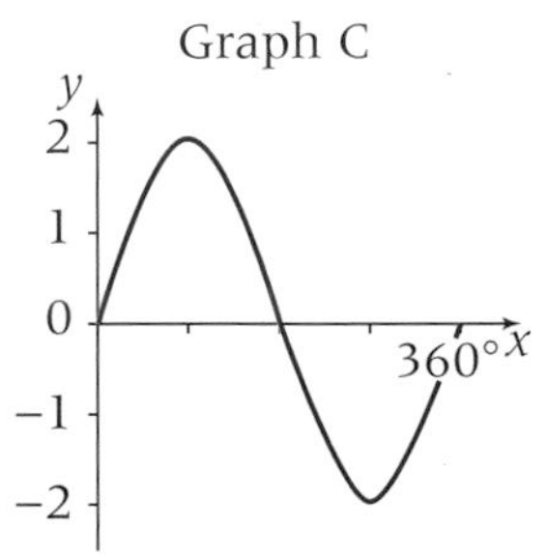

Graph D

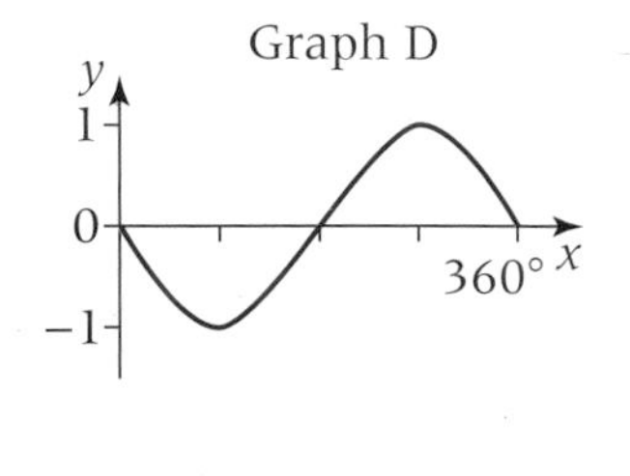

Graph E

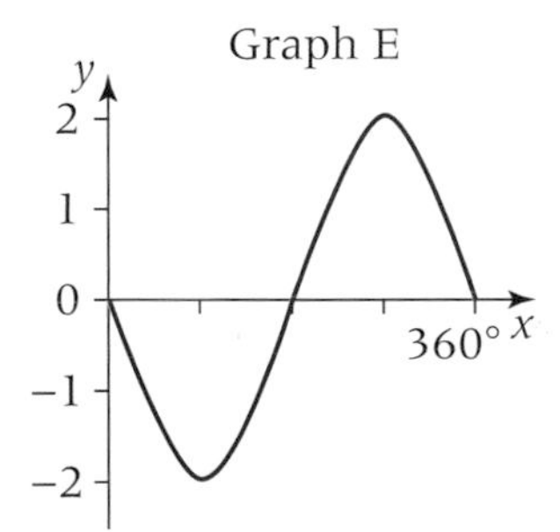

Graph F

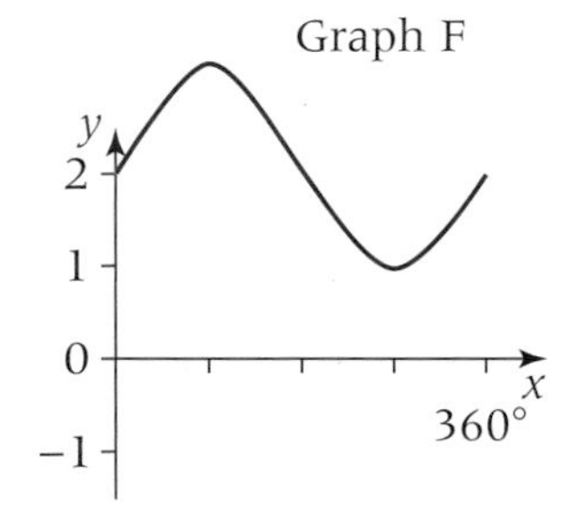

Match each of the graphs **A**, **B**, **C**, **D**, **E**, and **F** to the equations in the table.

Equation	Graph
$y = 2 \sin x$	
$y = -\sin x$	
$y = \sin 2x$	
$y = \sin x + 2$	
$y = \sin \frac{1}{2}x$	
$y = -2 \sin x$	

(4)

(Edexcel Limited 2007)

GCSE formulae

In your Edexcel GCSE examination you will be given a formula sheet like the one on this page.
You should use it as an aid to memory. It will be useful to become familiar with the information on this sheet.

Volume of a prism = area of cross section × length

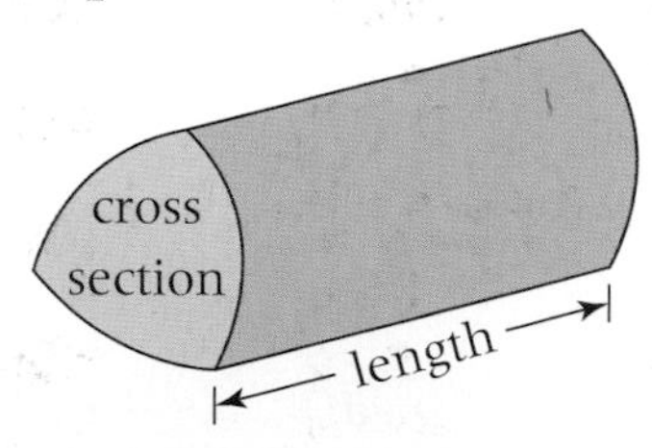

Volume of sphere $= \frac{4}{3}\pi r^3$

Surface area of sphere $= 4\pi r^2$

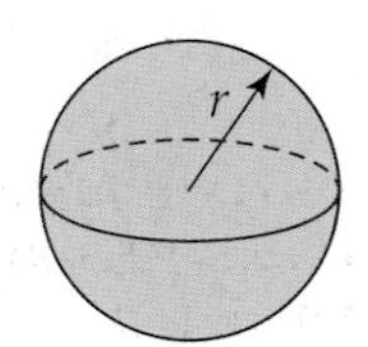

Volume of cone $= \frac{1}{3}\pi r^2 h$

Curved surface area of cone $= \pi r l$

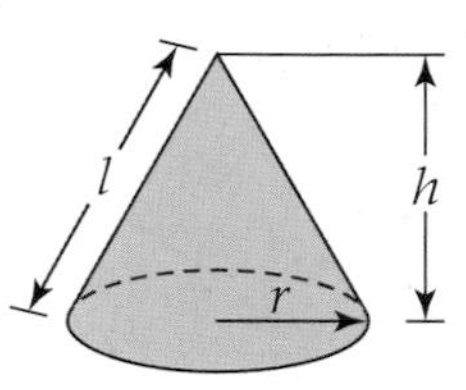

Area of trapezium $= \frac{1}{2}(a + b)h$

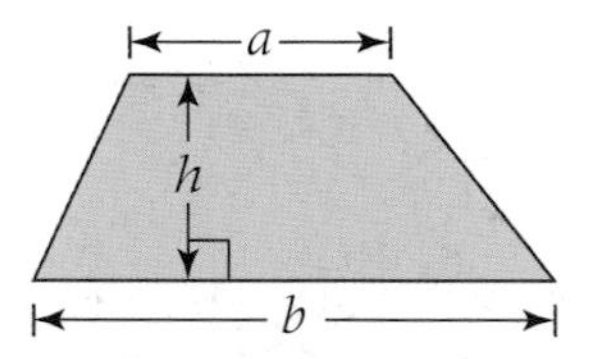

In any triangle ABC

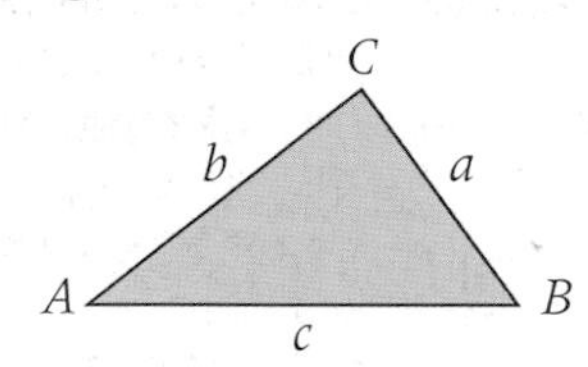

Sine rule $\frac{a}{\sin A} = \frac{b}{\sin B} = \frac{c}{\sin C}$

Cosine rule $a^2 = b^2 + c^2 - 2bc \cos A$

Area of triangle $= \frac{1}{2} ab \sin C$

The Quadratic Equation

The solutions of $ax^2 + bx + c = 0$ where $a \neq 0$, are given by

$$x = \frac{-b \pm \sqrt{(b^2 - 4ac)}}{2a}$$

D1 Check in

1 a A survey of an entire population.
  b In a census, everyone in the population is included. In a sample, only a selection of people from the population are included.
2 a Nicky collected primary data, Mike secondary data.
  b Primary data is data you collect yourself. Secondary data is taken from an existing source.
3 It suggests the correct answer is 'yes'.

D1.1

1 'Often' will be interpreted differently by different people, so he should give a choice of frequencies. People outside a bowling alley are likely to go bowling often, so they are not representative.
2 It is a leading question. People at a netball club are likely to find netball most exciting, so they are not representative.
3 a Year 11 students may have different views to students in other years.
  b Assign a number to each student, then use random numbers to choose 50 students.
  c List the students, pick a starting point at random, then pick every 20th student.
4 a Systematic sampling
  b He is only asking people who are prepared to travel to the match.
5 Systematic sampling because, for example, choosing every 20th rivet starting with the 19th rivet would give no substandard rivets.
6 a Assign a number to each student, then use random numbers to choose 90 students.
  b List the students, pick a starting point at random, then pick every fourth student.
  c The random sample could be biased if, for example, there is a higher proportion of girls in the sample than in the population as a whole. The systematic sample could be biased if the list used is ordered, for example alphabetically.

D1.2

1 The proportion of boys in the sample is much higher than the proportion of boys at the gym club.
2 Randomly select 550 women and 350 men.
3 For example, for a sample size of 100, randomly select 14 males 18–30, 14 males 31–50, 12 males over 50, 20 females 18–30, 19 females 31–50 and 21 females over 50.
4 a 10, 8 b 85
5 It would be more natural to choose a sample of size 61 with 9 adults, 26 boys and 26 girls.

D1.3

1 45.6
2 ai 6 aii 6 aiii 5.84
  aiv 4 av 3
  bi 5 bii 5 biii 5
  biv 4 bv 2
  ci 4 cii 6 ciii 6.63
  civ 6 cv 5
  di 6 dii 6 diii 5.68
  div 6 dv 3
3 1.5
4 7

D1.4

1 81 minutes
2 7.6 hours
3 26 lessons
4 The mean is higher for city dwellers than for small town dwellers, perhaps because they live closer to more museums.
5 The mean spend is higher for online customers than for in store customers, perhaps because they have to pay delivery costs.
6 $\frac{36c+42d}{78}$
7 $\frac{mx+fy}{m+f}$

D1.5

1 ai $10 < t \leqslant 15$ aii $10 < t \leqslant 15$
  aiii 15.5 minutes aiv 20 minutes
  bi $10 < t \leqslant 20$ bii $20 < t \leqslant 30$
  biii 24.2 minutes biv 40 minutes
  ci $10 < t \leqslant 15$ cii $10 < t \leqslant 15$
  ciii 14.6 minutes civ 25 minutes
  di $5 < t \leqslant 15$ dii $15 < t \leqslant 25$
  diii 26.5 minutes div 50 minutes
2 a $150 < B \leqslant 200$ b £171
3 a £88 b $8 < M \leqslant 120$
  c No, the 21st value is in the $8 < M \leqslant 120$ interval.

D1.6

1 The average suggests Sell-a-lot. (116 is a valid range for both data sets.)
2 The report was not valid. The mean and mode are greater for Year 7, but not double.

D1 Summary

1 a $35 \leqslant t < 40$
  b 34.75 minutes (accept 34.7, 34.8)

2 35
3 5 girls

N1 Check in

1 ai 38.5 aii 39
bi 16.1 bii 16
ci 103.9 cii 100
di 0.1 dii 0.082
ei 0.4 eii 0.38

2 a 954
b 337.415
c 48.99 (2dp)
d 22.37 (2dp)
e 105.59 (2dp)
f −45.19 (2dp)

N1.1

1 a 310 b 450 c 530 d 2170 e 56 690
2 a 43 b 1 c 23 d 32 e 44
3 ai 1300 aii 1000
bi 1100 bii 1000
ci 0 cii 0
di 500 dii 1000
ei 41 500 eii 41 000
4 a 0.3 b 0.74 c 0.151 d 0.68
5 a 0.056 b 3.2 c 15 d 950
6 a 0.52 b 34 600 c 72 700 d 0.0045
7 a 400 b 0.6 c 0.005 d 700 000
8 a 0.42 b 0.86 c 1800 d 4300
9 ai $500 \div 20$ aii $20 + 40 \div 4$
aiii $2000 + 30 \times 50$ aiv $2 + 3$
bi 25 bii 30 biii 3500 biv 5
ci 25.056 cii 26.5 ciii 3320
civ 5.1186
10 ai e.g. $250 + 350 = 600$
aii e.g. $60 \div 30 = 2$
aiii e.g. $60 \times 200 = 12\ 000$
aiv e.g. $100 - 60 = 40$
bi The example given produces an underestimate.
bii The example given produces an underestimate.
biii The example given produces an underestimate.
biv The example given produces an overestimate.
ci 620 cii 2.37 ciii 13 279.86 civ 31.9

N1.2

1 a 5.85 m, 5.75 m b 16.55 litres, 16.45 litres
c 0.95 kg, 0.85 kg d 6.35 N, 6.25 N
e 10.15 s, 10.05 s f 104.75 cm, 104.65 cm
g 16.05 km, 15.95 km h 9.35 m/s, 9.25 m/s
2 a 6.75 m, 6.65 m b 7.745 litres, 7.735 litres
c 0.8135 kg, 0.8125 kg d 6.5 N, 5.5 N
e 0.0015 s, 0.0005 s f 2.545 cm, 2.535 cm
g 1.1625 km, 1.1615 km h 15.5 m/s, 14.5 m/s
3 a 174 kg, 162 kg b 28.4 kg, 27.6 kg
4 a 37.5 mm, 32.5 mm b 42.5 mm, 37.5 mm
c 112.5 mm, 107.5 mm d 4.75 cm, 4.25 cm
5 The number of nails should be assumed to be exact so maximum = $38 \times 12.5 = 475$ g, minimum = $38 \times 11.5 = 437$ g.
6 Sum of upper bounds = 461 kg, which exceeds the limit.
7 526.75 cm^2, 481.75 cm^2
8 65 g, 63 g
9 49 crates

N1.3

1 a Larger b Smaller c Smaller d Smaller
e Larger f Larger g Smaller h Same
2 a $8 \div 2$ b $10 \div 5$ c $12 \div 4$ d $18 \div 10$
3 a $15 \div 5$ b $28 \div 4$ c $10 \div \frac{5}{2}$ d $72 \div 8$
4 a 18×2 b 24×4 c $8 \times \frac{3}{2}$ d 5.5×10
e 5.9×1 f $66 \times \frac{5}{3}$ g $7 \times \frac{10}{7}$ h 8×0.8
5 a 24 b 144 c 65 d 72
e 2 f 9 g 8 h 30
6 a 27 b 60 c 16 d 21
e 7 f 90 g 40 h 6
7 a 16 b 24 c 12 d 16
e 30 f 24 g 36 h 28
8 a 2.5 b 2 c 3.5 d 0
e 1 f −7 g 4.8 h −3.85
9 a False. If you divide a negative number by 2 the answer is bigger.
b True
c False. If you multiply a negative number by 5 the answer is smaller.
d False. Multiplying zero by 10 gives zero.
10 a Repeatedly multiplying a positive number by 0.9 will make it smaller and smaller, tending to zero. Repeatedly multiplying a negative number by 0.9 will make it bigger and bigger, tending to zero.
b The answer will alternate between being positive and negative, and tend to zero.

N1.4

1 a 8 b 500 c 0.008 d 10
e 4 f 50
2 a 1.4 b 2.8 c 3.2 d 3.9
e 4.5 f 5.1 g 5.7 h 6.7
i 8.4 j 9.2
3 a The denominator would become zero, so the result would become undefined.
b The approximation would be 1^8 instead of 1.49^8, and the large power would mean this was very inaccurate.
c This would simply give zero, which would not be very helpful.

4 **a** $\frac{300 \times 4}{0.2} = 6000$ **b** $\frac{4 \times 700}{0.8} = 3500$
c $\frac{5 \times 100}{500 \times 0.4} = 2.5$ **d** $5^2 - 9^2 = 44$
e $\frac{9-4}{0.04-0.02} = 250$ **f** $8 \times \frac{0.2 \times 200}{50-25} = 12.8$

5 **a** $\frac{50 \times 5}{100} = 2.5$ **b** $\frac{\sqrt{300^2}}{900} = 10$
c $\frac{\sqrt{100}}{0.1} = 100$ **d** $\frac{49+144}{\sqrt{50^2}} \approx \frac{200}{50} = 4$
e $\frac{(23-18)^2}{3+2} = 5$ **f** $\sqrt{\frac{3+6}{0.2^2}} = \frac{3}{0.2} = 15$

6 **a** $\frac{5 \times 10^3}{5 \times 10^2} = 10$ **b** $\frac{7 \times 10^3}{7 \times 10^{-2}} = 10^5$
c $\frac{8 \times 10^4 \times 4 \times 10^{-2}}{4 \times 10^3} = 0.8$ **d** $\frac{7 \times 10^2 \times 4 \times 10^{-1}}{7 \times 10^{-2} \times 4 \times 10^{-1}} = 10^4$
e $\frac{7 \times 10^2 + 9 \times 10^2}{6 \times 10^{-2} \times 8 \times 10^{-1}} = \frac{16}{48} \times 10^5 \approx 3 \times 10^4$
f $\frac{3 \times 10^3 \times 5 \times 10^2}{(30+40)^2} \approx \frac{15 \times 10^5}{5 \times 10^3} = 3 \times 10^2$

N1.5

1 **a** 1 **b** $\frac{3}{4}$ **c** $\frac{5}{6}$ **d** $\frac{3}{10}$ **e** $\frac{7}{12}$
2 **a** $\frac{1}{3}$ **b** $\frac{1}{3}$ **c** $\frac{1}{12}$ **d** $\frac{1}{15}$ **e** $\frac{5}{21}$
3 **a** $1\frac{3}{4}$ **b** $1\frac{2}{3}$ **c** $3\frac{3}{10}$ **d** $3\frac{1}{8}$
4 **a** $\frac{3}{2}$ **b** $\frac{4}{5}$ **c** $\frac{7}{16}$ **d** $\frac{5}{16}$
5 **a** $\frac{2}{9}$ **b** $\frac{5}{2}$ **c** $\frac{1}{9}$ **d** $\frac{9}{4}$
6 **a** $\frac{3}{5}$ **b** $\frac{3}{8}$ **c** $\frac{6}{35}$ **d** $\frac{8}{15}$
7 **a** 3 **b** $5\frac{3}{5}$ **c** $8\frac{1}{4}$ **d** $3\frac{3}{32}$
8 **a** $6\frac{1}{8}$ **b** $\frac{5}{8}$ **c** $10\frac{5}{16}$ **d** $\frac{2}{3}$
9 **a** $3\frac{1}{3}$ **b** $\frac{35}{64}$ **c** $\frac{15}{38}$ **d** $4\frac{57}{160}$
10 32
11 8

N1.6

1 **a** 0.5 **b** 2.4 **c** 0.15 **d** 0.04
2 **a** 0.375 **b** 0.8 **c** 0.0625 **d** 0.12
4 **a** $0.\dot{3}$ **b** $0.\dot{6}$ **c** $0.1\dot{6}$ **d** $0.\dot{1}$
e $0.8\dot{3}$
5 $0.\dot{1}4285\dot{7}$, $0.\dot{2}8571\dot{4}$, $0.\dot{4}2857\dot{1}$, $0.\dot{5}7142\dot{8}$, $0.\dot{7}1428\dot{5}$, $0.\dot{8}5714\dot{2}$
6 **a** $\frac{1}{2}$ **b** $\frac{3}{10}$ **c** $\frac{3}{4}$ **d** $\frac{19}{20}$ **e** $\frac{13}{20}$
7 **a** $\frac{1}{9}$ **b** $\frac{2}{9}$ **c** $\frac{5}{33}$ **d** $\frac{125}{999}$ **e** $\frac{8}{37}$
8 **a** $\frac{19}{90}$ **b** $\frac{13}{18}$ **c** $\frac{91}{110}$ **d** $\frac{421}{666}$ **e** $\frac{1349}{1650}$
9 **a** $\frac{8}{11}$ **b** $0.\dot{7}$ **c** $0.\dot{0}7\dot{4}$
10 **a** Let $x = 0.5757...$, then $100x = 57.5757...$ and $100x - x = 99x = 57$ so $x = \frac{57}{99} = \frac{19}{33}$
b $\frac{59}{165}$

N1.7

1 **a** 35% **b** 60.7% **c** 99.5% **d** 100%
e 215% **f** 0.056% **g** 1700% **h** 10.1%
2 **a** 55.6% **b** 34.4% **c** 75.8% **d** 51.3%
e 43.7% **f** 83.9% **g** 83.9% **h** 105.6%
3 **a** 0.22 **b** 0.185 **c** $0.\dot{5}$ **d** $0.3\dot{5}$
e $0.0\dot{6}\dot{5}$ **f** $0.6\dot{1}4\dot{9}$ **g** $0.54\dot{4}\dot{6}$ **h** $1.5\dot{2}$
4 **a** 80% **b** 15% **c** 87.5% **d** 75%
e 28% **f** 18.75% **g** 45% **h** 14%
5 **a** 66.7% **b** 6.7% **c** 42.9% **d** 83.3%
e 77.8% **f** 36.4% **g** 8.3% **h** 13.3%
6 **a** $\frac{1}{6}, \frac{1}{2}, \frac{3}{5}, \frac{2}{3}$ **b** $\frac{1}{12}, \frac{7}{24}, \frac{1}{3}, \frac{3}{8}, \frac{1}{2}$
c $\frac{1}{20}, \frac{7}{40}, \frac{3}{5}, \frac{5}{8}, \frac{3}{4}$ **d** $\frac{7}{36}, \frac{5}{18}, \frac{4}{9}, \frac{1}{2}, \frac{2}{3}, \frac{3}{4}$
7 **a** $\frac{2}{5}, \frac{4}{10}, \frac{3}{8}, \frac{1}{4}$ or $\frac{4}{10}, \frac{2}{5}, \frac{3}{8}, \frac{1}{4}$ **b** $\frac{2}{5}, \frac{1}{3}, \frac{3}{10}, \frac{3}{11}, \frac{2}{9}$
c $\frac{4}{9}, \frac{7}{20}, \frac{1}{3}, \frac{2}{7}, \frac{1}{5}$ **d** $\frac{5}{9}, \frac{3}{8}, \frac{1}{3}, \frac{4}{13}, \frac{2}{7}, \frac{3}{11}$
8 **a** 0.33, 33.3%, $33\frac{1}{3}$%
b $0.\dot{4}$, 44.5%, 0.45, 0.454
c 22.3%, 0.232, 23.22%, 0.233, $0.2\dot{3}$
d $0.\dot{6}\dot{5}$, 0.66, 66.6%, 0.6666, $\frac{2}{3}$
e 14%, $14.\dot{1}$%, 0.142, $\frac{1}{7}$, $\frac{51}{350}$
f $\frac{5}{6}$, $\frac{6}{7}$, 86%, 0.866, $0.8\dot{6}$
9 Abby is correct. Since $0.\dot{3}$ is exactly equal to $\frac{1}{3}$, $0.\dot{9}$ is exactly equal to $3 \times \frac{1}{3} = 1$.

N1 Summary

1 **a** 400 (Accept 386 to 420)
2 **a** 72.5 g
b 73.5 g
3 **a** 48 mph (Accept 0.8 miles per minute)
bi 22.5 litres
bii 23.5 litres
4 Let $x = 0.\dot{4}\dot{5}$
$100x = 45.\dot{4}\dot{5}$
$99x = 45$
$x = \frac{45}{99} = \frac{15}{33}$
5 $\frac{211}{900}$

D2 Check in

1 **a** 62 **b** 60 **c** 18.5 **d** 15 **e** 7
f 30 **g** 108 **h** 150
2 **a** £50 **b** £70 **c** 2 days

D2.1

1

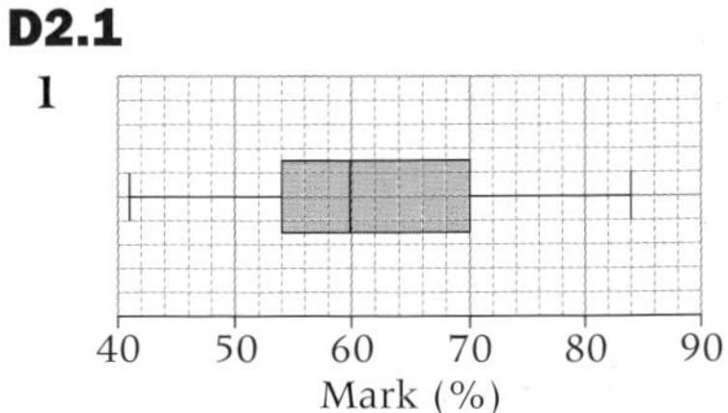

2

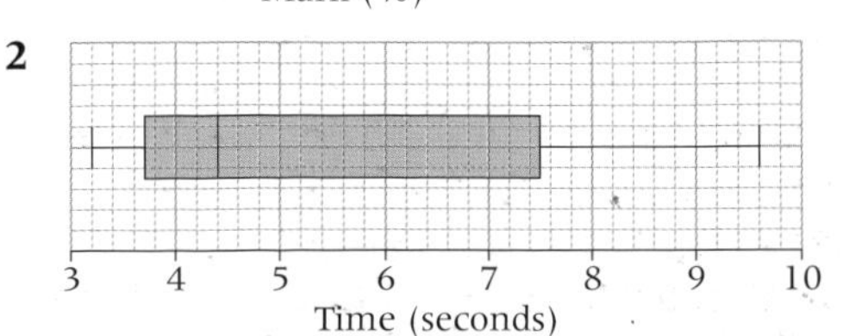

3

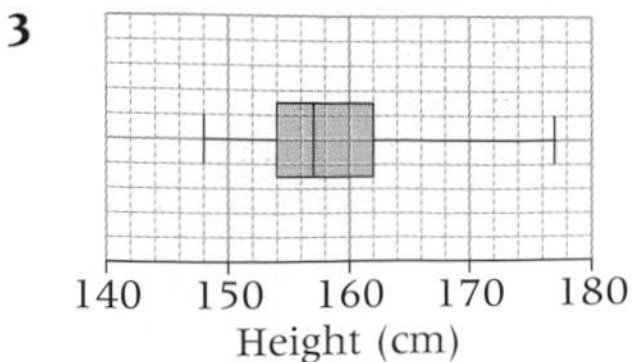

4 **a** 58 **b** 49 **c** 75

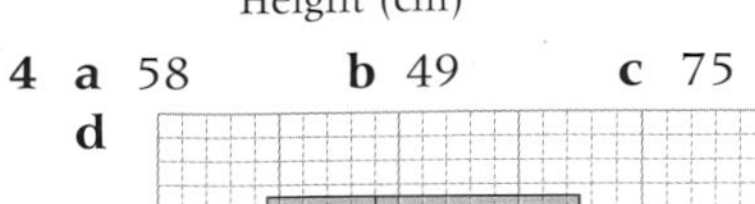

d

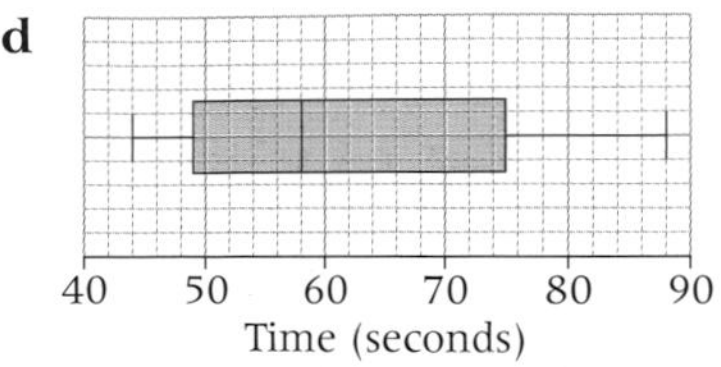

5 **a** Median = 44, LQ = 29, UQ = 56

b

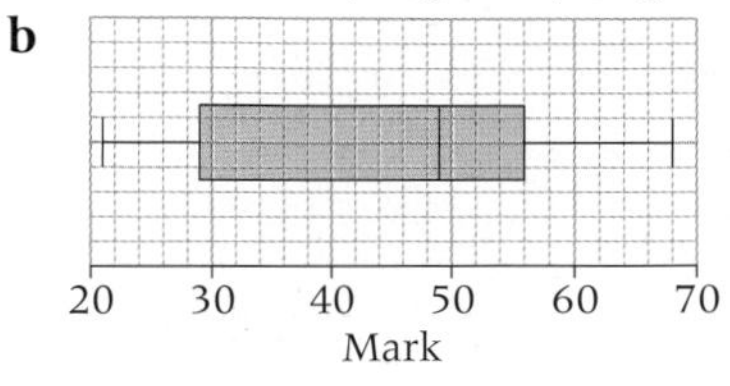

6

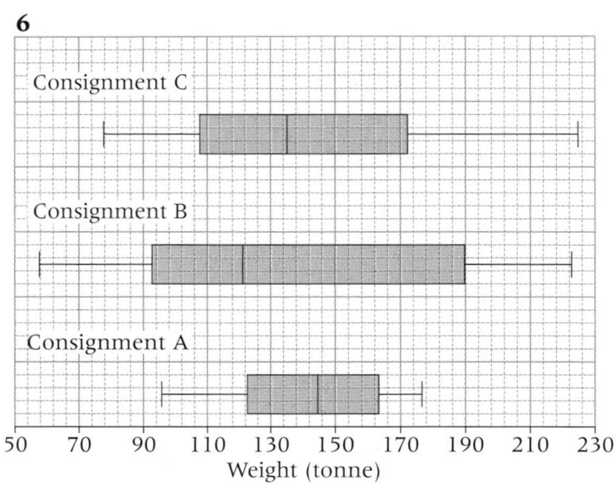

D2.2

1 **a**

Height, *h* cm	$h<$ 155	$h<$ 160	$h<$ 165	$h<$ 170	$h<$ 175
Cumulative frequency	9	36	81	97	100

b

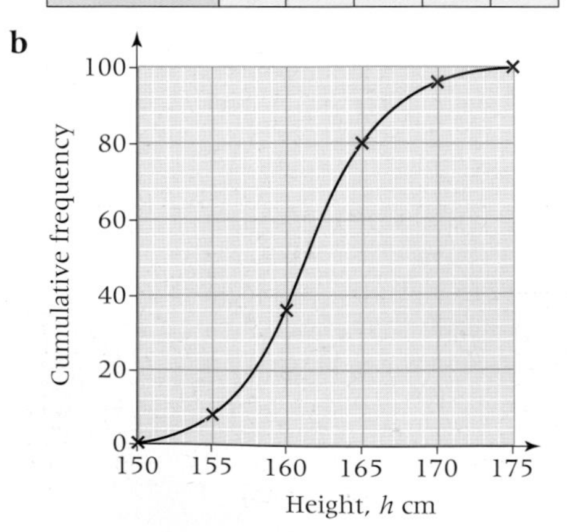

2 **a**

Age, *A*	$A<$ 30	$A<$ 40	$A<$ 50	$A<$ 60	$A<$ 70
Cumulative frequency	20	56	107	134	145

b

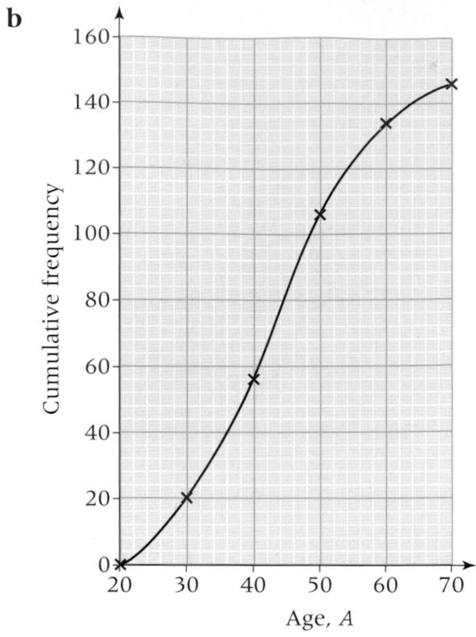

3 **a**

Time, *t* minutes	$t<$ 10	$t<$ 20	$t<$ 30	$t<$ 40	$t<$ 50	$t<$ 60
Cumulative frequency	6	24	53	88	109	120

b

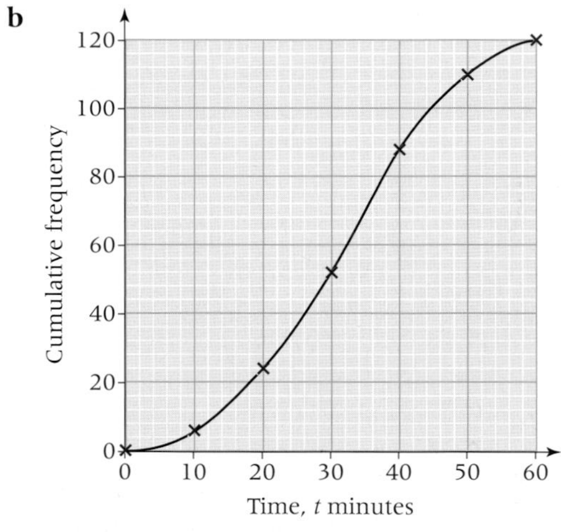

4 **a**

Weight, *w* minutes	$w<$ 3000	$w<$ 3500	$w<$ 4000	$w<$ 4500	$w<$ 5000
Cumulative frequency	7	30	64	89	100

b

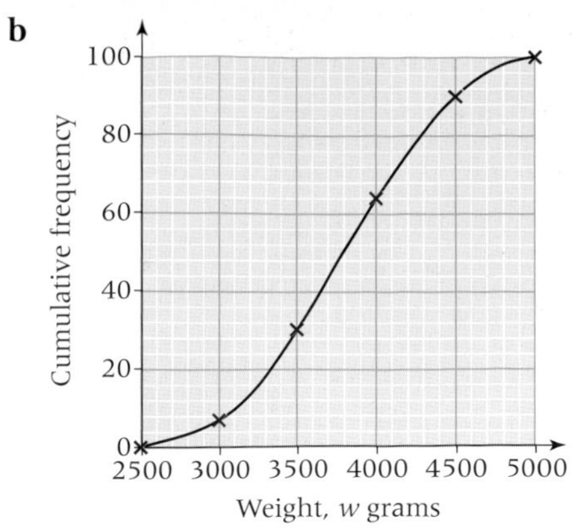

5 a

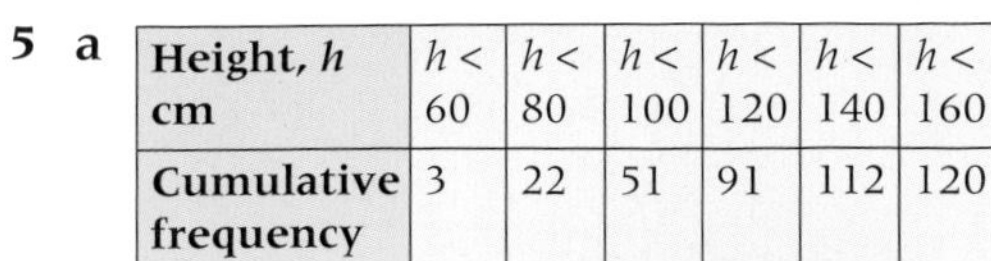

Height, *h* cm	$h <$ 60	$h <$ 80	$h <$ 100	$h <$ 120	$h <$ 140	$h <$ 160
Cumulative frequency	3	22	51	91	112	120

b

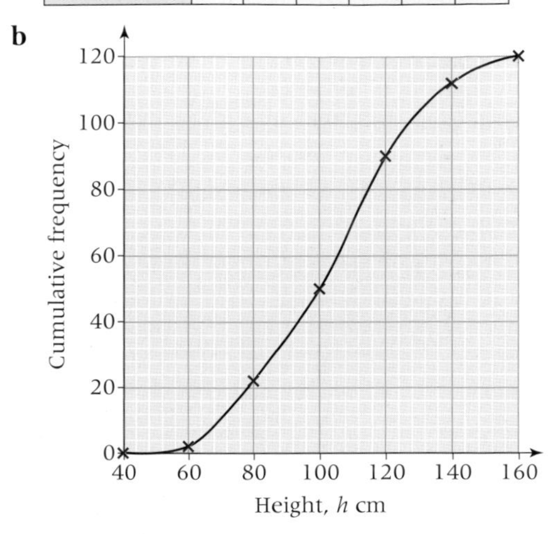

6 a

Test result, *t* %	$t <$ 40	$t <$ 50	$t <$ 60	$t <$ 70	$t <$ 80	$t <$ 90
Cumulative frequency	5	21	46	79	97	100

b

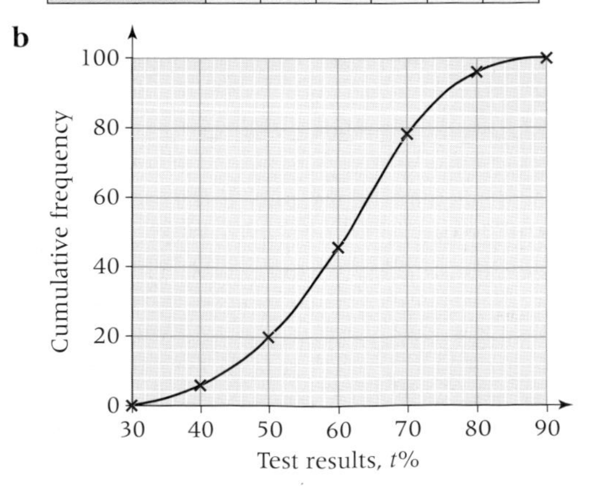

D2.3

1 a 162 **b** 7 **c** 35

2 a 43 **b** 15 **c** 10

3 a 32 **b** 19 **c** 56

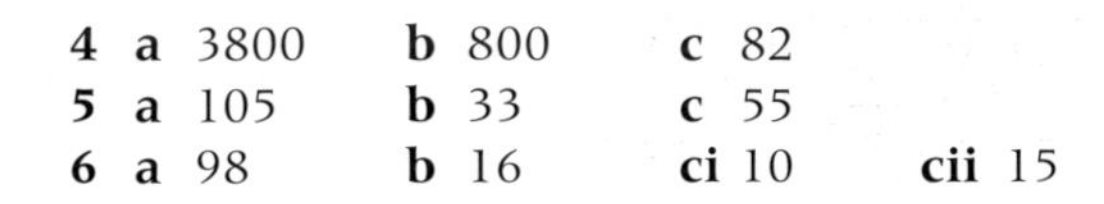

4 a 3800 **b** 800 **c** 82

5 a 105 **b** 33 **c** 55

6 a 98 **b** 16 **ci** 10 **cii** 15

D2.4

1

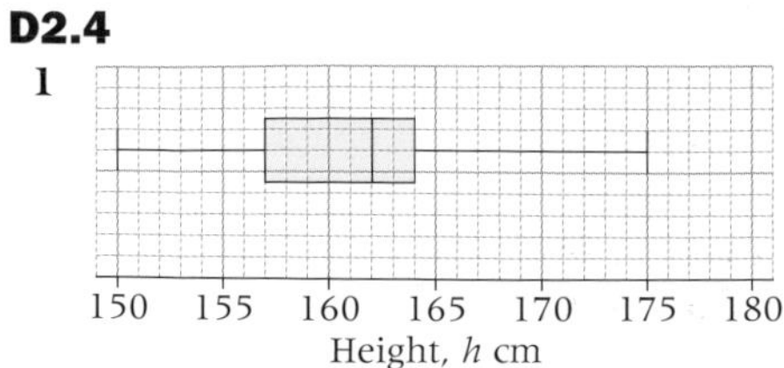

2

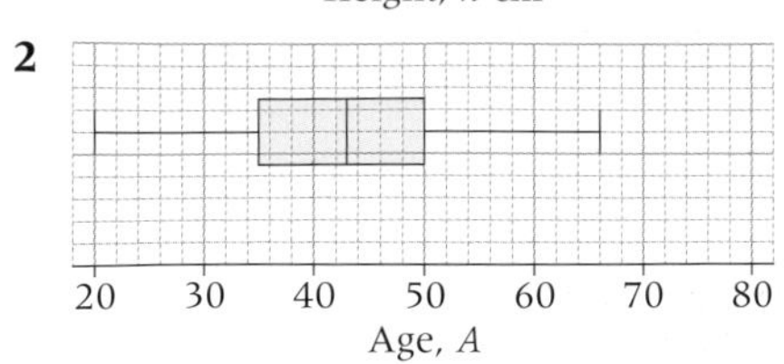

3

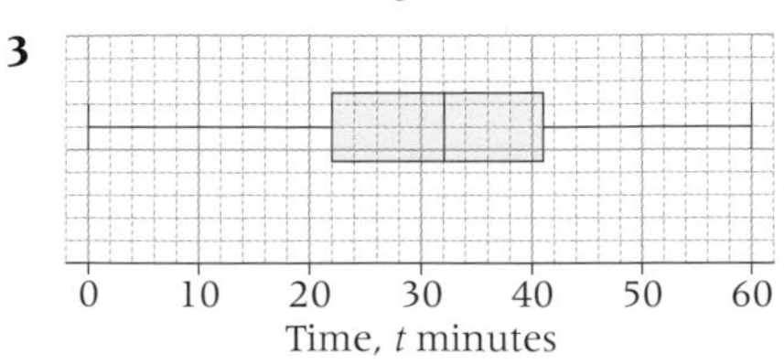

4

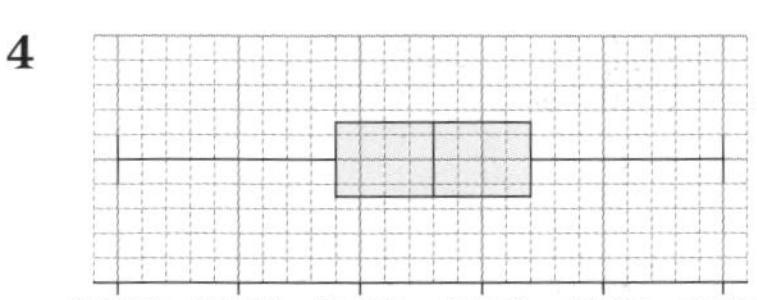

5

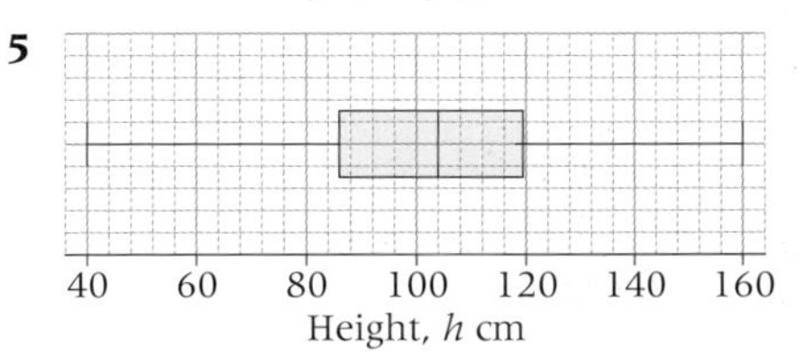

6

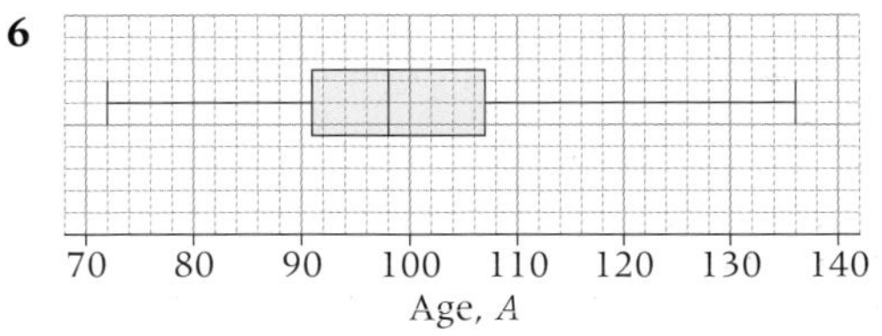

7 a

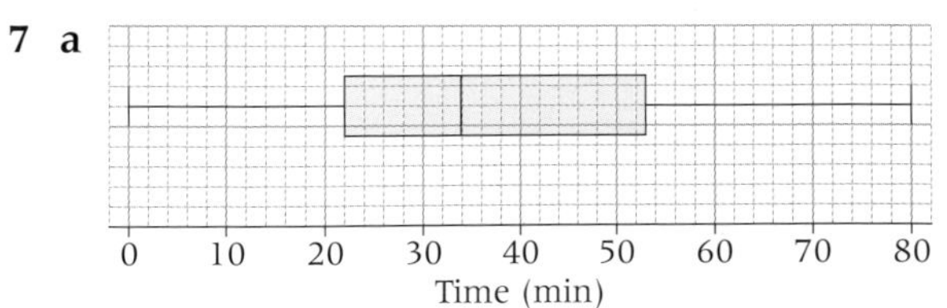

b

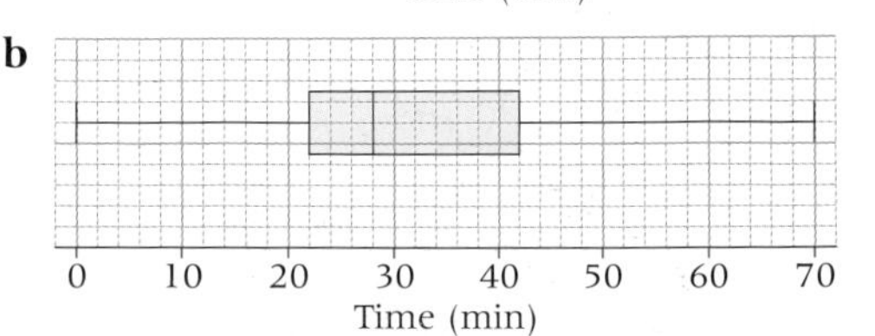

D2.5

1 On average, calls to the mobile phone helpline were longer. The length of calls to the computer helpline are more varied. The middle 50% of calls to the computer helpline vary in length more. More of the middle 50% of calls are less than the median for the mobile phone helpline.

2 On average the reaction times were the same. The reaction times of the teenagers are more varied. The middle 50% of reaction times of the over 60s is more varied. More of the middle 50% of times are less then the median for the over 60s.

3 On average, the girls got higher marks. The marks for the boys are more varied. The middle 50% of marks vary about the same for boys and girls.

4 On average, Farmer Green's sunflowers are taller. The heights of Farmer McArthur's sunflowers vary more. The middle 50% of Farmer McArthur's sunflowers vary more in height.

D2.6

1 a

8	9
9	0 2 3 5 9
10	2 4 5 6 7 7 8 9
11	0 1 3 5 5 6 7 8 9
12	0 0 6 7 8
13	0 2 2

Key: 13 | 2 means 132

b Most IQs are between 100 and 120

c Mode = 107, 115, 120, 132; median = 111; IQR = 16

d Median is best as there are several modes

2 a

0	4 5 8 9
1	0 1 1 2 5 5 6 8 8
2	1 2 3 3 3 5 6
3	1 2 3

Key: 1 | 1 means 11 minutes

b Most journeys take more than 10 minutes

c Mode = 23 minutes, median = 18 minutes, IQR = 12 minutes

d Median is best, as mode is quite high

3 a

14	5 7 8 8 8 9 9
15	1 3 3 4 5 6 6 7 7
16	2 5 6 7
17	0

Key: 14 | 5 means 14.5 kg

b Most are less than 15.7 kg

c Mode = 14.8 kg, median = 15.4 kg, IQR = 1.1 kg

d Median is best, as mode is quite low.

4 a

Women		Men
9 9 8	14	
8 7 7 7 6 4 3 2 2 1	15	7 8 9
8 8 7 6 3 2 0	16	4 6 8 8 9
7 5 2	17	2 2 3 3 4
	18	1 1 2 3 4

Key: 9 | 14 means 149 cm   Key: 18 | 1 means 18 cm

b Men: median = 173 cm, IQR = 13 cm, range = 32 cm
Women: median = 157 cm, IQR = 15 cm, range = 29 cm

c On average, men are taller. The variation in heights is similar.

5 a

B		A
9 7	9	1 4 6
6 5 4 4	10	1 1 2 3 8
7 6 2 2 1	11	0 0 1 5 6 7 9
9 9 8 4 3 2 1 0 0	12	1 3 5 8 9
1 1 0	13	1

Key: 9 | 12 means 129   Key: 13 | 1 means 131

b A: median = 111, IQR = 20.5, range = 40
B: median = 120, IQR = 22, range = 34

c On average, the marks in test B were higher. The middle 50% of marks in test A varied less than the middle 50% in test B (lower IQR), but overall the marks in test A varied more (higher range)

6 a

Y		X
	3	2 3 7 7 9 9
7 7 4 3 1	4	3 4 4 6
8 7 6 3 1	5	1 2 7 7 7 8
8 6 5 2 2 1	6	1 2 5 5 9
7 6 5 4 3	7	1 2
1 1	8	

Key: 7 | 4 means 4.7   Key: 7 | 1 means 7.1

b X: median = 5.2, IQR = 2.3, range = 4.0
Y: median = 6.2, IQR = 2.3, range = 4.0

c On average, the scores or club Y were higher. The variation in scores was the same.

D2 Summary

1 P and C
Q and D
R and B
S and A

Case study 1: Weather

16°C; 2.5°C; −8.9°C

England:
Range = 38.5 − −26.1 = 64.6°C

Wales:
Range = 35.2 − −23.3 = 58.5°C

Scotland:
Range = 32.9 − −27.2 = 60.1°C

N. Ireland:
Range = 30.8 − −17.5 = 48.3°C

D3 Check in

1 a $\frac{3}{5}$ b $\frac{3}{7}$ c $\frac{5}{8}$
d $\frac{5}{6}$ e $\frac{9}{20}$ f $\frac{23}{24}$

2 a $\frac{1}{6}$ b $\frac{3}{20}$ c $\frac{2}{15}$

3 a 0.7 b 0.75 c 0.375
d 0.4 e $0.\dot{3}$ f $0.\dot{1}4285\dot{7}$

D3.1

1 a $\frac{9}{25}$ b $\frac{16}{25}$ c $\frac{8}{25}$ d $\frac{17}{25}$ e $\frac{2}{5}$ f $\frac{19}{25}$

2 a The Hawks' chance of winning is twice the Jesters' chance of winning.
b 0.15

3 0.25

4 a $\frac{1}{9}$ b $\frac{8}{9}$ c $\frac{4}{9}$ d $\frac{1}{9}$ e $\frac{1}{3}$
f $\frac{2}{3}$ g $\frac{2}{3}$ h $\frac{1}{3}$ i $\frac{4}{9}$

5 a 30
bi $\frac{2}{3}$ bii $\frac{4}{15}$ biii $\frac{17}{30}$
biv $\frac{1}{3}$ bv $\frac{1}{6}$ bvi Can't tell

6 a $\frac{1}{32}$ b $\frac{5}{16}$ c $\frac{13}{16}$ d $\frac{15}{32}$

D3.2

1 120

2 $\frac{1}{5}$ or 0.2

3 ai 0.23 aii 0.33 aiii 0.44
aiv 70.72 bi 56 bii 112

4 a 152 b 126 c 215

5 a

	Canoeing	Abseiling	Potholing	Total
Male	13	4	5	22
Female	7	9	12	28
Total	20	13	17	50

bi $\frac{2}{5}$ bii $\frac{3}{5}$ biii $\frac{37}{50}$ biv $\frac{3}{5}$
ci 120 cii 168

D3.3

1 No, $\frac{124}{360}$ = 0.34 which is not close to 0.5.

2 Yes, as the probability of picking a black is close to 0.5.

3 Yes, all frequencies have similar values.

4 No, 3 occurs much more than other outcomes.

5 a 0.2, 0.3, 0.4, 0.1, 0.1, 0.2 0.3, 0.2, 0.1, 0.3
b 0.22

6 a 0.306, 0.222, 0.222, 0.25
b Probably not, as the relative frequencies are similar, but more trials are needed to be sure.

D3.4

1 a

	1	2	3	4	5	6
Head	H and 1	H and 2	H and 3	H and 4	H and 5	H and 6
Tail	T and 1	T and 2	T and 3	T and 4	T and 5	T and 6

bi $\frac{1}{12}$ bii $\frac{1}{12}$
The probabilities are the same as both coin and dice are fair.

2 a

Blue dice

Red dice	1	2	3	4	5	6
1	1, 1	1, 2	1, 3	1, 4	1, 5	1, 6
2	2, 1	2, 2	2, 3	2, 4	2, 5	2, 6
3	3, 1	3, 2	3, 3	3, 4	3, 5	3, 6
4	4, 1	4, 2	4, 3	4, 4	4, 5	4, 6
5	5, 1	5, 2	5, 3	5, 4	5, 5	5, 6
6	6, 1	6, 2	6, 3	6, 4	6, 5	6, 6

bi $\frac{1}{36}$ bii $\frac{1}{36}$ biii $\frac{1}{12}$ biv $\frac{1}{18}$

3 $\frac{1}{6}$

4 a $\frac{4}{25}$ b $\frac{1}{25}$ c $\frac{3}{50}$ d $\frac{1}{50}$ e $\frac{3}{100}$

5 a

2 pence coin

10 pence coin	Head	Tail
Head	Head, Head	Head, Tail
Tail	Tail, Head	Tail, Tail

b $\frac{1}{4}$

6 Yes, as 0.24 × 0.15 = 0.036

D3.5

1 0.99

2 $\frac{31}{45}$

3 $\frac{193}{288}$

4 a $\frac{99}{400}$ b $\frac{407}{1875}$

5 a $\frac{6}{23}$ b $\frac{65}{136}$

D3 Summary

1 £4

2 0.4712 (Accept 0.47, 0.471)

N2 Check in

1 a 60% b $\frac{7}{20}$ c $\frac{13}{20}$ d 0.35

2 a 80 b 21 c 19.6 d 168

N2.1

1 ai $\frac{7}{20}$ aii 35%
bi $\frac{2}{25}$ bii 8%
ci $\frac{3}{10}$ cii 30%

di $\frac{3}{40}$ dii 7.5%
ei $\frac{9}{40}$ eii 22.5%
fi $\frac{3}{20}$ fii 15%

2 a $\frac{3}{25}$ b 12%

3 a 16.8 g b 8.75 g c 9 g
d 67.2 g e 18.75 g f 48.72 g

4 a 45 kg b 84 g c 48 cm
d 285 m e 15.95 cc f 260 mm

5 a 135 kg b 32.085 m c £105
d 22.5 cm^3 e 510.15 g f £16.50

6 a £640 b €180 c 240 kg
d 304 cc e 1640 kg f 23.3 m

7 a £1500
bi $\frac{5}{12}$ bii 41.7%

N2.2

1 a £12 b 240 m c 52.2 kg
d 0.18 km e €70.40 f 18 seconds
g 86.1 cm h 34.2 g

2 a 0.25 b 0.4 c 0.9 d 0.05
e 1.1 f 0.3 g 1.05 h 0.95

3 a 0.3 b 0.32 c 0.325
d 0.0125 e 1.12 f 0.000 06

4 a 1.15 b 1.025 c 1.225
d 1.875 e 2.08 f 1.000 45

5 a 0.95 b 0.9175 c 0.7725
d 0.6175 e 0.02 f 0

6 a 1.1 b 1.21

7 a 1.113 b 0.972 c 1.014

8 It makes no difference which she does first.

9 2.13%

N2.3

1 a 2.5% b 3.33% c 55.2% d 1.25%

2 41.4%

3 16%, 12.7%, 36.6%, 22.8%

4 £8450

5 13 600

6 a £182.25 b £385.01

7 a £82.94 b £313.42 c £140.92

N2.4

1 188 mm, 42.75 litres, 229.4 cm^3, £177.01

2 a £330 b £246.50
c £805.35 d £342.42

3 a £182.32 b £299 250

4 a £167.60 b £339 023

5 a £213.75
b No, the compound interest would only be £178.30.
c No, the compound interest would be £390.07 but the simple interest would be £427.50.

6 6 years

7 Yes, as if it had exactly doubled she would have sold it for £122 500 × 1.035^{21} = £252 280.

8 a £2747.76 b £1560

N2.5

1 a 2 : 1 b 5 : 1 c 3 : 2
d 1 : 2 : 3 e 5 : 2 f 1 : 2 : 6

2 a £33 : £22 b 75 cm : 45 cm
c 48 seats : 36 seats : 12 seats
d 27 tickets : 9 tickets : 6 tickets
e 96 books : 36 books : 12 books
f 80 hours : 50 hours : 30 hours

3 £80 : £60, 51 cm : 34 cm, 10 h : 8 h, 7 cc : 42 cc, 21 min : 24 min, €570 : €150

4 Benny £222.22, Amber £177.78

5 Peggy £3200, Grant £3600, Mehmet £2800

6 Steven £6315.79, Will £3789.47, Phil £1894.74

7 141 cm^2 : 47 cm^2 : 12 cm^2,
19 cm : 12.7 cm : 6.3 cm,
208 m : 104 m : 138 m,
131 mm : 327 mm : 262 mm,
\$47.50 : \$17.81 : \$29.69

8 John £24.41, Janine £20.59

9 Con.170°, Lab.143°, L.D. 32°, other 15°

N2 Summary

1 a £21
b 47 kg OR 47.1 kg OR 47.12 kg

2 £348.48

3 8 years

4 £8400

D4 Check in

1 a 10 b 25 c 20 d 12.5 e 27.5

2 a 22 b 45

3 a 1.3 b 2.4 c 0.36 d 30

D4.1

1

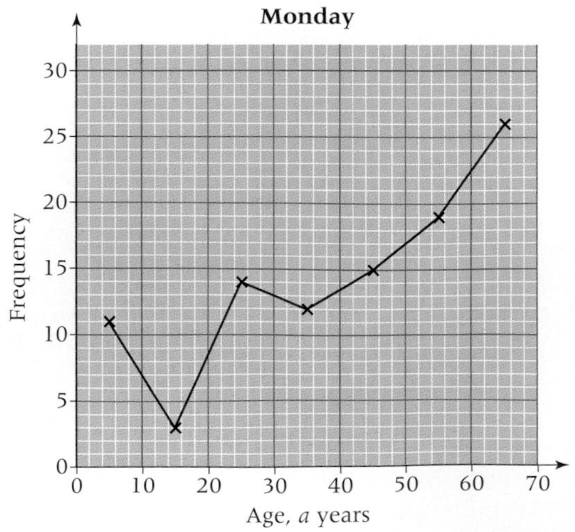

Monday

Frequency: 0, 5, 10, 15, 20, 25, 30

Age, *a* years: 0, 10, 20, 30, 40, 50, 60, 70

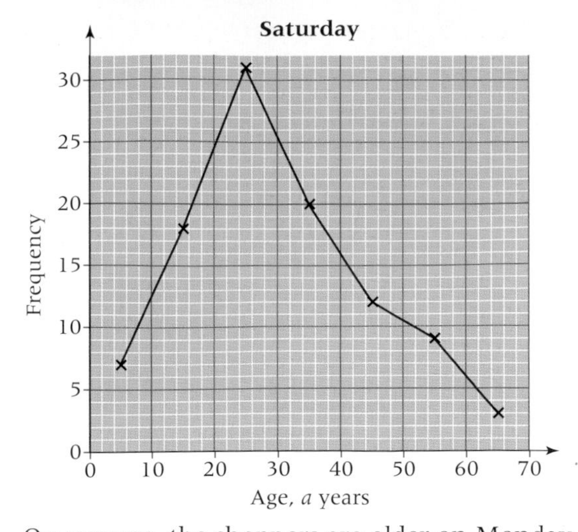

On average, the shoppers are older on Monday.

2 The modal journey time is greater for office workers. The highest frequency for teachers is in the range 20–30 minutes, for office workers it is in the range 30–40 minutes. The ranges are the same.

3 The modal daily mileages for January and December are in the same interval, 40–60 miles. The range for December is greater than the range for January, by 10 miles. January is likely to have a greater frequency of days where some mileage is attained, as December is likely to cover holiday time. The modal class is likely to be in the same range for the two months, as she will most likely travel to work to cover this mileage. The distribution for December is more skewed to the left (positive skew), as during the holidays Jayne is likely to spend more days driving smaller, local distances.

D4.2

1 a Class width: 2, 1, 1, 1, 3;
Frequency density: 6, 17, 19, 11, 6

b

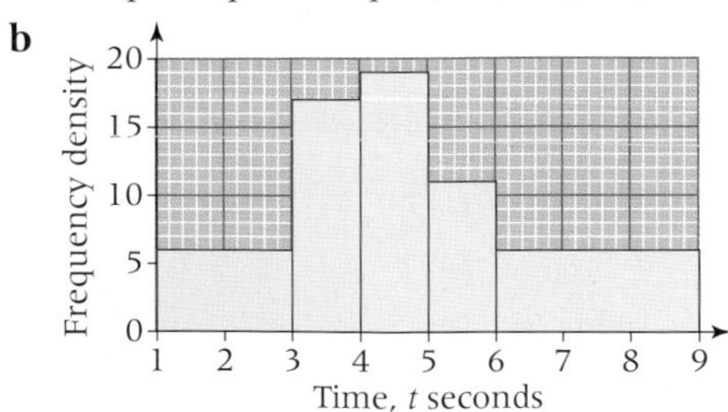

2 a Class width: 5, 5, 10, 20, 20, 40;
Frequency density: 1.2, 2, 2.3, 1.45, 1.2, 0.2

b

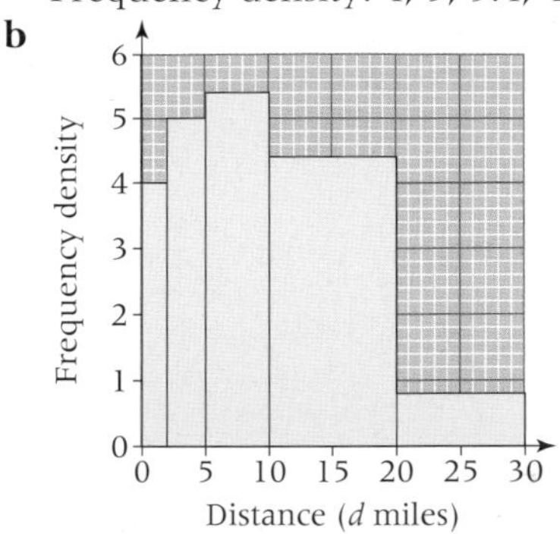

3 a Class width: 2, 3, 5, 10, 10;
Frequency density: 4, 5, 5.4, 4.4, 0.6

b

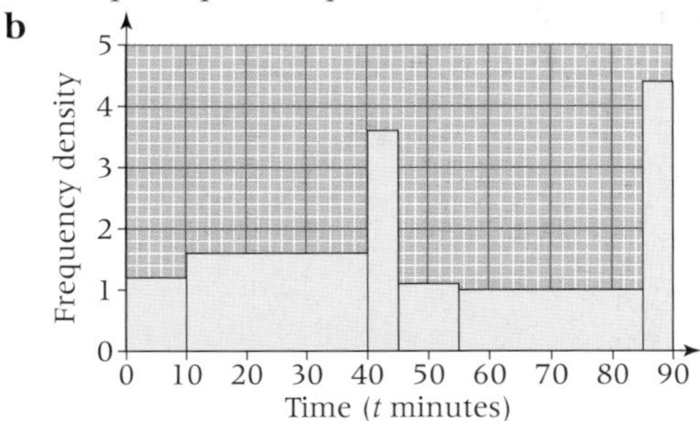

4 a Class width: 10, 30, 5, 10, 30, 5;
Frequency density: 1.2, 1.6, 3.6, 1.1, 1.0, 4.4

b

5 a Class width: 0.1, 0.3, 0.5, 1, 3;
Frequency density: 30, 40, 44, 25, 6

b

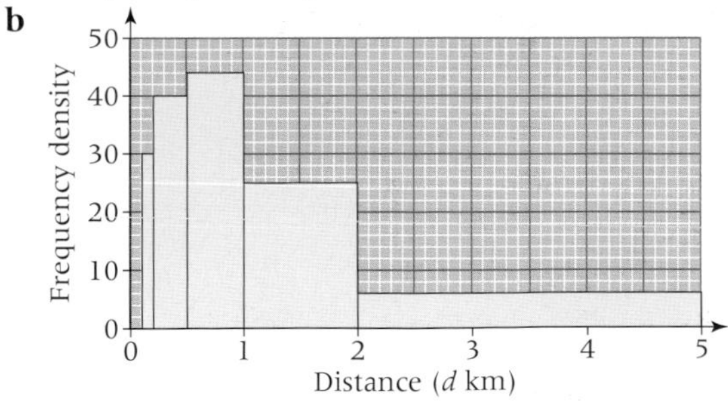

6 a Class width: 10, 20, 20, 30, 30;
Frequency 8, 16, 28, 39, 9
Frequency density: 0.8, 0.8, 1.4, 1.3, 0.3

b

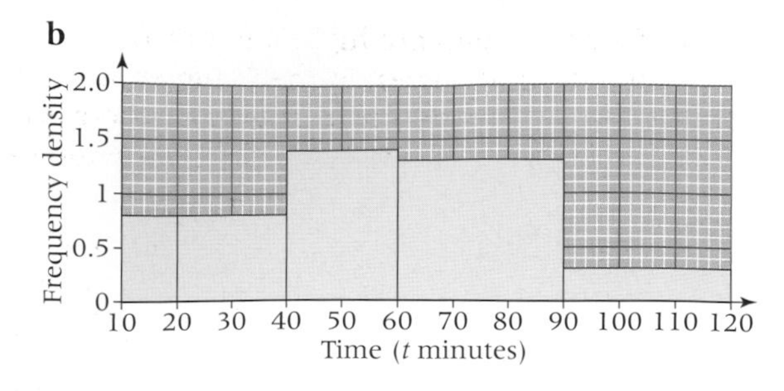

D4.3

1 a 10 b 4, 18, 53, 20, 10 c 105

2 a 48 b 3, 7, 28, 20, 8 c 66

3 a

Height (h cm)	Frequency
$110 \leqslant h < 115$	3
$115 \leqslant h < 120$	7
$120 \leqslant h < 125$	13
$125 \leqslant h < 135$	29
$135 \leqslant h < 145$	21
$145 \leqslant h < 150$	5
$150 \leqslant h < 170$	6

Total = 85

b

Height (h cm)	Frequency
$130 \leqslant h < 140$	1
$140 \leqslant h < 150$	9
$150 \leqslant h < 155$	9
$155 \leqslant h < 160$	12
$160 \leqslant h < 165$	14
$165 \leqslant h < 170$	13
$170 \leqslant h < 180$	15
$180 \leqslant h < 190$	1

Total = 74

D4.4

1 a 5, 8, 11

b

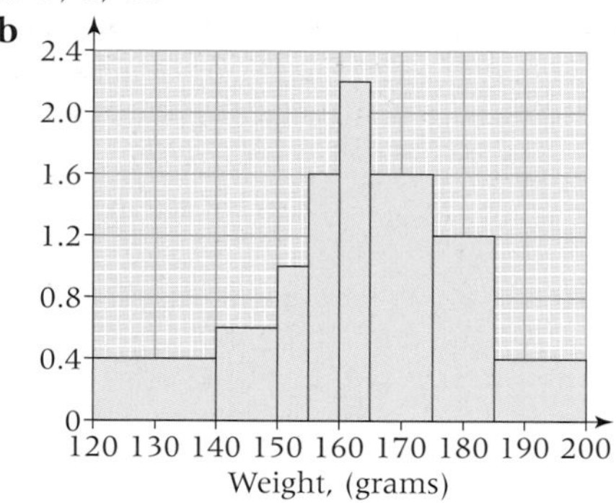

2 a 36, 42

b

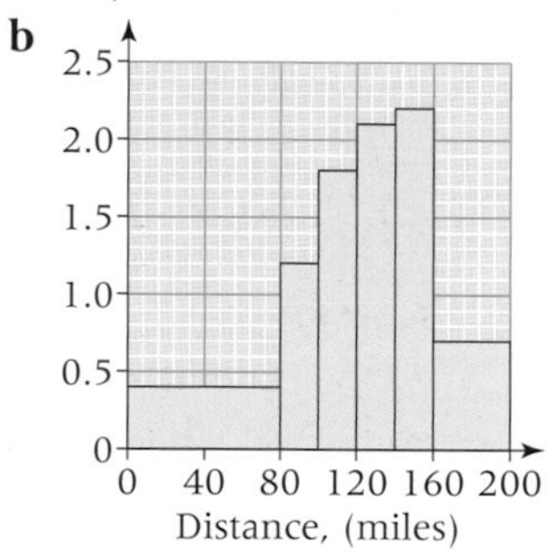

3 a Bar height = $\frac{2f}{15}$ cm
Frequency density = $\frac{f}{30}$

b Bar height = f
Frequency density = 0.4 and $\frac{f}{5}$

D4.5

1 The range for boys aged 11 is the same as the range for boys aged 16 (60 cm). The modal class for boys aged 11 is 125–135 cm, which is smaller than for 16-year-old boys, who have a modal class of 160–165 cm. The distribution for boys aged 11 is positively skewed, whereas for boys aged 16, the data is slightly negatively skewed.

2 The range for apples (80 g) is greater than the range for pears (70 g). The modal class for apples is 160–165 g, which is larger than for pears, having a modal class of 115–120 g. The distribution for apples is roughly symmetrical with no skew, whereas for pears, the data is slightly negatively skewed.

3 The range for girls (8 s) is greater than the range for boys (6 s). The modal class for girls is 6–6.5 s, which is the same as that for boys. Both distributions are slightly negatively skewed.

D4.6

1 a

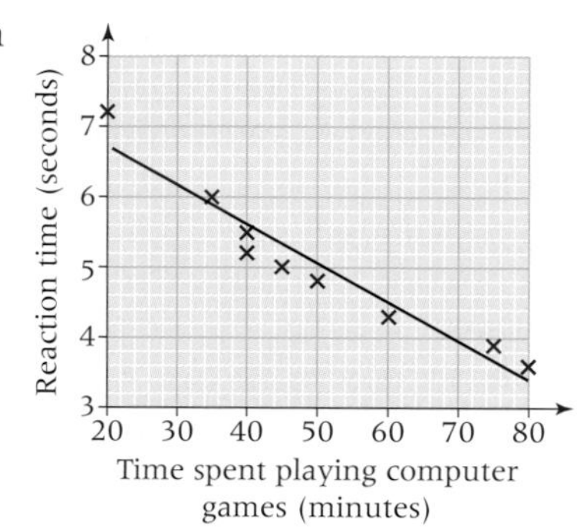

b Negative correlation

c As time spent playing computer grames increases, reaction time decreases

2 a

Length (cm)

Weight (grams)

b Positive correlation

c As weight of fish increases, so does length

3 a, c

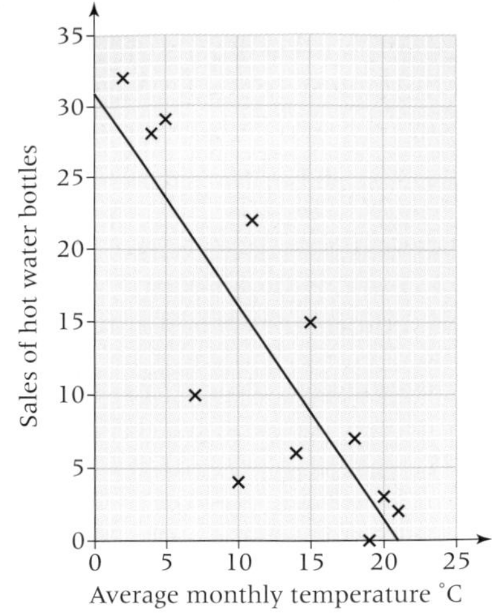

3 b Negative correlation: there will be less hot water bottles sold in the summer months when the temperature is generally higher.

d 7.5 °C

e By extending the line of best fit you can predict that 45 hot water bottles will be sold.

4 Let s = sales of hot water bottles

t = average monthly temperature in °C

Then $s = 31 - 1.5t$

At zero average temperature, expected sales are 31; for every 1°C rise in temperature monthly sales drop by an average of 1.5 bottles.

D4.7

1 a

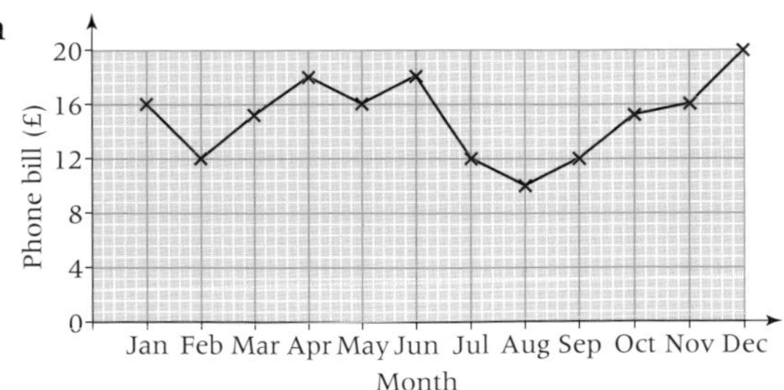

b Typical phone bills are about £15. These fall during the Summer months and peak in December.

2 a

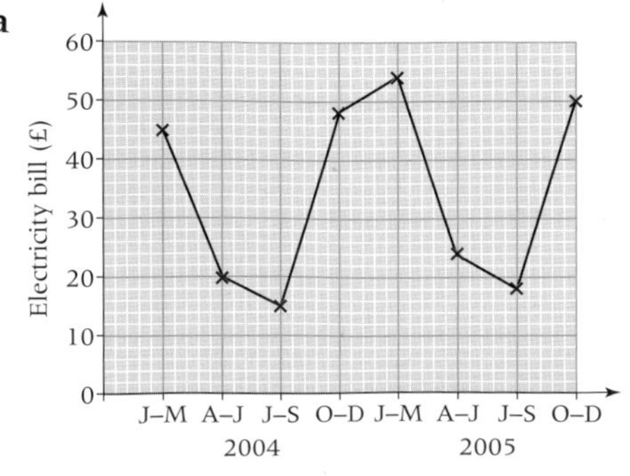

b Electricity bills are highest in the Winter months and lowest in the Summer months. This annual pattern repeats itself; there is a slight trend for bills to rise from year-to-year.

3 a

Sales (£)
0 10 20 30 40 50 60
Jan Feb Mar Apr May Jun Jul Aug Sep Oct Nov Dec
Month

b Icecream sales grow steadily during Spring and Summer but drop sharply in the Autumn. Sales are low during Autumn and Winter except for a peak in December.

4 a

Percentage
0 10 20 30
1998 1999 2000 2001 2002 2003 2004 2005
Year

b The percentage of students using the library grow steadily from about 15% in 1998 to 28% in 2001. It has since fallen back to around 20%.

5 a

Earnings (£)
0 10 20 30 40 50 60 70 80
J–A M–A S–D J–A M–A S–D J–A M–A S–D
2001 2002 2003
Month/year

b Christable's earnings have grown steadily during the three years. Her earnings peak in the Winter months.

6 a

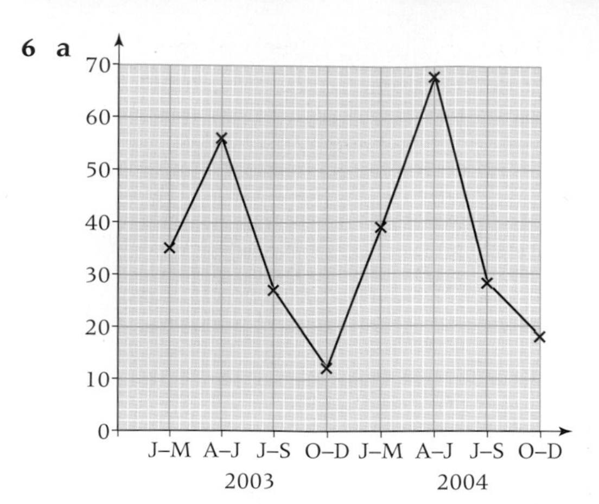

b Steve's expenses grow to a peak in Spring and fall back in Winter. The level of expenses appears to be growing from year-to-year.

D4 Summary

1 a 30 – 40 cups (Accept 31 – 40 cups)

b

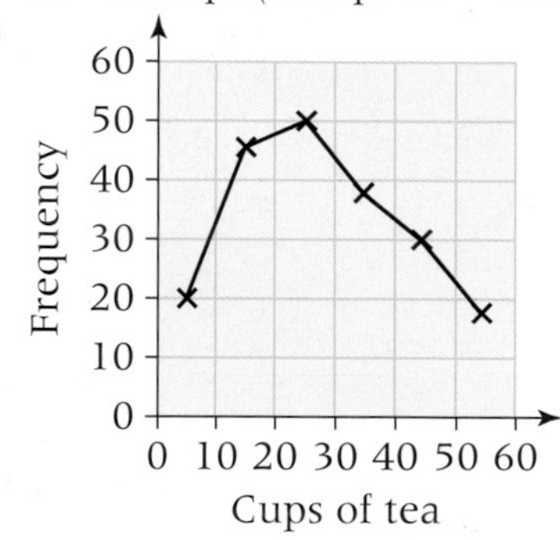

c No, because the maximum and minimum values are not known.

Case study 2: Recycling

Chart text, from DEFRA:

There has been a change in the composition of recycled waste over time. In 1997/8 paper and card was the largest component, making up 37% of the total, followed by compost (20 per cent) and glass (18 per cent). In 2007/08 compost was the largest component (36.1 per cent of the total) with the next largest being paper and card (18.1 per cent) followed by co-mingled (17.7 per cent). Co-mingled collections - the collection of a number of recyclable materials in the same box or bin, for example paper, cans, plastics - have become more widespread in recent years.'

'25.3 million tonnes of household waste was collected in England in 2007/08; 34.5% of this waste was collected for recycling or composting. The amount of household waste not re-used, recycled or composted was 16.6 million tonnes, a decrease of 7.0 per cent from 2006/07. This equates to 324 kg per person of residual household waste and shows progress towards the 2010 target, in the Waste Strategy 2007, of reducing this amount to 15.8 million tonnes.

New can weighs 55.8 g (to 3 s.f)

Weight has decreased by approximately 50%

Volume of container = 1296 cm^3

Volume of six tomatoes = 679 cm^3

48% of the available volume is empty.

D5 Check in

1 a	0.55	b	0.04	c	0.72	d	0.625
e	0.6	f	0.34	g	0.9	h	0.18
i	0.17	j	0.192	k	0.235	l	0.92
2 a	$\frac{1}{6}$	b	$\frac{4}{5}$	c	$\frac{2}{9}$	d	$\frac{13}{15}$
e	$\frac{11}{12}$	f	$\frac{5}{9}$	g	$\frac{8}{45}$	h	$\frac{2}{5}$

D5.1

1

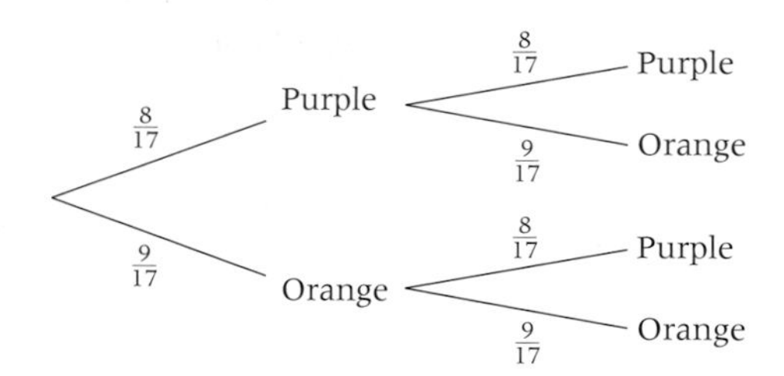

2 a

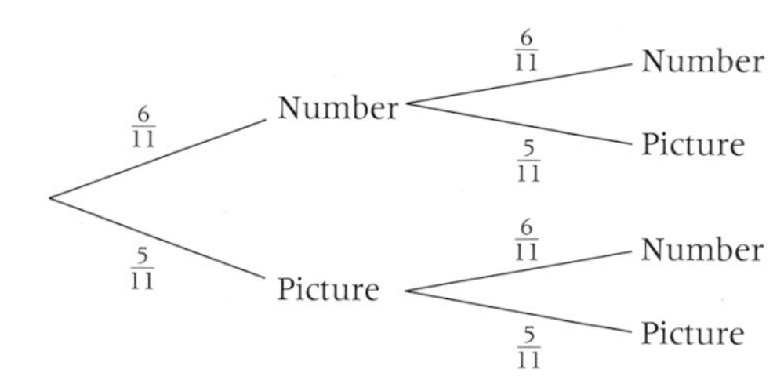

3

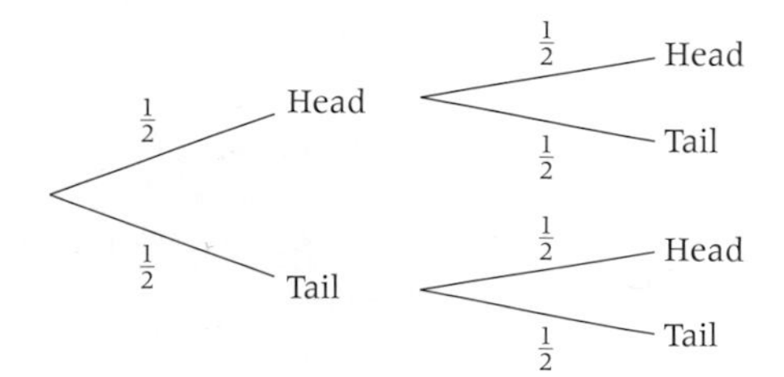

4 Marble Coin

Red: $\frac{7}{9}$ — Head $\frac{1}{2}$, Tail $\frac{1}{2}$

Blue: $\frac{2}{9}$ — Head $\frac{1}{2}$, Tail $\frac{1}{2}$

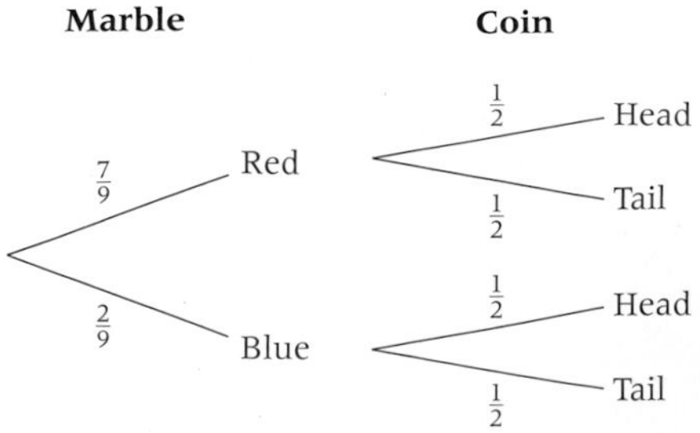

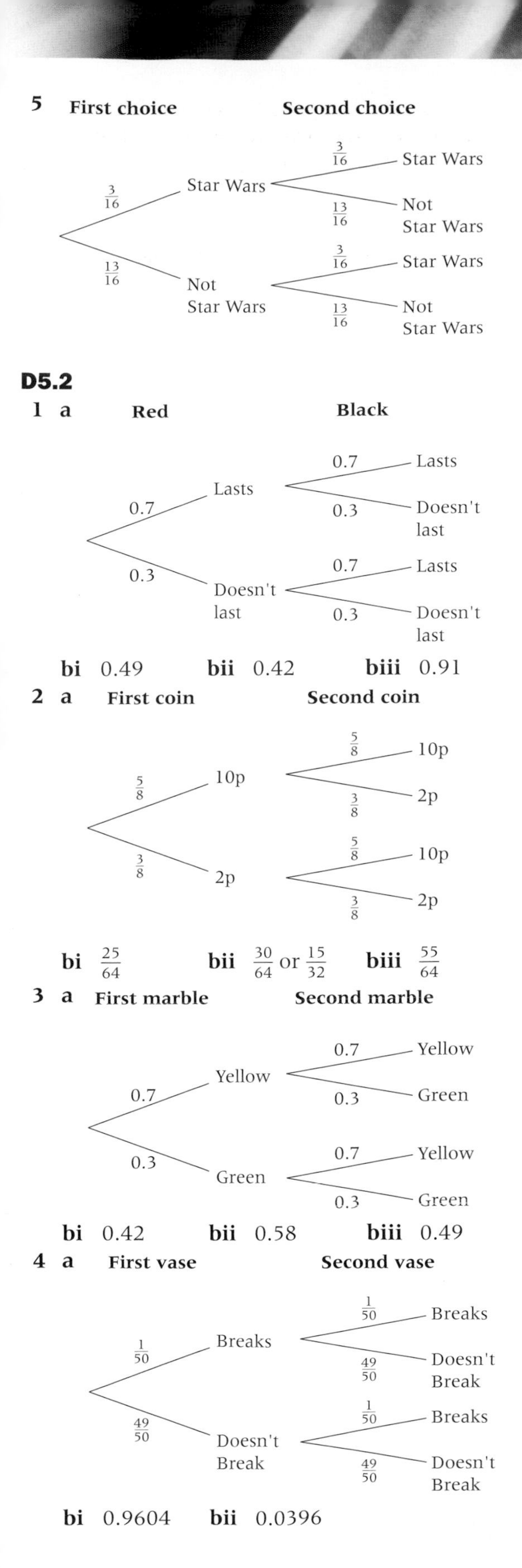
5 First choice
Second choice
3/16
Star Wars
3/16
Star Wars
13/16
Not Star Wars
13/16
Not Star Wars
3/16
Star Wars
13/16
Not Star Wars
D5.2
1 a Red
Black
0.7
Lasts
0.7
Lasts
0.3
Doesn't last
0.3
Doesn't last
0.7
Lasts
0.3
Doesn't last
bi 0.49
bii 0.42
biii 0.91
2 a First coin
Second coin
5/8
10p
5/8
10p
3/8
2p
3/8
2p
5/8
10p
3/8
2p
bi 25/64
bii 30/64 or 15/32
biii 55/64
3 a First marble
Second marble
0.7
Yellow
0.7
Yellow
0.3
Green
0.3
Green
0.7
Yellow
0.3
Green
bi 0.42
bii 0.58
biii 0.49
4 a First vase
Second vase
1/50
Breaks
1/50
Breaks
49/50
Doesn't Break
49/50
Doesn't Break
1/50
Breaks
49/50
Doesn't Break
bi 0.9604
bii 0.0396

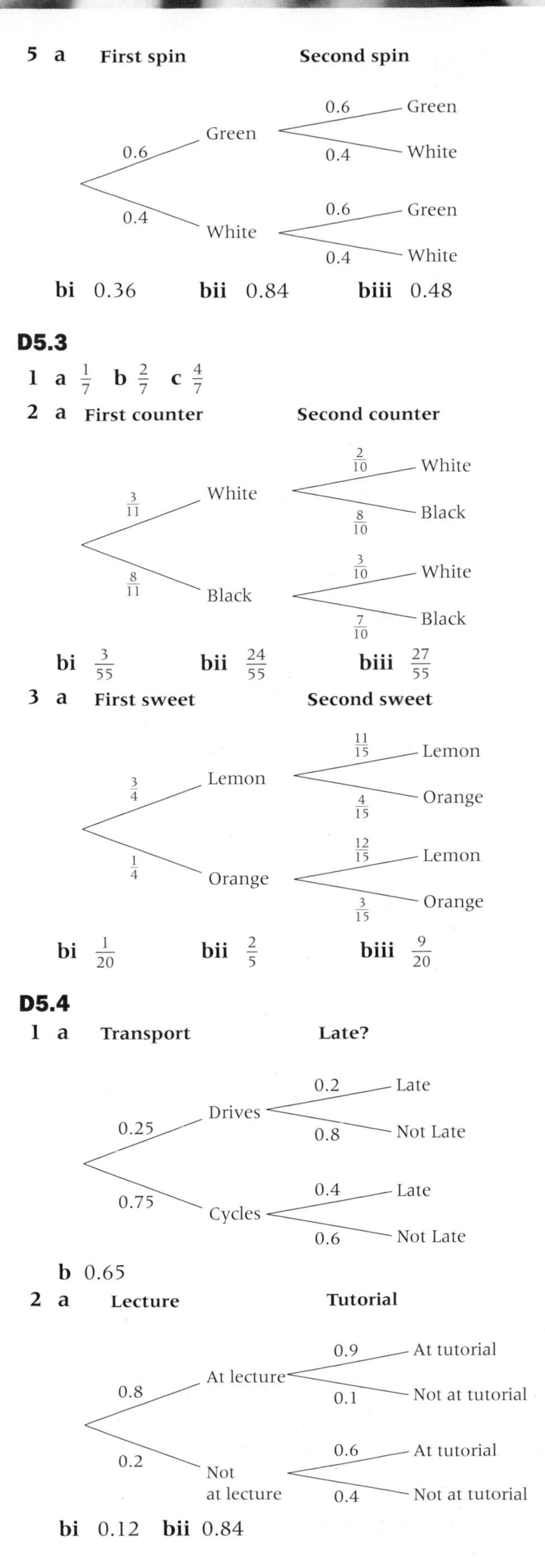
5 a First spin
Second spin
0.6
Green
0.6
Green
0.4
White
0.4
White
0.6
Green
0.4
White
bi 0.36
bii 0.84
biii 0.48
D5.3
1 a 1/7 b 2/7 c 4/7
2 a First counter
Second counter
3/11
White
2/10
White
8/10
Black
8/11
Black
3/10
White
7/10
Black
bi 3/55
bii 24/55
biii 27/55
3 a First sweet
Second sweet
3/4
Lemon
11/15
Lemon
4/15
Orange
1/4
Orange
12/15
Lemon
3/15
Orange
bi 1/20
bii 2/5
biii 9/20
D5.4
1 a Transport
Late?
0.25
Drives
0.2
Late
0.8
Not Late
0.75
Cycles
0.4
Late
0.6
Not Late
b 0.65
2 a Lecture
Tutorial
0.8
At lecture
0.9
At tutorial
0.1
Not at tutorial
0.2
Not at lecture
0.6
At tutorial
0.4
Not at tutorial
bi 0.12 bii 0.84

3 a

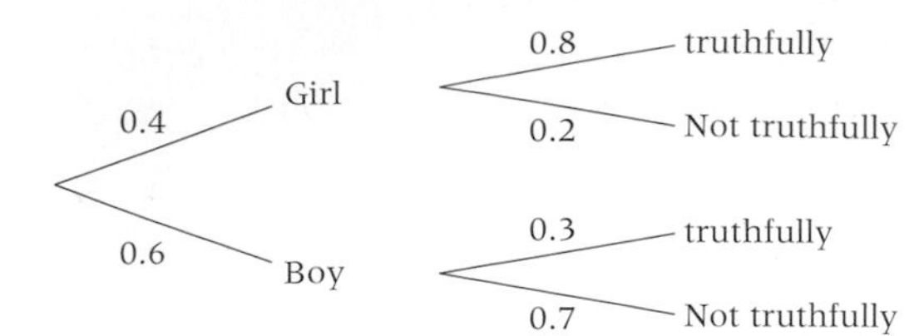

bi 0.5 **bii** 0.32

D5.5

1 a $\frac{103}{220}$ **b** $\frac{51}{103}$ **c** $\frac{17}{36}$

2 a $\frac{9}{25}$ **b** $\frac{4}{9}$ **c** $\frac{4}{11}$

3 a $\frac{1}{12}$ **b** $\frac{1}{17}$

4 a $\frac{2}{3}$ **b** $\frac{6}{7}$ **c** $\frac{3}{4}$

D5 Summary

1 a

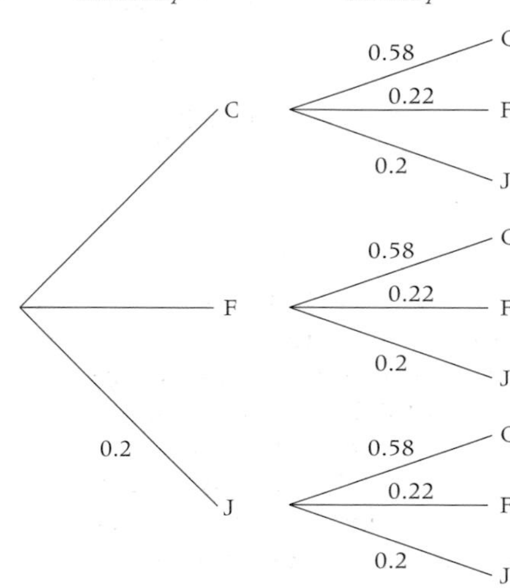

b 0.04

c 0.36

2 $\frac{129}{190}$ (or an equivalent fraction or decimal, not 129 out of 190 or a ratio)

A1 Check in

1 a 196 **b** 1 **c** −32 **d** $\frac{8}{27}$

e 0.000008 **f** 10 000 000 000 **g** −5 **h** $\frac{2}{5}$

2 a $4n + 7$ **b** $3(n - 6)$ **c** $10 - n^2$

d $\frac{3n-6}{2}$ **e** n^5

3 ai 72 **aii** 150 **aiii** 504

aiv 900 **av** 44 268

b x

A1.1

1 a y^{12} **b** k^4 **c** m^{12} **d** g^3 **e** h^{-6}

f b^{-12} **g** j^{-6} **h** t^{10} **i** n^{-2}

2 a $15h^{13}$ **b** $3p^2$ **c** $4p^{16}$ **d** $60r^{-1}$

e $27h^{-9}$ **f** $3b^8$ **g** $2m^{-9}$ **h** 2

3 $30w^{-9}$

4 $64p^{12} \div 64p^{14} = p^{-2}$

5 False, it simplifies to 3^{x+y}.

6 a 36 **b** 3 **c** 64 **d** −7 **e** 49

7 The first expression is the odd one out.

8 $x = 1$

9 a $9^x + 3^{x+1} = 3^{2x} + 3 \times 3^x = u^2 + 3u$

b $3^{3x} + 9 \times 3^x$ or $3^{3x} + 3^{x+2}$

c $3^{2x} - 3^{-x}$

d $u^4 - \frac{u^2}{9}$

A1.2

1 a $15x + 27$ **b** $8p^2 - 16p$

c $15m - 6m^2$ **d** $21y + 17$

e $10x^2 + 10xy - 45x$ **f** $57 - 2t$

g $4h + 16$ **h** $3x^3 + x^4$

2 a $x^2 + 13x + 42$ **b** $2x^2 - 2x - 24$

c $12p^2 + 23p + 10$ **d** $6m^2 - 32m + 42$

e $10y^2 + 17y - 63$ **f** $9t^2 - 12t + 4$

g $2x^2 + 4x - 15$ **h** $-25b^2 + 10b - 1$

3 a $(3y + 8)(2y - 1) = 6y^2 + 13y - 8$

b $\frac{1}{2}(2x - 3)(3x - 2) = 3x2 - 6.5x + 3$

4 $\sqrt{(3p-1)^2 + (2p+2)^2} = \sqrt{13p^2 + 2p + 5}$

5 $x(12 - x) = 12x - x^2$

6 $(2n + 1)^2 - (2n - 1)^2$

$= 4n^2 + 4n + 1 - 4n^2 + 4n - 1 = 8n$

7 a $x^3 + 6x^2 + 12x + 8$ **b** $2y^3 + 3y^2 - 32y + 15$

A1.3

1 a $3(x + 2y + 3z)$ **b** $5(2p - 3)$

c $x(5y + 7)$ **d** $3m(2n + 3t)$

e $4x(4x - 3y)$ **f** $3p(1 + 3q)$

g $7x(y - 8x)$ **h** $3x(x + 4x^2 - 2)$

i $(m + n)(3 + m + n)$ **j** $(p - q)[4 + (p - q)^2]$

k $(a + b)(x + y)$ **l** $(a - b)(c - d)$

2 a $(x + 2)(x + 5)$ **b** $(x + 3)(x + 5)$

c $(x + 2)(x + 6)$ **d** $(x + 5)(x + 7)$

e $(x - 5)(x + 2)$ **f** $(x - 7)(x + 5)$

g $(x - 5)(x - 3)$ **h** $(x - 5)(x + 4)$

i $(x - 20)(x + 12)$ **j** $(x - 9)(x + 12)$

k $(x - 5)(x + 5)$ **l** $(x - 3)(x + 2)$

3 a $(x + y)^2$ **b** 400

4 a $2(3x + 1)$ **b** $2(x - 6)(x - 4)$

5 $(x + 7)(x + 3) = 0$

6 a $2(x + 6)(x + 2)$ **b** $3(y + 3)(y + 12)$

c $4(m - 5)(m + 4)$ **d** $x(x + 5)(x + 3)$

e $x(y - 12)(y + 9)$ **f** $y(x - 4)(x + 4)$

7 a $(p - 4)(p + 3)$ **b** $3p(p + 2)$

c $10(x + 3)(x + 4)$ **d** $y(x - 9)(x + 7)$

e $a(a - b)^2$ **f** $3a(m + n + b + c)$

A1.4

1 a $(2x + 3)(x + 1)$ **b** $(3x + 2)(x + 2)$

c $(2x + 5)(x + 1)$ **d** $(2x + 3)(x + 4)$

e $(3x + 1)(x + 2)$ **f** $(2x + 1)(x + 3)$

g $(2x+7)(x-3)$ h $(3x+1)(x-2)$
i $(4x-3)(x-5)$ j $(6x-1)(x-3)$
k $(3x+2)(4x+5)$ l $(2x-3)(4x+1)$
m $3(2x-5)(x-2)$ n $(2x-3)(2x+3)$
o $(2x+3)(3x-1)$ p $(6x-1)(3x+4)$

2 You cannot find two numbers that multiply to 6 and add to 4.

3 a $9x(2x-1)$ b $4ab(1-2b)(1+2b)$
c $m(3n+8-m^2)$ d $(x-9)(x+2)$
e $(2x+5)(x-3)$ f $x(x+3)(x+4)$
g $p(2x+3)(x+4)$ h $50(x-5)(x+4)$
i $10(4p-3)(p-5)$ j $8(x+y)^2$

4 Total $= 8x^2 + 52x + 60$, so mean $= 2x^2 + 13x + 15$ $= (2x+3)(x+5)$

5 a $(2y+3)(1-3y)$ b $(3-2p)(4p+1)$
c $(3y-5)(2-y)$ d $3(2m-5)(2-m)$
e $y(3x+1)(2-x)$

A1.5

1 a $(x-10)(x+10)$ b $(y-4)(y+4)$
c $(m-12)(m+12)$ d $(p-8)(p+8)$
e $\left(x-\frac{1}{2}\right)\left(x+\frac{1}{2}\right)$ f $\left(k-\frac{5}{6}\right)\left(k+\frac{5}{6}\right)$
g $(w-50)(w+50)$ h $(7-b)(7+b)$
i $(2x-5)(2x+5)$ j $(3y-11)(3y+11)$
k $\left(4m-\frac{1}{2}\right)\left(4m+\frac{1}{2}\right)$ l $(20p-13)(20p+13)$
m $(x-y)(x+y)$ n $(2a-5b)(2a+5b)$
o $(3w-10v)(3w+10v)$ p $\left(5c-\frac{1}{2}d\right)\left(5c+\frac{1}{2}d\right)$
q $x(x-4)(x+4)$ r $2y(5-y)(5+y)$
s $\left(\frac{4}{7}x+\frac{8}{9}y\right)\left(\frac{4}{7}x-\frac{8}{9}y\right)$

2 a 400 b 240 000 c 199 d 157 000

3 a $30^2 - 7^2 = (30-7)(30+7) = 23 \times 37$
b $100^2 - 3^2 = (100-3)(100+3) = 97 \times 103$
c $34^2 - 23^2 = (34-23)(34+23) = 11 \times 57$
d $20^2 - 9^2 = (20-9)(20+9) = 11 \times 29$

4 $\sqrt{1200} = 20\sqrt{3}$

5 a $3(2x^2 - 5xy + 3y^2)$ b $(4a+3b)(4a-3b)$
c $(x-4)(x-7)$ d $(2x-3)(x+7)$
e $x(x-3)(x+6)$ f $5ab(1+2ab)$
g $(x+5)(2-x)$ h $10(1-x)(1+x)$
i $(y+9)(y-7)$ j $2x(x-8)(x+8)$
k $(3x-2)(2x-3)$ l $(x-y)(x+y)(x^2+y^2)$

6 $97^2 - 57^2 = (97-57)(97+57) = 40 \times 154$ $= 6160\text{ cm}^2$

A1.6

1 a $3w$ b $\frac{b}{3}$ c $2c$ d $\frac{4b}{d}$
e $4bd^2$ f $x+3$ g $x+1$ h $\frac{y-2}{3}$
i $x+2$ j $x-7$ k $\frac{1}{x-7}$ l $x-2$
m $2y-5$ n $\frac{1}{x+9}$ o $2x-5$ p $\frac{3x+4}{x-2}$
q $x-4$

2 The first fraction cannot be simplified as the numerator and denominator have no common factors.

3 $a(ab-1)$

4 a 3 b $\frac{12a}{7}$ c $\frac{2}{5}m$ d $\frac{15}{g^2}$
e $2\frac{2}{3}$ f $\frac{f}{2p^2}$ g $\frac{5}{y}$ h $3(x+7)$
i $\frac{2(x-2)(x-9)}{x^2-17x+18}$ j $\frac{2(x-5)}{x^2(x-6)}$

5 $\frac{15}{16}$ m by $\frac{4}{15}$ m

6 -3

A1.7

1 a $\frac{4p}{5}$ b $\frac{4y}{7}$ c $\frac{3}{p}$ d $\frac{11y}{8}$
e $\frac{p}{15}$ f $\frac{6y-7x}{xy}$ g $\frac{4x+2}{x^2}$

2 A–E, D–G, D–F

3 a $\frac{14x+3}{20}$ b $\frac{12x+13}{77}$ c $\frac{y-1}{15}$ d $\frac{41p-112}{35}$
e $\frac{5x-13}{(x-7)(x+4)}$ f $\frac{8x+9}{(x-2)(x+3)}$
g $\frac{11-y}{(y-2)(y+1)}$ h $\frac{-3p-17}{(p+3)(p-1)}$
i $\frac{12(w-2)}{w(w-8)}$ j $\frac{4x-3}{(x-2)^2}$

4 $\frac{2(8p-7)}{(p+1)(p-2)}$

A1.8

1 a 6.78 cm^2 b 113 mm^3
c 15.9 m d 1240 cm^3

2 a $F = 1.3m + 2.5$, where m = mileage
b $P = 0.39a + 0.61b + 1.28c + 4.45d$
where a = number of small letters
b = number of large letters
c = number of small packets
and d = number of large packets
c $B = d + \frac{cn}{100}$
where d = delivery charge
c = cost per 100 bricks
n = number of bricks
d $C = nc + md$
where c = cost of a can of cola
n = number of cans
d = cost of a bag of chips
m = number of bags

3 a $A = p^2 - \frac{1}{4}\pi p^2 = p^2\left(1 - \frac{\pi}{4}\right)$
b $A = (x-2)(x+3) + \frac{1}{2}(3+x+3)(x+6-(x-2))$
$= x^2 + 5x + 18$

A1 Summary

1 a $2a^2 - 3ab$ b $(a-7)(a+7)$

2 a $2x(2x-3y)$ b $(x-1)(x+6)$

3 a $x(x-3)$ b k^3
ci $7x-1$ cii $x^2 + 5xy + 6y^2$
d $(p+q)(p+q+5)$

4 a t^8 b m^5
c $8x^3$ d $12a^7h^5$

5 $\frac{x+3}{x-5}$

6 $\frac{x-3}{2x+3}$

7 $A = (4x-6)(6x-5) + 6(4x-6)$
$= 24x^2 - 20x - 36x + 30 + 24x - 36$
$= 24x^2 - 32x - 6$
OR
$A = 4x(4x-6) + (4x-6)(2x+1)$
$= 16x^2 - 24x + 8x^2 + 4x - 12x - 6$
$= 24x^2 - 32x - 6$

Working must be shown

N3 Check in

1 a 443 b 373 c 21.3 d 265
e 132.17 f 266.81 g 27.03 h 156.23
i 696 j 1104 k 722 l 13
m 15.54 n 2023 o $73\frac{6}{11}$

2 a $\frac{3}{4}$ b $\frac{1}{4}$ c $\frac{1}{2}$ d $1\frac{1}{6}$

3 a $\frac{1}{3}$ b $1\frac{3}{5}$ c $\frac{3}{10}$ d 8

N3.1

1 a $\frac{1}{2}$ b $\frac{3}{10}$ c $\frac{1}{4}$ d $\frac{2}{5}$
e $\frac{3}{5}$ f $\frac{1}{20}$ g $\frac{9}{20}$ h $\frac{3}{8}$

2 a 0.75 b 0.4 c 0.625 d 0.15
e 0.04 f 0.16 g 0.35 h 0.06

3 a $\frac{3}{4}$ b $\frac{3}{10}$ c $\frac{1}{2}$ d $\frac{9}{20}$
e $\frac{5}{8}$ f $\frac{5}{12}$ g $\frac{8}{15}$ h $\frac{5}{8}$

4 a $\frac{1}{4}$ b $\frac{3}{8}$ c $\frac{7}{10}$ d $\frac{7}{12}$
e $\frac{1}{9}$ f $\frac{1}{6}$ g $\frac{5}{24}$ h $\frac{19}{56}$

5 a $\frac{2}{15}$ b $\frac{3}{8}$ c $\frac{1}{6}$ d $\frac{1}{2}$
e $\frac{1}{6}$ f $\frac{1}{8}$ g $\frac{3}{7}$ h $\frac{1}{16}$

6 a 4 b 4 c 4 d $\frac{8}{9}$
e $\frac{5}{2}$ f $\frac{7}{5}$ g $\frac{2}{3}$ h $\frac{7}{5}$

7 a 0.7 b 1.2 c 2.1 d 0.2
e 0.9 f 6.5 g 0.72 h 0.33
i 1.85 j 16.23 k 10.214 l 7.883

8 a 0.96 b 31 c 101.6 d 1.29
e 1.12 f 7 g 0.45 h 15

9 a 48 b 51 c 152 d 246
e 24 f 392.2 g 58.5 h 3160

10 a £392 b £735 c £58.50 d £471.90
e £741 f £208.25

N3.2

1 a 126.741 b 17.29 c 58.32
d 62.04 e 62.641 f 50.73
g 282.2 h 9.768 i 67.31

3 a $\frac{31}{35}$ b $\frac{20}{27}$ c $\frac{5}{16}$
d $\frac{25}{56}$ e $\frac{37}{30}$ f $\frac{5}{18}$
g $\frac{1}{6}$ h $\frac{17}{40}$ i $\frac{4}{3}$

5 a 43.7 b 15.7 c 103
d 78.7 e 66.7 f 181
g 263 h 15.0 i 113

7 a 457.2 b 801.85 c 477.375 d 39.4

9 a 6.25% b 24% c 18.2%
d 1.3% e 54.2% f 14.9%

N3.3

1 a 38 456 b 29 412 c 67 887 d 159 803

2 a 156 b 177 c 209 d 319

3 a 90.78 b −5.797 c −10.6 d −10.67

4 a 2.55 b 33 c 2.50 d 4.09

5 a 216.32 b 37.24 c 555.03
d 91.63 e 146.775 f 26.0925
g 1.7556 h 0.0595 i 0.21665

6 a $\frac{13}{21}$ b $\frac{7}{20}$ c $\frac{1}{6}$ d $1\frac{1}{6}$
e $2\frac{1}{24}$ f $23\frac{11}{36}$ g $43\frac{7}{8}$ h $1\frac{13}{21}$

7 a 6.5×10^3 b 2.07×10^5
c 1.441×10^{-7} d 6.721×10^8
e 1.09×10^2 f 2.35×10^{-6}

8 a 1.62×10^8 b 6.85×10^5 c 1.17×10^{10}
d 4.84×10^{14} e 2.29×10^9 f 2.41×10^{12}

N3.4

1 Distributive law

2 $24 \times 19 = 24 \times (20 - 1) = 24 \times 20 - 24 = 480 - 24 = 456$

3 She has done $36 \div 6$ not $36 \div 5$. Division is not distributive over addition.

4 a e.g. $1 - 2 = -1$, $2 - 1 = 1$
b e.g. $1 \div 2 = 0.5$, $2 \div 1 = 2$
c e.g. $1 - (2 - 3) = 2$, $(1 - 2) - 3 = -4$
d e.g. $1 \div (2 \div 3) = 1\frac{1}{2}$, $(1 \div 2) \div 3 = \frac{1}{6}$

5 No, for example, $\sqrt{4} = \sqrt{(2+2)} \neq \sqrt{2} + \sqrt{2}$.

6 a 4600 b 28 c 36 400

7 a 1225 b 841 c 16.81

8 $5.7^2 = (4.7 + 1)^2 = 4.7^2 + 2 \times 4.7 + 1 = 22.09 + 9.4 + 1 = 32.49$

9 $14.6^2 = (15.6 - 1)^2 = 15.6^2 - 2 \times 15.6 + 1 = 243.36 - 31.2 + 1 = 213.16$

10 $(x + 0.5)^2 = x^2 + x + 0.25 = x(x + 1) + 0.25$

N3 Summary

1 a $1\frac{17}{20}$

2 24 cm

3 a 9.3×10^8 (Accept -9.3×10^8)
b $33\frac{1}{3}$%

Case study 3: Sandwich shop

1

Day	Number of Customers	
	Week 1	Week 2
Monday	50	54
Tuesday	68	60
Wednesday	47	53
Thursday	58	57
Friday	52	56
Saturday	76	70
TOTAL	351	350

a Daily average (mean) = 58.4
b Weekly average (mean) = 350.5
c Without Saturdays:
weekly average (mean) = 277.5
d With coach trip on second Wednesday:
weekly average (mean) = 362.5

From frequency polygon:
busiest time 1pm to 2pm;
quietest time 9am to 10am

Customer numbers vary from hour to hour.
There is a peak around lunchtime.
Customer numbers build at the start of the day and tail off towards the end of the day, with a lull in the early afternoon.
This seems to be quite a suitable pattern for an average day.

2 Sales for second week:
Ham 91, Cheese 63, Hummus 35, Tuna 52.5, Chicken 108.5

Product	Stock (packs)	Portions per pack	Portions left	Stock needed	Amount to order
Bread	6	20	120	351	12
Ham	2.5	10	25	91	7
Cheese	3	10	30	63	4
Hummus	2	8	16	35	3
Tuna	1.5	14	21	53	3
Chicken	1	10	10	109	10

Estimate of stock left on Wednesday morning:
Bread 3, Ham −6 (will have none left), Cheese 9, Hummus 4, Tuna 3, Chicken −27 (will have none left)

The shop will not be able to cater for the coach trip.
There would be enough tuna and cheese, but not enough of the other ingredients (including the bread).

The manager could calculate a percentage surplus to order so that unexpected customers can be catered for.
This value would need to be calculated so that the resulting waste was minimised.

G1 Check in

1 $a = 43°$, $b = 107°$, $c = 233°$, $d = 139°$
2 $e = 102°$, $f = 29°$, $g = 123°$, $h = 73°$

G1.1

1 66° **2** 43.5° **3** 126°
4 15° **5** 97° **6** 107°
7 131° **8** 29° **9** 133°
10 $i = j = k = 49°$ **11** $m = 34°$, $n = 67°$
12 $p = 72°$, $q = 28°$ **13** 90°
14 $s = 63°$, $t = 30°$ **15** $u = 35°$, $v = 53°$
16 180° **17** $x = 40°$, $y = 75°$
18 45°

G1.2

1 90° **2** 56° **3** 63°
4 128° **5** 59° **6** 32°
7 $g = 152°$, $h = 110°$ **8** $i = 134°$, $j = 67°$
9 $k = 48°$, $l = 132°$ **10** $m = 117°$, $n = 126°$
11 $p = q = 98°$ **12** 30° **13** 15°
14 $t = 86°$, $u = 94°$ **15** 122°
16 $w = x = 90°$ **17** 36° **18** 40°

G1.3

1 90° **2** $b = 44°$, $c = 46°$
3 $d = e = 28°$ **4** $f = g = 20°$
5 48° **6** 62°
7 $k = 69°$, $l = 111°$ **8** $m = 12°$, $n = 84°$
9 $p = 46°$, $q = 67°$ **10** $r = 73°$, $s = 53.5°$
11 $t = 112°$, $u = 56°$ **12** $v = 65°$, $w = 32.5°$
13 $x = 90°$, $y = 66°$ **14** 53°

G1.4

1 43° **2** 72° **3** 57°
4 19° **5** $e = 50°$, $f = 40°$
6 $g = 35°$, $h = 55°$ **7** 26°
8 42° **9** $k = l = 36°$
10 $m = n = 59°$ **11** 67°
12 $q = 44°$, $r = 92°$ **13** $s = 59°$, $t = 56°$
14 $u = 64°$, $v = 128°$, $w = 26°$
15 $x = 58°$, $y = 32°$ **16** 17°

G1.5

1 Angles PQR and RSP do not sum to 180°.
2 Rectangle

3 Since $120^2 + 64^2 = 136^2$ the triangle contains a right angle, hence DF is a diameter.
4 If C was the centre of the circle then the angle at C would be twice the angle at the circumference.
5 ∠XQP = ∠XPQ (isosceles triangle) so ∠RQP = ∠SPQ (angles on straight lines sum to 180°); ∠RQP + ∠RSP = 180° and ∠SPQ + ∠QRS = 180° (cyclic quadrilateral), so ∠RSP = ∠QRS (as ∠RQP = ∠SPQ); therefore RQPS is an isosceles trapezium.
6 Angle PXO = 100°, but it would be a right angle if PX were a tangent.

G1 Summary

1 **a** 60°
b 35°
c angle DAC + angle CAB = 65° + 25°
= 90°
Since the angle in a semicircle is 90°, BD is a diameter.
Yes, Ben is correct.
2 **a** AP = AP (Common line)
PB = PC (Tangents to a circle from the same point)
AB = AC (Triangle ABC is isosceles)
So triangle APB and triangle APC are congruent (SSS)
b 50°

A2 Check in

1 **a** 20, 27 **b** 32, 64 **c** $\frac{1}{25}, \frac{1}{36}$
d 125, 216 **e** 31, 50 **f** 243, −729
2 **a** 16 **b** 32 **c** 62
d 1 **e** 1 **f** $\frac{1}{16}$

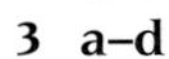
3 **a–d**

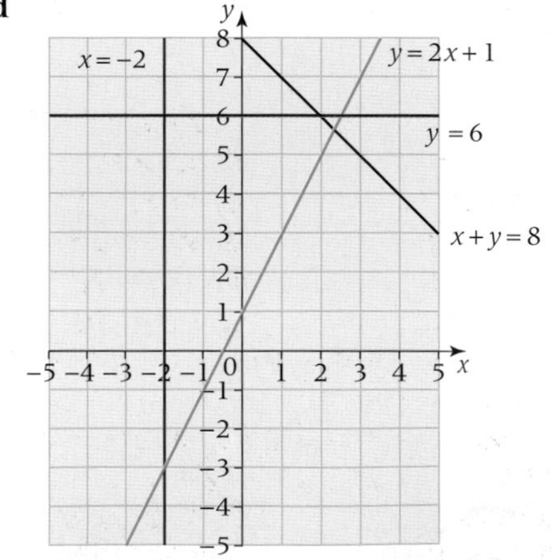

A2.1

1 **a** 5, 12, 19, 26, 33 **b** −3, 0, 5, 12, 21
c 1, 8, 27, 64, 125 **d** 8, 17, 32, 53, 80
e 4, 9, 16, 25, 36 **f** 7, 4, 1, −2, −5
g 1, 4, 9, 16, 25 **h** 6, 12, 20, 30, 42

2 **a** 1, 3, 5, 7, 9; odd numbers
b 7, 14, 21, 28, 35; multiples of 7
c 2, 4, 8, 16, 32; powers of 2
d 10, 100, 1000, 10 000, 100 000; powers of 10
e 1, 3, 6, 10, 15; triangular numbers
f 100, 81, 64, 49, 36; square numbers decreasing from 100
g 1, 2, 3, 4, 5; positive integers
h 1, 1, 2, 3, 5; Fibonacci sequence
3 **a** Convergent, has a limit of 1
b Divergent
c Convergent, has a limit of 0
d Convergent, has a limit of 0
e Oscillating
f Divergent
4 **a** D **b** A **c** C **d** B

A2.2

1 **a** $T_n = 3n + 10$ **b** $T_n = 3n - 1$
c $T_n = 5n + 20$ **d** $T_n = n + 3$
e $T_n = 0.2n + 0.8$ **f** $T_n = 53 - 3n$
g $T_n = 14 - 4n$
2 **a** Each cube has four faces showing, and the two end cubes each have an extra face showing.
b Area of rectangle = Height × Length, and the length is one more than the height.
c There is a blue bead above each red bead, and there is an extra blue bead on the left and one on the right. Each red bead has a link to the left and one above, and the last red bead also has a link to the right.
3 **a** $E = 5H + 1$ **b** $M = 2L(L + 1)$
4 3 × 3 × 3 cube: 1, 6, 12, 8; $n \times n \times n$ cube: $(n-2)^3$, $6(n-2)^2$, $12(n-2)$, 8;
$m \times n \times p$ cuboid: $(m-2)(n \times 2)(p-2)$,
$2(p-2)(n-2) + 2(p-2)(m-2) + 2(m-2)(n-2)$,
$4(p-2) + 4(m-2) + 4(n-2)$, 8

A2.3

1 **a** Gradient 3, y-intercept −2
b Gradient $\frac{1}{2}$, y-intercept 7
c gradient 3, y-intercept −2
d Gradient 2, y-intercept $2\frac{1}{2}$
2 **a** $y = 6x + 2$ **b** $y = -2x + 5$
c $y = -x + \frac{1}{2}$ **d** $y = -3x - 4$
3 **a** $y = 2x + c$ for any $c \neq -1$
b $y = -5x + c$ for any $c \neq 2$
c $y = -\frac{1}{4}x + c$ for any $c \neq 2$
d $y = c - 4x$ for any $c \neq 7$
e $y = c + \frac{3}{4}x$ for any $c \neq 6$
f $2y = 9x + c$ for any $c \neq -1$
4 **a** A, B and E
b C and F

c D
d A
e C and F

5 **a** $y = -4x - 2$ **b** $y = 1\frac{1}{2}x - 2$

6 **a** True **b** False **c** False **d** False

7 **a** (1, 5) **b** (3, 4) and (−1.4, −4.8)

A2.4

1 **a** 4 **b** 2 **c** $\frac{1}{3}$

2 **a–d**

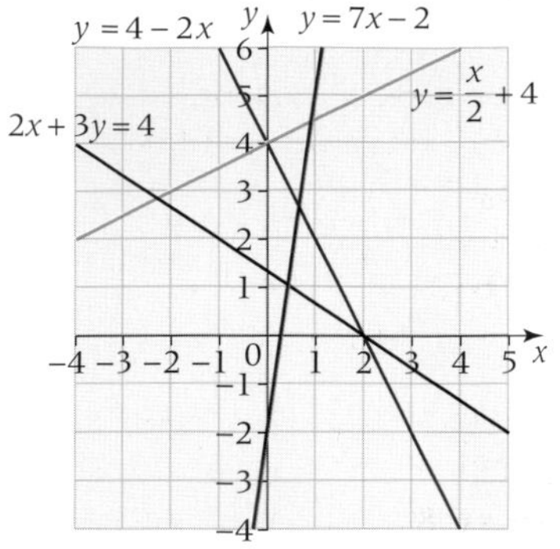

$y = 4 - 2x$, $y = 7x - 2$, $y = \frac{x}{2} + 4$, $2x + 3y = 4$

3 **a** $x = -3$ **b** $y = 4x - 2$ **c** $y = 3 - 2x$
d $y = -\frac{3}{4}x - 1$ **e** $y = \frac{1}{2}x + 1$

4 **a** $y = 4x + 5$ **b** $y = \frac{1}{2}x + 7$ **c** $y = -\frac{1}{3}x + 2$

5 **a** $(1\frac{1}{2}, 11)$ $11 = 4 \times 1\frac{1}{2} + 5$
b $(7, 10\frac{1}{2})$ $10\frac{1}{2} = \frac{1}{2} \times 7 + 7$

6 **a** $5\frac{1}{3}$ **b** 2

7 **a** $y = 2x - 8$ **b** $y = 1 - x$

A2.5

1 **a** $y = -\frac{1}{2}x + c$ for any c **b** $y = \frac{1}{5}x + c$ for any c
c $y = 4x + c$ for any c **d** $y = \frac{1}{4}x + c$ for any c
e $y = -1\frac{1}{3}x + c$ for any c **f** $y = -\frac{2}{9}x + c$ for any c

2 C (A–E, B–F, D–G)

3 **a** $y = -x + 19$ **b** $y = -\frac{1}{2}x + 8\frac{1}{2}$
c $y = 2x + 4$ **d** $y = 1\frac{1}{2}x + 5$

4 **a** $y = -2x + 21$ **b** $y = 1\frac{1}{2}x + 13\frac{3}{4}$
c $y = \frac{6}{17}x - 1\frac{29}{34}$

5 $\frac{t}{3}$

6 False

7 $y = 2x - 1$, $y = -\frac{1}{2}x - 4$

A2 Summary

1 **a** $n^2 - (n-1)(n+1)$ **b** 1

2 **a** $y = -x + 3$

3 **a** $y = \frac{1}{3}x + c$ with a given value of c, $c \neq 2$
b $y = mx + 2$ with a given value of m, $m \neq \frac{1}{3}$
c $y = -3x + 32$

N4 Check in

1 **a** 1, 2, 3, 4, 6, 12
b 1, 2, 3, 5, 6, 10, 15, 30
c 1, 2, 3, 4, 5, 6, 8, 10, 12, 15, 20, 24, 30, 40, 60, 120
d 1, 2, 3, 4, 5, 6, 8, 9, 10, 12, 15, 18, 20, 24, 30, 36, 40, 45, 60, 72, 90, 120, 180, 360

2 2, 3, 5, 7, 11, 13, 17, 19, 23, 29, 31, 37, 41, 43, 47, 53, 59, 61, 67, 71, 73, 79, 83, 89, 97

3 **a** 6 **b** 60

4 **a** 2^7 **b** 3^3 **c** 5^7 **d** 6^2
e 7^3 **f** 4^4

5 **a** 1 **b** 1 **c** 5 **d** $\frac{1}{2}$

6 **a** 7π **b** 13π **c** 36π **d** 4
e $3\sqrt{2}$ **f** $\sqrt{3}$

N4.1

1 **a** 10^2 **b** 10^1 **c** 10^3 **d** 10^0
e 10^4 **f** 10^6 **g** 10^5 **h** 10^8

2 **a** 10^{-2} **b** 10^{-1} **c** 10^{-3} **d** 10^{-5}
e 10^{-4} **f** 10^{-7} **g** 10^{-6} **h** 10^0

3 **a** 1000 **b** 1 000 000
c 100 000 **d** 1 000 000 000
e 10 000 **f** 10
g 100 **h** 10 000 000

4 **a** 1 **b** 0.01 **c** 0.000 01 **d** 0.001
e 0.000 000 1 **f** 0.1
g 0.0001 **h** 0.000 001

5 **a** 10^5 **b** 10^9 **c** 10^8 **d** 10^3
e 10^4 **f** 10^4

6 **a** 10^{-2} **b** 10^{-4} **c** 10^{-8} **d** 10^{-8}
e 10^{-8} **f** 10^6

7 **a** 10^2 **b** 10^{12} **c** 10^{-6} **d** 10^0

8 **a** 3 **b** 5 **c** 3 **d** 2.5

9 **a** 2 **b** 5 **c** −1 **d** 1.3

10 **a** 4.005 **b** 5 **c** 7 **d** −4

N4.2

1 **a** 1.375×10^3 **b** 2.0554×10^4
c $2.314\,55 \times 10^5$ **d** 5.8×10^9

2 **a** 3.4×10^{-4} **b** 1.067×10^{-1}
c 9.1×10^{-6} **d** 3.15×10^{-1}
e 5.05×10^{-5} **f** 1.82×10^{-2}
g 8.45×10^{-3} **h** 3.06×10^{-10}

3 9×10^3, 1.08×10^4, 3.898×10^4, 4.05×10^4, 4.55×10^4, 5×10^4

4 **a** 63 500 **b** 910 000 000 000 000 000
c 111 **d** 299 800 000

5 **a** 0.0045 **b** 0.000 031 7
c 0.00000109 **d** 0.000000979

6 **a** 2.15×10^4 **b** 7×10^{13}
c $1.225\,16 \times 10^{20}$ **d** 1.5×10^7

7 **a** 6×10^7 **b** 4×10^{12} **c** 1.5×10^9
d 7×10^0 **e** 1.5×10^{10} **f** 1.2×10^{11}

8 **a** 0.02 **b** 0.1 **c** 1.5×10^{-8} **d** 2000
e 0.000002 5 **f** 31 000

9 3.3 × 10^{-9} seconds

10 **a** 1.2 × 10^{-2} kg **b** 2.0 × 10^{-26} kg

11 5.1 × 10^4 km

N4.3

1 **a** Divisible by 3 **b** Divisible by 5
c Divisible by 3 **d** Divisible by 3
e Divisible by 2

2 **a** 3×7 **b** 2^3 **c** 3×5
d $2 \times 3^2 \times 5$ **e** $2^2 \times 31$

3 **a** $2^2 \times 3^2 \times 5^2$ **b** $2 \times 3^2 \times 5 \times 7$
c $7 \times 11 \times 13$ **d** $3^2 \times 5 \times 7^2$
e 3×457 **f** $3^4 \times 11$
g $2^2 \times 17 \times 41$ **h** $3^3 \times 53$
i 11×307 **j** $2 \times 3^3 \times 5 \times 13 \times 701$

4

Number	Prime? (Yes/No)	Prime factor decomposition
2000	No	$2^4 \times 5^3$
2001	No	$3 \times 23 \times 29$
2002	No	$2 \times 7 \times 11 \times 13$
2003	Yes	
2004	No	$2^2 \times 3 \times 167$
2005	No	5×401
2005	No	$2 \times 17 \times 59$
2007	No	$3^2 \times 223$
2008	No	$2^3 \times 251$
2009	No	$7^2 \times 41$
2010	No	$2 \times 3 \times 5 \times 67$

5 **ai** $2^6 \times 3^3$ **aii** $3^2 \times 47$
aiii $2^2 \times 7 \times 29$ **aiv** 23
bi 3 **bii** 47 **biii** 203 **biv** 23

6 2 × 42, 3 × 28, 7 × 12, 4 × 21, 6 × 14, 7 × 12

7 **a** $3 \times 5 \times 11^2$
b 3 × 5 × 121, 3 × 11 × 55, 5 × 11 × 33, 11 × 11 × 15

N4.4

1 **a** 5 **b** 16 **c** 3 **d** 5 **e** 14

2 **a** 48 **b** 800 **c** 66 **d** 416 **e** 280

3 **a** 60, 1260 **b** 7, 8085 **c** 48, 1680
d 2, 9216 **e** 2, 314 706

4 **a** $2^5 \times 3^2$ **b** $3^3 \times 5^2 \times 7$ **c** $5^3 \times 7^2 \times 11$

5 **ai** 1, 475 **aii** 1, 828
aiii 1, 323 **aiv** 1, 99
b The HCF is 1 (unless both numbers are the same)
c The LCM is the product of the two numbers (unless both numbers are the same)

6 **a** Not co-prime **b** Co-prime
c Co-prime **d** Not co-prime

7 **a** HCF = 3, LCM = 15 015
HCF = 1, LCM = 45 045
HCF = 1, LCM = 21 175
HCF = 7, LCM = 45 045
b HCF = 1, LCM = the product of the two numbers

8 **a** True

9 **a** 6, 1890 **b** 2, 34 650 **c** 1, 510 510

N4.5

1 **a** 64 **b** 625 **c** 64 **d** 27
e 81 **f** 2401 **g** 1 **h** 1

2 **a** 256 **b** 5 **c** 9
d $x = 4$, $y = 3$

3 **a** 3^4 **b** 5^5 **c** 6^7 **d** 7^9
e 2^{15} **f** 4^{10} **g** 10^{14} **h** 9^{13}

4

×	x	x^3	x^4	x^9
x^2	x^3	x^5	x^6	x^{11}
x^6	x^7	x^9	x^{10}	x^{15}
x^3	x^4	x^6	x^7	x^{12}
x^5	x^6	x^8	x^9	x^{14}

5 **a** 4^2 **b** 5^3 **c** 9^3 **d** 6^4
e 7^4 **f** 8^2 **g** 9^4 **h** 3^5

6 **a** 3^2 **b** 4^5 **c** 7^3 **d** 6^2
e 5^4 **f** 2^5 **g** 6^1 **h** 4^3

7 **a** 2^6 **b** 4^{10} **c** 7^4 **d** 5^{15}
e 3^{16} **f** 6^4 **g** 5^{21} **h** 10^{16}

8 **a** 4096 **b** 2 **c** 2 **d** 793

9 **a** 4^4 **b** 3^4 **c** 6^7 **d** 5^{12}
e 2^{10} **f** 7^{12} **g** 3^3 **h** 9^5

N4.6

1 **a** 4 **b** 3 **c** 3 **d** 0 **e** 10

2 **a** $\frac{1}{2}$ **b** $\frac{1}{3}$ **c** $\frac{1}{2}$ **d** $\frac{1}{4}$

3 **a** 5 **b** $\frac{1}{5}$ **c** 1 **d** 125 **e** $\frac{1}{25}$

4 **a** 8 **b** 4 **c** 243 **d** $\frac{1}{10}$ **e** $\frac{1}{64}$
f 100 **g** $\frac{1}{20}$ **h** 2197 **i** $\frac{1}{729}$ **j** $\frac{1}{8}$

5 **a** 4^{-1} **b** 4^2 **c** 4^{-2} **d** $4^{\frac{3}{2}}$ **e** $4^{\frac{5}{2}}$

6 **a** $16^{-\frac{1}{2}}$ **b** 16^1 **c** 16^{-1} **d** $16^{\frac{3}{4}}$ **e** $16^{\frac{5}{4}}$

7 **a** 10^2 **b** 10^{-1} **c** $10^{\frac{1}{2}}$ **d** $10^{\frac{3}{2}}$ **e** $10^{-\frac{5}{2}}$

8 **a** 5^{-2} **b** $5^{-\frac{1}{3}}$ **c** $5^{\frac{2}{3}}$ **d** $5^{-\frac{3}{2}}$ **e** $5^{-\frac{4}{3}}$

9 **a** $\frac{1}{9}$, $\frac{1}{3}$, 1, 3, 9
b, c

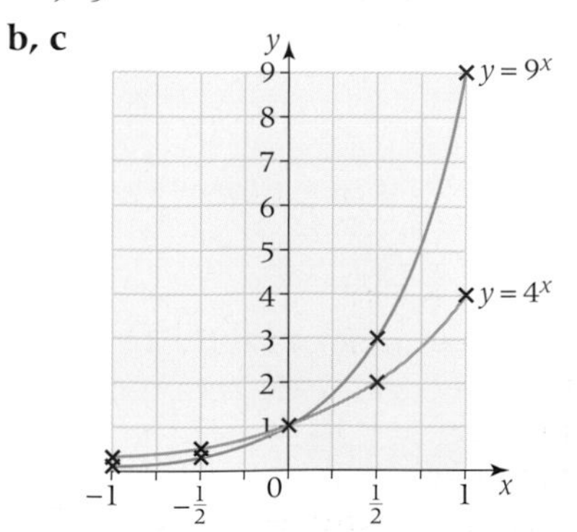

The graphs both pass through (0, 1) and have a similar shape, $y = 9^x$ is steeper than $y = 4^x$.

N4.7

1 a Rational b Irrational c Rational
 d Irrational e Rational f Irrational
2 All fractions are rational. The calculator does not have enough digits to show the repeating pattern.
3 a e.g. $\sqrt{3}$ b e.g. 25 c e.g. $\sqrt[3]{3}$ d e.g. π
4 64
5 Karla cannot be correct. If the number is rational, it can be written as a fraction $\frac{a}{b}$, and half of it can be written as the fraction $\frac{a}{2b}$.
6 Jim cannot be correct. If the final number is $\frac{a}{b}$, then the original number would have been $\frac{a-b}{b}$, which is rational.
7 Javier is correct.
8 a $\frac{\sqrt{2}}{2}$ b $\frac{\sqrt{3}}{3}$ c $\frac{\sqrt{7}}{7}$ d $\frac{\sqrt{6}}{6}$ e $\frac{\sqrt{5}}{5}$
9 a $\frac{\sqrt{8}}{4}$ b $\frac{\sqrt{10}}{5}$ c $\frac{\sqrt{12}}{4}$ d $\frac{\sqrt{30}}{6}$ e $\frac{\sqrt{40}}{5}$
10 Simone's number is rational, but Lisa's is irrational. For a repeating decimal to be rational, there must be a pattern that repeats exactly.

N4.8

1 a $\sqrt{6}$ b $\sqrt{15}$ c $\sqrt{231}$
2 a $\sqrt{2}\sqrt{7}$ b $\sqrt{3}\sqrt{11}$ c $\sqrt{3}\sqrt{7}$
3 a $2\sqrt{5}$ b $3\sqrt{3}$ c $7\sqrt{2}$
4 a $\sqrt{48}$ b $\sqrt{50}$ c $\sqrt{80}$
5 a 6 b 12 c 66
6 a $5\sqrt{5}$ b $7\sqrt{7}$ c $8\sqrt{3}$
7 a 11.180, 18.520, 13.856
 b 11.180, 18.520, 13.856
 c They are the same.
8 a $\sqrt{5}+5$ b $2\sqrt{3}-3$ c $7+3\sqrt{5}$
 d $1+\sqrt{3}$ e $38-14\sqrt{7}$ f 23
 g 16 h −38 i −26
9 a $\frac{\sqrt{11}}{11}$ b $\sqrt{2}-1$
 c $\frac{-1-2\sqrt{3}}{11}$ d $\frac{5-\sqrt{5}}{4}$
 e $2\sqrt{2}-3$ f $16+11\sqrt{2}$

N4 Summary

1 a 8.4×10^4 b 3.7×10^{-3}
2 3.02×10^{27}
3 a 42 b 144
4 a $2 \times 2 \times 3 \times 3 \times 7$ OR $2^2 \times 3^2 \times 7$
 b 9 and 15 OR 3 and 45
5 3 packs of buns
 5 packs of burgers
6 ai 1 aii 8 aiii $\frac{1}{16}$ b $n = 2.5$
7 a $\frac{\sqrt{5}}{5}$ b $8+4\sqrt{5}$
8 1

Case study 4: Sport

1 a 0.01s b 0.1s

Rank	Lower bound (s)	Time (s)	Upper bound (s)	Athlete
1	9.575	9.58	9.584	Bolt
2	9.685	9.69	9.694	Gay
3	9.715	9.72	9.724	Powell
4	9.785	9.79	9.794	Greene
5	9.835	9.84	9.844	Bailey
	9.835		9.844	Surin
7	9.845	9.85	9.854	Burrell
	9.845		9.854	Gatlin
	9.845		9.854	Fasuba
10	9.855	9.86	9.864	Lewis
	9.855		9.864	Fredericks
	9.855		9.864	Boldon
	9.855		9.864	Obikwelu

A3 Check in

1 a $x+16y$ b $9x^2+3x$ c $7ab$
 d $7p-9$ e $14x^3$ f $2x+3y-x^2$
2 a $24x-27$ b $5y+11$
 c $3x^2-2xy$ d $x^2+2x-63$
 e $6w^2-32w+32$ f p^2+q^2-2pq
3 a $\frac{2}{5}$ b $\frac{19}{45}$ c $\frac{1}{2}$ d $5\frac{17}{30}$ e $\frac{7}{12}$ f $3\frac{1}{8}$

A3.1

1 a $x=4\frac{1}{5}$ b $x=4\frac{5}{6}$ c $x=9$
 d $x=-\frac{1}{4}$ e $x=-4\frac{1}{3}$ f $y=-1\frac{1}{5}$
 g $x=9$ h $x=-3$ i $x=3$
2 a $x=3\frac{4}{5}$ b $x=2\frac{2}{7}$ c $y=1\frac{2}{5}$
 d $y=5$ e $w=-3\frac{2}{5}$ f $x=-\frac{1}{3}$
 g $y=-\frac{1}{3}$ h $z=1\frac{3}{16}$ i $p=2\frac{4}{5}$
 j $q=2\frac{3}{8}$ k $r=\frac{7}{9}$
3 a Length = 5 b $p=-5$
4 a 32 b Square length = 7.2

A3.2

1 a $x=47$ b $y=1\frac{13}{17}$ c $z=3$
 d $p=9\frac{1}{3}$ e $q=9\frac{9}{14}$ f $m=\frac{4}{23}$
2 False
3 A $9\frac{4}{9}$ mm², B $15\frac{1}{9}$ mm²
4 $1\frac{4}{11}$
5 35
6 3 cm, 4 cm; 1 cm, 4 cm

A3.3

2 a $x = 16\frac{9}{24}$ b $y = -16\frac{3}{4}$

3 The first two are possible, the third is not.

A3 Summary

1 a $b = 1.5$

2 $x = -3.5$

3 $x = 1, x = 9$

4 a $\frac{\sqrt{5}}{5}$ b $8 + 4\sqrt{5}$

5 1

G2 Check in

1

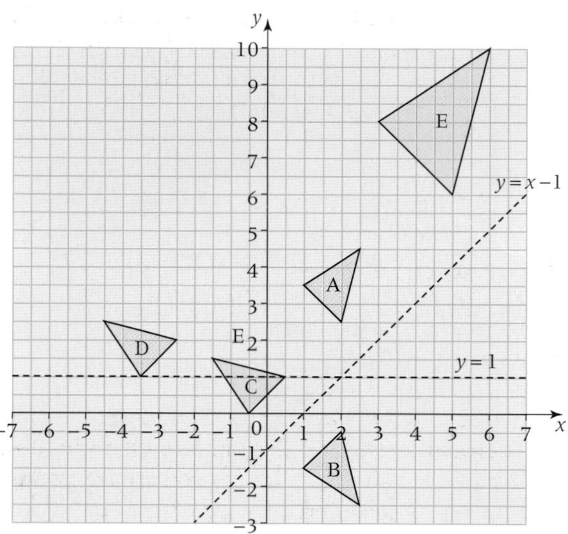

c Rotate 90° anticlockwise about the origin.

e Shapes A, B, C and D are congruent.

G2.1

1 a Reflection in y-axis

b Rotation by 180° about (0, 0)

c Reflection in x-axis

d Rotation by 180° about (0, 0)

e Reflection in y-axis

f Reflection in y-axis

2 a Translation by $\begin{pmatrix} 16 \\ 2 \end{pmatrix}$ b Translation by $\begin{pmatrix} 5 \\ 3 \end{pmatrix}$

c Translation by $\begin{pmatrix} 9 \\ 8 \end{pmatrix}$ d Translation by $\begin{pmatrix} -4 \\ -5 \end{pmatrix}$

e Translation by $\begin{pmatrix} 7 \\ -6 \end{pmatrix}$ f Translation by $\begin{pmatrix} -7 \\ 6 \end{pmatrix}$

3 a Rotation by 90° clockwise about (0, 0)

b Rotation by 180° about (0, 0)

c Rotation by 90° anticlockwise about (0, 0)

d Rotation by 90° clockwise about (0, 0)

e Rotation by 90° clockwise about (0, 0)

f Rotation by 180° about (0, 0)

G2.2

1

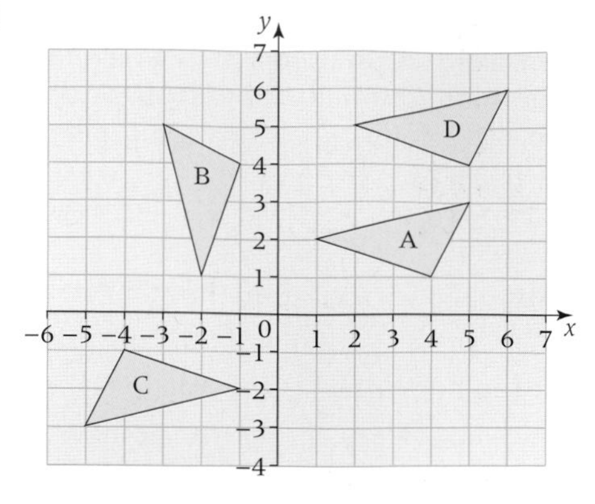

c Rotation 180° about (0, 0)

e Translation by $\begin{pmatrix} 1 \\ 3 \end{pmatrix}$

2

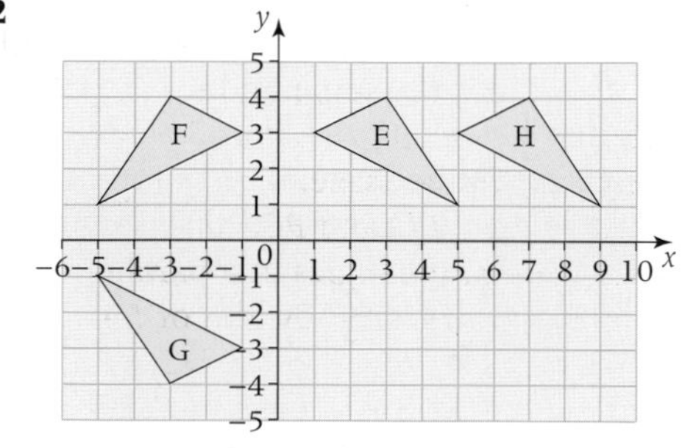

c Rotation 180° about (0, 0)

e Translation by $\begin{pmatrix} 4 \\ 0 \end{pmatrix}$

3

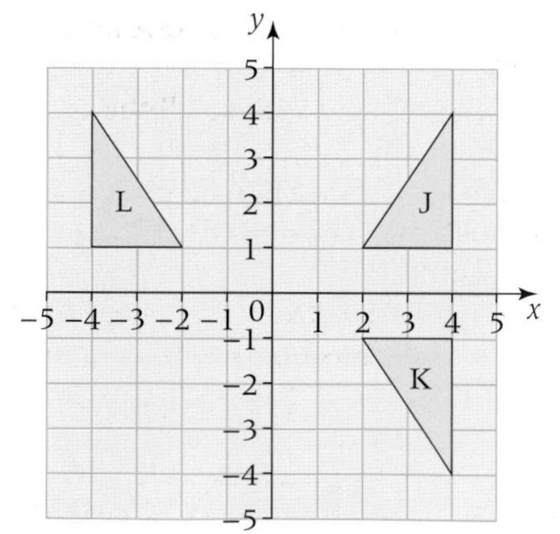

c Reflection in y-axis

4 **a** and **b**

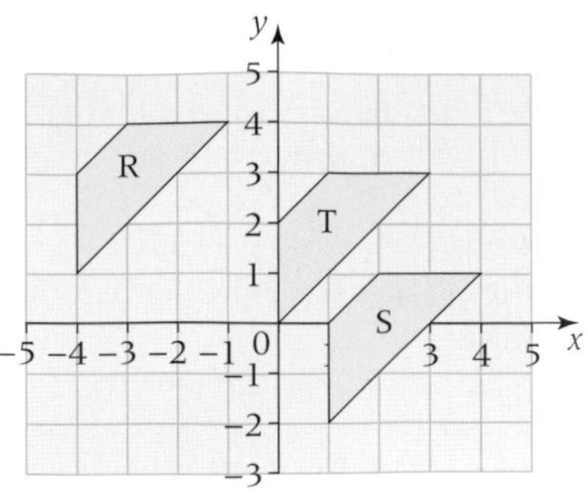

c Translation by $\begin{pmatrix} -4 \\ 1 \end{pmatrix}$

G2.3

1 **a** Congruent **b** Not congruent **c** Congruent

2 Congruent by RHS

3 WX = 5 cm by Pythagoras, so congruent by RHS

4 **a** Only one side is the same and no angles are the same.

b Three sides are the same.

5 **a** HFG and FHE, HEG and FGE

b Opposite sides and angles in a parallelogram are equal, so congruent by SSS or SAS.

G2.4

1 **a** ∠HGM = ∠MEF (alternate angles), ∠MHG = ∠MFE (alternate angles) and HG = EF so congruent by ASA

b MGH and MEF are congruent, so ME = MG, hence M is the midpoint of EG

2 **a** ORS is congruent to OTU by ASA and ORU is congruent to OTS by ASA, then OTU is congruent to OTS by SSS, so all the triangles are congruent.

b All angles at O are equal, so they must be 90°. OR = OT and UO = SO as ORS and OTU are congruent.

3 AX = AY = BX = BY so AXBY is a rhombus. The diagonals of a rhombus bisect each other at right angles so XY is the perpendicular bisector of AB.

4 OSX and OTX are congruent by SSS, so ∠OXS = ∠OXT which means OX bisects ∠SXT.

5 Using the diagram on page 152, the perpendicular bisector OD bisects angle O, so ∠AOD and ∠BOD are equal. The hypothenuse of both triangles is r, radius of the circle. Both triangles share line OD. Therefore, the triangles are congruent by SAS and AD = BD, so line OD bisects line AB.

G2.5

1 **b** Vertices of B: (1, −1), (−5, −2), (−2, −3)

c Vertices of C: (2, −2), (−10, −4), (−4, −6)

d They are congruent.

2 **b** Vertices of E: (3, −3), (−6, −6), (−12, 0)

c Vertices of F: (0.5, −0.5), (−1, −1), (−2, 0)

d 3

3 **b** Vertices of K: (5.5, −3.5), (5.5, −8), (−3.5, −3.5)

c Vertices of L: (2, 0), (2, −1), (0, 0)

d 3

4 **b** Vertices of Y: (2, −2), (0, −6), (−6, 4)

c Vertices of Z: (−3, 3), (0, 9), (9, 6)

d Enlargement by scale factor 3, centre (0, 0)

e Enlargement by scale factor $\frac{1}{3}$, centre (0, 0)

f Same centre, scale factors which are reciprocals of one another.

G2.6

1 $a = 6.4$ cm, $b = 4.5$ cm

2 $c = 9$ cm, $d = 1.9$ cm

3 RT = 5 cm, QR = 4 cm, QS = 12 cm

4 **a** XY = 6.6 cm, VY = 6 cm

b 15.5 cm

5 12.8 cm, 24.8 cm

6 **a** 1 : 3 **b** 1 : 3

c For any two circles, the ratio of the circumferences is the same as the ratio of the radii because each circumference is just the radius multiplied by a constant (2π).

G2 Summary

1 **a** 2 cm **b** $5\frac{1}{4}$ cm

2 Rotation of 180° about (0, −2)
OR
Enlargement scale factor −1 with centre of enlargement (0, −2)

A4 Check in

1 **a** $x(x-5)$ **b** $(x+3)(x+7)$
c $(x-5)(x+5)$ **d** $(y-3)^2$
e $(p-10)(p+10)$ **f** $3a(b-3a)$
g $(4m-7)(4m+7)$ **h** $(3x+1)(x+2)$

2 **a** −10 **b** 29

A4.1

1 **a** $x = 0$ or 3 **b** $x = 0$ or −8 **c** $x = 0$ or 4.5
d $x = 0$ or 3 **e** $x = 0$ or 5 **f** $x = 0$ or 7
g $x = 0$ or 12 **h** $x = 0$ or 2 **i** $x = 0$ or 6
j $y = 0$ or 3 **k** $w = 0$ or 7

2 **a** $x = -3$ or -4 **b** $x = -2$ or -6 **c** $x = -5$
d $x = -5$ or 3 **e** $x = -7$ or 2 **f** $x = 5$ or -1
g $x = 2$ or 3 **h** $x = 6$
i $x = -3$ or $-\frac{1}{2}$ **j** $x = -2$ or $-\frac{1}{3}$

k $x = -2$ or $-\frac{1}{2}$ **l** $y = -\frac{1}{2}$ or $-\frac{2}{3}$
m $x = 2$ or 6 **n** $x = 1.5$ or -5
o $x = 2$ or 3 **p** $x = -3$ or -7

3 **a** $x = 4$ or -4 **b** $x = 8$ or -8 **c** $y = 5$ or -5
d $x = \frac{2}{3}$ or $-\frac{2}{3}$ **e** $y = \frac{1}{2}$ or $-\frac{1}{2}$ **f** $x = 13$ or -13
g $x = 2.5$ or -2.5 **h** $y = 2$ or -2

4 **a** $x = 0$ or $\frac{1}{3}$ **b** $x = 5$ or -3 **c** $x = 3$ or $\frac{2}{3}$
d $y = 1\frac{1}{3}$ or $-1\frac{1}{3}$ **e** $x = 1\frac{1}{4}$ or $-1\frac{1}{4}$
f $x = 0$ or 1 **g** $x = -\frac{1}{4}$ or $\frac{3}{5}$ **h** $x = 2$ or 6

5 **a** $x = \frac{1}{2}$ or $-\frac{2}{3}$ **b** $x = 3$ **c** $x = -\frac{1}{2}$ or $\frac{3}{5}$
d $x = 1$ **e** $x = 2, -2, 3$ or -3

6 **a** $w(w + 7)$ **b** $w(w + 7) = 60 \Rightarrow w^2 + 7w - 60 = 0$
c 5 cm by 12 cm

A4.2

1 **a** $(x + 2)^2 + 2$ **b** $(x + 4)^2 - 1$
c $(x + 5)^2 + 1$ **d** $(x + 2)^2 - 4$
e $(x + 6)^2 - 26$ **f** $(x + 7)^2 - 24$
g $(x + 2)^2 - 14$ **h** $(x + 4)^2 - 19$
i $(x + 8)^2 - 65$ **j** $\left(x + \frac{3}{2}\right)^2 + \frac{7}{4}$
k $\left(x + \frac{5}{2}\right)^2 - \frac{1}{4}$ **l** $\left(x + \frac{7}{2}\right)^2 - \frac{9}{4}$
m $\left(x + \frac{9}{2}\right)^2 - \frac{81}{4}$ **n** $\left(x + \frac{5}{2}\right)^2 - \frac{33}{4}$
o $\left(x + \frac{11}{2}\right)^2 - \frac{137}{4}$

2 **a** $(x + 15)^2 - 135$ **b** $(x + 8)^2 - 66$
c $\left(x + \frac{7}{2}\right)^2 - \frac{49}{4}$ **d** $\left(x + \frac{17}{2}\right)^2 - \frac{297}{4}$
e $\left(x + \frac{1}{4}\right)^2 + \frac{15}{16}$ **f** $\left(x - \frac{9}{2}\right)^2 - \frac{125}{4}$

3 **a** $2\left(x + \frac{3}{2}\right)^2 - \frac{1}{2}$ **b** $3(x + 1)^2 + 6$
c $-(x - 3)^2 + 7$ **d** $5(x + 1)^2 + 10$
e $-(x - 3)^2 + 1$ **f** $2\left(x + \frac{7}{4}\right)^2 - \frac{73}{8}$

4 **a** Because there are different powers of x.
b $(x + b)^2 - b^2$
c $(x + b)^2 - b^2 = c \Rightarrow x = \pm\sqrt{c + b^2} - b$

5 **a** $x = \pm\sqrt{k + 4c^2} - 2c$ **b** $x = \pm\sqrt{9 + t^3} - 3$
c $x = \pm\sqrt{9g^2 - m} + 3g$ **d** $x = \pm\sqrt{\frac{p}{2} + c^2} - c$

6 $ax + bx + c = 0 \Rightarrow x^2 + \frac{b}{a}x + \frac{c}{a}$
$= 0 \Rightarrow \left(x + \frac{b}{2a}\right)^2 = \frac{b^2}{4a^2} - \frac{c}{a}$
$\Rightarrow \left(x + \frac{b}{2a}\right)^2 = \frac{b^2 - 4ac}{4a^2} \Rightarrow x$
$= \frac{-b \pm \sqrt{b^2 - 4ac}}{2a}$

A4.3

1 **a** $x = 2$ or 10 **b** $x = -5$ or 3
c $x = -1$ or 5 **d** $x = -1$
e $x = -9$ or 7 **f** $x = 7$
g $x = 0$ or 8 **h** $y = 0.0828$ or -12.1
i $p = 3.56$ or -0.562

2 **a** $(x + 4)^2 = -1$ is positive = negative, so there are no solutions.

3 **a** $y = 8$ or -11 **b** $x = -13$ or 11
c No solutions **d** $x = -0.634$ or -2.37
e $w = 3$

4 **a** $6(10 - x) = 2x(2x + 1) \Rightarrow x^2 + 2x - 15 = 0$
b $x = -5$ (reject) or 3 (accept); Painting A is 6 by 7, Painting B is 3 by 7

5 $x^2 + 4x + 10 = (x + 2)^2 + 6 > 6$, as the part in brackets is always positive.

6 **a** $(-4, -4)$ **b** $(-5, -30)$
c $(6, -40)$ **d** $\left(-1\frac{1}{2}, -1\frac{1}{4}\right)$
e There is no minimum.

7 $p^2 + 6p + 9 = (p + 3)^2 > 0$

A4.4

1 **a** $(0, 12)$ **b** $(0, 15)$ **c** $(0, 5)$
d $(0, 6)$ **e** $(0, 25)$

2 **a** $(-6, 0), (-2, 0)$ **b** $(3, 0), (5, 0)$
c $(-1, 0), (-5, 0)$ **d** $(-2, 0), (-3, 0)$
e $(-5, 0)$

3 **a** $(-4, -4)$ **b** $(4, -1)$ **c** $(-3, -4)$
d $\left(-2\frac{1}{2}, -\frac{1}{4}\right)$ **e** $(-5, 0)$

4 $y = (x + 4)^2 + 4 \geq 4$, so y never intersects the x-axis.

5

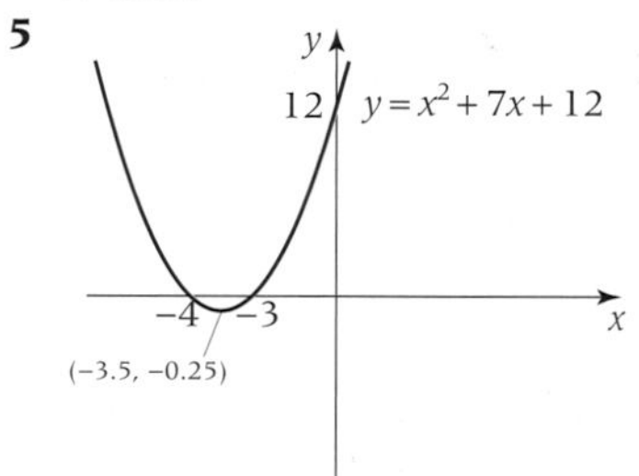

6 **a** **ii** $y = x^2 - x - 20$ **b** **iv** $y = x^2 + 2x - 24$
c **i** $y = (x - 4)^2$; graph of **iii** $y = (x - 6)(x + 4)$, has y-intercept $(0, -24)$ and intercepts the x-axis at $(6, 0)$ and $(-4, 0)$

7 **a** $y = x(x + 5)$ **b** $y = (x - 6)^2$
c $y = (x - 4)^2 - 2$

8 **a** e.g. $y = x^2 + 5$ **b** e.g. $y = (x + 12)^2$
c e.g. $y = 9 - (x - 4)^2$

9 **a** It is a cubic equation, not a quadratic equation, so the graph will be S-shaped.
b

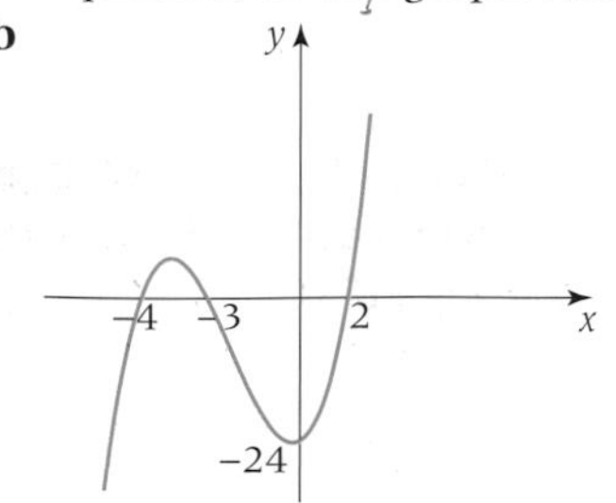

A4 Summary

1 a $(2x-3)(x+5)$
  b $x = 1.5, x = -5$
2 a $p = 3, q = 6$
  b

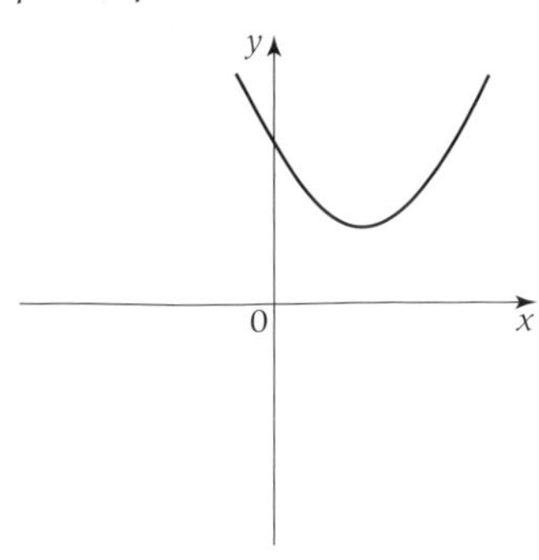

A U-shaped curve with the minimum point in the positive quadrant.
The points (3, 6) and (0, 15) do not have to be shown.

3 Minimum at (−2, −9)

Case study 5: Holiday

1 £32.63 (assuming 13 weeks)
2 £4 per hour, £130 (assuming 13 weeks)
3 CDs £2.50 each, DVDs £3.50 each
  £8.50
4 £1 : 1.31 Eur
  90.78 euros for £70 and 1% commission
  £1 : 10.37 FRF
  £1 : 3.07 DEM
  $°C = \frac{2}{5}°F - \frac{16}{5}$

G3 Check in

1 ai 78.5 cm² aii 31.4 cm
  bi 172.0 cm² bii 46.5 cm
2 a 600 cm³ b 351.9 cm³ c 45.92 cm³
3 a 500 cm² b 276.5 cm² c 105.2 cm²

G3.1

1 a 15.1 cm² b 15.7 cm² c 101.8 cm²
  d 15.7 cm² e 6.0 cm² f 12.3 cm²
2 a 7.5 cm b 10.5 cm c 33.9 cm
  d 6.3 cm e 5.0 cm f 7.0 cm
3 a 45.4 mm b 14.8 cm c 29.5 cm
4 23.6 cm²
5 a 11.1 m² b 13

G3.2

1 a 42 cm³ b 1.77 cm³ c 69.2 cm³
  d 32 cm³ e 263.9 cm³ f 87.1 cm³
2 a Y b 134 cm³
3 3
4 69.2 cm³

G3.3

1 a 443.4 mm² b 249.4 cm² c 46.8 cm²
  d 106.0 cm² e 138.5 cm² f 197.2 cm²
  g 85.1 cm² h 94.8 cm²
  i 9856.1 mm²
2 a 93.5 mm² b 4704.3 mm²

G3.4

1 a 128.8 cm² b 56.5 cm² c 4241.2 mm²
  d 23.6 cm² e 221.0 cm² f 784.6 cm²
2 a 188.5 cm² b 82.4 cm² c 316.5 cm²
3 a 15.7 cm² b 15.7 cm²
  c 6804.7 mm² d 184.4 cm²
  e 101.8 cm² f 40.2 m²
4 156.6 cm²

G3.5

1 a 1436.8 cm³ b 57 906 mm³
  c 1047.4 cm³ d 24 429 cm³
  e 70 276 cm³ f 10 306 000 mm³
2 a 615.8 cm² b 7238.2 mm²
  c 498.8 cm² d 4071.5 cm²
  e 8235.5 cm² f 229 022 mm²
3 7 cm
4 a 134.0 cm³ b 100.5 cm²
5 785.4 cm³
6 a Cube b Cube
7 ai 6 cm aii 12 cm
  b 452.4 cm² c 452.4 cm²
  di $4\pi r^2$ dii $2r \times 2\pi r$
  e They both equal $4\pi r^2$.

G3.6

1 a 32 m² b 4 mm c 20 000 cm³
  d 1.9 litres e 5.1 km f 0.58 m
  g 6.3 km h 24 000 mm³ i 630 000 cm³
  j 90 000 m² k 700 m² l 1000 mm²
2 a Length + Length + Length = Length
  b Length × Length × Length = Volume
3 a $2n + 2m\pi$ b $2mn + \pi m^2$
4 a $2ab + 2ac + 2bc$ b abc
5 $2.4ab + 0.6b^2$ cm³
6 $\frac{lwh}{10}$ cm³
7 There are four lengths multiplied together when there should only be three.

G3.7

1 a 107.5 cm² b 193.75 cm³
2 244 cm²
3 125 cm³
4 a Their sides are all the same length, so any scale factor will be constant on all dimensions.
  b The scale factor on one dimension could be different from that on another.
  ci 1 : 49 cii 1 : 343
5 a 1 : 1.5 or 2 : 3 b 34 cm² c 81 cm³
6 Deal A

G3.8

1 ai 490 cm³ — aii 659 cm²
bi 1186 cm³ — bii 707 cm²
ci 1057 cm³ — cii 704 cm²
di 1982 cm³ — dii 1152 cm²
ei 638 cm³ — eii 467 cm²
fi 159 cm³ — fii 252 cm²
gi 650 cm³ — gii 510 cm²
hi 1985 cm³ — hii 1020 cm²
i i 3146 cm³ — i ii 1528 cm²
ji 7242 cm³ — jii 3339 cm²
ki 3398 cm³ — kii 4610 cm²
li 16 040 cm³ — lii 4682 cm²

2 a $\frac{7\pi r^2 h}{3}$ cm³ **b** $5\pi r^2 + 3\pi r\sqrt{r^2 + h^2}$ cm²

G3 Summary

1 201 cm² (Accept 200 – 202)
2 20 000 cm² OR 2 m²
3 a 905 m³ (Accept 904 – 905)
b 4.92 m (Accept 4.915 – 4.925)

N5 Check in

1 672 g
2 3 hours
3 ai 10 — aii 5.5
bi 48 — bii ±5
ci $1\frac{1}{2}$ — cii $1\frac{1}{5}$

N5.1

1 16.6 m²
2 25.83 kg
3 **b**, **d** and **e**
4 Two variables that are in direct proportion have a straight-line graph through (0, 0).
5 a $w \propto l$ **b** $w = kl$ **c** 2.48 kg/m **d** 7.192 kg
6 €4.08
7 £17.12
8 Super

N5.2

1 a

x	1	2	3	4	5	6
x^2	1	4	9	16	25	36
y	4	16	36	64	100	144

b y is proportional to x^2. This means $y = kx^2$ for some constant, k. In this case, $k = 4$.
c $y = 4x^2$, $x = 14$ **d** Graph ii
2 a $R = 100s^2$ **b** 64 **c** 1.41
3 $A = \pi r^2$ so the area is directly proportional to the square of the radius, with constant of proportionality π. Also, $C = 2\pi r$ so the circumference is directly proportional to the radius, with constant of proportionality 2π.
4 $615\frac{2}{9}$ cm²

5 a $y = 2x^3$ **b** 4
6 50

N5.3

1 a Doubled **b** Halved
c Multiplied by 6 **d** Divided by 10
e Multiplied by a factor of 0.7
2 a Halved **b** Doubled
c Divided by 6 **d** Multiplied by 10
e Divided by 0.7
3 a 2 **b** 8 **c** 0.4 **d** 16 **e** 0.16
4 $y = \frac{500}{w}$
5 800
6 a $t = \frac{10}{n}$
bi 10 hours **bii** $2\frac{1}{2}$ hours **biii** $1\frac{1}{4}$ hours
biv 2 hours **bv** 1 hour 24 minutes
c Yes
7 a 2 amps **b** 4 amps **c** 6 amps
d 8 amps **e** 10 amps **f** 0.67 amps
g 0.5 amps **h** 2.4 amps
8 The graph tends to both axes: when R is small, I is big and when I is small, R is big.
9 4
10 a If $x = ky$, then $y = \frac{1}{k}x$. Since k is a constant, $\frac{1}{k}$ is a constant, and y is directly proportional to x.
b $y = \frac{k}{x}$ can be rearranged to $x = \frac{k}{y}$, so x is inversely proportional to y.

N5.4

1 a $F = 100/d^2$; when $d = 3$, $F = 11\frac{1}{9}$, when $d = 4$, $F = 6\frac{1}{4}$
b

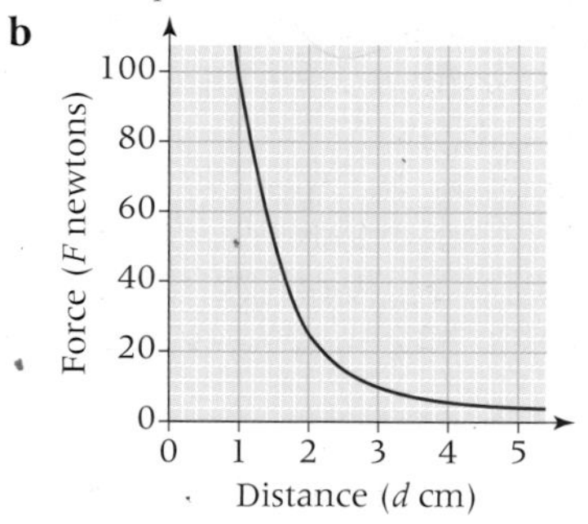

2 a ii $y \propto \frac{1}{x}$ **b iii** $y \propto x$
c i $y \propto \frac{1}{x^2}$ **d iv** $y \propto x^2$
3 1 hour
4 a $y = \frac{10}{\sqrt{x}}$ **bi** 1 **bii** $\frac{25}{36}$
5 a $P = \frac{32}{t^3}$ **bi** $1\frac{5}{27}$ **bii** 4

N5.5

1 a $\frac{8}{15}$ **b** $\frac{29}{35}$ **c** $\frac{5}{56}$ **d** $\frac{44}{45}$

e $4\frac{7}{8}$ f $\frac{59}{60}$ g $\frac{19}{40}$ h $9\frac{17}{36}$

2 a $\frac{1}{2}$ b $1\frac{2}{3}$ c $1\frac{9}{16}$ d $\frac{8}{15}$

e $1\frac{5}{9}$ f $14\frac{3}{10}$ g $2\frac{37}{55}$ h $26\frac{4}{9}$

3 Exact decimal representations are possible for:

1e 4.875 **1g** 0.475 **1j** 0.8

2a 0.5 **2c** 1.5625 **2f** 14.3

All of the other answers have recurring decimal representations, as the denominators have prime factors other than 2 or 5.

4 a $\frac{17}{30}$ b $1\frac{67}{87}$ c $\frac{1}{3}$

d $\frac{207}{1715}$ e $1\frac{133}{324}$ f $-\frac{8}{21}$

5 a $5\sqrt{5}$ b $3-\sqrt{3}$ c $\frac{3}{2}+\frac{3}{4}\sqrt{3}$

d $\frac{5}{3}-\sqrt{7}$ e $\frac{10}{3}+\frac{10}{9}\sqrt{2}$ f $\frac{20}{9}+\frac{7}{9}\sqrt{5}$

6 a $5+\sqrt{5}$ b $5\sqrt{3}-3$ c $9+5\sqrt{3}$

d $11-6\sqrt{2}$ e 18 f $9+14\sqrt{5}$

7 a $\frac{\sqrt{2}}{2}$ b $1+\frac{\sqrt{3}}{3}$ c $\sqrt{5}-2$ d $\frac{\sqrt{7}-1}{2}$

e $\frac{40+16\sqrt{3}}{13}$ f $\frac{(3-\sqrt{3})(4-\sqrt{5})}{11}$

8 a $\frac{-(1+\sqrt{5})(1-\sqrt{3})}{2}$ b $(2-\sqrt{3})(2+\sqrt{7})$

c $-3+\sqrt{11}$ d $8+5\sqrt{3}$

e $-9-\frac{16\sqrt{3}}{3}$ f $\frac{18-\sqrt{5}}{22}$

N5.6

1 0.455 m, 0.445 m

2 a 6.45 mm, 6.35 mm b 4.725 m, 4.715 m

c 18.5 s, 17.5 s d 0.3885 kg, 0.3875 kg

e 6.55 volts, 6.45 volts

3 20.968 m/s, 19.091 m/s

4 a 14.6 m b 13.7025 m^2

5 ai 60 601.5 cm^2, 59 401.5 cm^2

aii 1 015 075.125 cm^3, 985 074.875 cm^3

bi 6 006 001.5 mm^2, 5 994 001.5 mm^2

bii 1 001 500 750.125 m^3, 998 500 749.875 m^3

6 14 boxes

7 7 crates

8 No, because 5×87.5 kg = 437.5 kg < 440 kg

9 21.71 cm

10 a 86.5 cm, 85.5 cm

b Lower bound of distance travelled = 27.36 m. This is shorter than the upper bound of the length of the path, which is 28.5 m.

N5.7

1 a 7.74×10^{-3} b 9.63×10^{5}

c 4.38×10^{-5} d 2.55×10^{2}

e 3.4×10^{5} f 4.47×10^{-2}

2 a 2.173×10^{-10} b 1.74×10^{-7}

c 1.37×10^{-2} d 4.32×10^{-9}

e 6.46×10^{8} f 6.32×10^{3}

3 a 11 b $9\frac{5}{28}$ c 2.92×10^{3}

4 a 317.869 b 497.372 c 15 625

d 162.37 e 34.29 f 6.82

5 a 30.7 b 7.72 c 8.10

d 4.19 e 0.0633 f 0.486

6 a $\frac{22}{15}$ b $\frac{19}{12}$ c $\frac{1}{9}$ d $\frac{1}{12}$ e $\frac{6}{5}$

f $\frac{15}{8}$ g $\frac{97}{12}$ h $\frac{27}{10}$ i $\frac{15}{16}$ j $\frac{273}{20}$

7 a 0.322 b 18.4 c 0.312 d 0.0303

8 Amy 3.6% Beth 13.4%

N5 Summary

1 a $d = 5t^2$ b 45 metres

c 11 seconds

2 $f = 160$

3 a 9.75 b 30.7

A5 Check in

1 a–d

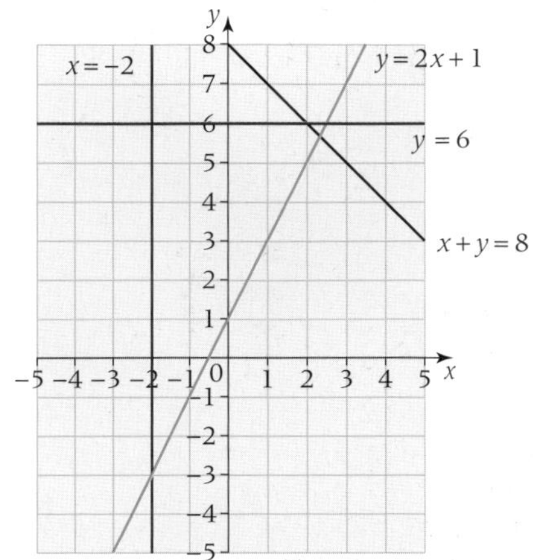

2 a < b > c < d =

3 a 5, 8 b 2, 7 c 3, 4

4 a 18 b 4.5 c 90 d 63

A5.1

1 a 31 b 10

2 False, $0 \leqslant x \leqslant 1$

3 a $x > 7\frac{2}{3}$ b $p \leqslant -32$ c $x \leqslant -2\frac{3}{4}$ d $y < -4$

e $q \leqslant 10$ f $z \leqslant -3$ g $y > -10\frac{4}{5}$

4 a $x > 7\frac{2}{3}$ b 8

5 a $-1 < x < 4$ b $-5 < x < 18$

6 These give $y \leqslant 6$ and $y > 6$, which do not intersect when combined.

7 a $-1 \leqslant y \leqslant 6$ b $-2 < z < 6$ c $\frac{1}{3} < p < 5$

8 a The solution is $-5 \leqslant x \leqslant 5$

bi $x < -4$, $x > 4$ bii $p < -2$, $p > 2$

9 a $30° < x < 150°$

b $0° < x < 45°$, $135° < x < 225°$, $315° < x < 360°$

A5.2

1 a

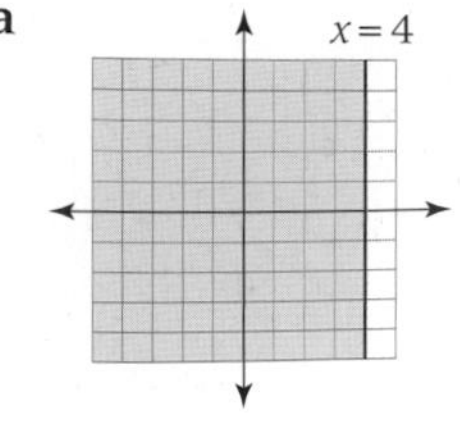

b

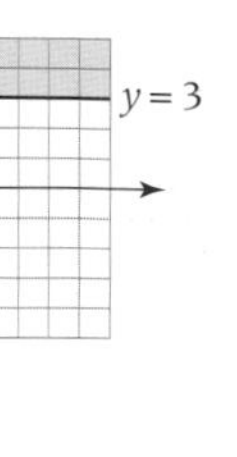

c $x = -2$

d $y = -3$

e

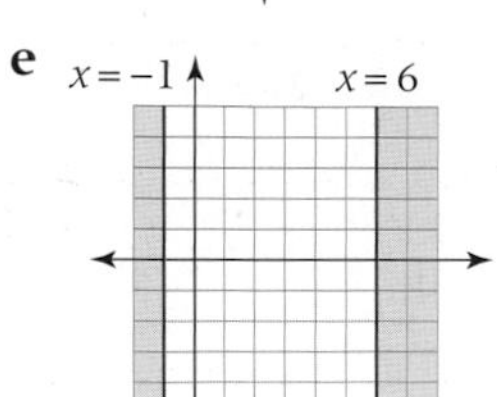

f

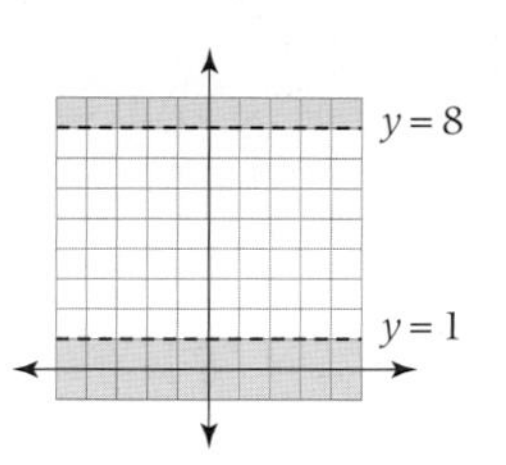

g

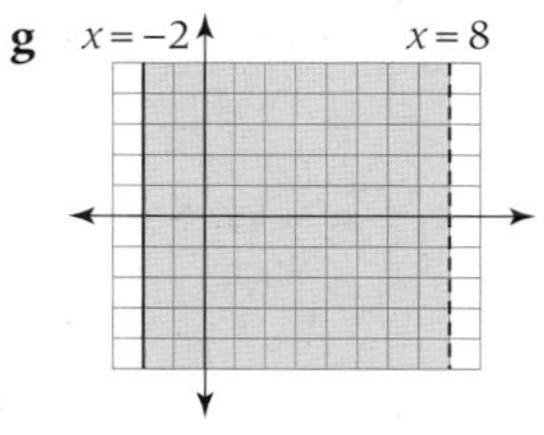

h

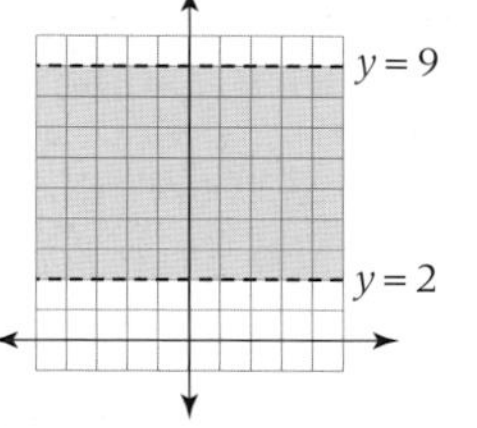

i

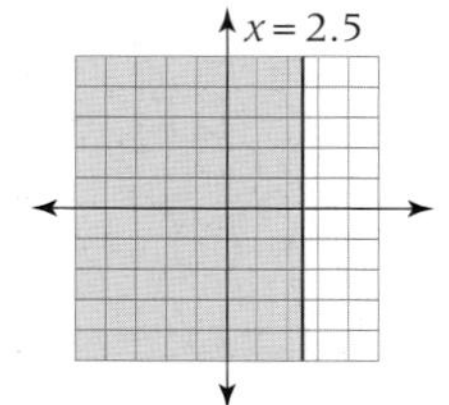

j $y = 8$

k

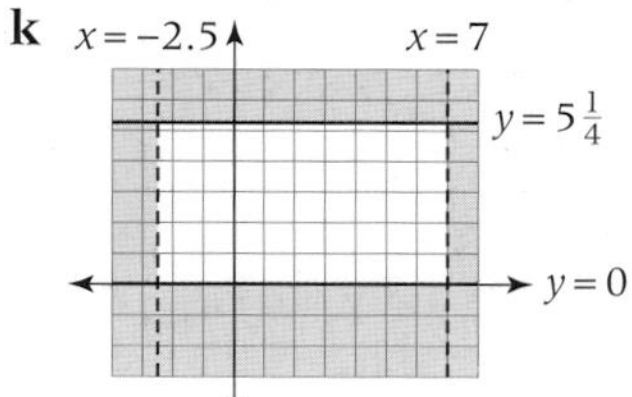
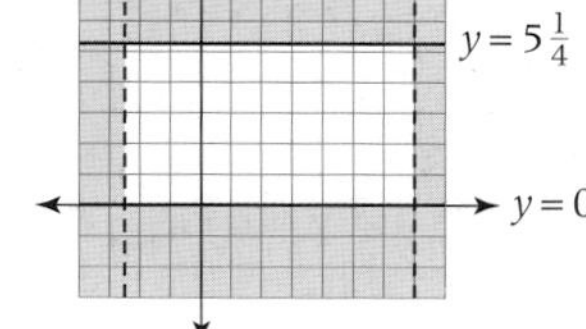

2 a $-2 < y \leqslant 5$ **b** $1 \leqslant x < 7.5$

3 $(-1, 0), (0, 0), (1, 0), (-1, 1), (0, 1), (1, 1)$

4 a e.g. $0 < x < 2$, $0 < y < 1$

b e.g. $x > 1$ or $x < 0$ and $y > 1$ or $y < 0$

c e.g. $0 < x < 2$, $0 < y < 1$

5 a $y \leqslant 9$, $y \geqslant x^2$ **b** $y < 5$, $y \geqslant (x-3)(x+2)$

A5.3

1 a

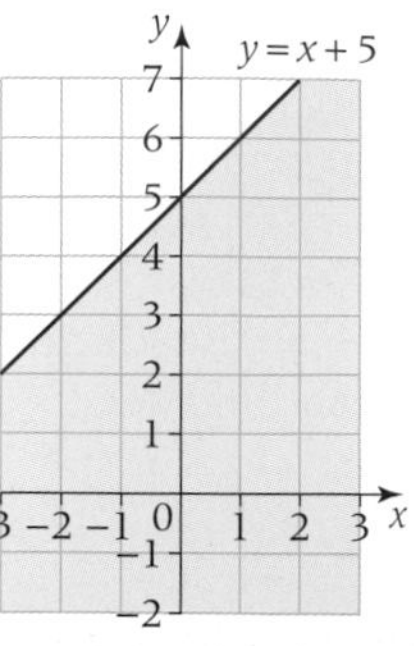

b

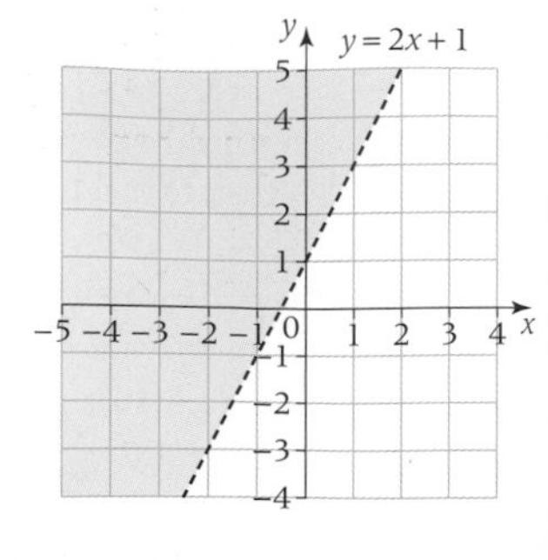

c

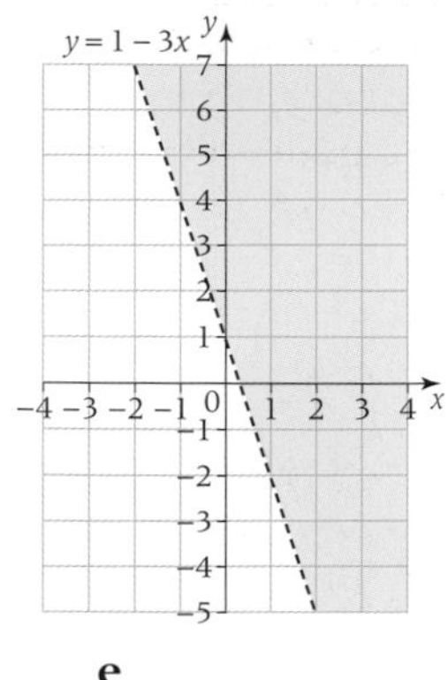

d

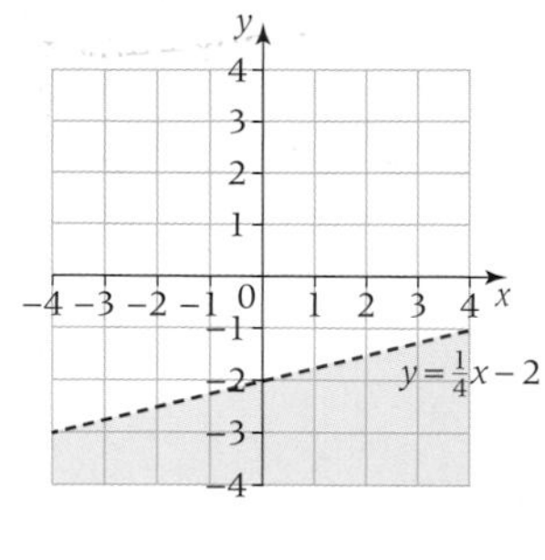

e

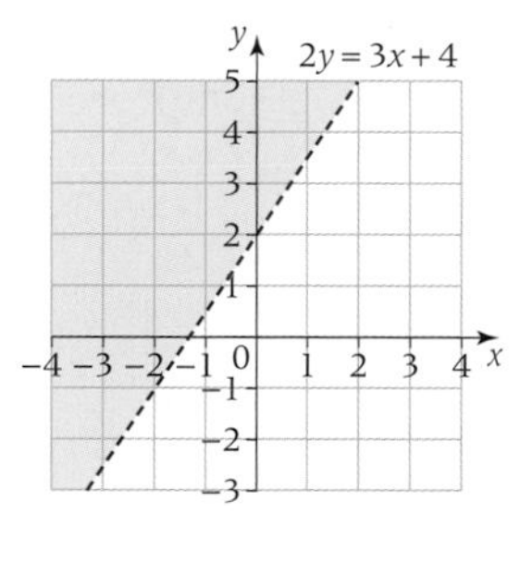

f

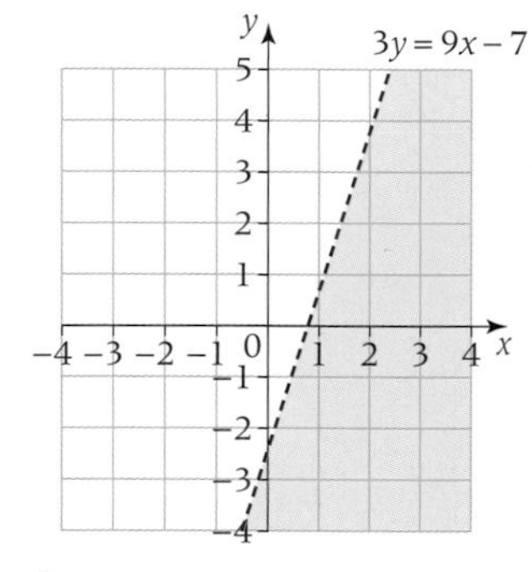

g

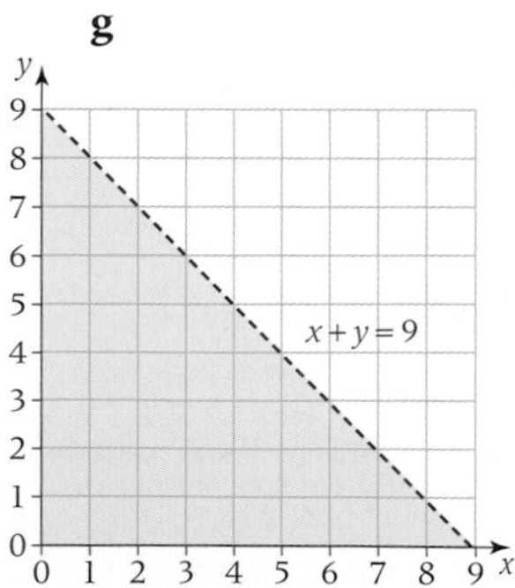

h

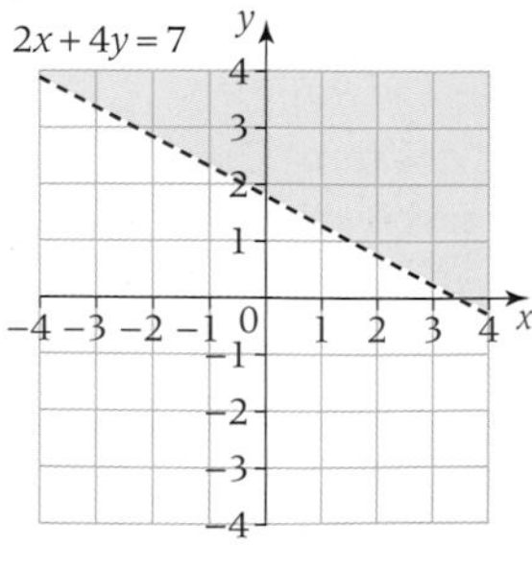

2 a $y < 2x + 4$ **b** $y \geqslant 2\frac{2}{3}x - 1$

3 a $(-1, 1), (-1, 0), (-1, -1), (-1, -2), (-1, -3), (-1, -4), (0, 1), (0, 0), (0, -1)$

b $(3, 1), (3, 0), (4, 0)$

4

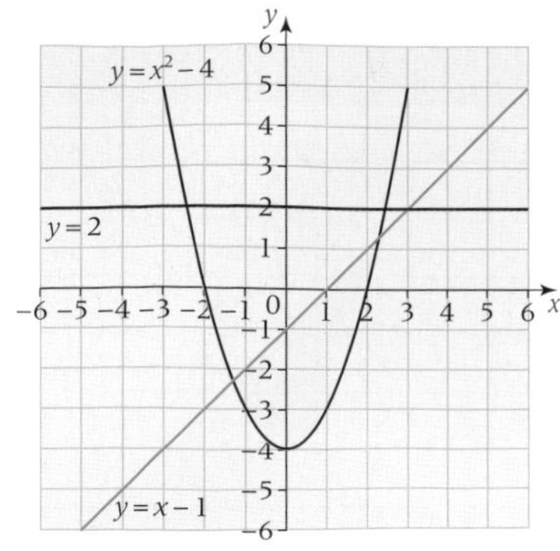

5 $x \geqslant 0$ and $y \geqslant 0$ because the number of each type of coach cannot be negative. The total number of seats is $20x + 48y$, which must be at least enough for 316 people, so $20x + 48y \geqslant 316$ or $5x + 12y \geqslant 79$. To have at least 2 adults per coach, there can be no more than 8 coaches, so $x + y \leqslant 8$.

6 **a** e.g. $x > 0$, $y > 0$, $x + y < 1$
b e.g. $y > 1$ or $y < 0$, $y > x$ or $y > 3 - x$

A5.4

1 **a** $x = 7$, $y = 4$ **b** $x = -4$, $y = -5$
c $c = 2$, $d = -1$ **d** $x = 2$, $y = 0.5$
e $a = -\frac{2}{3}$, $b = \frac{3}{4}$ **f** $p = 2.5$, $q = -3$

2 **a** 9, 14 **b** −2, 4

3 The star is worth 29 and the moon 17.

4 $y = -\frac{8}{9}x + 9\frac{2}{3}$

5 73

6 $y = \frac{8}{3}x^2 - \frac{19}{3}x + 7$

A5 Summary

1

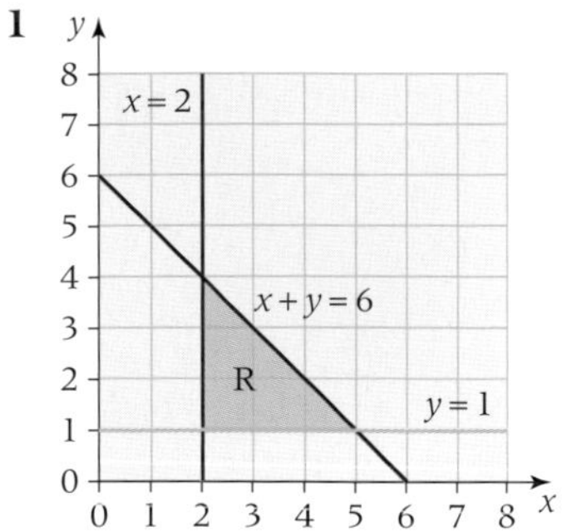

2 $x = 2\frac{1}{3}$, $y = -1$

3 **a** $x = 3$, $y = 2$
b Crosses at (4, 2), (5, 1), (5, 2), (5, 3)

Case study 6: Business

Original table:

	January (£)	February (£)	March (£)
Quantity of standard	18	16	24
Quantity of deluxe	12	14	9
Standard card sales	45.9	40.8	61.2
Deluxe card sales	43.2	50.4	32.4
TOTAL INCOME	89.1	91.2	93.6
Materials used	7.8	7.6	9
Wages	42	44	42
Craft fair fees	10	10	10
Advertising	5	5	5
TOTAL EXPENDITURE	64.8	66.6	66
NET CASH SURPLUS/DEFICIT	24.3	24.6	27.6
CASH BALANCE BROUGHT FORWARD		24.3	48.9
CASH BALANCE TO CARRY FORWARD	24.3	48.9	76.5

Fair fees £15:

	January (£)	February (£)	March (£)
Quantity of standard	18	16	24
Quantity of deluxe	12	14	9
Standard card sales	45.9	40.8	61.2
Deluxe card sales	43.2	50.4	32.4
TOTAL INCOME	89.1	91.2	93.6
Materials used	7.8	7.6	9
Wages	42	44	42
Craft fair fees	15	15	15
Advertising	5	5	5
TOTAL EXPENDITURE	69.8	71.6	71
NET CASH SURPLUS/DEFICIT	19.3	19.6	22.6
CASH BALANCE BROUGHT FORWARD		19.3	38.9
CASH BALANCE TO CARRY FORWARD	19.3	38.9	61.5

Material costs increase by 40%:

	January (£)	February (£)	March (£)
Quantity of standard	18	16	24
Quantity of deluxe	12	14	9
Standard card sales	45.9	40.8	61.2
Deluxe card sales	43.2	50.4	32.4
TOTAL INCOME	89.1	91.2	93.6
Materials used	10.92	10.64	12.6
Wages	42	44	42
Craft fair fees	10	10	10
Advertising	5	5	5
TOTAL EXPENDITURE	67.92	69.64	69.6
NET CASH SURPLUS/DEFICIT	21.18	21.56	24
CASH BALANCE BROUGHT FORWARD		21.18	42.74
CASH BALANCE TO CARRY FORWARD	21.18	42.74	66.74
Materials cost standard	0.42		
materials cost deluxe	0.28		

20% discount:

	January (£)	February (£)	March (£)
Quantity of standard	18	16	24
Quantity of deluxe	12	14	9
Standard card sales	36.72	32.64	48.96
Deluxe card sales	34.56	40.32	25.92
TOTAL INCOME	71.28	72.96	74.88
Materials used	7.8	7.6	9
Wages	42	44	42
Craft fair fees	10	10	10
Advertising	5	5	5
TOTAL EXPENDITURE	64.8	66.6	66
NET CASH SURPLUS/DEFICIT	6.48	6.36	8.88
CASH BALANCE BROUGHT FWD		6.48	12.84
CASH BALANCE TO CARRY FWD	6.48	12.84	21.72
discounted price standard	2.04		
discounted price deluxe	2.88		

Scenario with £8.00 for materials and £40 for wages.

Constraints:
$0.3s + 0.2d \leqslant 8$
$s + 2d \leqslant 40$
The aim is to maximise profit £P, where
$P = 1.25s + 1.40d - 15$
Gives maximum profit £24 when $s = 20$ and $d = 10$

G4 Check in

1 **a** 49 **b** 52 **c** 34
  d 48 **e** 45 **f** 8
2 **a** $x = 6y$ **b** $x = 5y$ **c** $x = 10y$
  d $x = \frac{2}{y}$ **e** $x = \frac{5}{y}$ **f** $x = \frac{8}{y}$
3 **a** $x = 5$ **b** $y = 77$
4 **a** $x = 4.93$ **b** $y = 2\frac{2}{3}$
5 **a** $x = 22.8$ **b** $y = 0.8$

G4.1

1 5.83 units
2 **a** 8.06 units **b** 10.30 units
3 **a** (1, 4.5) AM = 5.02 **b** (2.5, 2) AM = 5.41
4 14.14 cm
5 **a** 7.07 cm **b** 50 cm^2
6 12.9 cm
7 16.2 cm
8 19.5 cm

G4.2

1 a 3.75 cm b 8.63 cm c 11.3 cm
d 2.04 cm e 2.42 cm f 4.73 cm
g 2.40 cm h 8.48 cm i 68.4 cm
j 4.73 cm k 4 cm l 16.3 cm

2 Right-angled isosceles triangles. The missing side is equal to the given shorter side.

G4.3

1 5.14 cm 2 10.4 cm 3 5.26 cm
4 4.75 cm 5 9.50 cm 6 11.1 cm
7 8.20 cm 8 8.30 cm 9 51.8 cm
10 10.9 cm 11 8.27 cm 12 7.65 cm
13 13.3 cm 14 5.89 cm 15 15.1 cm

G4.4

1 15.8° 2 18.3° 3 65.3°
4 67.1° 5 24.7° 6 20.5°
7 55.7° 8 69.3° 9 22.2°
10 39.2° 11 69.8° 12 25.9°
13 74.5° 14 35.8° 15 45°
16 30° 17 60°

G4.5

1 a 18.9 cm b 20.6 cm c 4.54 cm
2 a 30.5° b 29.1° c 63.9°
3 31.8 cm²
4 36.7 cm²
5 89.8°
6 9.46 cm
7 50°, 50° and 80°
8 Edina's measurements make the height of the tree 25.6 m. Patsy's measurements make the height of the tree 27.0 m. They could both be correct as this type of measuring can be inaccurate (they may have taken different points as the top of the tree, the ground may not be level, ...).

G4.6

1 a $a = 4.5$ cm b $b = 5.4$ cm c $c = 3.7$ cm
d $d = 10.2$ cm e $e = 15.1$ cm f $f = 9.6$ cm
2 a $p = 32.6°$ b $q = 53.7°$ c $r = 55.6°$
d $s = 31.6°$ e $t = 42.5°$
3 a $AD^2 = 3^2 + 12^2 + 4^2$, so AD = 13 cm
b 18.2 cm
4 a 14.2 b 13.7 m

G4 Summary

1 a 39.8°
b 6.43 cm
2 13.5°

A6 Check in

1 a £31.50 b 54 mins
2 a 15 b $-2\frac{1}{2}$ c $\frac{5}{17}$ d $\frac{3}{7}$
3 a When $x = 0$, $y = 6$; when $y = 0$, $x = -2$ or -3.
b When $x = 0$, $y = 6$; when $y = 0$, $x = 2$ or 1.5.

A6.1

1 a $m = \frac{y-c}{x}$ b $m = k + wt$
c $m = \sqrt{p - kt}$ d $m = (l + k)^3$

2 a $x = 2(y - kw)$ b $x = \frac{4m + t^2}{a}$
c $x = 4y^2$ d $x = \frac{k - y^2}{a}$
e $x = \frac{(t - kh)^2}{a^2}$ f $x = \frac{p}{m + t}$
g $x = \frac{p}{w - c}$ h $x = \frac{y}{a(b + j)}$

3 Clare: In the second line the RHS should be $-ax$. Isla: In the second line the RHS should be $px - k$.

4 $a = b - \frac{c}{def}$, $a + \frac{c}{def} = b$, $\frac{c}{def} = b - a$, $\frac{c}{df(b - a)} = e$, $c = def(b - a)$

5 a $\frac{p(c - qt)}{m - x^2} = wr$, $pc - pqt = mwr - wrx^2$,
$x^2 = \frac{mwr + pqt - pc}{wr}$, $x = \sqrt{\frac{mwr + pqt - pc}{wr}}$

b $\frac{1}{a} + \frac{1}{b} = \frac{1}{c}$, $\frac{1}{b} = \frac{1}{c} - \frac{1}{a}$, $\frac{1}{b} = \frac{a - c}{ac}$, $b = \frac{ac}{a - c}$

c $\sqrt[4]{t - qx} = 2p$, $t - qx = 16p^4$, $t = qx + 16p^4$,
$x = \frac{t - 16p^4}{q}$

6 False, the last formula has no t.

7 a $f = \frac{uv}{u + v}$ b $v = \frac{fu}{u - f}$

A6.2

1 a $x = \sqrt{\frac{y - b}{a}}$ b $x = c + 2t$
c $x = \frac{c - k}{b}$ d $x = \frac{b}{p + q}$
e $x = \frac{b + 9t^2}{az}$ f $x = \sqrt[3]{k - t^2}$

2 Collecting the x-terms on one side gives $x^2 + 5x = -6$, so it is not possible to get x on its own.

3 a $w = \frac{t + r}{q - p}$ b $w = \frac{a - k}{c - l}$
c $w = \frac{py + qt}{p + q}$ d $w = \frac{t + r}{1 + t}$
e $w = \frac{c - rg}{r - 1}$ f $w = \frac{t + 5r}{x + r}$
g $w = \frac{t + 5k}{k - 1}$ h $w = \frac{3q - 4p}{7}$
i $w = \frac{-t - 25q}{24}$

4 a $A = 2\pi r^2 + 2\pi rh$
b $r = \frac{-2\pi h \pm \sqrt{4\pi^2h^2 + 8\pi A}}{4\pi}$

5 a $A = \pi r^2 + \pi rl$
b $r = \frac{-\pi l \pm \sqrt{\pi^2 l^2 + 4\pi A}}{2\pi}$

6 a $x = 2.098$ or -1.431
b $c = -ax^2 - bx$
c $b = \frac{-ax^2 - c}{x}$
d Because there are different powers of a.

7 $\sin(A + B) \neq \sin A + \sin B$

A6.3

1 a $x = -0.785$ or -2.55 b $x = 1$ or 0.2
c $x = 5$ or -1.5 d $x = -0.268$ or -3.73
e $x = 0.158$ or -3.16 f $y = 2.21$ or -0.377
g $x = 3$ or 0.333 h $x = 0.158$ or -3.16
i $x = 1.67$ or -4

2 a $x = 1.61$ or -5.61 b $x = 4.46$ or -2.46
c $x = 1.17$ or -0.284 d $x = 2.87$ or -4.87
e $x = 2.32$ or -0.323

3 a $(x-2)(x+7)$
b $(x-2)(x+7) = 20 \Rightarrow x^2 + 5x - 34 = 0$
c 1.84 cm by 10.84 cm

4 a $x = -4$ or -1.5; $x = -4.35$ or -1.15
b Only the first can be solved by factorization: $2x^2 + 11x + 12 = (2x+3)(1x+4)$
c $b^2 - 4ac$

5 $b^2 - 4ac$ is negative so you cannot take the square root.

A6.4

1 a $x = 3$ or 6 b $x = 10$ or -10 c $x = 0$ or 7

2 a $x = 8.24$ or -0.243 b $x = 1.42$ or -0.587
c $x = 1$ or -1.43

3 ai $x^2 - 4x - 117 = 0$
aii 9 and 13 or -9 and -13
bi $x^2 - 4x - 357 = 0$ bii 17 cm by 21 cm
ci $x^2 - 6x - 1 = 0$ cii 6.16 or -0.162
di $10x^2 - x - 3 = 0$ dii 0.6 or -0.5
ei $x^2 + 7x - 120 = 0$ eii 8 mm by 15 mm
fi $2x^2 - 9x + 4 = 0$ fii 10 units

4 a $2\pi r^2 + 2\pi r \times 5 = 100$, so $\pi r^2 + 5\pi r - 50 = 0$
b 4.42 cm

A6.5

1 a $x = -2$ or -5 b $x = 2$ or -6
c $x = -7$ or 7 d $x = 0$ or 8
e $x = 2$ or -6 f $x = \frac{1}{3}$ or 2

2 a $x = -0.838$ or -7.16 b $x = -0.227$ or -0.631
c $x = 2.41$ or -0.414 d $x = 0.657$ or -0.457
e $x = 0.425$ or -1.18 f $x = 2.65$ or 0.849

3 ai Factorises aii Two solutions
bi Factorises bii One solution
ci Does not factorise cii Two solutions
di Factorises dii Two solutions
ei Factorises eii Two solutions
fi Factorises fii Two solutions

4 a $y = 8$ or -11 b $x = -2.5$ or 2.5
c $x = -3$ or -0.5 d No solutions
e $y = 3$ f $p = 0.662$ or -22.7
g $x = 0.587$ or -0.730

5 False, as the discriminant $= -15 < 0$ there are no solutions.

6 8 cm by 15 cm

7 a $x = 0$, 2 or 3 b $x = 0$, 0.186 or -2.69
c $x = -3$, 0 or 2

A6.6

1 a $x = 2.28$ or 0.219 b $y = 9.47$ or 0.528
c $p = -3$ or -7 d $x = -1$ or 0.3
e $x = 0.762$ or -2.36

2 a 34 cm b 27 cm c (3, 4), (−1.4, −4.8)
d Volume of sphere = 19.0 units³, volume of cylinder = 51.7 units³, so cylinder has greater volume

3 a

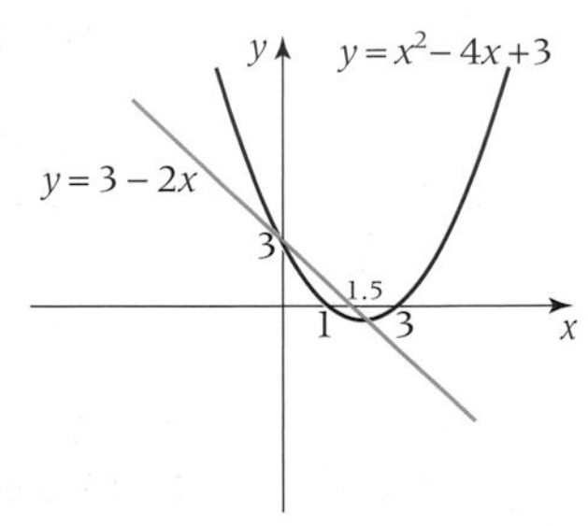

b (0, 3) c (2, −1)

4 2.8

A6 Summary

1 $b = \frac{2+5a}{a+7}$ $\left(\text{Accept } b = \frac{-2-5a}{-a-7}\right)$

2 a $f = 1\frac{3}{7}$ b $u = \frac{fv}{v-f}$

3 $x = 9.93$ (to 3 sf)

G5 Check in

1 a $a = 8.1$ units b $b = 9.8$ units

2 a $x = 2\frac{6}{7}$ b $x = 1\frac{7}{11}$ c $x = 5\frac{3}{5}$

3 a 5 b −38 c 6

4 ai 6.34 cm aii 38.06 cm²
bi 8.86 cm bii 57.61 cm²

G5.1

1 a 6.5 cm b 14.3 cm c 5.7 cm d 10.3 cm
e 9.5 cm f 18.1 cm g 28.9 cm h 9.8 cm

2 a 33.4° b 26.1° c 13.0° d 44.8°
e 18.3° f 59.8° g 35.6° h 25.2°

G5.2

1 a 6.2 cm b 13.9 cm c 12.0 cm d 6.6 cm
e 14 cm f 11.3 cm g 12.0 cm h 18.8 cm

2 a 33.0° b 34.8° c 85.3° d 55.8°
e 19.9° f 63.1° g 54.7° h 111°

3 B = 59.0°, C = 41.0°, $x = 10.4$ cm

G5.3

1 11.1 km, 191° 2 18.1 km, 261°
3 47.7 cm 4 41.3 cm
5 47.3 cm 6 a 13.4 cm b 10.5 cm
7 43.7 cm 8 12.7 cm
9 11.8 cm

G5.4

1 a 5.9 b 6.6 c 9.7 d 4.9

2 a 15.3 cm b 11.7 cm c 12.3 cm

3 a 51.8° b 43.1° c 18.9°
4 a 35.3° b 35.3°
5 12°
6 a 10.4 cm b 16.7°

G5.5

1 a 25.4 cm² b 36.1 cm² c 25.9 cm²
d 12.4 cm² e 40.7 cm² f 21.8 cm²
g 26.5 cm² h 37.1 cm² i 17.7 cm²
j 31.6 cm²
2 a 55.0 cm² b 34.9 cm² c 61.0 cm²
d 114 cm² e 49.0 cm² f 72.2 cm²
g 80.1 cm² h 74.8 cm²

G5.6

1 a 2.61 cm² b 0.916 cm² c 7.50 cm²
d 18.2 cm² e 17.6 cm² f 14.6 cm²
g 90.0 cm² h 1.70 cm²
2 a 39.1° b 60.1° c 17.0 cm²
3 a 22.1 cm² b 9.25 cm² c 18.0 cm²
4 a 56.4° b 10.2 cm²
5 a 19.9 cm² b 29.8 cm² c 36.9 cm²
6 1356 cm²

G5.7

1 $r = 10$ cm and $h = 2\sqrt{11}$ cm so $V = \frac{200}{3}\pi\sqrt{11}$ cm³
2 $r = 5$ cm and $h = 10\sqrt{2}$ so $V = \frac{250}{3}\pi\sqrt{2}$ cm³
3 $4\pi r^2 = 12\pi \Rightarrow r = \sqrt{3}$ so $V = \frac{4}{3}\pi(\sqrt{3})^3 = 4\pi\sqrt{3}$
4 103.1π cm²
5 48.17π cm²
6 $2\pi(2r)h = \pi rl \Rightarrow h = \frac{1}{4}l$
7 4 cm

G5 Summary

1 13 cm
2 5.22 cm
3 85.5°

Case study 7: Art

1 a 30
b 4.44 m
c length = 40 hands; height = 29.6 hands
d 30 cm squares
e area scale factor 900
2 ai $B = (2, 60)$ aii $A = (1, 210)$
aiii $C = (3, 300)$ b $D = (1.5, 0)$
3 a Circle radius = 4
b $a = 4$
c 9 units
d $k = 0.025$
5 a Centre is a circle – constant radius.
b Outer edge is a spiral – radius increases with angle. The two are joined by radial and tangent lines.

A7 Check in

1 a $x = 3$ or 4
b $y = 0$ or 8
c $x = -3.56$ or 0.56
d $x = -3$ or $-\frac{1}{2}$
e $y = 1.87$ or 13.37
f $x = -\frac{1}{3}$ or 1
2 a 40 b 6 c −80
3 a

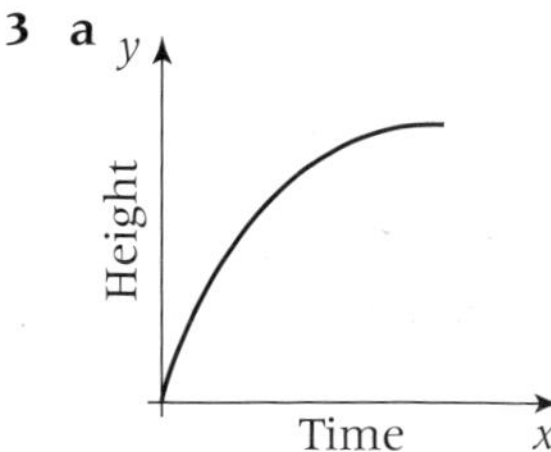

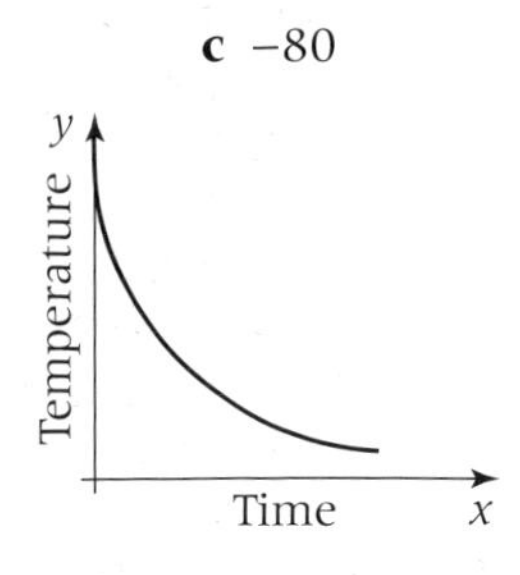

A7.1

1 a $x = 7, y = 6$ or $x = -7, y = 6$
b $x = 7, y = 5$ or $x = 7, y = -5$
c $x = 10, y = 0$ or $x = -10, y = 9$
2 a $x = 4, y = 8$ or $x = -1, y = 3$
b $x = 1, y = 0$
c $x = 32, y = 4$ or $x = \frac{1}{2}, y = \frac{1}{2}$
3 a $x = 3, y = 7$ or $x = 5, y = 17$
b $p = 2, q = 1$ or $p = -2, q = -1$
4 a $y = x^2 + x - 6$
b $y = x + 10$
c (4, 14) and (−4, 6)
5 $x = -3, y = 2$ or $x = 3, y = -2$
6 $p = -1, q = 6$

A7.2

1 a $x^2 + y^2 = 36$ b $4x^2 + 4y^2 = 1$
c $100x^2 + 100y^2 = 16$ d $x^2 + y^2 = 5$
2 a $x^2 + y^2 = 49$ b $x^2 + y^2 = 7$
3 (5, 12) and $\left(-\frac{33}{5}, -\frac{56}{5}\right)$
4 They do not intersect
5 ai Once aii Twice aiii Once b Tangent
6 a $(\sqrt{5}, 2\sqrt{5})$ and $(\sqrt{5}, -2\sqrt{5})$
b (1.38, 3.76) and (−2.18, −3.36)
7 a $(x - 3)^2 + (y - 5)^2 = 36$ b (6, 0)

A7.3

1 $x = 1.6, y = 2.6$ and $x = -1.6, y = -2.6$
2 a Two b Two c One d Two
e None f Two g Two h Three
3 a $x = 4, y = 3$ and $x = -4, y = 3$
b $x = 1.4, y = 1.4$ and $x = -1.4, y = -1.4$
c $x = 2.2, y = 5.6$ and $x = -1.6, y = -5.8$
d $x = 0.8, y = 0.6$ and $x = -0.8, y = 0.6$
4 a $x = 1.8, y = 3.6$ and $x = -1.8, y = -3.6$
b $x = 3.5, y = 3.5$ and $x = -3.5, y = -3.5$
c $x = 1.2, y = 2.7$ and $x = -0.6, y = -2.9$
d $x = 2.6, y = 6.5$ and $x = -2.6, y = 6.5$
5 Yes, as in question **2h**.

A7.4

1 a–i, b–iv, c–ii, d–iii

2 a

x	−4	−3	−2	−1	0	1	2	3	4
$2x^2$	32	18	8	2	0	2	8	18	32
$3x$	−12	−9	−6	−3	0	3	6	9	12
−6	−6	−6	−6	−6	−6	−6	−6	−6	−6
y	14	3	−4	−7	−6	−1	8	21	38

b

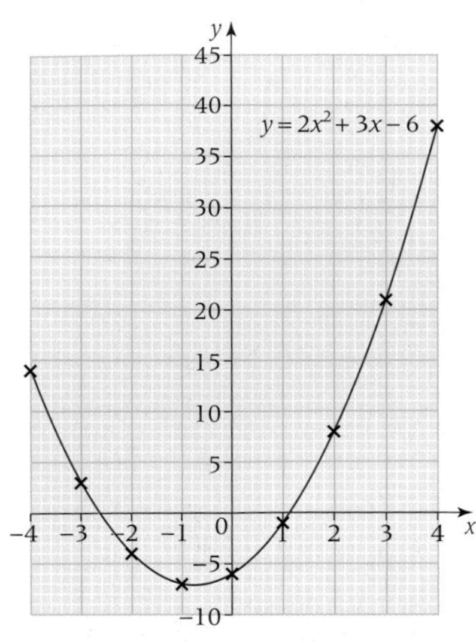

c $\left(-\frac{3}{4}, -7\frac{1}{8}\right)$

3 a

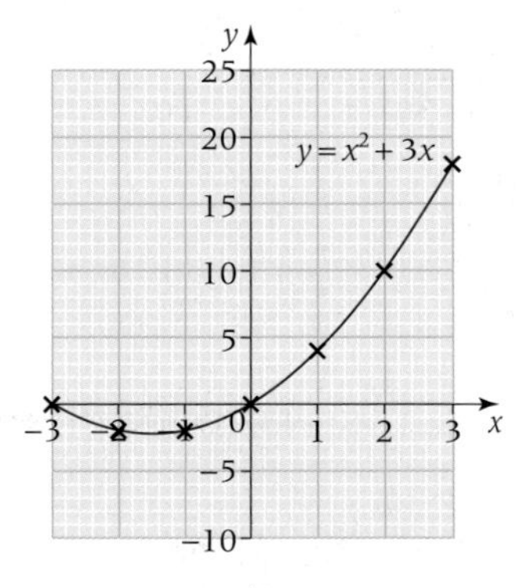

b

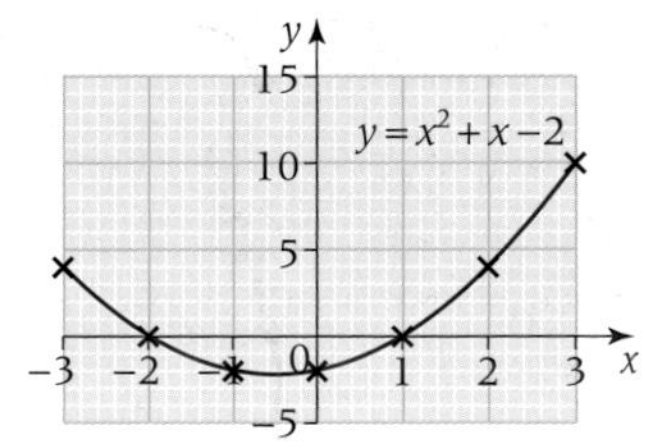

c

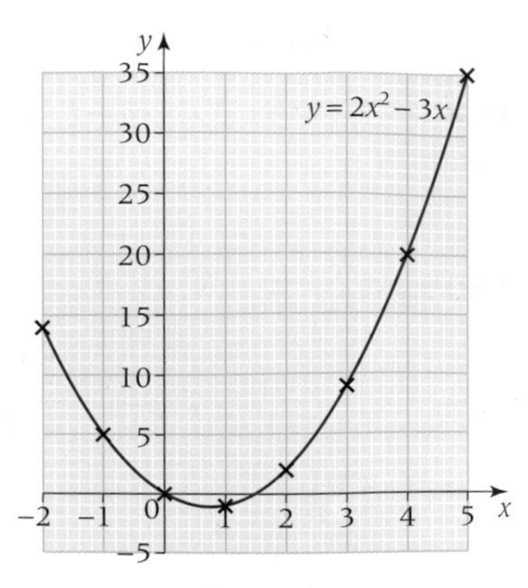

d

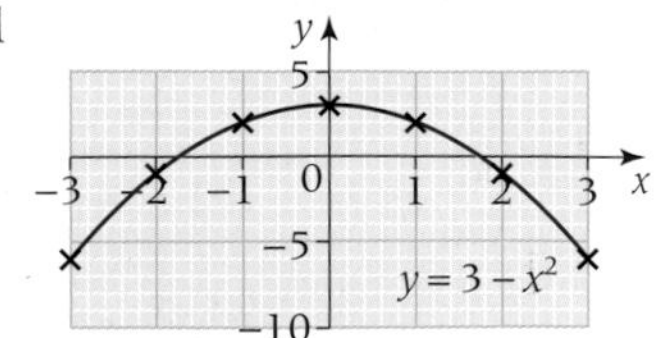

e **f**

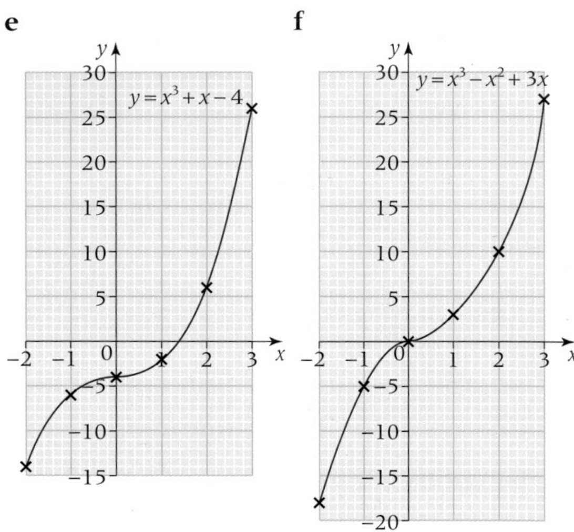

4 a

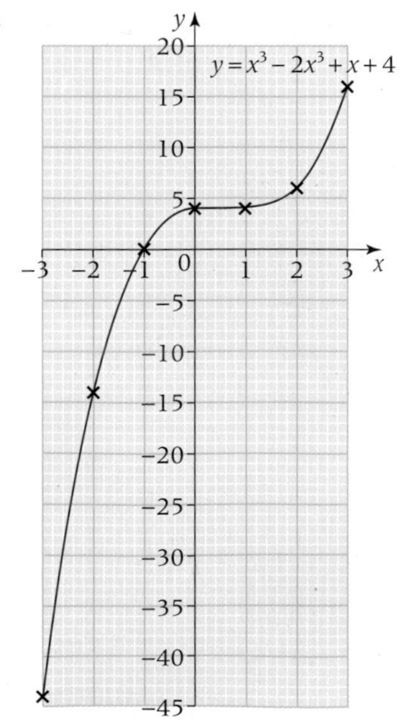

bi −2.3 **bii** 4.8

5 **ai** $l = 100 - 2w$ **aii** $A = w(100 - 2w)$

b 25 m by 50 m

A7.5

1 **a–iv, b–i, c–ii, d–iii**

2 **a**

x	−6	−5	−4	−3	−2	−1	0	1	2	3	4	5	6
f(x)	−2	−2.4	−3	−4	−6	−12	Asymptote	12	6	4	3	2.4	2

b

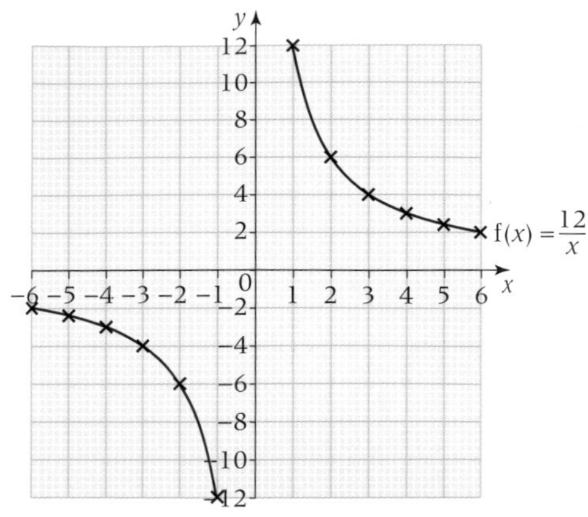

c The graph is a hyperbola. The axes are the asymptotes.

d 4.8

3 **a**

x	−4	−3	−2	−1	0	1	2	3	4
g(x)	$\frac{1}{8}$	$\frac{1}{4}$	$\frac{1}{2}$	1	2	4	8	16	32

b

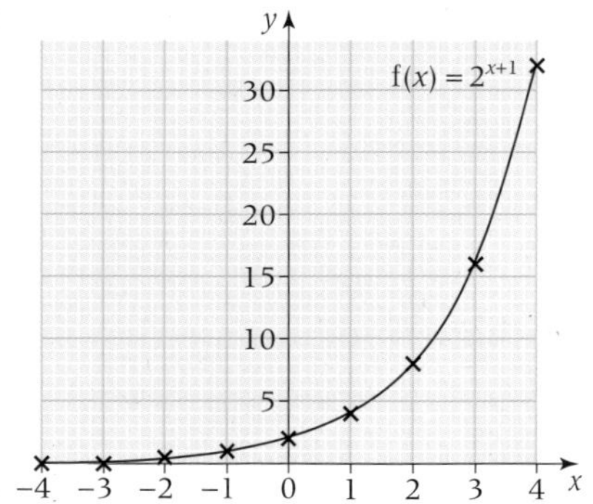

c 3.6

4

x	−4	−3	−2	−1	0	1	2	3	4	5
$\frac{20}{x}$	−5	$-6\frac{2}{3}$	−10	−20	–	20	10	$6\frac{2}{3}$	5	4
−5	−5	−5	−5	−5	−5	−5	−5	−5	−5	−5
y	−14	$-14\frac{2}{3}$	−17	−26	–	16	7	$4\frac{2}{3}$	4	4

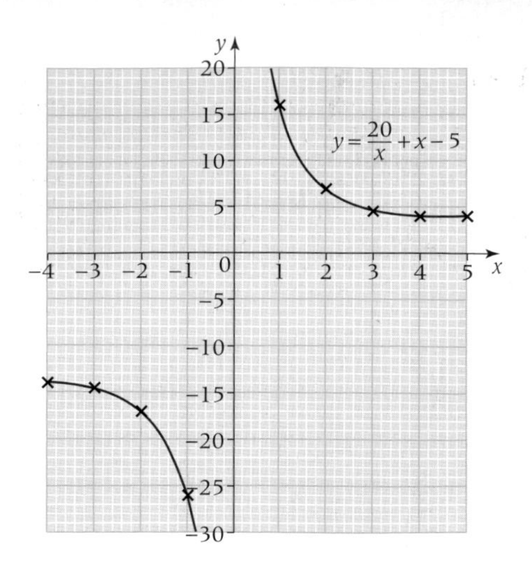

5 **a**

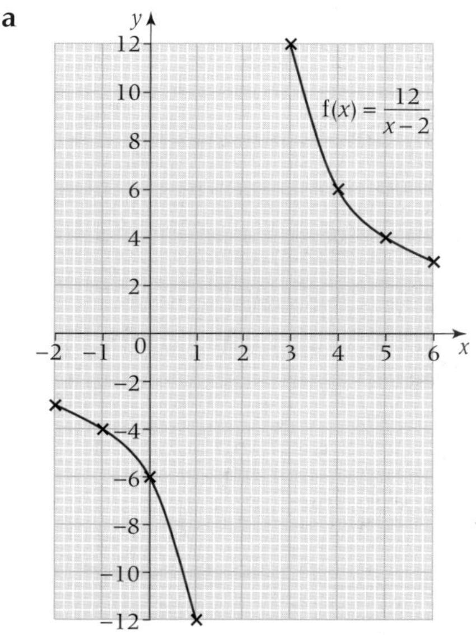

b

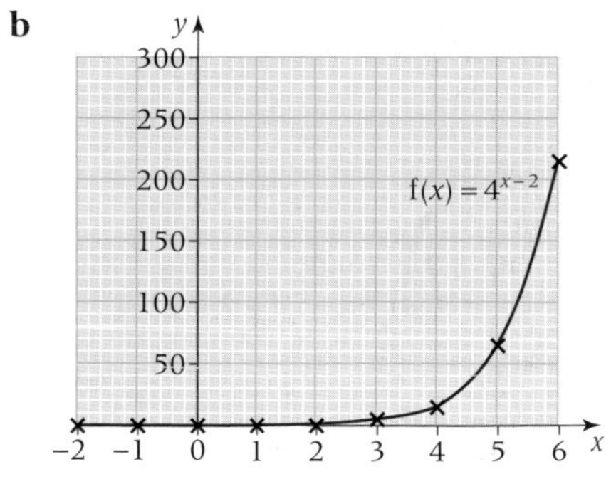

c

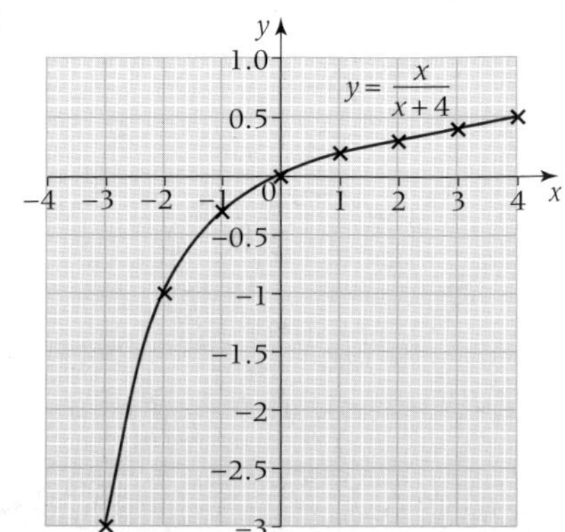

d

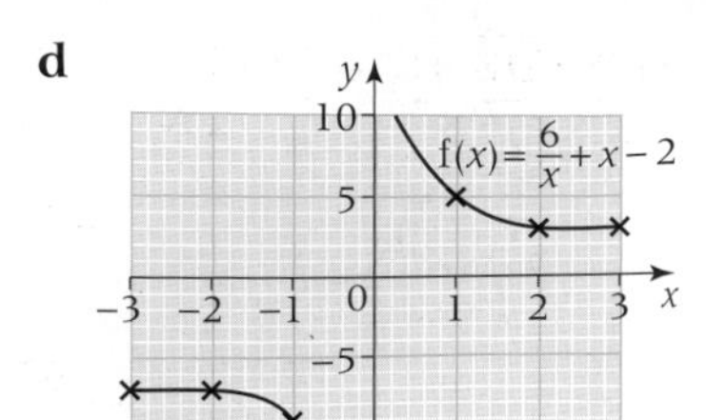

e

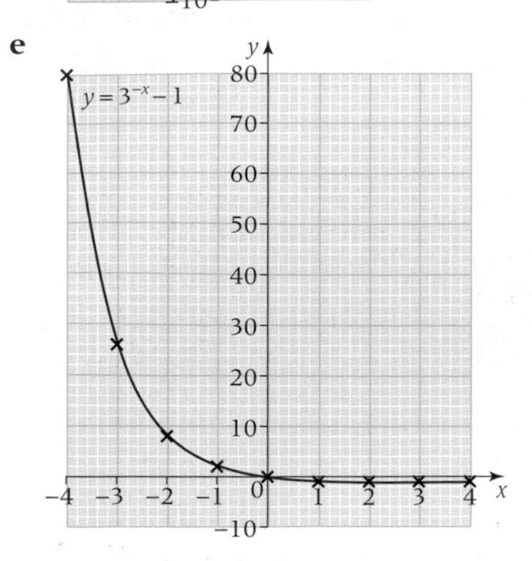

A7.6

1 **a** $x = 3, y = 7$ **b** $x = 1, y = 1$
c $x = 3, y = -1$

2 **ai** $x = -1.6$ or 2.6
aii $x = -0.6$ or 1.6
aiii $x = -1$ or 3
b On the x-axis; $x = -1$ or 2
c $y = 2 - x$

3 **a** $x = -0.7$ or 2.7 **b** $x = -1.2$ or 3.2
c $x = -0.8$ or 3.8 **d** $x = -1.8$ or 2.8
e $x = -0.9$ or 3.4 **f** $x = 1, y = 2$

4 **a** $x = 4.8$ **b** $x = 2.3$
c No solutions **d** $x = 3.2$

5 **a** None **b** None **c** One **d** One

A7.7

1 **a** $x = -3$ or 1 **b** $x = -2.6$ or 1.6
c $x = -3.4$ or 1.4 **d** $x = -1$ or 0

2 **a** $y = 0$ **b** $y = x + 2$
c $y = 2x - 3, y = 0$

3 **a** $y = 3$ **b** $y = 0$ **c** $y = 2x + 1$
d $y = 4$ **e** $y = 2$ **f** $y = 8x + 1$

4 **a** $y = 2x - 1$ **b** $y = 5x - 3$ **c** $y = 2x + 3$
d $y = 0$ **e** $y = x^2 - 1$ **f** $y = x^3 + x^2 - 1$

5 **bi** $x = 1.6$ **bii** $x = 0.2$ **biii** $x = -0.8, 2, 4$

6 **bi** $x = 8$ **bii** $x = 2.9$ **c** $y = x + 1$

A7.8

1 **a–i, b–ii, c–iii, d–v, e–iv**

2

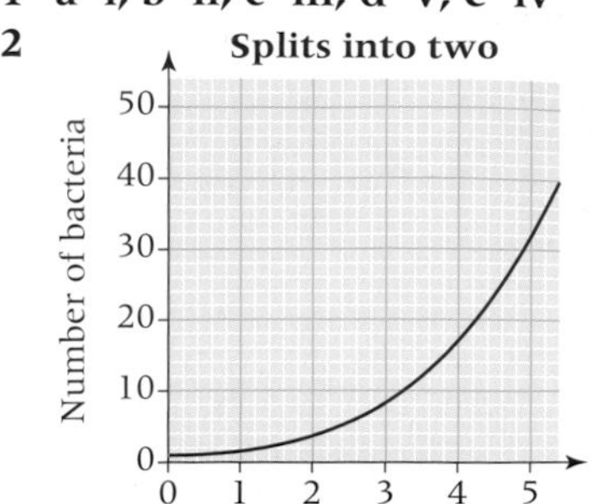

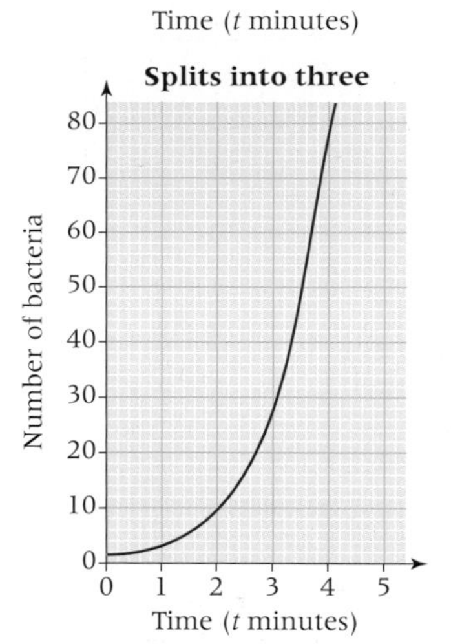

3 **a** 5 elephants
bi 8 elephants **bii** 11 elephants
c 4 years
d

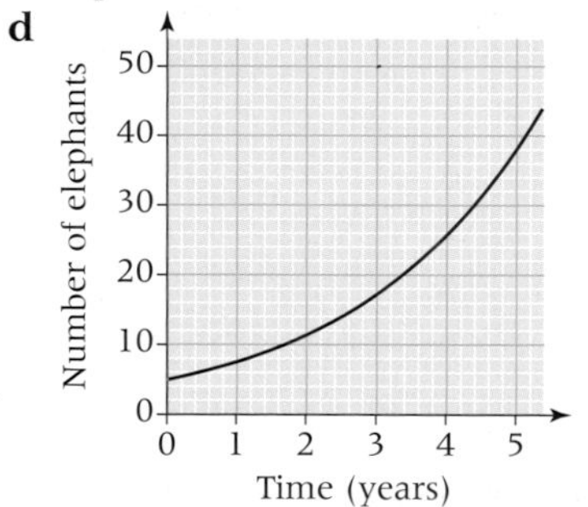

4 **a** $p = 3, q = 5$ **b** False, because $3 \times 5^2 \neq 100$.

5 **a** $m = 2000, n = 1.1$ **b** 10 years

A7 Summary

1 **i** E
ii A
iii I

2 $x = \frac{2}{5}, y = 2\frac{1}{5}$
$x = -1, y = -2$

3 **a** $x = -1, x = 2$
b $x = -2, x = 3$
c $x = 0, x = 2$

G6 Check in

1

	Square	Rhombus	Rectangle	Parallelogram	Trapezium	Kite
1 pair opposite sides parallel	✓	✓	✓	✓	✓	
2 pair opposite sides parallel	✓	✓	✓	✓		
Opposite sides equal	✓	✓	✓	✓		
All sides equal	✓	✓				
All angles equal	✓		✓			
Opposite angles equal	✓	✓	✓	✓		One pair
Diagonals equal	✓		✓			
Diagonals perpendicular	✓	✓		✓		✓
Diagonals bisect each other	✓	✓	✓			One bisects the other
Diagonals bisect the angle	✓	✓				One diagonal

2

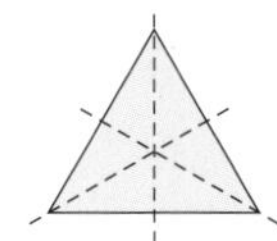

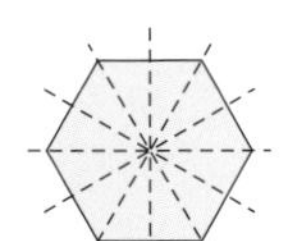

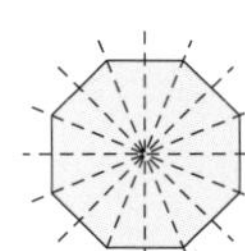

G6.1

1

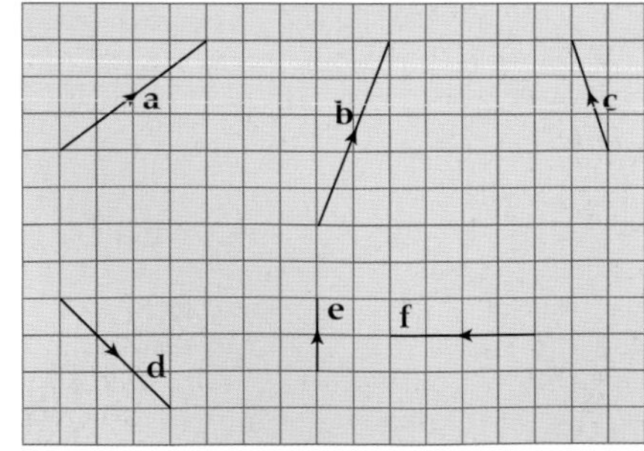

2 a

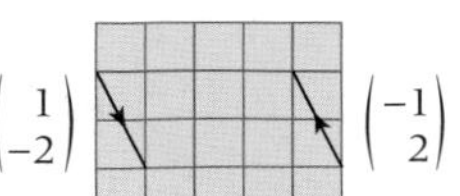

b They are parallel but in opposite directions.

3 $\overrightarrow{AB} = \overrightarrow{BC} = \overrightarrow{FH}$, $\overrightarrow{FG} = \overrightarrow{DE} = \overrightarrow{GH} = \overrightarrow{IJ}$,
$\overrightarrow{JH} = \overrightarrow{IG} = \overrightarrow{GE} = \overrightarrow{FD} = \overrightarrow{DB}$, $\overrightarrow{HC} = \overrightarrow{IE} = \overrightarrow{FB}$,
$\overrightarrow{IF} = \overrightarrow{JG} = \overrightarrow{GD} = \overrightarrow{HE} = \overrightarrow{EB}$, $\overrightarrow{FA} = \overrightarrow{JD} = \overrightarrow{HB}$,
$\overrightarrow{BA} = \overrightarrow{CB} = \overrightarrow{HF}$, $\overrightarrow{GF} = \overrightarrow{ED} = \overrightarrow{HG} = \overrightarrow{JI}$,
$\overrightarrow{HJ} = \overrightarrow{GI} = \overrightarrow{EG} = \overrightarrow{DF} = \overrightarrow{BD}$, $\overrightarrow{CH} = \overrightarrow{EI} = \overrightarrow{BF}$,
$\overrightarrow{FI} = \overrightarrow{GJ} = \overrightarrow{DG} = \overrightarrow{EH} = \overrightarrow{BE}$, $\overrightarrow{AF} = \overrightarrow{DJ} = \overrightarrow{BH}$

4 **ai** $\overrightarrow{XA}$, $\overrightarrow{CB}$, $\overrightarrow{DX}$, $\overrightarrow{EF}$ **aii** $\overrightarrow{AB}$, $\overrightarrow{FX}$, $\overrightarrow{XC}$, $\overrightarrow{ED}$
bi $\overrightarrow{AX}$, $\overrightarrow{BC}$, $\overrightarrow{XD}$, $\overrightarrow{FE}$ **bii** $\overrightarrow{BA}$, $\overrightarrow{XF}$, $\overrightarrow{CX}$, $\overrightarrow{DE}$

5 **ai** $\overrightarrow{OJ}$, $\overrightarrow{NK}$ **aii** $\overrightarrow{OM}$, $\overrightarrow{QK}$ **aiii** $\overrightarrow{OP}$, $\overrightarrow{LK}$
bi $\overrightarrow{JO}$, $\overrightarrow{KN}$ **bii** $\overrightarrow{MO}$, $\overrightarrow{KQ}$ **biii** $\overrightarrow{PO}$, $\overrightarrow{KL}$

G6.2

1

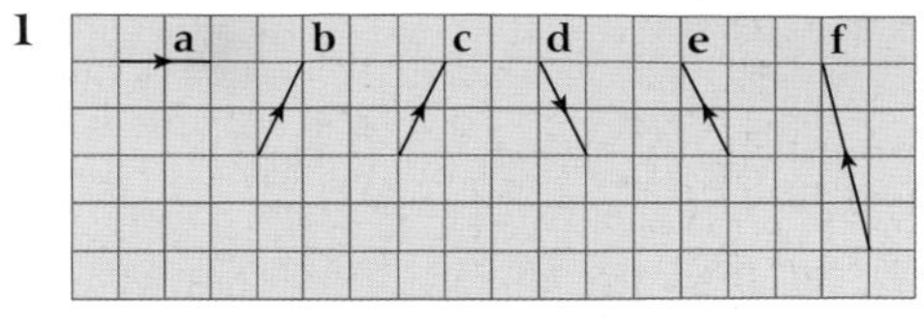

2

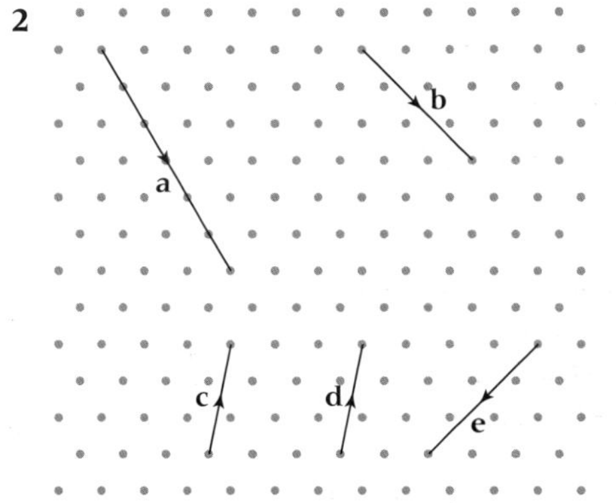

3 **a** r **b** p + r **c** −p − r **d** p − r
4 **a** a − b **b** b − a **c** −a **d** b − a
5 **a** l **b** l − j **c** −j − l **d** −j

G6.3

1 **a** 3p **b** 5p **c** 5p + 2q
d 5p + 6q **e** 4q **f** p + 3q
g p + 6q **h** 3p + 3q **i** p + 2q
j p + 6q **k** p + q **l** q − p
m −4q **n** −p − 4q **o** −3p − 4q
p 3q − p **q** p − 3q **r** p − 3q
s −4p − 4q **t** −p − 4q **u** −3p + 4q

2

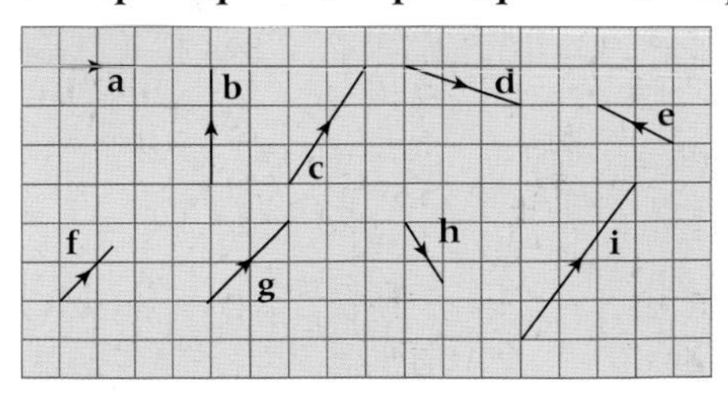

3 **a** a + d **b** 4d − a
4 **a** 1.5j **b** −1.5j

G6.4

1 a $\mathbf{z}$ b $\mathbf{w}+\mathbf{z}$ c $\frac{1}{2}(\mathbf{w}+\mathbf{z})$
d $\frac{1}{2}(\mathbf{w}+\mathbf{z})$ e $\frac{1}{2}(\mathbf{w}-\mathbf{z})$ f $\frac{1}{2}(\mathbf{z}+\mathbf{w})$

2 a $\mathbf{d}$ b $\mathbf{a}$ c $\mathbf{a}-\mathbf{d}$ d $\mathbf{a}-\mathbf{d}$

3 a $2\mathbf{c}$ b $-2\mathbf{b}$ c $\mathbf{c}-\mathbf{b}$ d $2(\mathbf{c}-\mathbf{b})$

4 a $\mathbf{t}$ b $\mathbf{r}+\mathbf{t}$ c $\frac{3}{4}(\mathbf{r}+\mathbf{t})$
d $-\frac{1}{4}(\mathbf{r}+\mathbf{t})$ e $\frac{3}{4}\mathbf{r}-\frac{1}{4}\mathbf{z}$

5 a $\mathbf{b}-\mathbf{a}$ b $\frac{4}{5}\mathbf{a}+\frac{1}{5}\mathbf{b}$

G6.5

1 $\overrightarrow{AB} = 2(\mathbf{b}-\mathbf{a})$, $\overrightarrow{CD} = 3(\mathbf{b}-\mathbf{a}) \Rightarrow \overrightarrow{CD} = 1.5\overrightarrow{AB} \Rightarrow$ $\overrightarrow{AB}$ is parallel to $\overrightarrow{CD}$.

2 ai $2\mathbf{r}+2\mathbf{a}$ aii $\mathbf{r}$ aiii $\mathbf{a}$ aiv $\mathbf{r}+\mathbf{a}$
b $\overrightarrow{RT} = 2\overrightarrow{MN}$ so $\overrightarrow{MN}$ is parallel to $\overrightarrow{RT}$.

3 ai $3\mathbf{p}$ aii $\mathbf{r}+3\mathbf{p}$ aiii $\mathbf{p}-\mathbf{r}$ aiv $2\mathbf{p}+\mathbf{r}$
b $\overrightarrow{OX} = \frac{1}{4}(3\mathbf{p}+\mathbf{r}) = \frac{1}{4}\overrightarrow{OQ}$, so $\overrightarrow{OX}$ and $\overrightarrow{OQ}$ are parallel. $\overrightarrow{OX}$ and $\overrightarrow{OQ}$ have the point O in common. So O, X and Q lie on the same straight line.

4 ai $\mathbf{j}+\mathbf{k}$ aii $\mathbf{k}$ aiii $\mathbf{k}-\mathbf{j}$
b $\overrightarrow{JX} = -\mathbf{j}\frac{1}{3}\mathbf{k}$ and $\overrightarrow{JM} = \mathbf{k} - 3\mathbf{j} = 3\overrightarrow{JX}$, so $\overrightarrow{JX}$ and $\overrightarrow{JM}$ are parallel. $\overrightarrow{JX}$ and $\overrightarrow{JM}$ have the point J in common. So J, X and M lie on the same straight line.

G6 Summary

1 a $\overrightarrow{MN} = 2\mathbf{a} - 2\mathbf{b}$ OR $\overrightarrow{MN} = 2(\mathbf{a}-\mathbf{b})$

b $\overrightarrow{MX} = \frac{1}{2}\overrightarrow{MN}$
$= \frac{1}{2}(2\mathbf{a}-2\mathbf{b})$
$= \mathbf{a}-\mathbf{b}$
$\overrightarrow{QR} = -2\mathbf{a} - 2\mathbf{b} + 6\mathbf{a}$
$= 4\mathbf{a} - 2\mathbf{b}$
$\overrightarrow{QY} = \frac{1}{2}\overrightarrow{QR}$
$= \frac{1}{2}(4\mathbf{a}-2\mathbf{b})$
$= 2\mathbf{a}-\mathbf{b}$
$\overrightarrow{XY} = -\overrightarrow{MX} + \overrightarrow{MQ} + \overrightarrow{QY}$
$= -(\mathbf{a}-\mathbf{b}) + \mathbf{a} + (2\mathbf{a} - \mathrm{b})$
$= -\mathbf{a} + \mathbf{b} + \mathbf{a} + 2\mathbf{a} - \mathbf{b}$
$= 2\mathbf{a}$
$\overrightarrow{OR} = 6\mathbf{a}$
So $3\overrightarrow{XY} = \overrightarrow{OR}$
ie XY and OR are parallel.

2 a $\overrightarrow{AB} = -\mathbf{a} + \mathbf{b}$ OR $\overrightarrow{AB} = \mathbf{b} - \mathbf{a}$

b $\overrightarrow{AP} = \frac{3}{5}\overrightarrow{AB}$
$= \frac{3}{5}(-\mathbf{a}+\mathbf{b})$
$= -\frac{3}{5}\mathbf{a} + \frac{3}{5}\mathbf{b}$
$\overrightarrow{OP} = \overrightarrow{OA} + \overrightarrow{AP}$
$= \mathbf{a} - \frac{3}{5}\mathbf{a} + \frac{3}{5}\mathbf{b}$
$= \frac{2}{5}\mathbf{a} + \frac{3}{5}\mathbf{b}$
$= \frac{1}{5}(2\mathbf{a} + 3\mathbf{b})$

A8 Check in

1 a Parabola b Straight line
c S-shape d Hyperbola

2 ai (0, 8) aii (−4, 0), (−2, 0)
bi (0, 24) bii (4, 0), (6, 0)
ci (0, 6) cii (−3, 0), (−2, 0)

3 a (−3, −1) b (5, −1) c $\left(-2.5, -\frac{1}{4}\right)$

4 a $\begin{pmatrix}2\\4\end{pmatrix}$ b $\begin{pmatrix}5\\-12\end{pmatrix}$ c $\begin{pmatrix}-3\\-3\end{pmatrix}$

A8.1

1 a A b C c D d B

2 a Graph translated 2 units up
b Graph translated 1 unit down
c Graph translated 4 units up
d Graph translated 3 units down

3 a A (−3, −2), B (0, 0), C (5, 5)
b A (−3, 3), B (0, 5), C (5, 10)
c A (−3, −5), B (0, −3), C (5, 2)
d A (−3, 10), B (0, 12), C (5, 17)
e A (−3, 2), B (0, 4), C (5, 9)

4 a 

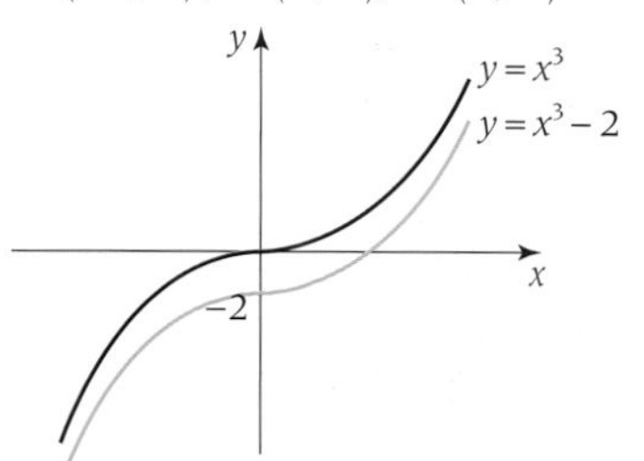

b

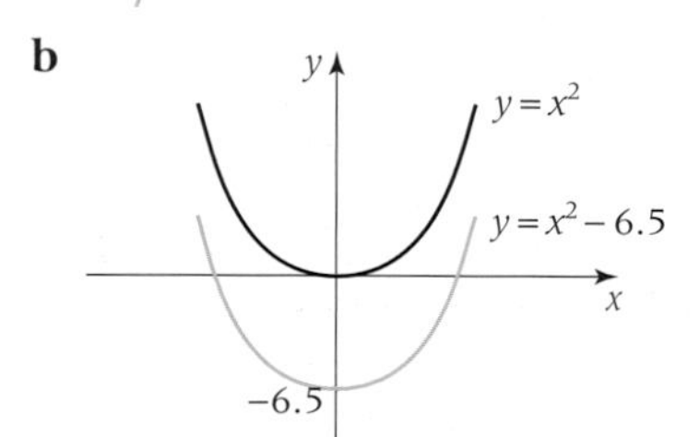

c

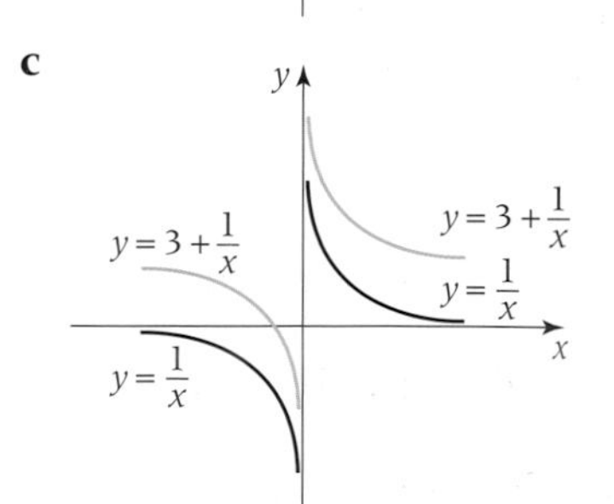

d

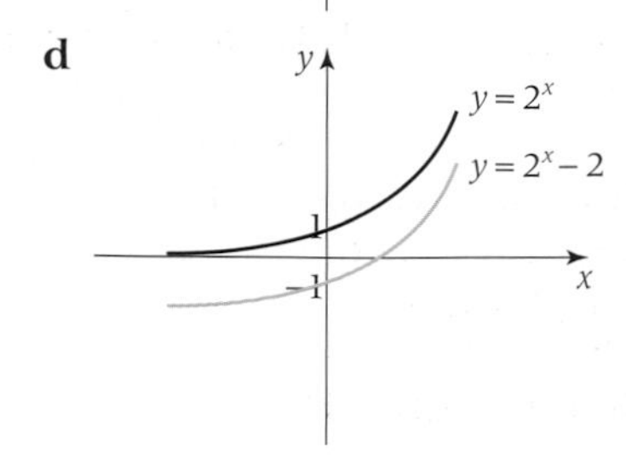

A8.2

1 a $y = f(x+2)$ b $y = f(x-1)$ c $y = f(x) - 5$
2 a Graph translated 3 units right
b Graph translated 2 units left
c Graph translated 2 units right
d Graph translated 1 unit left
3 a A (−1, 0), B (2, 6), C (6, 3)
b A (−6, 0), B (−3, 6), C (1, 3)
c A (−7, 0), B (−4, 6), C (0, 3)
d A (−3, −2), B (0, 4), C (4, 1)
e A (−3, 5), B (0, 11), C (4, 8)
4 a $y = f(x-2)$ b $y = f(x+1)$
c $y = f(x) + 3$ d $y = f(x+2) - 1$
5 $\begin{pmatrix} -4 \\ -2 \end{pmatrix}$

A8.3

1 a $y = -3f(x)$ b $y = 2f(x)$ c $y = \frac{1}{2}f(x)$
2 a Graph stretched 2 units parallel to y-axis
b Graph stretched 3 units parallel to y-axis
c Graph stretched $\frac{1}{2}$ unit parallel to y-axis
d Graph stretched 2 units parallel to y-axis and reflected in the x-axis
3 18, −1
4 a A(−2, 0), B(0, 8), C(5, 12)
b A(−2, 0), B(0, 40), C(5, 60)
c A(−2, 0), B(0, −12), C(5, −18)
d A(−2, 0), B(0, −2), C(5, −3)
e A(−2, 1), B(0, 9), C(5, 13)
f A(0, 0), B(2, 16), C(7, 24)
5 Reflection in the x-axis.
6 a y-intercept (0, 12), x-intercepts (−2, 0), (−6, 0), minimum point (−4, −4)
b y-intercept (0, 36), x-intercepts (−2, 0), (−6, 0), minimum point (−4, −12)
c y-intercept (0, 1), x-intercepts $(2-\sqrt{3}, 0)$, $(2+\sqrt{3}, 0)$, minimum point (2, −3)
d y-intercept (0, −1), x-intercepts $(2-\sqrt{3}, 0)$, $(2+\sqrt{3}, 0)$, maximum point (2, 3)

A8.4

1 a $f(2x)$ b $f(-x)$ c $f\left(\frac{1}{2}x\right)$
2 a Graph stretched by a scale factor of $\frac{1}{2}$ parallel to the x-axis
b Graph stretched by a scale factor of 4 parallel to the x-axis
c Graph reflected in the y-axis
d Graph stretched by a scale factor of $\frac{1}{3}$ parallel to the x-axis
3 1, −16
4 a A(−1, 0), B(0, 10), C(2, 6)
b A(−4, 0), B(0, 10), C(8, 6)
c A(−0.5, 0), B(0, 10), C(1, 6)
d A(2, 0), B(0, 10), C(−4, 6)
e A(−1, 5), B(0, 15), C(2, 11)
5 $f(-2x) = \left(\frac{1}{4}\right)^x$

A8.5

1 a e.g. take 3, 4, 5 then $(4+5) - (3+4) = 9 - 7 = 2$
bi $2n+1$ bii $2n+3$
c $(2n+3) - (2n+1) = 2$
2 a e.g. 1 is odd
b e.g. $2^3 - 1^3 = 8 - 1 = 7$, which is odd
c e.g. $1^2 = 1$, which is not greater than 1
d e.g. $(0+1)^2 = 0^2 + 1^2$
e e.g. $\sin(0° + 90°) = \sin 0° + \sin 90°$
f e.g. $x = 11$
3 a n, $n+1$ and $n+2$ are consecutive integers, $(n+1)^2 = n^2 + 2n + 1 = n(n+2) + 1$
b $2n-1$ and $2n+1$ are consecutive odd numbers, $(2n-1)(2n+1) = 4n^2 - 1$
c n, $n+1$ and $n+2$ are consecutive integers, $(n+1)^2 - n(n+2) = n^2 + 2n + 1 - n^2 - 2n = 1$
4 a Let n and m be the numbers on the top faces, then $7-n$ and $7-m$ are the numbers on the bottom faces, so sum $= nm + (7-n)(7-m) + n(7-m) + m(7-n) = 49$
5 a $n-2$, $n-1$, n, $n+1$ and $n+2$ are five consecutive integers, square of the mean $= n^2$, mean of squares $= n^2 + 2$

A8.6

1 Pythagoras: $(x+1)^2 + (x+3)^2 = (x+7)^2 \Rightarrow x^2 - 6x - 39 = 0$
2 $12n + (2n+1)^2 + 4n^2 + 3 = 8n^2 + 20n + 4 =$ which is even
3 AB $= \sqrt{53}$, AC $= \sqrt{29}$ and BC $= \sqrt{34}$, which are all different, so the triangle is scalene.
4 a $A = x(x+10)$
b $x(x+10) = 24 \Rightarrow (x+12)(x-2) = 0$ so $x = 2$ (reject $x = -12$ as lengths must be positive)
5 a $\frac{m}{m+6}$
b $\frac{6}{m+5}$
c P(both red) $= \frac{m}{m+6} \times \frac{m-1}{m+5} = \frac{2}{11} \Rightarrow 3m^2 - 11m = 20$
d 5
6 $(\sqrt{3})^2 + (\sqrt{2}+3)^2 = 14 + 6\sqrt{2} \neq (2+\sqrt{10})^2$ so Pythagoras' theorem does not hold, and the triangle is not right-angled.

A8 Summary

1 a

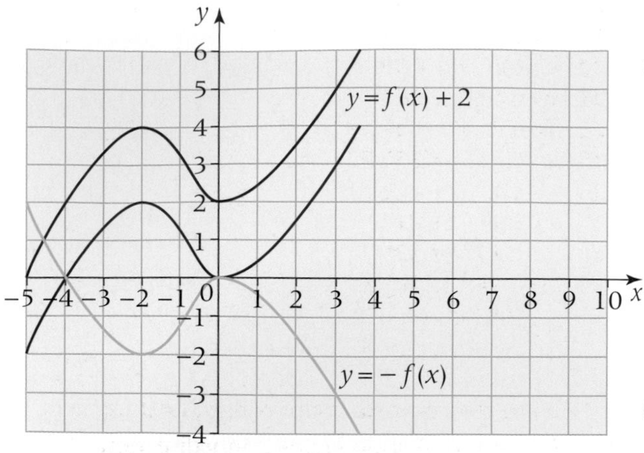

$y = f(x) + 2$ passes through (−4, 2), (−2, 4), (0, 2) and (3, 5).
$y = -f(x)$ passes through (−4, 0), (−2, −2), (0, 0) and (3, −3).

2 a When you add 2 to an even number the answer is always an even number.
OR
Even numbers go up in twos.
OR
Even numbers are two apart.

b $2n + 4$

c $2n + (2n + 2) + (2n + 4) = 2n + 2n + 2 + 2n + 4$
$= 6n + 6$
$= 6(n + 1)$

ie a multiple of 6.

Case study 8: Radio maths

1 a 35 GHz = 35 000 000 000 = 35 000 000 kHz $= 3.5 \times 10^7$ kHz

b 300 Hz = 0.000 300 MHz $= 3.0 \times 10^{-4}$ MHz

2 Waves **b** is loudest, followed by wave **a** and then wave **c**.
Waves **c** has the highest pitch, followed by wave **b** and then wave **a**.

3 ai 93.2 MHz

aii 93 200 000 Hz $= 9.32 \times 10^7$ Hz

aiii 0.0932 GHz $= 9.32 \times 10^{-2}$ GHz

b For Maths FM:
wave speed = frequency × wavelength
$= 9.32 \times 10^7 \times 3.22$
= 300104000 m/s

For Maths AM:
Wavelength $= \frac{\text{wave speed}}{\text{frequency}} = \frac{300\,104\,000}{930\,000} = 322.7$ m

c frequency (kHz) × wavelength = 300 104
so frequency = wavelength =
$\sqrt{300104} = 547.8$ kHz
This is in the AM frequency band.

4 ai 40 mins aii 20 mins

b Yes. A third of the show (33%) is speech based.

G7 Check in

1 ai 0.5 aii 0.5

bi 0.643 bii 0.643

c In a right-angled triangle, the sine of one of the smaller angles is equal to the cosine of the other smaller angle.

2 a Graph of $y = x^2$ translated 2 units up

b Graph of $y = x^2$ stretched by a scale factor of 2 parallel to the y-axis

c Graph of $y = x^2$ translated 2 units right

G7.1

1 a $x = 44°$ or $136°$ b $x = 53°$ or $127°$
c $x = 233°$ or $307°$ d $x = 210°$ or $330°$
e $x = 192°$ or $348°$ f $x = 24°$ or $156°$
g $x = 217°$ or $323°$ h $x = 17°$ or $163°$
i $x = 197°$ or $343°$ j $x = 270°$
k $x = 0°$ or $360°$ l $x = 30°$ or $150°$

2 a 0.5 b 0.5 c 0.71 d −0.71
e −0.71 f 1 g −1 h 0
i 0.97 j −0.87 k 0.98 l −0.87

3 a 130° b 112° c 40° d 323°
e 278° f 82° g 210° h 297°
i 168° j 231° k 8° l 253°

4 ai 72° aii 108°
bi 34.6° bii No second value possible
ci 60.3° cii No second value possible

5

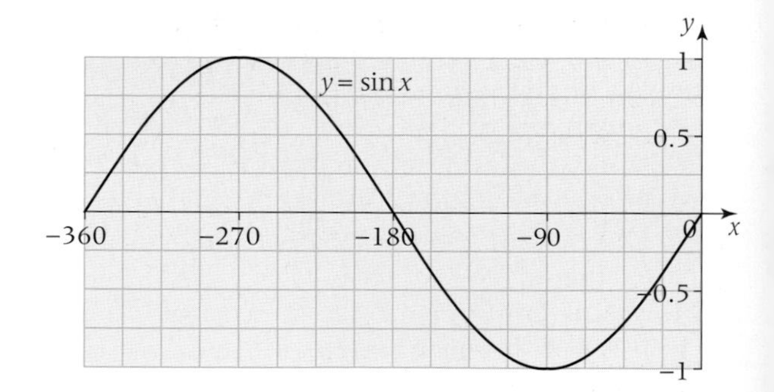

6 a Height of triangle $= \sqrt{3}$ so $\tan 30° = \frac{1}{\sqrt{3}} = \frac{\sqrt{3}}{3}$

b $\sqrt{3}$

ci 1 cii $\frac{1}{\sqrt{2}}$ or $\frac{\sqrt{2}}{2}$ ciii $\frac{1}{\sqrt{2}}$ or $\frac{\sqrt{2}}{2}$

G7.2

1 a $x = 46°$ or $314°$ b $x = 37°$ or $323°$
c $x = 143°$ or $217°$ d $x = 120°$ or $240°$
e $x = 102°$ or $258°$ f $x = 66°$ or $294°$
g $x = 127°$ or $233°$ h $x = 73°$ or $287°$
i $x = 107°$ or $253°$ j $x = 0°$ or $360°$
k $x = 90°$ or $270°$ l $x = 60°$ or $300°$

2 a 0.5 b −0.77 c 0.71 d −0.71

e 0.71	**f** 0	**g** 0	**h** −1
i 0.57	**j** −0.34	**k** 0.17	**l** 0.87

3 **a** 330° **b** 294° **c** 210° **d** 123°
e 196° **f** 282° **g** 235° **h** 126°
i 342° **j** 161° **k** 58° **l** 92°

4

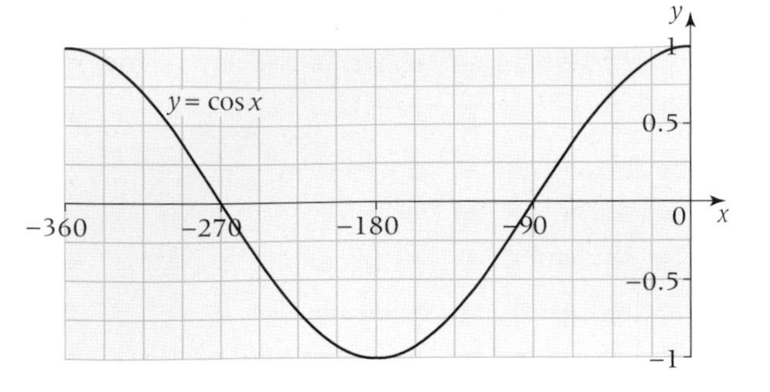

5 **b** $\cos\theta = \frac{a}{b} = \sin(90° - \theta)$

6

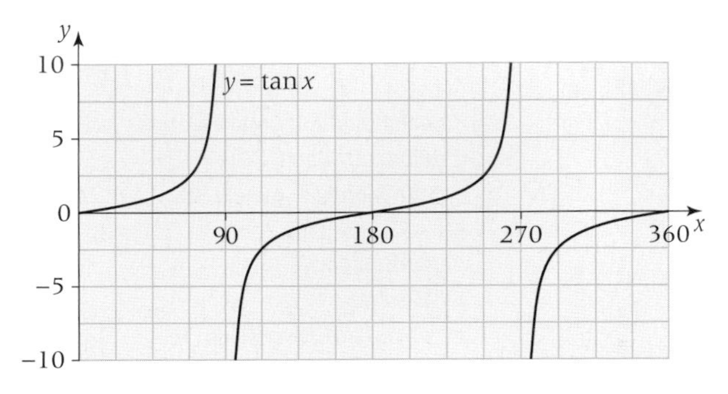

Exist values larger than 1, with vertical asymptotes at $x = 90°$ and 270°. The function is discontinuous at these points. The function has a period of 180°.

G7.3

1 $x = 32.3°$ or 147.7°
2 $x = 69.1°$ or 110.9°
3 $x = 265.6°$ or 274.4°
4 $x = 58.6°$ or 301.4°
5 $x = 82.9°$ or 277.1°
6 $x = 7.1°$ or 172.9°
7 $x = 38.3°$ or 141.7°
8 $x = 120.5°$ or 239.5°
9 $x = 60.0°$ or 120.0°
10 $x = 150.0°$ or 210.0°
11 $x = 47.1°$ or 132.9°
12 $x = 115.5°$ or 244.5°
13 $x = 136.5°$ or 223.5°
14 $x = 207.8°$ or 332.2°
15 $x = 18.7°$ or 161.3°
16 $x = 120.0°$ or 240.0°
17 $x = 30.0°$ or 150.0°
18 $x = 66.3°$ or 293.7°
19 $x = 134.0°$ or 226.0°
20 $x = 51.3°$ or 231.3°
21 $x = 69.0°$ or 249.0°
22 $x = 120.9°$ or 300.9°
23 $x = 62.1°$ or 242.1°
24 $x = 117.9°$ or 297.9°
25 $x = 77.5°$ or 257.5°
26 $x = 102.5°$ or 282.5°
27 $x = 38.9°$ or 218.9°
28 $x = 45.0°$ or 225.0°
29 $x = 83.4°$ or 263.4°
30 $x = 80.4$ or 260.4°

G7.4

1 **a** Stretch of scale factor $\frac{1}{2}$ in x direction
b Stretch of scale factor 2 in y direction
c Shift down 2 units
2 **a** Stretch of scale factor 2 in x direction
b Stretch of scale factor $\frac{1}{2}$ in y direction
c Shift 60° to the right
3 $a = 1$, $b = 3$
4 $p = 0.5$, $q = 1$
5 **a** Shift up 1 unit
b Shift 30° to the left
c Shift 30° to the left and up 1 unit

G7 Summary

1

$y = 2\sin x$	C
$y = -\sin x$	D
$y = \sin 2x$	A
$y = \sin x + 2$	F
$y = \sin\frac{1}{2}x$	B
$y = -2\sin x$	E

Index